LEARNING GUIDE

Kathleen Schmidt Prezbindowski
College of Mount St. Joseph

PRINCIPLES OF ANATOMY AND PHYSIOLOGY

Tenth Edition

Gerard J. Tortora
Bergen Community College

Sandra Reynolds Grabowski
Purdue University

John Wiley & Sons, Inc.

Cover photos: Photo of woman: ©PhotoDisc, Inc.
Photo of man: ©Ray Massey/Stone

To order books or for customer service call 1-800-CALL-WILEY (225-5945).

ISBN 0-471-43447-7

Printed in the United States of America

10 9 8 7 6 5 4 3

Printed and bound by Courier Kendallville, Inc.

Learning Guide
C O N T E N T S

Unit 1 Organization of the Human Body

Chapter 1 An Introduction to the Human Body 1

Chapter 2 The Chemical Level of Organization 19

Chapter 3 The Cellular Level of Organization 41

Chapter 4 The Tissue Level of Organization 65

Chapter 5 The Integumentary System 87

Unit 2 Principles of Support and Movement

Chapter 6 The Skeletal System: Bone Tissue 103

Chapter 7 The Skeletal System: The Axial Skeleton. 121

Chapter 8 The Skeletal System: The Appendicular Skeleton 143

Chapter 9 Joints 159

Chapter 10 Muscle Tissue 177

Chapter 11 The Muscular System 201

Unit 3 Control Systems of the Human Body

Chapter 12 Nervous Tissue 223

Chapter 13 The Spinal Cord and Spinal Nerves 245

Chapter 14 The Brain and Cranial Nerves 263

Chapter 15 Sensory, Motor, and Integrative Systems 287

Chapter 16 The Special Senses 311

Chapter 17 The Autonomic Nervous System 337

Chapter 18 The Endocrine System 357

Unit 4 Maintenance of the Human Body

Chapter 19 The Cardiovascular System: The Blood 391

Chapter 20 The Cardiovascular System: The Heart 417

Chapter 21 The Cardiovascular System: Blood Vessels and Hemodynamics 443

Chapter 22 The Lymphatic and Immune System and Resistance System to Disease 477

Chapter 23 The Respiratory System 507

Chapter 24 The Digestive System 535

Chapter 25 Metabolism 565

Chapter 26 The Urinary System 593

Chapter 27 Fluid, Electrolyte, and Acid–Base Homeostasis 625

Unit 5 Continuity

Chapter 28 The Reproductive Systems 645

Chapter 29 Development and Inheritance 675

P R E F A C E

This learning guide is intended to be used in conjunction with Tortora and Grabowski's *Principles of Anatomy and Physiology*, Tenth Edition. The emphasis is on active learning, not passive reading, and the student examines each concept through a variety of activities and exercises. By approaching each concept several times from different points of view, the student sees how the ideas of the text apply to real, clinical situations. In this way, the student may be helped not simply to study, but to learn.

The 29 chapters of the *Learning Guide* parallel those of the Tortora and Grabowski text. Students will find that the following features in each chapter enhance the effectiveness of an excellent text.

- Each chapter begins with an **Overview** that introduces content succinctly and a **Framework** that visually organizes the interrelationships among key concepts and terms.
- **Objectives** from the text are included for convenient previewing and reviewing. The Objectives are arranged according to major chapter divisions in the **Topic Outline**.
- **Wordbytes** present word roots, prefixes, and suffixes that facilitate understanding and recollection of key terms.
- **Checkpoints** offer a diversity of learning activities that follow the exact sequence of topics in the chapter, with cross-references to pages in the text. This format makes the Learning Guide especially "student-friendly." **Answers to the Selected Checkpoints** (those marked √) provide feedback to students and are found at the end of each chapter.
- **More Critical Thinking** provides students with opportunities for improving their critical thinking and writing skills by asking them to compose short essays on chapter concepts.
- A 25-question **Mastery Test** includes both objective and subjective questions as a final tool for student self-assessment of learning. Answers to the Mastery Test immediately follow.
- **The Big Picture: Looking Ahead** guides students into brief ventures in upcoming chapters where students will see further examples of concepts being introduced. These brief glimpses into the future are designed to reduce anxiety: when students reach these chapters later in the course, they will recognize familiar content. **The Big Picture: Looking Back** helps integrate concepts as students recognize that they are building upon information from previous chapters.

This learning guide is especially designed to fit the needs of people with different learning styles.

- Visual learners will identify the **Frameworks** as assets in organizing concepts and key terms. Visual and tactile learners will welcome the abundance of coloring and labeling exercises with color-code ovals. Further variety in Checkpoints is offered by short definitions, fill-ins, tables for completion, and multiple-choice, matching, and arrange-in-correct-sequence activities.
- Learning on multiple levels is made possible by **For Extra Review** sections that provide enrichment or assistance with particularly difficult topics.
- **Clinical Correlations** introduce application especially relevant for students in allied health. These are based on my own experiences as an R.N./M.S.N. and on discussions with colleagues. These challenging exercises enable students to apply abstract concepts to actual clinical settings.

I offer particular thanks to the students and faculty at the College of Mount St. Joseph for their ongoing sharing and support of teaching and learning. I thank my family (Amy, Joost, and Steffy; Laurie; Maureen and Alex) and close friends for their wisdom, encouragement and caring. I am thankful for the expertise of Mary O'Sullivan, Associate Editor and Rosa Bryant, Senior Production Editor of John Wiley & Sons.

Kathleen Schmidt Prezbindowski

TO THE STUDENTS

This learning guide is designed to help you do exactly what its title indicates – learn. It will serve as a step-by-step aid to help you bridge the gap between goals (objectives) and accomplishments (learning). The 29 chapters of the *Learning Guide* parallel the 29 chapters of Tortora and Grabowski's *Principles of Anatomy and Physiology*, Tenth Edition.

Take a moment to turn to Chapter 1 and look at the major headings in this learning guide. Now consider each one.

Overview. As you begin your study of each chapter, read this brief introduction, which presents the topics of the chapter. The overview is a look at the "forest" (or the "big picture") before examination of the individual "trees" (or details).

Framework. The Framework arranges key concepts and terms in an organizational chart that demonstrates interrelationships and lists key terms. Refer back to the Framework frequently to keep sight of the interrelationships.

Topic Outline and Objectives. Objectives that are identical to those in the text organize the chapter into "bite-size" pieces that will help you identify the principal subtopics that you must understand. Preview objectives as you start the chapter, and refer back to them as you complete each one. Objectives are arranged according to major chapter divisions in the Topic Outline.

Wordbytes present word roots, prefixes, and suffixes that will help you master terminology of the chapter and facilitate understanding and remembering of key terms. Study these; then check your knowledge. (Cover meanings and write them in the lines provided.) Try to think of additional terms in which these important bits of words are used.

Checkpoints offer a variety of learning activities arranged according to the sequence of topics in the chapter. At the start of each new Checkpoint section, note the related pages in the Tortora and Grabowski text. Read these pages carefully and then complete this group of Checkpoints. Checkpoints are designed to help you to "handle" each new concept, almost as if you were working with clay. Through a variety of exercises and activities, you can achieve a depth of understanding that comes only through active participation, not passive reading alone. You can challenge and verify your knowledge of key concepts by completing the Checkpoints along your path of learning.

A diversity of learning activities is included in Checkpoints. You will focus on specific facts in the chapter by doing exercises that require you to provide definitions and comparisons, and you will fill in blanks in paragraphs with key terms. You will also color and label figures. Fine-point, color, felt-tip pens or pencils will be helpful so that you can easily fill in related color-coded ovals. Your understanding will also be tested with matching exercises, multiple-choice questions, and arrange-in-correct-sequence activities.

Three special types of Checkpoints are **For Extra Review**, **Clinical Correlations**, and **The Big Picture**. For Extra Review provides additional help for difficult topics or refers you by page number to specific activities earlier in the *Learning Guide* so that you may better integrate related information. Applications for enrichment and interest in areas particularly relevant to nursing and other health science students are found in selected Clinical Correlation exercises. The Big Picture: Looking Ahead (or Looking Back) refers you to relevant sections of the Guide or text to help you integrate concepts within the entire text. When you come upon a Big Picture Checkpoint, take a few moments to turn to those sections of the book that may reinforce content you studied earlier and also reduce your anxiety about upcoming content.

When you have difficulty with an exercise, refer to the related text pages (listed with each section of the *Learning Guide*) before proceeding further. You will find specific references to many text figures and tables.

Answers to Selected Checkpoints. As you glance through the activities, you will notice that a number of questions are marked with a check mark (✓). Answers to these Checkpoints are given at the end of each chapter, in order to provide you with some immediate feedback about your progress. Incorrect answers alert you to review related objectives in the *Learning Guide* and corresponding text pages. Each *Learning Guide* (LG) figure included in Answers to Selected Checkpoints is identified with an "A", such as Figure LG 13.1A. Activities without answers are also purposely included in this book. The intention is to encourage you to verify some answers independently by consulting your textbook and also to stimulate discussion with students or faculty.

More Critical Thinking provides several suggestions for essays to improve critical thinking and application skills.

Mastery Test. This 25-question self-test provides an opportunity for a final review of the chapter as a whole. Its format will assist you in preparing for standardized tests or course exams that are objective in nature, because the first 20 questions are multiple-choice, true-false, and arrangement questions. The final five questions of each test help you to evaluate your learning with fill-in and short-answer questions. Answers for each mastery test are placed at the end of each chapter.

I wish you success and enjoyment in learning concepts relevant to your own anatomy and physiology and in applying this information in clinical settings.

Kathleen Schmidt Prezbindowski

A B O U T T H E A U T H O R

Kathleen Schmidt Prezbindowski, Ph.D., M.S.N., is a professor of biology at the College of Mount St. Joseph, where she was named Distinguished Teacher of the Year. She has taught anatomy and physiology, pathophysiology, and biology of aging for 30 years. Her Ph.D. is from Purdue University. Later, Dr. Prezbindowski earned a B.S.N. and then M.S.N.s in both Gerontological Nursing and Psychiatric/Mental Health Nursing. Through a grant from the U.S. Administration on Aging, she produced the first video series on the human body ever designed especially for older persons, as well as a training manual for health promotion and aging. Kathleen is a delegate to the Health Promotion Institute of the National Council on the Aging. She is the author of three other learning guides including one on pathophysiology.

DEDICATION
To Steffy

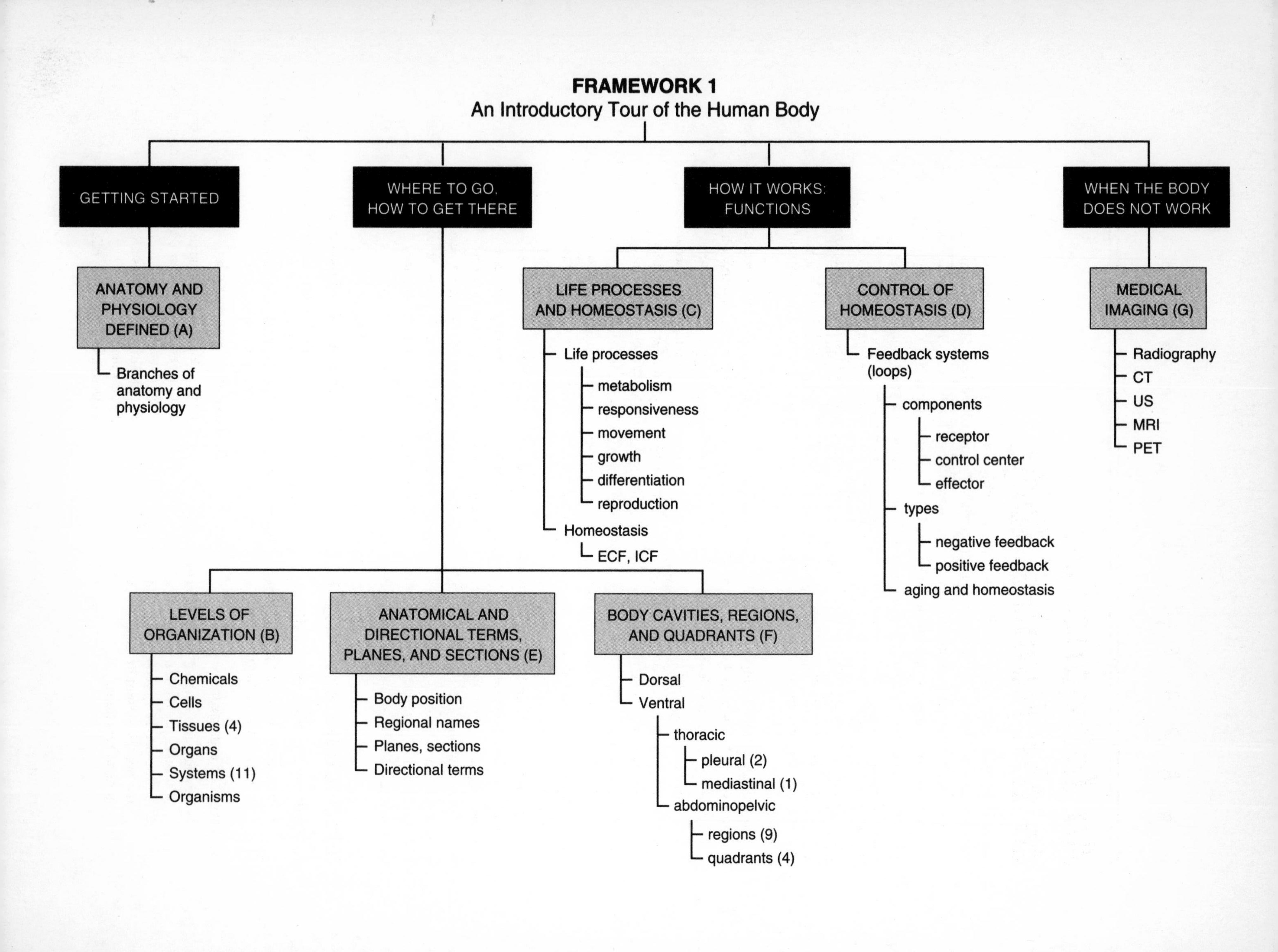

FRAMEWORK 1
An Introductory Tour of the Human Body
GETTING STARTED
ANATOMY AND PHYSIOLOGY DEFINED (A)
Branches of anatomy and physiology
WHERE TO GO, HOW TO GET THERE
LEVELS OF ORGANIZATION (B)
Chemicals
Cells
Tissues (4)
Organs
Systems (11)
Organisms
ANATOMICAL AND DIRECTIONAL TERMS, PLANES, AND SECTIONS (E)
Body position
Regional names
Planes, sections
Directional terms
BODY CAVITIES, REGIONS, AND QUADRANTS (F)
Dorsal
Ventral
thoracic
pleural (2)
mediastinal (1)
abdominopelvic
regions (9)
quadrants (4)
HOW IT WORKS: FUNCTIONS
LIFE PROCESSES AND HOMEOSTASIS (C)
Life processes
metabolism
responsiveness
movement
growth
differentiation
reproduction
Homeostasis
ECF, ICF
CONTROL OF HOMEOSTASIS (D)
Feedback systems (loops)
components
receptor
control center
effector
types
negative feedback
positive feedback
aging and homeostasis
WHEN THE BODY DOES NOT WORK
MEDICAL IMAGING (G)
Radiography
CT
US
MRI
PET

C H A P T E R 1

An Introduction to the Human Body

You are about to embark on a tour, one that you will enjoy for at least several months and hopefully for many years. Your destination: the far-reaching corners and fascinating treasures of the human body. Imagine the human body as an enormous, vibrant world, abundant in architectural wonders and bustling with activity. You are the traveler, the visitor, the learner on this tour.

Just as the world is composed of individual buildings, cities, countries, and continents, so the human body is organized into cells, tissues, organs, and systems. In fact, a close-up look at a cell (comparable to a detailed study of a building) will reveal structural intricacies—the array of chemicals that comprise cells. Any tour requires a sense of direction and knowledge of the language: directional and anatomical terms that guide the traveler through the body. A tour is sure to enlighten the visitor about the culture in each region—in the human body, the functions of each body part or system. By the time you leave each area (body part or system), you will have gained a sense of how that area contributes to overall stability and well-being (homeostasis). You will also learn about how feedback systems (loops) contribute to homeostasis.

In planning a trip, an itinerary serves as a useful guide. A kind of itinerary is found below in the objectives (same as in your text) listed according to a topic outline that can help you organize your study. It can also give you a well-deserved feeling of accomplishment as you complete each objective. Then, at the end of the trip/chapter, look back at this itinerary/outline to put all aspects of the chapter in context. Also look at the Framework for Chapter 1, study it carefully, and refer back to it as often as you wish. It contains the key terms of the chapter and serves as your map. Bon voyage!

TOPIC OUTLINE AND OBJECTIVES

A. Anatomy and physiology defined

1. Define anatomy and physiology and several subdisciplines of these sciences.

B. Levels of organization

1. Describe the levels of structural organization that make up the human body.
2. Define the 11 systems of the human body, the organs present in each, and their general functions.

C. Life processes and homeostasis

1. Define the important life processes of the human body.
2. Define homeostasis and explain its relation to interstitial fluid.

D. Control of homeostasis

1. Describe the components of a feedback system.
2. Contrast the operation of negative and positive feedback systems.
3. Explain why homeostatic imbalances cause disorders.
4. Describe some of the effects of aging.

E. Anatomical and directional terms, planes, and sections

1. Describe the orientation of the body in the anatomical position.
2. Relate the common names to the corresponding anatomical descriptive terms for various regions of the human body.
3. Define the anatomical planes and sections used to describe the human body.

F. Body cavities, regions, and quadrants

1. Describe the major body cavities, the organs they contain, and their associated linings.

G. Medical imaging

1. Describe the principles and importance of medical imaging procedures in the evaluation of organ functions and diagnosis of disease.

WORDBYTES

Now become familiar with the language of this chapter by studying each wordbyte, its meaning, and an example of its use within a term. After you study the entire list, self-check your understanding by writing the meaning of each wordbyte on the line. As you continue through the *Learning Guide,* identify and write in other examples of terms that contain the same wordbyte.

Wordbyte	Self-check	Meaning	Example(s)
ana-		apart	*ana*tomy
ante-		before	*ante*rior
dis-		part	*dis*section
homeo-		sameness	*homeo*stasis
inter-		between	*inter*cellular
intra-		within	*intra*cellular
-logy		study of	physio*logy*
-meter		measure	milli*meter*
pariet-		wall	*pariet*al
physio-		nature (or function) of	*physio*logy
post-		after	*post*erior
sagitta-		arrow	*sagitta*l
-section		act of cutting	dis*section*
-stasis		standing still	homeo*stasis*
-tomy		process of cutting	ana*tomy*
viscero-		body organs	*viscera*l

CHECKPOINTS

A. Anatomy and physiology defined (page 2)

✓ **A1.** Anatomy is the study of ____________ and physiology is the study of ____________.

✓ **A2.** Identity the subdivisions of anatomy and physiology by selecting the term that best fits the definitions provided below.

DA.	**Developmental anatomy**	**H.**	**Histology**
Embr.	**Embryology**	**I.**	**Immunology**
Endo.	**Endocrinology**	**PA.**	**Pathological anatomy**
EP.	**Exercise physiology**	**PP.**	**Pathophysiology**
GA.	**Gross anatomy**	**RP.**	**Renal physiology**

______ a. Study of tissues

______ b. Study of structures that can be examined without the aid of a microscope, such as study of an entire cadaver

______ c. Study of development during the first two months of life in utero

______ d. Study of function of kidneys

______ e. Study of functions of hormones

______ f. Study of functional changes associated with disease

______ g. Study of the defense mechanisms of the body

A3. An intimate relationship exists between structure and function: one determines the other. Explain how the structure of each of the following determines the functions for which it can be used.

a. Spoon/fork

b. Hand/foot

c. Incisors (front teeth)/molars

B. Levels of body organization (pages 3–6)

✓ **B1.** Arrange the terms listed in the left box from highest to lowest in level of organization by writing terms on lines provided at left below. One is done for you. Then match the examples in the right box with the related level of organization by writing those terms on lines at the right.

Cell	**Organism**
Chemical	**System**
Organ	**Tissue**

DNA	**Smooth muscle tissue**
Reproductive system	**Uterus**
Smooth muscle cell	**21-year-old woman**

Levels of organization	**Examples**
a. ______________ (highest)	a. ______________ (highest)
b. System	b. ______________
c. ______________	c. ______________
d. ______________	d. ______________
e. ______________	e. ______________
f. ______________ (lowest)	f. ______________ (lowest)

✓ **B2.** List the four basic tissue types.

______________ ______________

______________ ______________

B3. Complete Table LG 1.1 on the next page describing systems of the body. Name two or more organs in each system. Then list one or more functions of each system.

✓ **B4.** *The Big Picture: Looking Ahead.* Look ahead to upcoming chapters as you identify organs that function as parts of more than one system. Text page numbers in parentheses may guide you to answers.

a. The pancreas is part of the ______________ system because the pancreas secretes enzymes that digest foods (page 874). Because it produces hormones such as

insulin and glucagon, the pancreas. is also part of the ______________ system (page 614). List two other organs that are parts of these same two systems:

______________ and ______________ (pages 619 and 881)

b. Because the pituitary releases many hormones, such as human growth hormone

(hGH) and antidiuretic hormone (ADH), it is part of the ______________ system (page 599). Yet the connections of the pituitary

with the brain make it a component of the ______________ system (page 466).

c. Because ovaries produce female sex cells (ova) and also hormones (estrogens and

Table LG 1.1 Systems of the human body

System	Organs	Functions
a.	Skin, hair, nails	
b. Skeletal		
c. Muscular		
d. Nervous	Brain	Regulates body by nerve impulses
e. Endocrine	Glands that produce hormones	
f. Cardiovas	Blood, heart, blood vessels	
g. Lymphatic		
h. Resp		Supplies oxygen, removes carbon dioxide, regulates acid-base balance
i. Digest		Breaks down food and eliminates solid wastes
j. [illegible]	Kidney, ureters, urinary bladder, urethra	
k. Reproductive		

progesterone), ovaries are components of both the female __________________ system (page 616) and the __________________ system (page 1041).

d. The large bone of your thigh (the femur) provides strong support for your body and it also forms blood cells, so the femur is part of two systems:

__________________ (page 232) and __________________ (page 637).

e. The urethra of the male serves as a passageway for both urine (page 980) and semen, (page 1027) so it is part of both the __________________ system and the __________________ system.

✓ **B5.** *A clinical correlation.* Identify the procedures most likely to be used by a nurse in assessments of patients. Select from these answers:

Auscultation	**Palpation**	**Percussion**

a. Listening via stethoscope to hear "heart murmurs" caused by damaged heart valves: __________________

b. Pressing of fingers on breast tissue to check for the presence of hard, immobile masses or for softer, mobile masses: __________________

c. Examining for possible liver enlargement by pressing fingers just below the right side of the rib cage: __________________

d. While tapping finger over chest wall, listening for changes in sounds that reveal fluid-filled lungs: __________________

C. Characteristics of the living human organism (pages 7–8)

C1. Six characteristics distinguish you from nonliving things. List these characteristics below and give a brief definition of each. One is done for you.

a. ____

b. ____

c. ____

d. **Growth: increase in size and complexity**

e. ____

f. ____

✓ **C2.** Refer to the list of life processes in the box. Demonstrate your understanding of these processes by selecting the answer that best fits each of the activities in your own body, described below.

D.	**Differentiation**	**Move.**	**Movement**
G.	**Growth**	**Repro.**	**Reproduction**
Meta.	**Metabolism**	**Resp.**	**Responsiveness**

Resp. a. Your hunger at 8:00 A.M. prompts you to head towards breakfast.

Move. b. During breakfast you chew your toast with jelly and an egg; your stomach and intestine then contract to help break apart the food and propel it through the digestive tract.

Metab. c. Your body utilizes the starch, sugars, and proteins in your breakfast foods to provide building blocks for more muscle protein and to provide energy to your eyes and brain for studying anatomy and physiology.

grow d. After you work out four days a week for a month, you note that your arm and thigh muscles (biceps and quadriceps) are larger (two answers).

Diff e. As you work out, "stem cells" in your bone marrow are stimulated to undergo changes to become mature red blood cells (RBC) so you experience a healthy increase in your RBC count.

C3. Contrast *anabolism* with *catabolism*.

anabolism → syntesis of complex moleculles
catabolism → breakdown of complex moleculles

✓ **C4.** *Critical Thinking* about homeostasis.

a. Using your list of wordbytes, write a literal definition of homeostasis.

b. Explain why a broader definition of homeostasis is actually more accurate.

c. Besides control of blood glucose, what other aspects of your body need to be kept relatively constant, within a narrow range?

✓ **C5.** Describe the types of fluid in the body and their relationships to homeostasis by completing these statements and Figure LG 1.1.

a. Fluid inside of cells is known as ________________ fluid (ICF). Color areas containing this type of yellow fluid. Color yellow the corresponding color code oval.

b. Fluid in spaces between cells is called ________________. It surrounds and bathes cells and is one form of *(intracellular? extracellular?)* fluid. Color the spaces containing this fluid (as well as the related color code oval) light green.

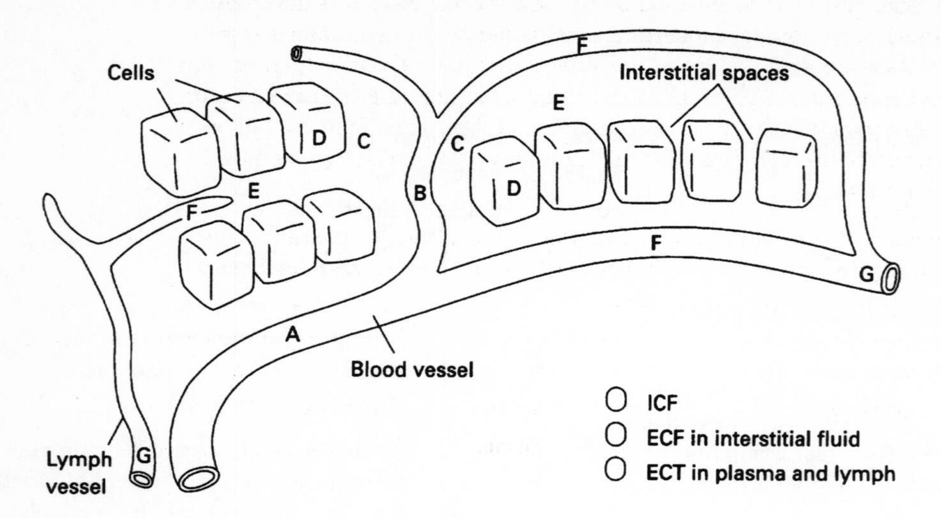

Figure LG 1.1 Internal environment of the body: types of fluid. Complete as directed. Areas labeled A to G refer to Checkpoint C6.

c. Another form of extracellular fluid is that located in ________________ vessels and ________________ vessels. Color these areas (and the color code oval) dark green.

d. The body's "internal environment" (that is, surrounding cells) is ________________ -cellular fluid (ECF) (all green areas in your figure). The condition of maintaining ECF in relative constancy is known as ________________.

✓ **C6.** Refer again to Figure LG 1.1. Show the pattern of circulation of body fluids by drawing arrows connecting letters in the figure in alphabetical order (A → B → C, etc.). Now fill in the blanks below describing this lettered pathway:

A (arteries and arterioles) → B (blood capillaries) →

C (________________) → D (________________) →

E (________________) → F (________________) →

G (________________).

D. Control of homeostasis; aging and homeostasis (pages 8–11)

D1. Think of a recent situation in which your mind and body have experienced stress. Identify the two body systems that are most often responsible for initiating changes that bring you back into homeostatic balance. (*Hint:* Refer to Checkpoint D5.)

✓ **D2.** Are you seated right now? For safety, hold on to the back of a chair or a desk and then stand up quickly. As you do, conceptualize what happens to your blood as you change position: As a response to gravity, blood tends to fall to the lower part of your body. If nothing happened to offset that action, your blood pressure (BP) would drop and you would likely faint because of insufficient blood supply to the brain. This usually does not happen because homeostatic mechanisms act to quickly elevate blood pressure back to normal.

Match the answers in the box with factors and functions that helped maintain your blood pressure when you stood up. (*Hint*: Figure 1.3 [page 10 your text] describes a situation that is just the opposite of this one.)

CCen.	**Control center**	**O.**	**Output**
CCond.	**Controlled condition**	**Recep.**	**Receptors**
E.	**Effector**	**Respo.**	**Response**
I.	**Input**	**S.**	**Stimulus**

_________ a. Your blood pressure (BP): a factor that must be maintained homeostatically

_________ b. Standing up; blood flowed by gravity to lower parts of your body

_________ c. Pressure-sensitive nerve cells in large arteries in your neck and chest

_________ d. Nerve impulses bearing the message "BP is too low" (since less blood remained in the upper part of your body as you were standing)

_________ e. Your brain

_________ f. Nerve impulses from your brain to your heart, bearing the message, "Beat faster!"

_________ g. Your heart (which beats faster)

_________ h. Elevated BP

✓ **D3.** *A clinical challenge.* In certain populations, such as malnourished or frail elderly, the mechanisms necessary for maintaining blood pressure occur more slowly than normal. As a result, blood pressure may drop as the person suddenly changes position (from lying to sitting or sitting to standing). This condition, known as postural hypotension, may lead to faintness and falls. What might a person subject to postural hypotension do to minimize the chance of falling?

✓ **D4.** Consider this example of a feedback system loop. Ten workers on an assembly line are producing handmade shoes. As shoes come off the assembly line, they pile up around the last worker, Worker Ten. When this happens, Worker Ten calls out, "We have hundreds of shoes piled up here." This information (input) is heard by (fed back to) Worker One, who determines when more shoes should be made. Worker One is therefore the ultimate controller of output. This controller could respond in either of two ways. In a *negative feedback system*, Worker One says, "We have an excess of unsold shoes. Let us slow down or stop production until the excess (stress) is relieved—until these shoes are sold." What might Worker One's response be if this were a *positive feedback system?*

✓ **D5.** Distinguish the two types of feedback in this exercise.

a. When blood pressure is abnormally low, the following mechanism should occur to raise blood pressure. Nerve impulses cause many tiny blood vessels that flow into stomach, intestine, and skin to become narrowed so that blood must stay in major arteries—raising blood pressure. This is an example of a *(negative? positive?)* feedback mechanism since the lowered blood pressure *(was raised? decreased even more?)*.

b. The increase of uterine contraction during labor as a response to increase of the hormone oxytocin is an example of a *(negative? positive?)* feedback mechanism because the uterine contractions are *(decreased? enhanced?)*.

c. Most feedback mechanisms in the human body are *(negative? positive?)*. Why do you think this might be advantageous and directed toward maintenance of homeostasis?

✓ **D6.** Circle the terms below that are classified as signs rather than as symptoms.

chest pain fever headache diarrhea swollen ankles skin rash toothache

D7. *A clinical challenge*. Contrast the two terms in each set.

a. *Local disease/systemic disease*

b. *Epidemiology/pharmacology*

c. *Medical history/diagnosis*

d. *Disease/aging*

D8. List 10 more normal aging changes.

E. Anatomical and directional terms, planes, and sections (page 12)

E1. How would you describe anatomical position? Assume that position yourself.

Contrast these body positions: *prone* and *supine*.

✓ **E2.** Complete the table relating common terms to anatomical terms. (See Figure 1.5, page 13 in your text, to check your answers.) For extra practice use common and anatomical terms to identify each region of your own body and that of a study partner.

Common Term	Anatomical Term
a. Armpit	Axillary
b. Fingers	phalanges
c. Arm	Brachial
d. Knee (back	Popliteal
e. head	Cephalic
f. Mouth	oral
g. groin	Inguinal
h. Chest	thoracic
i. neck chest	Cervical
j. backofarm	Antebrachial
k. Buttock	ass
l. heel	Calcaneal

✓ **E3.** Using your own body, a skeleton, and Exhibit 1.1 and Figure 1.6 determine relationships among body parts. Write correct directional terms to complete these statements.

a. The liver is ____________________ to the diaphragm.

b. Fingers (phalanges) are located ____________________ to wrist bones (carpals).

c. The skin on the dorsal surface of your body can also be said to be located on

your ______________________ surface.

d. The great (big) toe is ______________________ to the little toe.

e. The little toe is ______________________ to the great toe.

f. The skin on your leg is ______________________ to muscle tissue in your leg.

g. Muscles of your arm are ______________________ to skin on your arm.
h. When you float face down in a pool, you are lying on your surface.

i. The lungs and heart are located ______________________ to the abdominal organs.
j. Because the stomach and the spleen are both located on the left side of the

abdomen, they could be described as ______________________ -lateral.

✓ **E4.** Match each of the following planes with the phrase telling how the body would be divided by such a plane.

F.	**Frontal (coronal)**	**P.**	**Parasagittal**
M.	**Midsagittal (median)**	**T.**	**Transverse**

_______ a. Into superior and inferior portions

_______ b. Into equal right and left portions

_______ c. Into anterior and posterior portions

_______ d. Into unequal right and left portions

F. Body cavities, regions, and quadrants (pages 12–20)

✓ **Fl.** After you have studied Figures 1.9 through 1.13 in your text, complete this exercise about body cavities. Circle the correct answer in each statement.

a. The *(dorsal? ventral?)* cavity consists of the cranial cavity and the vertebral canal.
b. The viscera, including such structures as the liver, lungs, and stomach, are all located in the *(dorsal? ventral?)* cavity.
c. Of the two body cavities, the *(dorsal? ventral?)* appears to be better protected by bone.
d. Pleural, mediastinal, and pericardial are terms that refer to regions of the *(thorax? abdominopelvis?)*.
e. The *(heart? lungs? esophagus and trachea?)* are located in the pleural cavities.
f. Which organs are located in the mediastinum? *(Esophagus? Major airways such as the trachea? Heart and its pericardial covering? Stomach? Thyroid? Thymus?)*.
g. The stomach, pancreas, small intestine, and most of the large intestine are located in the *(abdomen? pelvis?)*.
h. The urinary bladder, rectum, and internal reproductive organs are located in the *(abdominal? pelvic?)* cavity.
i. Which is larger: the *(abdominal? pelvic?)* region of the abdominopelvic cavity.
j. Which membrane lines the inside of the thoracic cavity? *(Parietal? Visceral?) (peritoneum? pleura?)*. The *(parietal? visceral?)* peritoneum is attatched to the surface of the liver.

✓ **F2.** Complete Figure LG 1.2 according to the directions below.

a. Color each of the organs listed on the figure. Select different colors for each organ, and be sure to use the same color for the related color code oval ◯.

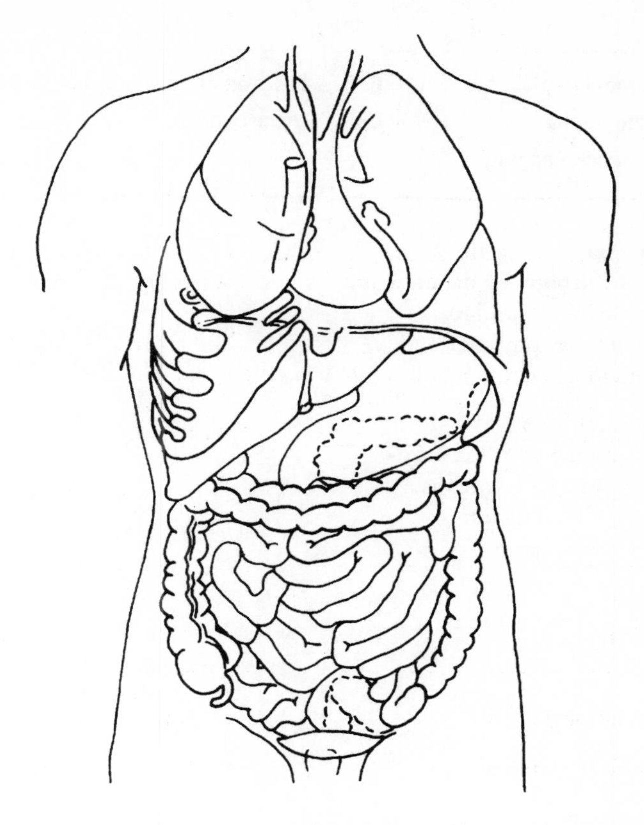

O Appendix	O Heart	O Lungs	O Spleen
O Diaphragm	O Large intestine	O Pancreas	O Stomach
O Gallbladder	O Liver	O Small intestine	

Figure LG 1.2 Regions of the ventral body cavity. Complete as directed in Checkpoint F2.

b. Next, label each organ on the figure.
c. *For extra review.* Draw lines dividing the abdomen into the nine regions, and note which organs are in each region.

✓ **F3.** Using the names of the nine abdominal regions and quadrants, complete these statements. Refer to the figure you just completed and to Figures 1.12 and 1.13 in the text.

a. From superior to inferior, the three abdominal regions on the right side are

______________________, ______________________ and ______________________.

b. The stomach is located primarily in the two regions named

______________________ and ______________________.

c. The navel (umbilicus) is located in the ______________________ region.
d. The region immediately superior to the urinary bladder is named the

______________________ region.

e. The liver and gallbladder are located in the ______________________ quadrant.

G. Medical imaging (pages 20–22)

✓ **G1.** Match the radiographic anatomy techniques listed in the box with the correct descriptions.

CR.	**Conventional radiograph**	**PET.**	**Positron emission tomography**
CT.	**Computed tomography**	**US.**	**Ultrasound**
MRI.	**Magnetic resonance imaging**		

_______ a. This method provides a cross-sectional picture of an area of the body by means of an x-ray source moving in an arc. Results are processed by a computer and displayed on a video monitor.

_______ b. This technique produces a two-dimensional image (radiograph) via a single barrage of x-rays. Images of organs overlap, making diagnosis difficult.

_______ c. Utilizes injected radioisotopes that assess function as well as structure.

_______ d. Two techniques that are noninvasive and do not utilize ionizing radiation (two answers).

ANSWERS TO SELECTED CHECKPOINTS: CHAPTER 1

A1. Structure, function.
A2. (a) H. (b) GA. (c) Embr (also a part of DA). (d) RP. (e) Endo. (f) PP. (g) I.
B1. (a) Organism; 21-year-old woman; (b) (System); reproductive system. (c) Organ; uterus. (d) Tissue; smooth muscle tissue. (e) Cell; smooth muscle cell. (f) Chemical; DNA.
B2. Epithelial tissue, connective tissue, muscle tissue, nervous tissue.
B4. (a) Digestive; endocrine; stomach, small intestine. (b) Endocrine; nervous. (c) Reproductive, endocrine. (d) Skeletal, cardiovascular. (e) Urinary, reproductive.
B5. (a) Auscultation. (b-c) Palpation. (d) Percussion.
C2. (a) Resp. (b) Move. (c) Meta. (d) G, Repro. (e) D.
C4. (a) Literal definition: standing still or staying the same. (b) Broader definition: maintaining a dynamic (ever-changing) state—yet over a narrow range. (c) Examples: blood levels of other chemicals such as potassium or calcium ions, water, fats; body temperature; heart rate and blood pressure.
C5. (a) Intracellular. (b) Interstitial (or intercellular or tissue) fluid, extracellular. (c) Blood, lymph. (d) Extra-; homeostasis.
C6. C, interstitial (intercellular) fluid; D, intracellular fluid; E, interstitial (intercellular) fluid again; F, blood or lymph capillaries; G, venules and veins or lymph vessels.
D2. (a) CCond. (b) S. (c) Recep. (d) I. (e) CCen. (f) O. (g) E. (h) Respo.
D3. Change position gradually, for example, by sitting on the side of the bed before getting out of bed. Support body weight on a nightstand or chair for a moment after standing. Wear support hose. Drink adequate fluid.
D4. "Hurray! We have produced hundreds of shoes; let us go for thousands! Do not worry that the shoes are not selling. Step up production even more."
D5. (a) Negative; was raised. (b) Positive, enhanced. (c) Negative; they permit continual fine-tuning so changes do not become excessive in one direction.
D6. Fever, diarrhea, swollen ankles, skin rash.
E2. (a) Armpit. (b) Phalangeal or digital. (c) Brachial. (d) Hollow area at back of knee. (e) Head. (f) Oral. (g) Groin. (h) Thoracic. (i) Neck. (j) Forearm. (k) Gluteal. (l) Heel.
E3. (a) Inferior (caudal). (b) Distal. (c) Posterior or dorsal (d) Medial. (e) Lateral. (f) Superficial. (g) Deep. (h) Anterior (ventral). (i) Superior (cephalic or cranial) (j) Ipsi-.
E4. (a) T. (b) M. (c) F. (d) P.
F1. (a) Dorsal. (b) Ventral. (c) Dorsal. (d) Thorax. (e) Lungs. (f) Esophagus, major airways such as the trachea, heart and its pericardial covering, thymus. (g) Abdomen. (h) Pelvic. (i) Abdominal. (j) Parietal, pleura; visceral.
F2. (a-c) See Figure LG 1.2A.

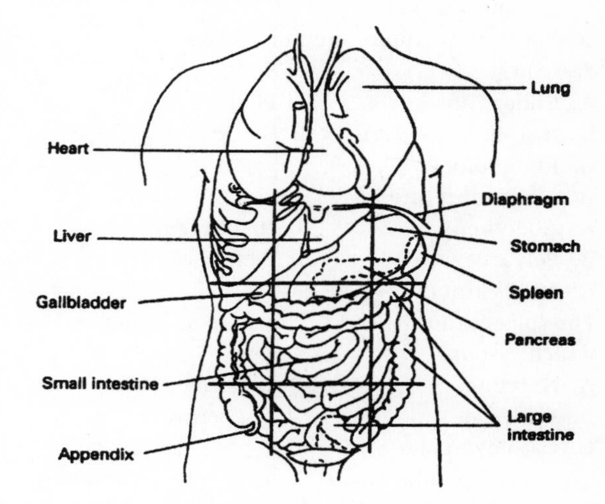

Figure LG 1.2A Regions of the ventral body cavity.

F3. (a) Right hypochondriac, right lumbar, right iliac (inguinal). (b) Left hypochondriac, epigastric. (c) Umbilical. (d) Hypogastric (pubic). (e) Right upper.
G1. (a) CT. (b) CR. (c) PET. (d) MRI, US.

MORE CRITICAL THINKING: CHAPTER 1

1. Think of a person you know who has a chronic (ongoing) illness. Explain how changes in anatomy (abnormal body structure) in that person are associated with changes in physiology (inadequate or altered body function).
2. Cancer is a disease of abnormal cell production. Explain how alterations at the cellular level affect higher levels of organization, such as specific organs, systems, and the entire organism.
3. Suppose you decide to run a quarter mile at your top speed. Tell how your body would respond to this stress created by exercise. List according to system the changes your body would probably make in order to maintain homeostasis.

MASTERY TEST: CHAPTER 1

Questions 1—11: Circle the letter preceding the one best answer to each question.

1. In a negative feedback system, when blood pressure decreases slightly, the body will respond by causing a number of changes which tend to:
   A. Lower blood pressure
   B. Raise blood pressure
2. The following structures are all located in the ventral cavity *except*:
   A. Spinal cord
   B. Urinary bladder
   C. Heart
   D. Gallbladder
   E. Esophagus
3. Which pair of common/anatomical terms is mismatched?
   A. Eye/ocular
   B. Skull/cranial
   C. Armpit/brachial
   D. Neck/cervical
   E. Buttock/gluteal
4. Which science studies the microscopic structure of tissues?
   A. Endocrinology
   B. Histology
   C. Embryology
   D. Pharmacology
   E. Epidemiology
5. Which is most inferiorly located?
   A. Abdomen
   B. Pelvic cavity
   C. Mediastinum
   D. Diaphragm
   E. Pleural cavity
6. The spleen, tonsils, and thymus are all organs in which system?
   A. Nervous
   B. Lymphatic
   C. Cardiovascular
   D. Digestive
   E. Endocrine
7. Which of the following structures is located totally outside of the upper right quadrant of the abdomen?
   A. Liver
   B. Gallbladder
   C. Transverse colon
   D. Spleen
   E. Pancreas
8. Choose the FALSE statement. (It may help you to mark each statement *T* for true or *F* for false as you read it.)
   A. Stress disturbs homeostasis.
   B. Homeostasis is a condition in which the body's internal environment remains relatively constant.
   C. The body's internal environment is best described as extracellular fluid.
   D. Extracellular fluid consists of plasma and intracellular fluid.
9. The system responsible for providing support, protection, and leverage; for storage of minerals; and for production of most blood cells is:
   A. Urinary
   B. Integumentary
   C. Muscular
   D. Reproductive
   E. Skeletal
10. Which is most proximally located?
    A. Ankle
    B. Hip
    C. Knee
    D. Toe
11. Which of the following describes pathological anatomy?
    A. Study of a specific region of the body such as the chest or arm
    B. Study of markings of the surface of the body
    C. Study of the body according to specific systems such as skeletal or respiratory
    D. Study of structural changes associated with disease

Questions 12—20: Circle T (true) or F (false). If the statement is false, change the underlined word or phrase so that the statement is correct.

T F 12. In order to assume the anatomical position, you should lie down with arms at your sides and palms turned backwards.

T F 13. The appendix is usually located in the left iliac region of the abdomen.

T F 14. Anatomy is the study of how structures function.

T F 15. In a negative feedback system, the body will respond to an increased blood glucose level by increasing blood glucose to an even higher level.

T F 16. A tissue is a group of cells and their intercellular substances that are similar in both origin and function.

T F 17. The right kidney is located mostly in the right iliac region of the abdomen.

T F 18. Most feedback mechanisms that help to maintain homeostasis are positive mechanisms.

T F 19. A parasagittal section is parallel to and lateral to a midsagittal section.

T F 20. The term homeostasis means that the body exists in a static state.

Questions 21—25: Fill-ins. Write the word that best fits the description.

_______________ 21. Region of thorax located between the lungs. Contains heart, thymus, esophagus.

_______________ 22. Name the fluid located within cells.

_______________ 23. Life process in which unspecialized cells, as in the early embryo, change into specialized cells such as nerve or bone.

_______________ 24. A painless diagnostic technique in which sound waves are generated and detected by a transducer, resulting in a sonogram.

_______________ 25. Name one or more extracellular fluids.

ANSWERS TO MASTERY TEST: CHAPTER 1

Multiple Choice

1. B
2. A
3. C
4. B
5. B
6. B
7. D
8. D
9. E
10. B
11. D

True—False

12. F. Stand erect facing observer, arms at sides, palms forward
13. F. Right iliac
14. F. Physiology
15. F. Decreasing blood glucose to a lower level
16. T
17. F. At the intersection of four regions of the abdomen: right hypochondriac, right lumbar, epigastric, and umbilical
18. F. Negative
19. T
20. F. In a state of dynamic equilibrium

Fill-ins

21. Mediastinum
22. Intracellular
23. Differentiation
24. Ultrasound (US)
25. Plasma, interstitial fluid, lymph, serous fluid, mucus, fluid within eyes or ears, cerebrospinal fluid (CSF)

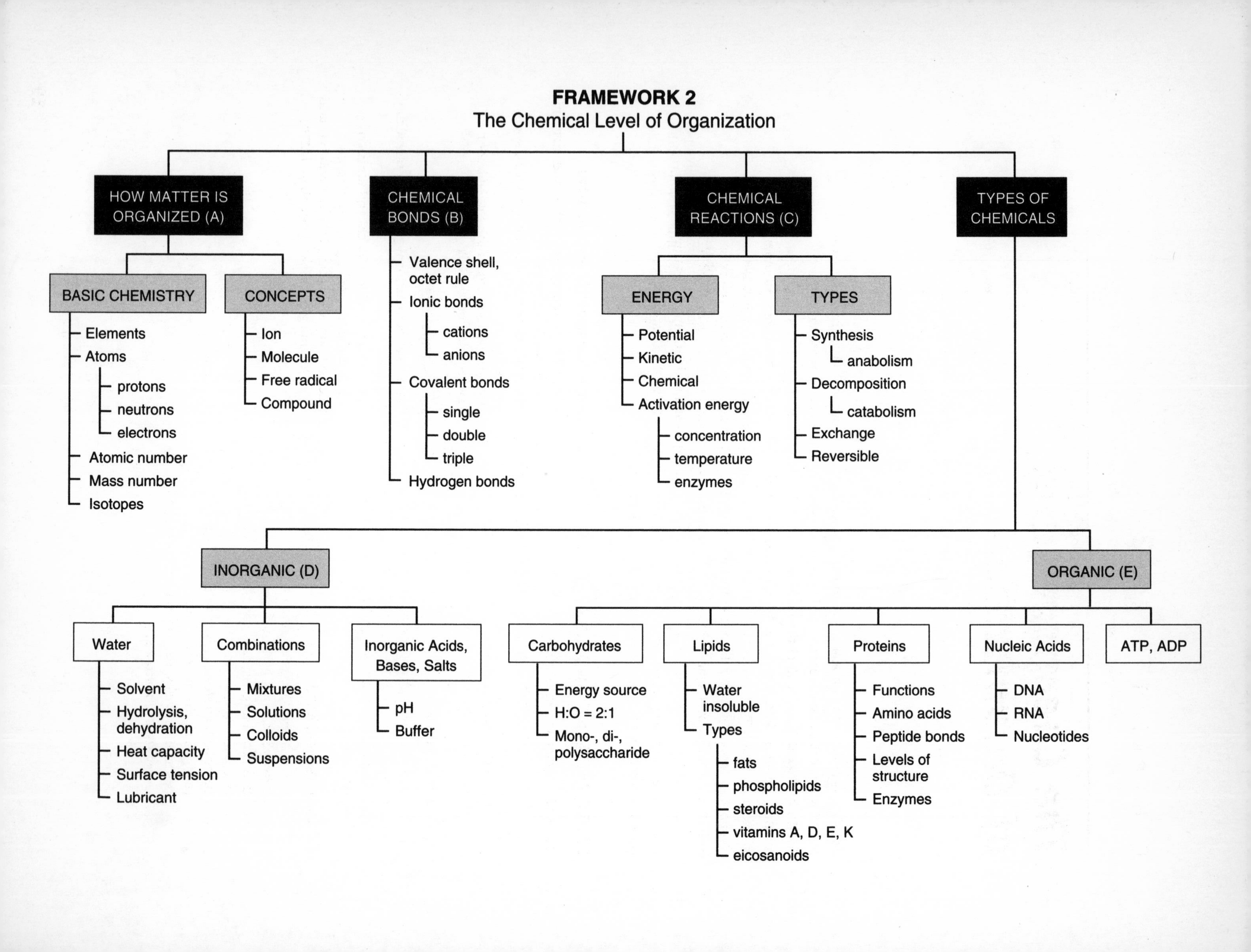
FRAMEWORK 2
The Chemical Level of Organization
HOW MATTER IS ORGANIZED (A)
BASIC CHEMISTRY
Elements
Atoms
protons
neutrons
electrons
Atomic number
Mass number
Isotopes
CONCEPTS
Ion
Molecule
Free radical
Compound
CHEMICAL BONDS (B)
Valence shell, octet rule
Ionic bonds
cations
anions
Covalent bonds
single
double
triple
Hydrogen bonds
CHEMICAL REACTIONS (C)
ENERGY
Potential
Kinetic
Chemical
Activation energy
concentration
temperature
enzymes
TYPES
Synthesis
anabolism
Decomposition
catabolism
Exchange
Reversible
TYPES OF CHEMICALS
INORGANIC (D)
Water
Solvent
Hydrolysis, dehydration
Heat capacity
Surface tension
Lubricant
Combinations
Mixtures
Solutions
Colloids
Suspensions
Inorganic Acids, Bases, Salts
pH
Buffer
ORGANIC (E)
Carbohydrates
Energy source
H:O = 2:1
Mono-, di-, polysaccharide
Lipids
Water insoluble
Types
fats
phospholipids
steroids
vitamins A, D, E, K
eicosanoids
Proteins
Functions
Amino acids
Peptide bonds
Levels of structure
Enzymes
Nucleic Acids
DNA
RNA
Nucleotides
ATP, ADP

C H A P T E R 2

The Chemical Level of Organization

Chemicals comprise the ultrastructure of the human body. They are the submicroscopic particles of which cells, tissues, organs, and systems are constructed. Chemicals are the minute entities that participate in all of the reactions that underlie bodily functions. Every complex compound that molds a living organism is composed of relatively simple atoms held together by chemical bonds. Two classes of chemicals contribute to structure and function: inorganic compounds, primarily water, but also inorganic acids, bases, and salts, and organic compounds including carbohydrates, lipids, proteins, nucleic acids, and the energy-storing compound ATP.

As you begin your study of the chemistry of life, carefully examine the Chapter 2 Framework and note the key terms associated with each section. Also refer to the Chapter 2 Topic Outline and check off objectives as you meet each one.

TOPIC OUTLINE AND OBJECTIVES

A. How matter is organized

1. Identify the main chemical elements of the human body.
2. Describe the structures of atoms, ions, molecules, free radicals, and compounds.

B. Chemical bonds

1. Describe how valence electrons form chemical bonds.
2. Distinguish among ionic, covalent, and hydrogen bonds.

C. Chemical reactions

1. Define a chemical reaction.
2. Describe the various forms of energy.
3. Compare exergonic and endergonic chemical reactions.
4. Describe the role of activation energy and catalysts in chemical reactions.
5. Describe synthesis, decomposition, exchange, and reversible reactions.

D. Inorganic compounds and solutions

1. Describe the properties of water and inorganic acids, bases, and salts.
2. Distinguish among solutions, colloids, and suspensions.
3. Define pH and explain the role of buffer systems in homeostasis.

E. Organic compounds

1. Describe the functional groups of organic molecules.
2. Identify the building blocks and functions of carbohydrates, lipids, proteins, enzymes, deoxyribonucleic acid (DNA), ribonucleic acid (RNA), and adenosine triphosphate (ATP).

WORDBYTES

Now become familiar with the language of this chapter by studying each wordbyte, its meaning, and an example of its use within a term. After you study the entire list, self-check your understanding by writing the meaning of each wordbyte on the line provided. As you continue through the *Learning Guide,* identify and write in additional terms that contain the same wordbyte.

Wordbyte	Self-check	Meaning	Example(s)
-ase	____________	enzyme	lip*ase*
de-	____________	remove	*de*hydrogenase
di-	____________	two	*di*saccharide
hex-	____________	six	*hex*ose
mono-	____________	one	*mono*saccharide
pent-	____________	five	*pent*ose
poly-	____________	many	*poly*peptide
prot-	____________	first	*prot*ein
-saccharide	____________	sugar	mono*saccharide*
tri-	____________	three	*tri*phosphate

CHECKPOINTS

A. How matter is organized (pages 27–30)

A1. Why is it important for you to understand the "language of chemistry" in order to communicate about life processes?

✓ **A2.** Complete this exercise about matter and chemical elements.

a. All matter is made up of building units called chemical ____________.

b. Of the total 112 different chemical elements found in nature, four elements make up 96% of the mass of the human body. These elements are ____________ (O), ____________ (C), hydrogen (______), and ____________ (______).

c. Circle the eight chemical symbols of elements listed here that together make up 3.8% of human body mass:
Ca Cl Co Cu Fe I K Mg Mn Na P S Si Sn Zn

d. Define this term: *trace elements.*

Now write on these lines the symbols of seven trace elements from the list of elements above (in d): ________ ________ ________ ________ ________ ________ ________

✓ **A3.** Refer to Table 2.1 (page 28) in your text. Match the elements listed in the box with descriptions of their functions in the body. The first one has been done for you.

C.	**Carbon**	**K.**	**Potassium**
Ca.	**Calcium**	**Na.**	**Sodium**
Cl.	**Chlorine**	**N.**	**Nitrogen**
Fe.	**Iron**	**O.**	**Oxygen**
F.	**Fluorine**	**P.**	**Phosphorus**
I.	**Iodine**		

__C__ a. Found in every organic molecule

______ b. Component of all protein molecules, as well as DNA and RNA

______ c. Vital to normal thyroid gland function

______ d. Essential component of hemoglobin, which gives red color to blood

______ e. Composes part of every water molecule; functions in cellular respiration

______ f. Contributes hardness to bone and teeth; required for blood clotting and for muscle contraction

______ g. Component of ATP, DNA, and RNA; also component of bone and teeth

______ h. Anion (negatively charged ion) in table salt (NaCl)

______ i. Cation (positively charged ion) in table salt (NaCl); helps control water balance in the body; important in nerve and muscle function

______ j. A trace element that contributes to tooth structure

______ k. Most abundant positively charged ion (cation) in intracellular fluid (inside of cells)

A4. Contrast *element* with *atom.*

✓ **A5.** Complete this exercise about atomic structure.

a. An atom consists of two main parts: ______________ and ______________.

b. Within the nucleus are positively charged particles called ______________ and uncharged (neutral) particles called ______________.

c. Electrons possess *(positive? negative?)* charges. The number of electrons forming a charged cloud around the nucleus is *(greater than? equal to? smaller than?)* the number of protons in the nucleus of the atom. Therefore each atom is said to be electrically *(charged? neutral?)* with a total charge of ______________.

d. Electrons are grouped in regions known as electron ______________. The shell closest to the nucleus can hold a maximum of ______ electrons; the next two shells outward may hold up to ______ and ______ electrons, respectively.

e. The largest element in the human body is *(iron? iodine?)*, which has ______ protons and ______ electrons. Fill in the number of electrons in its five shells from innermost to outermost: ______, ______, ______, ______, and ______.

✓ **A6.** Refer to Figure 2.2 in your text and complete this exercise about atomic structure.

a. The *atomic number* of potassium (K) is __________. Locate this number below the symbol K on the figure. This means that a potassium atom has ________ protons and ________ electrons.

b. Identify the *atomic mass* associated with potassium in the figure. This number is ___________, and it represents the average of the mass numbers of the potassium atom. This indicates the total number of protons and neutrons in the potassium nucleus.

✓ **A7.** Complete this exercise about isotopes.

a. If you subtract the *atomic number* from the *mass number*, you will identify the number of neutrons in an atom. Most potassium atoms have ________ neutrons. Different isotopes of potassium all have 19 protons, but have varying numbers of *(electrons? neutrons?)*.

b. The isotope 14C differs from 12C in that 14C has ________ neutrons.

c. Radioisotopes have a nuclear structure that is unstable and can decay, emitting ______________________. Each radioactive isotope has its own distinctive ______________________, which is the time required for the radioactive isotope to emit half of its original amount of radiation.

d. List one helpful and one harmful effect of radioactive isotopes.

A8. Complete this exercise about several different categories of chemicals. Select all answers from the box that fit descriptions below. All answers except two will be used.

Ca^{++}	Cl^-	$C_6H_{12}O_6$ (glucose)	H
H_2O	Na^+	O_2	O_2^- (superoxide)

______________ a. Molecule

______________ b. Compound

______________ c. Free radical

______________ d. Ion that has gained one electron

______________ e. Ion that has lost one electron

A9. A *clinical correlation.* Discuss roles of free radicals in disease and aging by completing this Checkpoint.

a. State how free radicals may form.

b. List six or more disorders or other changes associated with free radicals.

c. Name a category of chemicals that can slow effects of free radicals.

B. Chemical bonds (pages 31–34)

B1. The type of chemical reacting (or bonding) that occurs between atoms depends on the number of electrons possessed by the bonding atoms. Refer to Figure 2.2 in your text and to the atoms listed in the box below; then do this Checkpoint about electrons and bonding.

C	Cl	H	He	Na	Ne

a. Which of these atoms contain only one electron shell? _____ and _____. Write the maximum number of electrons in this shell: _____.

b. The other atoms in the box all have *(1? 2? 3?)* electron shells. As you refer to Figure 2.2, write the number of electrons present in the outermost electron shell of these atoms: Na: _____ Cl _____ C: _____. The maximum number of electrons that can be present in the outer shell of each of these atoms is _____. (For help, refer to Checkpoint A5d.)

c. The outer shell of any electron is called its ______________ shell. Atoms can achieve stability by filling this outer shell to the maximum; this process is normally achieved by bonding. The likelihood that an atom will form chemical bonds is directly related to the number of extra or deficient electrons in the valence shell of that atom. (Visualize that this valence shell is what atom A presents to nearby atom B, which may be a possible "bonding partner.")

d. Find sodium (Na) in Figure 2.2. Sodium has one *(extra? missing?)* electron in its valence shell. It is therefore more likely to become an electron *(donor? acceptor?).* When sodium gives up the electron (which is a negative entity), sodium becomes more *(positive? negative?).* In other words, sodium forms Na^+, which is a(n) *(anion? cation?).*

e. Chlorine has _____ electrons in its outer shell, so it has one *(extra? missing?)* electron compared to a full valence shell. As sodium chloride is formed, chlorine *(accepts? gives up?)* an electron and so becomes the *(anion? cation?)* Cl^-.

f. Ions are defined as atoms that have *(gained or lost? shared?)* one or more electrons into or from their valence shells. A cation, such as K^+, Na^+, or H^+, is an atom that has *(gained? lost?)* an electron (a negative entity) by ionic bonding. A(n) *(anion? cation?),* such as Cl^-, has gained an electron from another atom. Ions in solution, such as table salt (Na^+Cl^-) in water, are called _______________ because _______________.

g. The fact that most atoms interact in ways that permit them to achieve stable outer shells of eight electrons is known as the *(triplet? septet? octet?)* rule.

h. Some atoms exist in nature with the valence shell already filled; these atoms *(do? do not?)* not attempt to bond, and so are said to be ________________. Two examples of inert atoms are *(H? He?)* and *(Ne with the atomic number of 10? Na with the atomic number of 11?)*

✓ **B2.** Match the types of bonds listed in the box with the descriptions below.

C. Covalent	H. Hydrogen	I. Ionic

________ a. Atoms lose electrons if they have just one or two electrons in their valence shells; they gain electrons if they need just one or two electrons to complete the valence shell, as in NaCl, (see Activity B1 above).

________ b. This bond is a bridgelike, weak link between a hydrogen atom and another atom such as oxygen or nitrogen. The hydrogen atom has a partial positive charge (+) that attracts the partial negative charge (–) of the electronegative atom (such as oxygen or nitrogen).

________ c. A pair of electrons is *shared* in this type of bond, for example, in a C–H bond; the two electrons (one from C and one from H) spend most of their time in the region between the nuclei (of C and of H).

________ d. These bonds are only about 5% as strong as covalent bonds and so are easily formed and broken. They are vital for holding large molecules like protein or DNA in proper configurations.

________ e. Double bonds, such as O = O (O_2), or triple bonds, such as N ≡ N (N_2).

________ f. This type of bond is most easily formed by carbon (C) because its outer orbit is half-filled.

________ g. These bonds may be polar, such as H_2O (H — OH), or nonpolar, such as O = O (O_2).

✓ **B3.** Describe a property of water based on chemical bonds.
Each water molecule can link to *(2? 3? 4?)* neighboring water molecules by

________________ bonds (*Hint:* Refer to Figure 2.7 in your text). Consequently, water

molecules exhibit the property of ________________, defined as the tendency for similar particles to stay together. Such cohesion creates a *(high? low?)* surface tension. Each time you breathe, such high surface tension of water molecules lining

air sacs (alveoli) of lungs, ________-creases the work required to breathe. As you will discover in Chapter 23, surface tension-lowering agents (or surfactants) help to

________-crease the work required for breathing.

✓ **B4.** Refer to Figure LG 2.1 and complete this exercise about covalent bonding.

a. Write atomic numbers in the parentheses ( ) under each atom. (*Hint:* Count the number of electrons or protons = atomic number.) One is done for you.

b. Write the name or symbol for each atom on the line under that atom. (*Hint:* You may identify these by atomic number in Figure 2.2 in your text.)

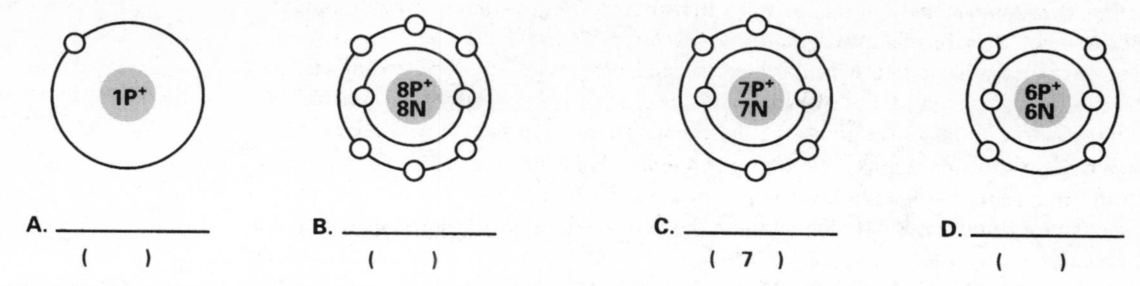

Figure LG 2.1 Atomic structure of four common atoms in the human body. Complete as directed in Checkpoint B4.

c. Based on the number of electrons that each of these atoms is missing from a complete outer electron shell, draw the number of covalent bonds that each of the atoms can be expected to form. Note that this is the valence, or combining capacity, of these atoms. One is done for you.

✓ **B5.** *Critical thinking* about covalently bonded molecules. Complete this exercise.

a. Which of the following molecules involves a triple (covalent) bond? (*Hint:* Refer to Figure 2.5 of your text.)

A. Oxygen molecule (O_2)	**B. Nitrogen molecule (N_2)**

b. In the margin (or on separate paper) draw the atomic structure of the covalently bonded molecule, ammonia (NH_3).

C. Chemical reactions (pages 34–37)

✓ **Cl.** Complete this Checkpoint about chemical reactions and energy.

a. When two reactants (such as hydrogen and oxygen) react, they form a product (such as ____________). The fact that the same number of atoms are in the reactants and the product is a statement of the law of conservation of *(mass? energy?)*.

b. The term ____________ refers to all of the chemical reactions in the body.

c. Chemical reactions involve changes in energy, which is defined as the capacity to do __________. Now identify three types of energy. The energy within chemical bonds is known as ____________ energy. Inactive or stored energy is known as *(kinetic? potential?)* energy, whereas the energy of motion is ____________ energy.

d. Chemical reactions in which more energy is released from chemical bonds than is used to break apart those bonds is called an *(endergonic? exergonic?)* reaction. Most reactions involved in breakdown of chemicals in food nutrients are *(endergonic? exergonic?)* reactions. Much of this released energy is stored in the chemical adenosine triphosphate (or)____________; this energy is stored for later use in fueling ____________ -gonic reactions, such as muscle building.

C2. Define *activation energy.*

Now write a short paragraph describing three factors facilitating collisions between atoms, increasing the likelihood that chemical reactions will take place. Which one of these three factors actually lowers the activation energy needed for a reaction to occur?

✓ **C3.** Identify the kind of chemical reaction described in each statement below.

D. Decomposition reaction	**R. Reversible reaction**
E. Exchange reaction	**S. Synthesis reaction**

_______ a. The end product can revert to the original combining molecules.

_______ b. Two or more reactants combine to form an end product; for example, many glucose molecules bond to form glycogen in an anabolic reaction.

_______ c. Such a reaction is partly synthesis and partly decomposition.

_______ d. All chemical reactions involve making or breaking of bonds. This type of reaction involves only breaking of bonds.

_______ e. This type of reaction is catabolic, as in digestion of foods, such as starch digestion to glucose.

D. Inorganic compounds and solutions (pages 37–41)

✓ **D1.** Write *I* next to phrases that describe inorganic compounds and write *O* next to phrases that describe organic compounds.

_______ a. Contain carbon and hydrogen; held together almost entirely by covalent bonds

_______ b. Tend to be very large molecules that serve as good building blocks for body structures

_______ c. The class that includes carbohydrates, proteins, fats, and nucleic acids

_______ d. The class that includes water, the most abundant compound in the body, and carbon dioxide (CO_2)

✓ **D2.** Do this exercise on roles of water in the body.

a. Water *(is? is not?)* a common solvent in the human body. This characteristic is due to two factors. One is related to the *(straight? bent?)* shape of water atoms, which provides positive regions (δ^+) present on the *(oxygen? hydrogen?)* atoms and negative regions (δ^-) present on the *(oxygen? hydrogen?)* atom. The other factor is based upon the *(polar covalent? nonpolar covalent? ionic?)* bonds of water. As a result, water will interact best with chemicals that are hydro-*(philic? phobic?)*, since those solutes also contain polar covalent or ionic bonds.

b. When added to water, salt is readily dissolved as the Na^+ ions are attracted to the *(positively charged hydrogens? negatively charged oxygens?)* and Cl^- ions are attracted to the *(positively charged hydrogens? negatively charged oxygens?)* of water.

c. Besides NaCl, name several other chemicals in the human body that are dissolved in the excellent solvent, water.

d. In organs of the digestive system, water breaks apart large food molecules via the breakdown process called ______________. In the opposite type of reaction, the hydrogen ion (H^+) and the hydroxide (_____) components of water are removed from two small molecules, which can then bond to form a larger chemical; this process is known as ______________. (See Figure LG 2.2)

e. Write two or more examples of how water functions as a lubricant in the human body.

✓ **D3.** Describe heat capacity of water in this exercise.

a. Water has a *(high? low?)* heat capacity, meaning that water can absorb much heat yet experience a *(large? small?)* increase in temperature. The reason for this phenomenon is that as water absorbs more heat, some of the energy is used to break ________________ bonds so less energy is left over to increase water temperature.

b. Since the composition of the human body is normally more than 50% water, this large amount of water *(enhances? lessens?)* the impact of high environmental temperatures. Also, evaporation of water from skin permits the body to release heat by the process of ________________.

✓ **D4.** Mixtures are combinations of *(chemically bonded? physically blended?)* elements or compounds. Name three common liquid mixtures: ________________, ________________, and ________________. Now take a closer look at mixtures in this Checkpoint.

a. An oil-based salad dressing is a *(solution? suspension?)* that is formed as the dressing is shaken. Upon sitting for a few seconds, the lighter weight oil rises to the top as the herbs settle out. Blood is another example of a *(solution? suspension?)* because blood cells are likely to settle out of blood plasma if the blood is centrifuged.

b. Coffee, tea, and tears are examples of *(solutions? suspensions?)*; water serves as the *(solute? solvent?)* in which ions and other *(solutes? solvents?)* (such as crystals of coffee or tea) are dissolved. Colloids are forms of solutions with very *(large? small?)* particles as the solutes. Plasma proteins in blood are examples of colloids.

✓ **D5.** Concentrations of solutions may be expressed in several ways. Describe them in this Checkpoint. (*Hint:* Refer to atomic masses for Na and Cl in Figure 2.2, in your text.)

a. A mole per liter (1 mol/liter) means that a number of *(milligrams? grams?)* equal to the molecular weight of the chemical should be dissolved in enough solvent to total one ________________ of solution. Because the atomic masses of Na and Cl are _______ and _______, then *(12.45? 58.44? 1000?)* g of NaCl would be needed to prepare a liter of a "one mole per liter" (1 mol/liter) solution of NaCl.

b. A one millimole per liter solution (1 mmol/liter) has *(1000 times the? 10 times the? the same? one-thousandth the?)* concentration of a one mole per liter (1 mol/liter) solution.

To prepare a liter of a 1 mmol/liter solution of NaCl, how much NaCl is needed?_____.

✓ **D6.** Fill in the blanks in this exercise about acids, bases, and salts.

a. NaOH (sodium hydroxide) is an example of a(n) ____________________, that is, a chemical that breaks apart into hydroxide ions (OH^-) (or proton acceptors) and one or more cations.

b. H_2SO_4 (hydrogen sulfate) is an example of a(n) ____________________, that is, a chemical that dissociates into hydrogen ions (H^+) (or proton __________) and one or more anions.

c. Salts, such as NaCl (sodium chloride), are chemicals that dissolve in water to form cations and ions, neither of which is ______ or ______.

D7. Draw a diagram showing the pH scale. Label the scale from 0 to 14. Indicate by arrows increasing acidity (H^+ concentration) and increasing alkalinity (OH^- concentration).

✓ **D8.** Choose the correct answers regarding pH.

a. Which pH is most acid?
A. 4 B. 7 C. 10

b. Which pH has the highest concentration of OH^- ions?
A. 4 B. 7 C. 10

c. Which solution has pH closest to neutral?
A. Gastric juice (digestive juices of the stomach)
B. Blood
C. Lemon juice
D. Milk of magnesia

d. A solution with pH 8 has 10 times *(more? fewer?)* H^+ ions than a solution with pH 7.

e. A solution with pH 5 has *(10? 20? 100?)* times *(more? fewer?)* H^+ ions than a solution with pH 7.

D9. Write a sentence or two explaining each of the following:

a. How pH is related to homeostasis

b. How buffers help to maintain homeostasis

D10. Complete this equation to show ionization of carboinc acid.

H_2CO_3 ⟶ ________________ + ________________.
(carbonic acid)

E. Organic compounds (pages 42–55)

E1. More than 50% of the human body is composed of the *(organic? inorganic?)*

compound named ________________. Organic compounds all contain carbon (C) and hydrogen (H). State two reasons why carbon is an ideal element to serve as the primary structural component for living systems.

✓ **E2.** Select the functional groups associated with carbon in organic componds that fit each description.

Amino	Carbonyl	Carboxyl	Ester	Phosphate	Sulfhydryl

a. -SH group helps to stabilize amino acids in proteins: ________________.

b. $-NH_2$ group found in amino acids: ________________.

c. -COOH group found in amino acids and fatty acids: ________________.

d. C=O found within ketones: ________________.

e. Three such groups are found in ATP: ________________.

✓ **E3.** Carbohydrates carry out important functions in your body.

a. State the principal function of carbohydrates.

b. State two secondary roles of carbohydrates.

E4. Complete these statements about carbohydrates.

a. The ratio of hydrogen to oxygen in all carbohydrates is ________:________. The general formula for carbohydrates is ____________________.

b. The chemical formula for a *hexose* is ____________________. Two common hexoses are ____________________ and ____________________.

c. One common pentose sugar found in DNA is ____________________. Its name and its formula ($C_5H_{10}O_4$) indicate that it lacks one oxygen atom from the typical pentose formula, which is ____________________.

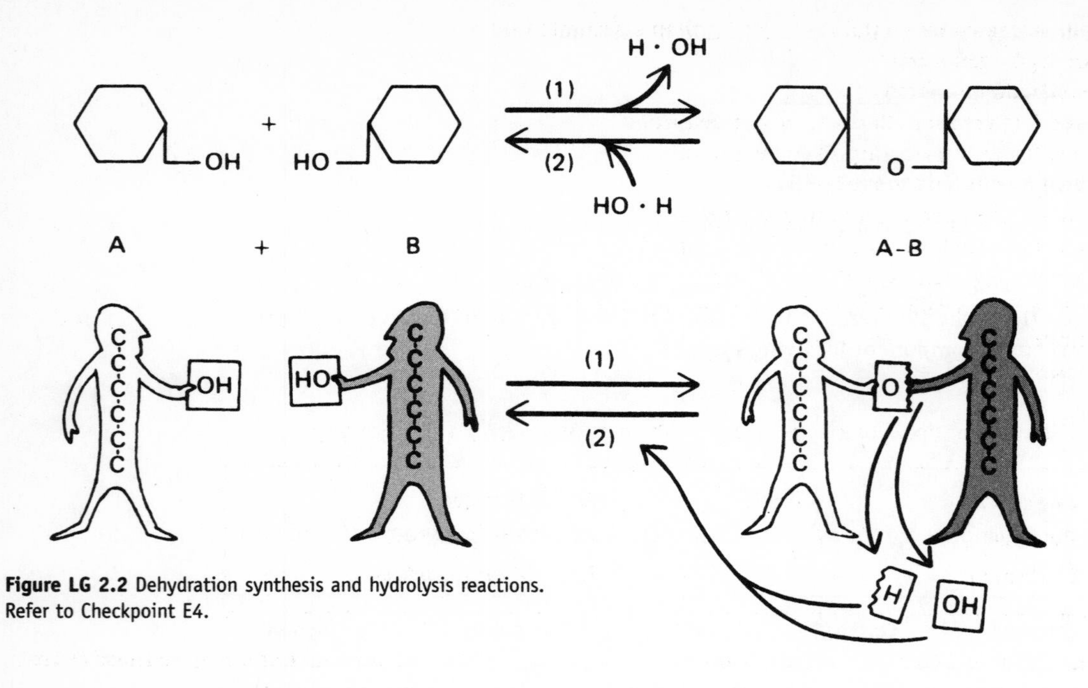

Figure LG 2.2 Dehydration synthesis and hydrolysis reactions. Refer to Checkpoint E4.

d. When two glucose molecules ($C_6H_{12}O_6$) combine, a disaccharide is formed. Its formula is ______________________, indicating that in a synthesis reaction such as this, a molecule of water is *(added? removed?)*. Refer to Figure LG 2.2. Identify which reaction *(1? 2?)* on that figure demonstrates synthesis.

e. Continued dehydration synthesis leads to enormous carbohydrates called ______________________. One such carbohydrate is ______________________.

f. When a disaccharide such as sucrose is broken, water is introduced to split the bond linking the two monosaccharides. Such a decomposition reaction is termed ______________________ which literally means "splitting using water." (See Figure 2.15 in your text.) On Figure LG 2.2, hydrolysis is shown by reaction *(1? 2?)*.

✓ **E5.** Contrast carbohydrates with lipids in this exercise. Write *C* next to any statement true of carbohydrates and write *L* next to any statement true of lipids.

________ a. These compounds are insoluble in water (hydrophobic). (*Hint:* Water alone is ineffective for washing such compounds out of clothes.)

________ b. These compounds are organic.

________ c. These substances have a hydrogen to oxygen ratio of about 2:1.

________ d. Very few oxygen atoms (compared to the numbers of carbon and hydrogen atoms) are contained in these compounds. (Look carefully at Figure 2.17 in your text.)

✓ **E6.** Circle the answer that correctly completes each statement.

a. A triglyceride consists of:
A. Three glucoses bonded together
B. Three fatty acids bonded to a glycerol

b. A typical fatty acid contains (see Figure 2.17 in your text):
A. About 3 carbons
B. About 16 carbons

c. A polyunsaturated fat contains ________ than a saturated fat.
   A. More hydrogen atoms
   B. Fewer hydrogen atoms
d. Saturated fats are more likely to be derived from:
   A. Plants, for example, corn oil or peanut oil
   B. Animals, as in beef or cheese
e. Oils that are classified as saturated fats include:
   A. Tropical oils such as coconut oil and palm oil
   B. Vegetable oils such as safflower oil, sunflower oil, and corn oil
f. A diet designed to reduce the risk for atherosclerosis is a diet low in:
   A. Saturated fats
   B. Polyunsaturated fats

✓ **E7.** Match the types of lipids listed in the box with their functions given below.

B.	**Bile salts**	**E.**	**Estrogens**
Car.	**Carotene**	**Pho.**	**Phospholipids**
Cho.	**Cholesterol**	**Pro.**	**Prostaglandins**

Car a. Present in egg yolk and carrots; leads to formation of vitamin A, a chemical necessary for good vision

Pho b. The major lipid in cell membranes

Cho c. Present in all cells, but infamous for its relationship to "hardening of the arteries"

B d. Substances that break up (emulsify) fats before their digestion

E e. Steroid sex hormones produced in large quantities by females

Pro f. Important regulatory compounds with many functions, including modifying responses to hormones and regulating body temperature

✓ **E8.** Complete this exercise about fat-soluble vitamins.

a. Name four fat-soluble vitamins (consult Table 2.7 in your text, for help):

________ ________ ________ ________

b. Because these four vitamins are absorbed into the body along with fats, any problem with fat absorption is likely to lead to:

c. The fat-soluble vitamin that helps with calcium absorption and so is necessary for normal bone growth is vitamin ________.

d. Vitamin ________ may be administered to a patient before surgery in order to prevent excessive bleeding because this vitamin helps in the formation of clotting factors.

e. Normal vision is associated with an adequate amount of vitamin ________.

f. Vitamin ________ has a variety of possible functions including the promotion of wound healing and prevention of scarring.

✓ **E9.** Match the types of proteins listed in the box with their functions given below.

Cat.	**Catalytic**	**R.**	**Regulatory**
Con.	**Contractile**	**S.**	**Structural**
I.	**Immunological**	**T.**	**Transport**

_______ a. Hemoglobin in red blood cells

_______ b. Proteins in muscle tissue, such as actin and myosin

_______ c. Keratin in skin and hair; collagen in bone

_______ d. Hormones such as insulin

_______ e. Defensive chemicals such as antibodies and interleukins

_______ f. Enzymes such as lipase, which digests lipids

✓ **E10.** Contrast proteins with the other organic compounds you have studied so far by completing this exercise.

a. Carbohydrates, lipids, and proteins all contain carbon, hydrogen, and oxygen. A fourth element, _______________, makes up a substantial portion of proteins.

b. Just as large carbohydrates are composed of repeating units (the _______________), proteins are composed of building blocks called _______________.

c. As in the synthesis of carbohydrates or fats, when two amino acids bond together, a water molecule must be *(added? removed?)*. This is another example of *(hydrolysis? dehydration?)* synthesis.

d. The product of such a reaction is called a ________-peptide. (See Figure 2.21 in your text.) When many amino acids are linked together in this way, a ________-peptide results. One or more polypeptide chains form a _______________.

e. Refer to Figure 2.20 of your text. Which part of the molecular structure varies among the 20 different kinds of amino acids to result in their unique identities? *(Amino group? Side chain = R group? Carboxyl group?)*

E11. There are _______ different kinds of amino acids forming the human body, much like a 20-letter alphabet that forms protein "words." It is the specific sequence of amino acids that determines the nature of the protein formed. In order to see the significance of this fact, do the following activity. Arrange the five letters listed below in several ways so that you form different five-letter words. Use all five letters in each word.

e i l m s

___ ___ ___ ___ ___ ___ ___ ___ ___ ___

___ ___ ___ ___ ___ ___ ___ ___ ___ ___

Note that although all of your words contain the same letters, the sequence of letters differs in each case. The resulting words have different meanings, or functions. Such is the case with proteins, too, except that protein "words" may be thousands of "letters" (amino acids) long, involving all or most of the 20-amino acid "alphabet." Proteins must be synthesized with accuracy. Think of the drastic consequences of omission or misplacement of even a single amino acid. (Think how different are *limes, miles, smile, and slime!*)

✓ **E12.** Match the description below with the four levels of protein organization in the box.

P.	**Primary structure**	**S.**	**Secondary structure**
Q.	**Quaternary structure**	**T.**	**Tertiary structure**

_______ a. Specific sequence of amino acids

_______ b. Twisted and folded arrangement (as in spirals or pleated sheets) due to hydrogen bonding

_______ c. Irregular three-dimensional shape; disulfide bonds play a role in this level of structure

_______ d. Relative arrangement of two or more polypeptide chains

✓ **E13.** Describe changes in protein structure by completing this activity.

a. Replacement of one amino acid (valine) with another (glutamate) in the protein portion of hemoglobin results in a condition known as ____________________.

b. Cooking an egg white results in destruction (or ____________________) of the protein albumin. Extreme increases in body temperature, for example, increases caused by ____________________, can denature body proteins such as those in skin or blood plasma.

✓ **E14.** Do this exercise about roles of enzymes related to chemical reactions.

a. Explain why it is necessary for chemical reactions to take place rapidly within the human body.

b. Increase in body temperature will affect rates of chemical reactions. Discuss pros and cons of increasing body temperature.

c. Explain how enzymes increase chemical reaction rates without requiring a raised body temperature.

d. Because enzymes speed up chemical reactions (yet are not altered themselves), they are known as ____________________.

e. Chemically, enzymes consist mostly of *(carbohydrates? lipids? proteins?)*. Each enzyme reacts with a specific molecule called a ____________________.

f. The substrate interacts with a specific region on the enzyme called the ____________________, producing an intermediate known as an ____________________–____________________ complex. Very quickly the *(enzyme? substrate?)* is transformed in some way (for example, is broken down or transferred to another substrate), while the (*enzyme? substrate?*) is recycled for

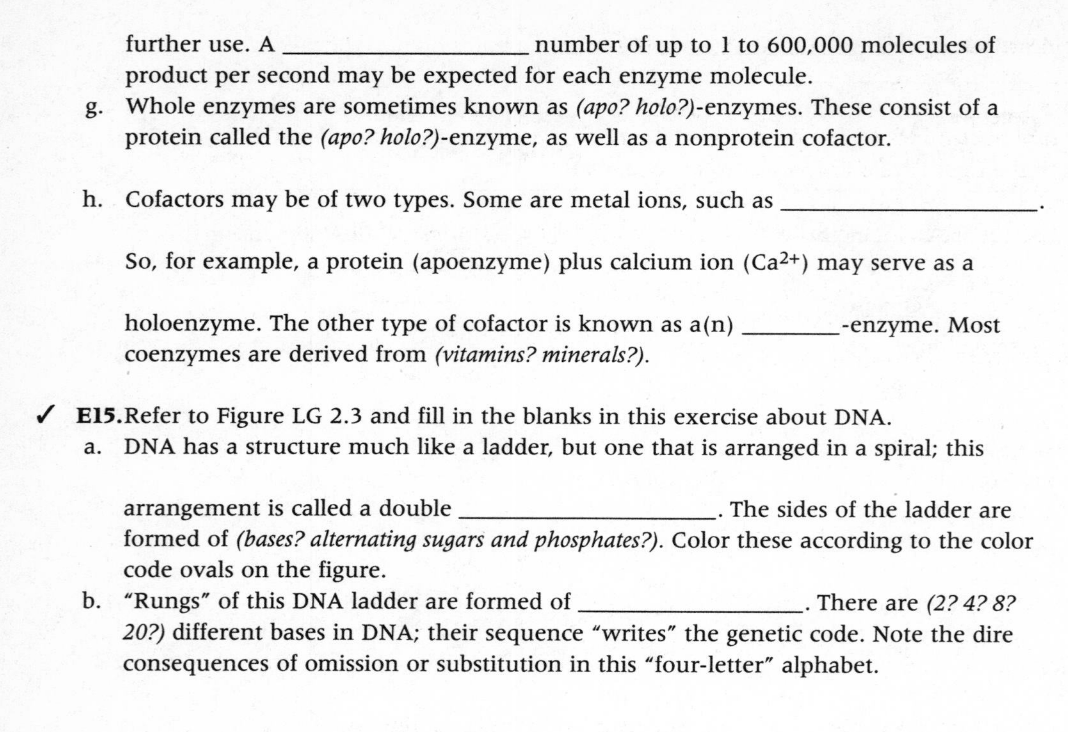

further use. A ____________________ number of up to 1 to 600,000 molecules of product per second may be expected for each enzyme molecule.

g. Whole enzymes are sometimes known as *(apo? holo?)*-enzymes. These consist of a protein called the *(apo? holo?)*-enzyme, as well as a nonprotein cofactor.

h. Cofactors may be of two types. Some are metal ions, such as ____________________.

So, for example, a protein (apoenzyme) plus calcium ion (Ca^{2+}) may serve as a

holoenzyme. The other type of cofactor is known as a(n) ________-enzyme. Most coenzymes are derived from *(vitamins? minerals?)*.

✓ **E15.** Refer to Figure LG 2.3 and fill in the blanks in this exercise about DNA.

a. DNA has a structure much like a ladder, but one that is arranged in a spiral; this

arrangement is called a double ____________________. The sides of the ladder are formed of *(bases? alternating sugars and phosphates?)*. Color these according to the color code ovals on the figure.

b. "Rungs" of this DNA ladder are formed of __________________. There are *(2? 4? 8? 20?)* different bases in DNA; their sequence "writes" the genetic code. Note the dire consequences of omission or substitution in this "four-letter" alphabet.

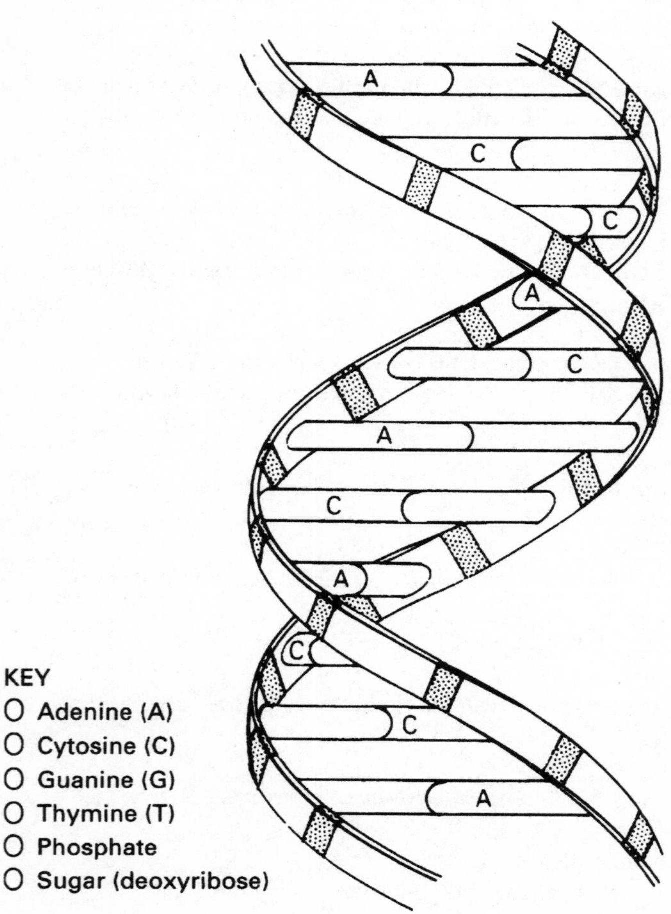

Figure LG 2.3 Structure of DNA. Complete as directed in Checkpoint E15.

c. Bases in DNA are paired ______________________ (A) to ______________________

(______) and ______________________ (C) to ______________________ (______). Label complementary nucleotides on the right side of the DNA in the figure. Then color all nucleotides in the figure according to the color code ovals.

d. A DNA ______________________ is one monomer of DNA, and it consists of three

parts: ______________________ (a sugar), a base, and a ______________________.
Circle one nucleotide in the figure.

e. Describe the structure and function of a gene.

E16. How does a strand of RNA differ from DNA?

E17. Describe the structure and significance of ATP by completing these statements.

a. ATP stands for adenosine triphosphate. Adenosine consists of a base that is a

component of DNA, that is, ______________________, along with the five-carbon sugar

named ______________________.

b. The "TP" of ATP stands for ______________________. The final *(one? two? three?)* phosphates are bonded to the molecule by high energy bonds.

c. When the terminal phosphate is broken, a great deal of energy is released as ATP is

split into ______________________ + ______________________.

d. ATP, the body's primary energy-storing molecule, is constantly reformed by the reverse of this reaction as energy is made available from foods you eat. Write this reversible reaction.

e. The anaerobic phase of cellular respiration yields ________ molecules of ATP, whereas

the aerobic phase yields ________ molecules of ATP.

ANSWERS TO SELECTED CHECKPOINTS: CHAPTER 2

A2. (a) Elements. (b) Oxygen, carbon, H, nitrogen (N). (c) Ca, Cl, Fe, K, Mg, Na, P, S. (d) Fourteen. elements found in very small amounts, together comprising only 0.2% of the total body mass; Co, Cu, I, Mn, Si, Sn, Zn.

A3. (b) N. (c) I. (d) Fe. (e) 0. (f) Ca. (g) P. (h) Cl. (i) Na. (j) F. (k) K.

A5. (a) Nucleus, electrons. (b) Protons (p^+), neutrons (n^0). (c) Negative; equal to; neutral; zero. (d) Shells; 2; 8, 18. (e) Iodine, 53, 53; 2, 8, 18, 18, 7.

A6. (a) 19; 19, 19. (b) 39.10 (rounded off to 39).

A7. (a) 20 (39–19); neutrons. *Note:* Varied masses of isotopes account for the uneven mass number, such as 39.10 for potassium. (b) 8. (c) Radiation; half-life. (d) Helpful: radioisotopes are used for medical imaging (as for PET, discussed in Chapter 1 of your text) and to treat cancer (such as cobalt treatments); harmful: radioisotopes can alter cells and lead to cancer.

A8. (a) $C_6H_{12}O_6$ (glucose), H_2O, O_2; O_2^-(superoxide). (b) $C_6H_{12}O_6$ (glucose), H_2O. (c) O_2^- (superoxide). (d) Cl^-. (e) Na^+.

B1. (a) H and He; 2. (b) 2 or 3; Na: 1, Cl: 7, C: 4; 8. (c) Valence. (d) Extra; donor; positive; cation. (e) 7, missing; accepts, anion. (f) Gained or lost; lost; anion; electrolytes (because) they can conduct an electrical charge. (g) Octet. (h) Do not, inert; He, Ne with atomic number of 10 (since Ne has 2 electrons in the inner shell and 8 electrons in the outer or valence shell).

B2. (a) I. (b) H. (c) C. (d) H. (e-g) C.

B3. 4, hydrogen; cohesion; high; in; de.

B4. (a, b: See Figure LG 2.1A) (c)

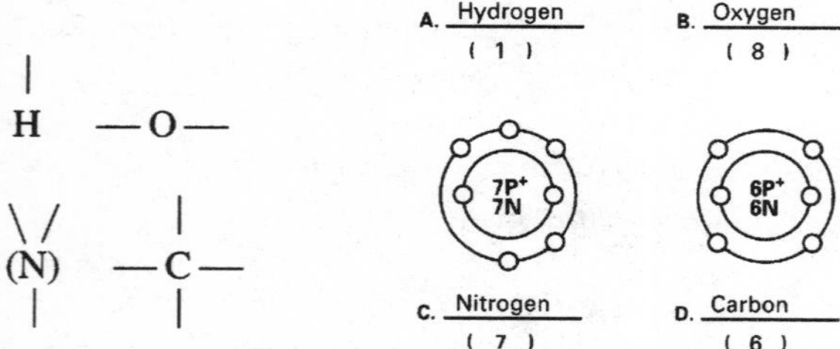

Figure LG 2.1A Atomic structure of four common atoms in the human body.

B5. (a) B. (b) See Figure at right: Atomic structure of ammonia (NH_3).

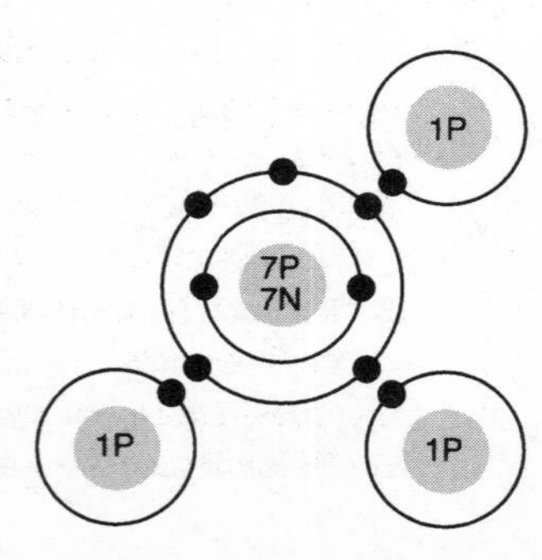

C1. (a) Water; mass. (b) Metabolism. (c) Work; chemical; potential, kinetic. (d) Exergonic; exergonic; ATP; ender-.

C3. (a) R. (b) S. (c) E. (d) D. (e) D.

D1. (a–c) O. (d) I.

D2. (a) Is; bent, hydrogen, oxygen; polar covalent; philic. (b) Negatively charged oxygen, positively charged hydrogens. (c) Some of the O_2 and CO_2 gas molecules in blood, wastes such as urea, and small sugars such as glucose. (d) Hydrolysis; OH, synthesis. (e) Water in mucus permits foods to slide along the digestive tract and wastes to be flushed out in the urinary tract; water in serous fluid and synovial fluids decreases friction.

D3. (a) High, small; hydrogen. (b) Lessens; sweating or perspiration.

D4. Physically blended; solutions, colloids, and suspensions. (a) Suspension; suspension. (b) Solutions, solvent, solutes; large.

D5. (a) Grams, liter; 22.99, 35.45, 58.44. (b) One thousandth the; 58.44 mg = 0.05844 g (or 0.058 g).

D6. (a) Base. (b) Acid, donors. (c) H^+ or OH^-.

D8. (a) A. (b) C. (c) B. (d) Fewer. (e) 100, more.

D10. HCO_3^- (bicarbonate) + H^+ (hydrogen ion).

E2. (a) Sulfhydryl. (b) Amino. (c) Carboxyl. (d) Carbonyl. (e) Phosphate.

E3. (a) Provide a readily available source of energy. (b) Converted to other substances; serve as food reserves.

E4. (a) 2:1; $(CH_2O)n$. (b) $C_6H_{12}O_6$ glucose, fructose, galactose. (c) Deoxyribose; $C_5H_{10}O_5$. (d) $C_{12}H_{22}O_{11}$, removed; 1. (e) Polysaccharides; starch, glycogen, cellulose. (f) Hydrolysis; 2.

E5. (a) L. (b) C, L. (c) C. (d) L.

E6. (a) B. (b) B. (c) B. (d) B. (e) A. (f) A.

E7. (a) Car. (b) Pho. (c) Cho. (d) B. (e) E. (f) Pro.

E8. (a) A, D, E, K. (b) Symptoms of deficiencies of fat-soluble vitamins. (c) D. (d) K. (e) A. (f) E.

E9. (a) T. (b) Con. (c) S. (d) R. (e) I. (f) Cat.

E10. (a) Nitrogen. (b) Monosaccharides, amino acids. (c) Removed; dehydration. (d) Di; poly; protein. (e) Side chain = R group.

E12. (a) P. (b) S. (c) T. (d) Q.

E13. (a) Sickle cell anemia. (b) Denaturation; burns or fever (hyperthermia).

E14. (a) Thousands of different types of chemical reactions are necessary daily to maintain homeostasis. (b) Chemical reactions occur more rapidly at higher temperature (one reason for increased catabolism and weight loss when you have a fever). However, very high temperatures will destroy body tissues. (c) Enzymes orient molecules so that they are more likely to react, lower activation energy needed for reactions, and increase frequency of collisions. (d) Catalysts. (e) Proteins; substrate. (f) Active site, enzyme-substrate; substrate, enzyme; turnover. (g) Holo; apo. (h) Magnesium (Mg^{2+}), zinc (Zn^{+}), or calcium (Ca^{2+}); co; vitamins.

E15. (a) Helix; alternating sugars and phosphates; see Figure LG 2.3A. (b) Bases; 4. (c) Adenine (A), thymine (T), cytosine (C), guanine (G); see Figure LG 2.3A. (d) Nucleotide, deoxyribose, phosphate; see Figure LG 2.3A. (e) Segment of DNA that determines one trait or regulates one activity.

E17. (a) Adenine, ribose. (b) Triphosphate; two. (c) ADP + phosphate (P). (d) ATP $\rightleftharpoons$ ADP + P + energy. (e) 2, 36–38.

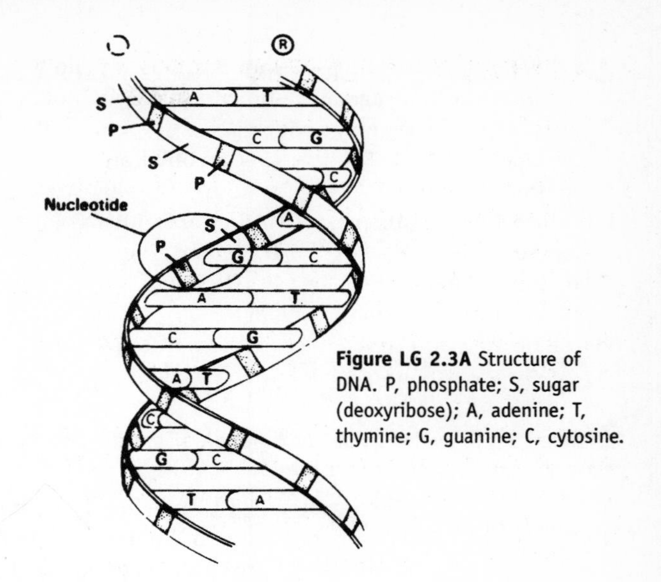

Figure LG 2.3A Structure of DNA. P, phosphate; S, sugar (deoxyribose); A, adenine; T, thymine; G, guanine; C, cytosine.

MORE CRITICAL THINKING: CHAPTER 2

1. Discuss the roles of the following in human anabolism and catabolism: (a) exergonic and endergonic reactions; (b) synthesis, decomposition, exchange, and reversible reactions.
2. How does an enzyme enhance the rate of chemical reactions in the human body? Exactly how does this "biological catalyst" do its job? Is the enzyme permanently altered during the chemical reaction?
3. Contrast the roles of carbohydrates and proteins in the human body.

MASTERY TEST: CHAPTER 2

Questions 1–12: Circle the letter preceding the one best answer to each question.

1. All of the following statements are true *except*:
   A. A reaction in which two amino acids join to form a dipeptide is called a dehydration synthesis reaction.
   B. A reaction in which a disaccharide is digested to form two monosaccharides is known as a hydrolysis reaction.
   C. About 65 to 75% of living matter consists of organic matter.
   D. Strong acids ionize more easily than weak acids do.
2. Choose the one *true* statement.
   A. A pH of 7.5 is more acidic than a pH of 6.5.
   B. Anabolism consists of a variety of decomposition reactions.
   C. An atom such as chlorine (Cl), with seven electrons in its outer orbit, is likely to be an electron donor (rather than electron receptor) in ionic bond formation.
   D. Polyunsaturated fats are more likely to reduce cholesterol level than saturated fats are.
3. In the formation of the ionically bonded salt NaCl, Na^+ has:
   A. Gained an electron from Cl^-
   B. Lost an electron to Cl^-
   C. Shared an electron with Cl
   D. Formed an isotope of Na^+
4. Over 99% of living cells consist of just six elements. Choose the element that is NOT one of these six.
   A. Calcium
   B. Hydrogen
   C. Carbon
   D. Iodine
   E. Nitrogen
   F. Oxygen
   G. Phosphorus
5. Which of the following describes the structure of a nucleotide?
   A. Base-base
   B. Phosphate-sugar-base
   C. Enzyme
   D. Dipeptide
   E. Adenine-ribose
6. Which of the following groups of chemicals includes only polysaccharides?
   A. Glycogen, starch
   B. Glycogen, glucose, galactose
   C. Glucose, fructose
   D. RNA, DNA
   E. Sucrose, polypeptide
7. $C_6H_{12}O_6$ is most likely the chemical formula for:
   A. Amino acid
   B. Fatty acid
   C. Hexose
   D. Polysaccharide
   E. Ribose
8. All of the following answers consist of correctly paired terms and descriptions related to enzymes *except:*
   A. Active site—a place on an enzyme that fits the substrate
   B. Substrate—molecule(s) upon which the enzyme acts
   C. -ase—ending of most enzyme names
   D. Holoenzyme—another name for a cofactor

9. Which of the following substances are used mainly for structure and regulatory functions and are not normally used as energy sources?
   A. Lipids
   B. Proteins
   C. Carbohydrates
10. Which is a component of DNA but not of RNA?
   A. Adenine
   B. Phosphate
   C. Guanine
   D. Ribose
   E. Thymine
11. All of the following answers consist of pairs of proteins and their correct functions except:
   A. Contractile—actin and myosin
   B. Immunological—collagen of connective tissue
   C. Catalytic—enzymes
   D. Transport—hemoglobin
12. All of the following answers consist of pairs of elements and their correct chemical symbols except:
   A. Nitrogen (N)
   B. Sodium (Na)
   C. Carbon (C)
   D. Calcium (Ca)
   E. Potassium (P)

Questions 13–20: Circle T (true) or F (false). If the statement is false, change the underlined word or phrase so that the statement is correct.

T F 13. Oxygen, water, NaCl, and glucose are inorganic compounds.
T F 14. There are four different kinds of amino acids found in human proteins.
T F 15. The number of protons always equals the number of neutrons in an atom.
T F 16. K^+ and Cl^- are both cations.
T F 17. Oxygen can form three bonds since it requires three electrons to fill its outer electron shell.
T F 18. Carbohydrates constitute about 18 to 25% of the total body weight.
T F 19. ATP contains more energy than ADP.
T F 20. Prostaglandins, ATP, steroids, and fats are all classified as lipids.

Questions 21–25: Fill-ins. Identify each organic compound below. Write the names of the compounds on the lines provided.

_______________ 21.

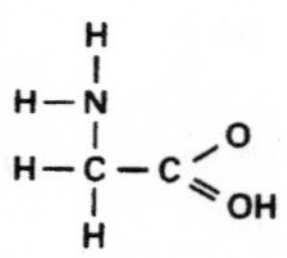

_______________ 22.

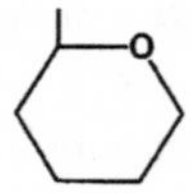

_______________ 23.

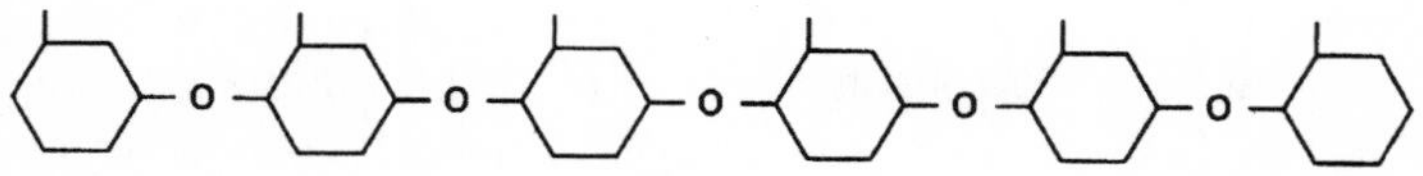

_______________ 24.

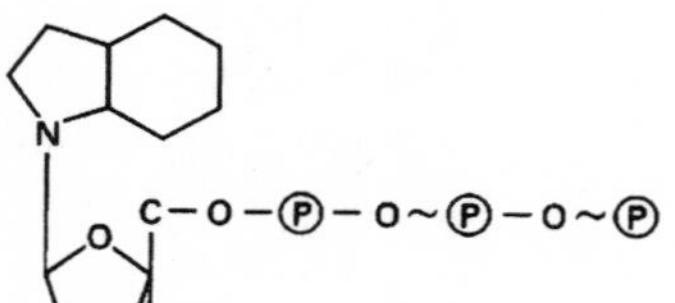

_______________ 25.

```
     H         H   H   H   H   H   H   H   H   H   H   H   H   H   H   H
     |         |   |   |   |   |   |   |   |   |   |   |   |   |   |   |
 H - C - O - C - C - C - C - C - C - C - C - C - C - C - C - C - C - C - C - H
     |       ||    |   |   |   |   |   |   |   |   |   |   |   |   |   |   |
     |       O     H   H   H   H   H   H   H   H   H   H   H   H   H   H   H
     |
     |         H   H   H   H   H   H   H   H   H   H   H   H   H   H   H
     |         |   |   |   |   |   |   |   |   |   |   |   |   |   |   |
 H - C - O - C - C - C - C - C - C - C - C - C - C - C - C - C - C - C - C - H
     |       ||    |   |   |   |   |   |   |   |   |   |   |   |   |   |   |
     |       O     H   H   H   H   H   H   H   H   H   H   H   H   H   H   H
     |
     |         H   H   H   H   H   H   H   H   H   H   H   H   H   H   H
     |         |   |   |   |   |   |   |   |   |   |   |   |   |   |   |
 H - C - O - C - C - C - C - C - C - C = C - C - C - C = C - C - C - C - C - H
     |       ||    |   |   |   |   |           |   |           |   |   |   |
     H       O     H   H   H   H   H           H   H           H   H   H   H
```

ANSWERS TO MASTERY TEST: CHAPTER 2

Multiple Choice

1. C
2. D
3. B
4. D
5. B
6. A
7. C
8. D
9. B
10. E
11. B
12. E

True–False

13. F. Water and NaCl (Oxygen is not a compound; glucose is organic.)
14. F. 20
15. F. Electrons
16. F. K^+ is a cation.
17. F. Two bonds because it requires two electrons
18. F. 2–3%
19. T.
20. F. Prostaglandins, steroids, and fats

Fill-ins

21. Amino acid
22. Monosaccharide or hexose or glucose
23. Polysaccharide or starch or glycogen
24. Adenosine triphosphate (ATP)
25. Fat or lipid or triglyceride (unsaturated)

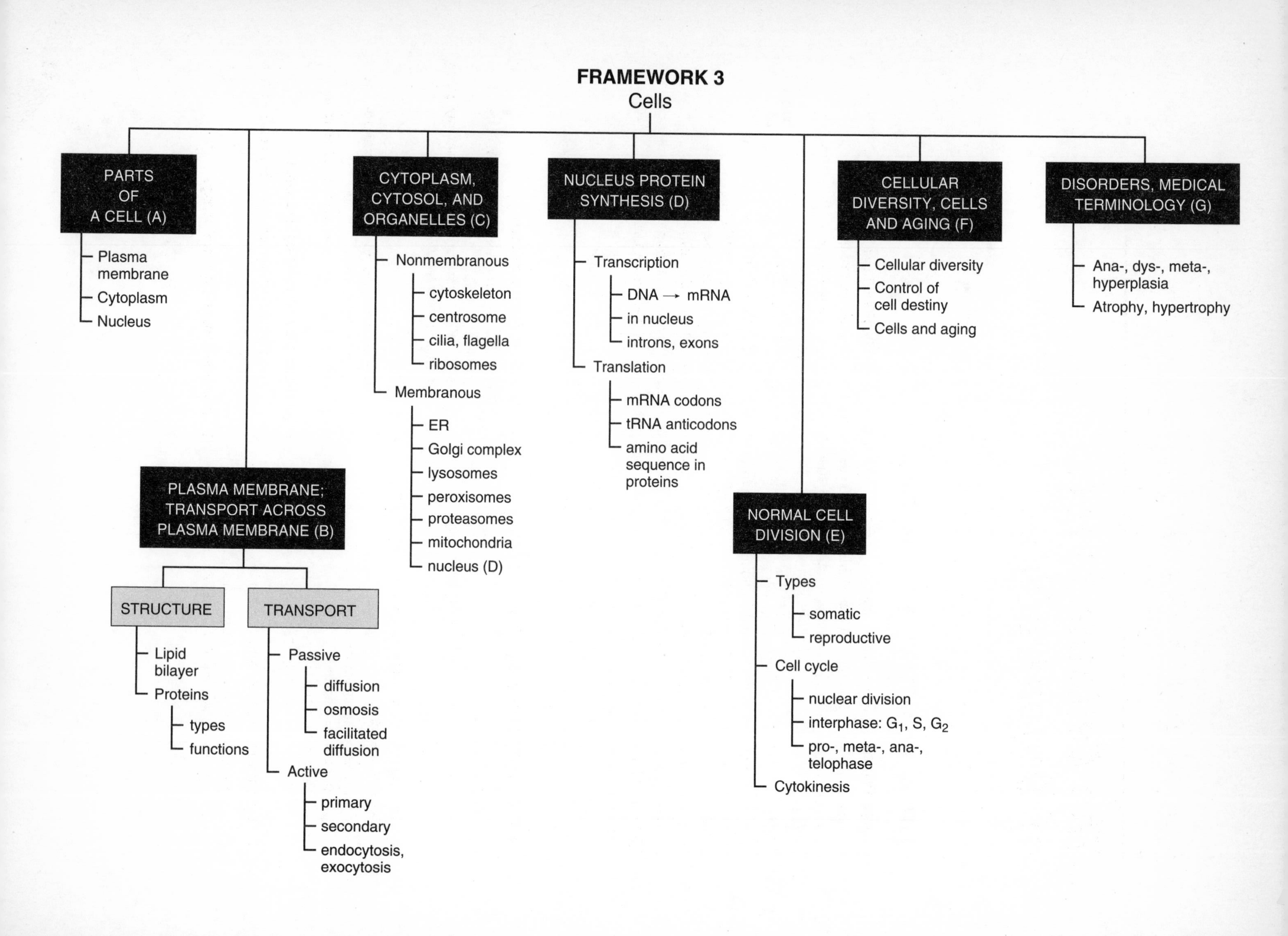
FRAMEWORK 3
Cells
PARTS OF A CELL (A)
Plasma membrane
Cytoplasm
Nucleus
PLASMA MEMBRANE; TRANSPORT ACROSS PLASMA MEMBRANE (B)
STRUCTURE
Lipid bilayer
Proteins
types
functions
TRANSPORT
Passive
diffusion
osmosis
facilitated diffusion
Active
primary
secondary
endocytosis, exocytosis
CYTOPLASM, CYTOSOL, AND ORGANELLES (C)
Nonmembranous
cytoskeleton
centrosome
cilia, flagella
ribosomes
Membranous
ER
Golgi complex
lysosomes
peroxisomes
proteasomes
mitochondria
nucleus (D)
NUCLEUS PROTEIN SYNTHESIS (D)
Transcription
DNA → mRNA
in nucleus
introns, exons
Translation
mRNA codons
tRNA anticodons
amino acid sequence in proteins
NORMAL CELL DIVISION (E)
Types
somatic
reproductive
Cell cycle
nuclear division
interphase: G_1, S, G_2
pro-, meta-, ana-, telophase
Cytokinesis
CELLULAR DIVERSITY, CELLS AND AGING (F)
Cellular diversity
Control of cell destiny
Cells and aging
DISORDERS, MEDICAL TERMINOLOGY (G)
Ana-, dys-, meta-, hyperplasia
Atrophy, hypertrophy

C H A P T E R **3**

The Cellular Level of Organization

Cells are the basic structural units of the human body, much as buildings comprise the cities of the world. Every cell possesses certain characteristics similar to those in all other cells of the body. For example, membranes surround and compartmentalize cells, just as walls support virtually all buildings. Yet individual cell types exhibit variations that make them uniquely designed to meet specific body needs. In an architectural tour, the structural components of a cathedral or mansion, lighthouse or fort, can be recognized as specific for activities of each type of building. So, too, the number and type of organelles differ in muscle or nerve, cartilage or bone cells. Cells carry out significant functions such as movement of substances into, out of, and throughout the cell; synthesis of the chemicals that cells need (like proteins, lipids, and carbohydrates); and cell division for growth, maintenance, repair, and formation of offspring. Just as buildings change with time and destructive forces, cells exhibit aging changes. In fact, sometimes their existence is haphazard and harmful, as in cancer.

As you begin your study of cells, carefully examine the Chapter 3 Topic Outline and Objectives and note relationships among concepts and key terms in the Framework.

TOPIC OUTLINE AND OBJECTIVES

A. Parts of a cell

1. Name and describe the three main parts of a cell.

B. The plasma membrane; transport across the plasma membrane

1. Describe the structure and functions of the plasma membrane.
2. Describe the processes that transport substances across the plasma membrane.

C. Cytoplasm, cytosol, and organelles

1. Describe the structure and function of cytoplasm, cytosol, and organelles.

D. Nucleus; protein synthesis

1. Describe the structure and function of the nucleus.
2. Describe the sequence of events that take place during protein synthesis.

E. Cell division

1. Discuss the stages, events, and significance of somatic cell division.
2. Discuss the signals that induce cell division.

F. Cellular diversity; aging and cells

1. Describe the cellular changes that occur with aging.

G. Disorders, medical terminology

WORDBYTES

Now become familiar with the language of this chapter by studying each wordbyte, its meaning, and an example of its use within a term. After you study the entire list, self-check your understanding by writing the meaning of each wordbyte on the line. As you continue through the *Learning Guide,* identify (and fill in) additional terms that contain the same wordbyte.

Wordbyte	Self-check	Meaning	Example(s)
a-	______________	without	*a*trophy
auto-	______________	self	*auto*phagy
chromo-	______________	colored	*chromo*some
cyt-	______________	cell	pino*cyt*osis
dys-	______________	abnormal	*dys*plasia
-elle	______________	small	organ*elle*
ger-	______________	old	*ger*iatrics
homo-	______________	same	*homo*logous
hydro-	______________	water	*hydro*static
hyper-	______________	above	*hyper*tonic
hypo-	______________	below	*hypo*tonic
-iatrics	______________	medicine	ger*iatrics*
iso-	______________	equal	*iso*tonic
lyso-	______________	dissolving	*lyso*some
meta-	______________	beyond	*meta*stasis
micro-	______________	small	*micro*villi
neo-	______________	new	*neo*plasm
-oma	______________	tumor	carcin*oma*
phag-	______________	eating	*phag*osome
phil-	______________	love	hydro*philic*
phob-	______________	fearing	hydro*phobic*
pino-	______________	drinking	*pino*cytosis
-plasia, -plasm-	______________	growth	dys*plasia*
pseudo-	______________	false	*pseudo*podia
reticulum	______________	network	endoplasmic *reticulum*
-some	______________	body	lyso*some*

Wordbyte	Self-check	Meaning	Example(s)
-stasis, -static	______________	stand, stay	meta*stasis*
-tonic	______________	pressure	hyper*tonic*
-trophy	______________	nourish	hyper*trophy*
villi-	______________	tufts of hair	micro*villi*

CHECKPOINTS

A. Parts of a cell (pages 60–61)

A1. Write principal functions next to these parts of a cell:

a. Plasma membrane (two functions)

b. Cytosol

c. Organelles

d. Nucleus

A2. Contrast terms in each pair:

a. *Cytoplasm* with *cytosol.*

b. *Chromosome* with *DNA*

B. The plasma membrane: transport across the plasma membrane (pages 61–75)

✓ **B1.** Refer to Figure LG 3.1 to complete the following exercise about the fluid mosaic model of the plasma membrane. First color each part of the membrane listed below (a-e) with a different color; then use the same colors to fill in the color code ovals next to the letters below. Next label these parts on the figure. Now use the lines provided next to the name of each membrane part to write one or more functions for each part.

○ a. Phospholipid polar "head" ______________

○ b. Phospholipid nonpolar "tail" ______________

○ c. Integral protein ______________

○ d. Glycoprotein ______________

○ e. Cholesterol ______________

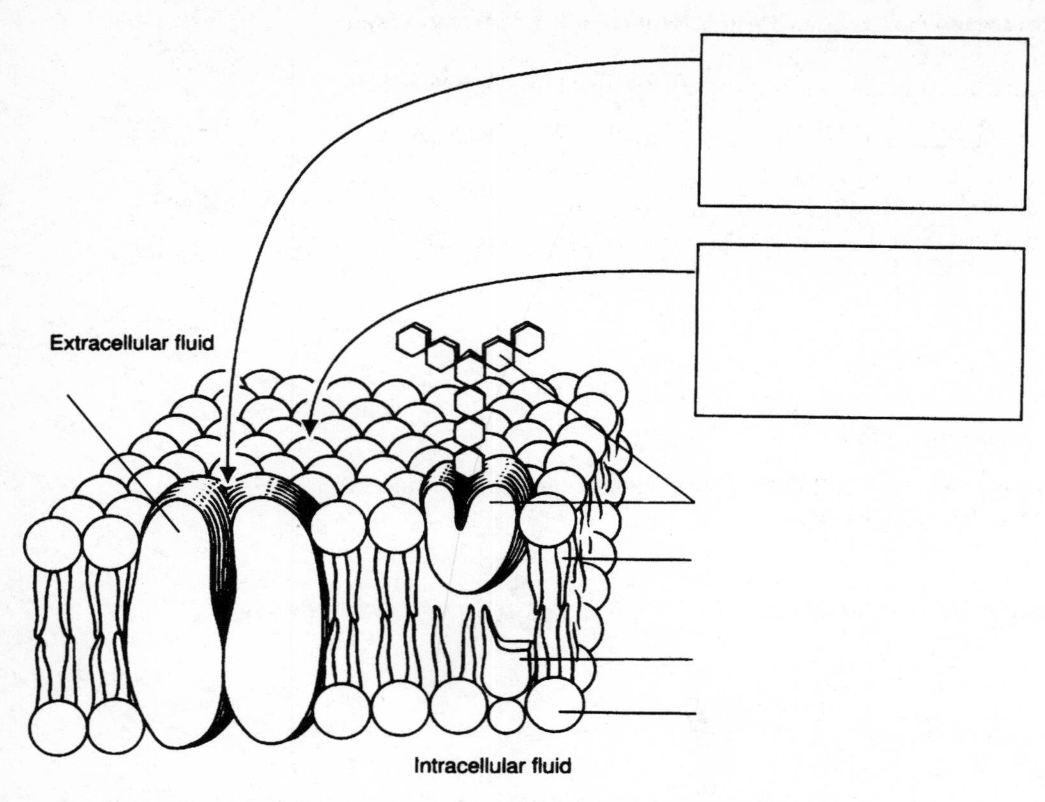

Figure LG 3.1 Plasma membrane, according to the fluid mosaic model. Color as directed in Checkpoint B1. Fill in answers in boxes according to Checkpoint B4.

✓ **B2.** *For extra review.* Complete this exercise about the plasma membrane.

a. Circle the two main chemical components of plasma membranes:
Carbohydrates Lipids Proteins

b. About 75% of the lipids are *(cholesterol? glycolipids? phospholipids?)*. Because phospholipids have polar and nonpolar parts, they are said to be ____________________. The polar part of a phospholipid molecule faces toward the *(interior? surface?)* of the membrane. The polar parts are hydro-*(philic? phobic?)*, meaning water-*(loving? fearing?)*.

c. About 20% of membrane lipids are *(cholesterol? glycolipids? phospholipids?)*. Glycolipids make up about _______% of membrane lipids, and appear only in the layer that faces *(extracellular? intracellular?)* fluid.

d. *(Lipid? Protein?)* exhibit more variety in structure and functions within different types of membranes. Which type of membrane protein extends all the way across the membrane so is likely to be amphipathic? *(Integral? Peripheral?)* Identify such ion channel proteins containing pores in Figure LG 3.1. Which type of proteins are present on surfaces or membranes? *(Integral? Peripheral?)*

e. Some integral proteins serve as receptors for chemicals such as ____________________ or ____________________. Other membrane proteins known as ____________________ form catalysts for chemical reactions. The fact that your body can accept only certain types of blood\transfusions is related to proteins called cell-identity ____________________. And finally, proteins called ____________________ help cells adhere to each other.

f. Glycoproteins are proteins associated with *(carbohydrates? lipids?)*. What forms *glycocalyx?* ______________________ List several functions of glycocalyx.

B3. *Critical thinking.* Discuss membrane fluidity in this exercise.

a. What structural features of cellular membranes account for fluidity?

b. What advantages are provided by membrane fluidity?

c. What factor accounts for the fact that the two halves of the bilayer seldom "flip-flop"?

d. What advantages are provided by the lack of "flip-flopping" of the two bilayers?

e. How does excessive cholesterol in the body lead to "hardening of the arteries"?

✓ **B4.** Complete this activity about selective permeability of membranes.

a. Plasma membranes are *(quite? not very?)* permeable to nonpolar chemicals such as steroids, O_2 and CO_2 because these molecules dissolve readily in the *(protein? phospholipid?)* portion of the membrane.

b. Charged chemicals (such as Na^+ K^+, or Ca^{2+}) and most polar chemicals pass through the *(protein? phospholipid?)* portions of the membrane. Two such proteins are ______________________ and ______________________.

c. Refer again to Figure LG 3.1. Write the names of the following chemicals in the correct boxes on the figure to differentiate chemicals that pass through the phospholipid bilayer from those that pass through protein channels:
(*Hint:* One of these types of chemicals will not normally pass across either part of healthy plasma membranes. Consult p. 69 of the text for information about transport of Ca^{2+}, Cl^-, vitamins A and D, and ammonia [NH_3].)

Ca^{++} Cl^- K^+ Na^+ H_2O protein

O_2 CO_2 vitamins A and D ammonia (NH3)

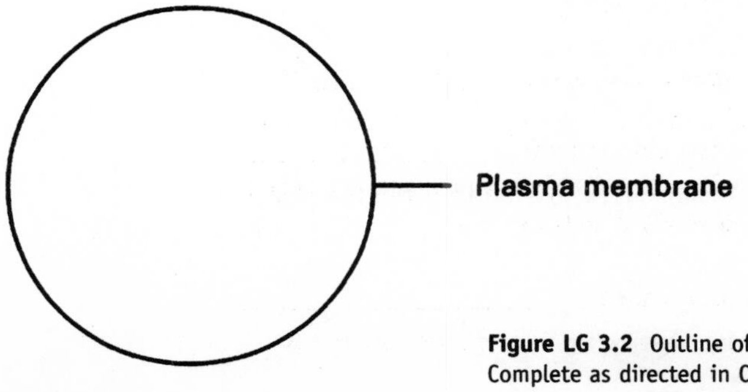

Plasma membrane

Figure LG 3.2 Outline of a cell.
Complete as directed in Checkpoint B5.

✓ **B5.** Complete this Checkpoint about how differences (or gradients) are maintained across plasma membranes.

a. Refer to Figure LG 3.2. Write inside the plasma membrane the symbols for the major chemicals concentrated in intracellular fluid (ICF). Write outside the plasma membrane the symbols for major ions or electrolytes concentrated in extracellular fluid (ECF). Use the following symbols. (*Hint:* proteins are made inside of cells and cannot easily pass through the plasma membrane. It may help to refer to Figure 27.6, page 997 in your text.)

K^+ Na^+ $Protein^-$ O_2 CO_2

b. Chemicals tend to move from where they are more concentrated to where they are less concentrated, in other words, *(up? down?)* the concentration gradient.

c. The plasma membrane normally exhibits a difference between the number of positively and negatively charged ions on either side of the membrane. This is known as a(n) *(chemical? electrical?)* gradient or a membrane ________________.
Which side of the membrane normally has a more negative charge *(Inside? Outside?)*

d. Combined, these two gradients across plasma membranes are known as the ________________ gradient.

B6. *Critical thinking.* Explain why substances must move across your plasma membranes in order for you to maintain homeostasis and survive. (What types of chemicals must be allowed into cells or kept out of cells?)

✓ **B7.** Contrast different categories of transport by circling the correct answers next to each description below.

a. A chemical is moved "down" its concentration gradient, a process that requires only the kinetic (or movement-related) energy of the chemicals. *(Active? Passive?)*

b. A chemical is moved "uphill," against its concentration gradient, a process that requires cellular energy. *(Active? Passive?)*

✓ **B8.** *Clinical correlation.* Circle which person in each pair of hospitalized clients is likely to have a *lower* rate of diffusion across plasma membranes, based upon different factors that influence diffusion rate.

a. Factor: surface area for diffusion of oxygen from lung alveoli into blood vessels there.
A. Mr. O'Loughlin: has had two-thirds of a lung removed due to lung cancer.
B. Ms. Yesberger: about the same size as Mr. O'Loughlin, but has both lungs intact.

b. Factor: thickness of membrane in lungs for diffusion of oxygen from lung alveoli into blood vessels there.
A. Mr. Claeys: has lungs free of infection.
B. Ms. Clemen: due to pneumonia, respiratory membranes of lungs are thickened by fluid.

c. Factor: concentration gradient of oxygen from blood into brain cells.
A. Mrs. Thielens has a normal level of oxygen (Po_2 = 96 mm Hg) in her blood.
B. Ms. Taylor has a low level of oxygen (Po_2 = 66 mm Hg) in her blood since she has a serious case of the flu.

d. Factor: body temperature affecting rate of diffusion of oxygen and nutrients from blood into muscle cells of the right thigh.
A. Six-year-old Amy: 98.6°F
B. Six-year-old Paula: 100.6°F

e. *For extra review.* Of the four factors listed above (a–d), three are *directly* related to diffusion rate. Which one is *inversely* related to diffusion rate? ______

✓ **B9.** Select from the following list of terms to identify passive transport processes described below.

FD. Facilitated diffusion	**O. Osmosis**	**D. Diffusion**

_______ a. Net movement of any substance (such as cocoa powder in hot milk) from region of higher concentration to region of lower concentration; membrane not required

_______ b. Same as (a) except movement across a semipermeable membrane with help of a transporter; ATP not required

_______ c. Net movement of water from region of high water concentration (such as 2% NaCl) to region of low water concentration (such as 10% NaCl) across semipermeable membrane; important in maintenance of normal cell size and shape

✓ **B10.** Complete the following exercise about osmosis in blood.

a. Human red blood cells (RBCs) contain intracellular fluid that is osmotically similar to _______ NaCl, which is known as normal saline.
A. 2.0% B. 0.90% C. 0% (pure water)

b. A solution that is hypertonic to RBCs contains *(more? fewer?)* solute particles and *(more? fewer?)* water molecules than blood.

c. Which of these solutions is hypertonic to RBCs?
A. 2.0% NaCl B. 0.90% NaCl

d. If RBCs are surrounded by hypertonic solution, water will tend to move *(into? out of?)* them, so they will *(crenate? hemolyze?)*.

e. A solution that is _______________ -tonic to RBCs will maintain the shape and size of the RBC. An example of such a solution is _______________.

f. Which solution will cause RBCs to hemolyze?
A. 2.0% NaCl B. 0.90% NaCl C. Pure water

g. Which solution has the highest osmotic pressure?
A. 2.0% NaCl B. 0.90% NaCl C. Pure water

✓ **B11.** Check your understanding of facilitated transport in this Checkpoint.

a. List two categories of chemicals that undergo facilitated transport.

b. This process *(does? does not?)* involve a transporter protein. This chemical *(moves across the membrane? undergoes a change in structure?)*, which allows the transported solute to reach the opposite side of the membrane.

c. Name the transporter protein for glucose: _______________. Name two other sugars transported by facilitated transport: _______________ and _______________.

d. The number of transporter molecules in a plasma membrane places an upper limit, or "transport _______________," on the rate at which this type of transport can occur. The hormone named _______________ promotes the insertion of many copies of a glucose transporter protein into plasma membranes of some cells, thereby permitting transport of glucose into those cells.

e. Once transported to the inside of the cell, the glucose is attached to a _______________ group with the help of an enzyme called a _______________. As a result, the intracellular concentration of "plain" glucose stays *(low? high?)*, a factor that favors additional transport of glucose into cells.

✓ **B12.** Describe active transport in this exercise.

a. Active transport may be either primary or ________________. The sodium pump, also known as the "_______ /_______ pump," is the most abundant form of *(primary? secondary?)* active transport. This pump requires energy derived from splitting ATP with an enzyme known as ________-ase. A typical cell may use up to ______% of its ATP to fuel primary active transport. When cells are energy-deprived, this pump is likely to stop working. As a result, cells ________________________________.

b. The sodium pump requires a transporter protein that changes shape when a ________________ group is added to it or removed from it. The effect of the pump is to concentrate sodium (Na^+) *(inside? outside?)* of the cell and to concentrate potassium (K^+) *(inside? outside?)* of the cell. (Refer to Figure 27.6.)

c. For every three Na^+ ions pumped out of the cell, *(two? three? six?)* K^+ ions are returned to the cell. This factor partially accounts for the negative electrical charge *(inside? outside?)* of the cell. (See Checkpoint B5c, page LG 46.) Concentration of these ions in ICF and ECF also contributes to ________________ pressure of these fluids.

d. Secondary active transport derives its energy (*directly from ATP? indirectly from ATP by utilizing concentration gradients [such as that for Na] developed by primary active transport?*). List four or more chemicals transported by secondary active transport.

e. Active transport utilizing a transporter that moves substances in opposite directions involves *(antiporters? symporters?)*.

f. Digitalis (or digoxin or Lanoxin) is a medication used for a failing heart. This chemical works by *(increasing? decreasing?)* the action of the sodium pump, which leads to a *(larger? smaller?)* concentration gradient of Na^+. As a result, Na^+/Ca^{2+} antiporters *(speed up? slow down?)*. Since more Ca^{2+} remains in heart cells, contractions of the heart are *(strengthened? weakened?)*.

B13. Endocytosis is a form of transport in which specific chemicals are transported *(into? out of?)* cells. Outline the main steps of receptor-mediated endocytosis, exemplified by HIV infection. Be sure to include these terms: *ligand, receptor (such as a CD4 receptor), vesicles and endosomes, lysosome, and HIV virus.*

B14. Contrast these types of active transport:

a. *Phagocytosis* and *pinocytosis*

b. *Exocytosis* and *transcytosis*

C. Cytoplasm, cytosol, and organelles (pages 75–82)

✓ **C1.** Cytosol is the *(fluid? solid?)* part of cytoplasm. Cytosol consists of about _______ to _______% water. Name six or more types of dissolved or suspended components of cytosol.

C2. Check your understanding of organelles in this Checkpoint.

a. Define *organelles.*

b. Of what advantage is compartmentalization offered by organelles?

✓ **C3.** Discuss the *cytoskeleton* in this activity.

a. State two categories of functions of the cytoskeleton.

b. Describe three types of movements provided by the cytoskeleton.

c. Arrange these components of the cytoskeleton according to thickness, from thickest to thinnest.

___ ___ ___ If. Intermediate filaments Mf. Microfilaments Mt. Microtubules

✓ **C4.** Match the nonmembranous organelles in the box with the best descriptions of their functions. Use each answer once.

Cen.	**Centrosomes**	**If.**	**Intermediate filaments**	**R.**	**Ribosomes**
Cil.	**Cilia**	**Mf.**	**Microfilaments**		
F.	**Flagella**	**Mt.**	**Microtubules**		

_______ a. Site of assembly of microtubules and organizing center for the mitotic spindle; consists of a pair of centrioles.

_______ b. Long, hairlike structures found only on sperm; like cilia, each is anchored to a basal body.

_______ c. Short, hairlike structures that move particles over cell surfaces, such as cells lining the uterine (Fallopian) tubes.

_______ d. Sites of protein synthesis; found free in the cytoplasm or bound to membranes (such as endoplasmic reticulum). Each is composed of a large and a small subunit.

_______ e. Composed of the protein tubulin; their movements are powered by motor proteins known as kinesins and dyneins.

_______ f. Composed of several kinds of proteins; are quite strong so help cells withstand mechanical stress.

_______ g. Form microvilli; assist in cell movements such as migration of skin cells during healing or of embryonic tissue during development, as well as invasion of tissues by white blood cells.

✓ **C5.** Match the membranous organelles in the box with the best descriptions of their functions. Use each answer once.

G.	**Golgi complex**	**Per.**	**Peroxisomes**
L.	**Lysosomes**	**Pro.**	**Proteasomes**
M.	**Mitochondria**	**RER.**	**Rough endoplasmic reticulum**
N.	**Nucleus**	**SER.**	**Smooth endoplasmic reticulum**

_______ a. Site of direction of cellular activities by means of DNA in chromosomes located here.

_______ b. Network of tubules that is extension of rough ER; site of synthesis of steroids and fatty acids.

_______ c. "Rough" appearance is due to presence of ribosomes on these flattened sacs; sites of protein and glycoprotein synthesis.

_______ d. Stacks of cisternae; involved in synthesis of glycolipids, glycoproteins, and lipoproteins.

_______ e. Vesicles that contain powerful enzymes that can digest bacteria, organelles, or entire cells by processes of phagocytosis, autophagy, or autolysis.

_______ f. Structurally similar to lysosomes; contain enzymes that help metabolize amino acids, fatty acids, and toxic chemicals such as alcohol.

_______ g. Cristae-containing structures, called "powerhouses of the cell" because ATP production takes place here.

_______ h. Minute organelles that contain protein-digesting enzymes.

✓ **C6.** *For extra review.* Choose the organelles that best fit descriptions below. Select answers from lists in Checkpoints C4 and C5. Those answers may be used once, more than once, or not at all.

_______ a. Abundant in white blood cells, which use these organelles to help in phagocytosis of bacteria

_______ b. Present in large numbers in muscle and liver cells, which require much energy

_______ c. Located on surface of cells of the respiratory tract; help to move mucus

_______ d. Absence of a single enzyme from these organelles causes Tay-Sachs, an inherited disease affecting nerve cells

_______ e. An acid pH must be present inside these organelles for enzymes to function effectively

_______ f. Sites of cellular respiration requiring oxygen (O_2); contain DNA located within 37 genes used for self-replication

_______ g. Known as sarcoplasmic reticulum in muscle cells; detoxification centers in liver cells

_______ h. Implicated in cystic fibrosis in which defective Cl^- transporter proteins are broken down leading to thick mucus that clogs airways

_______ i. Might be described as a "clean-up service" since enzymes released from these organelles help to remove debris within and around injured cells (two answers)

✓ **C7.** Describe the sequence of events in synthesis, secretion, and discharge of protein from the cell by listing these five steps in connect order: _____, _____, _____, _____, _____.

A. Proteins accumulate and are modified, sorted, and packaged in cisternae of Golgi complex.

B. Proteins pass within a transport vesicle to *cis* cisternae of Golgi complex.

C. Secretory vesicles move toward cell surface, where they are discharged.

D. Proteins are synthesized at ribosomes on rough ER.

E. Vesicles containing protein pinch off of Golgi complex *trans* cisternae to form secretory vesicles.

C8. Color and then label all parts of Figure LG 3.3 listed with color code ovals.

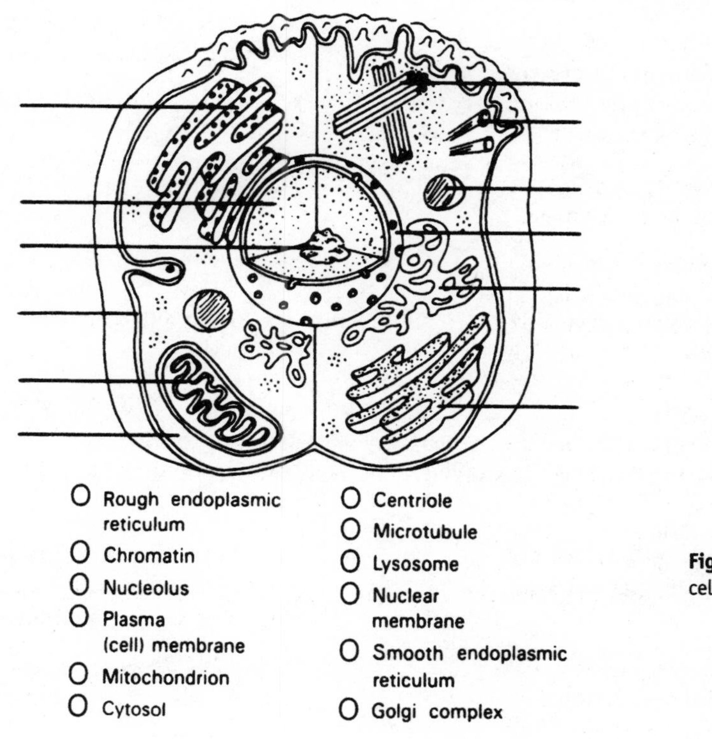

Figure LG 3.3 Typical structures found in body cells. Complete as directed in Checkpoint C8.

D. Nucleus; protein synthesis (pages 82–90)

✓ **D1.** Describe parts of the nucleus by completing this exercise.

a. The nuclear membrane has pores that are *(larger? smaller?)* than channels in the plasma membrane. Of what significance is this?

b. Ribosomes are assembled at sites known as ______________________, where protein, DNA, and RNA are stored.

c. Chromosomes consist of beadlike subunits known as ______________________ which are made of double-stranded ______________________ wrapped around core proteins called ______________________. Nucleosome "beads" are held together, necklace-fashion, by *(histones? linker DNA?)*. Nucleosomes are then held into chromatin fibers and folded into ______________________.

d. Before cell division, DNA duplicates and forms compact, coiled conformations known as *(chromatin? chromatids?)* that may be more easily moved around the cell during the division process.

e. The total genetic information carried in a human cell is known as the human ______________________, whereas the total set of proteins produced by human cells is known as the human ______________________. *(Almost all? Few?)* of the nucleotide bases are identical in different humans and contain the genetic code for producing chemicals common to humans. The average human gene contains *(30? 3,000? 3 million?)* nucleotides.

✓ **D2.** Define *gene expression.*

State the significance of protein synthesis in the cell.

✓ **D3.** Refer to Figure LG 3.4 and describe the process of protein synthesis in this exercise.

a. Write the names of the two major steps of protein synthesis on the two lines in Figure LG 3.4.

b. Transcription takes place in the *(nucleus? cytoplasm?)* with the help of the enzyme called ______________________. How does this enzyme know precisely where on DNA to start the transcription process? The "start" region consists of a specific nucleotide sequence known as a ______________________. At this site, ______________________ attaches to the DNA. Similarly, the "stop" section on the gene is a nucleotide sequence called a ______________________, at which point the enzyme detaches from the DNA.

c. In the process of transcription, a portion of one side of a double-stranded DNA serves as a mold or ______________________ and is called the *(sense? antisense?)* strand. Nucleotides of ______________________ line up in a complementary fashion next to DNA nucleotides (much as occurs between DNA nucleotides in replication of DNA prior to mitosis). Select different colors and color the components of DNA nucleotides (P, D. and bases T, A, C. and G).

d. Complete the strand of mRNA by writing letters of complementary bases. Then color the parts of mRNA (P. R, and bases U, A, C, and G).

e. The second step of protein synthesis is known as ______________________ because it involves translation of one "language" (the nucleotide base code of ______________________ RNA) into another "language" (the correct sequence of the 20 ______________________ to form a specific protein).

f. Translation occurs in the *(nucleus? cytoplasm?)*, specifically at a *(chromosome? lysosome? ribosome?)* to which mRNA attaches. The *(large? small?)* ribosomal subunit binds to one end of the mRNA and finds the point where translation will begin, known as *(ATP? the start codon?)*. The base sequence of the "start" codon is ___ ___ ___. *Note:* The "start" codon is shown on Figure LG 3.4.

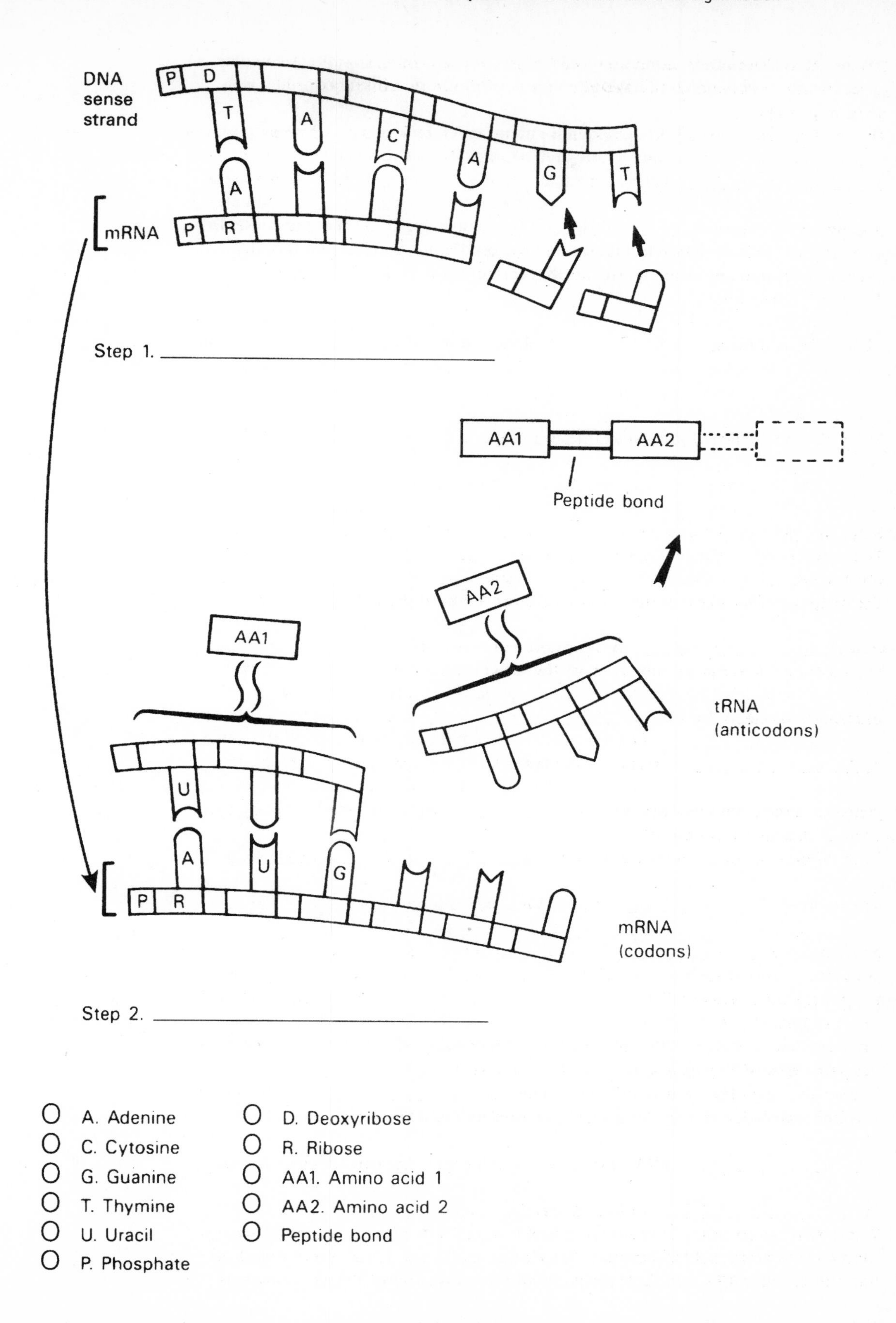

- O A. Adenine
- O C. Cytosine
- O G. Guanine
- O T. Thymine
- O U. Uracil
- O P. Phosphate
- O D. Deoxyribose
- O R. Ribose
- O AA1. Amino acid 1
- O AA2. Amino acid 2
- O Peptide bond

Figure LG 3.4 Protein synthesis. Label and color as directed in Checkpoint D3.

g. When the large ribosomal subunit attaches to the small subunit –mRNA complex, a functional ribosome results. Each amino acid is transported to the ribosome by

_______________________ RNA, which is characterized by a specific three-base unit or *(codon? anticodon?)*. For example, the anticodon triplet A G U would bind to the

complementary portion of mRNA (a codon) with the bases ____ ____ ____. In Step 2 of Figure LG 3.4, complete the labeling of bases and color all components of tRNA and mRNA. Be sure you select colors that correspond to those used in Step 1.

h. As the ribosome moves along each mRNA codon, additional amino acids are

transferred into place by ________ RNA. _______________________ bonds form between adjacent amino acids with the help of enzymes from the *(large? small?)* subunit of the ribosome. Now color the amino acids and peptide bonds that join them.

i. As the ribosome moves on to the next codon on mRNA, what happens to the "empty" tRNA?

j. When the protein is complete, synthesis is stopped by the _______________________ codon. As the protein is released, what happens to the ribosome?

k. When several ribosomes attach to the same strand of mRNA, a _______________________ is formed, permitting repeated synthesis of the identical protein.

✓ **D4.** Complete this additional activity about protein synthesis.

a. Sections of DNA that do code for parts of proteins are known as *(exons? introns?)*, whereas sections of DNA that do not code for parts of proteins are called

_______________________. The pre-mRNA formed initially by transcription contains *(only exons? only introns? both exons and introns?)*.

b. Removal of *(exons? introns)* occurs *(before? after?)* mRNA moves to the cytoplasm. This process is facilitated by enzymes known as "snurps," which refers to

_______________________ _______________________ _______________________.

c. Define the term *gene*.

d. A gene that contains 3000 nucleotides in an exon will code for a protein of approximately *(9000? 3000? 1000?)* amino acids.

D5. Explain how 40,000 genes in the human genome can code for up to a million human proteins.

D6. List six or more therapeutic substances that have been produced by bacteria containing recombinant DNA.

E. Cell division (pages 90–94)

✓ **E1.** Once you have reached adult size, do your cells continue to divide? *(Yes? No?)* Of what significance is this fact?

✓ **E2.** Describe aspects of cell division by completing this exercise.

a. Division of any cell in the body consists of two processes: division of the

____________________ and division of the ____________________.

b. In the formation of mature sperm and egg cells, nuclear division is known as

____________________. In the formation of all other body cells, that is, *somatic* cells,

nuclear division is called ____________________.

c. Cytoplasmic division in both somatic and reproductive division is known as

____________________. In *(meiosis? mitosis?)* the two daughter cells have the same hereditary material and genetic potential as the parent cell.

✓ **E3.** Describe interphase in this activity.

a. Interphase occurs *(during cell division? between cell divisions?)*.

b. Chromosomes are replicated during phase *(G_1? S? G_2?)* of interphase. In fact "S"

stands for ____________________ of new DNA. During this time, the DNA helix partially uncoils and the double DNA strand separates at points where

____________________ connect so that new complementary nucleotides can attach here. (See Figure LG 2.3.) In this manner the DNA of a single chromosome is

doubled to form two identical ____________________.

c. Two periods of growth occur during interphase, preparing the cell for cell division. The *(G_1? G_2?)* phase immediately precedes the S phase, whereas the *(G_1? G_2?)* phase follows the S portion of interphase. Which of these phases is usually longest? *(G_1? S? G_2?)*

d. Most nerve cells *(are? are not?)* capable of mitosis after birth; their development is

permanently arrested in the ________ phase, and are then said to be in the ________ phase.

✓ **E4.** Carefully study the phases of the cell cycle shown in Figure 3.33, page 92 in your text. Then check your understanding of the process by identifying major events in each phase.

A.	**Anaphase**	**P.**	**Prophase**
I.	**Interphase**	**T.**	**Telophase**
M.	**Metaphase**		

_______ a. This phase immediately follows interphase.

_______ b. Chromatin condenses into distinct chromosomes (chromatids) each held together by a centromere complexed with a kinetochore.

_______ c. Nucleoli and nuclear envelope break up; the two centrosomes are moved to opposite poles of the cell, and formation of the mitotic spindle begins.

_______ d. Chromatids line up with their centromeres along the equatorial plate.

_______ e. Centromeres split; chromatids (now called daughter chromosomes) are dragged by microtubules to opposite poles of the cell. Cytokinesis begins.

_______ f. Events of this phase are essentially a reversal of prophase; cytokinesis is completed.

E5. Describe roles of the cytoskeleton in each aspect of cell division:

a. Dragging of chromosomes to opposite poles of the cell

b. Formation of the contractile ring that contributes to cytokinesis

E6. Write the full names and a brief description of the roles of the following three chemicals that regulate cell division.

a. MPF

b. cdc2 proteins

c. Cyclin

E7. Which type of cell death is considered normal cell death? *(Apoptosis? Necrosis?)* Which type is followed by an inflammatory reaction? *(Apoptosis? Necrosis?)*

✓ **E8.** Tumor suppressor genes normally *(stimulate? inhibit?)* cell division. Loss or alteration of such a gene leads to *(cancer? atrophy?)*. The most common genetic change leading to cancer is a change in the tumor suppressor gene called *(cdc? cyclin? p53?)*: this gene is located on chromosome number _____.

F. Cellular diversity; aging and cells (pages 94–95)

✓ **F1.** Describe the diversity of human cells in this Checkpoint.

a. The human body is composed of about _______ different types of cells with a total of about _______ cells. These range in size from a red blood cell (_______ mu) to an ovum (_______ mu).

b. List several different shapes of cells. (More about this in Chapter 4.)

c. State the advantages afforded to cells by their own specific shapes:

Sperm with a long "tail" (flagellum) _______________________________________

Intestinal cells with microvilli _______________________________________

Thin, flat cells forming walls of the tiniest blood vessels _______________________

F2. Describe changes that normally occur in each of the following during the aging process.

a. Number of cells in the body. *Note:* Which three types of cells are permanently arrested in the G_0 phase?

b. Aging of cells is associated with *(shortening? elongation?)* of telomeres. Define *telomeres*.

c. Glucose cross-links

F3. In a sentence or two, summarize the main points of each of these theories of aging.

a. An "aging gene," affecting rate of mitosis

b. Free radicals

c. Immune system

F4. Contrast these two disorders: *progeria* and *Werner syndrome*.

G. Disorders, medical terminology (pages 95–97)

✓ **G1.** Complete this exercise about cancer.

a. A term that means "tumor or abnormal growth" is ___________________. A cancerous growth is a *(benign? malignant?)* neoplasm, whereas a noncancerous tumor is a *(benign? malignant?)* neoplasm. The term ___________________ refers to the study of cancer.

b. Tumors grow quickly by increasing their rate of cell division, a process known as ___________________. Cancer cells also compete with normal body cells by robbing those cells of ___________________. Tumors also favor their own growth by synthesizing chemicals known as TAFs. TAFs refer to ___________________ ___________________ ___________________, indicating that these chemicals stimulate growth of new ___________________ around tumor cells.

c. Malignant tumors are *(more? less?)* likely than benign tumors to spread and possibly cause death. A term that means "spread" of cancer cells is ___________________. By what routes do cancers reach distant parts of the body?

d. What may cause pain associated with cancer?

✓ **G2.** Match terms in the box with the definitions below.

C.	**Carcinoma**	**O.**	**Osteogenic sarcoma**
L.	**Lymphoma**	**S.**	**Sarcoma**

_______ a. General term for malignant tumor arising from any connective tissue

_______ b. Malignant tumor derived from bone (a connective tissue)

_______ c. General term for malignant tumor of epithelial tissue

_______ d. A malignancy of lymph tissue

G3. Relate the following terms to cancer:

a. Carcinogens

b. Proto-oncogenes

c. Oncogenes

d. Growth factors

e. Mutation

G4. *Clinical thinking*. Are all cancer cells alike within a single tumor? *(Yes? No?)* What significance is this in planning cancer treatment?

✓ **G5.** Match the terms in the box describing alterations in cells or tissues with the descriptions below.

A.	**Atrophy**	**HT.**	**Hypertrophy**
D.	**Dysplasia**	**M.**	**Metaplasia**
HP.	**Hyperplasia**		

_______ a. Increase in size of tissue or organ by increase in size (not number) of cells, such as growth of your biceps muscle with exercise

_______ b. Increase in size of tissue or organ due to increase in number of cells, such as a callus on your hand or growth of breast tissue during pregnancy

_______ c. Change of one cell type to another normal cell type, such as change from single row of tall (columnar) cells lining airways to multilayers of cells as response to constant irritation of smoking

_______ d. Abnormal change in cells in a tissue due to irritation or inflammation. May revert to normal if irritant removed, or may progress to neoplasia

_______ e. Decrease in size of cells with decrease in size of tissue or organ

ANSWERS TO SELECTED CHECKPOINTS: CHAPTER 3

B1. See Figure LG 3.1A. (a) Water-soluble and facing the membrane surface, so compatible with watery environment surrounding the membrane. (b) Forms a water-insoluble barrier between cell contents and extracellular fluid; nonpolar molecules such as lipids, oxygen, and carbon dioxide pass readily between "tails." (c) Channels for passage of water or ions; transporters; receptors; linkers (helping to bind cells to one another). (d) Form sugary coat (glycocalyx) that is important in cell recognition, immunity, protection, adhesion of cells; this coating also helps make red blood cells slippery so they can easily pass through tiny blood vessels. (e) Strengthens the membrane.

B2. (a) Lipids, proteins. (b) Phospholipids; amphipathic; surface; philic, loving. (c) Cholesterol; 5, extracellular. (d) Proteins; integral; peripheral. (e) Hormones such as insulin or nutrients such as glucose; enzymes; markers; linkers. (f) Carbohydrates; the carbohydrate portion of glycoproteins and glycolipids; see text page 62.

B4. (a) Quite, phospholipid. (b) Protein; ion channels with pores and transporters. (c) See Figure LG 3.1A.

B5. (a) See Figure LG 3.2A. (b) Down. (c) Electrical, potential; inside. (d) Electrochemical.

B7. (a) Passive. (b) Active.

B8. (a) A. (b) B. (c) B. (d) A. (e) b: membrane thickness.

B9. (a) D. (b) FD. (c) O.

B10. (a) B. (b) More, fewer. (c) A. (d) Out of, crenate. (e) Iso; 0.90% NaCl or 5% glucose. (f) C. (g) A.

B11. (a) Chemicals too polar or charged to go through the lipid bilayer, or too large to pass through membrane channels. (b) Does; undergoes a change in structure. (c) GluT; fructose and galactose. (d) Maximum; insulin. (e) Phosphate, kinase; low.

B12. (a) Secondary; Na^+/K^+, primary; ATP; 40; experience Na^+/K^+ pump failure and can die. (b) Phosphate; outside, inside. (c) Two, inside; osmotic. (d) Indirectly from ATP by utilizing concentration gradients [such as that for Na^+] developed by primary active transport; Ca^{2+}, H^+, glucose, and amino acids from foods. (e) Antiporters. (f) Decreasing, smaller; slow down; strengthened.

C1. Fluid; 75, 90; ions, glucose, amino acids, fatty acids, proteins, lipids, ATP, waste products, and temporary storage molecules such as lipid droplets and glycogen granules.

C3. (a) Provides a scaffold or structural framework that organizes the cell and gives it shape; permits movement. (b) Movement of the entire cell, movement of chromosomes during cell division, and movement of organelles and some chemicals within the cell. (c) Mt If Mf.

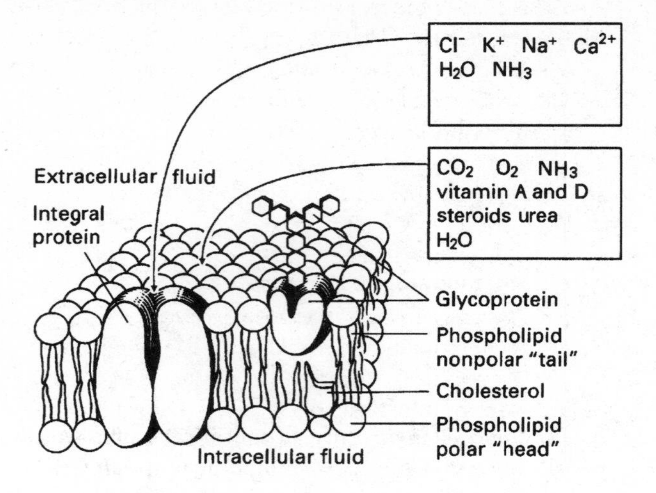

Figure LG 3.1A Plasma membrane

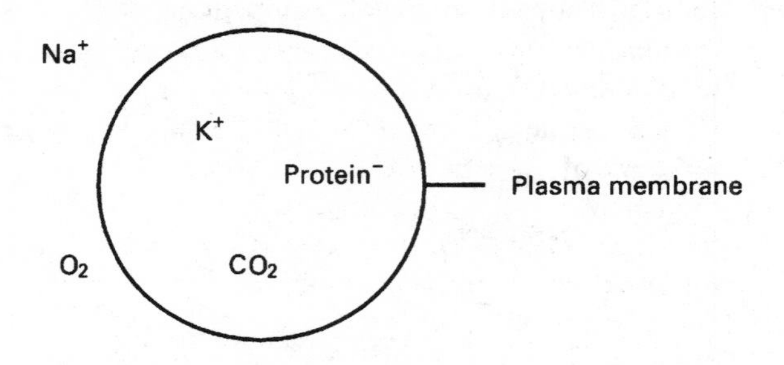

Figure LG 3.2A Major ions in intracellular fluid (ICF) and extracellular fluid (ECF).

C4. (a) Cen. (b) F. (c) Cil. (d) R. (e) Mt. (f) If. (g) Mf.

C5. (a) N. (b) SER. (c) RER. (d) G. (e) L. (f) Per. (g) M. (h) Pro.

C6. (a) L. (b) M. (c) Cil. (d) L. (e) L. (f) M. (g) SER. (h) Pro. (i) L, Per.

C7. D B A E C.

C8.

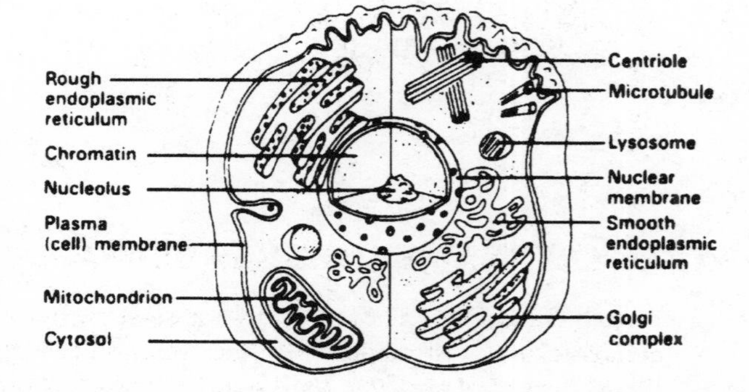

Figure LG 3.3A Typical structures found in body cells.

D1. (a) Larger; permit passage of large molecules such as RNA and proteins. (b) Nucleoli. (c) Nucleosomes. DNA, histones; linker DNA; loops. (d) Chromatids. (e) Genome, proteome; almost all; 3,000.

D2. The process whereby a gene's DNA is used to synthesize a specific protein. Proteins serve a number of significant functions (described in Chapter 2) including the production of all carbohydrates and lipids. Refer to Chapter 2, page 32 of the *Learning Guide*, Checkpoint E9.

D3. (a) See Figure LG 3.4A. (b) Nucleus, RNA polymerase; promotor; RNA polymerase; terminator. (c) Template, sense; messenger RNA (mRNA) (see Figure LG 3.4A). (d) See Figure LG 3.4A. (e) Translation, m (messenger), amino acids. (f) Cytoplasm, ribosome; small, the start codon; A U G. (g) t (transfer), anticodon; U C A, see Figure LG 3.4A. (h) t (transfer); peptide, large. (i) It is released and may be recycled to transfer another amino acid. (j) Stop; it splits into its large and small subunits. (k) Polyribosome.

D4. (a) Exons, introns; both exons and introns. (b) Introns, before; small nuclear ribonucleoproteins or snRNPs. (c) A group of nucleotides on a DNA molecule that codes for synthesis of a specific sequence of bases in mRNA (by transcription) and ultimately, a specific sequence of amino acids (by translation). (d) 1000 (because three RNA nucleotides code for one amino acid).

E1. Yes; cells can be replaced when damaged or destroyed, but overgrowth can lead to disorders such as cancer.

E2. (a) Nucleus, cytoplasm. (b) Meiosis; mitosis. (c) Cytokinesis; mitosis.

E3. (a) Between cell divisions. (b) S; synthesis; nitrogen bases; chromatids. (c) G_1, G_2, G_1. (d) Are not, G_1, G_0.

E4. (a) P. (b) P. (c) P. (d) M. (e) A. (f) T.

E7. Apoptosis; necrosis.

E8. Inhibit; cancer; p53; 17.

F1. (a) 200, 100 trillion; 8*u*m, 140*u*m. (b) Round, oval, flat, cuboidal, columnar, elongated, star-shaped, cylinrical, or disc-shaped. (c1) Mobility. (c2) Increased surface area. (c3) Permit diffusion.

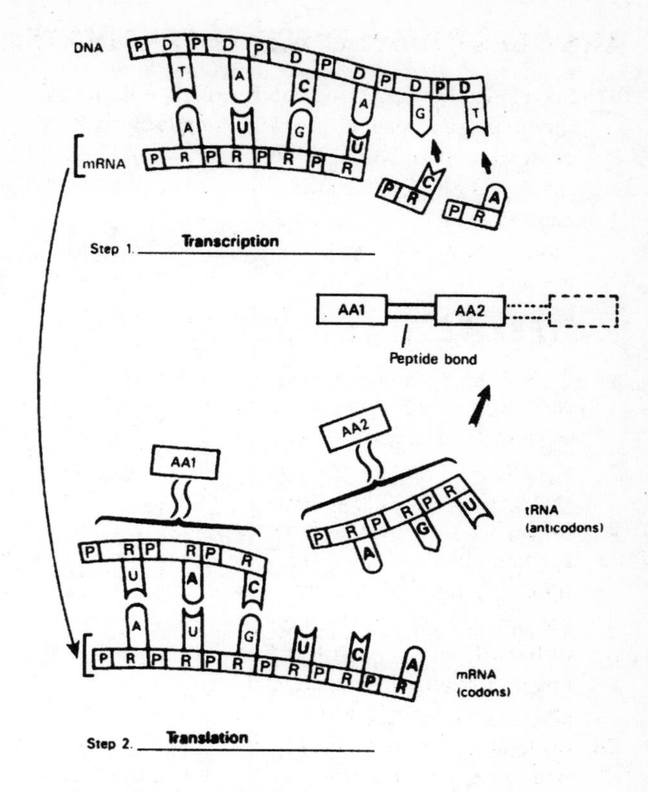

Figure LG 3.4A Protein synthesis.

F2. (a) Decrease; cardiac muscle, skeletal muscle, and neurons. (b) Shortening; sections of DNA at tips of each chromosome that should protect chromosomes. (c) Glucose added to cellular proteins leads to irreversible cross-links that alter aging tissues.

G1. (a) Neoplasm; malignant, benign; oncology. (b) Hyperplasia; nutrients and space; tumor angiogenesis factors, blood vessels (*angio* = vessel). (c) More; metastasis; blood or lymph, or by detaching from an organ, such as liver, and "seeding" into a cavity (as abdominal cavity) to reach other organs which border on that cavity. (d) Pressure on nerve, obstruction of passageway, loss of function of a vital organ.

G2. (a) S. (b) O. (c) C. (d) L.

G5. (a) HT. (b) HP. (c) M. (d) D. (e) A.

MORE CRITICAL THINKING: CHAPTER 3

1. Discuss clinical aspects of tonicity—for example, how would a client be affected by the accidental administration of 5% NaCl solution (which is hypertonic to human body fluids) instead of an 0.9% NaCl solution (which is isotonic to the human body)?
2. Inadequate dietary protein can rob the body of nutrients for critical functions. If plasma membranes lack adequate protein, what membrane functions are affected?
3. Explain how cells are compartmentalized and what advantages are provided to cells by compartmentalization.

4. Tell the "life history" of a digestive enzyme made within a pancreas cell. Remember that most of any enzyme consists of protein. Start with the "birth" (synthesis) of this protein and outline its "travels" from one part of the cell to another, including its exit from the cell via exocytosis.
5. In Chapter 2, you studied DNA and ATP. Describe the roles of each of these chemicals in cellular functions.
6. Since virtually all human cells contain the same type of DNA in their 46 chromosomes, explain what accounts for the difference between a muscle and a bone cell.

MASTERY TEST: CHAPTER 3

Questions 1-9: Circle the letter preceding the one best answer to each question.

1. The organelle that carries out the process of autophagy in which old organelles are digested so that their components can be recycled is:
   A. Peroxisome
   B. Mitochondrion
   C. Centrosome
   D. Lysosome
   E. Endoplasmic reticulum
2. Which statement about proteins is *false*?
   A. Proteins are synthesized on ribosomes.
   B. Integral proteins extend all the way across plasma membranes.
   C. Proteins are so large that they can be expected to create osmotic pressure important in maintaining fluid volume in blood.
   D. Proteins are small enough to pass through pores of typical plasma (cell) membranes.
3. Choose the *false* statement about prophase of mitosis.
   A. It occurs just after metaphase.
   B. The nuclear membrane breaks down then.
   C. Centrosomes move to opposite poles of the cell.
   D. The mitotic spindle forms then and moves centrosomes to the poles of the cell.
4. Choose the *false* statement about the Golgi complex.
   A. It consists of stacked membranous sacs known as cisternae.
   B. It is more extensive in cells that secrete proteins into ECF.
   C. It is involved in lipoprotein secretion.
   D. It pulls chromosomes towards poles of the cell during mitosis.
5. Choose the *false* statement about genes.
   A. Genes contain DNA.
   B. Genes contain information that controls heredity.
   C. Genes are located primarily in the nucleus of the cell.
   D. Genes are transcribed by messenger RNA during the translation step of protein synthesis.
6. Choose the *false* statement about protein synthesis.
   A. Translation occurs in the cytoplasm.
   B. Messenger RNA picks up and transports an amino acid during protein synthesis.
   C. Messenger RNA travels to a ribosome in the cytoplasm.
   D. Transfer RNA is attracted to mRNA due to their complementary bases.
7. All of the following structures are part of the nucleus *except*:
   A. Nucleolus
   B. Chromatin
   C. Chromosomes
   D. Centrosome
8. The Na^+/K^+ pump is an example of:
   A. Active transport
   B. Osmosis
   C. Diffusion
   D. Phagocytosis
9. Bone cancer is an example of what type of cancer?
   A. Sarcoma
   B. Lymphoma
   C. Carcinoma
   D. Leukemia

Questions 10-20: Circle T (true) or F (false). If the statement is false, change the underlined word or phrase so that the statement is correct.

T F 10. Plasma membranes consist of a double layer of carbohydrate molecules with proteins embedded in the bilayer.

T F 11. Free ribosomes in the cytosol are concerned primarily with synthesis of proteins for insertion in the plasma membrane or for export from the cell.

T F 12. Glucose in foods you eat is absorbed into cells lining the intestine by the process of facilitated transfusion.

T F 13. Cristae are folds of membrane in mitochondria.

T F 14. Sperm move by lashing their long tails named cilia.

T F 15. Hypertrophy refers to increase in size of a tissue related to increase in the number of cells.

T F 16. <u>Diffusion, osmosis, and facilitated diffusion</u> are all passive transport processes.

T F 17. A 5% glucose solution is <u>hypotonic</u> to a 10% glucose solution.

T F 18. <u>The free radical theory of aging</u> suggests that electrically charged, unstable, and highly reactive free radicals cause oxidative damage to chemical components of aging cells.

T F 19. The mitotic phase that follows metaphase is <u>anaphase</u>.

T F 20. <u>Cytokinesis</u> is another name for mitosis.

Questions 21-25: Fill-ins. Complete each sentence with the word which best fits.

_________________ 21. Cells of cardiac muscle, skeletal muscle, and nerves do not divide throughout a lifetime since they are arrested permanently in the _______ phase.

_________________ 22. Active processes involved in movement of substances across cell membranes are those that use energy from the splitting of ________.

_________________ 23. White blood cells engulf large solid particles by the process of ________.

_________________ 24. ________ DNA consists of DNA combined from several sources, such as from human cells and from bacteria.

_________________ 25. The sequence of bases of mRNA which would be complementary to DNA bases in the sequence A-T-T-C-A-C would be ________.

ANSWERS TO MASTERY TEST: CHAPTER 3

Multiple Choice

1. D
2. D
3. A
4. D
5. D
6. B
7. D
8. A
9. A

True-False

10. F. Phospholipid molecules
11. F. Proteins for use inside of the cell
12. F. Facilitated diffusion (or facilitated transport).
13. T
14. F. Flagella
15. F. Size
16. T
17. T
18. T
19. T
20. F. Nuclear division (Cytokinesis is cytoplasmic division.)

Fill-ins

21. G_0 or G_1
22. ATP
23. Phagocytosis
24. Recombinant
25. U-A-A-G-U-G

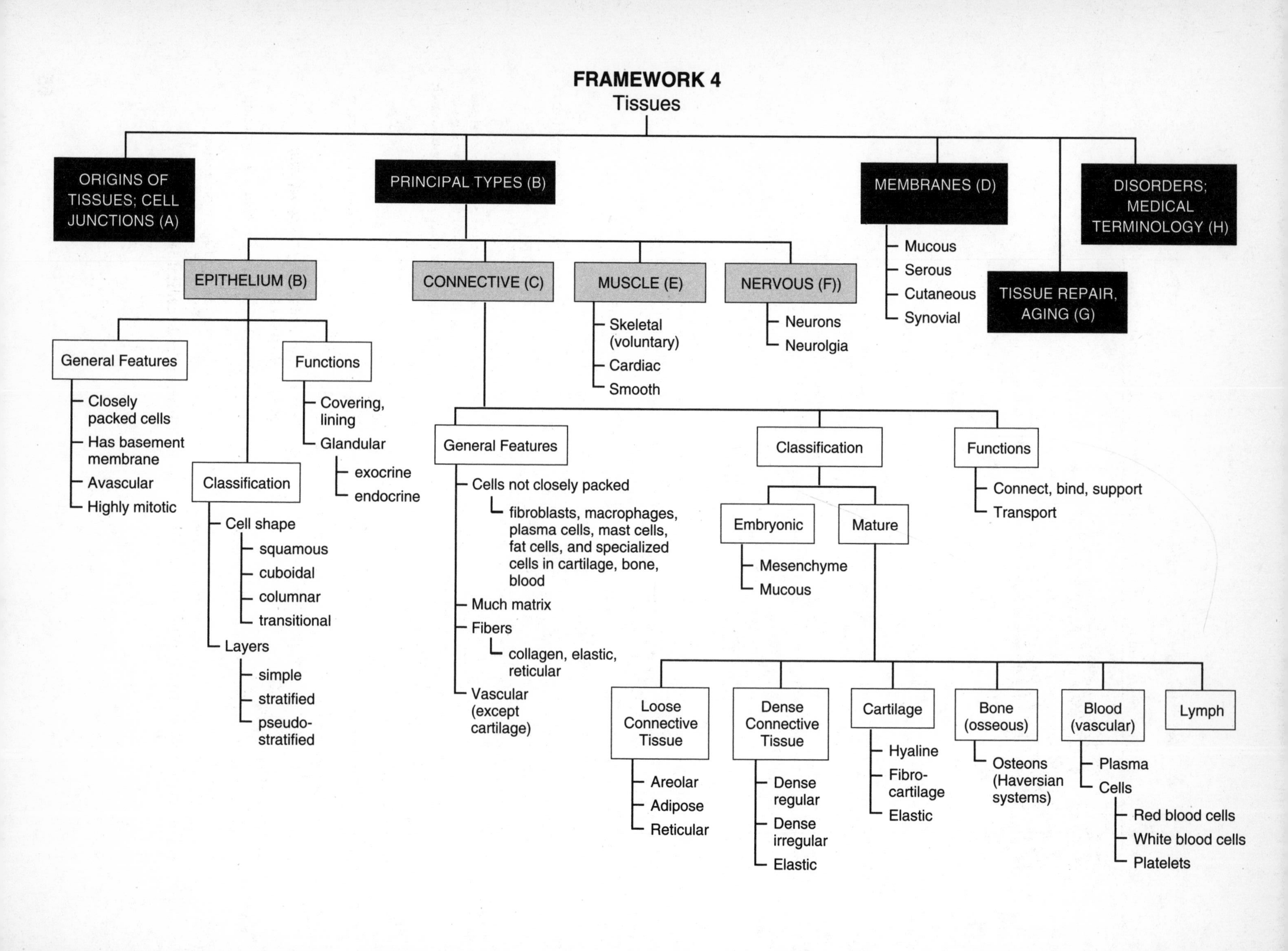
FRAMEWORK 4
Tissues
ORIGINS OF TISSUES; CELL JUNCTIONS (A)
PRINCIPAL TYPES (B)
MEMBRANES (D)
DISORDERS; MEDICAL TERMINOLOGY (H)
TISSUE REPAIR, AGING (G)
EPITHELIUM (B)
CONNECTIVE (C)
MUSCLE (E)
NERVOUS (F))
Mucous
Serous
Cutaneous
Synovial
Skeletal (voluntary)
Cardiac
Smooth
Neurons
Neurolgia
General Features
Closely packed cells
Has basement membrane
Avascular
Highly mitotic
Functions
Covering, lining
Glandular
exocrine
endocrine
Classification
Cell shape
squamous
cuboidal
columnar
transitional
Layers
simple
stratified
pseudo-stratified
General Features
Cells not closely packed
fibroblasts, macrophages, plasma cells, mast cells, fat cells, and specialized cells in cartilage, bone, blood
Much matrix
Fibers
collagen, elastic, reticular
Vascular (except cartilage)
Classification
Embryonic
Mesenchyme
Mucous
Mature
Functions
Connect, bind, support
Transport
Loose Connective Tissue
Areolar
Adipose
Reticular
Dense Connective Tissue
Dense regular
Dense irregular
Elastic
Cartilage
Hyaline
Fibro-cartilage
Elastic
Bone (osseous)
Osteons (Haversian systems)
Blood (vascular)
Plasma
Cells
Red blood cells
White blood cells
Platelets
Lymph

CHAPTER 4

The Tissue Level of Organization

Tissues consist of cells and the intercellular (between cells) materials secreted by these cells. All of the organs and systems of the body consist of tissues that may be categorized into four classes or types: epithelium, connective, muscle, and nerve. Each tissue exhibits structural characteristics (anatomy) that determine the function (physiology) of that tissue. For example, epithelial tissues consist of tightly packed cells that form the secretory glands and also provide protective barriers, as in the membranes that cover the surface of the body and line passageways or cavities. Tissues are constantly stressed by daily wear and tear and sometimes by trauma or infection. Tissue repair is therefore an ongoing human maintenance project.

As you begin your study of tissues, carefully examine the Chapter 4 Framework and note relationships among concepts and key terms there. Also refer to the Topic Outline and Objectives; you may want to check off each objective as you complete it.

TOPIC OUTLINE AND OBJECTIVES

A. Types and origins of tissues; cell junctions

1. Name the four basic types of tissues that make up the human body and describe the characteristics of each.
2. Describe the structure and function of the five main types of cell junctions.

B. Epithelial tissue

1. Describe the general features of epithelial tissue
2. For each different type of epithelium, list its location, structure and function.

C. Connective tissue

1. Describe the general features of connective tissues.
2. Describe the structure, location, and function, of the various types of connective tissues.

D. Membranes

1. Define a membrane, and describe the classification of membranes.

E. Muscle tissue

1. Describe the general features of muscle tissue.
2. Contrast the structure, location, and mode of control of skeletal, cardiac, and smooth muscle tissue.

F. Nervous tissue

1. Describe the structural features and function of nervous tissue.

G. Tissue repair: restoring homeostasis; aging and tissues

1. Describe the role of tissue repair in restoring homeostasis.
2. Describe the effects of aging on tissues.

H. Disorders and medical terminology

WORDBYTES

Now become familiar with the language of this chapter by studying each wordbyte, its meaning, and an example of its use within a term. After you study the entire list, self-check your understanding by writing the meaning of each wordbyte on the line. As you continue through the *Learning Guide*, identify (and fill in) additional terms that contain the same wordbyte.

Wordbyte	Self-check	Meaning	Example(s)
a-	____________	without	*a*vascular
acin	____________	berry	*acin*ar gland
adip(o)-	____________	fat	*adipo*se
apex, apic-	____________	tip	*api*cal
areol-	____________	small space	*areol*ar
ciliaris	____________	resembling an eyelash	*cilia*
-crine	____________	to secrete	endo*crine*
endo-	____________	within	*endo*crine
exo-	____________	outside	*exo*crine
hist -	____________	tissue	*hist*ology
lupus	____________	wolf	systemic *lupus* erythematosus
macro-	____________	large	*macro*phage
multi-	____________	many	*multi*cellular
path-	____________	disease	*path*ology
pseudo-	____________	false	*pseudo*stratified
squam-	____________	thin plate	*squam*ous
strat-	____________	layer	*strat*ified
uni-	____________	one	*uni*cellular
vasc-	____________	blood vessel	a*vasc*ular
xeno-	____________	strange, foreign	*xeno*trans-plantation

CHECKPOINTS

A. Types and origins of tissues; cell junctions (pages 104–106)

A1. Define the following terms.

a. Tissue

b. Histology

c. Pathologist

d. Biopsy

A2. Complete the table about the four main classes of tissues.

Tissue	General Functions
a. Epithelial tissue	
b. Connective tissue	
c.	Movement
d.	Initiates and transmits nerve impulses that help coordinate body activities

✓ **A3.** *The Big Picture: Looking Ahead*. Refer to Exhibit 29.1 in you text. Then do the following matching exercise on the three embryonic germ layers that form all body organs.

Ecto. Ectoderm	Endo. Endoderm	Meso. Mesoderm

________ a. Epithelial lining of the organs of digestion and of respiratory passageways

________ b. Brain and nerves

________ c. Muscles and bones

________ d. Outer layers of skin

________ e. Cartilage and fat tissue

✓ **A4.** Complete this exercise about extracellular materials and connections between cells.

a. Name one type of cell that typically moves around within extracellular areas.

______________________.

b. Most body cells stay in one place. They are held in place by points of contact between adjacent cell membranes; such points are known as cell ______________________. For example, *(anchoring gap? tight?)* junctions are helpful in forming a tight seal between

cells, such as that required in the lining of the stomach or the ____________________.

They consist of ____________________ proteins fused "zip-lock fashion."

c. Adherens junctions consist of plaque, which is a dense layer of *(lipid? protein?)* and

transmembrane glycoproteins known as ____________________ that extend to other cells to help resist separation of the tissue. In epithelial tissue, such junctions are

extensive and are called ____________________ belts.

d. Desmosomes help to prevent ____________________ cells from separating during contraction, and help to hold *(epidermal? dermal?)* cells together. Desmosomes *(do? do not?)* contain plaque and cadherins.

e. Resembling half demosomes, ____________________ contain transmembrane glycoproteins known as *(integrins? laminin?)* that anchor the basal surfaces of epithelial cells to the basement membrane.

f. A gap junction consists of proteins called ____________________ that form tunnels between cells. Passage of ions through these tunnels permits spread

of ____________________.

B. Epithelial tissue (pages 106–117)

✓ **B1.** Describe structural features of epithelium in this Checkpoint.

a. Epithelium consists of *(closely packed cells? intercellular material with few cells?) (Many? Few?)* cell junctions are present, securing these cells *(tightly? loosely?)* to each other at their *(apical? basal? lateral?)* surfaces.

b. Blood vessels *(do? do not?)* typically penetrate epithelium. A term meaning "lack of

blood vessels" is ____________________. State the significance of this feature of epithelium.

c. Epithelium *(has? lacks?)* a nerve supply.

d. Epithelium *(does? does not?)* have the capacity to undergo mitosis.

e. The apical surface area of some epithelial cells is modified by the presence of *(cilia? microvilli?)* for secretion or absorbtion, and ____________________ for moving substances along the cell surface.

✓ **B2.** Complete these sentences describing basement membrane.

a. The *(apical? basal?)* surfaces of epithelial cells attach by a ____________________ membrane to underlying connective tissue.

b. This membrane consists of *(cells? extracellular material?)*. The part of the membrane adjacent to epithelium is secreted by *(epithelium? connective tissue?)*. It consists of the protein named *(collagen? reticulin?)* along with proteoglycans (laminin) and is known as the *(reticular? basal?)* lamina.

c. The second underlying layer of the membrane is secreted by *(epithelium? connective tissue?)* and is called the ____________________ lamina.

B3. As you read the section of your text on types of epithelium, fill in the table below describing five or more locations and functions of epithelial tissue. One is done for you.

Location	Functions
a. Eyes, nose, outer layer of skin	Sensory reception
b.	
c.	
d.	
e.	

✓ **B4.** Name the two main subtypes of epithelial tissue based on location and function.

____________________ and ____________________.

✓ **B5.** Use each term in the box once to complete the sentences below.

Columnar	**Simple**	**Stratified**
Cuboidal	**Squamous**	**Transitional**

a. Epithelial cells arranged in one layer are known as ____________________ epithelium. ____________________ epithelium is composed of many layers of cells.

b. ____________________ epithelium consists of cells so flat that diffusion readily occurs across them. Tall, cylindrical or rectangular cells are called ____________________ epithelium. Intermediate between these two types of cells are the cells that form ____________________ epithelium.

c. A type of epithelium that can stretch; lines urinary bladder. ____________________

B6. Study diagrams of epithelial tissue types (A-F) in Figure LG 4.1 and then write the following information on the lines provided on the figure:

✓ a. The name of each tissue
b. One or more locations of each type of tissue
c. One or more functions of each type of tissue

Now color the following structures on each diagram:

O Nucleus O Intercellular material
O Cytoplasm O Basement membrane

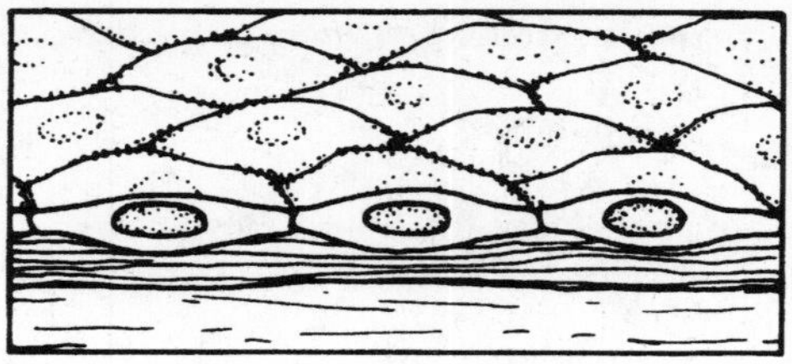

A Name

Location

Function

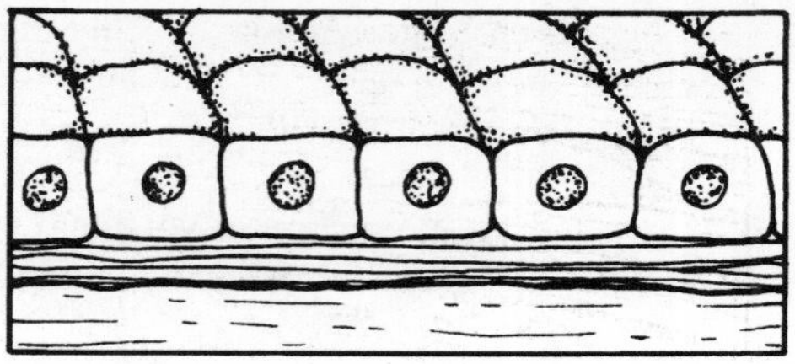

B Name

Location

Function

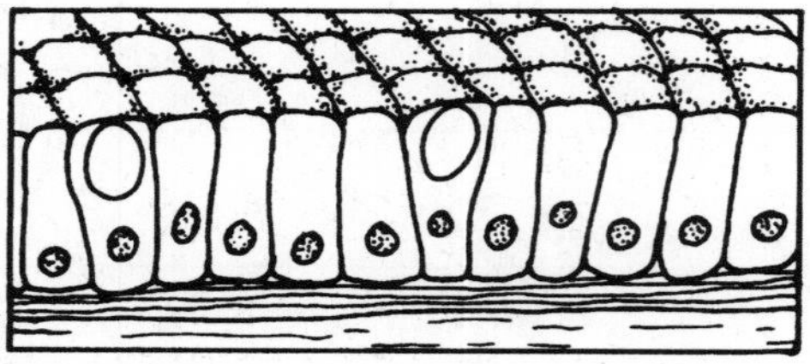

C Name

Location

Function

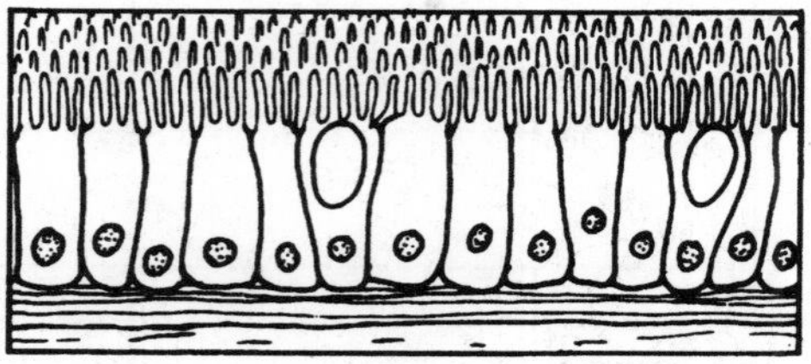

D Name

Location

Function

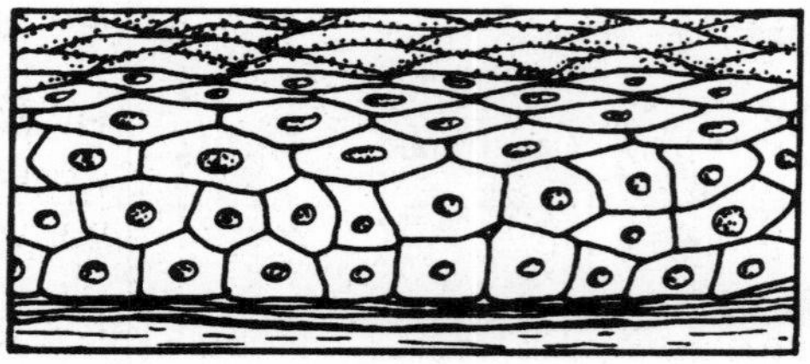

E Name

Location

Function

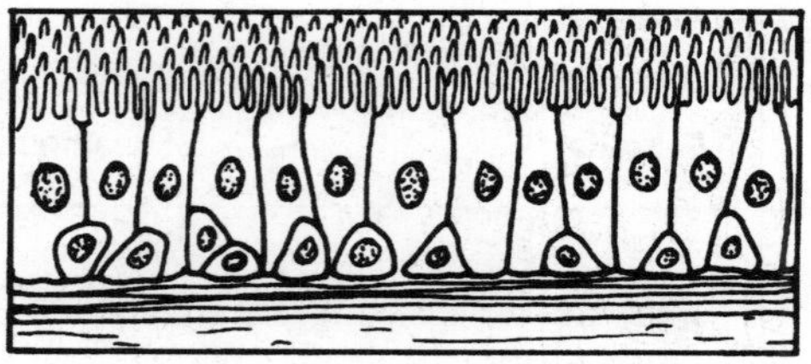

F Name

Location

Function

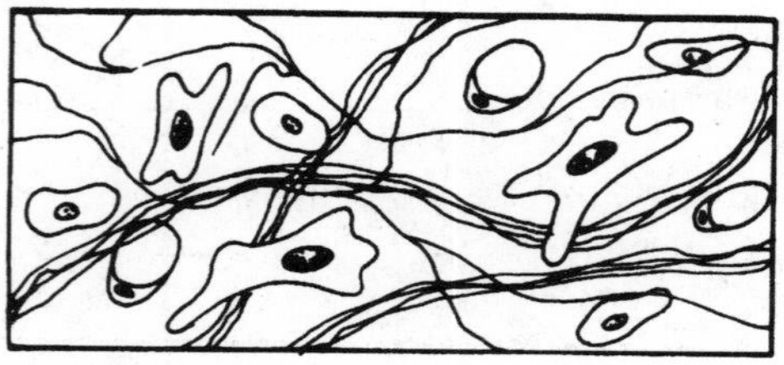

G Name

Location

Function

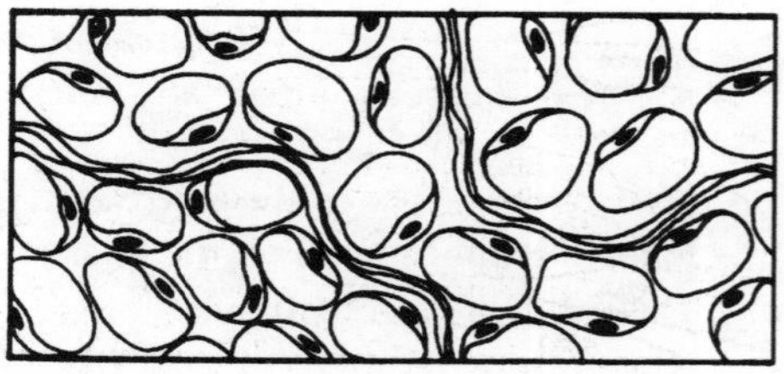

H Name

Location

Function

Figure LG 4.1 Diagrams of selected tissues types. Complete as directed.

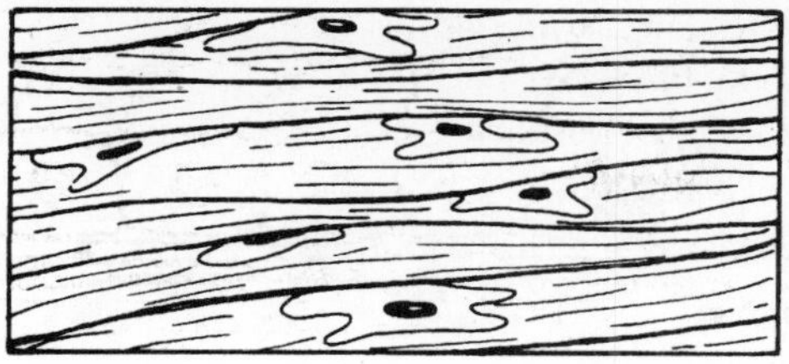

I Name ____________________

Location ____________________

Function ____________________

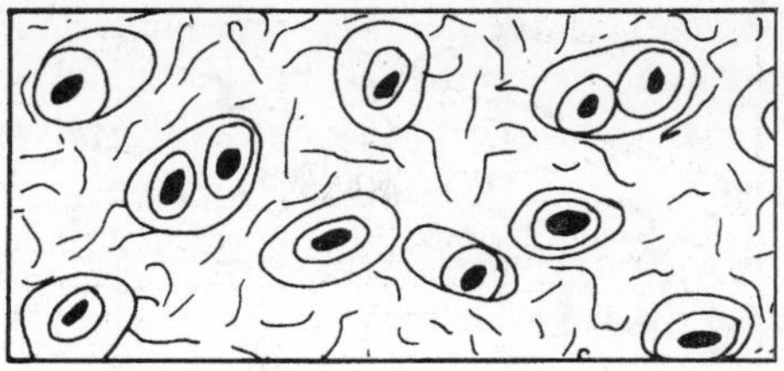

J Name ____________________

Location ____________________

Function ____________________

K Name ____________________

Location ____________________

Function ____________________

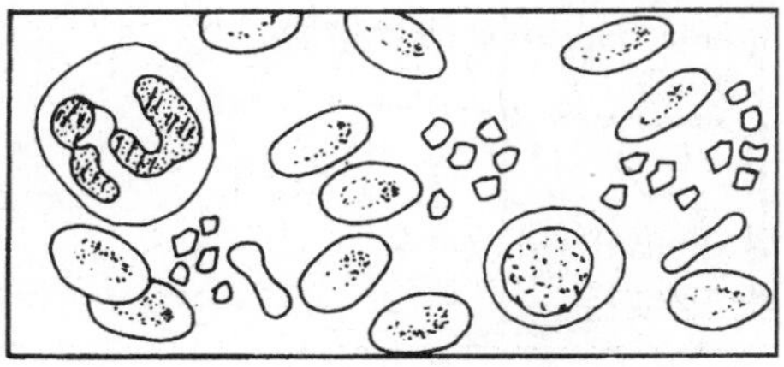

L Name ____________________

Location ____________________

Function ____________________

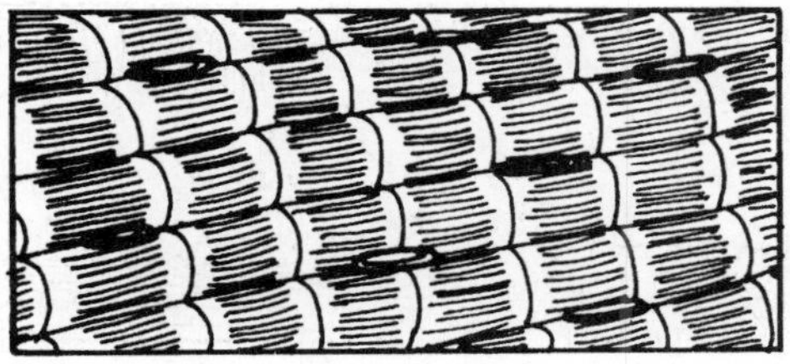

M Name ____________________

Location ____________________

Function ____________________

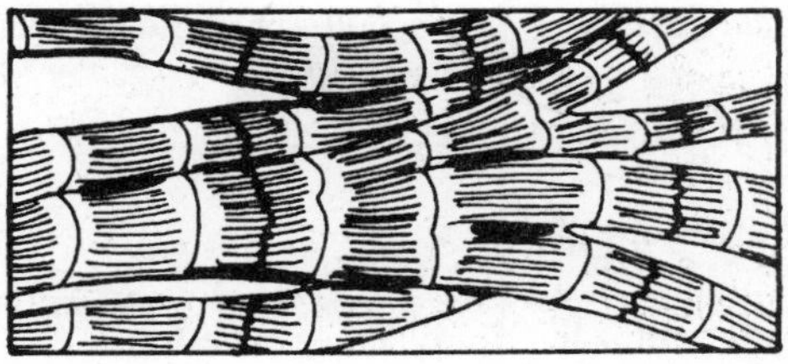

N Name ____________________

Location ____________________

Function ____________________

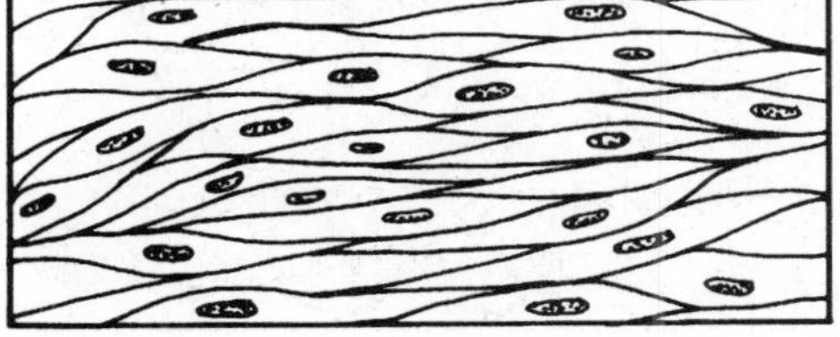

O Name ____________________

Location ____________________

Function ____________________

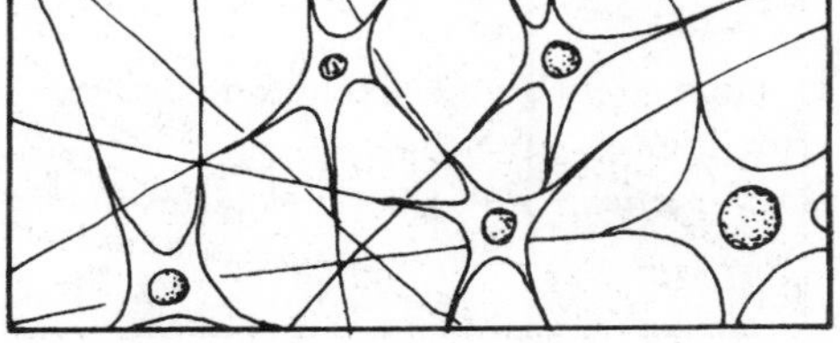

P Name ____________________

Location ____________________

Function ____________________

Figure LG 4.1 *(continued)*

✓ **B7.** *The Big Picture: Looking Ahead.* Check your understanding of the most common types of epithelium listed in the box by writing the name of the type after the phrase that describes it. To confirm your answers, refer to text figures or tables of upcoming chapters.

Pseudostratified ciliated columnar	**Simple squamous**
Simple columnar, ciliated	**Stratified squamous**
Simple columnar, nonciliated	**Transitional**
Simple cuboidal	

a. Lines the inner surface of the stomach (Figure 24.12) and intestine (Figure 24.23):

_____________________.

b. Forms the outer surface of skin (Figures 5.1, 5.3, 5.4, 5.6, and 5.7):

_____________________.

c. Single layer of cube-shaped cells; found in some glands, such as hormone-producing follicles of the thyroid gland (Figure 18.13c), as well as ducts of glands:

_____________________.

d. Endothelium, as in capillaries (Figure 21.1), as well as mesothelium lining serous

membranes: _____________________.

e. Lines air sacs of lungs where thin cells are required for diffusion of gases into blood

(Figure 23.12): _____________________.

f. Forms the delicate Bowman's capsule surrounding each filtering apparatus

(glomerulus) (Figure 26.8): _____________________.

g. Forms most of renal tubule walls (Table 26.1): _____________________.

h. Lines the urinary bladder; epithelial cells flatten as tissue is stretched (Figure 26.21):

_____________________.

✓ **B8.** *For extra review.* Circle the correct answers about epithelium.

a. Pseudostratified epithelium is *(one? many?)* layer(s) thick with *(all? only some?)* cells attached to the basement membrane and *(all? only some?)* cells reaching the apical surface.

b. Stratified epithelium is named after the cells that form the *(superficial? deepest?)* layer of that tissue.

c. Stratified squamous epithelium is likely to offer *(more? less?)* protection than simple squamous epithelium.

d. Stratified *(cuboidal? columnar?)* epithelium forms ducts of adult sweat glands.

B9. Complete the table describing distinguishing characteristics of specialized epithelial cells. Also fill in sites where such modified cells would likely be located.

Modification	Description	Location
a. Cilia		
b. Microvilli		Cells lining digestive tract
c. Keratin production	Waterproofing protein	
d. Goblet cells		

B10. *Clinical correlation.* State the purpose of a "Pap smear," and describe this procedure.

B11. Define glands and explain why glands are studied in this section on epithelium.

✓ **B12.** Write *EXO* before descriptions of *exocrine* glands, and *ENDO* before descriptions of *endocrine* glands. (Endocrine glands will be studied further in Chapter 18.)

_______ a. Their products are secreted into ducts that lead either directly or indirectly to the outside of the body.

_______ b. Their products are secreted into ECF and then the blood and so stay within the body; they are ductless glands.

_______ c. Examples are glands that secrete sweat, oil, mucus, and digestive enzymes.

_______ d. Examples are glands that secrete hormones.

✓ **B13.** Contrast classes of glands in this exercise. Circle the best answer in each case.

a. A gland with a duct that does branch is called a *(simple? compound?)* gland.
b. A gland consisting of one cell, such as a goblet cell, is a *(multi? uni?)*-cellular gland.
c. A gland with the secreting portions all shaped like flasks (or berries) is a(n) *(acinar? tubular? tubuloacinar?)* gland.
d. Most glands in the human body fit into the functional class of *(holo? mero?)*-crine gland. This means that *(only the secretion? the entire secretory cell, including the secretion?)* is released.
e. Name one location of each type of gland:

(1) Compound acinar: ______________________.

(2) Simple branched acinar: ______________________.

(3) Simple branched tubular: ______________________.

C. Connective tissue (pages 117–129)

✓ **C1.** Write *connective tissue* or *epithelial tissue* following the description below that correctly describes that tissue.

a. Consists of many cells with little intercellular substance (fibers and ground substance): ____________________

b. Penetrated by blood vessels (vascular): ____________________

c. Does not cover body surfaces or line passageways and cavities but is more internally located; binds, supports, protects: ____________________

✓ **C2.** List four or more categories of connective tissue matrix. (Note that *intercellular* means "between cells" or "extracellular.") What is the source of these materials?

✓ **C3.** Do this activity about connective tissue cells.

a. Cells in connective tissues derive from *(endo? meso? ecto?)*-derm cells in the embryo. These cells are called ____________________ cells.

b. Most matrix is secreted by *(immature? mature?)* connective tissue cells. Which type of cells have greater capacity for mitosis? *(Mature? Immature?)*

c. Names of immature cells end in *(-blast? -cyte?)*, whereas names of mature cells end in *(-blast? -cyte?)*. To check your knowledge further, write correct names for the following cells: immature bone cell: ____________________; immature cartilage cell: ____________________; mature bone cell: ____________________

✓ **C4.** Identify characteristics of each connective tissue cell type by writing the name of the correct cell type after its description. (Note that several of these cell types are shown in diagrams G, H, and I of Figure LG 4.1.)

Adipocyte	Fibroblast	Macrophage	Mast cell	Plasma cell	White blood cell

a. Derived from B lymphocyte; gives rise to antibodies; helpful in defense: ____________________

b. Phagocytic cell that engulfs bacteria and cleans up debris; important during infection; formed from monocyte: ____________________

c. Fat cell: ____________________

d. Forms collagenous and elastic fibers in injured tissue: ____________________

e. Abundant along walls of blood vessels; produces histamine, which dilates blood vessels: ____________________

f. Neutrophils and lymphocytes are examples: ____________________

✓ **C5.** The ground substance of connective tissue consists of *(large? small?)* molecules that are combinations of polysaccharides and *(lipids? proteins?)*. These are known as GAGs or ____________________. Identify chemical components of ground substance by selecting the answers listed in the box that best fit the descriptions below.

AP.	**Adhesion proteins**	**DS.**	**Dermatan sulfate**	**KS.**	**Keratan sulfate**
CS.	**Chondroitin sulfate**	**HA.**	**Hyaluronic acid**		

_______ a. Chemicals that stabilize position of cells; includes proteins such as fibronectin

_______ b. Viscous, slippery substance that binds cells together and lubricates joints

_______ c. Chemical component of bone, cartilage, and the cornea of the eye

_______ d. Jellylike; provides support and adhesiveness in bone, skin, blood vessels, and cartilage

_______ e. Found in skin, blood vessels, tendons, and heart valves

✓ **C6.** Select the type of fiber in the box that best fit descriptions below.

C.	**Collagen**	**E.**	**Elastic**	**R.**	**Reticular**

_______ a. Formed of collagen and glycoproteins; form branching networks that provide stroma for soft organs such as spleen and networks around fat, nerve, and muscle cells

_______ b. Strong fibers in bundles that provide great strength, as needed in bone and tendons; formed of the most abundant protein in the body

_______ c. Can be greatly stretched without breaking, an important quality of connective tissues forming skin, blood vessels, and lungs

_______ d. Composed of elastin surrounded by fibrillin (a glycoprotein that is defective in Marfan syndrome), a disorder affecting bones, large arteries, and eyes.

✓ **C7.** As described earlier, the embryonic connective tissue from which all of the connective tissues arise is called ____________________. *(Much? Little?)* of this type of tissue is found in the body after birth. Another embryonic tissue, known as mucous connective tissue or ____________________ jelly, gives support to the umbilical cord. Tissues such as areolar connective tissue, cartilage, or bone are known as *(embryonic? mature?)* connective tissues.

C8. After you study each mature connective tissue type in the text, refer to diagrams of connective tissues on Figure LG 4.1, G-L. Write the following information on the lines provided on the figure.

✓ a. The name of each tissue

b. One or more locations of the tissue

c. One or more functions of the tissue

Now color the following structures on each diagram, using the same colors as you did for diagrams A-F:

○ Nucleus ○ Matrix

○ Cytoplasm ○ Fat globule (a cellular inclusion), diagram H

✓ From this activity, you can conclude that connective tissues appear to consist mostly of *(cells? intercellular material [matrix]?)*.

C9. Label fibers in Figure LG 4.1, diagrams G, I, and J.
(Note that fibers *are* present in bone but not visible in diagram K.)

✓ **C10.** Which two types of tissue form subcutaneous tissue (superficial fascia)?

____________________, ____________________

✓ **C11.** Answer these questions about adipose tissue.

a. Explain what accounts for the fact that a fat cell appear like a class ring with a stone, as in Figure LG 4.1H.

b. List three or more functions of fat in the body.

c. Explain what gives the brown color to brown adiopse tissue (BAT), and state how this type of fat may be helpful to newborns.

✓ **C12.** *The Big Picture: Looking Ahead.* Refer to figures in your textbook and Table 4.2 (pages 121–127) to help you identify the dense connective tissues described below. Use the answers in the box. Use each answer once. Lines following descriptions are for Checkpoint C13.

A.	**Aponeurosis**	**P.**	**Periosteum**
C.	**Capsule**	**SL.**	**Suspensory ligament**
EWA.	**Elastic walls of arteries**	**T.**	**Tendon**
F.	**Fascia**	**TVC.**	**True vocal cords**
L.	**Ligament**	**V.**	**Valves**

________ a. Figure 6.3 shows the dense covering over bone which affords a pathway for blood to penetrate bone and also serves as a site of osteoblast cells that build new bone. _____

________ b. Strong but flexible bands that attach bone to bone are shown in Figure 9.11. _____

________ c. In Figure 9.14 a strong, fibrous band is shown attaching the quadriceps femoris muscle across the patella and eventually inserting it into bone (tibia). Similar fibrous bands attach forearm muscles into bones of the hand (Figure 11.17). _____

________ d. Figure 11.10 shows broad, sheetlike attachments of external and internal oblique muscles. _____

________ e. A broad sheet of fibrous tissue that provides support to the floor of the pelvic cavity (urogenital diaphragm) is shown in Figure 11.12. _____

________ f. Figure 20.4a and b demonstrates the strong connective tissue leaflets that direct the flow of blood within the heart. _____

________ g. Figure 21.1 shows the tissue that provides blood vessels' walls with the elasticity needed as the heart regularly pumps blood into these vessels. _____

________ h. Figure 23.6 demonstrates the structures that require strength yet flexibility as their continual alteration of tension changes vocal quality. _____

________ i. Connective tissue surrounds, supports, and protects each kidney (Figure 26.4). _____

________ j. Figure 28.3 shows the tissue that suspends and supports the penis. _____

✓ **C13.** Now identify the category to which each of the connective tissues listed in Activity C12 belongs. On the lines following each description above, write *DR* for dense regular, *DI* for dense irregular, or *E* for elastic.

✓ **C14.** Do this activity about cartilage.

a. In general, cartilage can endure *(more? less?)* stress than other connective tissues studied so far.

b. The strength of cartilage is due to its *(collagen? elastic? reticular?)* fibers, and its resilience is related to the presence of *(chondroitin? dermatan? keratan?)* sulfate.

c. Cartilage heals *(more? less?)* rapidly than bone. Explain why this is so.

d. Growth of cartilage occurs by two mechanisms. In childhood and adolescent years, cartilage increases in size by *(appositional? interstitial?)* growth by enlargement of existing cells within the cartilage.

e. Post adolescence, cartilage increases in thickness by *(appositional? interstitial?)*

growth. This process involves cells in the ____________________ that covers cartilage. Here cells known as *(chondroblasts? chondrocytes? fibroblasts?)* differentiate

into ____________________, which lay down matrix and eventually become mature chondrocytes.

✓ **C15.** Match the types of cartilage with the descriptions given.

E.	**Elastic cartilage**	**F.**	**Fibrouscartilage**	**H.**	**Hyaline cartilage**

________ a. Found where strength and rigidity are needed, as in discs between vertebrae and in symphysis pubis

________ b. White, glossy cartilage covering ends of bones (articular), covering ends of ribs (costal), and giving strength to nose, larynx, trachea, and bronchi

________ c. Provides strength and flexibility, as in external part of ear

________ d. Forms menisci (cartilage pads) of the knees

________ e. Forms most of the skeleton of the human embryo and is the most abundant type of cartilage in the adult

✓ **C16.** Complete this exercise about bone tissue.

a. Bone tissue is also known as ____________________ tissue. Compact bone consists of concentric rings or *(lamellae? canaliculi?)* with bone cells, called *(chondrocytes?*

osteocytes?), located in tiny spaces called ____________________. Nutrients in blood

reach osteocytes by vessels located in the ____________________ canal and then via

minute, radiating canals called ____________________ that extend out to lacunae. The basic unit of structure and function of compact bone is known as an

____________________ or Haversian system.

b. Spongy bone *(is? is not?)* arranged in osteons. Spongy bone consists of plates of bone

called ____________________.

c. List three or more functions of bone tissue.

✓ **C17.** List eight types of tissues or organs that are currently available (or are in progress) via tissue engineering.

✓ **C18.** Fill in the blanks about blood tissue.

a. Blood, or ________________ tissue, consists of a fluid called ________________

containing three types of formed elements: *(red? white?)* blood cells that transport

oxygen and carbon dioxide, ________________ blood cells that provide defense

for the body, and platelets that function in blood ________________.

b. *(Neutrophils? Lymphocytes?)* are white blood cells commonly found in lymph fluid that is leaving lymph nodes.

C19. Now label the following cell types found in diagrams J, K, or L of Figure LG 4.1: *chondrocytes, osteocytes, red blood cells, white blood cells,* and *platelets.*

D. Membranes (page 129)

D1. Complete the table about the four types of membranes in the body.

Type of Membrane	Location	Example of Specific Location	Function (s)
a.	Lines body cavity leading to exterior		
b. Serous			Allows organs to glide easily over each other
c.		Lines knee and hip joints	
d.	Covers body surface	Skin	

✓ **D2.** *The Big Picture: Looking Ahead.* Refer to figures in your textbook to help you identify the types of membranes described below. Use the answers in the box.

Cutaneous	**Pericardial**	**Pleural**
Mucous	**Peritoneal**	**Synovial**

a. A ____________________ membrane is another name for skin, shown in Figure 5.1.

b. Figure 9.3 shows a ____________________ membrane that secretes a lubricating fluid that reduces friction at joints. This type of membrane contains connective (rather than epithelial) tissue.

c. On Figure 20.2a, the serous membrane that covers the heart is shown. This is the ____________________ membrane. The same figure shows the ____________________ membrane covering the lungs and lining the chest.

d. Figure 24.2 contrasts the membrane lining the inside of digestive organs, known as a ____________________ membrane, with the ____________________ membrane that covers the outside of such organs. Figure 26.21 shows similar membranes of the urinary bladder: the pink ____________________ membrane with a rippled pattern (rugae) inside the organ and the ____________________ membrane over the outside of the bladder.

✓ **D3.** Describe specific layers of serous and mucous membranes in this exercise.

a. The layer of a serous membrane that adheres to an organ, such as the liver, is the *(parietal? visceral?)* layer, whereas the portion lining the body cavity wall is the *(parietal? visceral?)* layer.

b. The epithelial layer of mucous membranes varies, with *(simple columnar? pseudostratified columnar? stratified squamous?)* lining the mouth, ____________________ lining the small intestine, and ____________________ lining the large airways to the lungs.

c. The connective tissue of a mucous membrane is known as the lamina ____________________. Unlike the surface epithelial layer, the lamina propria *(is? is not?)* vascular.

✓ **D4.** *A clinical correlation.* Serous membranes are normally relatively free of microorganisms, whereas mucous and cutaneous membranes have many microorganisms on them. Based on this information, state a rationale for the administration of antibiotics to patients undergoing stomach or intestinal surgery.

E. Muscle tissue (pages 130–132)

✓ **El.** After you study each muscle tissue type in the text, refer to diagrams of muscle tissues in Figure LG 4.1, M-O. Write the following information on the lines provided on the figure.

✓ a. The name of each muscle tissue

b. One or more locations of the tissue

c. One or more functions of the tissue

Now color the following structures on each diagram, using the same colors as you did for diagrams A-L.

○ Nucleus ○ Cytoplasm ○ Intercellular material

✓ From this activity, you can conclude that muscle tissues appear to consist mostly of *(cells? intercellular material [matrix]?)*

✓ **E2.** Check your understanding of the three muscle types listed in the box by selecting types that best fit descriptions below.

C. **Cardiac**	**Sk.** **Skeletal**	**Sm.** **Smooth**

________ a. Tissue forming most of the wall of the heart

________ b. Attached to bones

________ c. Spindle-shaped cells with ends tapering to points

________ d. Contains intercalated discs

________ e. Found in walls of intestine, urinary bladder, and blood vessels

________ f. Voluntary muscle

F. Nervous tissue (page 132)

Fl. Refer to diagram P in Figure LG 4.1. Write the following information on the lines provided on the figure.

✓ a. The name of the tissue

b. One or more locations of the tissue

c. One or more functions of the tissue

F2. Describe cells of the nervous system in this exercise.

a. *(Neurons? Neuroglia?)* convert stimuli into nerve impulses and conduct impulses.

b. *(Axons? Dendrites?)* carry nerve impulses toward the cell body of a neuron, whereas

________________________ transmit nerve impulses away from the cell body.

G. Tissue repair: restoring homeostasis; aging and tissues (pages 132–133)

✓ **Gl.** Circle the type of tissue in each case that is more capable of repair:

a. Cardiac muscle - Smooth muscle

b. Bone - Cartilage

c. Nervous tissue - Epithelium

d. Blood - Tendon

✓ **G2.** Define a *stem cell.*

List three locations where stem cells serve as sources of new cells throughout life.

✓ **G3.** Choose the correct term to complete each sentence. Write *P* for parenchymal or *S* for stromal.

a. The part of an organ that consists of functioning cells, such as secreting epithelial

cells lining the intestine, is composed of ________ tissue.

b. The connective tissue cells that support the functional cells of the organ are called

________ cells.

c. If only the ________ cells are involved in the repair process, the repair will be close to perfect.

d. If ________ cells are involved in the repair, they will lay down fibrous tissue known as a fibrosis or a scar.

G4. Define each of the following terms related to tissue repair.

a. Regeneration

b. Granulation tissue

c. Adhesions

✓ **G5.** List three factors that affect the quality of tissue repair.

Describe two roles of vitamin C in tissue repair.

G6. Explain what changes in collagen and elastin result in aging of tissues.

H. Disorders and medical terminology (page 134)

✓ **H1.** Check your understanding of two connective tissue disorders, Sjögren's syndrome and systemic lupus erythematosus (SLE), by filling in the table in this Checkpoint.

Characteristic of disease	Sjögren's syndrome	SLE (lupus)
a. Autoimmune *(Yes? No?)*		
b. Type of tissue most affected?		
c. Gender most affected *(F? M?)*		
d. Distinguishing signs/symptoms		

H2. *Critical thinking.* An autograft refers to replacement of one part of a person's body (such as a traumatized bone or burned skin) with another part of the same person's body (such as a piece of hip bone or healthy skin). Contrast autograft with *xenotransplantation* (or *xenograft*) (see text page 134) and predict the success rate of the procedures.

ANSWERS TO SELECTED CHECKPOINTS: CHAPTER 4

A3. (a) Endo. (b) Ecto. (c) Meso. (d) Ecto. (e) Meso.

A4. (a) Phagocyte. (b) Junctions; tight, urinary bladder; transmembrane. (c) Protein; cadherins; adhesion. (d) Cardiac muscle, epidermal; do. (e) Hemidesmosomes, integrins. (f) Connexons; impulses.

B1. (a) Closely packed cells; many, tightly, lateral. (b) Do not; avascular; epithelial tissue is protective because it does not bleed when injured; however, the avascular epithelial cells must draw nutrients from adjacent tissues. (c) Has. (d) Does. (e) Microvilli, cilia.

B2. (a) Basal, basement. (b) Extracellular material; epithelium; collagen, basal. (c) Connective tissue, reticular.

B4. Covering or lining; glandular.

B5. (a) Simple; stratified. (b) Squamous; columnar; cuboidal. (c) Transitional.

B6. (A) Simple squamous epithelium. (B) Simple cuboidal epithelium. (C) Simple columnar epithelium (nonciliated). (D) Simple columnar epithelium (ciliated). (E) Stratified squamous epithelium. (F) Pseudostratified columnar epithelium, ciliated.

B7. (a) Simple columnar, nonciliated. (b) Stratified squamous. (c) Simple cuboidal. (d–f) Simple squamous. (g) Simple cuboidal. (h) Transitional.

B8. (a) One, all, only some. (b) Superficial. (c) More. (d) Cuboidal.

B12. (a) EXO. (b) ENDO. (c) EXO. (d) ENDO.

B13. (a) Compound. (b) Uni. (c) Acinar. (d) Mero; only the secretion. (e1) Mammary glands. (e2) Sebaceous (oil) glands. (e3) Gastric glands.

C1. (a) Epithelial tissue. (b) Connective tissue (except cartilage). (c) Connective tissue.

C2. Fluid, semifluid, gelatinous, fibrous, or calcified; usually secreted by connective tissue (or adjacent) cells.

C3. (a) Meso-; mesenchymal. (b) Immature; immature. (c) -blast, -cyte; osteoblast, chondroblast, osteocyte.

C4. (a) Plasma cell. (b) Macrophage. (c) Adipocyte. (d) Fibroblast. (e) Mast cell. (f) WBC (white blood cell).

C5. Large, proteins, glycosaminoglycans (a) AP. (b) HA. (c) KS. (d) CS. (e) DS.

C6. (a) R. (b) C. (c) E. (d) E.

C7. (a) Mesenchyme; little; Wharton's, mature.

C8. (G) Loose (areolar) connective tissue. (H) Adipose tissue. (I) Dense connective tissue. (J) Cartilage. (K) Osseous (bone) tissue. (L) Vascular (blood) tissue. Intercellular material (matrix).

C10. Loose (areolar) connective tissue; adipose tissue.

C11. (a) Most of the cell consists of a globule of stored triglycerides (fat); the nucleus and the rest of the cytoplasm are located in a narrow ring around the fat globule. (b) Fat insulates the body, supports and protects organs, and serves as a major source of energy. (c) This type of fat has an abundance of blood vessels as well as mitochondria (which appear brown due to the presence of copper-containing cytochromes). Brown fat may help to generate heat to maintain proper body temperature in newborns.

C12. 13. (a) P, DI. (b) L, DR. (c) T, DR. (d) A, DR. (e) F, DI. (f) V, DI. (g) EWA, E. (h) TVC, E. (i) C, DI. (j) SL, E.

C14. (a) More. (b) Collagen, chondroitin. (c) Less; cartilage is avascular, so chemicals needed for repair must reach cartilage by diffusion from the perichondrium. (d) Interstitial. (e) Appositional; perichondrium; fibroblasts, chondroblasts.

C15. (a) F. (b) H. (c) E. (d) F. (e) H.

C16. (a) Osseous; lamellae, osteocytes, lacunae; central (Haversian), canaliculi; osteon. (b) Is not; trabeculae. (c) Supports and protects, plays a role in movement, stores mineral salts, houses red and yellow marrow, including lipids.

C17. Skin, cartilage, bone, bone marrow, tendons, heart valves, dopamine-producing cells, insulin-producing cells, intestine, liver, or kidney.

C18. Vascular, plasma; red, white, clotting. (b) Lymphocytes.

D2. (a) Cutaneous. (b) Synovial. (c) Pericardial; pleural. (d) Mucous, peritoneal (or serous); mucous, peritoneal (or serous).

D3. (a) Visceral, parietal. (b) Stratified squamous, simple columnar, pseudostratified columnar. (c) Propria; is.

D4. Microorganisms from skin or from within the stomach or intestine may be introduced into the peritoneal cavity during surgery; the result may be an inflammation of the serous membrane (peritonitis).

E1. (M) Skeletal muscle. (N) Cardiac muscle.(O) Smooth muscle. Cells [known as muscle fibers].

E2. (a) C. (b) Sk. (c) Sm. (d) C. (e) Sm. (f) Sk.

F1. Nervous tissue (neurons).

F2. (a) Neurons. (b) Dendrites, axons. (*Hint:* Remember A for away from cell body.)

G1. (a) Smooth muscle. (b) Bone. (c) Epithelium. (d) Blood.

G2. Immature, undifferentiated cell; skin, lining of the GI tract, and bone marrow.

G3. (a) P. (b) S. (c) P. (d) S.

G5. Nutrition, circulation, and (younger) age; affects quality of collagen and blood vessels.

H1. (a) Yes, Yes. (b) Exocrine glands; diverse connective tissues. (c) F, F. (d) Sjögren's: dry mouth and dry eyes; also arthritis. SLE: butterfly rash, and other skin lesions, sensitivity to light, painful joints; possibly serious damage to organs such as kidneys.

MORE CRITICAL THINKING: CHAPTER 4

1. The outer part (epidermis) of skin is composed of epithelium. Explain what characteristics make epithelium suitable for this location.
2. Which tissue is likely to heal better: bone or cartilage? Explain why.
3. Discuss the process of differentiation in development of tissues by explaining the roles of mesenchyme and stem cells.
4. Surgical repair on fetuses (in utero) leaves no surgical scar. Explain why this is so.
5. Contrast hypertrophy and hyperplasia. (*Hint:* See Medical Terminology in Chapter 3.) Predict which types of tissues can grow in adults by hypertrophy but not by hyperplasia. State your rationale.

MASTERY TEST: CHAPTER 4

Questions 1-6: Circle the letter preceding the one best answer to each question.

1. A surgeon performing abdominal surgery will pass instruments through the skin, and then loose connective tissue (subcutaneous), then fascia covering muscle and

   muscle tissue itself to reach the _______ membrane lining the inside wall of the abdomen.
   A. Parietal pleura
   B. Parietal pericardium
   C. Parietal peritoneum
   D. Visceral pleura
   E. Visceral pericardium
   F. Visceral peritoneum
2. The type of tissue that covers body surfaces, lines body cavities, and forms glands is:
   A. Nervous
   B. Muscular
   C. Connective
   D. Epithelial
3. Which of the four types of tissue is the most abundant and most widely distributed tissue in the body?
   A. Nervous tissue
   B. Muscle tissue
   C. Connective tissue
   D. Epithelial tissue
4. Which statement about connective tissue is *false*?
   A. Cells are very closely packed together.
   B. Most connective tissues have an abundant blood supply.
   C. Intercellular substance is present in large amounts.
   D. Bone, cartilage, fat, and tendons are examples.
5. Modified columnar cells that are unicellular glands secreting mucus are known as:
   A. Cilia
   B. Microvilli
   C. Goblet cells
   D. Branched tubular glands
6. Which tissues are avascular?
   A. Skeletal muscle and cardiac muscle
   B. Bone and dense connective tissue
   C. Cartilage and epithelium
   D. Areolar and adipose tissues.

Questions 7-12: Match tissue types with correct descriptions

Adipose	**Simple columnar epithelium**
Cartilage	**Simple squamous epithelium**
Dense connective tissue	**Stratified squamous epithelium**

7. Contains lacunae and chondrocytes: ____________________
8. Forms fascia, tendons, and deeper layers of dermis of skin: ____________________
9. Forms thick surface layer of skin on hands and feet, providing extra protection:

   ____
10. Single layer of flat, scalelike cells: ____________________
11. Lines stomach and intestine: ____________________
12. Fat tissue: ____________________

Questions 13-20: Circle T (true) or F (false). If the statement is false, change the underlined word or phrase so that the statement is correct.

T F 13. Stratified transitional epithelium lines the urinary bladder.

T F 14. About 75% of lipids in plasma membranes are glycolipids.

T F 15. Hormones are classified as exocrine secretions.

T F 16. Elastic connective tissue, because it provides both stretch and strength, is found in walls of elastic arteries, in the vocal cords, and in some ligaments.

T F 17. The surface attachment between epithelium and connective tissue is called basement membrane.

T F 18. Simple squamous epithelium is most likely to line the areas of the body that are subject to wear and tear.

T F 19. Marfan syndrome is a muscle tissue disorder that results in persons being short with short extremities.

T F 20. Appositional growth of cartilage starts later in life than interstitial growth and continues throughout life.

Questions 21-25: Fill-ins. Complete each sentence with the word that best fits.

____________________ 21. The type of cell junction that forms a "zip-lock" is a ________ junction.

____________________ 22. ________ is a dense, irregular connective tissue covering over bone.

____________________ 23. The kind of tissue that lines alveoli (air sacs) of lungs is ________.

____________________ 24. Another term for intercellular is ________.

____________________ 25. All glands are formed of the tissue type named ________.

ANSWERS TO MASTERY TEST: CHAPTER 4

Multiple Choice

1. C
2. D
3. C
4. A
5. C
6. C

Matching

7. Cartilage
8. Dense connective tissue
9. Stratified squamous epithelium
10. Simple squamous epithelium
11. Simple columnar epithelium
12. Adipose

True-False

13. T
14. F. Phospholipids
15. F. Endocrine
16. T
17. T
18. F. Stratified
19. F. Connective tissue; tall stature with long extremities
20. T

Fill-ins

21. Tight
22. Periosteum
23. Simple squamous epithelium
24. Extracellular or interstitial
25. Epithelium

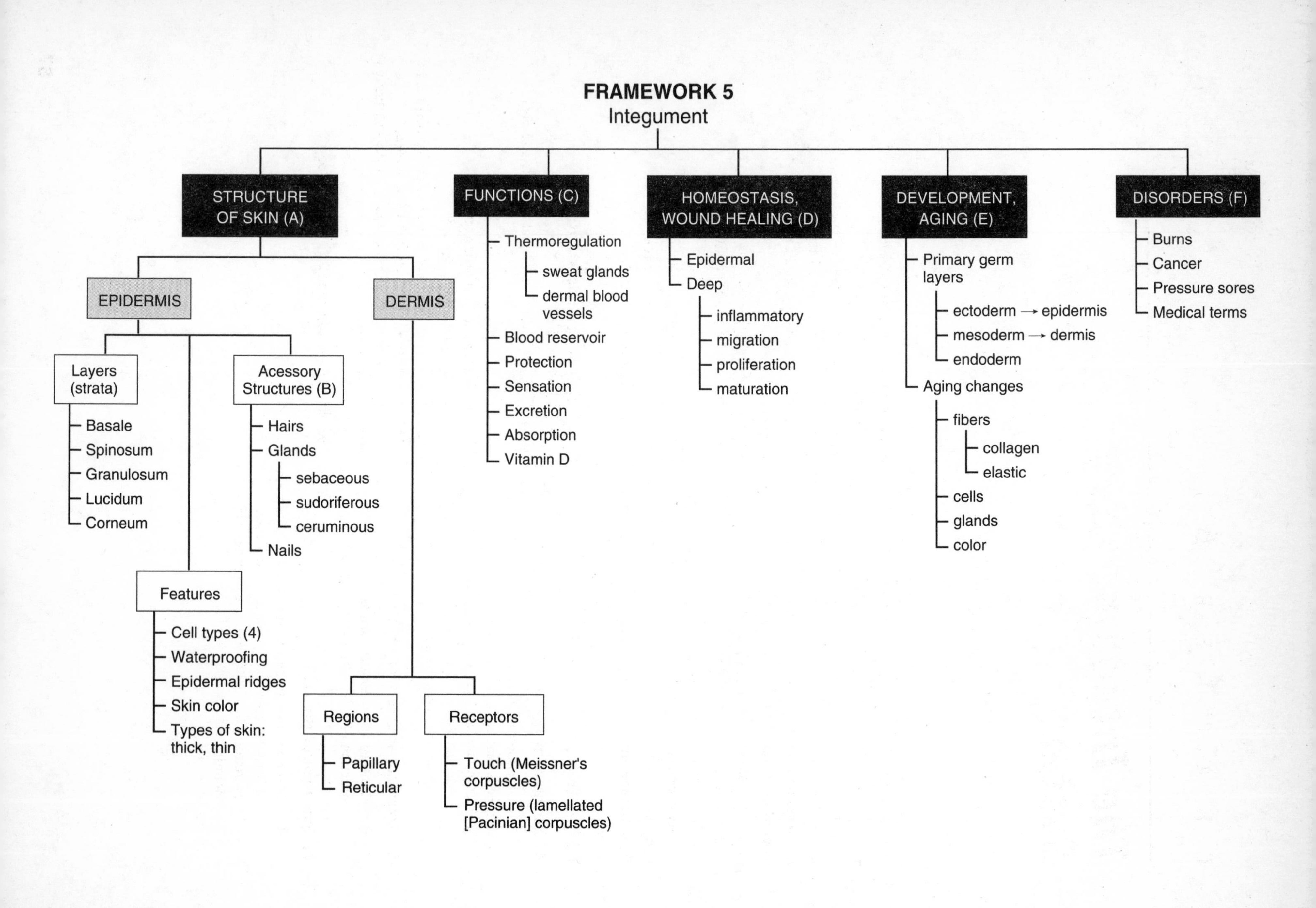
FRAMEWORK 5
Integument
STRUCTURE OF SKIN (A)
EPIDERMIS
Layers (strata)
Basale
Spinosum
Granulosum
Lucidum
Corneum
Acessory Structures (B)
Hairs
Glands
sebaceous
sudoriferous
ceruminous
Nails
Features
Cell types (4)
Waterproofing
Epidermal ridges
Skin color
Types of skin: thick, thin
DERMIS
Regions
Papillary
Reticular
Receptors
Touch (Meissner's corpuscles)
Pressure (lamellated [Pacinian] corpuscles)
FUNCTIONS (C)
Thermoregulation
sweat glands
dermal blood vessels
Blood reservoir
Protection
Sensation
Excretion
Absorption
Vitamin D
HOMEOSTASIS, WOUND HEALING (D)
Epidermal
Deep
inflammatory
migration
proliferation
maturation
DEVELOPMENT, AGING (E)
Primary germ layers
ectoderm → epidermis
mesoderm → dermis
endoderm
Aging changes
fibers
collagen
elastic
cells
glands
color
DISORDERS (F)
Burns
Cancer
Pressure sores
Medical terms

CHAPTER 5

The Integumentary System

In your human body tour you have now arrived at your first system, the integumentary system, which envelops the entire body. Composed of the skin, hair, nails, and glands, the integument serves as both a barrier and a link with the environment. The many waterproofed layers of cells, as well as the pigmentation of skin, afford protection to the body. Nails, glands, and even hairs offer additional fortification. Sense receptors and blood vessels in skin increase awareness of conditions around the body and facilitate appropriate responses. Even so, this microscopically thin surface cover may be traumatized, or invaded, or it may simply succumb to the passage of time. The integument then demands repair and healing for continued maintenance of homeostasis.

As you begin your study of the integumentary system, carefully examine the Chapter 5 Framework and note relationships among concepts and key terms there. Also refer to the Topic Outline and Objectives; you may want to check off each objective as you complete it.

TOPIC OUTLINE AND OBJECTIVES

A. Structure of the skin

1. Describe the layers of the epidermis and the cells that compose them.
2 Compare the composition of the papillary and reticular regions of the dermis.
3. Explain the basis for skin color.

B. Accessory structure of the skin

1. Contrast the structure, distribution, and function of hair, skin glands, and nails.

C. Functions of the skin

1. Describe how the skin contributes to regulation of body temperature, protection, sensation, excretion and absorption, and synthesis of vitamin D.

D. Maintaining homeostasis: skin wound healing

1. Explain how epidermal wounds and deep wounds heal.

E. Development and aging of the integumentary system

1. Describe the development of the epidermis, its accessory structures, and the dermis.
2. Describe the effects of aging on the integumentary system.

F. Disorders and medical terminology

WORDBYTES

Now become familiar with the language of this chapter by studying each wordbyte, its meaning, and an example of its use within a term. After you study the entire list, self-check your understanding by writing the meaning of each wordbyte on the line. As you continue through the *Learning Guide*, identify (and fill in) additional terms that contain the same wordbyte.

Wordbyte	Self-check	Meaning	Example(s)
albin-	____________	white	*albin*ism
basale	____________	base	stratum *basale*
cera-	____________	wax	*cer*umen
corneum	____________	horny	stratum *corneum*
cut-	____________	skin	*cut*aneous
derm-	____________	skin	*derm*atologist
epi-	____________	over	*epi*dermis
fer-	____________	carry	sudori*fer*ous
granulum	____________	little grain	stratum *granulosum*
lucidus	____________	clear	stratum *lucidum*
melan-	____________	black	*melan*oma
seb-	____________	fat	*seb*aceous
spinosum	____________	thornlike, prickly	stratum *spinosum*
sub-	____________	under	*sub*cutaneous
sudor-	____________	sweat	*sudor*iferous

CHECKPOINTS

A. Structure of the skin (pages 140–146)

A1. Name the structures included in the integumentary system.

A2. Skin may be one of the most underestimated organs in the body. What functions does your skin perform while it is "just lying there" covering your body? List some functions on the lines provided.

____________________ ____________________

____________________ ____________________

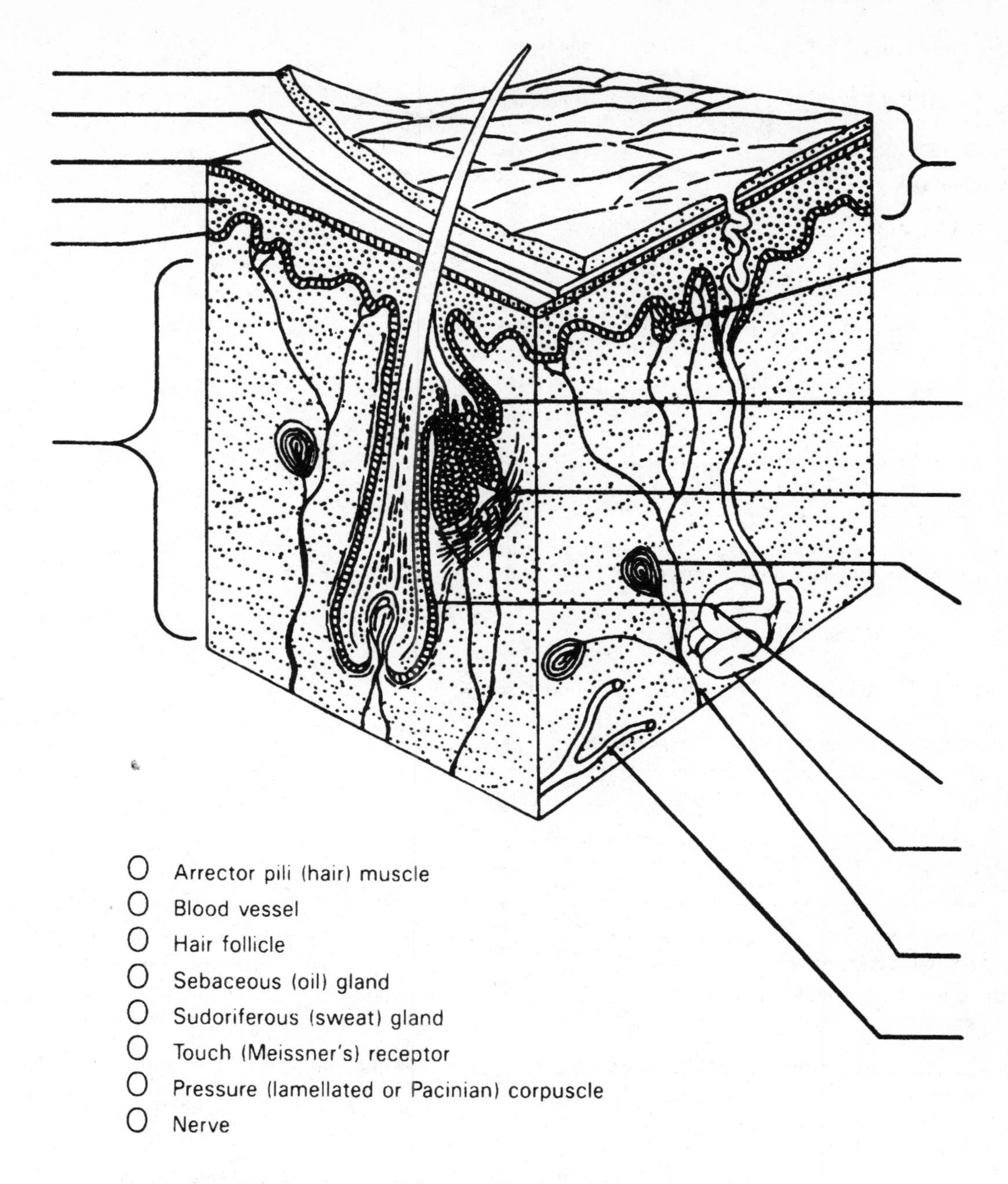

Figure LG 5.1 Structure of the skin. Label as directed in Checkpoints A4, A7, and A11.

✓ **A3.** Skin typically covers about ______ square meters in the adult body. The thickness of skin averages about ______ mm. In which part of the body is the thickest skin located? ________________

✓ **A4.** Answer these questions about the two regions of skin. Label them (bracketed regions) on Figure LG 5.1.

a. The outer layer is named the __________________. It is composed of *(connective tissue? epithelium?)*.

b. The inner portion of skin, called the __________________, is made of *(connective tissue? epithelium?)*. The dermis is *(thicker? thinner?)* than the epidermis.

✓ **A5.** The tissue underlying skin is called *subcutaneous,* meaning ____________________.

This layer is also called ____________________. It consists of two types of tissue,

____________________ and ____________________. What functions does subcutaneous tissue serve?

✓ **A6.** Epidermis contains four distinct cell types. Fill in the name of the cell type that fits each description.

a. The most numerous cell type, this cell produces the protein keratin, which helps to waterproof skin: ____________________.

b. This type of cell produces the pigments that give skin its color and protect against ultraviolet (UV) light: ____________________.

c. These cells, derived from macrophages, function in immunity: ____________________.

d. Located in the deepest layer of the epidermis, these cells that originate in bone marrow contribute to the sensation of touch: ____________________.

✓ **A7.** Label each of the five layers of the epidermis on the left side of Figure LG 5.1.

✓ **A8.** Select the names of the epidermal layers (strata) that fit descriptions below.

B.	**Basale**	**L.**	**Lucidum**
C.	**Corneum**	**S.**	**Spinosum**
G.	**Granulosum**		

________ a. Also known as the stratum germinativum because new cells arise from stem cells located here; attached to the basement membrane.

________ b. Attached by desmosomes to adjacent cells and by hemidesmosomes to basement membrane; intracellular tonofilaments here form keratin.

________ c. Stratum immediately superficial to the stratum basale; "prickly spines" help attach cells to adjacent cells, providing strength and flexibility.

________ d. Cells in this layer are undergoing apoptosis; they contain keratohyalin as well as lamellar granules—the latter releasing waterproofing lipids.

________ e. Cells begin to die in this stratum.

________ f. A clear, thin layer normally found only in thick layers of palms and soles.

________ g. Uppermost layer, consisting of 25 to 30 rows of flat, dead cells filled with keratin and embedded in waterproofing lipids.

A9. The journey of a cell from formation in the stratum basale to shoughing off in the stratum corneum normally requires four *(days? weeks? years?).* Psoriasis of skin including that of scalp (dandruff) involves a *(faster? slower?)* turnover of epithelial cells.

✓ **A10.** Describe the dermis in this exercise.

a. The outer one-fifth of the dermis is known as the *(papillary? reticular?)* region. It consists of *(loose? dense?)* connective tissue. Present in fingerlike projections known as dermal ____________________ are Meissner's corpuscles, sense receptors sensitive to *(pressure? touch?)*.

b. The remainder of the dermis is known as the ____________________ region, composed of *(loose? dense?)* connective tissue. Skin is strengthened by *(elastic ? collagenous?)* fibers in the reticular layer. Skin is extensible and elastic due to ____________________ fibers.

c. Epidermal ridges on palms, fingers, soles, and toes form the basis for ____________________ and ____________________. When are these ridges formed? ____________________. Do these patterns change during life? _____ What determines the patterns of these ridges?

✓ **A11.** Most sensory receptors, nerves, blood vessels, and glands are embedded in the *(epidermis? dermis?)*. Fill in all label lines on the right side of Figure LG 5.1. Then color those structures and their related color code ovals.

✓ **A12.** Explain what accounts for skin color by doing this exercise.

a. Dark skin is due primarily to *(a larger number of melanocytes? greater melanin production per melanocyte?)*. Melanocytes are in greatest abundance in the *(dermis? epidermis?)*. Melanin is produced in organelles called ____________________.

b. Melanin is derived from the amino acid ____________________. The enzyme that converts tyrosine to melanin is ____________________. This enzyme is activated by ____________________ light. In the inherited condition known as ____________________, melanocytes cannot synthesize tyrosinase, so skin, hair, and eyes lack melanin.

c. What are freckles and liver spots?

d. Carotene contributes a ________________ color to skin.

e. What accounts for the pinker color of skin during blushing and acts as a cooling mechanism during exercise?

This condition is known as *(cyanosis? erythema?)*.

f. Jaundiced skin has a more *(blue? red? yellow?)* hue, often due to ________________ problems. Cyanotic skin appears more *(blue? red? yellow?)* due to lack of oxygen and excessive carbon dioxide in the blood vessels of the skin.

A13. Contrast *thick skin* with *thin skin* by completing this table.

Characteristic	Thick skin	Thin skin
a. Locations	Palms, palmar surfaces of digits, and soles	
b. Structure of strata		
c. Epidermal ridges *(Yes? No?)*		
d. Sweat glands *(Many? Few?)*		
e. Sensory receptors *(Many? Few?)*		

B. Accessory structures of the skin (pages 146–150)

✓ **B1.** List the three types of accessory structures of skin. ________________

________________ ________________. These structures all develop from the embryonic *(dermis? epidermis?)*.

✓ **B2.** Complete this exercise about the structure of a hair and its follicle.

a. Arrange the parts of a hair from superficial to deep: _____ _____ _____
A. Shaft B. Bulb C. Root

b. Arrange the layers of a hair from outermost to innermost: _____ _____ _____
A. Medulla B. Cuticle C. Cortex

c. A hair is composed of *(cells? no cells, but only secretions of cells?)*. What accounts for the fact that hairs are waterproofed?

d. Surrounding the root of a hair is the hair ________________. The follicle consists of two parts. The *(external? internal?)* root sheath is an extension of deeper layers of the *(epidermis? dermis?)*. The internal root sheath consists of cells derived

from the ________________.

e. The ________________ is the part of a hair follicle where cells undergo mitosis, permitting growth of a new hair. What is the function of the papilla of the hair?

B3. Describe the relationship between the following pairs of terms related to hairs:

a. Round hair—oval hair

b. Sebaceous glands—root of hair

c. Arrector pili muscle—"goose bumps"

d. Growth stage—resting stage (in the growth cycle of a hair)

e. Lanugo—vellus hairs

B4. Complete this exercise about hair growth.

a. Scalp hair normally grows for about _____ years, and then rests for _____ months.

b. Normal hair loss is about ________ hairs/day. List several factors that can increase the rate of hair loss.

c. What is the meaning of the term *alopecia?*

B5. Describe the different chemicals in hairs that account for these colors:

a. Brown or black

b. Blonde or red

c. Graying

d. White

✓ **B6.** Complete this exercise about hair and hormones.

a. At puberty androgens are likely to _____-crease hair growth in men. Females with excessive androgen production may develop hirsutism which refers to *(deficient? excessive?)* hair growth.

b. Males who develop male-pattern baldness have *(inhibition? stimulation?)* of hair growth associated with androgens. Vasodilator medications improve hair growth in about *(33% 67% 100%)* of men who try it; it *(does? does not?)* work in men or women who are already bald.

✓ **B7.** Answer these questions about sebaceous glands.

a. In which parts of the body are these glands largest?

b. In which body parts are sebaceous glands absent?

c. What functions are attributed to these glands?

✓ **B8.** Contrast types of sudoriferous (sweat) glands in this activity. Use answers in the box.

Apo.	**Apocrine**	**Ecc.**	**Eccrine**

_______ a. Of apocrine and eccrine, the more common type of gland

_______ b. Responsible for "sweaty palms" because these glands are found in large numbers in palms

_______ c. Associated with foot odor because these glands are found in large numbers on soles

_______ d. Produce most of the sweat in armpits (axillae) and in pubic areas

_______ e. Of apocrine and eccrine glands, secrete more viscous (less watery) fluid

✓ **B9.** Write one major function and one minor function of sweat (perspiration).

✓ **B10.** *For extra review.* Write the answers (types of glands) listed in the box that best fit the descriptions below.

C.	**Ceruminous**	**Seb.**	**Sebaceous**	**Sud.**	**Sudoriferous**

_______ a. Sweat glands

_______ b. Inflamed in acne

_______ c. Ear wax

_______ d. Most numerous in skin of the palms and soles

_______ e. Usually most superficial in location

B11. Refer to Figure LG 5.2. Now look at one of your own fingernails. Label the following parts of the nail on the figure: *free edge, nail body, nail root, lunula,* and *eponychium (cuticle).* Check your accuracy by consulting Figure 5.5 page 150 in your text.

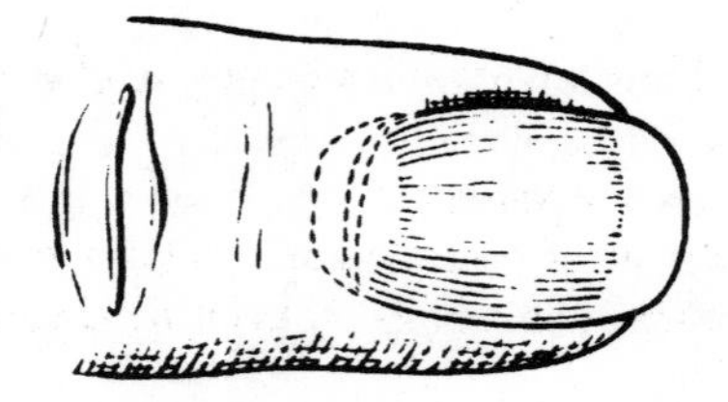

Figure LG 5.2 Structure of nails. Label according to Checkpoint B11.

✓ **B12.** Answer these questions about nails and hair.

a. Are nails formed of cells? *(Yes? No?)* Do nails contain keratin? *(Yes? No?)* What type of tissue forms nails? *(Dermis? Epidermis?)*

b. How is the function of the *nail matrix* similar to that of the *matrix of a hair?*

c. Why does the nail body appear pink, yet the lunula and free edge appear white?

C. Functions of the skin (pages 151–152)

✓ **C1.** Complete this exercise about temperature regulation.

a. When the body temperature is too hot, as during vigorous exercise, nerve messages from the brain inform sweat glands to *(in? de?)*-crease sweat production. In addition, blood vessels in the dermis are directed to *(dilate? constrict?)*. Skin blood vessels contain about ________% of the total body blood of a resting adult.

b. Temperature regulation is an example of a *(negative? positive?)* feedback system because the response (cooling) is *(the same as? opposite to?)* the stimulus (heating).

C2. List six or more ways in which skin protects the human body.

C3. List five or more types of sensations that arise in the skin.

C4. Check your understanding of the excretive and absorptive functions of skin in this Checkpoint.

a. How much water per day is likely to be lost from skin by evaporation?

_____ mL = _____ cup(s)

b. List several waste products excreted by skin.

c. *(Fat? Water?)*-soluble vitamins can be absorbed by skin. These include vitamins _____,

_____, _____, and ____. Sunlight on skin helps to activate vitamin _____, which is needed

for absorption of the mineral ______ from foods.

d. Administration of medications by a patch (or paste) on skin is known as __________ drug administration. Name four or more medications that can be absorbed in this way.

D. Maintaining homeostasis: skin wound healing (pages 152–153)

✓ **D1.** Describe the process of epidermal wound healing in this exercise.

a. State two examples of epidermal wounds.

b. Usually the deepest part of the wound is the *(central? peripheral?)* region.

c. In the process of repair, epidermal cells of the stratum *(corneum? basale?)* break contact from the basement membrane. These are cells at the *(center? periphery?)* of the wound.

d. These basal cells migrate toward the center of the wound, stopping when they meet other similar advancing cells. This cessation of migration is an example of the phenomenon known as _______________.

e. Both the migrated cells and the remaining epithelial cells at the periphery undergo _______________ to fill in the epithelium up to a normal (or close to normal) level.

Name a chemical that stimulates mitosis in the healing wound. _______________

D2. The process of deep wound healing involves four phases. List the three or four major events that occur in each phase.

a. Inflammatory

b. Migratory

c. Proliferative

d. Maturation

✓ **D3.** For *extra review* of deep wound healing, match the phases in the box with descriptions below.

I.	**Inflammation**	**Mig.**	**Migration**
Mat.	**Maturation**	**P.**	**Proliferation**

_______ a. Blood clot temporarily unites edges of wound; blood vessels dilate so neutrophils enter to clean up area.

_______ b. Clot forms a scab; epithelial cells migrate into scab; fibroblasts also migrate to start scar tissue; pink granulation tissue contains delicate new blood vessels.

_______ c. Epithelium and blood vessels grow; fibroblasts lay down many fibers.

_______ d. Scab sloughs off; epidermis grows to normal thickness; collagenous fibers give added strength to healing tissue; blood vessels are more normal.

D4. Contrast these two terms: *hypertrophied scar/keloid*

E. Development and aging of the integumentary system (pages 153–154)

✓ **E1.** *The Big Picture: Looking Ahead.* Do this exercise by referring to specific figures and exhibits in Chapter 29 in your text.

a. Identify and list the three primary germ layers (Figure 29.6b) developed from a fertilized egg: ______________-derm, ______________-derm, and ______________-derm.

b. The epidermis of skin develops from the ______________-derm (Table 29.1). All epidermal layers are formed by the *(second? fourth?)* month of the nine-month human gestation period (Table 29.2).

c. Nails form from the *(dermis? epidermis?)* and reach the ends of fingertips by the *(fourth? ninth?)* month of gestation (Table 29.2).

d. Delicate fetal hair called ______________ covers the body during the middle trimester of fetal development. It is normally *(present at? shed before?)* birth (Table 29.2), so its presence on a newborn is an indicator of prematurity.

e. The dermis and its associated blood vessels derive from the ______________-derm (Table 29.1).

f. What is the function of *vernix caseosa*?

✓ **E2.** Complete the table relating observable changes in aging of the integument to their causes.

Changes	Causes
a. Wrinkles; skin springs back less when gently pinched	
b.	Macrophages become less efficient; decrease in number of Langerhans cells
c. Dry, easily broken skin	
d.	Decrease in number and size of melanocytes

✓ **E3.** List and briefly describe two or more factors that can help keep your skin healthy throughout your lifetime.

F. Disorders and medical terminology (pages 154–157)

✓ **F1.** Circle the risk factor for cancer in each pair.

a. Skin type:
   A. Fair with red hair
   B. Tans well with dark hair

b. Sun exposure:
   A. Lives in Canada at sea level
   B. Lives in high-altitude mountainous area of South America

c. Age:
   A. 30 years old
   B. 75 years old

d. Immunological status:
   A. Taking immunosuppressive drugs for cancer or following organ transplant
   B. Normal immune system

✓ **F2.** Identify characteristics of the three different classes of burns by completing this exercise.

a. In a first-degree burn, only the superficial layers of the *(dermis? epidermis?)* are involved.

   The tissue appears ________________ in color. Blisters *(do? do not?)* form. Give one example of a first-degree burn.

b. Which parts of the skin are injured in a second-degree burn?

   Blisters usually *(do? do not?)* form. Epidermal derivatives, such as hair follicles and glands, *(are? are not ?)* injured. Healing usually occurs in about three to four *(days? weeks?)*.

c. Third-degree burns are called *(partial? full?)*-thickness burns. Such skin appears *(red and blistered? white, brown, or black and dry?)*. Such burned areas are usually *(painful? numb?)* because nerve endings are destroyed. Regeneration is usually *(rapid? slow?)*.

F3. Contrast systemic effects with local effects of burns.

✓ **F4.** *A clinical challenge.* Answer these questions about the Lund-Browder method of burns assessment.

a. What does the Lund-Browder method estimate? *(Depth of burn? Amount of body surface area burned?)* This method is based upon differences in body *(size? proportions?)* of age groups.

b. If an adult and a one-year-old each experience burns over the entire anterior surface of both legs, who is more burned? *(Adult? One-year-old?)* If both are burned over the entire anterior surface of the head, who is more burned? *(Adult? One-year-old?)*

✓ **F5.** Match the name of the disorder with the description given.

A. Acne	**M. Malignant melanoma**
B. Basal cell carcinoma	**N. Nevus**
D. Decubitus ulcers	**P. Pruritus**
I. Impetigo	

_______ a. Pressure sores

_______ b. Mole

_______ c. Staphylococcal or streptococcal infection that may become epidemic in nurseries

_______ d. Inflammation of sebaceous glands especially in chin area; occurs under hormonal influence

_______ e. Rapidly metastasizing form of cancer

_______ f. The most common form of skin cancer

_______ g. Itching

F6. Contrast the following pairs of terms.

a. *Topical/intradermal*

b. *Corn/wart*

ANSWERS TO SELECTED CHECKPOINTS: CHAPTER 5

A3. 2; 1-2; heels.

A4. See Figure LG 5.1A. (a) Epidermis; epithelium. (b) Dermis, connective tissue; thicker.

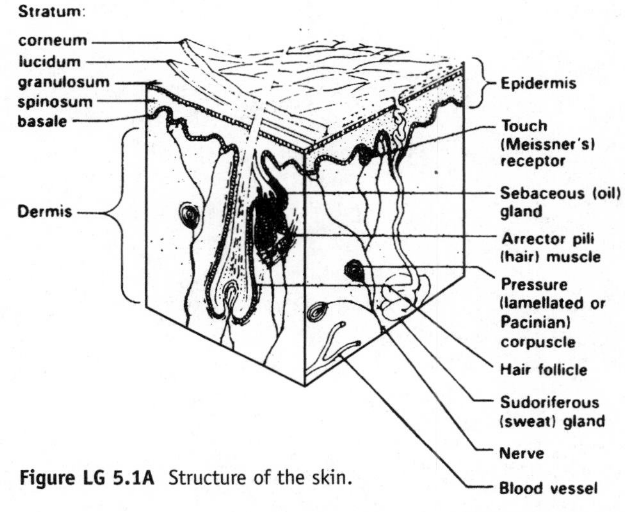

Figure LG 5.1A Structure of the skin.

A5. Under the skin; superficial fascia or hypodermis; areolar, adipose; anchors skin to underlying tissues and organs, insulates by fat here, contains blood vessels and pressure receptors.

A6. (a) Keratinocyte. (b) Melanocyte. (c) Langerhans cell. (d) Merkel cell.

A7. See Figure LG 5.1A above.

A8. (a) B. (b) B. (c) S. (d) G. (e) G. (f) L. (g) C.

A9. Weeks; faster.

A10. (a) Papillary; loose; papillae, touch. (b) Reticular, dense; collagenous; elastic. (c) Fingerprints and footprints; third and fourth fetal months; no; patterns of dermal papillae.

A11. Dermis. See Figure LG 5.1A.

A12. (a) Greater melanin production per melanocyte; epidermis; melanosomes. (b) Tyrosine; tyrosinase; ultraviolet (UV); albinism. (c) Accumulations of melanin. (d) Yellow. (e) Widening (dilation) of blood vessels; erythema. (f) Yellow, liver; blue.

B1. Hair, nails, and glands; epidermis.

B2. (a) A C B. (b) B C A. (c) Cells; keratin (especially in cuticle). (d) Follicle; external, epidermis; matrix. (e) Matrix (of bulb of hair follicle); contains blood vessels that nourish the hair.

B6. (a) In; excessive. (b) inhibition; 33, does not.

B7. (a) In skin of the breast, face, neck, and upper chest. (b) Palms and soles. (c) Keep hair and skin from drying out, lower risk of dehydration and infection of skin. (d) Enlarged sebaceous gland with accumulated sebum; no, it is due to the presence of melanin and oxidation of the oil within sebum.

B8. (a-c) Ecc. (d-e) Apo.

B9. Major: regulate body temperature; minor: eliminate wastes such as urea, uric acid, ammonia, lactic acid, and excess salts.

B10. (a) Sud. (b) Seb. (c) Cer. (d) Sud. (e) Seb.

B12. (a) Yes; yes; epidermis. (b) They bring about growth of nails and hairs, respectively. (c) The pink color is related to visibility of blood vessels deep to the nail body; the free edge has no tissue (so no blood vessels) deep to it, whereas the thickened stratum basale obscures blood vessels deep to the lunula.

C1. (a) In; dilate; 8–10(%). (c) Negative, opposite to.

C4. (a) 400mL. (b) Salts, carbon dioxide, ammonia and urea. (c) Fat; A,D,E, and K; D, Ca^{2+}. (d) Transdermal; corticosteroids, nitroglycerin, scopolamine, estradiol, nicotine.

D1. (a) Skinned knee, first- or second-degree burn. (b) Central. (c) Basale; periphery. (d) Contact inhibition. (e) Mitosis; epidermal growth factor (EGF).

D3. (a) I. (b) Mig. (c) P. (d) Mat.

E1. (a) Endo, meso, ecto. (b) Ecto; fourth (c) Epidermis; ninth. (d) Lanugo; shed before. (e) Meso. (f) This cheesy substance covers and protects fetal skin from amniotic fluid, and its slippery nature facilitates the birth process.

E2.

Changes	Causes
a. Wrinkles; skin springs back less when gently pinched	Elastic fibers thicken into clumps and fray
b. Increased susceptibility to skin infections and skin breakdown	Phagocytes become less efficient; decrease in number of Langerhans cells
c. Dry, easily broken skin	Decreased secretion of sebum by sebeceous glands; decreased sweat production
d. Gray or white hair; atypical skin pigmentation	Decrease in number and size of melanocytes

E3. Good nutrition, decreased stress, balance of rest and exercise, not smoking, protection from the sun.

F1. (a) A. (b) B. (c) B. (d) A.

F2. (a) Epidermis; redder; do not; typical sunburn. (b) All of epidermis and possibly upper regions of dermis; do; are not; weeks. (c) Full; white, brown, or black and dry; numb; slow.

F4. (a) Amount of body surface burned; proportions. (b) Adult; one-year-old.

F5. (a) D. (b) N. (c) I. (d) A. (e) M. (f) B. (g) P.

MORE CRITICAL THINKING: CHAPTER 5

1. Recall (from Chapter 1) the definitions of an organ and a system. Is skin an organ or a system? State your rationale.
2. Skin may provide clues about the health of the body; in this case at least, you can tell a lot about a book by its cover. Discuss how a person's physical or emotional health state may be detected by inspection of skin.
3. Explain how skin helps you to regain and maintain homeostasis. State as many examples as you can think of.
4. Compare and contrast the structure of a hair with its follicle and a nail with its nail root.
5. Give scientific rationales for four readily observed aging changes of skin.

MASTERY TEST: CHAPTER 5

Questions 1-4: Circle the letter preceding the one best answer to each question.

1. "Goose bumps" occur as a result of:
   A. Contraction of arrector pili muscles
   B. Secretion of sebum
   C. Contraction of elastic fibers in the bulb of the hair follicle
   D. Contraction of papillae
2. Select the one *false* statement about the stratum basale.
   A. It is the one layer of cells that can undergo cell division.
   B. It consists of a single layer of squamous epithelial cells.
   C. It is the stratum germinativum.
   D. It is the deepest layer of the epidermis.
3. Select the one *false* statement.
   A. Epidermis is composed of epithelium.
   B. Dermis is composed of connective tissue.
   C. Pressure-sensitive Pacinian corpuscles are normally more superficial in location than Meissner's touch receptors.
   D. The amino acid tyrosine is necessary for production of the skin pigment melanin.
4. At about what age do epidermal ridges which cause fingerprints develop?
   A. Months 3–4 in fetal development
   B. Age 3–4 months
   C. Age 3–4 years
   D. Age 30–40 years

Questions 5–11: Arrange the answers in correct sequence.

_______ _______ _______ 5. From most serious to least serious type of burn:
A. First-degree
B. Second-degree
C. Third-degree

_______ _______ _______ 6. Deep wound healing involves four phases. List in order the phases following the inflammatory phase.
A. Migration
B. Maturation
C. Proliferation

_______ _______ _______ 7. From outside of hair to inside of hair:
A. Medulla
B. Cortex
C. Cuticle

_______ _______ _______ 8. From most superficial to deepest:
A. Dermis
B. Epidermis
C. Superficial fascia

_______ _______ _______ 9. From most superficial to deepest:
A. Stratum lucidum
B. Stratum corneum
C. Stratum germinativum

_______ _______ _______ 10. From most to least abundant in epidermis:
A. Melanocytes
B. Merkel cells
C. Keratinocytes

_______ _______ _______ 11. From greatest to least in incidence (frequency of occurrence):
A. Malignant melanoma
B. Basal cell carcinoma
C. Squamous cell carcinoma

Questions 12–20: Circle T (true) or F (false). If the statement is false, change the underlined word or phrase so that the statement is correct.

T F 12. The color of skin is due primarily to a pigment named keratin.
T F 13. The outermost layers of epidermis are composed of dead cells.
T F 14. Eccrine sweat glands are more numerous than apocrine sweat glands and are especially dense on palms and soles.
T F 15. The dermis consists of two regions; the papillary region is more superficial and the reticular region is deeper.
T F 16. Temperature regulation is a positive feedback system.
T F 17. The internal root sheath is a downward continuation of the epidermis.
T F 18. Both epidermis and dermis contain blood vessels (are vascular).
T F 19. Hair, glands, and nails are all derived from the dermis.
T F 20. Hairs are noncellular structures composed entirely of nonliving substances secreted by follicle cells.

Questions 21–25: Fill-ins. Complete each sentence with the word or phrase that best fits.

_______________ 21. Skin contains a chemical that, under the influence of ultraviolet radiation, leads to formation of vitamin _______.

_______________ 22. The cells that are sloughed off as skin cells age and undergo keratinization are those of the stratum _______.

_______________ 23. Fingerprints are the result of a series of grooves called _______.

_______________ 24. The oily glandular secretion that keeps skin and hairs from drying is called _______.

_______________ 25. When the temperature of the body increases, nerve messages from brain to skin will decrease body temperature by _______.

ANSWERS TO MASTERY TEST: CHAPTER 5

Multiple Choice

1. A
2. B
3. C
4. A

Arrange

5. C B A
6. A C B
7. C B A
8. B A C
9. B A C
10. C A B
11. B C A

True-False

12. F. Melanin
13. T
14. T
15. T
16. F. Negative
17. F. External root sheath
18. F. Only the dermis contains, is vascular
19. F. Epidermis
20. F. Composed of different kinds of cells

Fill-ins

21. D
22. Corneum
23. Epidermal ridges or grooves
24. Sebum
25. Stimulating sweat glands to secrete and blood vessels to dilate

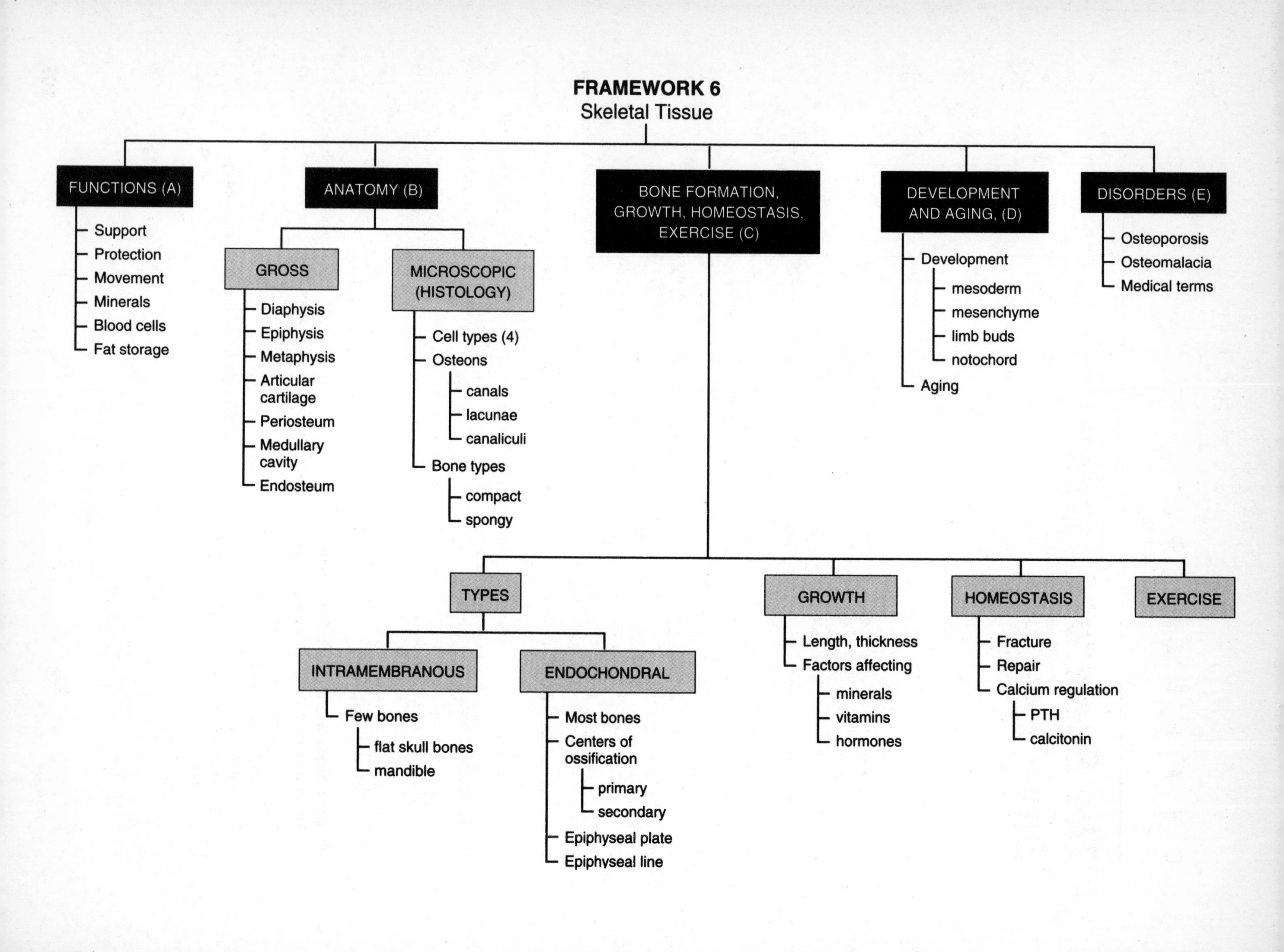

FRAMEWORK 6
Skeletal Tissue
FUNCTIONS (A)
Support
Protection
Movement
Minerals
Blood cells
Fat storage
ANATOMY (B)
GROSS
Diaphysis
Epiphysis
Metaphysis
Articular cartilage
Periosteum
Medullary cavity
Endosteum
MICROSCOPIC (HISTOLOGY)
Cell types (4)
Osteons
canals
lacunae
canaliculi
Bone types
compact
spongy
BONE FORMATION, GROWTH, HOMEOSTASIS, EXERCISE (C)
TYPES
INTRAMEMBRANOUS
Few bones
flat skull bones
mandible
ENDOCHONDRAL
Most bones
Centers of ossification
primary
secondary
Epiphyseal plate
Epiphyseal line
GROWTH
Length, thickness
Factors affecting
minerals
vitamins
hormones
HOMEOSTASIS
Fracture
Repair
Calcium regulation
PTH
calcitonin
EXERCISE
DEVELOPMENT AND AGING, (D)
Development
mesoderm
mesenchyme
limb buds
notochord
Aging
DISORDERS (E)
Osteoporosis
Osteomalacia
Medical terms

C H A P T E R **6**

The Skeletal System: Bone Tissue

The skeletal system provides the framework for the entire body, affording strength, support, and firm anchorage for the muscles that move the body. In this chapter, the first of Unit II, you will explore tissue that forms the skeleton. Bones provide rigid levers covered with connective tissue designed for joint formation and muscle attachment. Microscopically, bones present a variety of cell types intricately arranged in an osseous sea of calcified intercellular material. Bone growth or ossification occurs throughout life. Essential nutrients, hormones, and exercise regulate the growth and maintenance of the skeleton. Fractures and other disorders may result from a lack of such normal regulation.

As you begin your study of bone tissue, carefully examine the Chapter 6 Framework and note relationships among concepts and key terms there. Also refer to the Topic Outline and Objectives; you may want to check off each objective as you complete it.

TOPIC OUTLINE AND OBJECTIVES

A. Functions of the skeletal system

1. Discuss the functions of the skeletal system.

B. Structure of bone

1. Describe the parts of a long bone.
2. Describe the histological features of bone tissue.
3. Describe the blood and nerve supply of bone.

C. Bone formation and growth; bone and homeostasis; exercise and bone tissue

1. Describe the steps involved in intramembranous and endochondral ossification.
2. Describe how bone grows in length and thickness.
3. Explain the role of nutrients and hormones in regulating bone growth.
4. Describe the processes involved in bone remodeling.
5. Describe the sequence of events in repair of a fracture.
6. Describe the role of bone in calcium homeostasis.
7. Describe how exercise and mechanical stress affect bone tissue.

D. Developmental anatomy and aging of the skeletal system

1. Describe the development of the skeletal system and the limbs.
2. Describe the effects of aging on bone tissue.

E. Disorders and medical terminology

WORDBYTES

Now become familiar with the language of this chapter by studying each wordbyte, its meaning, and an example of its use within a term. After you study the entire list, self-check your understanding by writing the meaning of each wordbyte on the line. As you continue through the *Learning Guide*, identify (and fill in) additional terms that contain the same wordbyte.

Wordbyte	Self-check	Meaning	Example(s)
-blast	______________	germ, to form	osteo*blast*
chondr-	______________	cartilage	peri*chondr*ium
-clast	______________	to break	osteo*clast*
dia-	______________	through	*dia*physis
endo-	______________	within	*endo*steum
epi-	______________	over	*epi*physis
intra-	______________	within	*intra*membranous
lacuna	______________	little lake	*lacunae*
meta-	______________	after, beyond	*meta*physis
os-, osteo-	______________	bone	*os*sification, *osteo*cyte
peri-	______________	around	*peri*osteum
-physis	______________	growing	dia*physis*

CHECKPOINTS

A. Functions of the skeletal system (page 162)

✓ **A1.** List six functions of the skeletal system.

______________ ______________

______________ ______________

______________ ______________

✓ **A2.** What other systems of the body depend on a healthy skeletal system? Explain why in each case.

A3. Define *osteology.*

B. Structure of bone (pages 162–168)

✓ **B1.** Name several types of tissue making up the skeletal system.

A single bone, such as one of your finger bones, is considered a(n) *(organ? system? tissue?).*

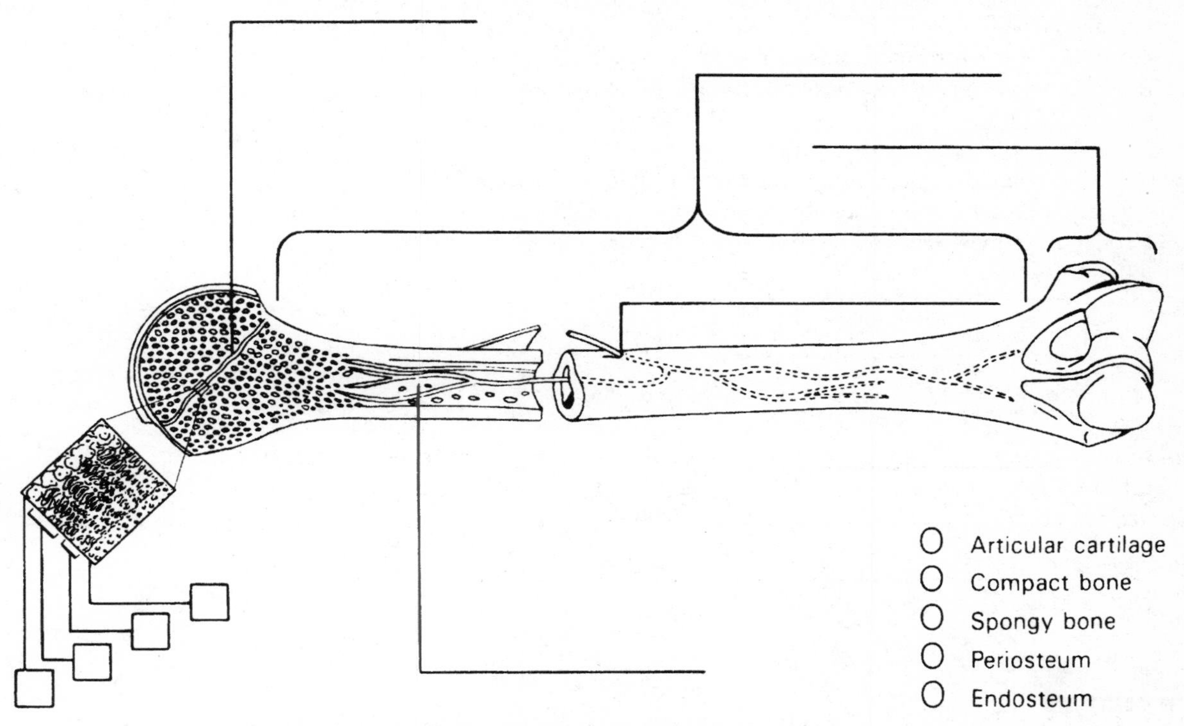

Figure LG 6.1 Diagram of a long bone that has been partially sectioned lengthwise. Color and label as directed in Checkpoints B2 and C5.

✓ **B2.** On Figure LG 6.1, label the *diaphysis, epiphysis, medullary (marrow) cavity,* and *nutrient foramen.* Then color the parts of a long bone using color code ovals on the figure.

✓ **B3.** Match the names of parts of a long bone listed in the box with the descriptions below.

A. Articular cartilage	**M. Metaphysis**
E. Endosteum	**P. Periosteum**

________ a. Thin layer of hyaline cartilage at end of long bone

________ b. Region of mature bone where diaphysis joins epiphysis

________ c. Fibrous covering over bone into which tendons and ligaments attach; inner portion is site of osteogenic cells

________ d. Layer of bone cells lining the marrow cavity

✓ **B4.** Describe the components of bone by doing this exercise.

a. Typical of all connective tissues, bone consists mainly of *(cells? intercellular material?).*

b. The intercellular substance of bone is unique among connective tissues. Protein fibers form about *(25? 50?)* % of the weight of bone, whereas mineral salts account for

about __________% of the weight of bone.

c. The two main salts present in bone are __________________ and __________________.

d. The *(hardness? tensile strength?)* of bones is related to inorganic chemicals in bone,

namely the mineral salts deposited around __________________ fibers. Resistance to being stretched or torn apart, a property of bone known as *(hardness? tensile strength?)* is afforded by *(mineral salts? collagen fibers?).*

e. Bone *(is completely solid? contains some spaces?).* Write two advantages of this structural feature of bone.

✓ **B5.** In the following exercise, bone cells are described in the sequence in which they form, and later destroy, bone. Write the name of the correct bone type after its description.

a. Derived from mesenchyme, these cells can undergo mitosis and differentiate into osteoblasts: __________________

↓

b. Form bone initially by secreting mineral salts and fibers; not mitotic. __________________

↓

c. Mature bone cells; maintain bone tissue; not mitotic. __________________

d. Huge cells in the endosteum; powerful lysosomal enzymes permit these cells to resorb (degrade) bone in bone repair, remodeling, and aging. __________________

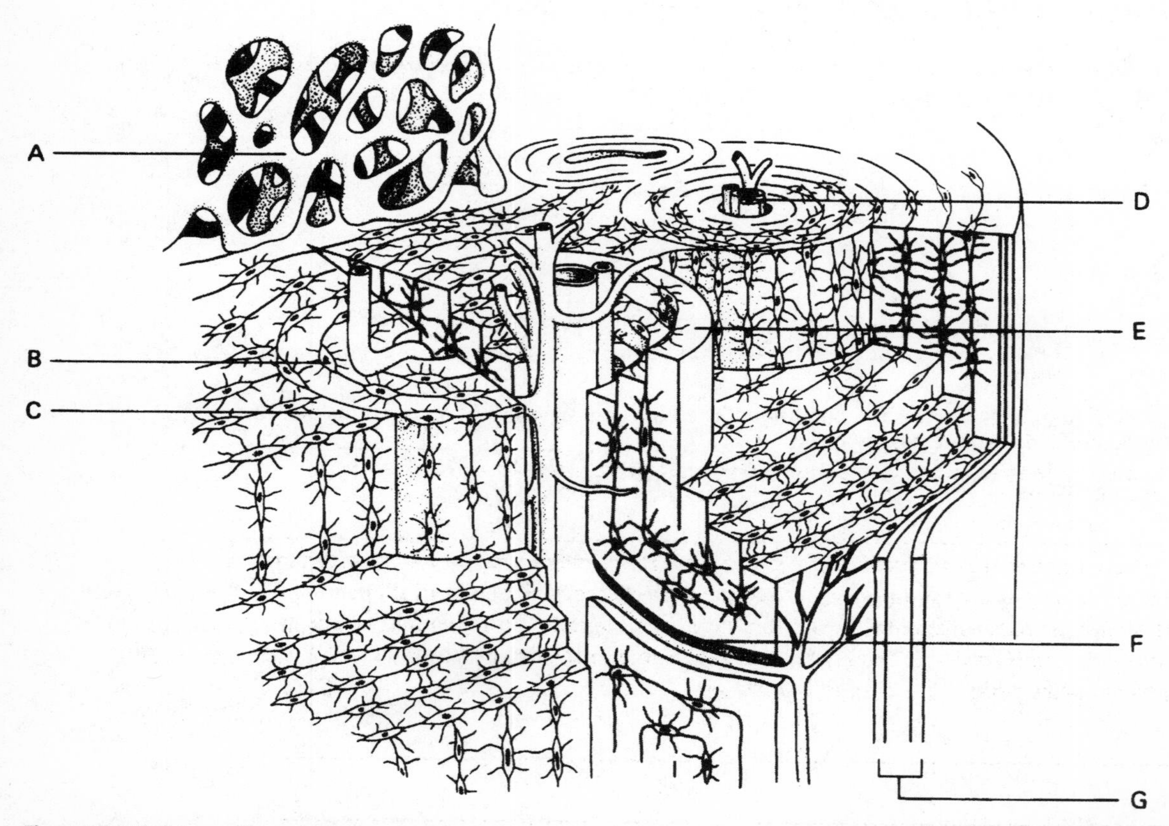

Figure LG 6.2 Osteons (Haversian systems) of compact bone. Identify lettered structures and color as directed in Checkpoint B6.

✓ **B6.** Refer to Figure LG 6.2 and complete this exercise about bone structure.

a. Compact bone is arranged in concentric circle patterns known as ____________________. Each individual concentric layer of bone is known as a ____________________ labeled with letter _______ in the figure.

b. The osteon pattern of compact bone permits blood vessels and nerves to supply bone cells trapped in hard bone tissue. Blood vessels and nerves penetrate bone from the periosteum, labeled with letter _______ in the figure. These structures then pass through horizontal canals, labeled _______ and known as perforating (Volkmann's) canals. These vessels and nerves finally pass into microscopic channels, labeled ______, in the center of each osteon. Color blood vessels red and blue in the figure.

c. Mature bone cells, known as ____________________, are located relatively far apart in bone tissue. These are present in "little lakes," or ____________________, labeled _______ in the figure. Color ten lacunae green.

d. Minute canals, known as canaliculi, are labeled with letter ______. What are the functions of these channels?

e. Compact bone is strengthened by the arrangement of osteons *(perpendicular to? along the same direction as?)* lines of stress. These lines of stress (and therefore the arrangement of osteons) *(remain constant? change?)* over a lifetime.

B7. Contrast these four types of lamellae: *concentric, interstitial, outer circumferential, inner circumferential.* Which type is found within osteons?

✓ **B8.** Do this exercise about spongy bone.

a. Spongy bone is arranged in *(osteons? trabeculae?)*, which may be defined as:

____

b. Spongy bone makes up most of the *(diaphyses? epiphyses?)* of long bones. Spongy bone *(is not? is also)* located within bones that are short, flat, and irregular in shape.

c. *A clinical correlation.* Red blood cell formation (hemopoiesis) normally takes place in *(all? only certain areas of?)* spongy bone tissue. Name four or more bones in which this process takes place.

______________________________ ______________________________

______________________________ ______________________________

Of what clinical importance is this information?

B9. Describe a *bone scan* in this exercise.

a. Contrast the diagnostic interpretation of a "hot spot" versus a "cold spot."

b. What diagnostic advantages do bone scans offer over traditional x-rays?

B10. Discuss the blood and nerve supply to bones in this activity.

a. Contrast the blood supply *to* bones (arteries) with that *from* bones (veins).

b. Explain why cancer of the bone is particularly painful.

C. Bone formation and growth; bone and homeostasis; exercise and bone tissue (pages 168–178)

✓ **C1.** Complete this overview of bone formation.

a. Bone formation is also called ________________.

b. All bones develop from preexisting *(epithelial? connective? muscle?)* tissue,

specifically from ________________. This tissue then differentiates into either bone or cartilage, beginning at the sixth or seventh *(week? month?)* of development.

c. Two methods of bone formation occur in the human body. Name the bones that are formed by each of these methods of ossification:

intramembranous ____

endochondral ____

d. These two methods of bone formation described above *(do? do not?)* lead to different types of bone structurally.

✓ **C2.** Refer to Figure 7.4 in your text. Locate the parietal bone, a typical "flat bone" forming in the fetal skull. Describe the intramembranous formation of this bone by using each of the terms in the box once.

Compact	**Mesenchyme**	**Osteogenic**
Fontanel	**Ossification**	**Protection**
Growth	**Osteoblast**	**Spongy**
Marrow	**Osteocyte**	

a. Fibrous membranes surrounding the head of the developing baby contain

embryonic ____________________ cells. These cells cluster in areas known as centers of ____________________, and these mark areas where bone tissue will form.

b. Differentiation of these cells leads to ____________________ cells and then into

____________________ and ____________________ cells that surround themselves with organic and inorganic matrix forming spongy bone. (For help, look back at Checkpoint B5, page LG 106.)

c. Eventually, mesenchyme on the outside of the bone develops into a periosteal covering over the skull bone. This lays down some compact bone over spongy bone. The resulting typical flat bone is analogous to a flattened jelly sandwich: two firm (flat)

"slices of bread" comparable to ____________________ bone with a "jelly" of

____________________ bone filled with red ____________________.

d. By birth, fetal skull bones are still not completed. Some fibrous membranes still remain

at points where skull bones meet, known as ____________________. Of what advan-

tage are these "soft spots"? They permit additional ____________________ of skull

bones as well as safer passage (____________________) of the fetal head during birth.

✓ **C3.** Endochondral ossification refers to bone formation from an initial model made of *(fibrous membranes? cartilage ?)*. Increase your understanding of this process by building your own outline in this exercise.

a. First identify the four major steps in endochondral ossification by filling in the five long lines (I–V) below with the correct terms selected from those in the box below.

Articular cartilage and epiphyseal plate	**Primary ossification center**
Development	**Secondary ossification center**
Growth	

I. ____________________ of the cartilage model ____ ____

II. ____________________ of the cartilage model ____ ____

III. Development of the ____________________ ____ ____ ____

IV. Development of the ____________________ ____ ____ ____

V. Formation of the ____________________ ____

b. Now fill in details of the process by placing the following statements of events in the correct sequence. Write the letters on the short lines to the right of the correct phase (I–V) above. One is done for you.

A. The cartilage model grows in length as chondrocytes divide and secrete more cartilage matrix; it grows in thickness as new chondroblasts develop within the perichondrium.

B. Cartilage cells in the center of the diaphysis accumulate glycogen, enlarge, and burst, releasing chemicals that alter pH and trigger calcification of cartilage.

C. Cartilage cells die because they are deprived of nutrients; in this way spaces are formed within the cartilage model.

D. Blood vessels penetrate the perichondrium, stimulating perichondrial cells to form osteoblasts. A collar of bone forms and gradually thickens around the diaphysis. The membrane covering the developing bone is now called the periosteum.

E. A hyaline cartilage model of future bone is laid down by differentiation of mesenchyme into chondroblasts.
F. A perichondrium develops around the cartilage model.
G. Secondary ossification centers develop in epiphyses, forming spongy bone there about the time of birth.
H. Capillaries grow into spaces, and osteoblasts deposit bone matrix over disintegrating calcified cartilage. In this way spongy bone is forming within the diaphysis at the primary ossification center.
I. Osteoclasts break down the newly formed spongy bone in the very center of the bone, thereby leaving the medullary (marrow) cavity.
J. The original hyaline cartilage remains as articular cartilage at the ends of the new bone and also as the epiphyseal plate for as long as the bone grows.

✓ **C4.** In growth of the cartilage model (Step A in Checkpoint C3), cartilage grows lengthwise by *(appositional? interstitial?)* growth; this means growth *(from within? on the cartilage surface?)*. Cartilage grows in thickness by *(appositional? interstitial?)* growth, which means growth *(from within? on the cartilage surface?)*.

✓ **C5.** On Figure LG 6.1, label the *epiphyseal plate*. This plate of growing bone consists of four zones. First identify each region by placing the letter of the name of that zone next to the correct description below. Next place these same four letters in the boxes in the inset on Figure LG 6.1 to indicate the locations of these cartilage zones within the epiphyseal plate of the developing bone.

C. Calcified	**P. Proliferating**
H. Hypertrophic	**R. Resting**

_______ a. Zone of cartilage that is not involved in bone growth but anchors the epiphyseal plate (site of bone growth) to the bone of the epiphysis

_______ b. New cartilage cells form here by mitosis and are arranged like stacks of coins

_______ c. Cartilage cells mature and die here as matrix around them calcifies

_______ d. Osteoblasts and capillaries from the diaphysis invade this region to lay down bone upon the calcified cartilage remnants here

✓ **C6.** Complete this summary statement about bone growth at the epiphyseal plate. Cartilage cells multiply on the *(epiphysis? diaphysis?)* side of the epiphyseal plate, providing temporary new tissue. Cartilage cells then die and are replaced by bone cells on the *(epiphysis? diaphysis?)* side of the epiphyseal plate. In other words, as the bone grows in length, the *(diaphysis? epiphyses?)* grow(s) longer. If the epiphyseal plate of a growing bone is damaged, for example by fracture, the bone is likely to be *(shorter than normal? of normal length? longer than normal?)*.

✓ **C7.** Complete this activity about maturation of the skeleton.

a. The epiphyseal *(line? plate?)* is the bony region in bones of adults that marks the original cartilaginous epiphyseal *(line? plate?)*.
b. Ossification of bones is completed by age *(15? 25?)*. If the epiphyseal plate of a growing bone is damaged by fracture or by sex steroids (especially estrogens), the epiphyseal line is likely to form *(earlier? later?)* in development. This accounts for the shorter than normal bone length.
c. Bones must grow in diameter to match growth in length. *(Osteoblasts? Osteoclasts?)* are cells that destroy bone to enlarge the medullary cavity, whereas *(osteoblasts? osteoclasts?)* are cells from the periosteum that increase the thickness of compact bone.

✓ **C8.** Fill in the names of hormones involved in bone growth in this exercise. Use the answers in the box.

hGH.	**Human growth hormone**	**INS.**	**Insulin**
IGFs.	**Insulinlike growth factor**	**SH.**	**Sex hormones**

_______ a. Hormone made by the pituitary. In excessive amounts, results in giantism; if deficient, leads to dwarfism.

_______ b. Produced by bone cells and also by the liver, these chemicals stimulate protein synthesis and growth of bones.

_______ c. Made by the pancreas; promotes normal bone growth.

_______ d. Produced by ovaries and testes, these hormones stimulate osteoblast activity and lead to growth spurt and changes in skeletal structure at puberty.

✓ **C9.** List 10 factors that are necessary for normal bone growth, remodeling, and repair of fractured bone.

a. Six minerals: _______ _______ _______ _______ _______ _______

b. Four vitamins: _______ _______ _______ _______

C10. Defend or dispute this statement: "Once a bone, such as your thighbone, is formed, the bone tissue is never replaced unless the bone is broken."

✓ **C11.** Discuss bone remodeling in this Checkpoint.

a. Explain the roles of lysosomes and acids in osteoclastic activity.

b. What fills in the tunnels formed by osteoclasts?

c. Explain one mechanism by which estrogen helps to prevent bone breakdown.

C12. Contrast the types of fractures in each pair.

a. *Simple/compound*

b. *Pott's/Colles'*

c. *Comminuted/greenstick*

✓ **C13.** Describe the four main steps in repair of a fracture by filling in terms from this list in the spaces below. One answer will be used twice.

Cell Type	Process or Stage	Time Period
Chondroblasts	Fracture hematoma	Hours
Fibroblasts	Inflammation and cell death	Months
Osteoblasts	Procallus	Weeks
Osteoclasts	Remodeling	
Phagocytes		

a. Blood flows from torn blood vessels, forming a clot known as a

hematoma ________ within 6 to 8 hours ________ of the injury.

Because of the trauma and lack of circulation, inflam + death ________ occurs in

the area. As a result, cells known as phago ________ and

osteoclas ________ enter the area and remove debris over a period of several

weeks ________.

b. Next, granulation tissue known as a procallus ________ forms. Then cells

called fibro ________ produce collagen fibers that connect broken ends

of the bone. Other cells, the chondro ________, form a soft callus at a distance from the remaining healthy, vascular bone. This stage lasts about 3

~~months~~ weeks ________.

c. The hard callus stage occurs next as osteoblasts ________ cells form bone adjacent to the remaining healthy, vascular bone. This stage lasts 3 to 4

~~week~~ months ________.

d. Callus is reshaped during the final (or Remodel ________) phase.

C14. *A clinical correlation.* What is the major difference between these two methods of setting a fracture: *closed reduction and open reduction?*

✓ **C15.** *The Big Picture: Looking Ahead.* Refer to Figure 18.14 and other sections of your text, and answer these questions about regulation of calcium homeostasis.

a. Figures 1.2 and 1.3 in your text introduced the concept of negative feedback systems. Using that reference and Figure 6.11 in your text, complete Figure LG 6.3. Fill in the six boxes and the two lines using the list of answers provided in the figure.

b. Figure LG 6.3 shows that parathyroid gland cells respond to a lowered blood level of

Ca^{2+} by signaling increased production of cyclic ________. This INPUT then *("turns on"? "turns off"?)* the PTH cell gene in parathyroid glands. The increased blood level of PTH acts as OUTPUT to "tell" effectors to respond. Effectors include kidneys as well

as ________________ which *(raise? lower?)* the blood level of Ca^{2+} in several ways.

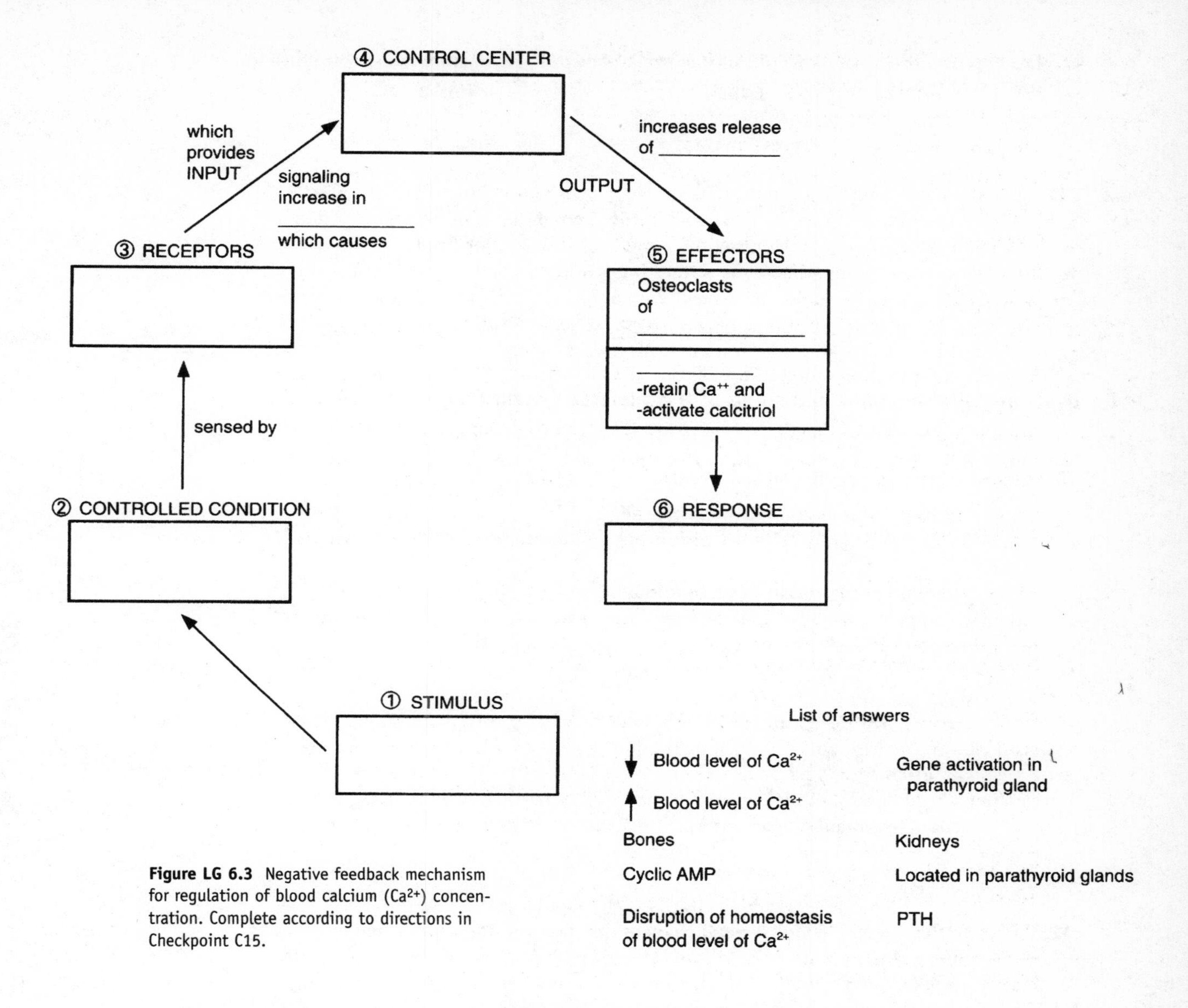

Figure LG 6.3 Negative feedback mechanism for regulation of blood calcium (Ca^{2+}) concentration. Complete according to directions in Checkpoint C15.

c. PTH signals kidneys to *(eliminate? retain?)* calcium in blood so that less is lost in urine. PTH also activates kidneys to produce *(calcitriol? calcitonin?)*, which is vitamin

________. This vitamin also raises blood levels of Ca^{2+} by increasing absorption of Ca^{2+} from foods in the intestine.

d. PTH activates osteo-*(blasts? clasts?)* of bone cells so that calcium is *(deposited in? resorbed from?)* bone.

e. Regulation of blood calcium (Ca^{2+}) exemplifies a *(positive? negative?)* feedback cycle

because a decrease in blood Ca^{2+} ultimately causes PTH to ________-crease blood Ca^{2+}.

f. The effect of PTH is *(the same as? opposite to?)* that of calcitonin (Figure 18.14, page 608). Calcitonin causes calcium to be *(deposited in? resorbed from?)* bone. In other words, calcitonin, which is released by the *(thyroid? parathyroid?)* gland, causes the

blood level of Ca^{2+} to ________-crease.

g. Chapter 27 (page 1000) points out that *(hypo? hyper?)*-parathyroidism refers to a deficiency of PTH. This condition is likely to lead to *(high? low?)* blood calcium levels, known as *(hypo? hyper?)*-calcemia. Symptoms of this disorder include increased *(hardness? fragility?)* of bones because bones have *(lost? gained?)* calcium and of *(hyper? hypo?)*-active muscles because low blood levels of calcium increase activity of nerves.

h. List two or more other functions of the human body affected by the blood concentration of calcium.

✓ **C16.** Circle the correct answers about the effects of exercise upon bones.

a. Exercise *(strengthens? weakens?)* bones. A fractured bone that is not exercised is likely to become *(stronger? weaker?)* during the period that it is immobilized.

b. The pull of muscles upon bones, as well as the tension on bones as they support body weight during exercise, causes *(increased? decreased?)* production of the protein collagen.

c. The stress of exercise also stimulates *(osteoblasts? osteoclasts?)* and increases production of the hormone calcitonin, which inhibits *(osteoblasts? osteoclasts?)*.

D. Developmental anatomy and aging of the skeletal system (pages 178–180)

✓ **D1.** Complete this learning activity about development of the skeleton.

a. Bones and cartilage are formed from *(ecto? meso? endo?)*-derm, which later differentiates into the embryonic connective tissue called ____________________. Some of these cells become chondroblasts, which eventually form *(bone? cartilage?)*, whereas ____________________-blasts develop into bone.

b. Limb buds appear during the *(fifth? seventh? tenth?)* week of development. At this point the skeleton in these buds consists of *(bone? cartilage?)*. Bone begins to form during week ________.

c. By week *(six? seven? eight?)* the upper extremity is evident, with defined shoulder, arm, elbow, forearm, wrist, and hand. The lower extremity is also developing but at a slightly *(faster? slower?)* pace.

d. The notochord develops in the region of the *(head? vertebral column? pelvis?)*. Most of the notochord eventually *(forms vertebrae? disappears?)*. Parts of it persist in the ________________________.

✓ **D2.** Complete this exercise about skeletal changes that occur in the normal aging process.

a. The amount of calcium in bones *(de? in?)*-creases with age. As a result, bones of the elderly are likely to be *(stronger? weaker?)* than bones of younger persons. This change occurs at a younger age in *(men? women?)*.

This loss of calcium typically begins at about age ___ in females, and by age 70, about ____% of calcium in bones is lost. In males, calcium loss does not usually begin until after age _____.

b. Another component of bones that decreases with age is __________________. What is the significance of this change?

c. Write two or more health practices that can minimize skeletal changes associated with aging.

E. Disorders and medical terminology (pages 180–181)

✓ **E1.** Describe osteoporosis in this exercise.

a. Osteoporotic bones exhibit _____-creased bone mass and _____-creased susceptibility to fractures. This disorder is associated with a loss of the hormone _______________ which normally stimulates osteo-____________________ activity.

b. Circle the category of persons at higher risk for osteoporosis in each pair:

Younger persons/older persons — Men/women

Short, thin persons/tall, large-build persons — Black persons/white persons

c. List three factors in daily life that will help to prevent osteoporosis. (*Hint:* Focus on diet and activities.)

✓ **E2.** Check your understanding of bone disorders by matching the terms in the box with related descriptions.

Osteoarthritis	**Osteosarcoma**	**Osteomyelitis**

a. Malignant bone tumor: ____________

b. Bone infection, for example, caused by *Staphylococcus:* ____________

c. Degenerative joint disease caused by destruction of articular cartilage: ____________

ANSWERS TO SELECTED CHECKPOINTS: CHAPTER 6

A1. Support, protection, assistance in movement, homeostasis of minerals, hemopoiesis, and fat storage.

A2. Essentially all do. For example, muscles need intact bones for movement to occur; bones are sites of blood formation; bones provide protection for viscera of nervous, digestive, urinary, reproductive, cardiovascular, respiratory, and endocrine systems; broken bones can injure integument.

B1. Bone, cartilage, dense connective tissues, epithelium, blood-forming tissues, adipose tissue, and nervous tissue; organ.

B2. Figure LG 6.1A at right.

B3. (a) A. (b) M. (c) P. (d) E.

B4. (a) Intercellular material. (b) 25, 50. (c) Calcium phosphate, calcium carbonate. (d) Hardness, collagen; tensile strength, collagen fibers. (e) Contains some spaces; provides channels for blood vessels and makes bones lighter weight.

B5. (a) Osteogenic cells. (b) Osteoblast. (c) Osteocyte. (d) Osteoclast.

B6. (a) Osteons (Haversian systems); lamella, E. (b) G; F; D. (c) Osteocytes; lacunae, B. (d) C; contain extensions of osteocytes bathed in extra-cellular fluid (ECF); provide routes for oxygen and nutrients to reach osteocytes and for

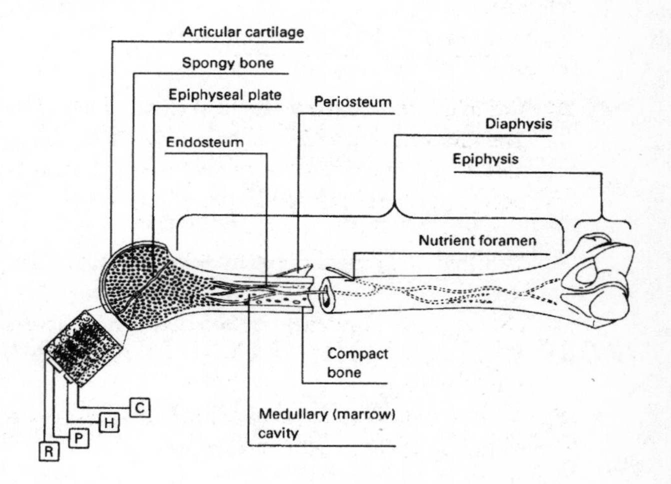

Figure LG 6.1A Diagram of a developing long bone that has been partially sectioned lengthwise.

wastes to diffuse away. (e) Along the same direction as; change.

B8. (a) Trabeculae, irregular latticework of bony lamellae containing osteocytes surrounded by lacunae and blood vessels. (b) Epiphyses; is also. (c) Only certain areas of; hipbones, ribs, sternum, vertebrae, skull bones, and epiphyses of long bones such as femurs. These areas (such as the sternum) may be biopsied to examine for aplastic anemia or for response of bone marrow to anti-anemia medications. These sites (such as hip-bones) may be utilized for marrow transplant.

C1. (a) Ossification or osteogenesis. (b) Connective, mesenchyme; week. (c) Intramembranous: flat bones of the cranium and the mandibles (lower jaw bones); endochondral: all other bones of the body. (d) Do not.

C2. (a) Mesenchyme; ossification. (b) Osteogenic, osteoblast, osteocyte. (c) Compact, spongy, marrow. (d) Fontanels; growth, protection.

C3. Cartilage. (I) Development: E F. (II) Growth: A B C. (Ill) Primary ossification center: D H I. (IV) Secondary ossification center: G. (V) Articular cartilage and epiphyseal plate: J.

C4. Interstitial, from within; appositional, on the cartilage surface.

C5. See Figure LG 6.1A. (a) R. (b) P. (c) H. (d) C.

C6. Epiphysis; diaphysis; diaphysis; shorter than normal.

C7. (a) Line, plate. (b) 25; earlier. (c) Osteoclasts, osteoblasts.

C8. (a) hGH. (b) IGFs. (c) INS. (d) SH.

C9. (a) Calcium, phosphorus, fluoride, magnesium, iron, and manganese. (b) C, K, B_{12}., and A.

C11. (a) Lysosomes release enzymes that digest the protein collagen; acids dissolve minerals of bone. (b) Osteoblasts lay down new bone. (c) By promoting death of osteoclasts.

C13. (a) Fracture hematoma, hours; inflammation and cell death; osteoclasts, phagocytes, weeks. (b) Procallus; fibroblasts; chondroblasts; weeks. (c) Osteoblast; months. (d) Remodeling.

C15. (a) See Figure LG 6.3A. (b) AMP; "turns on"; bones, raise. (c) Retain; calcitriol, D. (d) Clasts, resorbed from. (e) Negative, in. (f) Opposite to; deposited in; thyroid, de. (g) Hypo; low, hypo; fragility, lost, hyper. (h) Blood clotting, heart activity, breathing, enzyme function.

C16. (a) Strengthens; weaker. (b) Increased. (c) Osteoblasts, osteoclasts.

D1. (a) Meso, mesenchyme; cartilage, osteo. (b) Fifth; cartilage; six or seven. (c) Eight; slower. (d) Vertebral column; disappears; intervertebral discs.

D2. (a) De; weaker; women; 30, 30; 60. (b) Protein; bones are more brittle and vulnerable to fracture. (c) Weight-bearing exercise; good diet with adequate calcium, vitamin D, and protein; not smoking because smoking constricts blood vessels.

E1. (a) De, in; estrogen; blast. (b) Older; women; short, thin persons; white persons. (c) Adequate intake of calcium, vitamin D, weight-bearing exercise, and not smoking; estrogen replacement therapy (HRT) may be advised for some women.

E2. (a) Osteosarcoma. (b) Osteomyelitis. (c) Osteoarthritis.

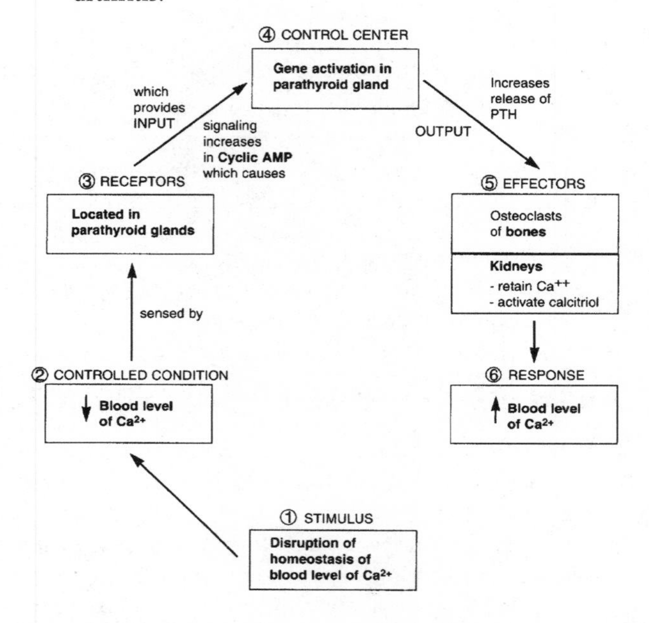

Figure LG 6.3A Negative feedback regulation of calcium.

MORE CRITICAL THINKING: CHAPTER 6

1. Explain how the skeletal system is absolutely vital for basic functions that keep you alive, for example, for breathing, eating, movement, and elimination of wastes.
2. Do you think bone is dynamic or "dead"? State your rationale.
3. Contrast compact bone with spongy bone according to structure and locations.
4. Describe mesenchyme and osteogenic cell roles in formation, growth, and maintenance of skeletal tissues.
5. One process of bone formation involves development of bone from fibrous connective tissue membranes rather than from cartilage. Tell which bones form by this process, and explain how this might be advantageous to the developing human body.
6. Discuss probable effects of these hormonal changes upon bone development and health:
   (a) abnormally high production of sex steroids in a nine-year-old boy
   (b) removal of both ovaries in a 26-year-old woman
   (c) excessive production of parathyroid hormone (PTH) in a 55-year-old woman

MASTERY TEST: CHAPTER 6

Questions 1–15: Circle T (true) or F (false). If the statement is false, change the underlined word or phrase so that the statement is correct.

T F 1. Greenstick fractures occur only in adults.

T F 2. Appositional growth of cartilage or bone means growth in thickness due to action of cells from the perichondrium or periosteum.

T F 3. Calcitriol is an active form of calcium.

T F 4. The epiphyseal plate appears earlier in life than the epiphyseal line.

T F 5. Osteons (Haversian systems) are found in compact bone, but not in spongy bone.

T F 6. Another name for the epiphysis is the shaft of the bone.

T F 7. A compound fracture is defined as one in which the bone is broken into many pieces.

T F 8. Haversian canals run longitudinally (lengthwise) through bone, but perforating (Volkmann's) canals run horizontally across bone.

T F 9. Canaliculi are tiny canals containing blood that nourishes bone cells in lacunae.

T F 10. Compact bone that is of intramembranous origin differs structurally from compact bone developed from cartilage.

T F 11. In a long bone the primary ossification center is located in the diaphysis whereas the secondary center of ossification is in the epiphysis.

T F 12. Osteoblasts are bone-destroying cells.

T F 13. Most bones start out in embryonic life as hyaline cartilage.

T F 14. The layer of compact bone is thicker in the diaphysis than in the epiphysis.

T F 15. Lamellae are small spaces containing bone cells.

Questions 16–20: Arrange the answers in correct sequence.

_______ _______ _______ 16. Steps in marrow formation during endochondral bone formation, from first to last:

A. Cartilage cells burst, causing intercellular pH to become more alkaline.
B. pH changes cause calcification, death of cartilage cells; eventually spaces left by dead cells are penetrated by blood vessels.
C. Cartilage cells hypertrophy with accumulated glycogen.

_______ _______ _______ 17. From most superficial to deepest:

A. Endosteum
B. Periosteum
C. Compact bone

_______ _______ _______ 18. Phases in repair of a fracture, in chronological order:

A. Fracture hematoma formation
B. Remodeling
C. Callus formation

_______ _______ _______ _______ 19. Phases in formation of bone in chronological order:

A. Mesenchyme cells
B. Osteocytes
C. Osteoblasts
D. Osteogenic

_______ _______ _______ _______ 20. Portions of the epiphyseal plate, from closest to epiphysis to closest to diaphysis:

A. Zone of resting cartilage
B. Zone of proliferating cartilage
C. Zone of calcified cartilage
D. Zone of hypertrophic cartilage

Questions 21–25: Fill-ins. Write the word or phrase that best completes the statement.

_____________________ 21. _______ is a term that refers to the shaft of the bone.

_____________________ 22. _______ is a disorder primarily of older women associated with decreased estrogen level; it is characterized by weakened bones.

_____________________ 23. _______ is the connective tissue covering over cartilage in adults and also over embryonic cartilaginous skeleton.

_____________________ 24. The majority of bones formed by intramembranous ossification are located in the _______.

_____________________ 25. Vitamin _______ is a vitamin that is necessary for absorption of calcium from the gastrointestinal tract, and so it is important for bone growth and maintenance.

ANSWERS TO MASTERY TEST: CHAPTER 6

True-False

1. F. Children
2. T
3. F. Active form of vitamin D
4. T
5. T
6. F. End of the bone (often bulbous)
7. F. Protrudes through the skin
8. T
9. F. Parts of osteocytes and fluid from blood vessels in Haversian canals, but not blood itself
10. F. Only in origin
11. T
12. F. Osteoclasts
13. T
14. T
15. F. Lacunae

Arrange

16. C A B
17. B C A
18. A C B
19. A D C B
20. A B D C

Fill-ins

21. Diaphysis
22. Osteoporosis
23. Perichondrium
24. Skull or cranium
25. D (or D3)

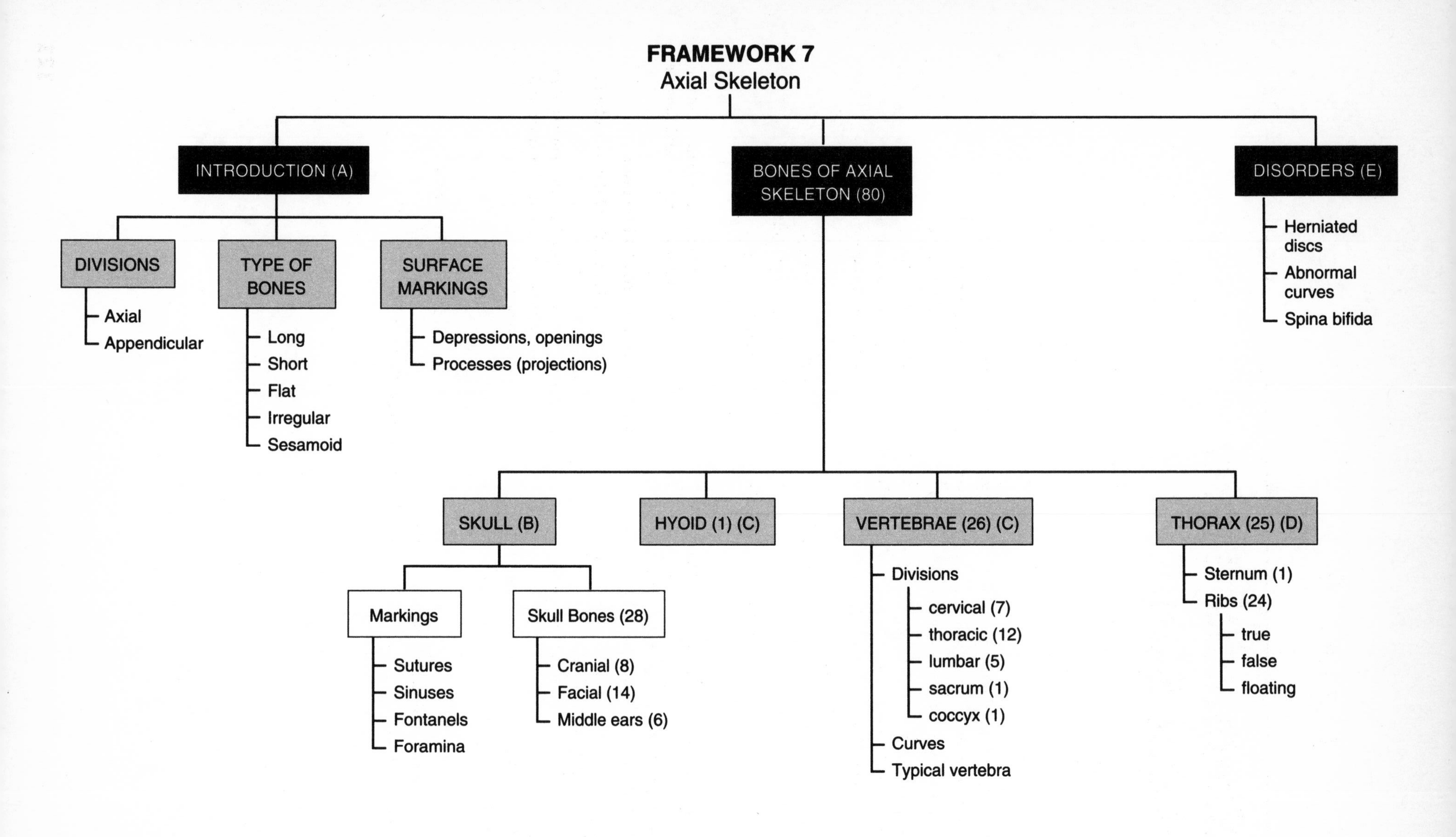

Note: numbers in ( ) refer to number of each bone in the body.

CHAPTER 7

The Skeletal System: The Axial Skeleton

The 206 bones of the human skeleton are classified into two divisions—axial and appendicular—based on their locations. The axial skeleton is composed of the portion of the skeleton immediately surrounding the axis of the skeleton, primarily the skull bones, vertebral column, and bones surrounding the thorax. Bones of the appendages, or extremities, comprise the appendicular skeleton. Chapter 7 begins with an introduction to types and surface characteristics (markings) of all bones and then focuses on the 80 bones of the axial skeleton. Disorders of the vertebral column are also included.

As you begin your study of the axial skeleton, carefully examine the Chapter 7 Framework and note relationships among concepts and key terms there. Also refer to the Topic Outline and Objectives; you may want to check off each objective as you complete it.

TOPIC OUTLINE AND OBJECTIVES

A. Divisions of the skeletal system, types of bones, and bone surface markings

1. Describe how the skeleton is divided into axial and appendicular divisions.
2. Classify bones based on their shape and location.
3. Describe the principal surface markings on bones and the functions of each.

B. Skull

1. Name the cranial and facial bones and indicate the number of each.
2. Describe the following special features of the skull: sutures, paranasal sinuses, and fontanels.

C. Hyoid bone and vertebral column

1. Describe the relationship of the hyoid bone to the skull.
2. Identify the regions and normal curves of the vertebral column and describe its structural and functional features.

D. Thorax

1. Identify the bones of the thorax.

E. Disorders

WORDBYTES

Now become familiar with the language of this chapter by studying each wordbyte, its meaning, and an example of its use within a term. After you study the entire list, self-check your understanding by writing the meaning of each wordbyte on the line. As you continue through the *Learning Guide,* identify (and fill in) additional terms that contain the same wordbyte.

Wordbyte	Self-check	Meaning	Example(s)
annulus	______________________	ring	*annulus* fibrosus
appendic-	______________________	to hang onto	*appendic*ular
cervic-	______________________	neck	*cervic*al collar
concha	______________________	shell	inferior nasal *concha*
costa	______________________	rib	*costa*l cartilage
cribr-	______________________	sieve	*cribr*iform plate
crist-	______________________	crest	*crist*a galli; *crist*ae
ethm-	______________________	sieve	*ethm*oid bone
infra-	______________________	below	*infra*orbital
lambd-	______________________	L-shaped	*lambd*oidal suture
lumb-	______________________	loin	*lumb*osacral
optic	______________________	eye	*optic* foramen
pulp	______________________	soft, flesh	nucleus *pulp*osus
sacrum-	______________________	sacred bone	*sacro*iliac
sphenoid-	______________________	wedge-shaped	*spheno*id bone
supra-	______________________	above	*supra*orbital

CHECKPOINTS

A. Divisions of the skeletal system, types of bones, and bone surface markings (pages 186–189)

A1. Explain how the skeletal system is absolutely necessary for vital life functions of eating, breathing, and movement.

✓ **A2.** Describe the bones in the two principal divisions of the skeletal system by completing this exercise. It may help to refer to Figure LG 8.1, page 146.

a. Bones that lie along the axis of the body are included in the *(axial? appendicular?)* skeleton.

b. The axial skeleton includes the following groups of bones. Indicate how many bones are in each category.

_______ Skull (cranium, face) _______ Vertebrae

_______ Earbones (ossicles) _______ Sternum

_______ Hyoid _______ Ribs

c. The total number of bones in the axial skeleton is _______.

d. The appendicular skeleton consists of bones in which parts of the body?

e. Write the number of bones in each category. Note that you are counting bones on one side of the body only.

_______ Left shoulder girdle

_______ Left upper extremity (arm, forearm, wrist, hand)

_______ Left hipbone

_______ Left lower extremity (thigh, kneecap, leg, foot)

f. There are _______ bones in the appendicular skeleton.

g. In the entire human body there are _______ bones.

A3. Complete the table about the four major and two minor types of bones.

Type of Bone	Structural Features	Examples
a. Long	Slightly curved to absorb stress better	
b.		Bones in wrist and back of feet
c.	Composed of two thin plates of bone	
d. Irregular		
e. Sutural		
f. Small bones in tendons		

A4. In general, what are the functions of surface markings of bones?

A5. Contrast the bone markings in each of the following pairs.

a. *Tubercle/tuberosity*

b. *Crest/line*

c. *Fossa/foramen*

d. *Condyle/epicondyle*

✓ **A6.** Write the correct answer from the following list of specific bone markings on the line next to its description. The first one has been done for you.

Articular *facet* on vertebra	**Maxillary *sinus***
External auditory *meatus*	**Optic *foramen***
Greater *trochanter*	**Styloid *process***
***Head* of humerus**	**Superior orbital *fissure***

a. Air-filled cavity within a bone, connected to nasal cavity **Maxillary sinus**

b. Narrow, cleftlike opening between adjacent parts of bone; passageway for blood vessels and nerves: ______

c. Rounded hole, passageway for blood vessels and nerves: ______

d. Tubelike passageway through bone: ______

e. Large, rounded projection above constricted neck: ______

f. Large, blunt projection: ______

g. Sharp, slender projection: ______

h. Smooth, flat surface: ______

B. Skull (pages 189–202)

✓ **B1.** The skull consists of two major regions. Describe these in this checkpoint.

a. The ____________________ consisting of *(6? 8? 14? 16?)* bones. These bones surround the ____________________. Located in one pair of the cranial bones (the temporals) are the tiny bones of the middle ear. Each middle ear contains *(2? 3? 6? 8?)* bones.

b. The remaining bones of the skull are comprised of _____ bones that provide a structural framework for the face. The total number of bones in the cranium, middle ears, and face is _____ bones.

✓ **B2.** Answer these questions on the cranium.

a. What is the primary function of the cranium? ____________________
List several other functions of the cranium.

b. Which bones make up the cranium (rather than the face)? Write the number (1 or 2) of each bone you name.

O Ethmoid bone
O Frontal bone
O Lacrimal bone
O Mandible bone
O Maxilla bone
O Nasal bone
O Occipital bone
O Parietal bone
O Sphenoid bone
O Temporal bone
O Zygomatic bone

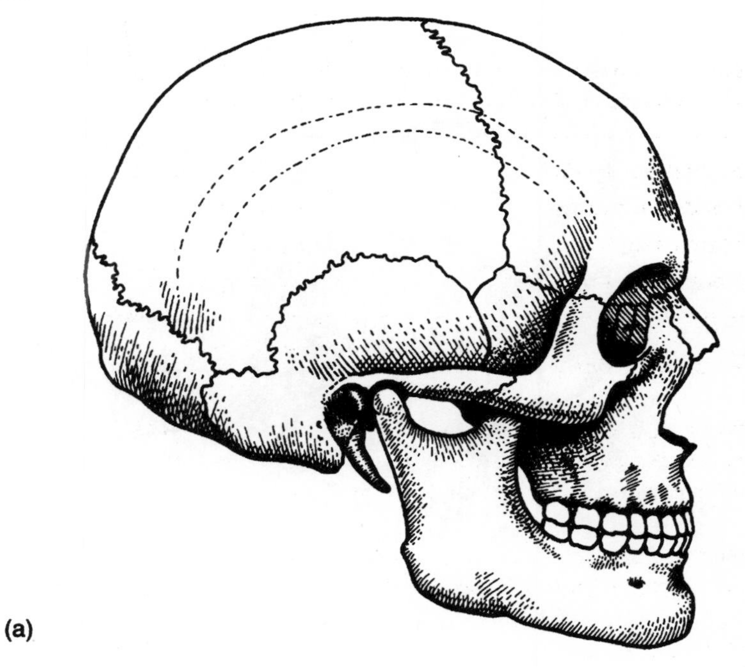

(a)

Figure LG 7.1 Skull bones. (a) Skull viewed from right side. Color and label as directed in Checkpoints B3, B6, and B9.

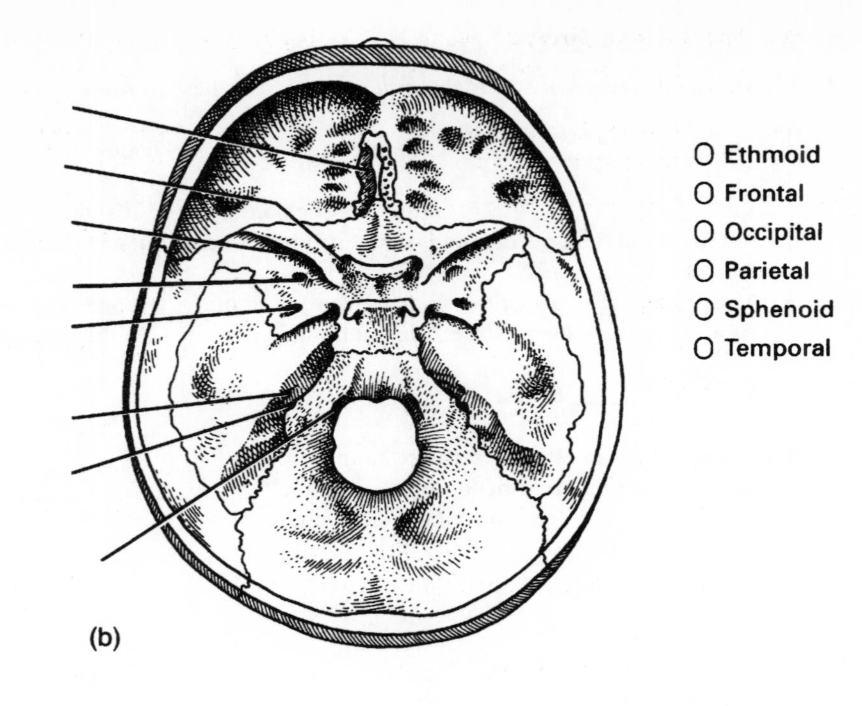

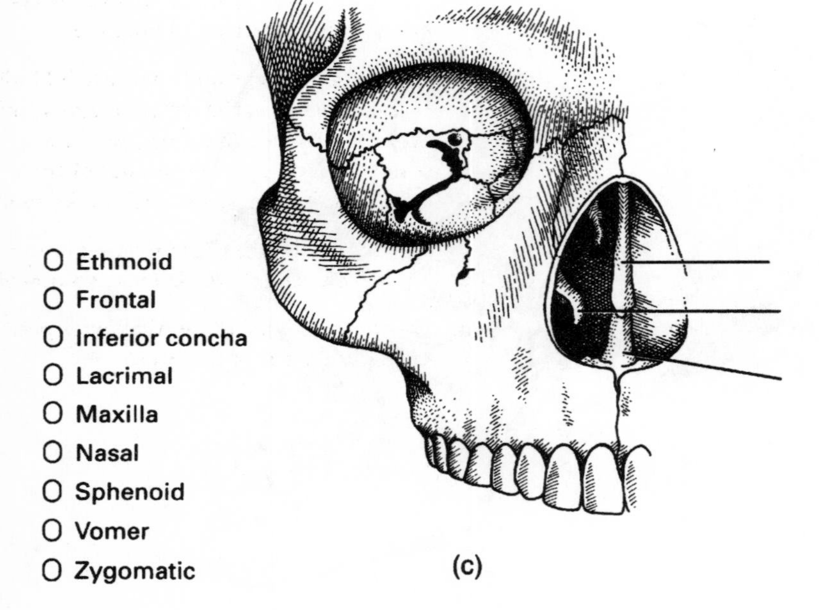

Figure LG 7.1 Skull bones. (b) Floor of the cranial cavity. (c) Anterior view of bones that form the right orbit and the nose. Color and label as directed in Checkpoints B3, B6, B7, B9, and B10.

✓ **B3.** Color the skull bones on Figure LG 7.1a, b, and c. Be sure to color the corresponding color code oval for each bone listed on the figure.

✓ **B4.** Check your understanding of locations and functions of skull bones by matching names of bones in the box with their descriptions below. Use each bone only once.

Ethmoid	**Maxilla**	**Sphenoid**
Frontal	**Nasal**	**Temporal**
Inferior concha	**Occipital**	**Vomer**
Lacrimal	**Palatine**	**Zygomatic (malar)**
Mandible	**Parietal**	

________ a. This bone forms the lower jaw, including the chin.

________ b. These are the cheek bones; they also form lateral walls of the orbit of the eye.

________ c. Tears pass through tiny foramina in these bones; they are the smallest bones in the face.

________ d. The bridge of the nose is formed by these bones.

________ e. Organs of hearing (internal part of ears) and mastoid air cells are located in and protected by these bones.

________ f. This bone sits directly over the spinal column; it contains the foramen through which the spinal cord connects to the brain.

________ g. The name means "wall." The bones form most of the roof and much of the side walls of the skull.

________ h. These bones form most of the roof of the mouth (hard palate) and contain the sockets into which upper teeth are set.

________ i. L-shaped bones form the posterior parts of the hard palate and nose.

________ j. Commonly called the forehead, it provides protection for the anterior portion of the brain.

________ k. A fragile bone, it forms much of the roof and internal structure of the nose.

________ l. It serves as a "keystone" because it binds together many of the other bones of the skull. It is shaped like a bat, with the wings forming part of the sides of the skull and the legs at the back of the nose.

________ m. This bone forms the inferior part of the septum dividing the nose into two nostrils.

________ n. Two delicate bones form the lower parts of the side walls of the nose.

✓ **B5.** Complete the table describing major markings of the skull.

Marking	Bone	Function
a. Greater wings	Sphenoid	
b.		Forms superior portion of septum of nose
c.		Site of pituitary gland
d. Petrous portion		
e.		Largest hole in skull; passageway for spinal cord
f.	(2)	Bony sockets for teeth
g.		Passageway for sound waves to enter ear
h. Condylar process	Mandible	

✓ **B6.** Now label each of the markings listed in the above table on Figure LG 7.1.

✓ **B7.** *The Big Picture: Looking Ahead.* Refer to figures on text pages listed in parentheses as you do this activity. The 12 pairs of nerves attached to the brain are called cranial nerves (Figure 14.5 in your text). Holes in the cranium permit passage of these nerves to and from the brain. These nerves are numbered according to the order in which they attach to the brain (and leave the cranium) from I (most anterior) to XII (most posterior). To help you visualize their sequence, label the foramina for cranial nerves in order on the left side of Figure LG 7.1b. Complete the table summarizing these foramina. The first one is done for you.

Number and Name of Cranial Nerve	Location of Opening for Nerve
a. I Olfactory (Figure 16.1)	Cribriform plate of ethmoid bone
b. II Optic (Figure 16.6)	
c. III Oculomotor (Figure 17.3) IV Trochlear V Trigeminal (ophthalmic branch) VI Abducens	
d. V Trigeminal (maxillary branch)	
e. V Trigeminal (mandibular branch)	
f. VII Facial (Figure 17.3) VIII Vestibulocochlear (Figure 16.17)	
g. IX Glossopharyngeal (Figure 17.3) X Vagus (Figure 17.3) XI Accessory	
h. XII Hypoglossal	

✓ **B8.** *For extra review.* Write the names of the markings listed in the box next to the bones in which those markings are located. Write the name of one marking on each line.

Carotid canal	**Mental foramina**	**Superior nasal conchae**
Condylar process	**Olfactory foramina**	**Superior nuchal line**
Crista galli	**Palatine process**	**Superior orbital fissure**
Inferior orbital fissure	**Perpendicular plate**	**Supraorbital margin**
Lesser wings	**Pterygoid processes**	**Zygomatic process**
Mastoid air cells		

a. Ethmoid: ______________ ______________

______________ ______________

b. Frontal: ______________

c. Mandible: ______________ ______________

d. Maxilla: ______________ ______________

e. Occipital: ______________

f. Sphenoid: ______________ ______________

____

g. Temporal: ______________ ______________

____

✓ **B9.** Check your understanding of special features of the skull in this Checkpoint.

a. *Define sutures.* Between which two bones is the sagittal suture

located? ____________________

On Figure LG 7.1a, label the following sutures: *coronal, squamous, lambdoid.*

b. "Soft spots" of a newborn baby's head are known as ____________________.

What is the location of the largest one? ____________________

c. List three functions of fontanels.

d. A good way to test your ability to visualize locations of important skull bones is to try to identify bones that form the orbits surrounding and protecting the eyes. On Figure LG 7.1c, locate six bones that form the orbit. (*Note:* The tiny portion of the seventh bone, the superior tip of the palatine bone, is not visible in the figure.)

✓ **B10.** Do this exercise about bony structures related to the nose.

a. Name four bones that contain paranasal sinuses.

Practice identifying their locations the next time you have a cold!

b. List three functions of paranasal sinuses.

c. Most internal portions of the nose are formed by a bone that includes these markings: paranasal sinuses, conchae, and part of the nasal septum. Name the bone.

d. What functions do nasal conchae perform?

e. On Figure LG 7.1c, label the bone that forms the inferior portion of the nasal septum.

The anterior portion of the nasal septum consists of flexible ____________________.

✓ **B11.** *The Big Picture: Looking Ahead* at details of markings of the skull, circle the correct answers in each statement. Refer to figures in Chapters 11, 16, and 23 as you complete this Checkpoint.

a. The zygomatic arch (Figure 11.6) is formed of two bones; these are the *(parietal? sphenoid? temporal? zygomatic?)* bones.

b. The styloid process (Figure 11.7) is a penlike projection from the *(parietal? sphenoid? temporal?)* bone.

c. The tear duct (Figure 16.5b) passes through a canal in the *(lacrimal? mental?)* foramen.
d. The frontal sinus (Figure 23.2b) is located *(superior? inferior?)* to the nasal cavity.
e. The nasal conchae (Figure 23.2b) are located *(superior to? within?)* the nasal cavity.
f. The hard palate (Figure 23.2b) is formed anteriorly of the *(hyoid? mandible? maxilla? palatine?)* bone and posteriorly of the *(mandible? maxilla? palatine?)* bone.

✓ **B12.** *Clinical correlation.* Name the skull bone(s) and specific markings likely to be involved in each clinical case.

a. Ten-year-old Chris suffered from a "black eye" after she was hit by a ball.

______________________________ ______________________________

b. An auto accident left Miji, age 24, with a "deviated septum."

______________________________ ______________________________

c. Susanna, 8 months, was born with a "cleft palate."

______________________________ ______________________________

d. Leonardo, age 40, consulted a pain clinic for chronic "TMJ" pain.

______________________________ ______________________________

C. Hyoid bone and vertebral column (pages 202–211)

✓ **C1.** Identify the location of the hyoid bone on yourself. Place your hand on your throat and swallow. Feel your larynx move upward? The hyoid sits just *(superior? inferior?)* to the larynx at the level of the mandible.

✓ **C2.** In what way is the hyoid unique among all the bones of the body?

✓ **C3.** The five regions of the vertebral column are grouped (A–E) in Figure LG 7.2. Color vertebrae in each region and be sure to select the same color for the corresponding color code oval. Now write on lines next to A–E the number of vertebrae in each region. One is done for you. Also label the first two cervical vertebrae.

✓ **C4.** Note which regions of the vertebral column in Figure 7.2 normally retain an anteriorly concave curvature in the adult: ________________ and ________________. These are considered *(primary? secondary?)* curvatures. This classification is based upon the fact that these curves *(were present originally during fetal life? are more important?).*

✓ **C5.** On Figure LG 7.3, color the parts of vertebrae and corresponding color code ovals. As you do this, notice differences in size and shape among the three vertebral types in the figure.

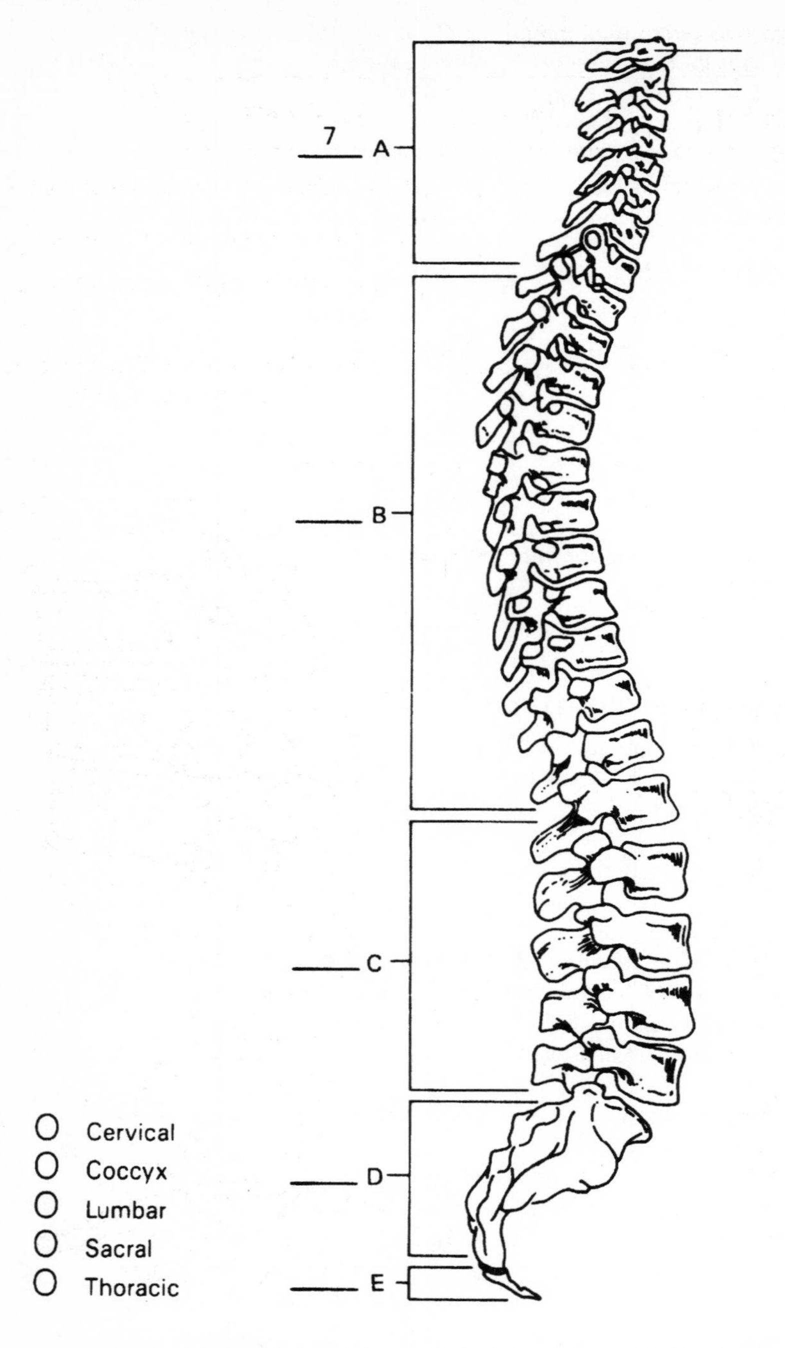

Figure LG 7.2 Right lateral view of the vertebral column. Color and label as directed in Checkpoint C3.

✓ **C6.** Describe bones and markings involved in each joint.

C. Cervical	**L.** Lumbar	**S.** Sacral	**T.** Thoracic

_______ a. Small body, foramina for vertebral blood vessels in transverse processes

_______ b. Only vertebrae that articulate with ribs

_______ c. Massive body, blunt spinous process and articular processes directed medially or laterally

_______ d. Long spinous processes that point inferiorly

_______ e. Articulate with the two hipbones

_______ f. The atlas and axis are vertebrae in this region

_______ g. Spinous processes in this region include several that are normally bifid

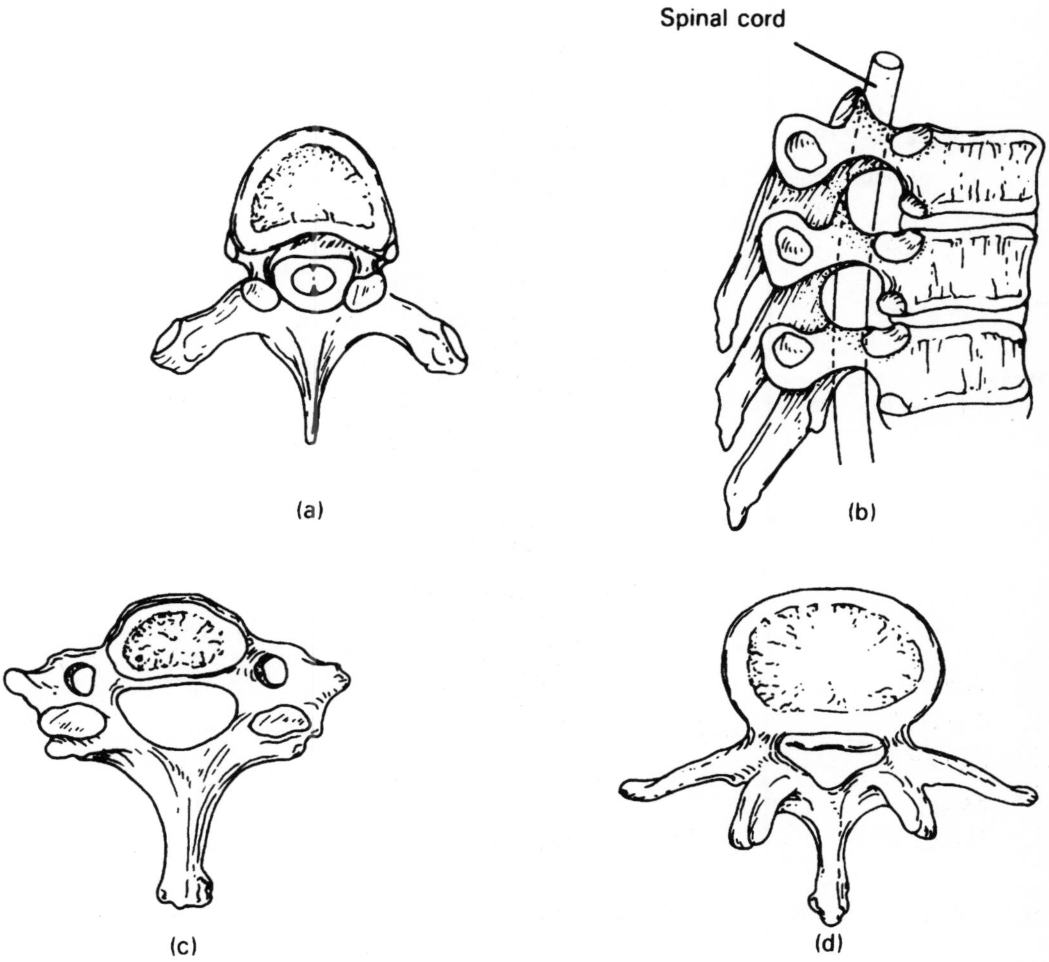

Spinal cord

(a) (b)

(c) (d)

- O Body
- O Facets for articulation with rib
- O Intervertebral disc
- O Intervertebral foramen
- O Lamina
- O Pedicle
- O Spinous process (spine)
- O Superior articular process
- O Transverse foramen
- O Transverse process
- O Vertebral foramen

Figure LG 7.3 Typical vertebrae. (a) Thoracic vertebra, superior view. (b) Thoracic vertebrae, right lateral view (c) Cervical vertebra, superior view. (d) Lumbar vertebra, superior view. Color and label as directed in Checkpoint C5.

✓ **C7.** Describe bones and markings involved in each joint.

a. Atlanto-occipital joint: ______________________________

b. Atlanto-axial (two different joint types): ______________________________

____

✓ **C8.** *A clinical correlation.* State the clinical significance of these markings.

a. Sacral hiatus

b. Sacral promontory

c. Dens

D. Thorax (pages 211–213)

D1. Name the structures that compose the thoracic cage. (Why is it called a "cage"?)

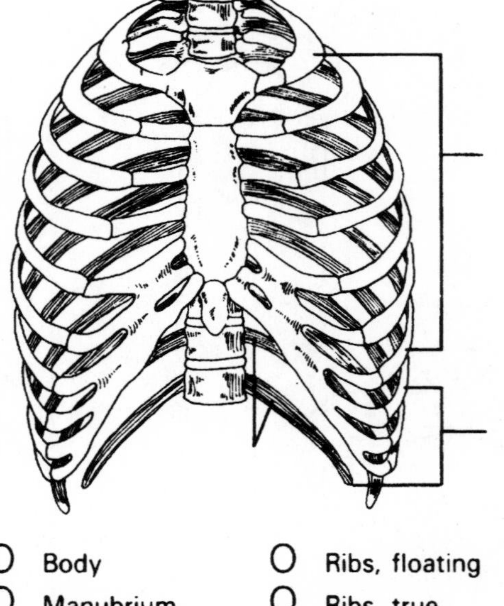

Figure LG 7.4 Anterior view of the thorax. Color as directed in Checkpoints D2 and D3.

O Body
O Manubrium
O Ribs, false
O Ribs, floating
O Ribs, true
O Xiphoid process

✓ **D2.** On Figure LG 7.4, color the three parts of the sternum, and label these parts: *sternal angle, suprasternal notch,* and *clavicular notch.*

✓ **D3.** Complete this exercise about ribs. Refer again to Figure LG 7.4.

a. There is a total of _____ ribs (_____ pairs) in the human skeleton.
b. Ribs slant in such a way that the anterior portion of the rib is *(superior? inferior?)* to the posterior end of the rib.

c. Posteriorly, all ribs articulate with ________________. Ribs also pass *(anterior? posterior?)* to and articulate with transverse processes of vertebrae. What is the functional advantage of such an arrangement?

d. Anteriorly, ribs numbered _____ to _____ attach to the sternum directly by means of

strips of hyaline cartilage, called ________________ cartilage. These ribs are called *(true? false?)* ribs. Color these ribs on Figure LG 7.4, leaving the costal cartilages white.

e. Ribs 8 to 12 are called ________________. Color these ribs a different color, again leaving the costal cartilages white. Do these ribs attach to the sternum? *(Yes? No?)* If so, in what manner?

f. Ribs _____ and _____ are called "floating ribs." Use a third color for these ribs. Why are they so named?

g. What function is served by the costal groove?

h. What occupies intercostal spaces?

✓ **D4.** At what point are ribs most commonly fractured?

E. Disorders (page 214)

✓ **E1.** Complete this exercise about slipped discs.

a. The normal intervertebral disc consists of two parts: an outer ring of *(hyaline? elastic?*

fibro-?) cartilage called ________________ and a soft, elastic, inner portion called

the ________________.

b. Ligaments normally keep discs in alignment with vertebral bodies. What may happen if these ligaments weaken?

c. Why might pain result from a slipped disc?

d. In what part of the vertebral column are slipped discs most common?

________________. What symptoms may result from a slipped disc in this region?

✓ **E2.** Match types of abnormal curvatures of the vertebral column with descriptions below.

K. Kyphosis	**L. Lordosis**	**S. Scoliosis**

_______ a. Exaggerated lumbar curvature; "swayback"

_______ b. Exaggerated thoracic curvature; "hunchback"

_______ c. S- or C-shaped lateral bending; most common abnormal curvature

E3. Imperfect union of the vertebral arches at the midline is the condition known as

_________________. Why is it crucial that the vertebral foramen be completely surrounded by bone? What problems may result from incomplete closure?

ANSWERS TO SELECTED CHECKPOINTS: CHAPTER 7

A2. (a) Axial. (b) 22 skull, 6 earbones (studied in Chapter 16), 1 hyoid, 26 vertebrae, 1 sternum, 24 ribs. (c) 80. (d) Shoulder girdles, upper extremities, hipbones, lower extremities. (e) 2 left shoulder girdle, 30 left upper extremity, 1 left hipbone, 30 left lower extremity. (f) 126. (g) 206.

A6. (b) Superior orbital fissure. (c) Optic foramen. (d) External auditory meatus. (e) Head of humerus. (f) Greater trochanter. (g) Styloid process. (h) Articular facet on vertebra.

B1. (a) Cranium, 8; brain; 3. (b) 14; 28.

B2. (a) Protection of the brain; attachment for meninges and muscles that move the head; openings for blood vessels and nerves. (b) Frontal (1), parietals (2), occipital (1), temporals (2), ethmoid (1), and sphenoid (1).

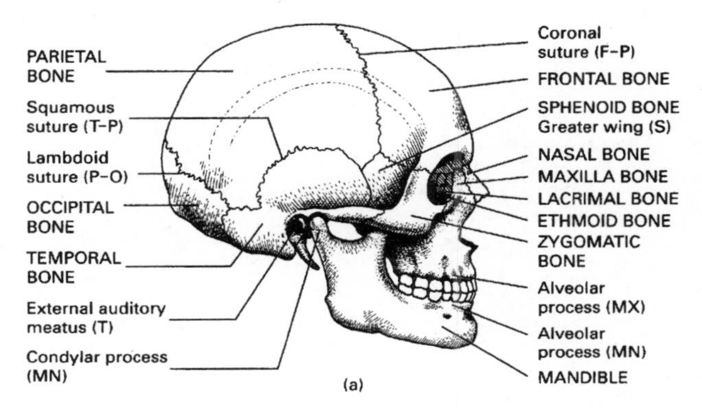

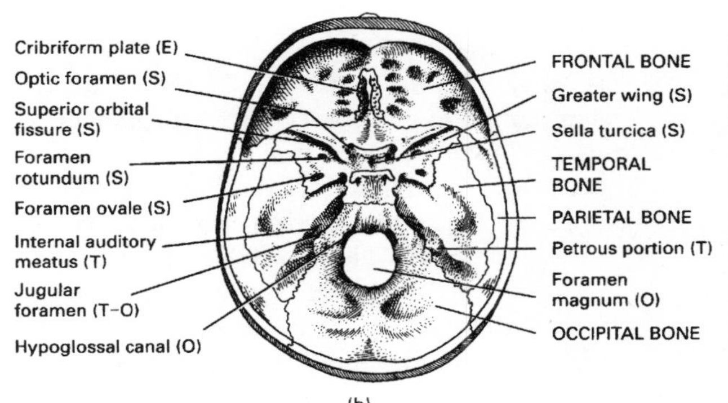

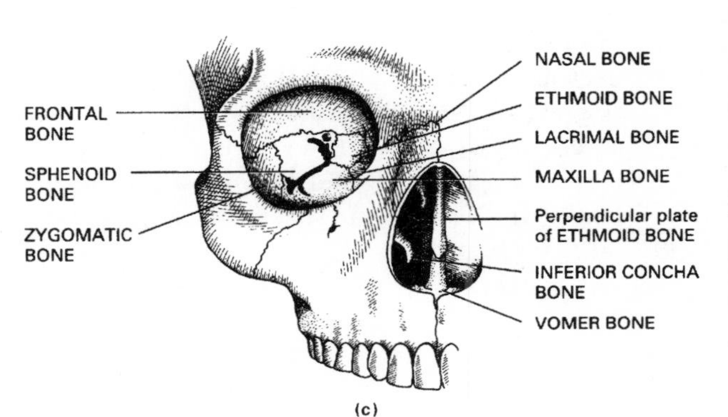

Figure LG 7.1A Skull bones. (a) Skull viewed from right side. (b) Floor of the cranial cavity. (c) Anterior view of bones that form the right orbit and the nose. Bone labels are capitalized; markings are in lowercase with related bone initial in parentheses.

B3. See Figure LG 7.1A.

B4. (a) Mandible. (b) Zygomatic (malar). (c) Lacrimal. (d) Nasal. (e) Temporal. (f) Occipital. (g) Parietal. (h) Maxilla. (I) Palatine. (j) Frontal. (k) Ethmoid. (l) Sphenoid. (m) Vomer. (n) Inferior concha.

B5.

Marking	Bone	Function
a. Greater wings	Sphenoid	**Form part of side walls of skull**
b. **Perpendicular plate**	**Ethmoid**	Forms superior portion of septum of nose
c. **Sella Turcica**	**Sphenoid**	Site of pituitary gland
d. Petrous portion	**Temporal**	**Houses middle ear and inner ear**
e. **Foramen magnum**	**Occipital**	Largest hole in skull; passageway for spinal cord
f. **Alveolar processess**	(2) **Maxillae and mandible**	Bony sockets for teeth
g. **External auditory meatus**	**Temporal**	Passageway for sound waves to enter ear
h. Condylar process	Mandible	**Articulates with temporal bone (in TMJ)**

B6. See Figure LG 7.1A

B7. See Figure LG 7.1A

Number and Name of Cranial Nerve	Location of Opening for Nerve
a. I Olfactory (Figure 16.1)	Cribriform plate of ethmoid bone
b. II Optic (Figure 16.6)	**Optic foramen of sphenoid bone**
c. III Oculomotor (Figure 17.3) IV Trochlear V Trigeminal (ophthalmic branch) VI Abducens	**Superior orbital fissure of sphenoid bone**
d. V Trigeminal (maxillary branch)	**Foramen rotundum of sphenoid bone**
e. V Trigeminal (mandibular branch)	**Foramen ovale of sphenoid bone**
f. VII Facial (Figure 17.3) VIII Vestibulocochlear (Figure 16.17)	**Internal auditory meatus**
g. IX Glossopharyngeal (Figure 17.3) X Vagus (Figure 17.3) XI Accessory	**Jugular foramen**
h. XII Hypoglossal	**Hypoglossal canal**

B8. (a) Crista galli, olfactory foramina, perpendicular plate, superior nasal conchae. (b) Supraorbital margin. (c) Condylar process, mental foramina. (d) Inferior orbital fissure (next to sphenoid), palatine process. (e) Superior nuchal line. (f) Lesser wings, pterygoid processes, superior orbital fissure. (g) Carotid canal, mastoid air cells, zygomatic process.

B9. (a) Immovable (fibrous) joints between skull bones; between parietal bones; see Figure LG 7.1A. (b) Fontanels; between frontal and parietal bones. (c) They permit change in size and shape of the fetal skull during the birth process and they permit rapid growth of the skull during infancy; the anterior fontanel serves as a clinical landmark. (d) See on Figure LG 7.1cA: frontal, zygomatic, maxilla, lacrimal, ethmoid, and sphenoid.

B10. (a) Frontal, ethmoid, sphenoid, maxilla. (b) Warm and humidify air because air sinuses are lined with mucous membrane; serve as resonant chambers for speech and other sounds; make the skull lighter weight. (c) Ethmoid. (d) Like paranasal sinuses, they are covered with mucous membrane, which warms, humidifies, and cleanses air entering the nose. (e) Vomer; cartilage.

B11. (a) Temporal, zygomatic. (b) Temporal. (c) Lacrimal. (d) Superior. (e) Within. (f) Maxilla, palatine.

B12. (a) Frontal: supraorbital margin. (b) Ethmoid: perpendicular plate; vomer; cartilage (anteriorly). (c) Maxilla bones: palatine processes; palatine bones: horizontal plates. (d) Temporal bone: condylar fossa; mandible: condyle.

Cl. Superior.

C2. It articulates (forms a joint) with no other bones.

C3. A, cervical (7); B, thoracic (12); C, lumbar (5); D, sacrum (1); E, coccyx (1); Cl, atlas; C2, axis.

C4. B (thoracic) and D (sacral); primary; were present originally during fetal life.

C5. See Figure LG 7.3A

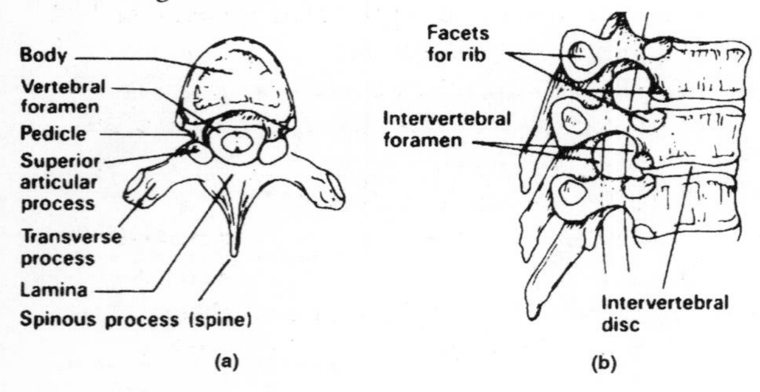

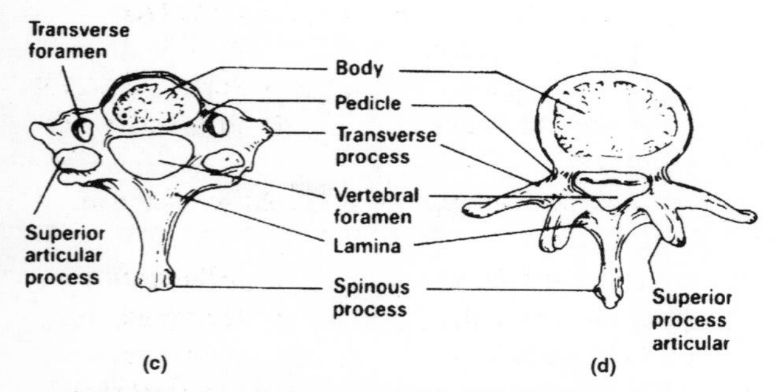

Figure LG 7.3A Typical vertebrae. (a) Thoracic vertebra, superior view. (b) Thoracic vertebrae, right lateral view. (c) Cervical vertebra, superior view. (d) Lumbar vertebra, superior view.

C6. (a) C. (b) T. (c) L. (d) T. (e) S. (f) C. (g) C.

C7. (a) Superior articular facets of atlas-occipital condyles. (b) Inferior articular facets of atlas—superior articular facets of axis; also anterior arch of atlas—dens of axis.

C8. (a) Site used for administration of epidural anesthesia. (b) Obstetrical landmark for measurement of size of the pelvis. (c) In whiplash, it may injure the brainstem.

D2. See Figure LG 7.4A

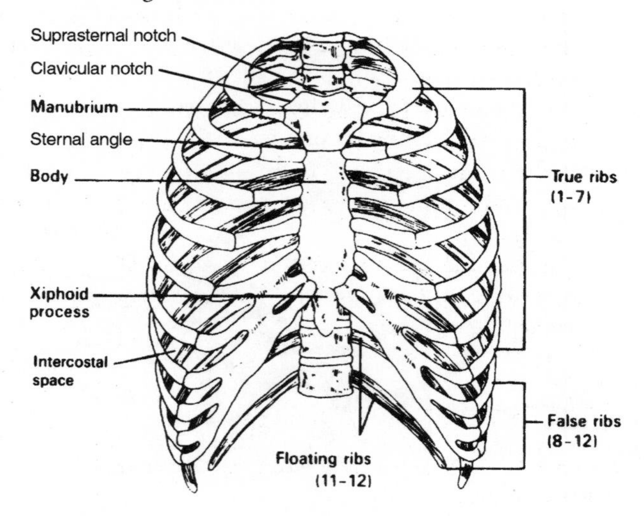

Figure LG 7.4A Anterior view of the thorax.

D3. (a) 24, 12. (b) Inferior. (c) Bodies of thoracic vertebrae; anterior; prevents ribs from slipping posteriorward. (d) 1, 7, costal; true. (e) False; ribs 8–10 attach to sternum indirectly via seventh costal cartilage. (f) 11, 12; they have no anterior attachment to sternum. (g) Provides protective channel for intercostal nerve, artery, and vein. (h) Intercostal muscles. Refer also to Figure LG 7.4A.

D4. Just anterior to the costal angle, especially involving ribs 3 to 10.

El. (a) Fibro-, annulus fibrosus, nucleus pulposus. (b) Annulus fibrosus may rupture, allowing the nucleus pulposus to protrude (herniate), causing a "slipped disc." (c) Disc tissue may press on spinal nerves. (d) L4 to L5 or L5 to sacrum; pain in posterior of thigh and leg(s) due to pressure on sciatic nerve, which attaches to spinal cord at L4 to S3.

E2. (a) L. (b) K. (c) S.

MORE CRITICAL THINKING: CHAPTER 7

1. Contrast the effects of the following injuries to the skull. Consider what functions of each traumatized bone would be lost, and deduce the relative seriousness in each case.
   a. The right "greater wing" of the sphenoid; the left "lesser wing" of the sphenoid.
   b. The orbit; the foramen magnum. Hint: Which (how many) bone(s) form(s) each?
   c. The mandible; the right maxilla bone
2. One surgical approach to the pituitary gland involves passage of an instrument though the nose, up through the cribriform plate. Which sensation is likely to be impaired after this surgery?
3. The ethmoid bone could be considered as delicate as a French pastry. Look closely at a skull and notice why each of the following parts of this bone is particularly vulnerable:
   (a) Orbital surface
   (b) Superior and middle turbinates (conchae)
   (c) Perpendicular plate
4. Contrast the vertebral column with the spinal column. Then describe functions of the vertebral column, and explain what structural features make the vertebral column strong yet flexible.
5. Consider the structure and functions of the axis and atlas. How do these bones help you to move your head and neck? What parts of these bones (and related joints) are most vulnerable to injury?

MASTERY TEST: CHAPTER 7

Questions 1–10: Circle the letter preceding the one best answer to each question.

1. All of these bones contain paranasal sinuses except:
   A. Frontal
   B. Maxilla
   C. Nasal
   D. Sphenoid
   E. Ethmoid
2. Choose the one false statement.
   A. There are seven vertebrae in the cervical region.
   B. The cervical region normally exhibits a curve that is slightly concave anteriorly.
   C. The lumbar vertebrae are superior to the sacrum.
   D. Intervertebral discs are located between bodies of vertebrae.
3. The hard palate is composed of _____ bones.
   A. Two maxilla and two mandible
   B. Two maxilla and two palatine
   C. Two maxilla
   D. Two palatine
   E. Vomer, ethmoid, and two temporal
4. The lateral wall of the orbit is formed mostly by which two bones?
   A. Zygomatic and maxilla
   B. Zygomatic and sphenoid
   C. Sphenoid and ethmoid
   D. Lacrimal and ethmoid
   E. Zygomatic and ethmoid
5. Choose the one true statement.
   A. All of the ribs articulate anteriorly with the sternum.
   B. Ribs 8 to 10 are called true ribs.
   C. There are 23 ribs in the male skeleton and 24 in the female skeleton.
   D. Rib 7 is larger than rib 3.
   E. Cartilage discs between vertebrae are called costal cartilages.
6. Which is the largest fontanel?
   A. Frontal
   B. Occipital
   C. Sphenoid
   D. Mastoid
7. Immovable joints of the skull are called:
   A. Sesamoid bones
   B. Sutures
   C. Conchae
   D. Sinuses
   E. Fontanels
8. _____ articulate with every bone of the face except the mandible.
   A. Lacrimal bones
   B. Zygomatic bones
   C. Maxillae
   D. Sphenoid bones
   E. Ethmoid bones
9. All of these markings are parts of the sphenoid bone except:
   A. Lesser wings
   B. Optic foramen
   C. Crista galli
   D. Sella turcica
   E. Pterygoid processes
10. All of these bones are included in the axial skeleton except:
   A. Rib
   B. Sternum
   C. Clavicle
   D. Hyoid
   E. Ethmoid

Questions 11–15: Arrange the answers in correct sequence.

______ ______ ______ 11. From anterior to posterior:
A. Ethmoid bone
B. Sphenoid bone
C. Occipital bone

______ ______ ______ 12. Parts of vertebra from anterior to posterior:
A. Vertebral foramen
B. Body
C. Lamina

______ ______ ______ 13. Vertebral regions, from superior to inferior:
A. Lumbar
B. Thoracic
C. Cervical

______ ______ ______ 14. From superior to inferior:
A. Atlas
B. Axis
C. Occipital bone

______ ______ ______ 15. From superior to inferior:
A. Atlas
B. Manubrium of sternum
C. Hyoid

Questions 16–20: Circle T (true) or F (false). If the statement is false, change the underlined word or phrase so that the statement is correct.

T F 16. The space between two ribs is called the <u>costal groove</u>.

T F 17. The thoracic and sacral curves are called <u>primary</u> curves, meaning that they retain the original curve of the fetal vertebral column.

T F 18. The jugular vein passes through the same foramen as cranial nerves <u>V, VI, and VII</u>.

T F 19. In general, <u>foramina, meati, and fissures</u> in the skull serve as openings for nerves and blood vessels.

T F 20. The annulus fibrosus portion of an intervertebral disc is a <u>firm ring of fibrocartilage</u> surrounding the nucleus pulposus.

Questions 21–25: Fill-ins. Write the word or phrase that best completes the statement.

________________ 21. A finger- or toothlike projection called the dens is part of the ________________ bone.

________________ 22. The ramus, angle, mental foramen, and alveolar processes are all markings on the ________________ bone.

________________ 23. The squamous, petrous, and zygomatic portions are markings on the ________________ bone.

________________ 24. The perpendicular plate, crista galli, and superior and middle conchae are markings found on the ________________ bone.

________________ 25. The most common type of abnormal curvature of the spine is ________________.

ANSWERS TO MASTERY TEST: CHAPTER 7

Multiple Choice

1. C
2. B
3. B
4. B
5. D
6. A
7. B
8. C
9. C
10. C

Arrange

11. A B C
12. B A C
13. C B A
14. C A B
15. A C B

True-False

16. F. Intercostal space
17. T
18. F. IX, X, and XI
19. T
20. T

Fill-Ins

21. Axis (second cervical vertebra)
22. Mandible
23. Temporal
24. Ethmoid
25. Scoliosis

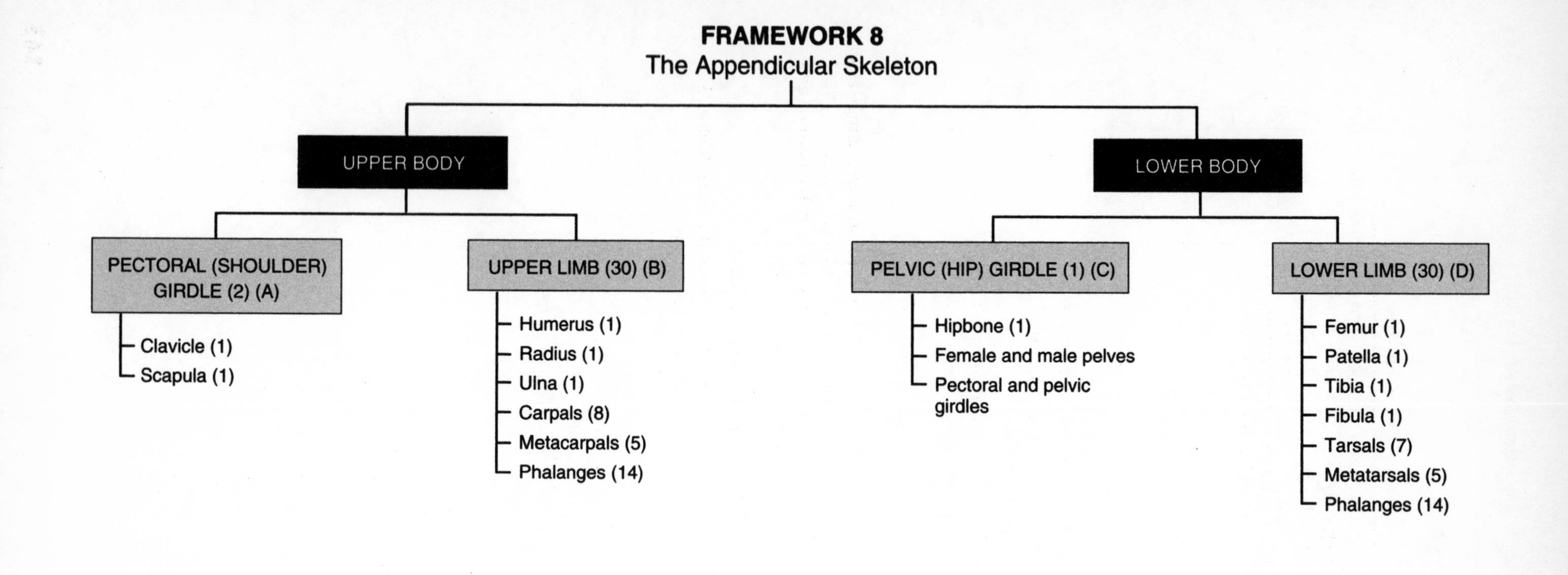

Note: numbers in ( ) refer to number of each bone on *one* side of the body.

CHAPTER 8

The Skeletal System: The Appendicular Skeleton

The appendicular skeleton includes the bones of the limbs, or extremities, as well as the supportive bones of the shoulder and hip girdles. Male and female skeletons exhibit some differences in skeletal structure, noted particularly in bones of the pelvis.

As you begin your study of the appendicular skeleton, carefully examine the Chapter 8 Topic Outline and Objectives and note relationships among concepts and key terms in the Framework.

TOPIC OUTLINE AND OBJECTIVES

A. Pectoral (shoulder) girdle

1. Identify the bones of the pectoral (shoulder) girdle and their principal markings.

B. Upper limb (extremity)

1. Identify the bones of the upper limb and their principal markings.

C. Pelvic (hip) girdle; comparison of female and male pelves

1. Identify the bones of the pelvic girdle and their principal markings.
2. Compare the principal structural differences between female and male pelves.
3. Describe the differences in the pectoral and pelvic girdles.

D. Lower limb (extremity)

1. Identify the bones of the lower limb and their principal markings.

WORDBYTES

Now become familiar with the language of this chapter by studying each wordbyte, its meaning, and an example of its use within a term. After you study the entire list, self-check your understanding by writing the meaning of each wordbyte on the line. As you continue through the *Learning Guide*, identify (and fill in) additional terms that contain the same wordbyte.

Wordbyte	Self-check	Meaning	Example(s)
acrom-	______________	topmost	*acromi*on process
acetabulum	______________	vinegar cup	*acetabul*ar notch
artho-	______________	joint	*artho*plasty
auric-	______________	ear-shaped	*auric*ular surface
capit-	______________	head	*capit*ulum, *capit*ate
coracoid	______________	like a crow's beak	*coracoid* process
coronoid	______________	crown-shaped	*coronoid* process
cox-	______________	hip	*cox*al bone
genu	______________	knee	*genu* valgum
glut-	______________	buttock	anterior *glut*eal line
hallux	______________	big toe	extensor *halluc*is
ilium	______________	flank	*iliac* crest
lun-	______________	moon	*lunate* bone
meta-	______________	after, beyond	*meta*carpal
obtur-	______________	closed up	*obtur*ator foramen
olecranon	______________	elbow	*olecranon* process
pelv-	______________	basin	true *pelv*is
-plasty	______________	molding, repair	anthro*plasty*
ram-	______________	branch	superior *ram*us
scaphoid	______________	boatlike	*scaphoid* bone
tri-	______________	three	*tri*quetrum
valgum	______________	bent outwards	genu *valgum*
varum	______________	bent towards midline	genu *varum*

CHECKPOINTS

A. Pectoral (shoulder) girdle (pages 219–221)

✓ **A1.** Do this exercise about the pectoral girdle.

a. Which bones form the pectoral girdle?

b. Do these bones articulate (form a joint) with vertebrae or ribs? ________

c. The pectoral girdle is part of the *(axial? appendicular?)* skeleton. Identify the point (marked by *) on Figure LG 8.1 at which the shoulder girdle articulates with the axial skeleton. Name the two bones forming that joint. Palpate (press and feel) the bones at this joint on yourself.

✓ **A2.** *Critical thinking* about the clavicle.
State two or more functions of the clavicle.

If the clavicle is fractured, this bone is not set in a cast. Why not?

✓ **A3.** Write the name of the bone that articulates with each of these markings on the scapula.

a. Acromion process: ____________________________

b. Glenoid cavity: ____________________________

c. Coracoid process: ____________________________

✓ **A4.** Study a scapula carefully, using a skeleton or Figure 8.3 in your text. Then match the markings in the box with the descriptions given.

A. Axillary border	**M. Medial border**
I. Infraspinatus fossa	**S. Spine**

________ a. Sharp ridge on the posterior surface

________ b. Depression inferior to the spine; location of infraspinatus muscle

________ c. Edge closest to the vertebral column

________ d. Thick edge closest to the arm

B. Upper limb (extremity) (pages 222–227)

✓ **B1.** List the bone (or groups of bones) in the upper limb from proximal to distal. Indicate how many of each bone there are in one limb. Two are done for you. Refer to Figure LG 8.1 to check your answers.

a. **Humerus** **(1)**

b. ____________________ ( )

c. ____________________ ( )

d. ____________________ ( )

e. **Metacarpals** **(5)**

f. ____________________ ( )

✓ **B2.** On Figure LG 8.2, select different colors and color each of the markings indicated by ○. Where possible, color markings on both the anterior and posterior views.

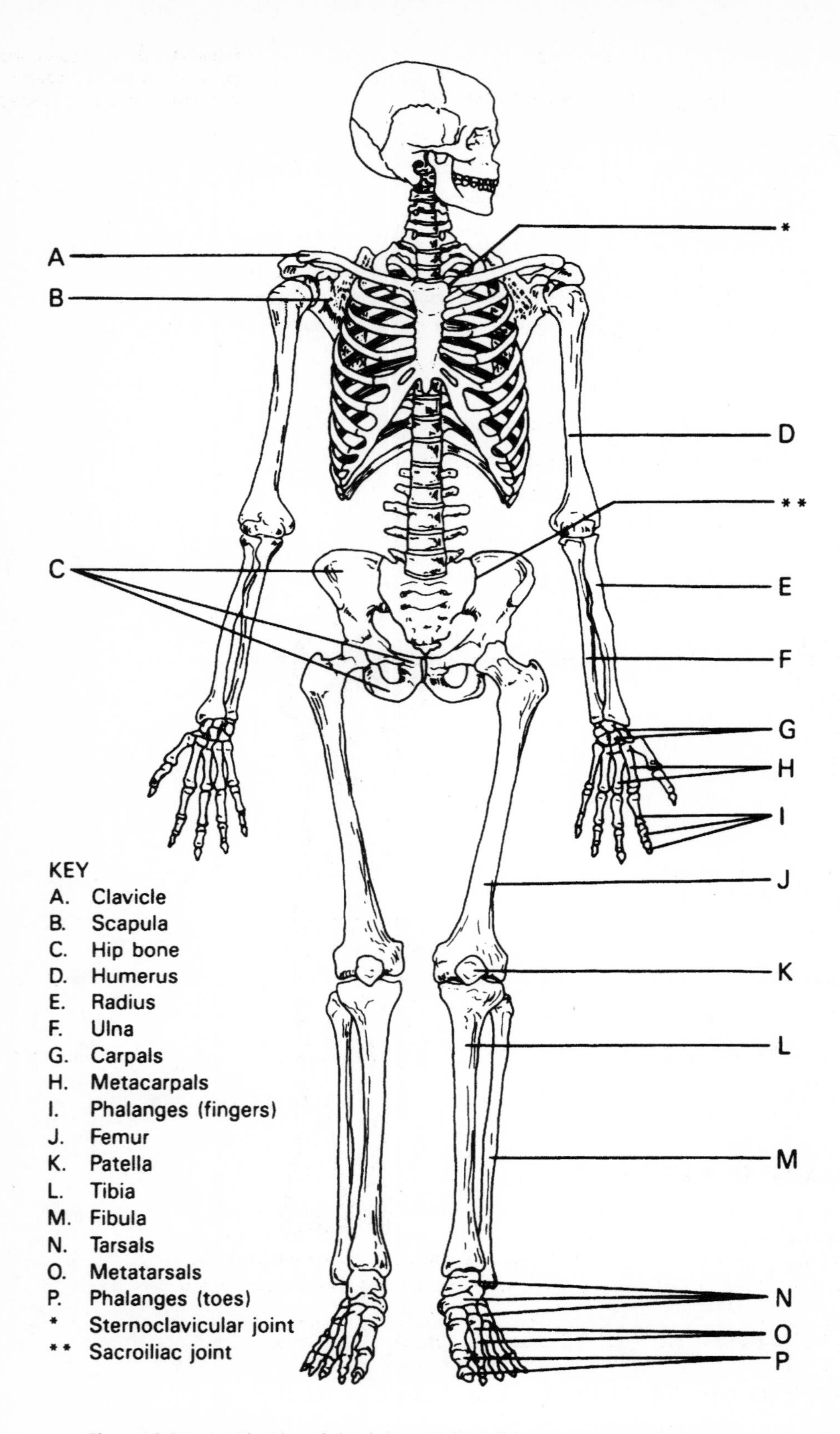

Figure LG 8.1 Anterior view of the skeleton. Color, label, and answer questions as directed in Checkpoints A1, B1, C1, D1, D2, and D9.

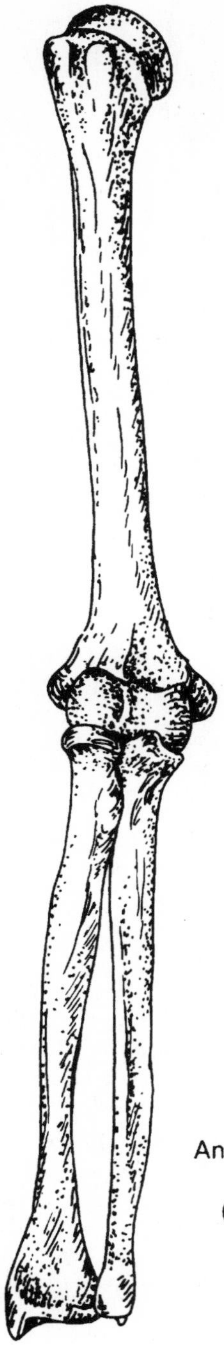

Humerus:

- O Anatomical neck
- O Capitulum
- O Coronoid fossa
- O Deltoid tuberosity
- O Greater tubercle
- O Head
- O Medial epicondyle
- O Trochlea

Ulna

- O Coronoid process
- O Olecranon process
- O Styloid process

Radius

- O Head
- O Radial tuberosity
- O Styloid process

Anterior

(a)

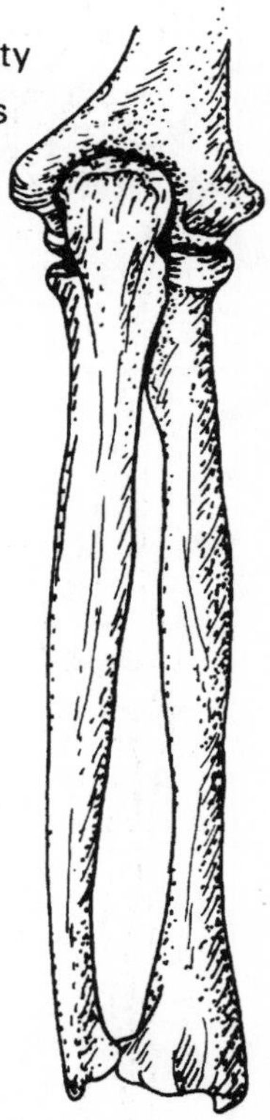

Posterior

(b)

Figure LG 8.2 Right upper limb. (a) Anterior view. (b) Posterior view. Color and label as directed in Checkpoint B2.

✓ **B3.** Name the marking that fits each description. Write H, U, or R to indicate whether the marking is part of the humerus (H), ulna (U), or radius (R). One has been done for you.

a. Articulates with glenoid fossa: ____**Head**____ (__**H**__)

b. Rounded head that articulates with radius: ____________________ (_____)

c. Posterior depression that receives the olecranon process: ____________________ (_____)

d. Halfmoon-shaped curved area that articulates with trochlea: ____________________ (_____)

e. Slight depression in which the head of the radius pivots (at the proximal radioulnar joint): ____________________ (_____)

f. Tendon of the biceps brachii muscle attaches here: ____________________ (_____)

g. Easily palpated (felt) projection on the posterior side of the distal end of the forearm on the little finger side. ____________________ (_____)

h. Projection on the medial side of the elbow; sensitive to touch since ulnar nerve passes through here: ____________________ (_____)

✓ **B4.** Answer these questions about the wrist.

a. Wrist bones are called ____________________. There are *(5? 7? 8? 14?)* of them in each wrist.

b. The wrist bone most subject to fracture is the bone named ____________________ located just distal to the *(radius? ulna?)*.

c. *Carpal tunnel syndrome,* due to repetitive wrist movements, involves inflammation of hand muscle tendons and the ____________________ that are wrapped by a connective tissue sheathe, the flexor ____________________, in a "tunnel" formed by four carpal bones.

✓ **B5.** If you wear a ring on your "ring finger," the ring surrounds the *(proximal? distal? middle?)* phalanx number *(II? IV?)*. The *pollex* is a term that refers to ____________________.

B6. Trace an outline of your hand. Draw in and label all bones.

C. Pelvic (hip) girdle; comparison of female and male pelves (pages 227–232)

✓ **C1.** Describe the pelvic bones in this exercise.

a. Name the bones that form the pelvic girdle. ______________________________

Which bones form the pelvis? ______________________________

b. Which of these bones is/are part of the axial skeleton?

c. Locate the point (**) on Figure LG 8.1 at which the pelvic girdle portion of the appendicular skeleton articulates with the axial skeleton. Name the bones and markings involved in that joint. ____________ and ____________

✓ **C2.** Answer the following questions about coxal bones.

a. Each coxal (hip) bone originates as three bones that fuse early in life. These bones are the ____________, ____________, and ____________.

At what location do the bones fuse? ____________

b. The largest of the three bones is the ____________.
A ridge along the superior border is called the iliac crest. Locate this on yourself.

c. The iliac crest ends anteriorly as the ____________ spine (known by the acronym ASIS). A posterior marking known as the ____________ spine (or PSIS) causes a dimpling of the skin just lateral to the sacrum, which can be used as a landmark for administering hip injections accurately.

✓ **C3.** Complete the table about markings of the coxal bones.

Marking	Location on Coxal Bone	Function
a. Greater sciatic notch		
b.		Supports most of body weight in sitting position
c.		Fibrocartilaginous joint between two coxal bones
d.		Socket for head of femur
e. Obturator oramen	Large foramen surrounded by pubic and ischial rami and acetabulum	

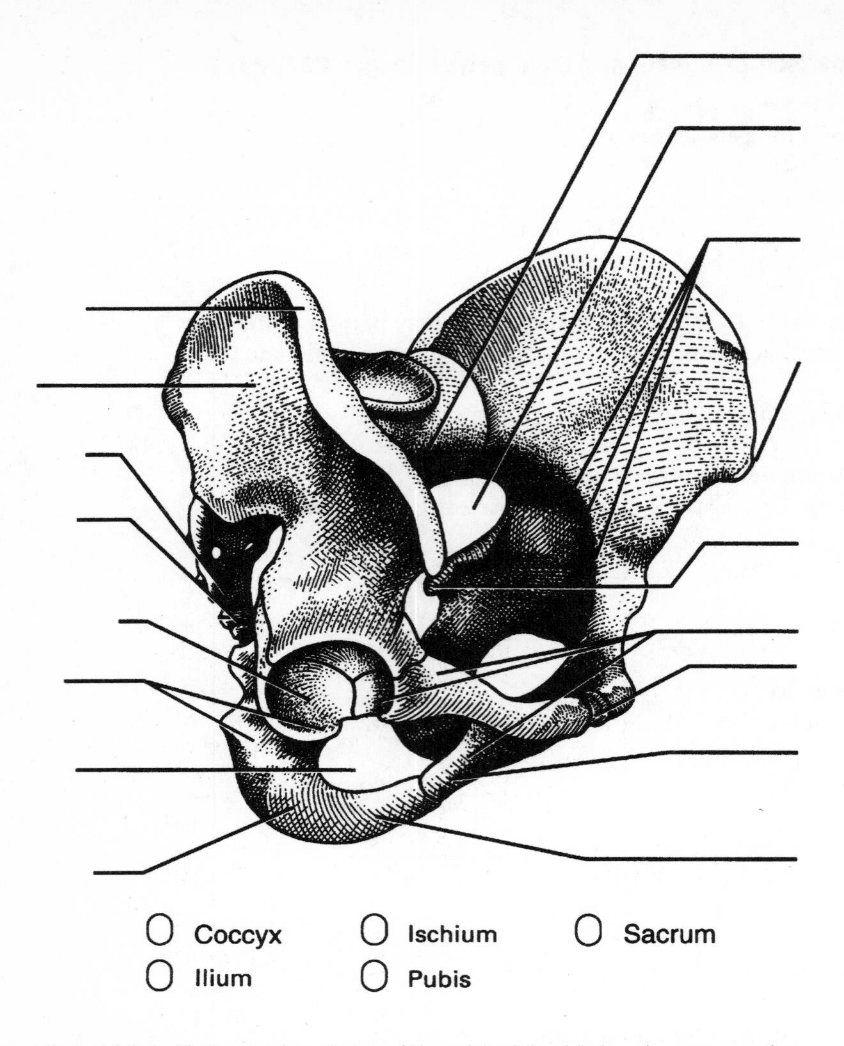

Figure LG 8.3 Right anterolateral view of the pelvis. Color, label, and answer questions as directed in Checkpoints C4 and C6.

✓ **C4.** Refer to Figure LG 8.3 and answer these questions about pelvic markings.

a. Trace with your pencil, and then label, the brim of the pelvis on the figure. It is a relatively *(smooth? irregular?)* line that demarcates the *(superior? inferior?)* border of the lesser (true) pelvis. The pelvic brim encircles the pelvic *(inlet? outlet?)*.

b. The bony border of the pelvic outlet is *(smooth? irregular?)*. Why is the pelvic outlet so named?

C5. Contrast the two principal parts of the pelvis. Describe their locations and name the structures that compose them.

a. Greater (false) pelvis

b. Lesser (true) pelvis

✓ **C6.** Color the pelvic structures indicated by color code ovals on Figure LG 8.3. Also label the following markings: *acetabulum, anterior superior iliac spine, greater sciatic notch, iliac crest, inferior pubic ramus, ischial spine, ischial tuberosity, obturator foramen, sacral promontory,* and *symphysis pubis.*

C7. State several characteristics of the female pelvis that make it more suitable for childbirth than the male pelvis.

✓ **C8.** Identify specific differences in pelvic structure in the two sexes by placing M before characteristics of the male pelvis and F before structural descriptions of the female pelvis.

_______ a. Shallow greater pelvis

_______ b. Heart-shaped inlet

_______ c. Pubic arch greater than 90° angle

_______ d. Acetabulum small

_______ e. Pelvic inlet comparatively small

✓ **C9.** *The Big Picture: Looking Ahead.* Refer to figures in Chapter 11 of your text as you check your understanding of the appendicular skeleton in this Checkpoint.

a. Contrast Figures 11.12 and 11.13. In each figure, find the coccyx and the ischial tuberosities. Which pelvis has a relatively wider distance between ischial tuberosities? *(Female? Male?)* Also notice that *(bone? muscle?)* forms most of the floor of the human pelvis.

b. The ischial tuberosities are points of attachment of several large muscles (Figure 11.20c and Exhibit 11.18). Name two or more muscles attached to these bony markings of the hip bone.

c. Which muscles cover the posterior surface of the scapula, padding this thin, hard bone (Figure 11.15c)?

The ___________________-spinatus muscle lies superior to the spine of the scapula,

whereas the ___________________-spinatus and teres ___________________ are located inferior to the spine.

d. Name two muscles that are anchored into the comma-shaped coracoid process of the scapula: the pectoralis *(major? minor?)* (Figure 11.14a) and the _______________ brachii (Figure 11.15a). These two muscles both lie closer to the *(anterior? posterior?)* surface of the body.

e. Figure 11.16a shows that the distal tendon of the biceps brachii attaches to the anterior of the *(humerus? radius? ulna?)*.

C10. *Critical thinking.* Explain how structural differences in the pectoral and pelvic girdles account for differences in support and mobility of the body.

D. Lower limb (extremity) (pages 232–239)

✓ **D1.** Refer to Figure LG 8.1 and list the bones (or groups of bones) in the lower limb from proximal to distal. Indicate how many of each bone there are. One is done for you.

a. **Femur** **(1)**

b. ______________ ( )

c. ______________ ( )

d. ______________ ( )

e. ______________ ( )

f. ______________ ( )

g. ______________ ( )

D2. Contrast the size, location, and names of the bones of the upper and lower limbs by coloring the bones on Figure LG 8.1 as follows. Color on one side of the figure only.

Humerus and femur (red)
Patella (brown)
Ulna and tibia (green)
Radius and fibula (yellow)
Carpals and tarsals (blue)
Metacarpals and metatarsals (orange)
Phalanges (purple)

✓ **D3.** Circle the term that correctly indicates the location and structure of these parts of the lower limb.

a. The head is the *(proximal? distal?)* epiphysis of the femur.
b. The greater trochanter is *(lateral? medial?)* to the lesser trochanter.
c. The intercondylar fossa is on the *(anterior? posterior?)* surface of the femur.
d. The tibial condyles are more *(concave? convex?)* than the femoral condyles.
e. The lateral condyle of the femur articulates with the *(fibula? lateral condyle of the tibia?)*.
f. The tibial tuberosity is *(proximal? distal?)* to the patella.
g. The tibia is *(medial? lateral?)* to the fibula.
h. The outer portion of the ankle is the *(lateral? medial?)* malleolus, which is part of the *(tibia? fibula?)*.

✓ **D4.** Complete the Checkpoint about the patella.

a. This bone is located on the *(anterior? posterior?)* aspect of the knee joint.
b. The patella is a sesamoid bone. Define sesamoid bone.

c. What are its functions?

d. What function does the patellar ligament serve?

e. Does the patella articulate with the fibula? ______

f. "Runner's knee" is a common term for ________________ ________________ syndrome, in which the patella fails to track (glide) in the groove between the ________________ of the femur.

✓ **D5.** *A clinical correlation* about the lower extremity.

a. The angle of the femur (between hip and knee joints) is known as the angle of ________________ since the knees are normally more *(medial? lateral?)* than the hips. This angle is normally *(larger? smaller?)* in females because hips are broader in females.

b. Contrast *genu valgum* with *genu varum*.

✓ **D6.** The tarsal bone that is most superior in location (and that articulates with the tibia and fibula) is the ________________. The largest and strongest of the tarsals is the ________________.

D7. Answer these questions about the arch of the foot.

a. How is the foot maintained in an arched position?

b. Locate each of these arches on your own foot. Refer to Figure 8.17 of your text. *(For extra review:* List the bones that form each arch.)

Longitudinal: medial side

Longitudinal: lateral side

Transverse

c. What causes flatfoot?

✓ **D8.** Now that you have seen all of the bones of the appendicular skeleton, complete this table relating common and anatomical names of bones.

Common name	Anatomical term
a. Shoulder blade	
b.	Pollex
c. Collarbone	
d. Heel bone	
e.	Olecranon process
f. Kneecap	
g.	Tibial crest
h. Toes	
i. Palm of hand	
j. Wrist bones	

D9. *For extra review.* Label all bones marked with label lines on Figure LG 8.1.

ANSWERS TO SELECTED CHECKPOINTS: CHAPTER 8

A1. (a) Two clavicles and two scapulas. (b) No. (c) Appendicular; clavicles, manubrium of sternum.

A2. The clavicle serves as a thick pencil-like wedge that prevents the arm from collapsing inward towards the sternum; serves as attachment point for muscles. Totally immobilizing the clavicle would limit movements of breathing.

A3. Clavicle. (b) Humerus. (c) None; tendons attach here.

A4. (a) S. (b) I. (c) M. (d) A.

B1. (a) Humerus, 1. (b) Ulna, 1. (c) Radius, 1. (d) Carpals, 8. (e) Metacarpals, 5. (f) Phalanges, 14.

B2. See Figure LG 8.2A

B3. (b) Capitulum (H). (c) Olecranon fossa (H). (d) Trochlear notch (U). (e) Radial notch (U). (f) Radial tuberosity (R). (g). Styloid process (U). (h) Medial epicondyle (H).

B4. (a) Carpals, 8. (b) Scaphoid, radius. (c) Median nerve, retinaculum.

B5. Proximal, IV; thumb.

C1. (a) Two coxal (hip) bones; hip bones plus sacrum

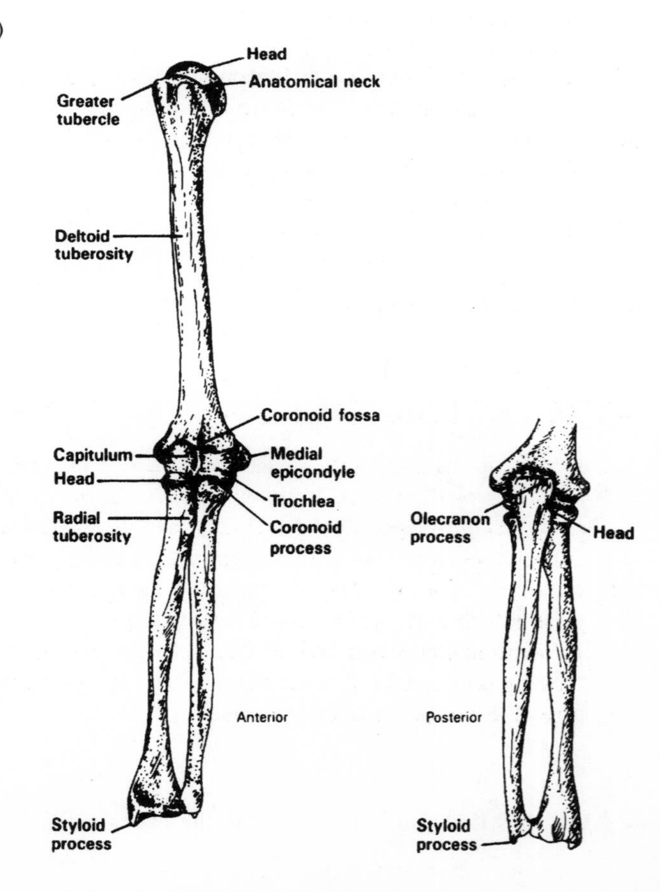

Figure LG 8.2A Right upper limb. (a) Anterior view. (b) Posterior view.

and coccyx. (b) Sacrum and coccyx. (c) Sacrum (right and left auricular surfaces) and hip bone (auricular surface).

C2. (a) Ilium, ischium, pubis; acetabulum (socket for the head of the femur). (b) Ilium. (c) Anterior superior iliac spine (ASIS); posterior superior iliac spine.

C3. See Table at right.

C4. See Figure LG 8.3A. (a) Smooth, superior, inlet. (b) Irregular; feces, urine, semen, menstrual flow and baby (at birth) exit via openings in the muscular floor attached to bony outlet.

C6. See Figure LG 8.3A.

C8. (a) F (b) M. (c) F. (d) F. (e) M.

C9. (a) Female; muscle. (b) The three heads of the hamstrings (semimembranosus, semitendinosus, and biceps femoris), as well as the adductor magnus, quadratus femoris, and inferior gemellus. (c) Supra, infra, minor. (d) Minor, biceps: anterior. (e) Radius.

Dl. (a) Femur, 1. (b) Patella, 1. (c) Tibia, 1. (d) Fibula, 1. (e) Tarsals, 7. (f) Metatarsals, 5. (g) Phalanges, 14.

D3. (a) Proximal. (b) Lateral. (c) Posterior. (d) Concave. (e) Lateral condyle of the tibia. (f) Distal. (g) Medial. (h) Lateral, fibula.

D4. (a) Anterior. (b) Sesame seed-shaped bone that develops within a tendon. (c) Protects the joint and its tendons from wear and tear; changes the direction of pull of a tendon, which provides mechanical advantage and some stability at the joint. (d) Attaches the patella to the tibial tuberosity. (e) No. (f) Patellofemoral stress; condyles.

D5. (a) Convergence, medial, larger. (b) Genu valgum is knock-knee, and genu varum is bowleg.

D6. Talus; calcaneus.

D8. (a) Scapula. (b) Thumb. (c) Clavicle. (d) Calcaneus. (e) Elbow. (f) Patella. (g) Shinbone. (h) Phalanges. (i) Metacarpals. (j) Carpals.

Marking	Location on Coxal Bone	Function
a. Greater sciatic notch	**Inferior to posterior inferior iliac spine**	**Sciatic nerve passes inferior to notch**
b. **Ischial tuberosity**	**Posterior and inferior to obturator foramen**	Supports most of body weight in sitting position
c. **Symphysis pubis**	**Most anterior portion of pelvis**	Fibrocartilaginous joint between two coxal bones
d. **Acetabulum**	**At junction of ilium, ischium, and pubis**	Socket for head of femur
e. Obturator oramen	Large foramen surrounded by pubic and ischial rami and acetabulum	**Blood vessels and nerves pass through**

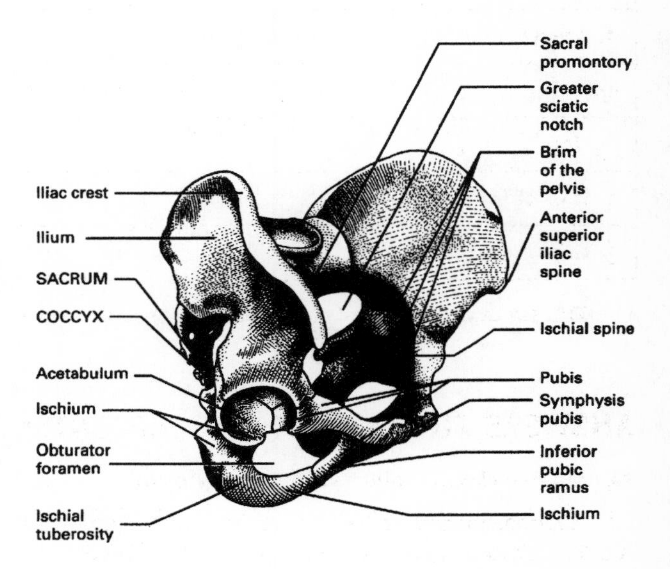

Figure LG 8.3A Right anterolateral view of the pelvis. Bone labels are capitalized; markings are in lowercase.

MORE CRITICAL THINKING: CHAPTER 8

1. Which joint is the point (bilaterally) at which the entire upper appendage attaches to the axial skeleton? At which joint (bilaterally) does the entire lower extremity attach to the axial skeleton? Speculate about what additional structures help to prevent the appendages from falling away from the axial skeleton.
2. Name the bones with which the scapula articulates. Based on the shape of bones in each joint, which joint appears to be most stable?
3. A fractured elbow can involve disruption of any/all of three joints. Describe these.
4. Identify functional differences that can be attributed to differences in location and structure of the carpal and tarsal bones.

MASTERY TEST: CHAPTER 8

Questions 1–6: Circle the letter preceding the one best answer to each question.

1. The point at which the upper part of the appendicular skeleton is joined to (articulates with) the axial skeleton is at the joint between:
   A. Sternum and ribs
   B. Humerus and clavicle
   C. Scapula and clavicle
   D. Scapula and humerus
   E. Sternum and clavicle
2. All of the following are markings on the femur *except.*
   A. Acetabulum
   B. Head
   C. Condyles
   D. Greater trochanter
   E. Intercondylar fossa
3. All of the following are bones in the lower limb *except:*
   A. Talus
   B. Tibia
   C. Calcaneus
   D. Ulna
   E. Fibula
4. Which structures are on the posterior surface of the upper limb (in anatomical position)?
   A. Radial tuberosity and lesser tubercle
   B. Trochlea and capitulum
   C. Coronoid process and coronoid fossa
   D. Olecranon process and olecranon fossa
5. The humerus articulates with all of these bones *except:*
   A. Ulna
   B. Radius
   C. Clavicle
   D. Scapula
6. Choose the *false* statement.
   A. The capitulum articulates with the head of the radius.
   B. The medial and lateral epicondyles are located at the distal ends of the tibia and fibula.
   C. The coronoid fossa articulates with the ulna when the forearm is flexed (elbow joint).
   D. The trochlea articulates with the trochlear notch of the ulna.

Questions 7–11: Arrange the answers in correct sequence.

_______ _______ _______ 7. According to size of the bones, from largest to smallest:
A. Femur
B. Ulna
C. Humerus

_______ _______ _______ 8. Parts of the humerus, from proximal to distal:
A. Anatomical neck
B. Surgical neck
C. Head

_______ _______ _______ 9. From proximal to distal:
A. Phalanges
B. Metacarpals
C. Carpals

_______ _______ _______ 10. From superior to inferior:
A. Lesser pelvis
B. Greater pelvis
C. Pelvic brim

_______ _______ _______ 11. Markings on hipbones in anatomical position, from superior to inferior:
A. Acetabulum
B. Ischial tuberosity
C. Iliac crest

Questions 12–20: Circle T (true) or F (false). If the statement is false, change the underlined word or phrase so that the statement is correct.

T F 12. Another name for the true pelvis is the greater pelvis.

T F 13. The scapulae do articulate with the vertebrae.

T F 14. The olecranon process is a marking on the ulna, and the olecranon fossa is a marking on the humerus.

T F 15. The female pelvis is deeper and more heart-shaped than the male pelvis.

T F 16. The greater tubercle of the humerus is lateral to the lesser tubercle.

T F 17. There are 14 phalanges in each hand and also in each foot.

T F 18. The fibula articulates with the femur, tibia, talus, and calcaneus.

T F 19. The organs contained within the right iliac, hypogastric, and left iliac portions (ninths) of the abdomen are located in the true pelvis.

T F 20. The total number of bones in one upper limb (including arm, forearm, wrist, hand, and fingers, but excluding shoulder girdle) is 29.

Questions 21–25: Fill-ins. Write the word or phrase that best completes the statement.

__________________ 21. In about three-fourths of all carpal bone fractures, only the __________________ bone is involved.

__________________ 22. The __________________ is the thinnest bone in the body compared to its length.

__________________ 23. The point of fusion of the three bones forming the hip bone is the __________________.

__________________ 24. The kneecap is the common name for the __________________.

__________________ 25. The bone that forms the heel is the __________________.

ANSWERS TO MASTERY TEST: CHAPTER 8

Multiple Choice

1. E
2. A
3. D
4. D
5. C
6. B

Arrange

7. A C B
8. C A B
9. C B A
10. B C A
11. C A B

True-False

12. F. Lesser
13. F. Do not
14. T
15. F. Shallower and more oval
16. T
17. T
18. F. Tibia and talus only
19. F. False
20. F. 30

Fill-ins

21. Scaphoid
22. Fibula
23. Acetabulum
24. Patella
25. Calcaneus

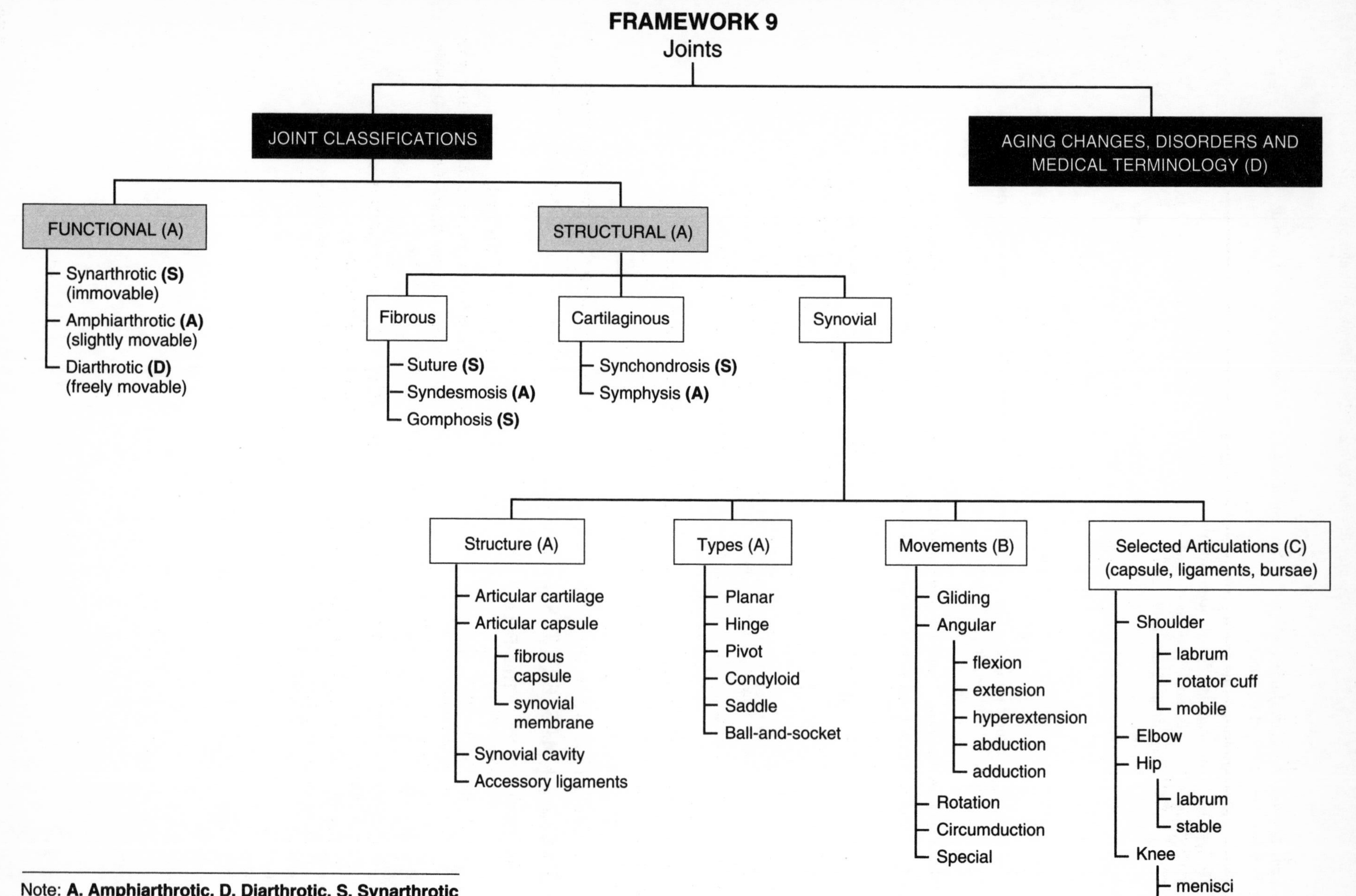

FRAMEWORK 9
Joints
JOINT CLASSIFICATIONS
AGING CHANGES, DISORDERS AND MEDICAL TERMINOLOGY (D)
FUNCTIONAL (A)
Synarthrotic (S) (immovable)
Amphiarthrotic (A) (slightly movable)
Diarthrotic (D) (freely movable)
STRUCTURAL (A)
Fibrous
Suture (S)
Syndesmosis (A)
Gomphosis (S)
Cartilaginous
Synchondrosis (S)
Symphysis (A)
Synovial
Structure (A)
Articular cartilage
Articular capsule
fibrous capsule
synovial membrane
Synovial cavity
Accessory ligaments
Types (A)
Planar
Hinge
Pivot
Condyloid
Saddle
Ball-and-socket
Movements (B)
Gliding
Angular
flexion
extension
hyperextension
abduction
adduction
Rotation
Circumduction
Special
Selected Articulations (C) (capsule, ligaments, bursae)
Shoulder
labrum
rotator cuff
mobile
Elbow
Hip
labrum
stable
Knee
menisci
unstable
Note: A. Amphiarthrotic, D. Diarthrotic, S. Synarthrotic

CHAPTER 9

Joints

In the last three chapters you have learned a great deal about the 206 bones in the body. Separated, or disarticulated, these bones would constitute a pile as disorganized as the rubble of a ravaged city deprived of its structural integrity. Fortunately, bones are arranged in precise order and held together in specific conformations—articulations or joints—that permit bones to function effectively. Joints may be classified by function (how movable) or by structure (the type of tissue forming the joint). Synovial joints are most common; their structure, movements, and types will be discussed. A detailed examination of four important synovial joints (shoulder, elbow, hip, and knee) is also included in this chapter. Joint disorders such as arthritis are usually not life threatening but plague much of the population, particularly the elderly. An introduction to joint disorders completes this chapter.

As you begin your study of articulations, carefully examine the Chapter 9 Topic Outline and Objectives and note relationships among concepts and key terms in the Framework.

TOPIC OUTLINE AND OBJECTIVES

A. Joint classifications

1. Describe the structural and functional classifications of joints.
2. Describe the structure and functions of the three types of fibrous joints.
3. Describe the structure and functions of the two types of cartilaginous joints.
4. Describe the structure of synovial joints.
5. Describe the six subtypes of synovial joints.
6. Describe the structure and function of bursae and tendon sheaths.

B. Types of movements at synovial joints

1. Describe the types of movements that can occur at synovial joints.

C. Selected joints of the body

1. Describe five factors that influence the type of movement and range of motion possible at a synovial joint.

D. Aging and joints; disorders and medical terminology

1. Explain the effects of aging on joints.

WORDBYTES

Now become familiar with the language of this chapter by studying each wordbyte, its meaning, and an example of its use within a term. After you study the entire list, self-check your understanding by writing the meaning of each wordbyte on the line. As you continue through the *Learning Guide*, identify (and fill in) additional terms that contain the same wordbyte.

Wordbyte	Self-check	Meaning	Example(s)
amphi-	______________	both	*amphi*arthrotic
arthr-	______________	joint	osteo*arthr*itis
articulat-	______________	joint	*articulat*ion
bursa-	______________	purse	*burs*itis
cruci-	______________	cross	*cruci*ate
gompho-	______________	bolt, nail	*gompho*sis
-itis	______________	inflammation	osteoarthr*itis*
ov-	______________	egg	syn*ov*ial fluid
-scopy	______________	observation	arthro*scopy*
sutur-	______________	seam	coronal *suture*
syn-	______________	together	*syn*arthrotic
syndesmo-	______________	band or ligament	*syndesmo*sis

CHECKPOINTS

A. Joint classifications (pages 244–250)

✓ **A1.** Define the term *joint (articulation).*

State two criteria used to classify joints according to structure.

✓ **A2.** Name three classes of joints based on structure.

✓ **A3.** Fill in the blanks below to name three classes of joints according to the amount of movement they permit.

a. Synarthrosis: ______________

b. ______________: slightly movable

c. ______________: freely movable

✓ **A4.** Describe fibrous joints by completing this exercise.

a. Fibrous joints *(have? lack?)* a joint cavity. They are held together by

_______________________ connective tissue.

b. One type of fibrous joint is a _______________________ found between skull bones.

Such joints are *(freely? slightly? im-?)* movable, or _________-arthrotic. Replacement of a fibrous suture with bony fusion is known as a *(synchondrosis? synostosis?)*.

c. Which of the following is a site of a gomphosis?

A. Joint at distal ends of tibia and fibula
B. Epiphyseal plate
C. Attachment of tooth by periodontal ligament to tooth socket in maxilla or mandible

d. The distal end of the tibia/fibula joint is a fibrous joint. It is *(more? less?)* mobile

than a suture and is therefore _________ -arthrotic. It is called a ___________________.

✓ **A5.** Describe cartilaginous joints by completing this exercise.

a. Synchondroses involve *(hyaline? fibrous?)* cartilage between regions of bone. An

example is the _______________________ cartilage between diaphysis and epiphysis of a growing bone. These joints are *(immovable? somewhat movable?)*. This cartilage *(persists through life? is replaced by bone during adult life?)*, forming a *(symphysis? syndesmosis? synostosis?)*.

b. Fibrocartilage is present in the type of joint known as a _________________________.

These joints permit some movement and so are called ________________-arthrotic.

Two locations of symphyses are _____________________ and ____________________.

✓ **A6.** What structural features of synovial joints make them more freely movable than fibrous or cartilaginous joints?

✓ **A7.** On Figure LG 9.1, color the indicated structures.

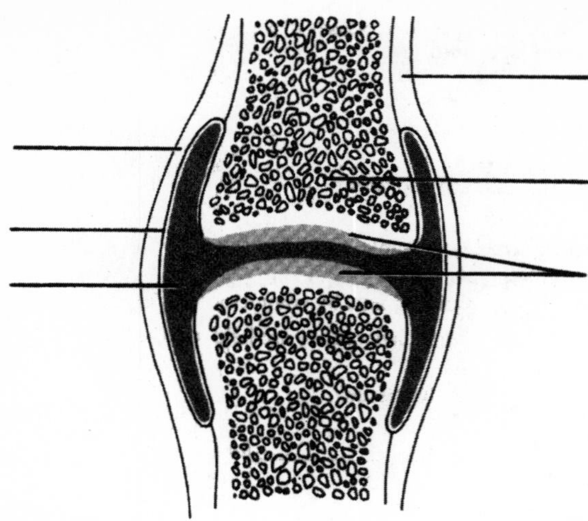

- O Articular cartilage
- O Articulating bone
- O Fibrous capsule
- O Periosteum
- O Synovial (joint) cavity
- O Synovial membrane

Figure LG 9.1 Structure of a generalized synovial joint. Color as directed in Checkpoint A7.

✓ **A8.** Select the parts of a synovial joint (listed in the box) that fit the descriptions below.

A. Articular cartilage	**SF. Synovial fluid**
F. Fibrous capsule	**SM. Synovial membrane**
L. Ligaments	

_________ a. Hyaline cartilage that covers ends of articulating bones but does not bind them together

_________ b. With the consistency of uncooked egg white or oil, it lubricates the joint and nourishes the avascular articular cartilage

_________ c. Connective tissue membrane that lines synovial cavity and secretes synovial fluid

_________ d. Parallel fibers in some fibrous capsules; bind bones together

_________ e. Together these form the articular capsule (two answers)

_________ f. Provides both flexibility and tensile strength

_________ g. Reduces friction at the joint

_________ h. Contains phagocytes that remove debris and microbes from joint

A9. Describe the structure, function, and location of the following structures that are associated with synovial joints.

a. Intracapsular ligaments

b. Extracapsular ligaments

c. Articular discs (menisci)

A10. Do this activity about "torn cartilage" of the knee.

a. In this condition, the injured tissues are *(articular cartilages? menisci?)*.

b. What functions are normally served by these cartilages?

c. How may these cartilages be repaired with minimal destruction to other tissues?

✓ **A11.** A *(sprain? strain?)* is an overstretching of a muscle, whereas a *(sprain? strain?)* involves a more serious injury to joint structures. The joint most often sprained is the *(ankle? knee?)* joint.

A12. Contrast *bursae* and *tendon sheathes* according to structure and locations.

✓ **A13.** Complete Table LG 9.1 on six subtypes of synovial joints. Select answers from lists in the box. *Note: omit the third column (types of movement) until Checkpoint B4.*

Subtype of synovial joint	Planes of movement	Types of movement	Examples of joints
Ball-and-socket	Biaxial	ABD. Abduction	List name of joint,
Condyloid (ellipsoidal)	Monoaxial	ADD. Adduction	articulating bones
Hinge	Triaxial	CIR. Circumduction	with specific
Pivot	Nonaxial	E. Extension	markings at points
Planar		F. Flexion	of articulation.
Saddle		ROT. Rotation	

Table LG 9.1 Subtypes of synovial joints.

Subtype of Synovial Joint	Planes of Movement	Types of Movement	Examples
a. Ball-and-socket			1. Shoulder: scapula (glenoid cavity)-humerus (head) 2. Hip: hip bone (acetabulum)-femur (head)
b. Condyloid (ellipsoidal)			
c.			Carpal (trapezium)-first metacarpal (base)
d. Hinge		F, E	1. Knee: femur (condyles)-tibia (condyles)
e.	Monoaxial	ROT	
f.		Gliding	

Complete according to directions in Checkpoints A13 and B4.

✓ **A14.** *For extra review* of types of synovial joints, choose the type of joint that fits the description. (Answers may be used more than once.)

B.	**Ball-and-socket**	**H.**	**Hinge**	**Pla.**	**Planar**
C.	**Condyloid**	**Piv.**	**Pivot**	**S.**	**Saddle**

________ a. Monaxial joint; only rotation possible

________ b. Examples include atlas-axis joint and joint between head of radius and radial notch at proximal end of ulna

________ c. Triaxial joint, allowing movement in all three planes

________ d. Hip and shoulder joints

________ e. Spoollike (convex) surface articulated with concave surface, for example, elbow, ankle, and joints between phalanges

________ f. One type of monoaxial joint in which only flexion and extension are possible

________ g. Found in joints at sternoclavicular and claviculoscapular joints

________ h. Thumb joint located between metacarpal proximal to thumb and carpal bone (trapezium)

________ i. Biaxial joints (two answers)

B. Types of movements at synovial joints (pages 251–256)

✓ **B1.** From the terms listed in the box, choose the one that fits the type of movement in each case. Not all answers will be used.

Abd.	**Abduction**	**E.**	**Extension**	**LF.**	**Lateral flexion**
Add.	**Adduction**	**F.**	**Flexion**	**O.**	**Opposition**
C.	**Circumduction**	**G.**	**Gliding**	**P.**	**Plantar flexion**
D.	**Dorsiflexion**	**I.**	**Inversion**	**R.**	**Rotation**

_______ a. Decrease in angle between anterior surfaces of bones (or between posterior surfaces at knee and toe joints)

_______ b. Simplest kind of movement that can occur at a joint; no angular or rotary motion involved; example: ribs moving against vertebrae

_______ c. State of entire body when it is in anatomical position

_______ d. Movement away from the midline of the body

_______ e. Movement of a bone around its own axis

_______ f. Position of foot when heel is on the floor and rest of foot is raised

✓ **B2.** Perform the action described. Then write in the name of the type of movement.

a. Describe a cone with your arm, as if you were winding up to pitch a ball.

The movement at your shoulder joint is called ____________________.

b. Stand in anatomical position (palms forward). Turn your palms backward. This action is called ____________________.

c. Move your fingers from "fingers together" to "fingers apart" position. This action is ____________________ of fingers.

d. Raise your shoulders, as if to shrug them. This movement is called ____________________ of the shoulders.

e. Stand on your toes. This action at the ankle joint is called ____________________.

f. Grasp a ball in your hand. Your fingers are performing the type of movement called ____________________.

g. Sit with the soles of your feet pressed against each other. In this position, your feet are performing the action called ____________________.

h. Thrust your jaw outward (gently!). This action is ________________ of the mandible.

i. Move your thumb across your palm to touch your fingertips. ____________________

j. Move to the right or left at the waist, as if to slide your fingertips down the lateral aspect of your leg. ____________________

✓ **B3.** Identify the kinds of movements shown in Figure LG 9.2. Write the name of the movement below each figure. Use the following terms: *abduction, adduction, extension, flexion,* and *hyperextension.*

✓ **B4.** Now complete the "types of movement" column in Table LG 9.1. (See page LG 164).

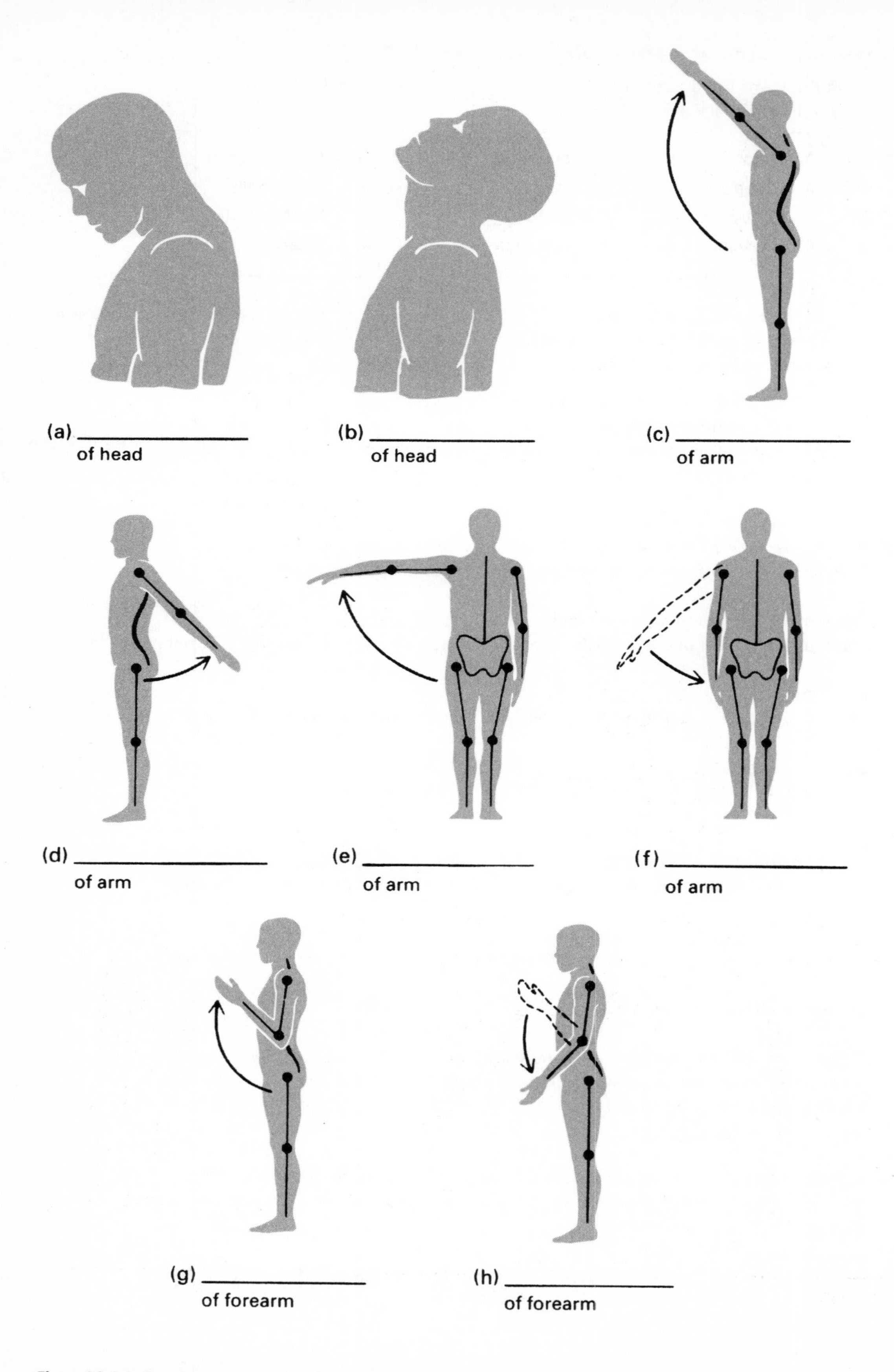

Figure LG 9.2 Movements at synovial joints. Answer questions as directed in Checkpoint B3 and in Chapter 11.

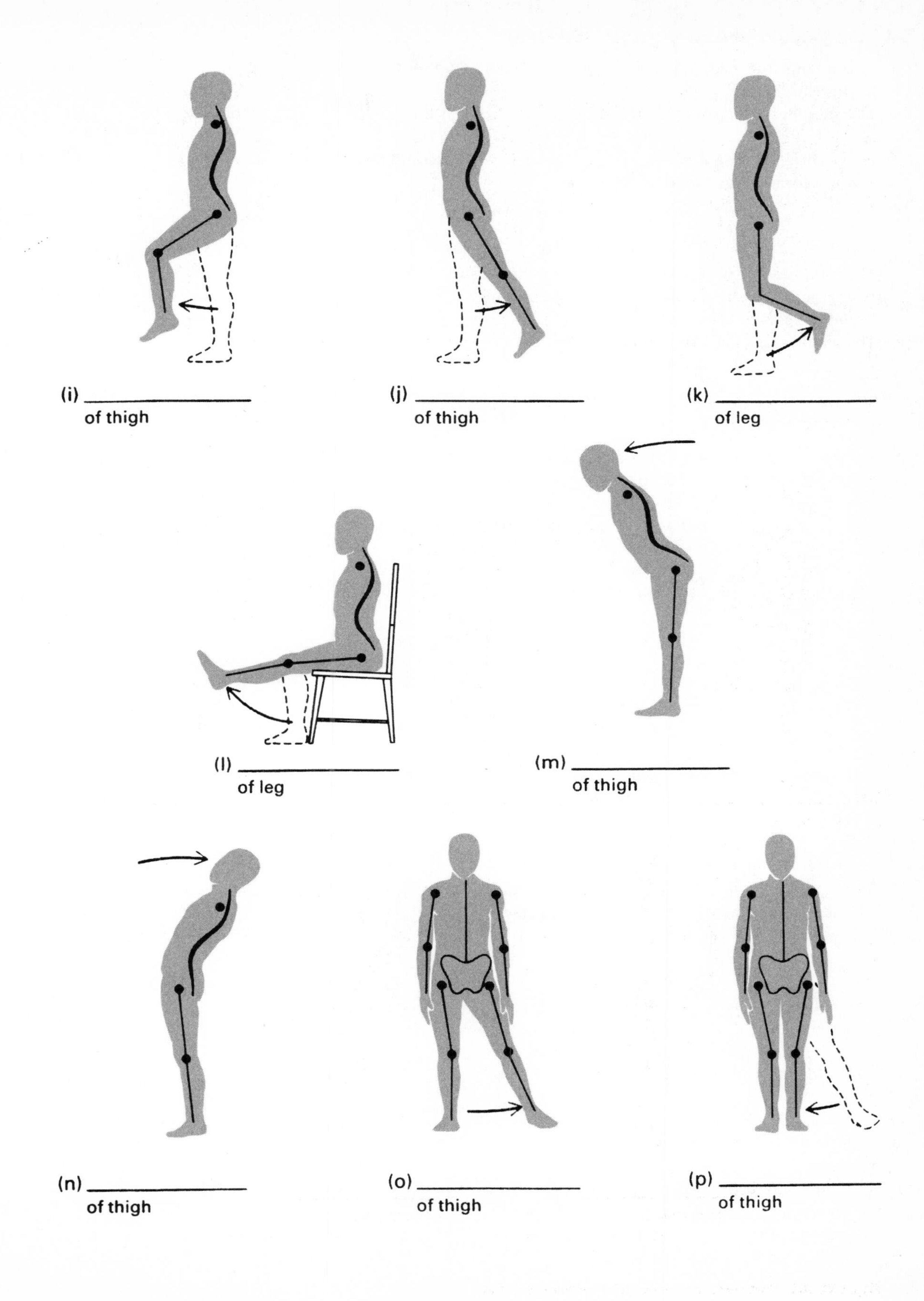
(i) ________________
of thigh
(j) ________________
of thigh
(k) ________________
of leg
(l) ________________
of leg
(m) ________________
of thigh
(n) ________________
of thigh
(o) ________________
of thigh
(p) ________________
of thigh

C. Selected joints of the body (pages 256–266)

✓ **C1.** Check your understanding of structure, function, and disorders of the shoulder joint in this Checkpoint.

a. The shoulder has *(greater? less?)* freedom of movement than any other joint in the body. This joint is said to be _________-axial. What structural features of the joint account for this freedom of movement?

b. The joint has several ligaments supporting the capsule. Which ligament is broad and strengthens the upper part of the articular capsule? *(Coraco-? Gleno-? Transverse)* humeral ligament.

c. The glenoid labrum is composed of *(bone? fat? fibrocartilage?)*. How does the labrum stabilize the shoulder joint?

d. Four bursae provide some cushioning and reduce friction at the shoulder joint. These bursae are located deep to bone or muscle tissue in this joint. Name them:

sub-_______________, sub-_______________, sub-_______________,

and sub-_______________.

e. What is the function of the rotator cuff?

What happens if the rotator cuff fails to do its job?

What tissues comprise the rotator cuff?

✓ **C2.** Contrast the hip joint with the shoulder joint in this exercise.

a. The hip joint exhibits *(more? less?)* stability and *(more? less?)* range of motion than the shoulder joint. State three factors that contribute to the strength of this joint.

b. Name three ligaments that limit hyperextension at the hip joint:

___________-femoral, ___________-femoral, and ___________-femoral. What limits adduction at this joint?

✓ **C3.** Complete this exercise describing the knee (tibiofemoral) joint. Consult Figure 9.14 in your text.

a. The knee joint actually consists of three joints:
   1. Between the femur and the *(patella? fibula?)*
   2. Between the lateral condyle of the femur, *(medial? lateral?)* meniscus, and the lateral condyle of the *(tibia? fibula?)*

3. Between the medial condyle of the femur, *(medial? lateral?)* meniscus, and the medial condyle of the *(tibia? fibula?)*

b. The first joint (1) described above is a *(modified hinge? pivot? planar?)* joint. The other two joints (2 and 3) are ________________ joints.

c. The knee (tibiofemoral) joint *(does? does not?)* include a complete capsule uniting the two bones. This factor contributes to the relative *(strength? weakness?)* of this joint.

d. The medial and lateral patellar retinacula and patellar ligament are tissues that serve as insertions of the *(hamstrings? quadriceps femoris?)* muscles.

e. A number of other ligaments provide strength and support. Identify whether the following ligaments are extracapsular (E) or intracapsular (I).

   1. Arcuate popliteal ligament: ______
   2. Anterior and posterior cruciate ligaments: ______
   3. Tibial (medial) and fibular (lateral) collateral ligaments: ______

f. Two cartilages, much like two C-shaped stadiums facing one another, each making an incomplete circle, are located between condyles of the femur and tibia. These articular discs are called ________________. What is their function?

g. There are *(no? several?)* bursae associated with the knee joint. Inflammation of bursae is called ________________.

✓ **C4.** Match the names of joints with descriptions provided.

A.	**Ankle joint**	**S.**	**Shoulder**
Ao.	**Atlanto-occipital joint**	**TMJ.**	**Temporomandibular joint**
E.	**Elbow joint**	**W.**	**Wrist**
H.	**Hip (coxal) joint**		

_______ a. Talocrural joint; capable of plantar flexion and dorsiflexion

_______ b. Glenohumoral joint

_______ c. A synovial joint with hinge and gliding actions; moves the lower (but not upper) jaw bone within the mandibular fossa of the temporal bone

_______ d. Radiocarpal joint

_______ e. Joint between the skull and the first cervical vertebra; condyloid joint

_______ f. Joint between the humerus (trochlea) and the trochlear notch of the ulna as well as the capitulum of humerus with head of the radius

_______ g. Joint between the head of the femur and acetabulum of the coxal bone

✓ **C5.** *The Big Picture: Looking Back and Looking Ahead.* The design of synovial joints permits free movement of bones. However, if bones moved too freely, they would potentially move right out of their joint cavities (dislocation). A number of factors modify movement at synovial joints. Identify these factors as you refer to figures and text in your textbook. Select from key factors listed in the box as you fill in lines with *.

Bones	**Hormones**	**Muscles**	**Ligaments**	**Soft parts**

a. Shape of articulating *________________ affects movement at joint, for example, at shoulder (Figure 8.4) and hip joints (Figure 8.10). Which joint has a deeper socket, contributing to increased stability and decreased mobility? *(Shoulder? Hip?)*

b. Position and tautness of *__________________, such as the anterior cruciate (Figure 9.14), strengthen the joint and limit excessive knee movement. In about _______% of all serious knee injuries, this ligament is damaged. It normally forms a cross ("*cruci*") with the __________________ cruciate ligament.

c. Figure 11.16, page 350 shows __________________ located on the anterior and posterior of the arm. The *(biceps? triceps?)* lies on the anterior of the arm. When this muscle contracts, the arm moves forward (flexion). The triceps, located on the *(anterior? posterior?)* of the arm, must stretch during this movement; because it has a limited ability to stretch, the triceps *(enhances? limits and opposes?)* flexion of the arm. In fact, the triceps is said to be an *antagonist* to the biceps.

d. Bend (flex) your own forearm (Figure 11.1 page 310). Try to place your wrist directly on your shoulder. If this is not possible for you, it is probably due to the meeting (apposition) of __________________ of your forearm against those covering your arm. Muscles are "soft parts." Name several others.

e. Another factor affecting movement at joints is *__________________ such as relaxin (Figure 28.25, page 1042). Made by ovaries and the __________________ toward the end of pregnancy, this hormone helps to relax the anterior joint between hipbones, known as the __________________ __________________.

D. Aging and joints; disorders and medical terminology (pages 266–268).

✓ **D1.** Fill in blanks to address these question about aging effects on joints.

a. With aging, the volume of synovial fluid tends to _____-crease.

b. Articular cartilage tends to _____-crease in thickness.

c. Ligaments are likely to _____-crease in length and flexibility.

d. One type of arthritis, namely __________________, is at least partly age-related.

✓ **D2.** How are *arthritis* and *rheumatism* related? Choose the correct answer.

A. Arthritis is a form of rheumatism.

B. Rheumatism is a form of arthritis.

✓ **D3.** List several forms of arthritis. Name one symptom common to all forms of this ailment.

✓ **D4.** Contrast *rheumatoid arthritis* with *osteoarthritis* in Table LG 9.2. Write *Yes, No, Larger,* or *Smaller* for answers.

Table LG 9.2 Comparison of major types of arthritis.

	Rheumatoid Arthritis	Osteoarthritis
a. Is known as "wear-and-tear" arthritis		
b. Which types of joints are most affected?		
c. Is this an inflammatory, autoimmune condition?		
d. Is synovial membrane affected? Does pannus form?		
e. Does articular cartilage degenerate?		
f. Does fibrous tissue join bone ends?		
g. Is movement limited?		

Complete as indicated in Checkpoint D4.

✓ **D5.** Complete this exercise describing disorders involving articulations.

a. Gouty arthritis is a condition due to an excess of ________________ in blood leading to deposit of ________________ in the joints. Gout can also involve damage to joints or to organs such as the ________________ because crystals may be deposited there also.

b. An acute chronic inflammation of a bursa is called ________________.

c. *Luxation*, or ________________, is displacement of a bone from its joint with tearing of ligaments.

d. Pain in a joint is known as ________________.

e. Named after a town in Connecticut where it was first reported, this is a systemic, bacterial disorder transmitted by tick bites. ________________

ANSWERS TO SELECTED CHECKPOINTS: CHAPTER 9

Al. Point of contact between bones, between bone and cartilage, and between teeth and bone; the presence or absence of a space (synovial cavity) between bones, and the type of connective tissue that binds the bones together.

A2. Fibrous, cartilage, and synovial.

A3. (a) Immovable. (b) Amphiarthroses. (c) Diarthroses.

A4. (a) Lack; fibrous. (b) Suture; im-, syn; synostosis. (c) C. (d) More, amphi; syndesmosis.

A5. (a) Hyaline; epiphyseal; immovable; is replaced by bone during adult life, synostosis. (b) Symphysis; amphi; discs between vertebrae, symphysis pubis between hipbones.

A6. The space (synovial cavity) between the articulating bones and the absence of tissue between those bones (which might restrict movement) make the joints more freely movable. The articular capsule and ligaments also contribute to free movement.

A7.

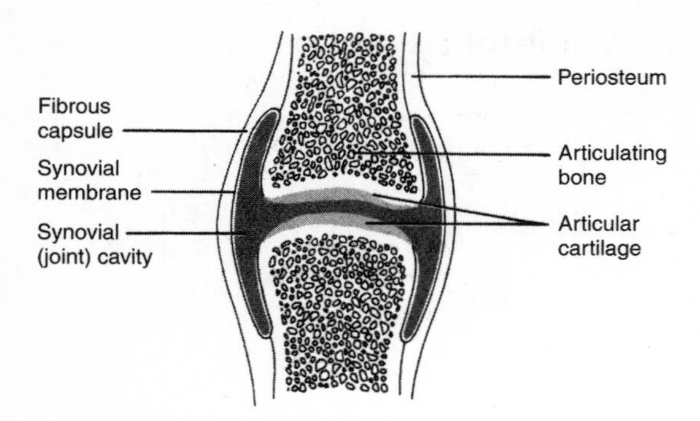

Figure LG 9.1A Structure of a generalized synovial joint.

A8. (a) A. (b) SF. (c) SM. (d) L. (e) SM and F. (f) F (and L, somewhat). (g) SF. (h) SF.

A11. Strain, sprain; ankle.

A13. See Table LG 9.1A.

Table LG 9.1A Subtypes of synovial joints.

Subtype of Synovial Joint	Planes of Movement	Types of Movement	Examples
a. Ball-and-socket	**Triaxial**	**F, E, ADD, ABD, CIR, ROT**	1. Shoulder: scapula (glenoid cavity)-humerus (head) 2. Hip: hip bone (acetabulum)-femur (head)
b. Condyloid (ellipsoidal)	**Biaxial**	**F, E, ADD, ABD, CIR**	**Wrist (radius-carpals)** **Metacarpal (head)-phalangeal (base)**
c. **Saddle**	**Biaxial**	**F, E, ABD, ADD, CIR**	Carpal (trapezium)-first metacarpal (base)
d. Hinge	**Monoaxial**	F, E	1. Knee: femur (condyles)-tibia (condyles) **2. Elbow: humerus (trochlea)-ulna (trochlear [semilunar] notch)** **3. Ankle: tibia and fibula-talus** **4. Phalanges** **5. Atlanto-occipital: atlas-occipital bone (condyles)**
e. **Pivot**	Monoaxial	ROT	**1. Atlas-axis (dens)** **2. Radius (head)-ulna (radial notch)**
f. **Planar**	Nonaxial	Gliding **Some ROT**	**1. Ribs (heads, tubercles)-vertebrae (transverse processes)** **2. Clavicle-sternum (manubrium), clavicle-scapula (acromiom)** **3. Intercarpal and intertarsal joints**

A14. (a) Piv. (b) Piv. (c) B. (d) B. (e) H. (f) H. (g) Pla. (h) S (i) C, S.

B1. (a) F. (b) G. (c) E. (d) Abd. (e) R. (f) D.

B2. (a) Circumduction. (b) Pronation. (c) Abduction. (d) Elevation. (e) Plantar flexion. (f) Flexion. (g) Inversion. (h) Protraction. (i) Opposition. (j) Lateral flexion.

B3. (a) Flexion. (b) Hyperextension. (c) Flexion. (d) Hyperextension. (e) Abduction. (f) Adduction. (g) Flexion. (h) Extension. (i) Flexion, while leg also slightly flexed. (j) Hyperextension, while leg extended. (k) Flexion, while thigh extended. (1) Extension, with thigh flexed. (m) Flexion. (n) Hyperextension. (o) Abduction. (p) Adduction.

B4. See Table LG 9.1A.

Cl. (a) Greater; tri; loose articular capsule and large, shallow glenoid cavity/socket for the head of the humerus. (b) Coraco-. (c) Fibrocartilage; increases the depth of the glenoid cavity. (d) Acromi-, coracoid, deltoid, scapular. (e) Helps to hold the humerus in the glenoid cavity; subluxation (dislocation) of the humerus out of the shoulder socket; four muscles and their tendons: supraspinatus, infraspinatus, teres minor, and subscapularis.

C2. (a) More, less; strong accessory ligaments and muscles around the joint capsule, as well as the fit of the femur into the deep acetabular socket. (b) llio, pubo, ischio; contact with the opposite thigh, as well as ligamentous tension.

C3. (al) Patella, (a2) lateral, tibia, (a3) medial, tibia. (Note that neither the femur nor the patella articulates with the fibula.) (b) Planar; modified hinge. (c) Does not; weakness. (d) Quadriceps femoris. (el) E; (e2) I, (e3) E. (f) Menisci; provide some stability to an otherwise unstable joint. (g) Several; bursitis.

C4. (a) A. (b) S. (c) TMJ. (d) W. (e) Ao. (f) E. (g) H.

C5. (a) Bones; hip. (b) Ligaments; 70; posterior. (c) Muscles; biceps; posterior; limits and opposes. (d) Soft parts; skin, fat, bursae, muscles and blood vessels. (e) Hormones; placenta, pubic symphysis.

Dl. (a) De. (b) De. (c) De. (d) Osteoarthritis.

D2. A.

D3. Rheumatoid arthritis (RA), osteoarthritis (OA), and gouty arthritis; pain.

D4.

Table LG 9.2A Comparison of major types of arthritis.

	Rheumatoid Arthritis	Osteoarthritis
a. Is known as "wear-and-tear" arthritis	**No**	**Yes**
b. Which types of joints are affected first?	**Smaller**	**Larger**
c. Is this an inflammatory autoimmune condition?	**Yes**	**No**
d. Is synovial membrane affected? Does pannus form?	**Yes**	**No**
e. Does articular cartilage degenerate?	**Yes**	**Yes**
f. Does fibrous tissue join bone ends?	**Yes**	**No**
g. Is movement limited?	**Yes**	**Yes**

D5. (a) Uric acid, sodium urate; kidneys. (b) Bursitis. (c) Dislocation. (d) Arthralgia. (e) Lyme.

MORE CRITICAL THINKING: CHAPTER 9

1. Synovial joints do contain cartilage. Exactly where is it located? Is it vascular? How does synovial fluid help with this problem?
2. Explain what prevents hyperextension at joints such as the elbow and knee. Speculate about what causes "double-jointedness." Which of these joints (elbow and knee) do exhibit some hyperextension?
3. What is the rotator cuff? What is likely to cause a "torn rotator cuff," and what are likely to be the consequences of that injury?
4. Which joint is usually more mobile? Hip or shoulder? Describe several anatomical factors that account for this difference.

MASTERY TEST: CHAPTER 9

Questions 1–13: Circle T (true) or F (false). If the statement is false, change the underlined word or phrase so that the statement is correct.

T F 1. A fibrous joint is one in which there is no joint cavity and bones are held together by fibrous connective tissue.

T F 2. Sutures, syndesmoses, and symphyses are examples of fibrous joints.

T F 3. Flexion and extension are movements that both occur in the sagittal plane.

T F 4. Ball-and-socket, planar, pivot, and ellipsoidal joints are all diarthrotic joints.

T F 5. All fibrous joints are synarthrotic and all cartilaginous joints are amphiarthrotic.

T F 6. In synovial joints, synovial membranes cover the surfaces of articular cartilages.

T F 7. Bursae are saclike structures that reduce friction at joints.

T F 8. Abduction of fingers occurs around an imaginary line through the middle finger, whereas abduction of the toes occurs around an imaginary line through the second toe.

T F 9. When your arm is in the supine position, your radius and ulna are parallel (not crossed).

T F 10. When you touch your toes, the major action you perform at your hip joint is called hyperextension.

T F 11. Abduction is movement away from the midline of the body.

T F 12. The elbow, knee, and ankle joints are all hinge joints.

T F 13. Joints that are relatively stable (such as hip joints) tend to have more mobility than joints that are less stable (such as shoulder joints).

Questions 14–15: Arrange the answers in correct sequence.

_______ _______ _______ 14. From most mobile to least mobile:

A. Amphiarthrotic
B. Diarthrotic
C. Synarthrotic

_______ _______ _______ 15. Stages in rheumatoid arthritis, in chronological order:

A. Articular cartilage is destroyed and fibrous tissue joins exposed bones.
B. The synovial membrane produces pannus which adheres to articular cartilage.
C. Synovial membrane becomes inflamed and thickened, and synovial fluid accumulates.

Questions 16–20: Circle the letter preceding the one best answer to each question.

16. The movement of circumscribing a cone with the humerus in the glenoid fossa is an example of the type of movement called:
    A. Supination
    B. Pronation
    C. Circumduction
    D. Internal rotation
17. Which structure is extracapsular in location?
    A. Meniscus
    B. Posterior cruciate ligament
    C. Synovial membrane
    D. Tibial collateral ligament
18. Which joint is amphiarthrotic and cartilaginous?
    A. Symphysis
    B. Synchondrosis
    C. Syndesmosis
    D. Synostosis
19. All of these structures are associated with the knee joint *except:*
    A. Glenoid labrum
    B. Patellar ligament
    C. Infrapatellar bursa
    D. Medial meniscus
    E. Fibular collateral ligament
20. A suture is found between:
    A. The two pubic bones
    B. The two parietal bones
    C. Radius and ulna
    D. Diaphysis and epiphysis
    E. Tibia and fibula (distal ends)

Questions 21–25: Fill-ins. Write the word or phrase that best completes the statement.

____________________ 21. ______________ is the forcible wrenching or twisting of a joint with partial rupture of it, but without dislocation.

____________________ 22. The action of pulling the jaw back from a thrust-out position so that it becomes in line with the upper jaw is the movement called ______________.

____________________ 23. Another name for a freely movable joint is ______________.

____________________ 24. The type of joint between the atlas and axis and also between proximal ends of the radius and ulna is a ______________ joint.

____________________ 25. Synovial fluid normally has the consistency resembling ______________.

ANSWERS TO MASTERY TEST: CHAPTER 9

True-False

1. T
2. F. Sutures, syndesmoses, and gomphoses
3. T
4. T
5. F. Either synarthrotic or amphiarthrotic, and the same is true of cartilaginous
6. F. Do not cover surfaces of articular cartilages. which may be visualized as floor and ceiling of a room; but synovial membranes do line the rest of the inside of the joint cavity (much like wallpaper covering the four walls of the room).
7. T
8. T
9. T
10. F. Flexion
11. T
12. T. (Or modified hinge)
13. F. Less

Arrange

14. B A C
15. C B A

Multiple Choice

16. C
17. D
18. A
19. A
20. B

Fill-ins

21. Sprain
22. Retraction
23. Diarthrotic
24. Pivot (or synovial or diarthrotic)
25. Egg white

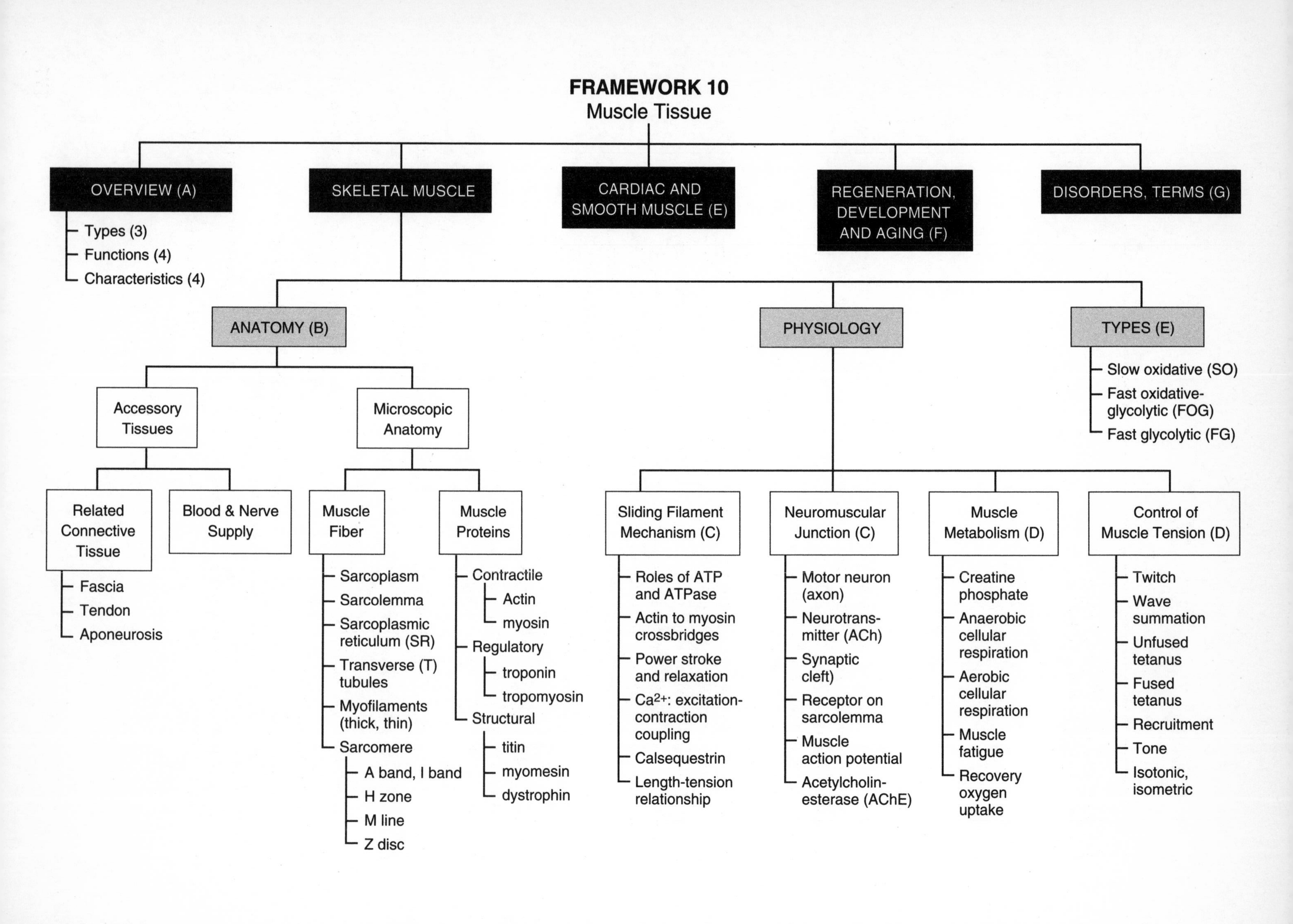

FRAMEWORK 10
Muscle Tissue
OVERVIEW (A)
Types (3)
Functions (4)
Characteristics (4)
SKELETAL MUSCLE
CARDIAC AND SMOOTH MUSCLE (E)
REGENERATION, DEVELOPMENT AND AGING (F)
DISORDERS, TERMS (G)
ANATOMY (B)
PHYSIOLOGY
TYPES (E)
Slow oxidative (SO)
Fast oxidative-glycolytic (FOG)
Fast glycolytic (FG)
Accessory Tissues
Microscopic Anatomy
Related Connective Tissue
Fascia
Tendon
Aponeurosis
Blood & Nerve Supply
Muscle Fiber
Sarcoplasm
Sarcolemma
Sarcoplasmic reticulum (SR)
Transverse (T) tubules
Myofilaments (thick, thin)
Sarcomere
A band, I band
H zone
M line
Z disc
Muscle Proteins
Contractile
Actin
myosin
Regulatory
troponin
tropomyosin
Structural
titin
myomesin
dystrophin
Sliding Filament Mechanism (C)
Roles of ATP and ATPase
Actin to myosin crossbridges
Power stroke and relaxation
Ca^{2+}: excitation-contraction coupling
Calsequestrin
Length-tension relationship
Neuromuscular Junction (C)
Motor neuron (axon)
Neurotrans-mitter (ACh)
Synaptic cleft)
Receptor on sarcolemma
Muscle action potential
Acetylcholin-esterase (AChE)
Muscle Metabolism (D)
Creatine phosphate
Anaerobic cellular respiration
Aerobic cellular respiration
Muscle fatigue
Recovery oxygen uptake
Control of Muscle Tension (D)
Twitch
Wave summation
Unfused tetanus
Fused tetanus
Recruitment
Tone
Isotonic, isometric

CHAPTER 10

Muscle Tissue

Muscles are making it possible for you to read this paragraph: facilitating movement of your eyes and head, maintaining your posture, producing heat to keep you comfortable. Muscles do not work in isolation, however. Connective tissue binds muscle cells into bundles and attaches muscle to bones. Blood delivers the oxygen and nutrients for muscle work, and nerves initiate a cascade of events that culminate in muscle contraction.

A tour of the mechanisms of muscle action finds the visitor caught in the web of muscle protein filaments sliding back and forth as the muscle contracts and relaxes. Neurotransmitters, as well as calcium and power-packing ATP, serve as chief regulators. Muscles are observed contracting in a multitude of modes, from twitch to treppe and isotonic to isometric. It may be surprising to hear a tour guide's announcement that most muscle cells present now were actually around at birth. Very few muscle cells can multiply, but fortunately muscle cells can greatly enlarge, as with exercise. With aging, cell numbers decline and so do muscle strength and endurance; however, regular exercise can minimize those changes. Certain disorders decrease muscle function and debilitate even in very early years of life.

As you begin your study of muscle tissue, carefully examine the Chapter 10 Topic Outline and Objectives; check off each one as you complete it. To organize your study of muscle tissue, glance over the Chapter 10 Framework now. Be sure to refer to the Framework frequently and note relationships among key terms in each section.

TOPIC OUTLINE AND OBJECTIVES

A. Overview of muscle tissue

1. Correlate the three types of muscle tissue with their functions and special properties.

B. Skeletal muscle tissue

1. Explain the relation of connective tissue components, blood vessels, and nerves to skeletal muscles.
2. Describe the microscopic anatomy of a skeletal muscle fiber.

C. Contraction and relaxation of skeletal muscle fibers

1. Outline the steps involved in the sliding filament mechanism of muscle contraction.
2. Describe how muscle action potentials arise at neuromuscular junctions.

D. Muscle metabolism; control of muscle tension

1. Describe the reactions by which muscle fibers produce ATP.
2. Describe the structure and function of a motor unit.

3. Explain the phases of a twitch contraction.
4. Describe how frequency of stimulation affects muscle tension.

E. Types of skeletal muscle fibers; cardiac and smooth muscle

1. Compare the structure and function of the three types of skeletal muscle fibers.
2. Describe the effects of exercise on different types of skeletal muscle fibers.
3. Describe the main structural and functional characteristics of cardiac muscle tissue.
4. Describe the main structural and functional characteristics of smooth muscle tissue.

F. Regeneration of muscle tissue; developmental anatomy and aging of muscles

1. Explain how muscle fibers regenerate.
2. Describe the development of muscles.
3. Explain how aging affects skeletal muscle.

G. Disorders and medical terminology

WORDBYTES

Now become familiar with the language of this chapter by studying each wordbyte, its meaning, and an example of its use within a term. After you study the entire list, self-check your understanding by writing the meaning of each wordbyte on the line. As you continue through the *Learning Guide*, identify (and fill in) additional terms that contain the same wordbyte.

Wordbyte	Self-check	Meaning	Example(s)
a-	______________	not	*a*trophy
-algia	______________	pain	fibromy*algia*
apo-	______________	from	*apo*neurosis
dys-	______________	bad, difficult	muscular *dys*trophy
endo-	______________	within	*endo*mysium
epi-	______________	upon	*epi*mysium
fascia	______________	bandage	superficial *fascia*
hyper-	______________	above, excessive	*hyper*trophy
intercal-	______________	to insert between	*intercal*ated disc
iso-	______________	equal	*iso*metric, *iso*tonic
-lemma	______________	sheath	sarco*lemma*
-mere	______________	part	sarco*mere*
-metric	______________	measure, length	iso*metric*
myo-	______________	muscle	*myo*fiber, *myo*sin
mys-	______________	muscle	epi*mys*ium
peri-	______________	around	*peri*mysium
sarc-	______________	flesh, muscle	*sarc*olemma

Wordbyte	Self-check	Meaning	Example(s)
tetan-	________________	rigid, tense	*tetan*us
titan	________________	gigantic	*tit*in
-tonic	________________	tension	iso*tonic*
tri-	________________	three	*tri*ad
troph-	________________	nourishment	muscular dys*trophy*

CHECKPOINTS

A. Overview of muscle tissue (pages 274–275)

✓ **A1.** Match the muscle types listed in the box with descriptions below.

C. Cardiac	Sk. Skeletal	Sm. Smooth

________ a. Involuntary muscle found in blood vessels and intestines

________ b. Involuntary striated muscle

________ c. Striated muscle attached to bones

________ d. The only type of muscle that is voluntary

________ e. Involuntary muscle in arrector pili muscles that causes "goose bumps"

✓ **A2.** Describe functions of muscles that are important for maintenance of homeostasis.

a. ________________ due to action of muscles pulling on bones.

b. Movement of substances within the body, such as:

1. ________________ within blood vessels due to pumping of heart muscle,
2. ________________ within the digestive system,
3. ________________ through the urinary system, and
4. ________________ within the reproductive system.

c. ________________ of body position, and maintenance of stored fluids within organs such as ________________.

d. Thermogenesis, which is generation of ________________ by muscles; an example is the warming effect of ________________ during cold weather.

e. Ringlike bands of smooth muscle known as ________________ control outflow of substances from organs, such as urine from the urinary bladder, or feces from the rectum.

✓ **A3.** Match the characteristics of muscles listed in the box with descriptions below.

Contractility	Elasticity	Excitability	Extensibility

a. Ability of muscle to stretch without damaging the tissue ____________________

b. Tendency of stretched or contracted muscle to return to its original shape

____

c. Ability of muscle cells and nerve cells (neurons) to respond to stimuli by producing action potentials; irritability

____

d. Ability of muscle tissue to shorten and thicken in response to action potential

____

✓ **A4.** Contrast isometric and isotonic contractions by doing this exercise.

a. A contraction in which a muscle shortens while tension (tone) of the muscle remains constant is known as an *(isometric? isotonic?)* contraction.

b. In an isometric contraction the muscle length *(shortens? stays about the same?)* and tension of the muscle *(increases? stays the same?)*.

B. Skeletal muscle tissue (pages 275–281)

✓ **B1.** Contrast two kinds of fascia by indicating which of the following are characteristics of superficial fascia (S) or deep fascia (D).

________ a. Located immediately under the skin (subcutaneous)

________ b. Composed of dense connective tissue that extends inward to surround and compartmentalize muscles

________ c. A route for nerves and blood vessels to enter muscles; contains much fat, so provides insulation and protection

✓ **B2.** Arrange the following terms (connective tissue) in correct sequence according to the amount of muscle surrounded: *endomysium, epimysium, perimysium.*

____________________ → ____________________ → ____________________

(entire muscle) (bundle [fascicle] of muscle fibers) (individual muscle fiber)

For extra review. Color these three connective tissues on Figure LG 10.la. Label a *fascicle.*

✓ **B3.** The connective tissues surrounding and compartmentalizing muscle fascicles join together to attach the muscle into the ____________________ covering over bone. A(n) *(aponeurosis? tendon?)* resembles a cord attaching the tapered end of a muscle into the covering over bone, whereas a(n) ____________________ is a broad, flat structure attaching a broad muscle into periosteum.

✓ **B4.** Summarize the nerve and blood supply to skeletal muscle in this Checkpoint.

a. Skeletal muscle receives its nerve supply by cells known as ____________________. Branches (called ____________________) of these nerve cells supply nerve impulses to several muscle fibers (cells). The ends (or synaptic end ____________________) of

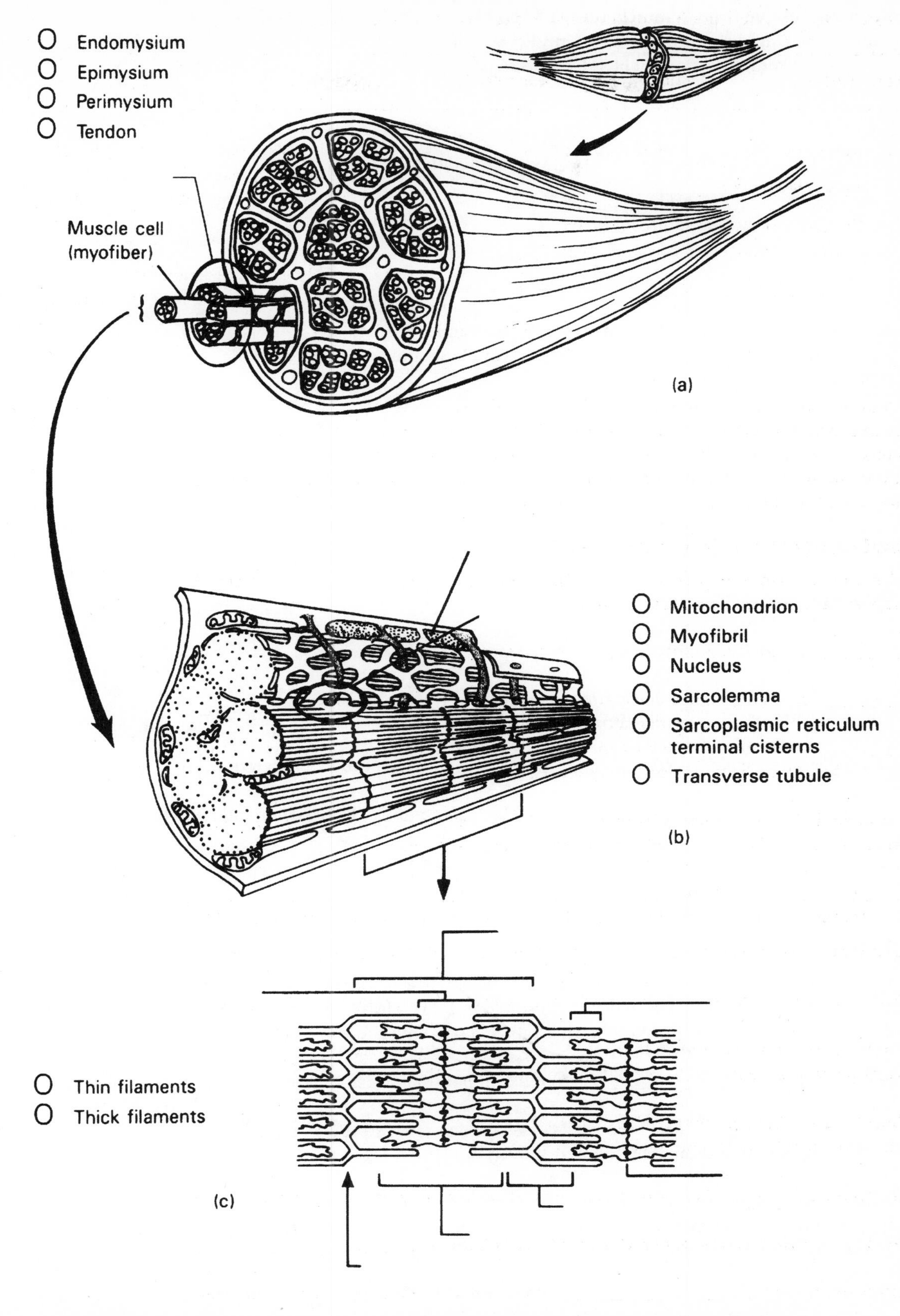

Figure LG 10.1 Diagram of skeletal muscle. (a) Skeletal muscle cut to show cross section and longitudinal section with connective tissues. (b) Section of one muscle cell (myofiber). (c) Detail of sarcomere of muscle cell. Color and label as indicated in Checkpoints B2, B6, and B10.

the nerve cells lie close to muscle cells at NMJs or ____________________ junctions. One will be considered in more detail in Checkpoint C6.

b. State two or more functions of blood vessels that supply muscles.

B5. Describe the roles of *myoblasts* in muscle cell development.

Do myoblasts multiply after birth? ________

✓ **B6.** Refer to Figure LG 10.1, and do this exercise about muscle structure.

a. Arrange the following terms in correct order from largest to smallest in size: *filaments (thick or thin), myofibrils, muscle fiber (cell).*

____________________ → ______________________ → ______________________
(largest) (smallest)

b. The plasma membrane of the skeletal muscle fiber (cell) is known as the ____________________. Continuous with this membrane are thousands of tiny structures called ______ (____________________) tubules. What major function do these structures serve?

c. The cytoplasm of the skeletal muscle cell is called its ____________________. Two major chemicals within it are ____________________, which can be broken down to form glucose, and ____________________, which gives muscle its red color and binds the gas ____________________. Both of these chemicals are needed for ____________________ production, which occurs in the myriad of organelles named ____________________ within the muscle.

d. Sarcoplasmic reticulum (SR) is comparable to *(endoplasmic reticulum? ribosomes? mitochondria?)* in nonmuscle cells. *(Ca^{2+}? K^{+}?)* stored in SR is released to sarcoplasm as the trigger for muscle contraction. Each T tubule along with the dilated ends (terminal ____________________) of SR on both sides of it are known as a ____________________. Label the triad circled on Figure LG 10.1b.

e. Color all structures indicated by color code ovals in Figured LG 10.1.

✓ **B7.** *A clinical correlation.* Complete this activity on changes in muscle size.

a. Hypertrophy refers to increase in *(number? size)* of muscle fibers. State two or more factors that may lead to hypertrophy.

b. When the nerve supply of a muscle is disrupted, the muscle is likely to experience

____________________ atrophy. Within 6 months to 2 years, the muscle is likely to be reduced in size by *(25% 50% 75%)*. What type of tissue replaces the muscle?

____________________.

c. State two examples of causes of disuse atrophy.

✓ **B8.** Match the correct term from the list in the box with its description below.

A band	**M line**	**Sarcoplasm**	**Zone of overlap**
I band	**Sarcomere**	**Z disc**	

a. Cytoplasm of muscle cell ____________________

b. Extends from Z disc to Z disc ____________________

c. Dark area in striated muscle; contains thick and thin filaments ____________________

d. Light area on either side of Z disc; location of thin filaments only ____________________

e. Located in the center of the H zone, which is the exact middle of the sarcomere.

f. Located in the center of the I band; appear damaged in electron micrographs of severely exercised muscle ____________________.

g. Located at both ends of each A band; consists of six thin filaments arranged around each thick filament ____________________.

✓ **B9.** *A clinical correlation.* Describe possible effects of intensive exercise upon muscles by filling in the blanks in this activity.

a. The acronym DOMS refers to D ____________________ O ____________________ M ____________________ S ____________________.

b. ____________________ and ____________________ are two proteins normally present within healthy muscle cells. Increased blood levels of these proteins may be correlated to the degree of muscle cell damage, as these chemicals move out of injured cells and into blood.

c. DOMS is likely to occur within ______ to ______ hours after strenuous exercise.

✓ **B10.** After you study Figures 10.4 and 10.6 in your text, describe the arrangement of the three categories of proteins in myofibrils.

a. Using the leader lines provided on Figure LG 10.1, label the following structures: *sarcomere, A band, I band, H zone, M line, Z disc, zone of overlap.*

b. Actin and myosin are the major components of thin and thick filaments, and they are both *(contractile? regulatory? structural?)* proteins. ____________________ and ____________________ are two regulatory proteins.

c. Each *(thick? thin?)* filament is composed of about 300 myosin molecules. Each molecule is shaped like two intertwined *(footballs? golf clubs?)*. The ends of the "handles" point toward the *(M lines? Z discs?)*. The rounded head of the "club" is called a

_________________ and it attaches to *(actin? troponin?)* in one of the six thin filaments as muscles begin contraction.

d. Thin filaments are anchored at *(M lines? Z discs?)* and *(do? do not?)* extend into H zones. These filaments are composed mostly of *(actin? myosin?)* molecules that are twisted into a helix. Thin filaments also contain two other proteins. The protein that covers myosin-binding sites in relaxed muscle is called *(tropomyosin? troponin?)*,

whereas tropomyosin is held in place by the regulatory protein _________________.

e. A third category of muscle proteins includes about 12 _________________ proteins. One of these proteins, titin, is named so because of its *(huge? tiny?)* size. Titin functions to anchor *(thick? thin?)* filaments between two parts of the sarcomere,

namely _________________ and _________________. Titin contributes greatly to the *(tensile strength? elasticity?)* of muscles.

f. Which structural protein forms the M line? _________________ Which one reinforces connections between the thin filaments, sarcolemma, and fibrous connective tissue surrounding muscles? _________________

C. Contraction and relaxation of skeletal muscle fibers (pages 282–288)

✓ **C1.** Summarize one theory of muscle contraction in this Checkpoint.

a. In order to effect muscle shortening (or contraction), heads (cross bridges) of

_________________ filaments pull on molecules of thin filaments, moving them towards the *(M line? Z discs?)*.

b. As a result, *(thick? thin?)* filaments move toward the center of the sarcomere. Because the thick and thin filaments *(shorten? slide?)* to decrease the length of the

sarcomere, this theory of muscle contraction is known as the _________________ theory. During this sliding action, the lengths of the thick and thin filaments *(do? do not?)* shorten.

✓ **C2.** Check your understanding of events that occur in muscle contraction and relaxation by completing this Checkpoint.

a. To initiate a muscle contraction, calcium ions (Ca^{2+}) are released from storage areas within the *(mitochondria? sarcoplasmic reticulum? T tubules?)*. (More about this process below in Checkpoint C3.)

b. The role of calcium ions is to bind to the protein named _______________ and therefore to cause the troponin-tropomyosin complex to pull away from sites on *(actin? myosin?)* where "heads" of the other contractile protein (_______________) must bind.

c. Myosin heads have ATP attached to them at this point. The enzyme

_________________ on myosin breaks down ATP into its products:

_________________ and _________________, which are still attached to myosin. The energy released by this reaction permits myosin heads to attach to

binding sites on _________________.

d. Release of the phosphate group triggers the "_________________ stroke," in which the myosin head rotates, releases *(ATP? ADP?)*, and slides the thin filament towards the center of the _________________.
e. Recall that each thick filament includes _____ myosin molecules (*Hint:* Look back at Checkpoint B10c). Each of these myosin molecules undergoes power strokes at a rate of about _____ times per second.
f. More ATP is then needed to bind to the myosin head and detach it from _________________. Splitting of ATP by ATPase located on *(actin? myosin?)* again activates myosin and returns it to its original position, where it is ready to bind to another _________________-binding site on an actin further along the thin filament.
g. Repeated power strokes slide actin filaments *(toward or even across? away from?)* the H zone and M line and so shorten the sarcomere (and entire muscle). If this process is compared to running on a treadmill, *(actin? myosin?)* plays the role of the runner staying in one place, whereas the treadmill that slides backward (toward H zones) is comparable to the *(thick? thin?)* sliding filament.

✓ **C3.** Describe the significance of intracellular locations of calcium and also the availability of ATP in regulation of muscle contraction and relaxation.

a. In a relaxed muscle, the concentration of calcium ions (Ca^{2+}) is *(high? low?)* in sarcoplasm but very high inside of SR. This concentration gradient is maintained in relaxed muscle by two mechanisms:
   1) *(Diffusion? Active transport?)* across the SR membrane.
   2) To enhance this effect, proteins named _________________ in SR membranes bind to the Ca^{2+} that enters SR, concentrating it inside SR *(10? 100? 10,000)* times over the level in the sarcoplasm. As a consequence, enormous amounts of calcium ions stand waiting in the SR for the next signal for the muscle to contract.
b. The signal to begin a muscle contraction is stimulation of a muscle by a nerve, which spreads an action potential across the sarcolemma and across the _________________ to SR. As a result, Ca^{2+} release _________________ in the SR membrane will open up, allowing Ca^{2+} to flow out of SR into _________________ surrounding thick and thin filaments. These ions are then available to exert their impact on regulatory proteins (troponin-tropomyosin) to initiate muscle contraction (review Checkpoint C2b).
c. When nerve stimulation ceases, the muscle relaxes as SR channels are no longer signaled to release Ca^{2+} into the cytosol. In fact, Ca^{2+} are sequestered back into SR by mechanisms just described. With such a low level of Ca^{2+} now in the sarcoplasm surrounding myofilaments, the troponin-tropomyosin complex once again blocks binding sites on _________________. As a result, thick and thin filaments detach, slip back into normal position, and the muscle is said to _________________.
d. The movement of Ca^{2+} back into sarcoplasmic reticulum (C3a) is *(an active? a passive?)* process. After death, a supply of ATP *(is? is not?)* available, so an active transport process cannot occur. Explain why the condition of rigor mortis results.

C4. *For extra review.* Sequence the events in excitation-contraction coupling by writing the letters of the five answers in correct chronological order.

_____ _____ _____ _____ _____

A. The troponin-tropomyosin complex is moved away from actin; myosin's ATP is hydrolyzed to ATP and phosphate, which stay attached to myosin.
B. The thin filament slides towards the center of the sarcomere as the muscle shortens; another ATP binds to myosin, releasing myosin from actin so the cycle can repeat.
C. Nerve impulse occurs; action potential spreads across sarcolemma and T tubules.
D. Calcium release channels release Ca^{2+} from SR to cytoplasm; Ca^{2+} complexes with troponin.
E. The energized myosin head can now attach to actin; phosphate is released from myosin, triggering rotation of myosin with release of ADP and the "power stroke."

✓ **C5.** Do this activity about changes in the force of muscle contraction.

a. When a muscle is stretched to its optimal length, there is *(much? little or no?)* overlap of myosin cross bridges on thick filaments with actin on thin filaments. In this case, a muscle demonstrates *(minimal? maximal?)* force of contraction. When a skeletal muscle is stretched excessively (such as to 170% of its optimal length), then *(many? no?)* myosin cross bridges can bind to actin.

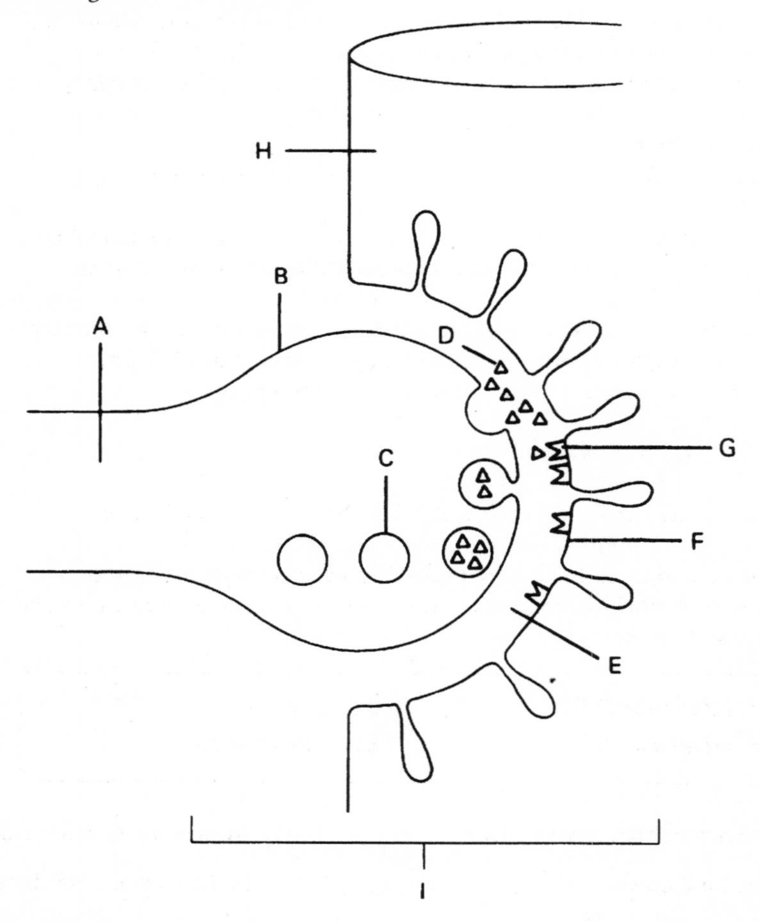

O Motor neuron (axon with terminal)
O Synaptic vesicle
O Neurotransmitter
O Skeletal muscle myofiber (cell)

Figure LG 10.2 Diagram of a neuromuscular junction. Refer to Checkpoints C6 and C7.

b. As a general rule, the more a muscle is stretched (within limits), the *(stronger? weaker?)* the contraction.

✓ **C6.** Do this activity about the nerve supply of skeletal muscles.

a. Refer to Figure LG 10.2, in which we zoom in at the point at which a branch of one neuron stimulates a single muscle fiber (cell). A nerve impulse travels along an axon terminal (at letter ________) toward one muscle fiber (letter ________). The axon is enlarged at its end into a synaptic ____________________ (at letter B).

b. The nerve impulse causes synaptic vesicles (letter ________) to fuse with the plasma membrane of the axon. Next the vesicles release the neurotransmitter (letter ________) named ____________________ into the synaptic cleft (letter ________)

c. The region of the muscle fiber membrane (sarcolemma) close to the axon terminals is called a *(motor end plate? neuromuscular junction [NMJ]?)*. This site contains specific ____________________ (letter G) that recognize and bind to ACh. A typical motor end plate contains about *(30-40? 3000-4000? 30-40 million?)* ACh receptors.

d. The effect of ACh is to cause Na^+ channels in the sarcolemma to *(open? close?)* so that Na^+ enters the muscle fiber. As a result, an action potential is initiated, leading to *(contraction? relaxation?)* of the muscle fiber. The diagnostic technique of recording such electrical activity in muscle cells is called ____________________.

e. The combination of the axon terminals and the motor end plate is known as a ____________________. At only one site along a muscle fiber (at about its middle) does a nerve approach a muscle cell to innervate it. In other words, there is/are usually *(only one? many?)* neuromuscular junction(s), labeled ________ on the figure, for each muscle fiber (cell). (In Chapter 12 we will consider regions similar to NMJ's, but where one neuron meets another neuron. These sites are known as ____________________.)

C7. *For extra review.* Color the parts of Figure LG 10.2 indicated by color code ovals.

✓ **C8.** Complete this exercise about pharmacology of the neuromuscular junction. Match the names of chemicals in the box with descriptions below. Short lines following each description will be used in Checkpoint C9.

AChE. Acetylcholinesterase (AChE)	**CUR. Curare**
CBT. *Clostridium botulinum* toxin	**NEO. Neostigmine**

________ a. Synaptic cleft enzyme that breaks down ACh into acetyl and choline. _____

________ b. Muscle relaxant that acts by blocking ACh receptors on motor end plate. _____

________ c. Anticholinesterase medication used for patients with myasthenia gravis. _____

________ d. Blocks release of ACh. _____

✓ **C9.** Now write arrows on lines in Checkpoint C8 to indicate whether each of the chemicals described there increases (↑) or decreases (↓) muscle stimulation.

D. Muscle metabolism; control of muscle tension (pages 289–294)

✓ **D1.** Complete this exercise about energy sources for muscle contraction.

a. Breakdown of *(ADP? ATP?)* provides the energy muscles use for contraction. Recall from Checkpoint C2c that ATP is attached to *(actin? myosin?)* cross bridges and so is available to energize the power stroke. Complete the chemical reaction showing ATP breakdown.

ATP →

b. ATP must be regenerated constantly. One method involves use of ADP and energy from food sources. Complete that chemical reaction.

ADP +

In essence, ADP is serving as a transport vehicle that can pick up and drop off energy stored in an extra high energy phosphate bond (~ P).

c. But ATP is used for other cell activities such as ___________________. To assure adequate energy for muscle work, muscle cells contain an additional molecule for

transporting high energy phosphate; this is ___________________. Complete the reaction in Figure LG 10.3 showing how the creatine phosphate and ADP transport "vehicles" can meet and transfer the high phosphate "trailer" so that more ATP is formed for muscle work.

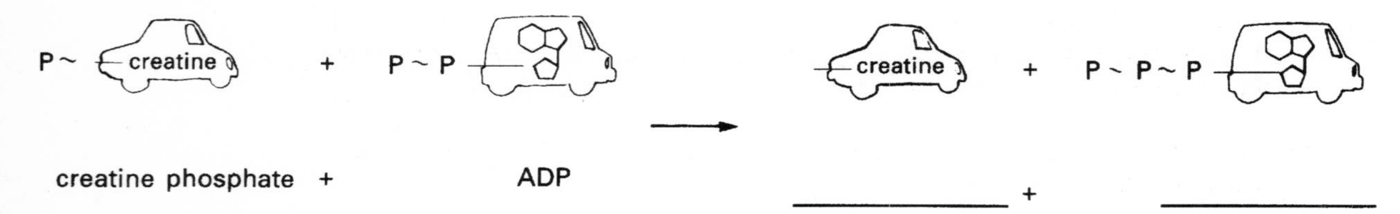

Figure LG 10.3 High energy molecules of the creatine phosphate system: the (.-P) "trailer" tradeoff. Complete figure as indicated in Checkpoint D1c, d.

d. How is creatine phosphate regenerated?
Show this on Figure LG 10.3. What enzyme catalyzes this reaction?_______________.
Which is more plentiful in the sarcoplasm? *(ATP? Creatine phosphate?)*.

e. ATP and creatine phosphate together provide only enough energy to power muscle activity for about *(an hour? 10 minutes? 15 seconds?)*. After that, muscles turn first to *(aerobic? anaerobic?)* pathways and later to *(aerobic? anaerobic?)* pathways. Complete Figure 10.4 to show how muscles get energy via these pathways.

f. During strenuous exercise an adequate supply of oxygen may not be available to

muscles. They must convert pyruvic acid to ___________________ acid by an *(aerobic? anaerobic?)* process. Excessive amounts of lactic acid in muscle tissue con-

tribute to some muscle ___________________. Name several types of cells that can use lactic acid to form ATP.

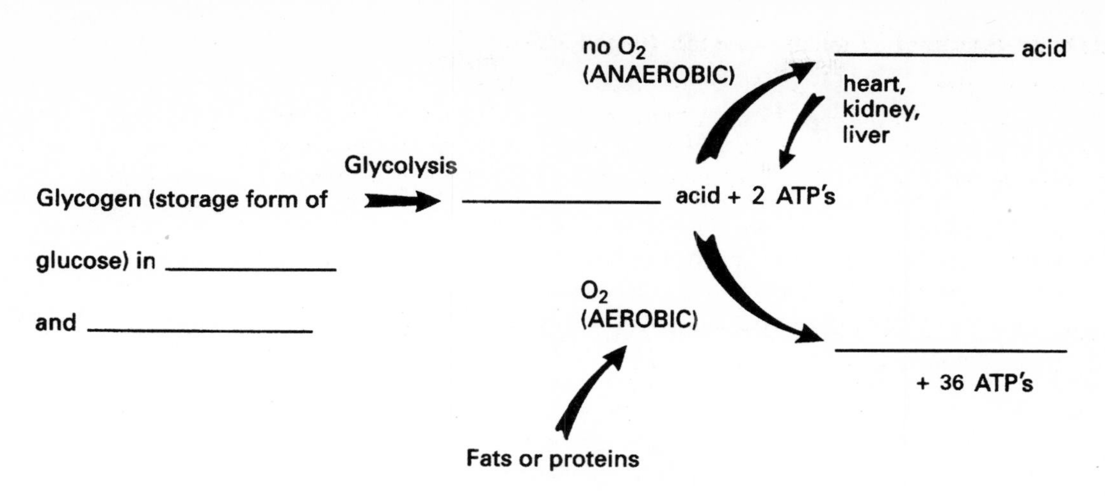

Figure LG 10.4 Anaerobic and aerobic energy sources for muscles. Complete figure as indicated in Checkpoint D1e.

g. One substance in muscle that stores oxygen until oxygen is needed by mitochondria is ________________. This protein is structurally somewhat like the ________________ -globin molecule in blood that also binds to and stores oxygen. Both of these molecules have a ________________ color that accounts for the color of blood and also of red muscle.

h. During exercise lasting more than 10 minutes, more than 90% of ATP is provided by the *(anaerobic? aerobic?)* breakdown of pyruvic acid (See Figure LG 10.4.).

D2. List several other factors besides the buildup of lactic acid that contribute to the following effects of exercise:

a. Muscle fatigue

b. Need for recovery oxygen uptake

✓ **D3.** *For extra review.* Complete this Checkpoint to identify metabolic processes used to supply energy for maximal exercise of varying durations. Select answers from those in the box.

AER.	**Aerobic cellular respiration**	**CP.**	**Creatine phosphate**
ANAER.	**Anaerobic cellular respiration**		

________ a. 15 sec (100-meter dash)

________ b. 30-40 sec (300-meter race); carbohydrate loading prior to an athletic event helps prepare muscles by maximizing glycogen storage there; does not require O_2

________ c. Longer than 10 minutes (2 miles to 26-mile marathon); "fuel" sources include glucose, fatty acids, and amino acids; does require O_2

✓ **D4.** List three or more factors that can decrease the strength of a muscle contraction.

✓ **D5.** In most cases a single nerve cell (neuron) innervates *(one muscle fiber [cell]? an average of 150 muscle fibers [cells]?).* An example of a motor unit that is likely to consist of just two or three muscle fibers (cells) precisely innervated by one neuron is *(laryngeal muscles controlling speech? the calf [gastrocnemius] muscles controlling walking?)*

✓ **D6.** Match the terms in the box with definitions below. One answer will be used twice; one will not be used.

C.	**Contraction period**	**Ref.**	**Refractory period**
L.	**Latent period**	**Rel.**	**Relaxation period**
M.	**Myogram**	**T.**	**Twitch**

________ a. Rapid, jerky response to a single action potential by a motor neuron

________ b. Recording of a muscle contraction

________ c. Period between application of a stimulus and start of a contraction; Ca^{2+} is being released from the SR during this time

________ d. Period when a muscle is not responsive to a stimulus

________ e. Active transport of Ca^{2+} back into the SR is occurring

________ f. A short period (about 0.005 second) in skeletal muscle and a long one (about 0.30 second) in cardiac muscle

✓ **D7.** Choose the type of contraction from the box that fits each descriptive phrase.

F.	**Fused tetanus**	**U.**	**Unfused tetanus**	**W.**	**Wave summation**

________ a. Sustained contraction due to stimulation at a rate of 20–30 stimuli per second

________ b. Sustained contraction due to stimulation at a rate of 80–100 stimuli per second

________ c. Stimuli arriving at different times cause larger contractions

________ d. Smooth, sustained voluntary muscle contractions are due primarily to out-of-synchrony ______ in different motor units

✓ **D8.** Contrast effects of different types of training in this Checkpoint.

a. Repeated isometric contractions (such as weight lifting) rely on *(aerobic? anaerobic?)* ATP production. Repeated short bouts of exercise such as sprints also depend on *(aerobic? anaerobic?)* ATP production.

b. Repeated isotonic contractions (such as jogging or dancing) rely on *(aerobic? anaerobic?)* ATP production. *(Aerobic? Anaerobic?)* workouts as in endurance training or prolonged jogging are more likely to increase blood vessels to skeletal and heart muscle to provide needed oxygen.

D9. Contrast concentric isotonic contraction with eccentric isotonic contraction.

E. Types of skeletal muscle fibers; cardiac and smooth muscle (pages 294–299)

✓ **E1.** Refer to the table below and contrast the three types of skeletal muscle fibers by doing this activity. You may find it helpful to refer back to Checkpoints D3 and D8 also.

Type and Rate of Contraction	Color	Myoglobin Concentration	Mitochondria and Blood Vessels	Source of ATP	Fatigue Easily?
SO. Slow oxidative	Red	High	Many	Aerobic	No
FOG. Fast oxidative-glycolytic	Red	Very high	Many	Aerobic	Moderate
FG. Fast glycolytic	White	Low	Few	Anaerobic	Yes

a. Postural muscles (as in neck and back) are used *(constantly? mostly for short bursts of energy?)*. Therefore it is appropriate that they contract at a *(fast? slow?)* rate. Such muscle tissue consists mostly of slow red fibers that *(do? do not?)* fatigue easily. These appear red because they have *(much? little?)* myoglobin and *(many? few?)* blood vessels. They have *(many? few?)* mitochondria and therefore can depend on *(aerobic? anaerobic?)* metabolism for energy. These muscles are classified as type *(SO? FOG? FG?)*.

b. Muscles of the arms are used *(constantly? mostly for short bursts of energy?)* as in lifting and throwing. Therefore they must contract at a *(fast? slow?)* rate. The arms consist mainly of fast twitch white fibers. These respond rapidly to nerve impulses and contract *(fast? slowly?)*. They *(do? do not?)* fatigue easily because they are designed for the relatively inefficient processes of *(aerobic? anaerobic?)* metabolism. These muscles are classified as *(SO? FOG? FG?)*.

c. Endurance exercises tend to enhance development of the more efficient (aerobic)

____________________ fibers, whereas weight lifting, which requires short bursts of

energy, tends to develop ____________________ fibers.

d. Most skeletal muscles, such as the biceps brachii, are *(all of one type? mixtures of types SO, FOG, and FG?)*. The muscles of any one motor unit are *(all of one type? mixtures of types SO, FOG, and FG?)*.

e. Arrange the three types of fibers in the sequence in which they are "called to action":

1. ____________________ → 2. ____________________ → 3. ____________________

(weak contraction) (stronger contraction) (maximal contraction)

f. Endurance training (such as running or swimming) tends to cause transformation

of some ________.

A. Fast oxidative-glycolytic (FOG) into fast glycolytic (FG) fibers
B. Fast glycolytic (FG) into fast oxidative-glycolytic (FOG) fibers

E2. State the desired effects as well as disadvantages of uses of anabolic steroids.

✓ **E3.** Compare the structure and function of skeletal and cardiac muscle by completing this table. State the significance of characteristics that have asterisks (*).

Characteristic	Skeletal Muscle	Cardiac Muscle
a. Number of nuclei per fiber	A hundred or more	
b. Number of mitochondria per fiber (More or Fewer)*		
c. Striated appearance due to alternated actin and myosin myofilaments (Yes or No)	Yes	
d. Presence of intercalated discs (Yes or No)*		
e. Nerve stimulation with release of ACh required for contraction (Yes or No)*		
f. Duration of muscle contraction (Long or Short)*		

E4. Compare the two types of smooth muscle.

Muscle Type	Structure	Spread of Stimulus	Locations
a. Visceral		Impulse spreads and causes contraction of adjacent fibers	
b.			Large blood vessels and airways, iris and ciliary muscles of eye, arrector pili muscles

✓ **E5.** Contrast smooth muscle with the muscle tissue you have already studied by circling the correct answers in this paragraph.

Smooth muscle fibers are *(cylinder shaped? tapered at each end?)* with *(several nuclei? one nucleus?)* per cell. They *(do? do not?)* contain actin and myosin. However, due to the irregular arrangement of these filaments, smooth muscle tissue appears *(striated? nonstriated or "smooth"?)*. In general, smooth muscle contracts and relaxes more *(rapidly? slowly?)* than skeletal muscle does, and smooth muscle holds the contraction for a *(shorter? longer?)* period of time than skeletal muscle does. This difference in smooth muscle is due at least partly to the fact that smooth muscle has *(no? many?)* transverse tubules to transmit calcium into the muscle fiber and has a *(large? small?)* amount of SR.

✓ **E6.** Smooth muscle *(is? is not?)* normally under voluntary control. List three chemicals released in the body or other factors that can also lead to smooth muscle contraction or relaxation.

✓ **E7.** *For extra review.* Of the three muscle types, select the correct muscle type from the box that matches the description below.

C. Cardiac	**Sk. Skeletal**	**Sm. Smooth**

_______ a. Starts most slowly and lasts much longer

_______ b. Has no transverse tubules and only scanty sarcoplasmic reticulum

_______ c. Has no striations because thick and thin filaments are not arranged in a regular pattern

_______ d. Has intermediate filaments attached to dense bodies (comparable to Z discs)

_______ e. Calmodulin, the regulatory protein in this muscle, activates an enzyme called myosin light chain kinase

_______ f. These muscles demonstrate the stress-relaxation response

_______ g. Spindle-shaped fibers with one nucleus per cell

_______ h. Unbranched, cylindrical cells; multi-nucleate

_______ i. Has fastest speed of contraction

_______ j. Stimulated only by the neurotransmitter acetylcholine (ACh)

_______ k. Includes three types: SO, FOG, and FG

_______ l. Branched, cylindrical cells; usually have one nucleus per cell

_______ m. Found only in the heart

_______ n. Striated, involuntary muscle

_______ o. Contains intercalated discs with desmosomes and gap junctions

E8. *The Big Picture: Looking Back.* For additional review of the three muscle types, look back at LG Chapter 4, Checkpoints El-E2, and Figure LG 4.1, M-O, page LG 71.

F. Regeneration of muscle tissue; developmental anatomy and aging of muscles (pages 299–301)

✓ **F1.** Arrange the three types of your muscle tissue in correct sequence according to ability (from most to least) to regenerate during your lifetime.

_____ _____ _____ A. Heart muscle

B. Biceps muscle

C. Muscle of an artery or intestine

✓ **F2.** Both skeletal and cardiac muscle tissue increases in size by *(hyperplasia? hypertrophy?)*, which means increase in *(number? size?)* of cells. If either of these tissues is damaged,

fibrosis, which means replacement of muscle tissue with _____________________ tissue, is likely to occur.

F3. Describe the roles of the following cells in regeneration of muscle tissue:

a. Satellite cells

b. Pericytes

✓ **F4.** Complete this exercise about the development of muscles.

a. Which of the three types of muscle tissue develop from mesoderm? *(Skeletal? Cardiac? Smooth?)*

b. Part of the mesoderm forms columns on either side of the developing nervous system. This tissue segments into blocks of tissue called ________________. The first pair of somites forms on day *(10? 20? 30?)* of gestation. By day 35 a total of ______ pairs of somites are present.

c. Which part of a somite develops into vertebrae? *(Myo-? Derma-? Sclero-?)* derm. The dermis of the skin and other connective tissues is formed from __________________ -tomes, while most skeletal muscles develop from __________________-tomes.

F5. With aging, muscles experience a relative increase in *(SO? FOG? FG?)* fibers. Discuss possible reasons for this change.

G. Disorders and medical terminology (pages 302–303)

G1. Write the correct medical term related to muscles after its description.

a. Muscle or tendon pain and stiffness: ____________________

b. Increased muscle tone, rigidity, or spasticity: ____________________

c. Inherited, muscle-destroying disease causing atrophy of muscles: __________________

d. A muscle tumor: ____________________

e. An involuntary, brief twitch of a muscle that is ordinarily under voluntary control, such as in the eyelid: ____________________

f. An involuntary, brief twitch of an entire motor unit that is visible under the skin; may be seen in MS or ALS patients: ____________________

✓ **G2.** Describe myasthenia gravis in this exercise.

a. In order for skeletal muscle to contract, a nerve must release the chemical ____________________ at the myoneural junction. Normally ACh binds to ____________________ on the muscle fiber membrane.

b. It is believed that a person with myasthenia gravis produces ________________ that bind to these receptors, making them unavailable for ACh binding. Therefore, ACh *(can? cannot?)* stimulate the muscle, and it is weakened.

c. One treatment for this condition employs ________________ drugs, which enhance muscle contraction by permitting the ACh molecules that *do* bind to act longer (and not be destroyed by AChE).

d. Other treatments include ________________-suppressants, which decrease the

patient's antibody production, or ________________, which removes the patient's harmful antibodies.

G3. Define each of these types of abnormal muscle contractions.

a. Spasm

b. Fibrillation

c. Tremor

ANSWERS TO SELECTED CHECKPOINTS

Al. (a) Sm. (b) C. (c) Sk. (d) Sk. (e) Sm.

A2. (a) Motion or movement. (b1) Blood, (b2) food, bile, enzymes, or wastes, (b3) urine, (b4) ova and sperm or eggs. (c) Stabilization, stomach or bladder. (d) Heat, shivering. (e) Sphincters.

A3. (a) Extensibility. (b) Elasticity. (c) Excitability. (d) Contractility.

A4. (a) Isotonic. (b) Stays about the same, increases.

B1. (a) S. (b) D. (c) S.

B2. Epimysium → perimysium → endomysium. See Figure LG 10.1A

B3. Periosteum; tendon, aponeurosis.

B4. (a) Somatic motor neurons; axons; bulbs, neuromuscular. (b) Provide nutrients and oxygen for ATP production, and remove heat and wastes.

B6. (a) Muscle fiber (cell) → myofibrils → filaments (thick or thin). (b) Sarcolemma; T (transverse); they transmit action potentials from the sarcolemma throughout the muscle fiber so that all parts of it contract in unison. (c) Sarcoplasm; glycogen, myoglobin, oxygen; ATP, mitochondria. (d) Endoplasmic reticulum; Ca^{2+}; cisternae, triad; see Figure LG 10.1A. (e) See Figure LG 10.1A.

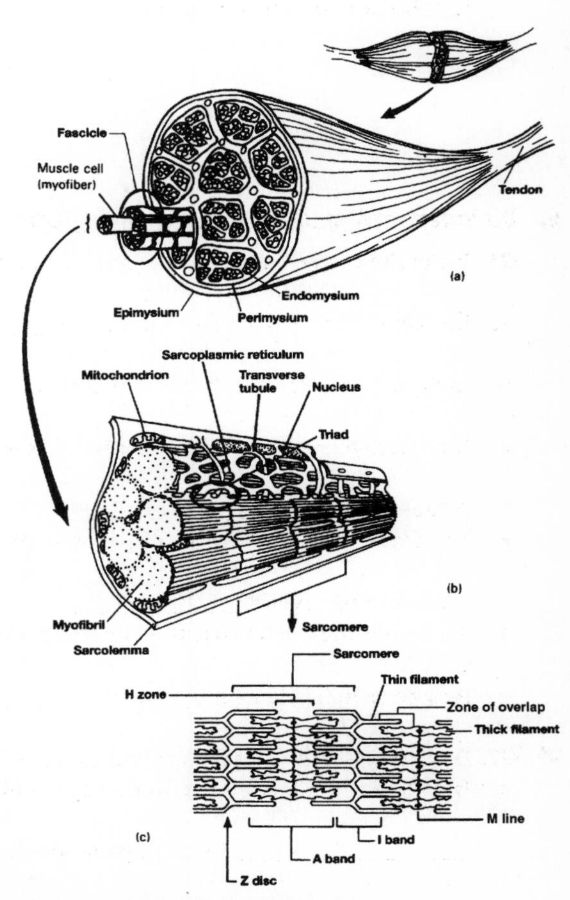

Figure LG 10.1A Diagram of skeletal muscle. (a) Skeletal muscle cut to show cross section and longitudinal section with connective tissues. (b) Section of one muscle cell (myofiber). (c) Detail of sarcomere of muscle cell.

B7. (a) Size; exercise and good nutrition; the heart can also hypertrophy when it must pump hard to move blood out into a high pressure system (in a person with high blood pressure).(b) Denervation; 75; fibrous connective. (c) Immobilization of the extremity by a cast or other immobilizer, lack of exercise; being bedridden (for example, in a coma).
B8. (a) Sarcoplasm. (b) Sarcomere. (c) A band. (d) I band. (e) M line. (f) Z disc. (g) Zone of overlap.
B9. (a) Delayed onset muscle soreness. (b) Myoglobin, creatine kinase. (c) 12, 48.
B10. (a) See Figure LG 10.1A. (b) Contractile; troponin, tropomyosin. (c) Thick; golf clubs; M lines; myosin head or crossbridge, actin. (d) Z discs, do not; actin; tropomyosin, troponin. (e) Structural; huge (titanic); thick, Z disc, M line; elasticity. (f) Myomesin; dystrophin.
C1. (a) Myosin, actin, M line. (b) Thin; slide, sliding-filament; do not.
C2. (a) Sarcoplasmic reticulum (SRI). (b) Troponin, actin, myosin. (c) ATPase, ADP, phosphate; actin. (d) Power, ADP, sarcomere. (e) 600 (2 per myosin x 300 myosin molecules per thick filament); five. (f) Actin; myosin, myosin. (g) Toward or even across; myosin, thin.
C3. (a) Low; (1) Active transport; (2) Calsequestrin, 10,000. (b) T tubules; channels, sarcoplasm. (c) Actin; relax. (d) An active; is not; myosin cross bridges stay attached to actin and muscles remain in a state of partial contraction (rigor mortis) for about a day.
C4. C D A E B.
C5. (a) Much; maximal; no. (b) Stronger.
C6. (a) A, H, end bulb. (b) C, D, acetylcholine (ACh), E. (c) Motor end plate; receptors; 30-40 million. (d) Open; contraction. (e) Neuromuscular junction (NMJ); only one, I; synapses.
C8-9. (a) AChE; ↓. (b) CUR; ↓. (c) NEO; ↑. (d) CBT; ↓.
D1. (a) ATP; myosin; ATP → ADP + P + energy. (b) ADP + P + energy (from foods) → ATP. (c) Active transport; creatine (that combines with phosphate to form creatine phosphate; creatine phosphate + ADP → creatine + ATP. (d) Creatine phosphate + ADP ← creatine + ATP (draw arrow to LEFT in Figure LG 10.3); creatine kinase (CK); creatine phosphate. (e) 15 seconds; anaerobic; aerobic; see Figure LG 10.4A. (f) Lactic, anaerobic; fatigue; heart, liver, kidney; see Figure LG 10.4A. (g) Myoglobin; hemo; red. (h) Aerobic.
D3. (a) CP. (b) ANAER. (c) AER.
D4. Decrease in frequency of stimulation by neurons, shorter length muscle fibers just prior to contraction, fewer and smaller motor units recruited, lack of nutrients, lack of oxygen. and muscle fatigue.
D5. An average of 150 muscle fibers [cells]; laryngeal muscles controlling speech.
D6. (a) T. (b) M. (c) L. (d) Ref. (e) Rel. (f) Ref.
D7. (a) U. (b) F. (c) W. (d) U.
D8. (a) Anaerobic; anaerobic. (b) Aerobic; aerobic.
E1. (a) Constantly; slow; do not; much, many; many, aerobic; SO. (b) Mostly for short bursts of energy; fast; fast; do, anaerobic; FG. (c) FOG, FG. (d) Mixtures of types SO, FOG, and FG; all of one type. (e1) SO, (e2) FOG, (e3) FG. (f) B.
E3.

Characteristic	Skeletal Muscle	Cardiac Muscle
a. Number of nuclei per fiber	A hundred or more	**One or two (see Chapter 20)**
b. Number of mitochondria per fiber (More or Fewer)*	**Fewer**	**More, because heart muscle requires constant generation of energy**
c. Striated appearance due to alternated actin and myosin myofilaments (Yes or No)	Yes	**Yes**
d. Presence of intercalated discs (Yes or No)*	**No**	**Yes, allow muscle action potentials to spread from one cardiac muscle fiber to another**
e. Nerve stimulation with release of ACh required for contraction (Yes or No)*	**Yes**	**No, so heart can contract without nerve stimulation, but nerves can increase or decrease heart rate**
f. Duration of muscle contraction (Long or Short)*	**Short (0.01 sec.–0.100 sec)**	**10–15 times longer so heart can pump blood out of heart**

E5. Tapered at each end, one nucleus; do; nonstriated or "smooth"; slowly, longer; no, small.
E6. Is not; hormones, pH or temperature changes, O_2 and CO_2 levels, and certain ions.
E7. (a–g) Sm. (h–k) Sk. (l–o) C.
F1. C B A
F2. Hypertrophy, size; fibrous (or connective or scar).
F4. (a) All three: skeletal, cardiac, and smooth. (b) Somites; 20; 44. (c) Sclero-; derma-, myo-.
G1. (a) Myalgia. (b) Hypertonia. (c) Muscular dystrophy. (d) Myoma. (e) Tic. (f) Fasciculation.

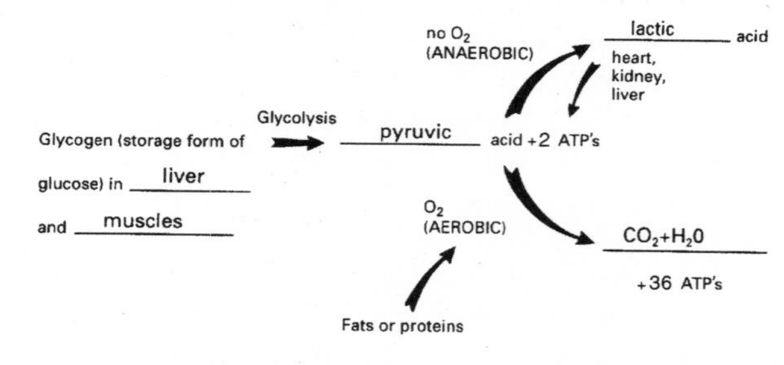

Figure LG 10.4A Anaerobic and aerobic energy sources for muscles.

G2. (a) Acetylcholine (ACh); receptors. (b) Antibodies; cannot. (c) Anticholinesterase (antiAChE). (d) Immuno-, plasmapheresis.

MORE CRITICAL THINKING: CHAPTER 10

1. Explain why stretching a muscle before exercising can maximize muscle contraction. Then explain why *over*stretched heart muscle (as in the condition heart failure) may decrease the effectiveness of the heart as a pump.
2. Discuss results of research on the potential value of using creatine as a performance-enhancing supplement.
3. Discuss the pros and cons of anabolic steroids as you would with a friend who is considering using these chemicals.
4. Explain how the microscopic structure of cardiac muscle fibers accounts for its continuous, rhythmic activity.
5. Describe several reasons why smooth muscle contracts more slowly than other muscle types.
6. Write a rationale for why endurance and strength training programs are effective for maintaining healthy musculature in elderly persons.

MASTERY TEST: CHAPTER 10

Questions 1–9: Circle T (true) or F (false). If the statement is false, change the underlined word or phrase so that the statement is correct.

T F 1. Tendons are cords of connective tissue, whereas aponeuroses are broad, flat bands of connective tissue.

T F 2. During contraction of muscle both A bands and I bands get shorter.

T F 3. In general, red (rather than white) skeletal muscle is more suited for endurance exercise than for short bursts of energy.

T F 4. Elasticity is the ability of a muscle to be stretched or extended.

T F 5. Myasthenia gravis is caused by an excess of acetylcholine production at the myoneural junction.

T F 6. Muscle fibers remain relaxed if there are few calcium ions in the sarcoplasm.

T F 7. Most myoblasts do continue to divide to form new skeletal muscle cells throughout life.

T F 8. Muscles that are used mostly for quick bursts of energy (such as those in the arms) contain large numbers of fast glycolytic (FG) fibers.

T F 9. In a relaxed stretched muscle fiber thin and thick filaments overlap to form the zone of overlap.

Questions 10–12: Arrange the answers in correct sequence.

______ ______ ______ 10. Number of transverse tubules from greatest to least per muscle cell in these types of muscle:

A. Smooth
B. Skeletal
C. Cardiac

______ ______ ______ 11. According to the amount of muscle tissue they surround, from most to least:

A. Perimysium
B. Endomysium
C. Epimysium

______ ______ ______ 12. From largest to smallest:

A. Myofibril
B. Filament
C. Muscle fiber (myofiber or cell)

Questions 13–20: Circle the letter preceding the one best answer to each question.

13. All of the following molecules are parts of thin filaments *except:*
    A. Actin
    B. Myosin
    C. Tropomyosin
    D. Troponin

14. Which of the following answers includes only structural (not contractile or regulatory) proteins of muscle?
    A. Actin and myosin
    B. Titin, dystrophin, and myomesin
    C. Troponin and tropomyosin
    D. Actin, troponin, and tropomyosin

15. Choose the one statement that is *false:*
    A. The A band is darker than the I band.
    B. Thick filaments reach the Z disc in relaxed muscle.
    C. The H zone contains thick filaments, but not thin ones.
    D. The I band is shorter in contracted muscle than in relaxed muscle.

16. All of the following terms are correctly matched with descriptions *except:*
    A. Denervation atrophy—wasting of a muscle to one-quarter of its size within two years of loss of nerve supply to a muscle
    B. Disuse atrophy—decrease of muscle mass in a person who is bedridden or who has a cast on
    C. Muscular hypertrophy—increase in size of a muscle by increase in the number of muscle cells
    D. Cramp—painful, spasmodic contraction

17. Which statement about muscle physiology in the relaxed state is *false?*
    A. Myosin cross bridges are bound to ATP.
    B. Calcium ions are stored in sarcoplasmic reticulum.
    C. Myosin cross bridges are bound to actin.
    D. Tropomyosin-troponin complex is bound to actin.

18. All of the following answers match a chemical with its correct description *except*:
    A. Neurotransmitter released by nerve at a neuromuscular junction: ACh
    B. Enzyme that destroys ACh within the synaptic cleft: AChE
    C. A contractile protein: tropomyosin
    D. Chemical that binds to and changes shape of troponin-tropomyosin complex, exposing myosin-binding site on actin: Ca^{2+}

19. Choose the *false* statement about cardiac muscle.
    A. Cardiac muscle is involuntary.
    B. Cardiac muscle contain fewer mitochondria than skeletal muscles contain.
    C. Cardiac muscle cells are called cardiac muscle fibers.
    D. Cardiac fibers are separated by intercalated discs.
    E. Cardiac fibers are tapered at each end with no striations.

20. Most voluntary movements of the body are results of ________ contractions.
    A. Isometric
    B. Fibrillation
    C. Twitch
    D. Unfused tetanus
    E. Fused tetanus

Questions 21–25: Fill-ins. Complete each sentence with the word or phrase that best fits.

________________ 21. ________________ muscle cells are nonstriated and tapered at each end with one nucleus per cell.

________________ 22. Cardiac muscle remains contracted longer than skeletal muscle does because ________________ is slower in cardiac than in skeletal muscle.

________________ 23. DOMS is an acronym that means ________________.

________________ 24. ________________ is the transmitter released from synaptic vesicles of axons supplying skeletal muscle.

________________ 25. ADP + creatine phosphate → ________________ (Write the products.)

ANSWERS TO MASTERY TEST: CHAPTER 10

True-False

1. T
2. F. I bands but not A bands
3. T
4. F. Extensibility
5. F. An autoimmune disorder in which antibodies are produced that bind to and block receptors on the sarcolemma
6. T
7. Few myoblasts (those called satellite cells) do
8. T
9. T

Arrange

10. B C A
11. C A B
12. C A B

Multiple Choice

13. B
14. B
15. B
16. C
17. C
18. C
19. B
20. D

Fill-Ins

21. Smooth
22. Passage of calcium ions from the extracellular fluid through sarcolemma
23. Delayed onset muscle soreness
24. Acetylcholine
25. ATP + creatine

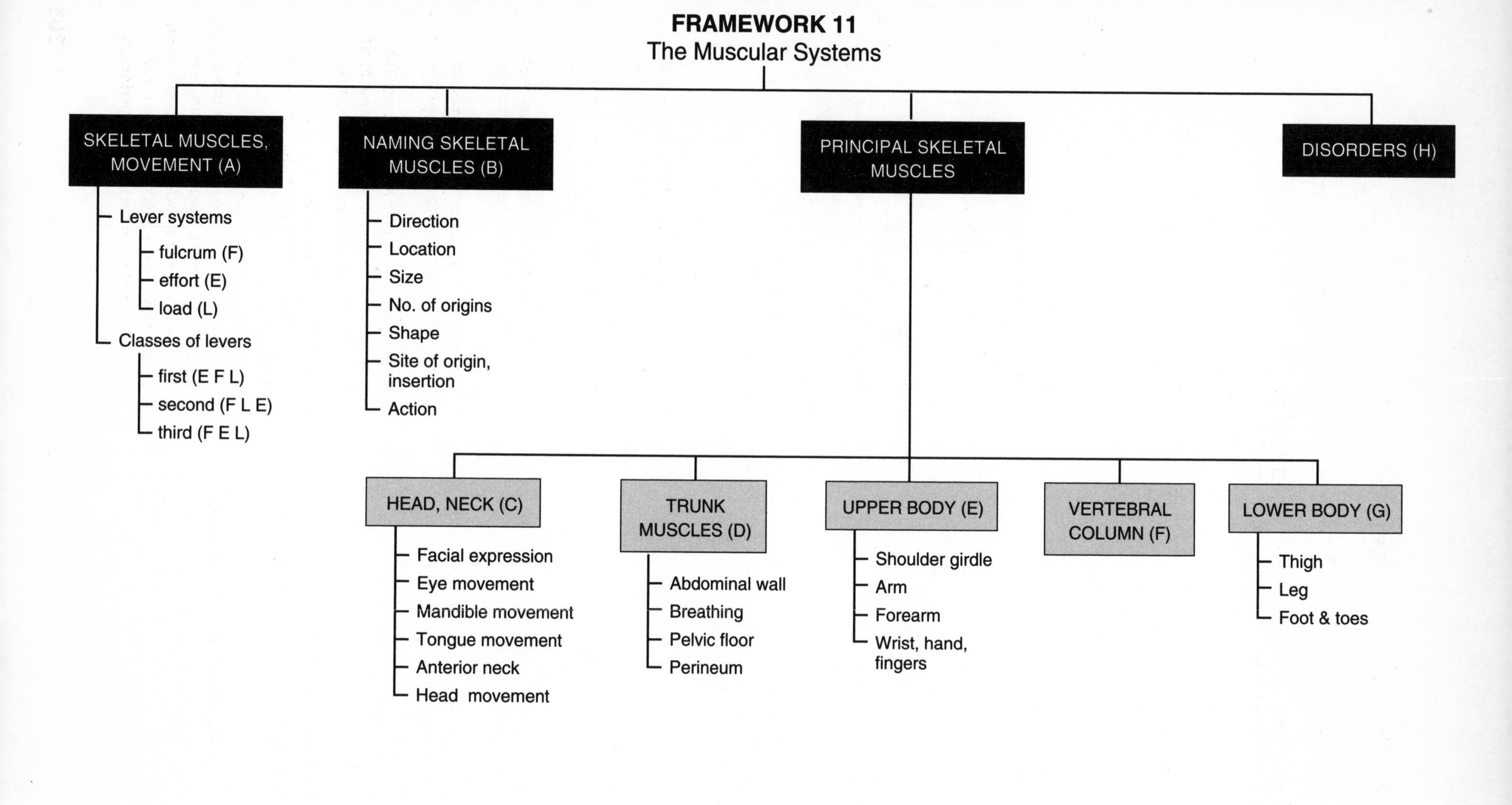
FRAMEWORK 11
The Muscular Systems
SKELETAL MUSCLES, MOVEMENT (A)
Lever systems
fulcrum (F)
effort (E)
load (L)
Classes of levers
first (E F L)
second (F L E)
third (F E L)
NAMING SKELETAL MUSCLES (B)
Direction
Location
Size
No. of origins
Shape
Site of origin, insertion
Action
PRINCIPAL SKELETAL MUSCLES
HEAD, NECK (C)
Facial expression
Eye movement
Mandible movement
Tongue movement
Anterior neck
Head movement
TRUNK MUSCLES (D)
Abdominal wall
Breathing
Pelvic floor
Perineum
UPPER BODY (E)
Shoulder girdle
Arm
Forearm
Wrist, hand, fingers
VERTEBRAL COLUMN (F)
LOWER BODY (G)
Thigh
Leg
Foot & toes
DISORDERS (H)

C H A P T E R **11**

The Muscular System

More than 600 different muscles attach to the bones of the skeleton. Muscles are arranged in groups that work much as a symphony, with certain muscles quiet while others are performing. The results are smooth, harmonious movements, rather than erratic, haphazard discord.

Much information can be gained by a careful initial examination of two aspects of each muscle: its name and location. A muscle name may offer such clues as size or shape, direction of fibers, or points of attachment of the muscle. A look at the precise location of the muscle and the joint it crosses—combined with logic—can usually lead to a correct understanding of action(s) of that muscle.

As you begin your study of the muscular system, carefully examine the Chapter 11 Topic Outline and Objectives; check off each one as you complete it. To organize your study of the muscular system, glance over the Chapter 11 Framework now. Be sure to refer to the Framework frequently and note relationships among key terms in each section.

TOPIC OUTLINE AND OBJECTIVES

A. How skeletal muscles produce movement

1. Describe the relationship between bones and skeletal muscles in producing body movements.
2. Define lever and fulcrum, and compare the three types of levers based on location of the fulcrum, effort, and load.
3. Identify the types of fascicle arrangements in a skeletal muscle, and relate the arrangements to strength of contraction and range of motion.
4. Explain how the prime mover, antagonist, synergist, and fixator in a muscle group work together to produce movement.

B. How skeletal muscles are named

1. Explain seven features used in naming skeletal muscles.

C. Principal skeletal muscles of the head and neck

1. Describe the origin, insertion, action, and innervation of the muscles of facial expression.
2. Describe the origin, insertion, action, and innervation of the extrinsic eye muscles.
3. Describe the origin, insertion, action, and innervation of the muscles that move the mandible.
4. Describe the origin, insertion, action, and innervation of the extrinsic muscles of the tongue.
5. Describe the origin, insertion, action, and innervation of the muscles of the anterior neck.
6. Describe the origin, insertion, action, and innervation of the muscles that move the head.

D. Principal skeletal muscles that act on the abdominal wall, muscles used in breathing, muscles of the pelvic floor and perineum

1. Describe the origin, insertion, action, and innervation of the muscles that act on the abdominal wall.
2. Describe the origin, insertion, action, and innervation of the muscles used in breathing.
3. Describe the origin, insertion, action, and innervation of the muscles of the pelvic floor.
4. Describe the origin, insertion, action, and innervation of the muscles of the perineum.

E. Principal skeletal muscles that move the shoulder girdle and upper limb

1. Describe the origin, insertion, action, and innervation of the muscles that move the pectoral girdle.
2. Describe the origin, insertion, action, and innervation of the muscles that move the humerus.
3. Describe the origin, insertion, action, and innervation of the muscles that move the radius and ulna.
4. Describe the origin, insertion, action, and innervation of the muscles that move the wrist, hand, thumb and fingers.
5. Describe the origin, insertion, action, and innervation of the intrinsic muscles of the hand.

F. Principal skeletal muscles that move the vertebral column

1. Describe the origin, insertion, action, and innervation of the muscles that move the vertebral column.

G. Principal skeletal muscles that move the lower limb

1. Describe the origin, insertion, action, and innervation of the muscles of the femur.
2. Describe the origin, insertion, action, and innervation of the muscles that act on the femur, tibia, and fibula.
3. Describe the origin, insertion, action, and innervation of the muscles that move the foot and toes.
4. Describe the origin, insertion, action, and innervation of the intrinsic muscles of the foot.

H. Disorders

WORDBYTES

Now become familiar with the language of this chapter by studying each wordbyte, its meaning, and an example of its use within a term. After you study the entire list, self-check your understanding by writing the meaning of each wordbyte on the line. As you continue through the *Learning Guide,* identify (and fill in) additional terms that contain the same wordbyte.

Wordbyte	Self-check	Meaning	Example(s)
bi-	______________	two	*bi*ceps
brachi-	______________	arm	*brachi*alis
brev-	______________	short	flexor digitorum *brev*is
bucc-	______________	mouth, cheek	*bucc*inator
cap-, -ceps	______________	head	*cap*itis, tri*ceps*
-cnem-	______________	leg	gastro*cnem*ius
delt-	______________	Greek D (△)	*delt*oid
ergon-	______________	work	syn*erg*ist
gastro-	______________	stomach, belly	*gastro*cnemius
genio-	______________	chin	*genio*glossus
glossus-	______________	tongue	*gloss*ary, stylo*glossus*
glute-	______________	buttock	*glute*us medius
grac-	______________	slender	*grac*ilis
-issimus	______________	the most	lat*issimus* dorsi
lat-	______________	wide	*lat*issimus dorsi

Wordbyte	**Self-check**	**Meaning**	**Example(s)**
maxi-	______________________	large	gluteus *maxi*mus
mini-	______________________	small	gluteus *mini*mus
or-	______________________	mouth	orbicularis *or*is
quad-	______________________	four	*quad*riceps femoris
rect-	______________________	straight	*rect*us femoris
sartor-	______________________	tailor	*sartor*ius
serra-	______________________	saw toothed	*serra*tus anterior
teres-	______________________	round	*teres* major
tri-	______________________	three	*tri*ceps femoris
vast-	______________________	great	*vast*us lateralis

CHECKPOINTS

A. How skeletal muscles produce movement (pages 309–313)

✓ **A1.** What structures constitute the *muscular system*?

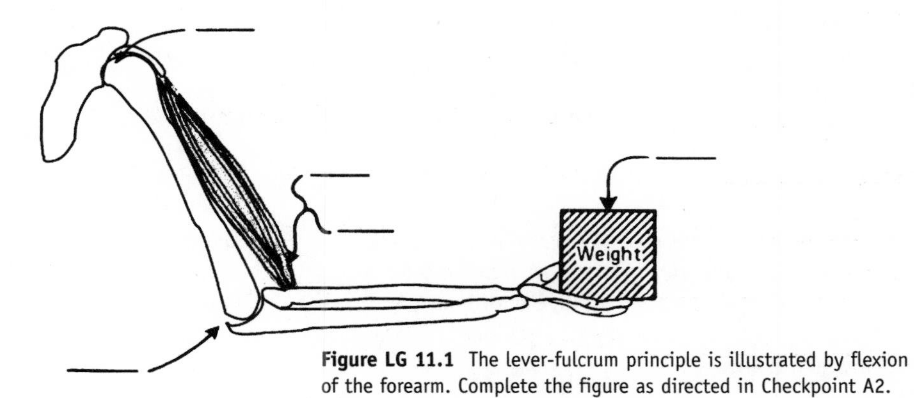

Figure LG 11.1 The lever-fulcrum principle is illustrated by flexion of the forearm. Complete the figure as directed in Checkpoint A2.

✓ **A2.** Refer to Figure LG 11.1 and consider flexion of your own forearm as you do this learning activity.

a. In flexion your forearm serves as a rigid rod, or ________________, which moves about a fixed point, called a ________________ (your elbow joint, in this case).

b. Hold a weight in your hand as you flex your forearm. The weight plus your forearm serve as the *(effort? fulcrum? load?)* during this movement.

c. The effort to move this load is provided by contraction of a ________________. Note that if you held a heavy telephone book in your hand while your forearm were flexed, much more ________________ by your arm muscles would be required.

d. In Figure LG 11.1 identify the exact point at which the muscle causing flexion attaches to the forearm. It is the *(proximal? distal?)* end of the *(humerus? radius? ulna?)*. Write an E and an I on the two lines next to the arrow at that point in the figure. This indicates that this is the site where the muscle exerts its effort (E) in the lever system, and it is also the insertion (I) end of the muscle. (More about insertions in a minute.)

e. Each skeletal muscle is attached to at least two bones. As the muscle shortens, one bone stays in place and so is called the *(origin? insertion?)* end of the muscle. What bone in the figure appears to serve as the origin bone? ______________________ Write O on that line at that point in the figure.

f. Now label the remaining arrows in Figure LG 11.1: F at fulcrum and L at load. This is an example of a *(first? second? third?)* -class lever.

A3. Complete the table about levers. KEY: E. Effort, F. Fulcrum, and L. load.

Class	Analogy of EFL	Arrangement	Mechanical advantage: (Yes, No). Explain.	Example in human body
a. First				Head and neck: atlanto-occipital joint
b.	Wheelbarrow	F L E		
c.	(Omit)		No since E is closer (than L) to F	

✓ **A4.** *For extra review.* Check your understanding of lever systems in the body in this Checkpoint.

a. Refer to Figure 11.2a in your text. Now hyperextend your head as if to look at the sky. The weight of your face and jaw serves as *(E? F? L?)*, while your neck muscles provide *(E? F? L?)*. The fulcrum is the joint between the ________________ and the ________________ bones. This is an example of a ________________ class lever.

b. Mechanical advantage (leverage) means that a *(more? less?)* forceful effort can move a *(more? less?)* forceful load. Mechanical advantage is provided by having the load always *(closer to? farther from?)* the fulcrum than the effort. Which class of lever provides more mechanical advantage?
A. Second-class lever B. Third-class lever

✓ **A5.** Correlate fascicular arrangement with muscle power and range of motion of muscles.

a. A muscle with *(many? long?)* fibers will tend to have great strength. An example is the *(parallel? pennate?)* arrangement.

b. A muscle with *(many? long?)* fibers will tend to have great range of motion (but less power). An example is the *(parallel? pennate?)* arrangement.

✓ **A6.** Refer again to Figure LG 11.1 and do this exercise about how muscles of the body work in groups.

a. The muscle that contracts to cause flexion of the forearm is called a ______________.
An example of a prime mover in this action would be the ______________ muscle.

b. The triceps brachii must relax as the biceps brachii flexes the forearm. The triceps is an extensor. Because its action is opposite to that of the biceps, the triceps is called *(a synergist? an agonist? an antagonist?)* of the biceps.

c. What would happen if the flexors of your forearm were functional, but not the antagonistic extensors?

d. What action would occur if both the flexors and extensors contracted simultaneously?

e. Muscles that assist or cooperate with the prime mover to cause a given action are known as ____________________.

B. How skeletal muscles are named (pages 313–317)

✓ **B1.** Review the Wordbyte section above. Match the names of the following muscles with their meanings.

A.	**Large muscle of the buttock region**	**D.**	**The broadest muscle of the back**
B.	**Belly-shaped muscle in leg**	**E.**	**Large muscle in medial thigh area**
C.	**Thigh muscle with four origins**	**F.**	**Muscle that raises the upper lip**

________ a. Latissimus dorsi

________ b. Vastus medialis

________ c. Quadriceps femoris

________ d. Gastrocnemius

________ e. Gluteus maximus

________ f. Levator labii superioris

✓ **B2.** As you study the names of muscles, you will find that most of them provide a good description of the muscle. For each of the following, indicate the type of clue that each part of the name gives. The first one is done for you.

A.	**Action**	**N.**	**Number of heads or origins**
D.	**Direction of fibers**	**P.**	**Points of attachment of origin and insertion**
L.	**Location**	**S.**	**Size or shape**

__**D, L**__ a. Rectus abdominis

________ b. Flexor carpi ulnaris

________ c. Biceps brachii

________ d. Sternocleidomastoid

________ e. Adductor longus

C. Principal skeletal muscles of the head and neck (pages 317–331)

✓ **C1.** After studying Exhibit 11.1 (pages 318–321) in your text, check your understanding of the muscles of facial expression. Write the name of the muscle that answers each description. Locate muscles in Figure LG 11.2a O-R. Cover the key and write the name of each facial muscle next to its lettered leader line.

a. Allows you to show surprise by raising your eyebrows and forming horizontal forehead wrinkle: ______________________________

b. Muscle surrounding opening of your mouth; allows you to use your lips in kissing and in speech: ______________________________

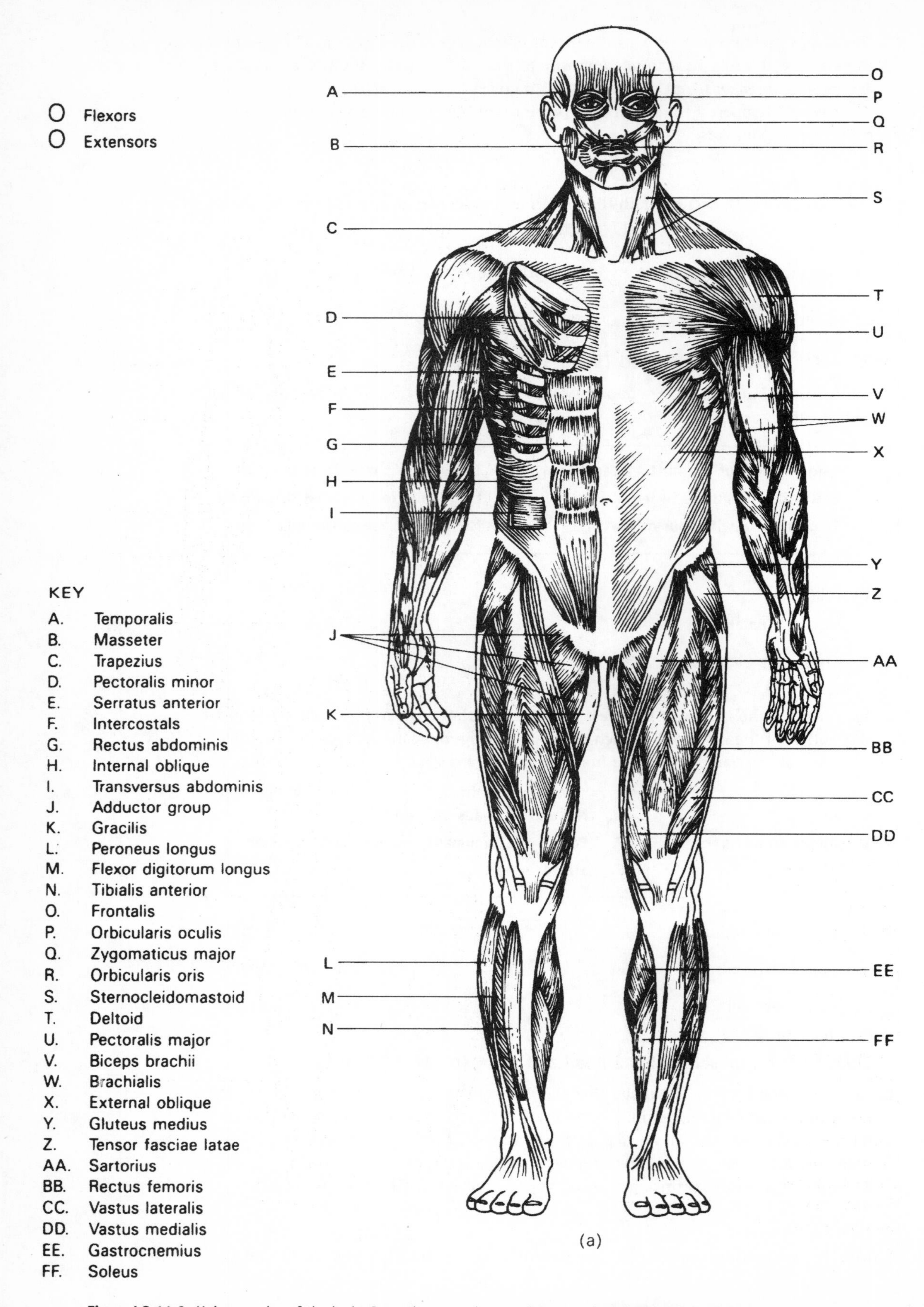

Figure LG 11.2 Major muscles of the body. Some deep muscles are shown on the left side of the figure. All muscles on the right side are superficial. Label and color as directed. (a) Anterior view. (b) Posterior view.

O Flexors
O Extensors

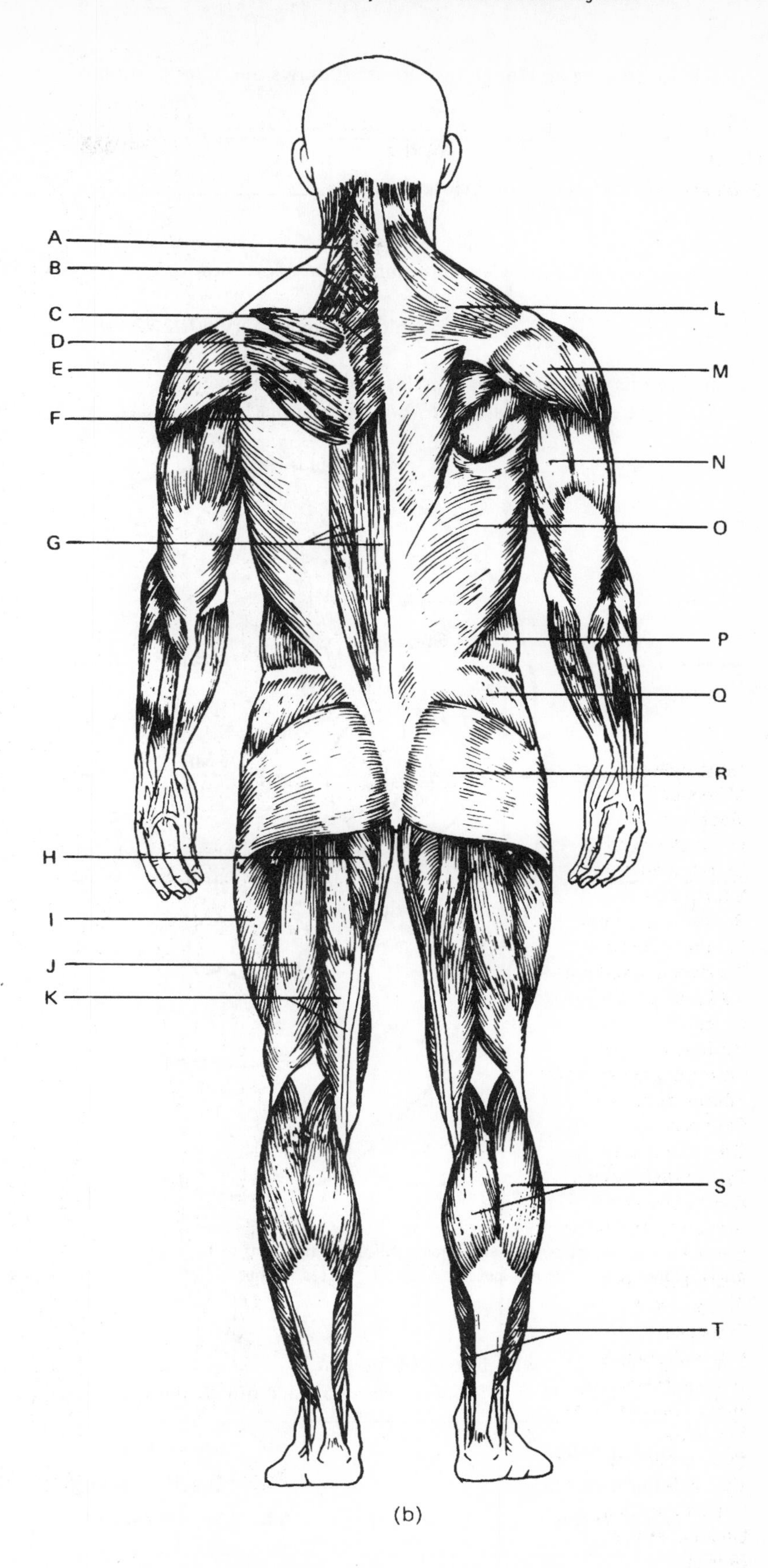

(b)

KEY

A. Levator scapulae
B. Rhomboideus minor
C. Rhomboideus major
D. Supraspinatus
E. Infraspinatus
F. Teres major
G. Erector spinae
H. Adductors
I. Vastus lateralis
J. Biceps femoris
K. Semimembranosus and semitendinosus
L. Trapezius
M. Deltoid
N. Triceps brachii
O. Latissimus dorsi
P. External oblique
Q. Gluteus medius
R. Gluteus maximus
S. Gastrocnemius
T. Soleus

c. Muscle for smiling and laughing because it draws the outer portion of the mouth upward and outward: ______________________________

d. Circular muscle around eye; closes eye: ______________________________

✓ **C2.** Essentially all of the muscles controlling facial expression receive nerve impulses via the ________________ nerve, which is cranial nerve *(III? V? VII?)*.

✓ **C3.** Complete this exercise about muscles that move the eyeballs.

a. Why are these muscles called *extrinsic* eyeball muscles?

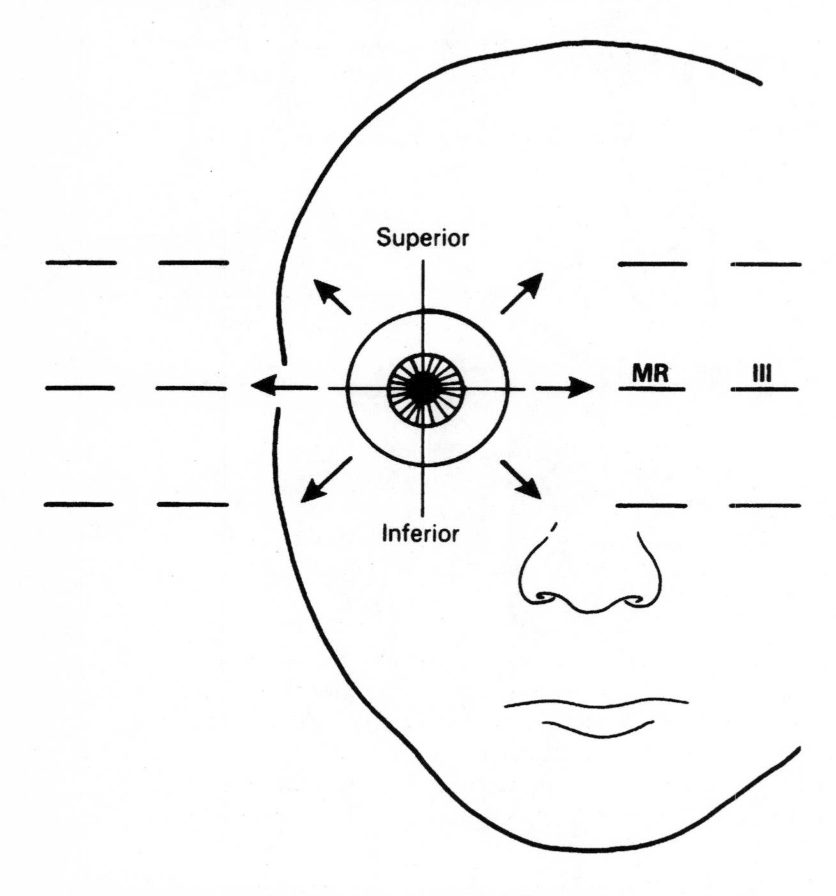

Figure LG 11.3 Right eyeball with arrows indicating directions of eye movements produced by the extrinsic eye muscles. Label as directed in Checkpoint C3.

b. Label Figure LG 11.3 with initials of the names of the six extrinsic eye muscles listed in the box to indicate *direction of movement* of each muscle. One has been done for you.

IO.	**Inferior oblique**	**MR.**	**Medial rectus**
IR.	**Inferior rectus**	**SO.**	**Superior oblique**
LR.	**Lateral rectus**	**SR.**	**Superior rectus**

c. Now write next to each muscle the number of the cranial nerve innervating that muscle. One has been done for you.

d. Note that most of the extrinsic eye muscles are innervated by cranial nerve *(III? IV? VI?)*, as indicated by the name of that nerve *(oculomotor)*. Cranial nerve VI, the *abducens*, supplies the muscles that cause most *(medial? lateral?)* movement, or abduction, of the eye. [*Hint:* Remember the chemical symbol for sulfate (SO_4) to remind yourself that the remaining muscle, superior oblique (SO), is supplied by the fourth cranial nerve.]

✓ **C4.** *For extra review* of actions of eye muscles, work with a study partner. One person moves the eyes in a particular direction; the partner then names the eye muscles used for that action. *Note:* Will both eyes use muscles of the same name? For example, as you look to your right, will the lateral rectus muscles attached to both eyes contract?

✓ **C5.** Place your index finger and thumb on the origin and insertion of each of the muscles that move your lower jaw. (Refer to Exhibit 11.3 and Figure 11.6 in the text, for help.) Then do this learning activity.

a. Two large muscles help you to close your mouth forcefully, as in chewing. Both of these act by *(lowering the maxilla? elevating the mandible?)*. The

_____________________ covers your temple, and the _____________________ covers the ramus of the mandible.

b. Most of the muscles involved in chewing are ones that help you to *(open? close?)* your mouth. Think about this the next time you go to the dentist and try to hold

your mouth open for a long time, with only your _____________________ muscles to force your mouth wide open.

c. The muscles that move the lower jaw and so aid in chewing are innervated by

cranial nerve *(III? V? VII?)*, known as the _____________________ nerve.

d. Refer to Figure LG 11.2a. Muscles *A* and *B* are used primarily for *(facial expressions? chewing?)*, whereas muscles *0, P, Q,* and *R* are used mainly for *(facial expressions? chewing?)*.

✓ **C6.** Actions such as opening your mouth for eating or speaking and also swallowing require integrated action of a number of muscles. You have already studied some that allow you to open your mouth (such as masseter and temporalis). We will now consider muscles of the tongue and the floor of the oral cavity.
Identify locations of three bony structures relative to your tongue. Tell whether each is anterior or posterior and superior or inferior to your tongue. (For help refer to Figure 11.7.)

a. Styloid process of temporal bone _____________________________________

b. Hyoid bone _____________________________________

c. Chin (anterior portion) _____________________________________

✓ **C7.** Remembering the locations of these bony points and that muscles move a structure (tongue) by pulling on it, determine the direction that the tongue is pulled by each of these muscles.

a. Styloglossus: _____________________________________

b. Hyoglossus: _____________________________________

c. Genioglossus: _____________________________________

✓ **C8.** The three muscles listed in Checkpoint C7 are innervated by cranial nerve ________;

the name ____________________ nerve indicates that this pair of cranial nerves supplies the region "under the tongue."

✓ **C9.** Match groups of muscles in each answer with the descriptions below.

A.	**Styloglossus, hyoglossus, and genioglossus**
B.	**Stylohyoid, mylohyoid, and geniohyoid**

________ a. Located superior to the hyoid (suprahyoid), these form the floor of the oral cavity. (Recall that the hyoid is directly posterior to the inferior border of the U-shaped mandible.)

________ b. Muscles permitting tongue movements

✓ **C10.** Using a mirror, find the origin and insertion of your left sternocleidomastoid muscle. (See Figure 11.9 in your text.) The muscle contracts when you pull your chin down and to the right; this diagonal muscle of your neck will then be readily located. Note that the left sternocleidomastoid pulls your face toward the *(same? opposite?)* side. It also *(flexes? extends?)* the head.

✓ **C11.** Alternately flex and extend vertebrae of your neck by looking at your toes and then toward the sky. As you look at the sky, you are *(flexing? extending and then hyperextending?)* your head and cervical vertebrae. On which surface of the neck would you expect to find extensors of the head and neck *(Anterior? Posterior?)*? Note that these muscles are the most superior muscles of the columns of extensors of the vertebrae. Find them on Figure 11.19, page 362. Now name three of these muscles.

✓ **C12.** *The Big Picture: Looking Ahead.* Refer to Figure 14.5 of your text, as well as Table 14.3, pages 459 and 485–489, and summarize head and neck muscle innervation by cranial nerves. Write the correct cranial nerve numbers next to the names of muscles innervated. Select from answers in the box. All except one answer will be used.

III.	**Oculomotor**	**VII.**	**Facial**
IV.	**Trochlear**	**IX.**	**Glossopharyngeal**
V(mand).	**Trigeminal (mandibular branch)**	**XI.**	**Spinal accessory**
VI.	**Abducens**	**XII.**	**Hypoglossal**

________ a. Most extrinsic eye muscles

________ b. Superior oblique eye muscle

________ c. Lateral rectus eye muscle

________ d. Upper eyelid (raises it)

________ e. Muscles of mastication (chewing), such as temporalis and masseter

________ f. Muscles that move the tongue

________ g. Muscles of facial expression, such as smiling, laughing, and frowning, as well as muscles used in speaking, pouting, or sucking

________ h. Sternocleidomastoid muscle that causes flexion of the head

D. Principal skeletal muscles that act on the abdominal wall, muscles used in breathing, muscles of the pelvic floor and perineum (pages 332–341)

✓ **Dl.** Each half of the abdominal wall is composed of *(two? three ? four?)* muscles. Describe these in the exercise below.

a. Just lateral to the midline is the rectus abdominis muscle. Its fibers are *(vertical? horizontal?)*, attached inferiorly to the ____________________ and superiorly to the ____________________. Contraction of this muscle permits *(flexion? extension?)* of the vertebral column.

b. List the remaining abdominal muscles that form the sides of the abdominal wall from most superficial to deepest.

____________________ → ____________________ → ____________________
most superficial deepest

c. Do all three of these muscles have fibers running in the same direction? *(Yes? No?)* Of what advantage is this?

✓ **D2.** Answer these questions about muscles used for breathing.

a. The diaphragm is ____________________-shaped. Its oval origin is located ____________________ ____________________. Its insertion is not into bone, but rather into dense connective tissue forming the roof of the diaphragm; this tissue is called the ____________________.

b. Contraction of the diaphragm flattens the dome, causing the size of the thorax to *(increase? decrease?)*, as occurs during *(inspiration? expiration?)*.

c. The name intercostals indicates that these muscles are located ____________________. Which set is used during expiration? *(Internal? External?)*

D3. *For extra review.* Cover the key to Figure LG 11.2a and write labels for muscles F, G, H, I, and X.

D4. Look at the inferior of the human pelvic bones on a skeleton (or refer to Figure 8.9 in your text). Note that a gaping hole (outlet) is present. Pelvic floor muscles attach to the bony pelvic outlet. Name these muscles and state functions of the pelvic floor.

✓ **D5.** Fill in the blanks in this paragraph. *Diaphragm* means literally ____________________ *(dia-)* ____________________ *(-phragm)*. You are familiar with the diaphragm that separates the ____________________ from the abdomen. The *pelvic diaphragm* consists of all of the muscles of the ____________________ floor plus their fasciae. It is a "wall" or barrier between the inside and outside of the body. Name the largest muscle of the pelvic floor. ____________________

E. Principal skeletal muscles that move the shoulder girdle and upper limb (pages 342–358)

✓ **E1.** From the list in the box, select the names of all muscles that move the shoulder girdle as indicated. Write one answer on each line provided.

LS.	**Levator scapulae**	**SA.**	**Serratus anterior**
RMM.	**Rhomboideus major and minor**	**T.**	**Trapezius**
PM.	**Pectoralis minor**		

a. Superiorly (elevation) __________ __________ __________

b. Inferiorly (depression) __________ __________

c. Toward vertebrae (adduction) __________ __________

d. Away from vertebrae (abduction) __________

✓ **E2.** Note on Figure LG 11.2a and b that the only two muscles listed in Checkpoint E1 that are superficial are the __________ and a small portion of the __________. The others are all deep muscles.

✓ **E3.** On Figure LG 11.2a and b identify the pectoralis major, deltoid, and latissimus dorsi muscles. All three of these muscles are *(superficial? deep?)*. They are all directly involved with movement of the *(shoulder girdle? humerus? radius/ulna?)*

Points of origin or insertion		**Actions (of humerus)**	
C.	Clavicle	Ab.	Abducts
H.	Humerus	Ad.	Adducts
I.	Ilium	EH.	Extension, hyperextension
RC.	Ribs or costal cartilages	F.	Flexion
Sc.	Scapula		
St.	Sternum		
VS.	Vertebrae and sacrum		

✓ **E4.** Fill in letters for *all points of origin and insertion* and *all actions* that apply for each muscle. Numbers in parentheses indicate the number of answers required in each case.

Muscles	**Points of Origin or Insertion**	**Actions (of Humerus)**
a. Pectoralis major	__________ (4)	__________ (2)
b. Deltoid	__________ (3)	__________ (3)
c. Latissimus dorsi	__________ (4)	__________ (2)

✓ **E5.** *The Big Picture: Looking Back.* Combine your knowledge of muscle actions with your knowledge of movements at joints from Chapter 9. Return to Figure LG 9.2, page LG 166. Write the name(s) of one or two muscles that produce each of the actions (a)-(h). Write muscle names next to each figure.

✓ **E6.** Complete the table describing three muscles that move the forearm.

Muscle name	Origin	Insertion	Action on Forearm
a.		Radial tuberosity	
b. Brachialis			
c.			Extension

✓ **E7.** Complete this exercise about muscles that move the wrist and fingers.

a. Examine your own forearm, palm, and fingers. There is more muscle mass on the *(anterior? posterior?)* surface. You therefore have more muscles that can *(flex? extend?)* your wrist and fingers.

b. Locate the flexor carpi ulnaris muscle on Figure 11.17a your text. What action does it have other than flexion of the wrist? ____________________ What muscles would you expect to abduct the wrist?

c. What is the difference in location between flexor digitorum superficialis and flexor digitorum profundus?

d. What muscle helps you to point (extend) your index finger? ____________________

✓ **E8.** For review of more details of the structures of the hand, do this exercise.

a. The *(extensor retinaculum? flexor retinaculum?)* is located over the palmar surface of the hand. This ligament secures the position of a number of long tendons of the wrist and digits, as well as the ____________________ nerve. Inflammation of these tissues by repetitive wrist flexion may lead to carpal tunnel syndrome.

b. The flexor pollicis longus is a muscle that flexes the *(thumb? middle finger? little finger?)*.

c. Which action of the thumb is movement of the thumb medially across the palm? (*Hint:* Refer to Figure 11.18.) *(Abduction? Adduction? Extension? Flexion? Opposition?)* Which action allows you to touch your thumb to the tip your of your little finger?

____

d. Extensor digitorum muscles extend the *(wrist? fingers?)*.

e. Thenar and hypothenar muscles are *(extrinsic? intrinsic?)* muscles of the hand. Thenar muscles form the pronounced muscle mass of the hand that moves the *(thumb? little finger?)*, whereas hypothenar muscles move the ____________________.

✓ **E9.** *For extra review* of muscles that move the upper limbs, write the name of one or more muscles that fit these descriptions.

a. Covers most of the posterior of the humerus: ______

b. Turns your hand from palm down to palm up position: ______

c. Originates from upper eight or nine ribs; inserts on scapula; moves scapula laterally:

______

d. Used when a baseball is grasped: ______

e. Antagonist to serratus anterior: ______

f. Largest muscle of the chest region; used to throw a ball in the air (flex humerus) and to adduct arm: ______

g. Raises or lowers scapula, depending on which portion of the muscle contracts:

______

h. Controls action at the elbow for a movement such as the downstroke in hammering a nail ______

i. Hyperextends the humerus, as in doing the "crawl" stroke in swimming or exerting a downward blow; also adducts the humerus: ______

F. Principal skeletal muscles that move the vertebral column (pages 359–363)

✓ **F1.** Describe the muscles that comprise the sacrospinalis. Locate and label on Figure LG 11.2b.

a. The sacrospinalis muscle is also called the ______.

b. The muscle consists of three groups: ______ (lateral), ______ (intermediate), and ______ (medial).

c. In general, these muscles have attachments between ______.

d. They are *(flexors? extensors?)* of the vertebral column, and so are *(synergists? antagonists?)* of the rectus abdominis muscles.

✓ **F2.** *Critical thinking.* Explain why it is common for women in their final weeks of pregnancy to experience frequent back pains.

✓ **F3.** Circle the correct origin and insertion of the scalene muscles:

A. Ribs—iliac crest

B. Cervical vertebrae—occipital and temporal bones

C. Cervical vertebrae—first two ribs

D. Thoracic vertebrae—sacrum

G. Principal skeletal muscles that move the lower limb (pages 364–379)

G1. Cover the key in Figure LG 11.2a and b and identify by size, shape, and location the major muscles that move the lower limb.

✓ **G2.** Now match muscle names in the box with their descriptions below.

Ad.	**Adductor group**	**Ham.**	**Hamstrings**
Gas.	**Gastrocnemius**	**Il.**	**Iliopsoas**
GMax.	**Gluteus maximus**	**QF.**	**Quadriceps femoris**
GMed.	**Gluteus medius**	**Sar.**	**Sartorius**

________ a. Consists of four heads: rectus femoris and three vastus muscles (lateralis, medialis, and intermedius)

________ b. This muscle mass lies in the posterior (flexor) compartment of the thigh; antagonist to quadriceps femoris

________ c. Attached to lumbar vertebrae, anterior of ilium, and lesser trochanter, it crosses anterior to hip joint

________ d. Large muscle mass of the buttocks; antagonist to the iliopsoas

________ e. The only one of these muscles located in the leg (between knee and ankle), it forms the "calf;" attaches to calcaneus by "Achilles tendon"

________ f. Forms the medial compartment of the thigh; moves femur medially

________ g. Crossing the femur obliquely, it moves lower extremity into "tailor position"

________ h. Located posterior to upper, outer portion of the ilium, it forms a preferred site for intramuscular (IM) injections

✓ **G3.** Complete the table by marking an X below each action produced by contraction of the muscles listed. (Some muscles will have two or three answers.)

Key to actions in table: Ab. Abduct; Ad. Adduct; EH, extend or hyperextend; F, flex

	Movements of Thigh (Hip Joint)				**Movement of Leg (Knee)**	
	Ab	Ad	EH	F	EH	F
a. Iliopsoas						
b. Gluteus maximus and medius						
c. Adductor mass						
d. Tensor fasciae latae						
e. Quadriceps femoris						
f. Hamstrings						
g. Gracilis						
h. Sartorius						

✓ **G4.** Refer to Figure LG 11.4.

a. Locate the major muscle groups of the right thigh in this cross section. Color the groups as indicated by color code ovals on the figure.

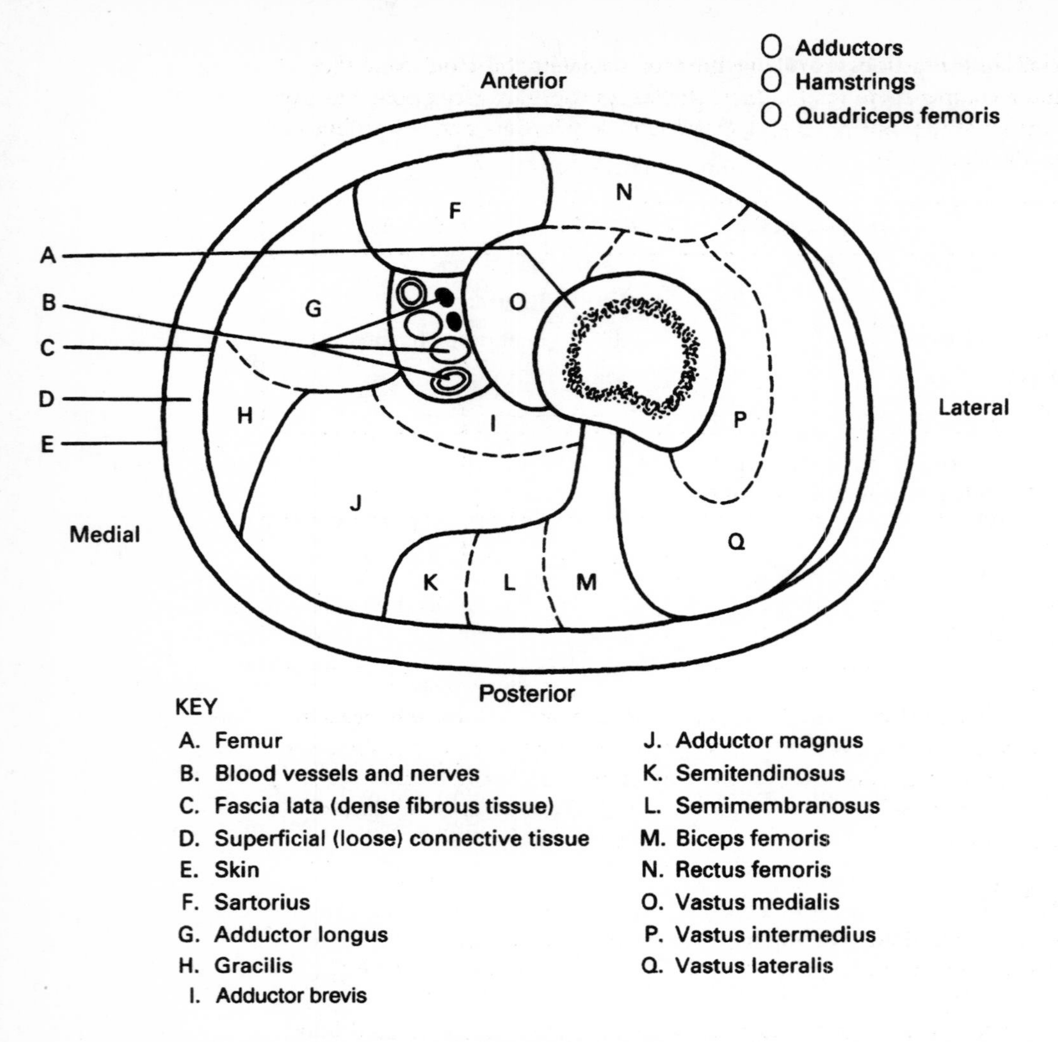

Figure LG 11.4 Cross section of the right thigh midway between hip and knee joints. Structures A to E are nonmuscle; F to Q are skeletal muscles of the thigh. Color as directed in Checkpoint G4.

b. Cover the key to Figure LG 11.4 and identify each muscle. Relate position of muscles in this figure to views of muscles in Figure LG 11.2a and b.

✓ **G5.** *The Big Picture: Looking Back.* What muscles cause the actions (i)-(p) shown in Figure LG 9.2 on page LG 167. Write the muscle names next to each diagram.

✓ **G6.** Perform these actions of your foot and toes. Feel which muscles are contracting. Then match names of actions with descriptions.

DF.	**Dorsiflex**	**F.**	**Flex toes**
Ev.	**Evert foot**	**In.**	**Invert foot**
Ex.	**Extend toes**	**PF.**	**Plantar flex**

_________ a. Jump, as if to touch ceiling

_________ b. Walk around on your heels

_________ c. Curl toes down

_________ d. Lift toes upward away from floor

_________ e. Move sole of foot medially

_________ f. Move sole of foot laterally

✓ **G7.** To review details of leg muscles that move the foot, complete this table. Note that muscles within a compartment tend to have similar functions. Muscles with * flex or extend only the great toe (not all toes). Use the same key for foot and toe actions as in the box for Checkpoint G6.

	DF PF	**In Ev**	**F Ex**
Posterior compartment: a. Gastrocnemius and soleus b. Tibialis posterior c. Flexor digitorum longus d. Flexor hallucis longus*			
Lateral compartment: e. Peroneus (longus and brevis)			
Anterior compartment: f. Tibialis anterior g. Extensor digitorum longus h. Extensor hallucis longus*			

✓ **G8.** *For extra review* of all muscles, color flexors and extensors using color code ovals on Figure LG 11.2a and b. This activity will allow you to see on which sides of the body most muscles with those actions are located. Omit plantar flexors and dorsiflexors. Note that some muscles are flexors *and* extensors-at different joints. (See Mastery Test question 23 also.)

H. Disorders (page 380)

✓ **H1.** Answer these questions about running injuries.

a. The most common site of injury for runners is the *(calcaneal tendon? groin? hip? knee?)*.

b. *Patellofemoral stress syndrome* is a technical term for ____________________. Briefly describe this problem.

c. *Shinsplint syndrome* refers to soreness along the *(patella? tibia? fibula?)*.

d. Write several suggestions you might make to a beginning runner to help to avoid runners' injuries.

e. Initial treatment of sports injuries usually calls for *RICE* therapy. To what does RICE therapy refer? R ____________________ I ____________________

C ____________________ E ____________________

H2. Write three or more possible causes of compartment syndrome. Then explain how this syndrome may lead to contractures.

ANSWERS TO SELECTED CHECKPOINTS: CHAPTER 11

A1. Skeletal muscle tissues and connective tissues.

A2. (a) Lever, fulcrum. (b) Load. (c) Muscle; effort. (d) Proximal, radius. (e) Origin, scapula. (f) See Figure LG 11.1A; third.

A3.

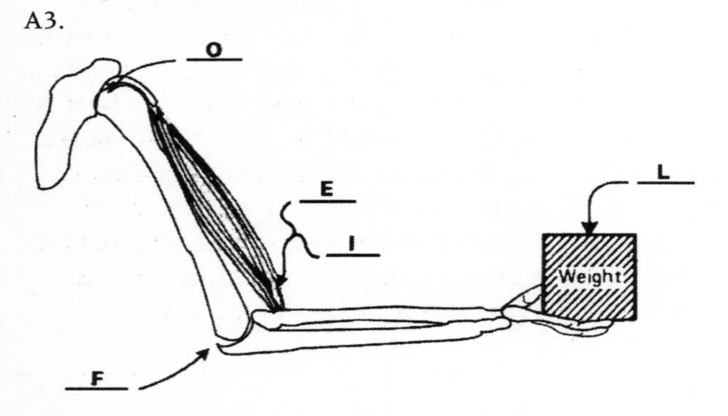

Figure LG 11.1A The lever-fulcrum principle is illustrated by flexion of the forearm.

Class	Analogy of EFL	Arrangement	Mechanical advantage: (Yes, No). Explain.	Example in human body
a. First	Seesaw	E F L	Yes if E is closer to F No if E is closer to F	Head and neck: atlanto-occipital joint
b. Second	Wheelbarrow	F L E	Yes since E is closer (than E) to F	None
c. Third	(Omit)	F E L	No since E is closer (than E) to F	Elbow joint: biceps brachii

A4. (a) L, E; atlas, occipital; first. (b) Less, more, closer to; A.

A5. (a) Many; pennate. (b) Long; parallel.

A6. (a) Prime mover (agonist); biceps brachii. (b) An antagonist. (c) Your forearm would stay in the flexed position. (d) None: each opposing muscle would negate the action of the other. (e) Synergists.

B1. (a) D. (b) E. (c) C. (d) B. (e) A. (f) F.

B2. (b) A, P, L. (c) N, L. (d) P. (e) A, S.

Cl. (a) Frontalis portion of epicranius. (b) Orbicularis oris. (c) Zygomaticus major. (d) Orbicularis oculi.

C2. Facial, VII.

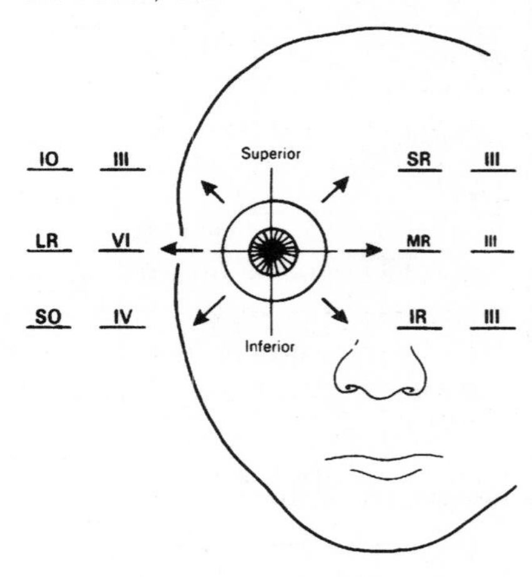

Figure LG 11.3A Right eyeball with arrows indicating directions of eye movements produced by the extrinsic eye muscles.

C3. (a) They are outside of the eyeballs, not intrinsic like the iris. (b) and (c) See Figure LG 11.3A. (d) III; lateral.

C4. No. The left eye contracts its medial rectus, while the right eye uses its lateral rectus and exerts some tension upon both oblique muscles.

C5. (a) Elevating the mandible; temporalis, masseter. (b) Close; lateral pterygoid. (c) V, trigeminal (d) Chewing, facial expression.

C6. (a) Posterior and superior. (b) Inferior. (c) Anterior and inferior.

C7. Same answers as for C6.

C8. XII, hypoglossal.

C9. (a) B. (b) A.

C10. Opposite; flexes.

C11. Extending and then hyperextending; posterior; longissimus capitis, semispinalis capitis, and splenius capitis.

C12. (a) III. (b) IV: remember SO_4. (c) VI. (d) III. (e) V (mand). (f) XII. (g) VII. (h) XI.

Dl. Four. (a) Vertical, pubic crest and symphysis pubis, ribs 5 to 7; flexion. (b) External oblique → internal oblique → transversus abdominis. (c) No; strength is provided by the three different directions.

D2. (a) Dome; around the bottom of the rib cage and on lumbar vertebrae; central tendon. (b) Increase, inspiration. (c) Between ribs (costa); internal.

D5. Across, wall; thorax; pelvic; levator ani (pubococcygeus and iliococcygeus).

El. (a) LS, RMM, upper fibers of T. (b) PM, lower fibers of T. (c) RMM, T. (d) SA.

E2. Trapezius, serratus anterior.

E3. Superficial; humerus.

E4. (a) C, St, RC, H; Ad, F. (b) C, H, Sc; Ab, EH (posterior fibers), F (anterior fibers). (c) I, RC, VS, H; Ad, EH.

E5. (a) Sternocleidomastoid. (b) Three capitis muscles. (c) Pectoralis major, coracobrachialis, deltoid (anterior fibers), biceps brachii. (d) Latissimus dorsi, teres major, deltoid (posterior fibers), triceps brachii. (e) Deltoid and supraspinatus. (f) Pectoralis major, latissimus dorsi, teres major. (g) Biceps brachii, brachialis, brachioradialis. (h) Triceps brachii.

E6.

Muscle name	Origin	Insertion	Action on Forearm
a. Biceps brachii	Scapula (2 sites)	Radial tuberosity (Anterior)	Flexion, supination (*Note:* also flexes humerus)
b. Brachialis	Anterior humerus	Ulna (coronoid process)	Flexion
c. Triceps brachii	Scapula and 2 sites on posterior of humerus	Posterior of ulna (olecranon)	Extension (also extension of humerus)

E7. (a) Anterior; flex. (b) Adducts wrist; those lying over radius, such as flexor and extensor carpi radialis. (c) Superficialis is more superficial, and profundus lies deep. (d) Extensor indicis.

E8. (a) Flexor retinaculum; median. (b) Thumb. (c) Flexion; opposition. (d) Fingers (digits). (e) Intrinsic; thumb, little finger.

E9. (a) Triceps brachii. (b) Supinator and biceps brachii. (c) Serratus anterior. (d) Flexor digitorum superficialis and profundus. (e) Trapezius. (f) Pectoralis major. (g) Trapezius. (h) Triceps brachii. (i) Latissimus dorsi.

Fl. (a) Erector spinae. (b) Iliocostalis, longissimus, spinalis. (c) Hipbone (ilium), ribs and vertebrae. (d) Extensors, antagonists.

F2. Extra weight of the abdomen demands extra support (contraction) by sacrospinalis muscles.

F3. C.

G2. (a) QF. (b) Ham. (c) Il. (d) GMax. (e) Gas. (f) Ad. (g) Sar. (h) Gmed.

G3.

	Movements of Thigh (Hip Joint)				Movement of Leg (Knee)	
	Ab	Ad	EH	F	EH	F
a. Iliopsoas				X		
b. Gluteus maximus and medius			X			
c. Adductor mass		X		X		
d. Tensor fasciae latae	X			X		
e. Quadriceps femoris				X	X	
f. Hamstrings			X			X
g. Gracilis		X				X
h. Sartorius	X			X		X

G4. (a) Adductors: G H I J; Hamstrings: K L M; Quadriceps: N O P Q. (b) See key to Figure LG 11.4.

G5. (i) Iliacus + psoas (iliopsoas), rectus femoris, adductors, sartorius; (j) Gluteus maximus, hamstrings, adductor magnus (posterior portion); (k) Hamstrings, gracilis, sartorius, gastrocnemius; (1) Quadriceps femoris; (m) Same as i but bilateral; (n) Same as j but bilateral; (o) Tensor fasciae latae, gluteus (medius and minimus), piriformis, superior and inferior gemellus, and obturator internus; (p) Adductors (longus, magnus, and brevis), pectineus, quadratus femoris, and gracilis.

G6. (a) PF. (b) DF. (c) F. (d) Ex. (e) In. (f) Ev.

G7.

	DF	PF	In	Ev	F	Ex
Posterior compartment:						
a. Gastrocnemius and soleus		X				
b. Tibialis posterior		X	X			
c. Flexor digitorum longus		X	X		X	
d. Flexor hallucis longus*		X	X		X	
Lateral compartment:						
e. Peroneus (longus and brevis)		X		X		
Anterior compartment:						
f. Tibialis anterior	X		X			
g. Extensor digitorum longus	X			X		X
h. Extensor hallucis longus*	X		X			X

G8. Figure LG 11.2a: flexors: G, J, K, S, T (anterior portion), U, V, W, AA, BB, EE; extensors: BB, CC, DD. Figure LG 11.2b: flexors: J, K, S; extensors: F, G, H, I, J, K, L, M (posterior fibers), N, O, R.

H1. (a) Knee. (b) "Runner's knee"; the patella tracks (glides) laterally, causing pain. (c) Tibia. (d) Replace worn shoes with new, supportive ones; do stretching and strengthening exercises; build up gradually; get proper rest; consider other exercise if prone to leg injuries because most (70%) runners do experience some injuries. (e) Rest, ice, compression (such as by elastic bandage), and elevation (of the injured part).

MORE CRITICAL THINKING: CHAPTER 11

1. Explain the kinds of information conveyed about muscles by examining muscle names. Give at least six examples of muscle names that provide clues to muscle locations, actions, size or shape, direction of fibers, numbers of origins, or points of attachment.
2. Identify types of movements required at shoulder, elbow, and wrist as you perform the action of tossing a tennis ball upward. Indicate groups of muscles that must contract and relax to accomplish this action.
3. Perform the actions of abducting and adducting your right thigh. Contrast the amount of muscle mass contracting to effect each of these movements. Name several muscles contracting as you carry out each action.

MASTERY TEST: CHAPTER 11

Questions 1–2: Arrange the answers in correct sequence.

______ ______ ______ 1. Abdominal wall muscles, from superficial to deep:
A. Transversus abdominis
B. External oblique
C. Internal oblique

______ ______ ______ 2. From superior to inferior in location:
A. Sternocleidomastoid
B. Pelvic diaphragm
C. Diaphragm and intercostal muscles

Questions 3–12: Circle T (true) or F (false). If the statement is false, change the underlined word or phrase so that the statement is correct

T F 3. In extension of the thigh the hip joint serves as the <u>fulcrum (F)</u>, while the <u>hamstrings and gluteus maximus</u> serve as the effort (E).

T F 4. A scissors and the capitis muscles are both examples of the action of <u>second</u>-class levers.

T F 5. The hamstrings are <u>antagonists</u> to the quadriceps femoris.

T F 6. The name deltoid is based on the <u>action</u> of that muscle.

T F 7. The most important muscle used for normal breathing is the <u>diaphragm</u>.

T F 8. The insertion end of a muscle is the attachment to the bone that <u>does move</u>.

T F 9. In general, adductors (of the arm and thigh) are located more on the <u>medial</u> than on the <u>lateral</u> surface of the body.

T F 10. Both the <u>pectoralis major and latissimus dorsi muscles extend</u> the humerus.

T F 11. The capitis muscles (such as splenius capitis) are <u>extensors</u> of the head and neck.

T F 12. The biceps brachii and biceps femoris are both muscles with <u>two heads of origin located on the arm</u>.

Questions 13–20: Circle the letter preceding the one best answer to each question.

13. All of these muscles are located in the lower limb *except:*
A. Hamstrings
B. Gracilis
C. Tensor fasciae latae
D. Deltoid
E. Peroneus longus

14. All of these muscles are located on the anterior of the body *except:*
A. Tibialis anterior
B. Rectus femoris
C. Sacrospinalis
D. Pectoralis major
E. Rectus abdominis

15. Contraction of all of the following muscles causes extension of the leg *except:*
A. Rectus femoris
B. Biceps femoris
C. Vastus medialis
D. Vastus intermedius
E. Vastus lateralis

16. All of these muscles are attached to ribs *except:*
A. Serratus anterior
B. Intercostals
C. Trapezius
D. Iliocostalis
E. External oblique

17. All of these muscles are directly involved with movement of the scapulae except:
    A. Levator scapulae
    B. Pectoralis major
    C. Pectoralis minor
    D. Rhomboideus major
    E. Serratus anterior
18. All of these muscles have attachments to the hip bones *except:*
    A. Adductor muscles (longus, magnus, brevis)
    B. Biceps femoris
    C. Rectus femoris
    D. Vastus medialis
    E. Latissimus dorsi
19. The masseter and temporalis muscles are used for:
    A. Chewing
    B. Pouting
    C. Frowning
    D. Depressing tongue
    E. Elevating tongue
20. All of these muscles are used for facial expression *except:*
    A. Zygomaticus major
    B. Orbicularis oculi
    C. Platysma
    D. Rectus abdominis
    E. Mentalis

Questions 21–25: Fill-ins. Refer to Figure LG 11.2a and b on pages LG 206–207. Write the word or phrase or key letters of muscles that best complete the statement or answer the questions.

__________________ 21. Muscles G, S, U, V, and W on Figure LG 11.2a all have in common the fact that they carry out the action of ______.

__________________ 22. Muscles G, N, 0, and R on Figure LG 11.2b all have in common the fact that they carry out the action ______.

__________________ 23. If you colored all flexors red and all extensors blue on these two figures, the view of the ______ surface of the body would appear more blue.

__________________ 24. Choose the letters of all of the muscles listed below that would contract as you raise your left arm straight in front of you, as if to point toward a distant mountain: Figure LG 11.2a: D T U V; Figure LG 11.2b: N O P.

__________________ 25. Choose the letters of all of the muscles listed below that would contract as you raise your knee and extend your leg straight out in front of you, as if you are starting to march off to the distant mountain: Figure LG 11.2a: AA BB CC DD; Figure LG 11.2b; I J K R.

ANSWERS TO MASTERY TEST: CHAPTER 11

Arrange

1. B C A
2. A C B

True-False

3. T
4. F. *First*
5. T
6. F. Shape
7. T
8. T
9. T
10. F. The latissimus dorsi extends, but the pectoralis major flexes.
11. T
12. F. Two heads of origin; but origins of biceps brachii are on the scapula, and origins of biceps femoris are on ischium and femur.

Multiple Choice

13. D
14. C
15. B
16. C
17. B
18. D
19. A
20. D

Fill-ins

21. Flexion
22. Extension
23. Posterior
24. Figure LG 11.2a: T (anterior fibers) U V; Figure LG 11.2b: none.
25. Figure LG 11.2a: AA BB CC DD; Figure LG 11.2b: I.

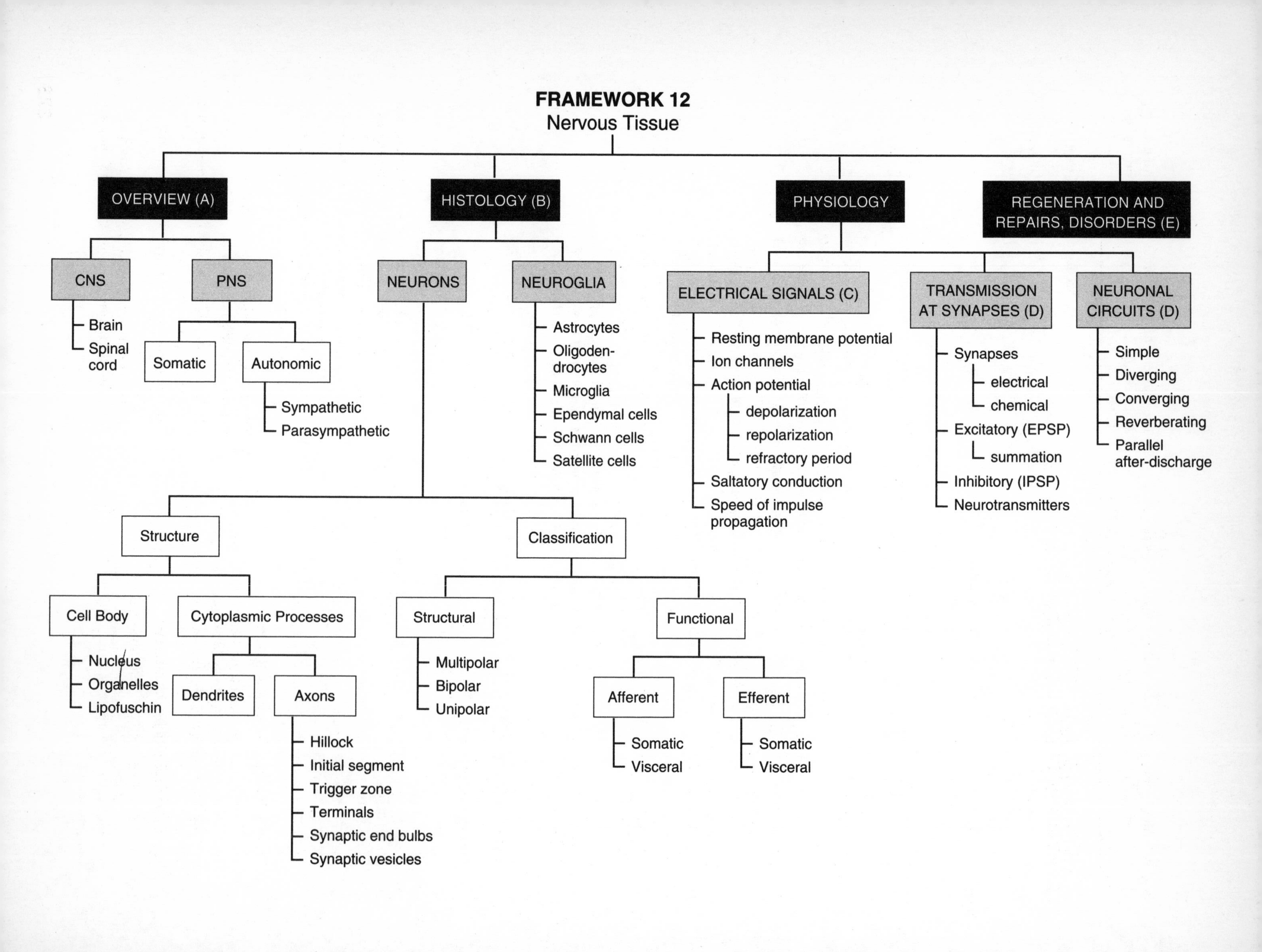

FRAMEWORK 12
Nervous Tissue
OVERVIEW (A)
CNS
Brain
Spinal cord
PNS
Somatic
Autonomic
Sympathetic
Parasympathetic
HISTOLOGY (B)
NEURONS
Structure
Cell Body
Nucleus
Organelles
Lipofuschin
Cytoplasmic Processes
Dendrites
Axons
Hillock
Initial segment
Trigger zone
Terminals
Synaptic end bulbs
Synaptic vesicles
Classification
Structural
Multipolar
Bipolar
Unipolar
Functional
Afferent
Somatic
Visceral
Efferent
Somatic
Visceral
NEUROGLIA
Astrocytes
Oligoden-drocytes
Microglia
Ependymal cells
Schwann cells
Satellite cells
PHYSIOLOGY
ELECTRICAL SIGNALS (C)
Resting membrane potential
Ion channels
Action potential
depolarization
repolarization
refractory period
Saltatory conduction
Speed of impulse propagation
TRANSMISSION AT SYNAPSES (D)
Synapses
electrical
chemical
Excitatory (EPSP)
summation
Inhibitory (IPSP)
Neurotransmitters
NEURONAL CIRCUITS (D)
Simple
Diverging
Converging
Reverberating
Parallel after-discharge
REGENERATION AND REPAIRS, DISORDERS (E)

C H A P T E R 12

Nervous Tissue

Two systems—nervous and endocrine—are responsible for regulating our diverse body functions. Each system exerts its control with the help of specific chemicals, namely neurotransmitters (nervous system) and hormones (endocrine system). In Unit III both systems of regulation will be considered, starting with the tissue of the nervous system.

Nervous tissue consists of two types of cells: neurons and neuroglia. The name neuroglia (*glia* = glue) offers a clue as to function of these cells: They bind, support, and protect neurons. Neurons perform the work of transmitting nerve impulses. These long and microscopically slender cells sometimes convey information several feet along a single neuron. Their function relies on an intricate balance between ions (Na^+ and K^+) found in and around nerve cells. Neurons release neurotransmitters that bridge the gaps between adjacent neurons and at nerve-muscle or nerve-gland junctions. Analogous to a complex global telephone system, the nervous tissue of the human body boasts a design and organization that permits accurate communication, coordination, and integration of virtually all thoughts, sensations, and movements.

As you begin your study of nervous tissue, carefully examine the Chapter 12 Topic Outline and Objectives; check off each one as you complete it. To organize your study of nervous tissue, glance over the Chapter 12 Framework now. Be sure to refer to the Framework frequently and note relationships among key terms in each section.

TOPIC OUTLINE AND OBJECTIVES

A. Overview of the nervous system

1. List the structures and basic functions of the nervous system.
2. Describe the organization of the nervous system.

B. Histology of nervous tissue

1. Contrast the histological characteristics and the functions of neurons and neuroglia.
2. Distinguish between gray matter and white matter.

C. Electrical signals in neurons

1. Describe the cellular properties that permit communication among neurons and effectors.
2. Compare the basic types of ion channels, and explain how they relate to action potentials and graded potentials.
3. Describe the factors that maintain a resting membrane potential.
4. List the sequence of events that generate an action potential.

D. Signal transmission at synapses; neurotransmitters; circuits

1. Explain the events of signal transmission at a chemical synapse.
2. Distinguish between spatial and temporal summation.
3. Give examples of excitatory and inhibitory neurotransmitters and describe how they act.

4. Describe the classes and functions of neurotransmitters.
5. Identify the various types of neuronal circuits in the nervous system.

E. Regeneration and repair of nervous tissue; disorders

1. Describe plasticity and neurogenesis.
2. Describe the events in damage and repair of peripheral nerves.

WORDBYTES

Now become familiar with the language of this chapter by studying each wordbyte, its meaning, and an example of its use within a term. After you study the entire list, self-check your understanding by writing the meaning of each wordbyte on the line. As you continue through the *Learning Guide*, identify (and fill in) additional terms that contain the same wordbyte.

Wordbyte	Self-check	Meaning	Example(s)
af-	____________	toward	*af*ferent
astro-	____________	star	*astro*glia
auto-	____________	self	*auto*matic
dendr-	____________	little tree	*dendr*ite
ef-	____________	away from	*ef*ferent
enter-	____________	intestines	*enter*ic nervous system
-ferent	____________	carried	ef*ferent*
ganglia	____________	swelling, knot	*ganglia*
-glia	____________	glue	neuro*glia*
-lemm	____________	sheath	neuro*lemma*
neuro-	____________	nerve	*neuro*n
olig-	____________	few	*olig*odendrocytes
saltat-	____________	leaping	*saltat*ory
sclera-	____________	hard	multiple *scler*osis
-soma-	____________	body	axo*soma*tic
syn-	____________	together	*syn*apse

CHECKPOINTS

A. Overview of the nervous system (pages 386–388)

A1. Contrast functions of the nervous system and the endocrine system in maintaining homeostasis.

A2. Define *neurology.*

✓ **A3.** Match the components of the nervous system in the box with descriptions below.

CN.	**Cranial nerves**	**SC.**	**Spinal cord**
B.	**Brain**	**SN.**	**Spinal nerves**
EP.	**Enteric plexuses**	**SR.**	**Sensory receptors**
G.	**Ganglia**		

________ a. Largest part of the nervous system; housed within the cranium

________ b. Twelve pairs of nerves attached to the base of the brain

________ c. Thirty-one pairs of nerves attached to the spinal cord

________ d. Structure that extends inferiorly from foramen magnum within the vertebral canal

________ e. Clusters of neuron cell bodies located outside the brain and spinal cord

________ f. Networks of neurons within walls of GI organs

________ g. Dendrites of neurons or complex clusters of cells that detect changes in environment

✓ **A4.** List the three basic functions of the nervous system.

✓ **A5.** Check your understanding of the organization of the nervous system by selecting answers that best fit descriptions below.

Aff.	**Afferent**	**ENS.**	**Enteric nervous system**
ANS.	**Autonomic nervous system**	**PNS.**	**Peripheral nervous system**
CNS.	**Central nervous system**	**SNS.**	**Somatic nervous system**
Eff.	**Efferent**		

________ a. Brain and spinal cord

________ b. Sensory nerves

________ c. Carry information from CNS to skeletal muscles

________ d. Consists of sympathetic and parasympathetic divisions

________ e. Nerves that convey impulses to smooth muscle, cardiac muscle, and glands; involuntary

________ f. Cranial nerves, spinal nerves, ganglia, and sensory receptors

________ g. "Brain of the gut"

B. Histology of nervous tissue (pages 388–395)

✓ **B1.** On Figure LG 12.1, label all structures with leader lines. Next, draw arrows beside the figure to indicate direction of nerve impulses. Then color structures to match color code ovals.

✓ **B2.** Match the parts of a neuron listed in the box with the descriptions below.

A.	**Axon**	**M.**	**Mitochondria**
CB.	**Cell body (soma)**	**NB.**	**Nissl bodies**
D.	**Dendrite**	**NF.**	**Neurofibrils**
L.	**Lipofuscin**	**T.**	**Trigger zone**

_________ a. Contains nucleus; cannot regenerate

_________ b. Yellowish brown pigment that increases with age; a product of lysosomes

_________ c. Provide energy for neurons

_________ d. Long, thin filaments that provide support and shape for the cell

_________ e. Orderly arrangement of rough ER; site of protein synthesis

_________ f. Conducts impulses toward cell body

_________ g. Conducts impulses away from cell body; has synaptic end bulbs that secrete neurotransmitter

_________ h. Located at junction of axon hillock and initial segment; nerve impulses arise here in many neurons

B3. *Clinical correlation.* Contrast the two types of axonal transport and tell which type is involved in the spread of the toxin that causes tetanus.

B4. Name and give a brief description of three types of neurons based on structural characteristics and three types of neurons classified according to functional differences.

Structural **Description**

_________________-polar ____

_________________-polar ____

_________________-polar ____

Functional **Description**

_________________-fferent ____

_________________-fferent ____

Interneurons ____

✓ **B5.** Refer to Checkpoint B4 above and place an asterisk next to the most common structural type of neuron in the brain and spinal cord.

✓ **B6.** Write *neurons* or *neuroglia* after descriptions of these cells.

a. Conduct impulses from one part of the nervous system to another: _______________

b. Provide support and protection for the nervous system: _______________

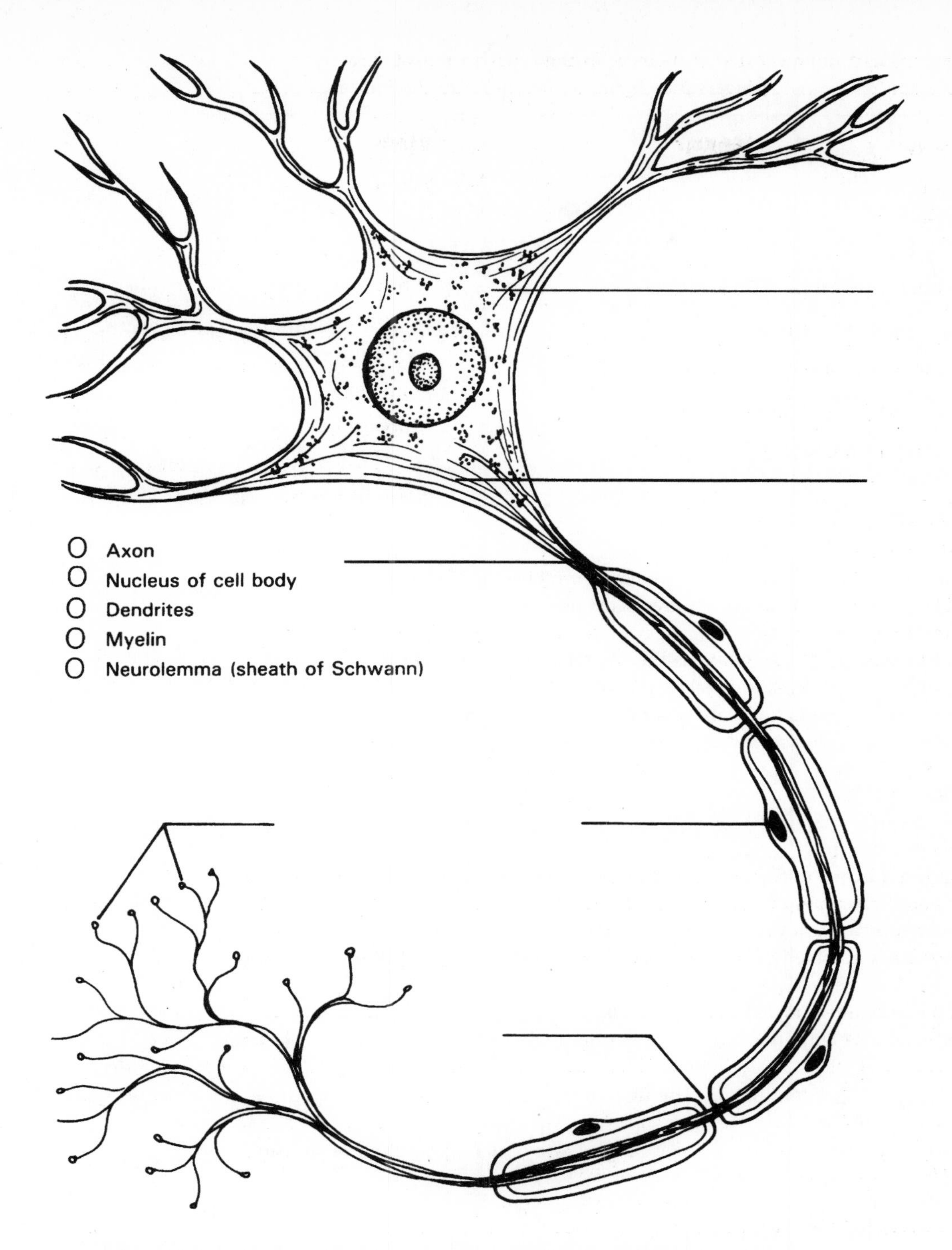

Figure LG 12.1 Structure of a typical neuron as exemplified by an efferent (motor) neuron. Complete the figure as directed in Checkpoint B1.

c. Bind nervous tissue to blood vessels, form myelin, and serve phagocytic functions:

____

d. Smaller in size, but much more abundant in number: ________________________

e. Can undergo cell division in the mature nervous system, possibly forming tumors:

____

✓ **B7.** Check your understanding of neuroglia by selecting types of neuroglia in the box that fit descriptions below. Answers may be used once or more than once.

A.	**Astrocytes**	**O.**	**Oligodendrocytes**
E.	**Ependymal cells**	**SAT.**	**Satellite cells**
M.	**Microglia**	**SCH.**	**Schwann**

_______ a. Neuroglia found in the PNS (two answers)

_______ b. Form myelin (two answers)

_______ c. Maintain proper balance of K^+ and Ca^{2+} needed for nerve impulses

_______ d. Help form the blood–brain barrier and provide nutrients to neurons

_______ e. Star-shaped cells

_______ f. Protective because they are phagocytic

_______ g. Line the fluid-filled spaces (ventricles) within the brain, and form cerebrospinal fluid (CSF) there

_______ h. Support neurons in clusters (ganglia) within the PNS

_______ i. Help regenerate PNS axons

✓ **B8.** Check your understanding of coverings over nerve fibers in this Checkpoint.

a. In the *(CNS? PNS?)*, myelination requires Schwann cells.

b. Myelin is formed of _______, whereas the neurolemma is formed of _______.
   A. Up to 100 layers of Schwann cell membrane
   B. The outer nucleated cytoplasmic layer of the Schwann cell

c. Arrange these layers from outermost to innermost: _______ _______ _______
   A. Axon
   B. Neurolemma
   C. Myelin sheath

d. A neurolemma is found _______.
   A. Only around axons in the CNS
   B. Only around axons in the PNS
   C. Around axons in both the CNS and the PNS

e. A neurolemma is involved in the process of _________________ of injured axons.

f. Nodes of Ranvier are sections of axons that lack _________________. Consequently, nerve impulses must "jump" from node to node; this process is called

_________________ conduction. The ultimate effect is that myelin arranged

between nodes causes a(n) ______-crease in the rate of nerve impulse transmission. (Hint. See page 402 in your text.)

g. Do any CNS axons have a neurolemma? *(Yes? No?)* State one consequence of this fact. (*Hint:* See page 412 your text.)

Do any CNS axons have a myelin sheath? *(Yes? No?)* What cells produce this?

____

✓ **B9.** *The Big Picture: Looking Ahead.* Refer to figures in your text to complete this Checkpoint. Write *G* for gray matter or *W* for white matter next to related descriptions.

_______ a. Consists mostly of myelinated nerve fibers (axons or dendrites) (Figures 12.3 and 13.10b.)

_______ b. Consists mostly of cell bodies containing gray Nissl bodies and unmyelinated nerve fibers, such as in ganglia (Figure 17.5)

_______ c. Forms the H-shaped inner portion of the spinal cord as seen in cross section (Figure 13.3a)

_______ d. Forms the outer part of the brain known as the cerebral cortex (Figure 14.11a)

_______ e. Forms nuclei (clusters of neuron cell bodies within the CNS) (Figures 14.9 and 14.10)

_______ f. Forms spinal nerves (Figure 13.16) and cranial nerves (Figure 14.5) in the PNS

C. Electrical signs in neurons (pages 395–404)

✓ **C1.** Complete this Checkpoint describing the properties of plasma membranes of cells, such as those of muscles or glands, that may be stimulated (excited) by neurons.

a. Excitable cells exhibit a membrane _______________ along which an electric charge (or _______________) may pass. In living cells, this current is usually carried by *(electrons? ions?)*.

b. Two types of potential may be developed: _______________ potential and _______________ potential. These potentials can develop because of the presence of _______________ channels that open or close to control ion flow. These channels provide the main path for flow of current across the membrane because the phospholipid bilayer normally permits *(little? much?)* passage of ions.

✓ **C2.** Do this exercise about roles of ion channels in development of membrane potentials.

a. There are two kinds of ion channels: *(gated? leakage?)* channels that randomly open and _______________ channels that require a stimulus to activate opening or closure. The plasma cell of a membrane is more permeable to *(K^+? Na^+?)* because there are more leakage channels for *(K^+? Na^+?)*.

b. Identify the type of gated channel that is responsive in each case described here. Choose from these answers: *(voltage? ligand? mechanically?)* gated channels.

1. Channels in plasma membranes of neurons and muscle cells: _______________
2. Channels in auditory receptors stimulated by sound waves in ears: _______________
3. Channels responsive to the neurotransmitter acetylcholine: _______________
4. Channels in neurons responsive to touch (in skin): _______________

✓ **C3.** Contrast direct vs. indirect change in membrane permeability to ions in this Checkpoint.

a. The neurotransmitter acetylcholine *(directly? indirectly?)* alters membrane permeability to ions such as Na^+, Ca^{2+}, or K^+.

b. Name chemicals involved in an *indirect* method to change membrane permeability.

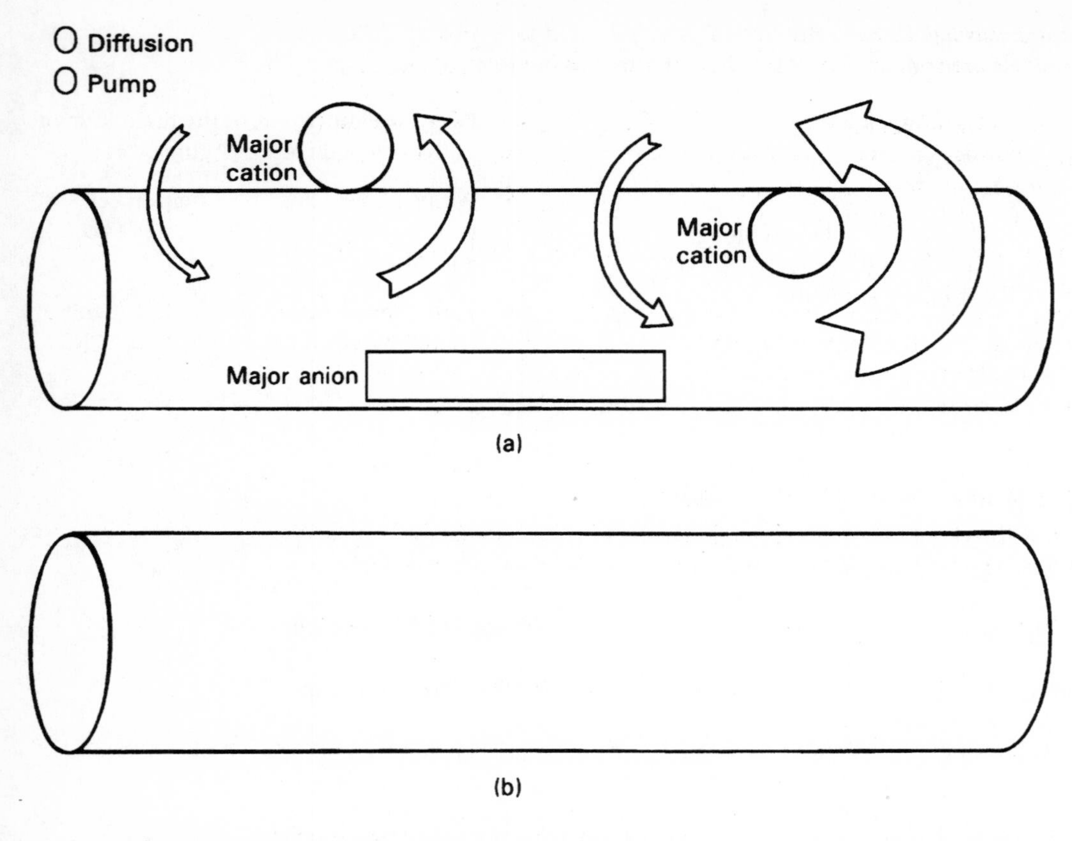

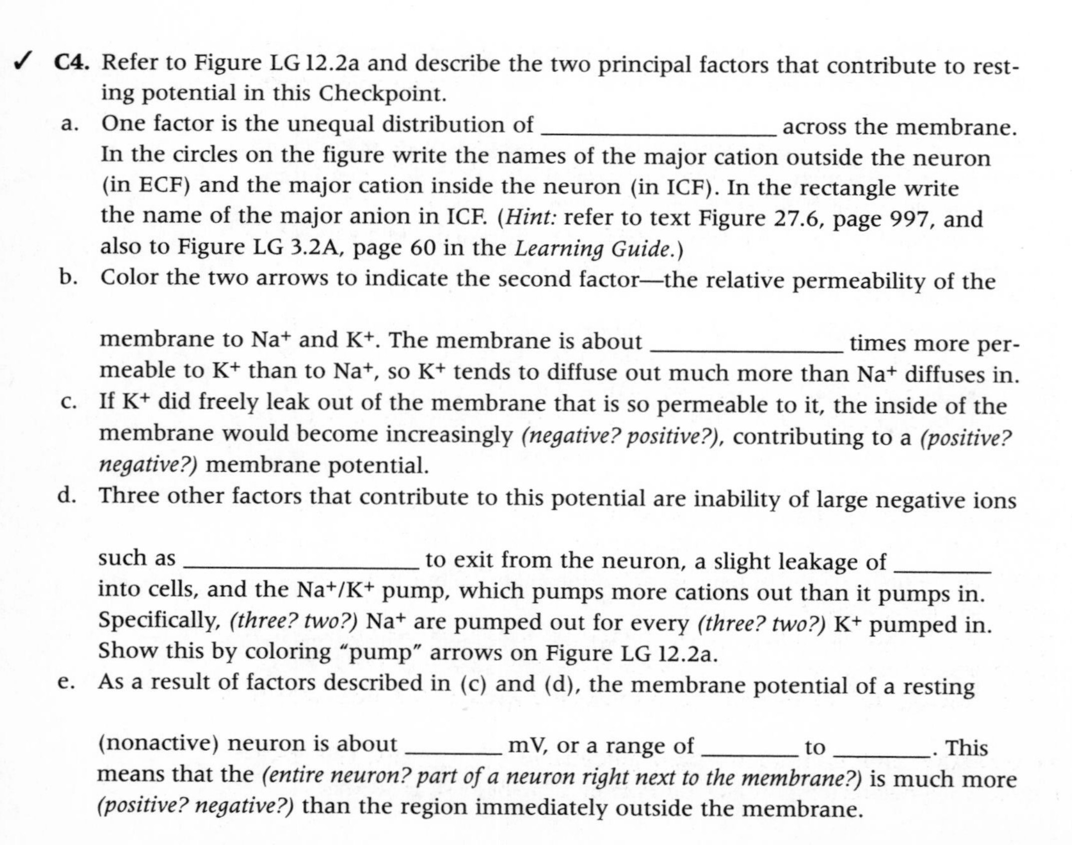

Figure LG 12.2 Diagrams of a nerve cell. (a) Factors that contribute to membrane potential of a resting neuron. Label and color as directed in checkpoint C4. (b) Complete as directed in Checkpoint C8. ECF–Extracellular fluid; ICF–Intracellular fluid.

✓ **C4.** Refer to Figure LG 12.2a and describe the two principal factors that contribute to resting potential in this Checkpoint.

a. One factor is the unequal distribution of ____________________ across the membrane. In the circles on the figure write the names of the major cation outside the neuron (in ECF) and the major cation inside the neuron (in ICF). In the rectangle write the name of the major anion in ICF. (*Hint:* refer to text Figure 27.6, page 997, and also to Figure LG 3.2A, page 60 in the *Learning Guide.*)

b. Color the two arrows to indicate the second factor—the relative permeability of the membrane to Na^+ and K^+. The membrane is about ________________ times more permeable to K^+ than to Na^+, so K^+ tends to diffuse out much more than Na^+ diffuses in.

c. If K^+ did freely leak out of the membrane that is so permeable to it, the inside of the membrane would become increasingly *(negative? positive?)*, contributing to a *(positive? negative?)* membrane potential.

d. Three other factors that contribute to this potential are inability of large negative ions such as ____________________ to exit from the neuron, a slight leakage of ________ into cells, and the Na^+/K^+ pump, which pumps more cations out than it pumps in. Specifically, *(three? two?)* Na^+ are pumped out for every *(three? two?)* K^+ pumped in. Show this by coloring "pump" arrows on Figure LG 12.2a.

e. As a result of factors described in (c) and (d), the membrane potential of a resting (nonactive) neuron is about ________ mV, or a range of ________ to ________. This means that the *(entire neuron? part of a neuron right next to the membrane?)* is much more *(positive? negative?)* than the region immediately outside the membrane.

✓ **C5.** Describe *graded potentials* in this exercise.

a. Graded potentials are so named because they vary in amplitude (________________)

depending on the strength of the ________________. The size of the potentials varies *(directly? indirectly?)* with the number of gated ion channels open or closed as well as the amount of time they are open.

b. These channels alter flow of ________________ across the membrane. If the flow of ions makes the inside of the membrane more positive (such as from –70 to –60 mV), it is said to be *(de? hyper?)*-polarized. If the potential becomes positive enough (such as from –70 to –55 mV) to reach threshold (described below), an action potential *(is? is not?)* likely to occur. If the membrane potential becomes more negative (such as from –70 to –80 mV),

the membrane is said to be ________________ polarized, a point farther away from starting an action potential (nerve impulse).

c. Graded potentials have names based on the type of stimulus applied. For example, neurotransmitters affecting receptors of chemically gated ion channels at synapses

lead to a post- ________________ potential. Sensory receptors respond to stimuli

by producing graded potentials called ________________ potentials, or

________________ potentials.

d. Graded potentials most often occur in *(dendrites or cell bodies? axons?)*. These potentials produce currents along *(just a few micrometers? the entire length?)* of a neuron, whereas action potentials produce currents along *(short distances? short or long distances?)* of a neuron such as along the entire length of axons.

✓ **C6.** Check your understanding of an action potential (nerve impulse) in this Checkpoint.

a. A stimulus causes the nerve cell membrane to become *(more? less?)* permeable to Na^+.

Na^+ can then enter the cell as voltage-gated Na^+ ________________ become activated and open. Notice from Figure 12.12 in your text that Na^+ channels have two gates, much like a double door system. Which gate is located in the outer portion of the neuron cell membrane? *(Activation? Inactivation?)* gate. This gate opens in response to the stimulus.

b. At rest, the membrane had a potential of ________ mV. As Na^+ enters the cell, the inside of the membrane becomes more *(positive? negative?)*. The potential will tend to go toward *(–80? –55?)* mV. The process of *(polarization? depolarization?)* is occurring. This process causes changes in more Na^+ channels so that even more Na^+ enters. This is an example of a *(positive? negative?)* feedback mechanism. The result is a nerve

impulse or nerve ________________.

c. The membrane is completely depolarized at exactly *(–50? 0? +30?)* mV. Na^+ channels stay open until the inside of the membrane potential is *(reversed? repolarized?)* at +30 mV. Then inactivation gates close just milliseconds after the activation gates had opened, preventing more inward flow of Na^+.

d. After a fraction of a second, K^+ voltage-gated channels at the site of the original stimulus open. K^+ is more concentrated *(outside? inside?)* the cell (as you showed on Figure 12.2a); therefore; K^+ diffuses *(in? out?)*. This causes the inside of the membrane to

become more negative and return to its resting potential of ________ mV. The process is known as *(de? re ?)*-polarization. In fact, outflow of K^+ may be so great that

____________________-polarization occurs in which membrane potential becomes closer to *(–50? –90?)* mV.

e. During depolarization, the nerve cannot be stimulated at all. This period is known as the *(absolute? relative?)* refractory period. *(Large? Small?)*-diameter axons have a longer absolute refractory period, with this about *(0.4? 4? 40?)* msec. In other words, slower impulses are likely to occur along *(large? small?)*-diameter neurons. Only a stronger-than-normal stimulus will result in an action potential during the __________________ refractory period, which corresponds roughly with *(de? re?)*-polarization.

f. The nerve impulse is propagated (or ____________________) along the nerve, as adjacent areas are depolarized, causing more channels to be activated and more *(Na^+? K^+?)* to enter.

g. What effect do anesthetics such as procaine (Novocaine) have on action potentials?

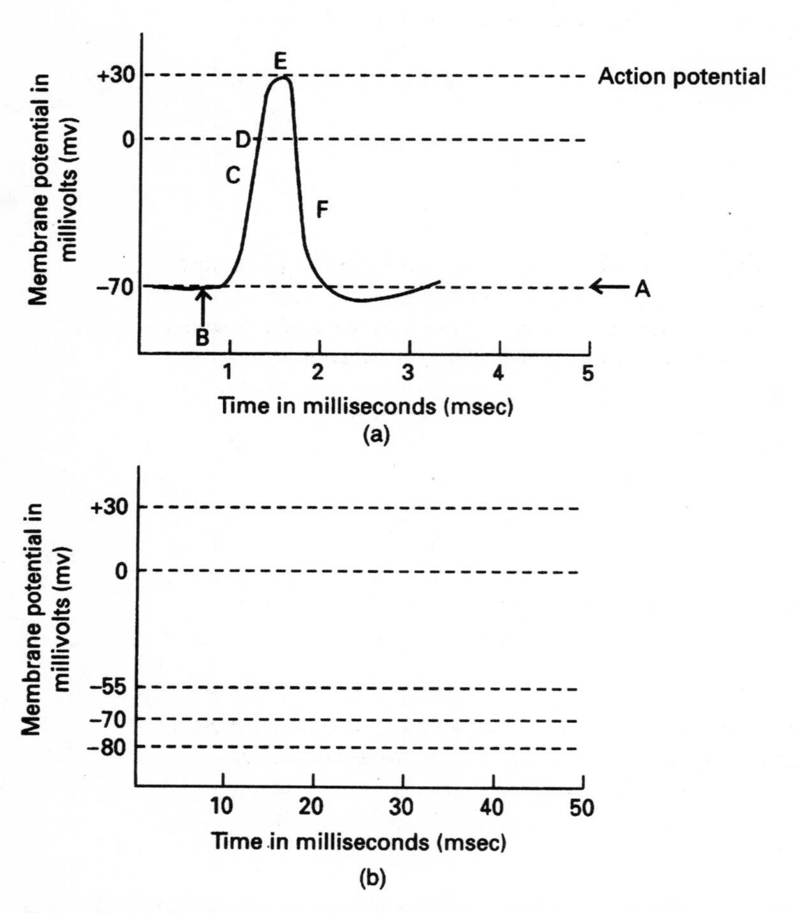

Figure LG 12.3 Diagrams for showing nerve action potentials. (a) Identify letter labels in Checkpoint C7. (b) Complete as directed in Checkpoint D2(b) and (d).

✓ **C7.** Write the correct letter label from Figure LG 12.3a next to each description.

_______ a. The stimulus is applied at this point.

_______ b. Resting membrane potential is at this level.

_______ c. Membrane becomes so permeable to K^+ that K^+ diffuses rapidly out through K^+ channels.

_______ d. The membrane is becoming more positive inside as Na^+ enters; its potential is –30 mV. The process of depolarization is occurring.

_______ e. The membrane is completely depolarized at this point.

_______ f. The membrane is repolarizing at this point.

_______ g. Reversed polarization occurs; enough Na^+ has entered so that the region just inside the membrane at this part of the cell is more positive inside than outside.

C8. *For extra review.* Show the events that occur during initiation and propagation of a nerve impulse on Figure LG 12.2b. Compare your diagram to Figures 12.12 and 12.13 in the text.

C9. Explain how the all-or-none principle resembles a domino effect.

✓ **C10.** Describe how myelination, fiber thickness, and temperature affect speed of impulse propagation.

a. Saltatory conduction occurs along *(myelinated? unmyelinated?)* nerve fibers. Saltatory transmission is *(faster? slower?)* and takes *(more? less?)* energy than continuous conduction. Explain why.

b. Type A fibers have the *(largest? smallest?)* diameter and *(are? are not?)* myelinated. These fibers conduct impulses *(rapidly? slowly?)*. Give two examples of A fibers.

c. Type C fibers have the *(largest? smallest?)* diameter and *(are? are not?)* myelinated. These fibers conduct impulses *(rapidly? slowly?)*. Give two examples of C fibers.

d. *A clinical correlation. (Warm? Cool?)* nerve fibers conduct impulses faster. How can this information be applied clinically?

✓ **C11.** Contrast action potentials of nerve and muscle by completing this table.

Tissue	Typical Resting Membrane Potential	Duration of Nerve impulse
a. Neuron		
b. Muscle		

C12. In Checkpoint C5 we considered graded potentials. Contrast *graded potentials* with *action potentials* by completing this table.

Characteristic	Graded Potential	Action Potential
a. Amplitude		All-or-none potential, usually about 100 mV
b. Duration	Long; may last several minutes	
c. Types of ion channels		Only voltage-gated
d. Location where potential arises	Usually dendrites or cell bodies	
e. Propagation		Long distances
f. Refractory period	None, so summation may occur	

D. Signal transmission at synapses; neurotransmitters; circuits (pages 404–411)

✓ **D1.** Do this activity about types of synapses.

a. In Chapter 10, you studied the point at which a neuron comes close to contacting a muscle. This is known as a ____________________ junction, shown in Figure LG 10.2, page LG 186. The minute space between two neurons is known as a ____________________.

b. There are two types of synapses. A(n) *(chemical? electrical?)* synapse is designed to allow spread of an ionic current from one neuron to the next. These synapses consist of ____________________ junctions made of hundreds of proteinaceous tunnels called ____________________. Gap junctions are common in *(skeletal? smooth* and *cardiac?)* muscle where a group of muscle fibers works in synchrony.

c. In chemical synapses membranes of the two cells *(do? do not?)* touch. Does a nerve impulse actually "jump" across the synaptic cleft? Explain.

d. Now summarize the sequence of events at a chemical synapse by placing these

events in order from first to last: ________ → ________ → ________

A. Electrical signal: postsynaptic potential (graded potential; if threshold level, action potential results)
B. Electrical signal: presynaptic potential (nerve impulse)
C. Chemical signal: release of neurotransmitter into synaptic cleft

e. Write these structures in order to summarize the anatomical pathway of a chemical synapse similar to the NMS in Figure LG 10.2.

EB.	**End bulb of presynaptic neuron**	**SC.**	**Synaptic cleft**
N.	**Neurotransmitter**	**SV.**	**Synaptic vesicle**
NR.	**Neurotransmitter receptor**		

________ → ________ → ________ → ________ → ________
First Last

f. Explain the role of calcium ions (Ca^{2+}) in nerve transmission at chemical synapses.

g. The rationale for *one-way information transfer* at chemical synapses is that only synaptic end bulbs of *(pre? post?)*-synaptic neurons release neurotransmitters, and only *(pre? post?)*-synaptic neurons have neurotransmitter receptors.

h. Neurotransmitter receptors are parts of *(ligand? voltage?)*-gated channels on postsynaptic neurons.

✓ **D2.** Complete the following exercise about postsynaptic potentials.

a. If a neurotransmitter causes depolarization of a postsynaptic neuron, for example, from –70 to –65 mV, the postsynaptic potential (PSP) is called *(excitatory? inhibitory?)*; in other words, it is called an *(EPSP? IPSP?)*. EPSPs result primarily from inflow of *(K^+? Na^+? Ca^{2+}?)* through chemically gated chemicals.

b. Usually a single EPSP within a single neuron *(is? is not?)* sufficient to cause a threshold potential and initiate a nerve impulse. Instead, a single EPSP can cause *(partial? total?)* depolarization. Diagram a partial depolarization from –70 to –62 mV in green on Figure LG 12.3b.

c. If neurotransmitters are released from a number of presynaptic end bulbs at one time, their combined effect may produce threshold EPSP of about ______ mV. This phenomenon is known as (*spatial? temporal?*) summation. Temporal summation is that due to accumulation of transmitters from (*one? many?*) presynaptic end bulb(s) over a period of time.

d. If a neurotransmitter causes inhibition of the postsynaptic neuron, the process is known as (*de? hyper?*)-polarization, and the PSP is (*excitatory? inhibitory?*). IPSPs often result from influx of (*Cl^-? K^+? Na^+?*) and/or outflow of (*Cl^-? K^+? Na^+?*) through gated channels. An example of an IPSP would be change in the membrane PSP from –70 to (*–60? –80?*) my. Diagram such an IPSP in red on Figure 12.3b.

e. In other words, a neuron with an IPSP is (*closer? farther?*) from threshold than a neuron at resting membrane potential (RMP), so a cell with an IPSP is less likely to have an action potential. An example of use of an IPSP is inhibition of your triceps brachii muscle as the ________________ is stimulated (with an EPSP) to contract, or innervation of the heart by the vagus nerve (CN X) which (*in? de?*)-creases heart rate.

✓ **D3.** *A clinical correlation.* Once a neurotransmitter completes its job, it must be removed from the synaptic cleft. Describe two mechanisms for getting rid of these chemicals. Notice the consequences of alterations of these mechanisms.

a. Inactivation of a neurotransmitter may occur, such as breakdown of ________________ (ACh) by the enzyme named ________________. Certain drugs (such as physostigmine) destroy this enzyme, so postsynaptic neurons (or muscles) (*remain? are less?*) activated. Predict the effects of such a drug.

b. A neurotransmitter such as norepinephrine (NE) can be removed from the synapse by a different mechanism. Describe this method.

c. Explain how the antidepressant Prozac alters transmission at synapses.

✓ **D4.** Answer these questions about different types of neurotransmitters (NTs).

a. The number of known (or strongly suspected) NTs is about *(5? 30? 100?)*. A *(large? tiny?)* amount of NT is released at a single synapse.

b. Which type of NT acts more quickly at a synapse? One that (*opens or closes ion channels in the neuron membrane? acts by a second messenger system that influences reactions inside the postsynaptic cell?*).

c. (*All? Not all?*) NTs are made by neurons. Name one type of cell other than a neuron that can synthesize NTs. ____________ Name one type of cell within the brain that can make hormones. ____________

d. We have already discussed acetylcholine (______), an NT released at synapses and neuromuscular junctions. ACh is (*excitatory? inhibitory?*) toward skeletal muscle, but ACh released from the vagus nerve is (*excitatory? inhibitory?*) toward cardiac muscle. So when the vagus nerve sends impulses to your heart, your pulse (heart rate) becomes (*faster? slower?*).

e. GABA and glycine are both (*excitatory? inhibitory?*) NTs. They act by opening (*Cl^-? Na^+?*) channels, leading to IPSPs. The most common inhibitory transmitter in the brain is (*GABA? glycine?*), whereas ____________ is more commonly released by neurons in the spinal cord.

f. Strychnine is a chemical that blocks (*GABA? glycine?*) receptors so that muscles are not properly inhibited (relaxed). Strychnine poisoning is likely to lead to death because the muscles of the ____________ cannot relax so that air that is high in carbon dioxide cannot be exhaled.

g. List three NTs that are categorized as catecholamines. These NTs are removed from the synapse by (*enzymatic breakdown as by COMT or MAO? recycling into synaptic vesicles? both of these mechanisms?*).

h. Pharmaceuticals that mimic the body's NTs (or cause NTs to linger in the synapse) are called (*agonists? antagonists?*), whereas drugs that block the action of a NT are called (*agonists? antagonists?*).

✓ **D5.** Match the neurotransmitter with the description that best fits. The first one is done for you. Lines at the end of parts *a-f* are for Checkpoint D6

ACh.	**Acetylcholine**	**No.**	**Nitric oxide**
DA.	**Dopamine**	**P.**	**Purines**
EED.	**Enkephalins, endorphins, and dynorphins**	**Ser.**	**Serotonin**
GABA.	**Gamma aminobutyric acid**	**SP.**	**Substance P**

GABA a. The most common inhibitory neurotransmitter in the brain; enhanced by antianxiety drugs such as diazepam (Valium) AA

______ b. Transmits pain-related impulses ______

______ c. Morphinelike chemicals that are the body's natural painkillers; suppress release of substance P ______

______ d. Concentrated in brainstem neurons, this chemical helps regulate sleep, temperature, and mood ______

______ e. Made by neurons involved with emotional responses and automatic movements of skeletal muscles ______

______ f. Neurons that degenerate in Parkinson's disease fail to produce this chemical ______

______ g. This chemical is not stored but is formed on demand and acts immediately; dilates blood vessels; Sildenafil (Viagra) enhances its function

______ h. Adenosine (which is present in ATP and ADP) is an example of this classs of excitatory neurotransmitters

✓ **D6.** Select answers in the box below to identify the correct category of each of the neurotransmitters and neuromodulators listed in Checkpoint D5*a-f*. Write answers on lines after each of those descriptions. The first one is done for you.

ACh.	**Acetylcholine**	**B.**	**Biogenic amines**
AA.	**Amino acids**	**N.**	**Neuropeptides**

D7. Answer these questions about opioid peptides such as enkephalins.

a. Briefly explain the relationship between these chemicals and acupuncture.

b. Besides serving as painkillers, enkephalins have several other functions. List four or more of these functions.

✓ **D8.** *Clinical correlation.* Discuss roles of neurotransmitters in these clinical situations.

ACh.	**Acetylcholine**	**DA.**	**Dopamine**	**E/NE.**	**Epinephrine or norepinephrine**

a. Paula has a form of food poisoning known as botulism; the involved toxin prevents her motor neurons from releasing ________

b. Drew's cocaine "high" is related to interference of ________ reuptake so that this neurotransmitter remains active for a longer time at his synapses

c. Mr. Danielson's Parkinson's disease is treated with L-DOPA which enhances production of ________

d. Tiffany uses an inhaler with chemicals similar to ________ during an asthma attack.

✓ **D9.** Match the types of circuits in the box with related descriptions. Answers may be used more than once.

C.	**Converging**	**P.**	**Parallel after-discharge**
D.	**Diverging**	**R.**	**Reverberating**

________ a. Impulse from a single presynaptic neuron causes stimulation of increasing numbers of cells along the circuit.

________ b. An example is a single motor neuron in the brain that stimulates many motor neurons in the spinal cord, therefore activating many muscle fibers.

________ c. Branches from a second and third neuron in a pathway may send impulses back to the first, so the signal may last for hours, as in coordinated muscle activities.

________ d. One postsynaptic neuron receives impulses from several nerve fibers.

________ e. A single presynaptic neuron stimulates intermediate neurons which synapse with a common postsynaptic neuron, allowing this neuron to send out a stream of impulses, as in precise mathematical calculations.

E. Regeneration and repair of nervous tissue; disorders (pages 412–414)

E1. Define *plasticity* and list several factors that account for this.

✓ **E2.** Which of the following can regenerate if destroyed? Briefly explain why in each case.

a. Neuron cell body

b. CNS nerve fiber

c. PNS nerve fiber

E3. Discuss methods currently being investigated to promote regrowth of damaged neurons.

✓ **E4.** Answer these questions about nerve regeneration.

a. In order for a damaged neuron to be repaired, it must have an intact cell body and also a ____________________.

b. Can axons in the CNS regenerate? Explain.

c. When a nerve fiber (axon or dendrite) is injured, the changes that follow in the cell body are called *(chromatolysis? Wallerian degeneration?)*. Those that occur in the portion of the fiber distal to the injury are known as ____________________.

✓ **E5.** Discuss multiple sclerosis (MS) in this Checkpoint.

a. This condition involves destruction of ____________________ surrounding axons. This condition most commonly manifests initial symptoms in *(women? men?)* in their *(20s–30s? 50s–60s?)*.

b. The name multiple sclerosis refers to the many locations in which myelin is replaced by ____________________. MS *(is? is not?)* progressive and *(does? does not?)* involve remissions and relapses.

E6. Check your understanding of epilepsy in this Checkpoint.

a. Define the term epileptic seizure.

b. Most epileptic seizures are *(caused by changes in blood chemistry? idiopathic [of unknown cause]?)*. Epilepsy *(usually? almost never?)* affects intelligence.

ANSWERS TO SELECTED CHECKPOINTS: CHAPTER 12

A3. (a) B. (b) CN. (c) SN. (d) SC. (e) G. (f) EP. (g) SR.
A4. Sensation, integration, and motor or glandular response.
A5. (a) CNS. (b) Aff. (c) SNS or Eff (*Hint:* remember S A M E: Sensory = Afferent; *M*otor = *E*fferent). (d) ANS. (e) ANS. (f) PNS. (g) ENS.
B1. See Figure LG 12.1A to the right.
B2. (a) CB. (b) L. (c) M. (d) NF. (e) NB. (f) D. (g) A. (h) T.
B5. Multipolar.
B6. (a) Neurons. (b-e) Neuroglia.
B7. (a) SAT, SCH. (b) O, SCH. (c-e) A. (f) M. (g) E. (h) SAT. (i) SCH.
B8. (a) PNS. (b) A, B. (c) B C A. (d) B. (e) Regeneration or growth. (f) Myelin; saltatory; in. (g) No; injured CNS nerve fibers do not heal well; yes; oligodendrocytes.
B9. (a) W. (b–e) G. (f) W.
C1. (a) Potential, current; ions. (b) Graded, action; ion; little.
C2. (a) Leakage, gated; K^+, K^+. (b1) Voltage. (b2) Mechanically. (b3) Ligand. (b4) Mechanically.
C3. (a) Directly. (b) G protein and a second messenger system using molecules in cytosol.

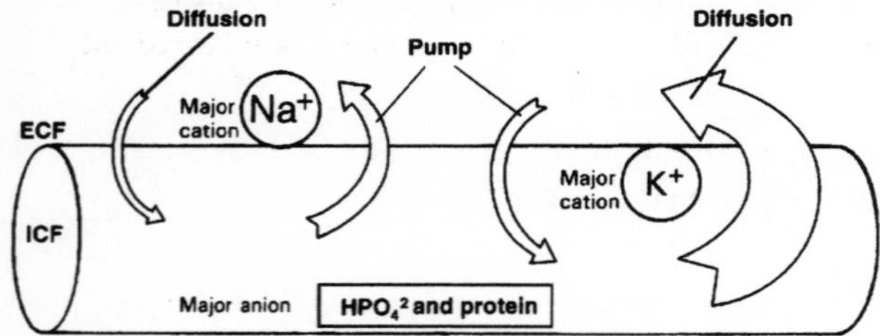

Figure 12.2aA Diagram of a nerve cell.

C4. (a) Ions; see Figure LG 12.2aA. (b) 50 to 100. (c) Negative, negative. (d) Anions of proteins, Cl^-; three, two; see Figure LG 12.2aA (e) –70, –40 to –90; part of a neuron right next to the membrane, negative.
C5. (a) Size, stimulus; directly. (b) Ions; de; is; hyper. (c) Synaptic; receptor, generator. (d) Dendrites or cell bodies; just a few micrometers, short or long distances.

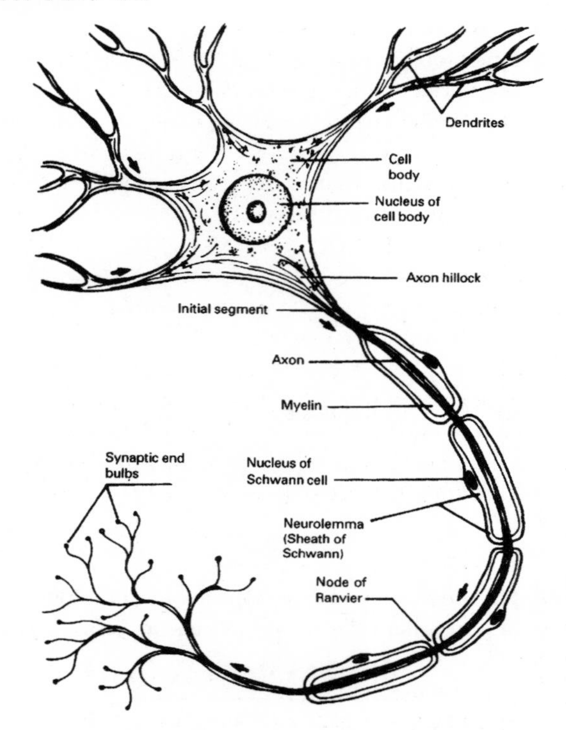

Figure LG 12.1A Structure of a typical neuron as exemplified by an efferent (motor) neuron.

C6. (a) More; channels; activation. (b) –70; positive; –55; depolarization; positive; action potential. (c) 0; reversed. (d) Inside, out; –70; re; after-hyper, –90. (e) Absolute; small, 4; small; relative, re. (f) Transmitted, Nat. (g) Such anesthetics prevent opening of voltage-gated Na^+ channels in pain fibers.
C7. (a) B. (b) A. (c) F. (d) C. (e) D. (f) F. (g) E.
C10. (a) Myelinated; faster because the nodes of Ranvier have a high density of voltage-gated Na^+ channels; less energy because only small regions (the nodes) have inflow of Na^+, which then must be pumped out by Na^+/K^+ pumps. (b) Largest, are; rapidly; large, sensory neurons, such as for touch, position of joints, and temperature, as well as nerves to skeletal muscles for quick reactions.

(c) Smallest, are not; slowly; nerves that carry impulses to and from viscera. (d) Warm; ice or other cold applications can slow conduction of pain impulses.

Tissue	Typical Resting Membrane Potential	Duration of Nerve impulse
a. Neuron	Higher (–70 mV)	Shorter (0.5–2 msec)
b. Muscle	Lower (–90 mV)	Longer (1.0–5.0 msec: skeletal muscle; 10–300 msec: cardiac and smooth muscle)

C11.

Dl. (a) Neuromuscular; synapse. (b) Electrical; gap, connexons; smooth and cardiac. (c) Do not; no; The nerve impulse travels along the presynaptic axon to the end of the axon, where synaptic vesicles are triggered to release neurotransmitters. These chemicals enter the synaptic cleft and then affect the postsynaptic neuron (either excite it or inhibit it). (d) B C A. (e) EB SV N SC NR. (f) The nerve impulse in the presynaptic neuron opens voltage-gated Ca^{2+} channels, creating an influx of Ca^{2+} into the synaptic end bulbs. The Ca^{2+} here triggers exocytosis of synaptic vesicles with release of neurotransmitters. (g) Pre, post. (h) Lingand.

D2. (a) Excitatory, EPSP; Na^+. (b) Is not; partial; see Figure LG 12.3bA. (c) –55; spatial; one. (d) Hyper, inhibitory; Cl^-, K^+ –80; see Figure LG 12.3bA.

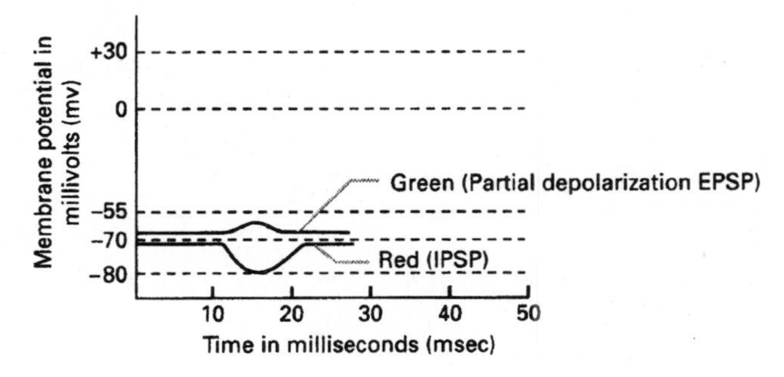

Figure LG 12.3bA Diagram of a facilitation EPSP and an IPSP.

(e) Farther; biceps, de.

D3. (a) Acetylcholine, acetylcholinesterase (AChE); remain. Such drugs may help activate muscles of persons with myasthenia gravis because many of their neurotransmitter receptors are unavailable; the extra activation of ACh can enhance response of remaining receptors. On the other hand, use of some powerful anticholinesterase drugs (as "nerve gases") cause profound and lethal effects. (b) NE can be recycled into the presynaptic neuron. (c) Prozac inhibits recycling (or reuptake) of the neurotransmitter serotonin, which stays longer in the synapse, affecting postsynaptic action potentials that lower depression.

D4. (a) 100: tiny. (b) Opens or closes ion channels in the neuron membrane. (c) Not all; endocrine cells; neurosecretory. (d) ACh: excitatory, inhibitory; slower. (e) Inhibitory; Cl^-; GABA, glycine. (f) Glycine; diaphragm. (g) Norepinephrine (NE), epinephrine (epi), and dopamine (DA); both of these mechanisms. (h) Agonists; antagonists.

D5, D6. (b) SP (N). (c) EED (N). (d) Ser (B). (e) DA (B). (f) DA (B). (g) NO. (h) P.

D8. (a) ACh. (b) DA. (c) DA. (d) E/NE.

D9. (a) D. (b) D. (c) R. (d) C. (e) P.

E2. (a) No, related to inhibition from neuroglia (especially oligodendrocytes) as well as absence of growth-stimulating cues such as EGF that trigger mitosis present during fetal development. (b) No. CNS nerve fibers lack the neurolemma necessary for regeneration, scar tissue builds up by proliferation of astrocytes, and myelin also inhibits regeneration of CNS axons. (c) Yes. PNS fibers do have a neurolemma.

E4. (a) Neurolemma. (b) No. They lack a neurolemma. (c) Chromatolysis; Wallerian degeneration.

E5. (a) Myelin; women, 20s and 30s. (b) Hardened scars or plaques; is, does.

MORE CRITICAL THINKING: CHAPTER 12

1. Contrast axons and dendrites with regard to structure and function.
2. If you cut a nerve in your finger, it can regenerate. However, severed axons in the brain or spinal cord cannot heal. Explain why.
3. Contrast the process of myelination of neurons in the CNS and in the PNS.
4. Explain how the structure (thickness and presence or absence of myelination) affects the rate of conduction along "life-or-death" neurons (such as those that detect heat or pain in fingertips) versus the rate along neurons that innervate the intestine.
5. Explain how your sensory systems can detect the difference between a weak handshake and a firm one.
6. Contrast effects of the following neurotransmitters: glutamate, glycine, and catecholamines such as norepinephrine.
7. Describe the steps in repair of peripheral neurons.
8. Describe causes and symptoms of and methods for alleviating epileptic seizures.

MASTERY TEST: CHAPTER 12

Questions 1–2: Arrange the answers in correct sequence.

______ ______ ______ 1. In order of transmission across synapse, from first structure to last:
A. Presynaptic end bulb
B. Postsynaptic neuron
C. Synaptic cleft

______ ______ ______ 2. Membrane potential values, from most negative to zero:
A. Resting membrane potential
B. Depolarized membrane potential
C. Threshold potential

Questions 3–10: Circle the letter preceding the one best answer to each question.

3. Choose the one *false* statement.
   A. The membrane of a resting neuron has a membrane potential of –70 mV.
   B. In a resting membrane, permeability to K^+ ions is about 100 times less than permeability to Na^+ ions.
   C. C fibers are thin, unmyelinated fibers with a relatively slow rate of nerve transmission.
   D. C fibers are more likely to innervate the heart and bladder than structures (such as skeletal muscles) that must make instantaneous responses.
4. All of the following are listed with a correct function *except:*
   A. Schwann cell—myelination of neurons in the PNS
   B. Oligodendrocytes–myelination of neurons in the CNS
   C. Astrocytes—form an epithelial lining of the ventricles of the brain
   D. Microglia—phagocytic
5. Synaptic end bulbs are located:
   A. At ends of axon terminals
   B. On axon hillocks
   C. On neuron cell bodies
   D. At ends of dendrites
   E. At ends of both axons and dendrites
6. Which of these is equivalent to a nerve fiber?
   A. A neuron
   B. A ganglion
   C. An axon or dendrite
   D. A nerve, such as sciatic nerve
7. ACh is an abbreviation for:
   A. Acetylcholine
   B. Norepinephrine
   C. Acetylcholinesterase
   D. Serotonin
   E. Inhibitory postsynaptic potential
8. A term that means the same thing as *afferent* is:
   A. Autonomic
   B. Somatic
   C. Peripheral
   D. Motor
   E. Sensory
9. An excitatory transmitter substance that changes the membrane potential from –70 to –65 mV causes:
   A. Impulse conduction
   B. Partial depolarization
   C. Inhibition
   D. Hyperpolarization
10. All of these statements are true EXCEPT:
    A. Neuroglia are more numerous than neurons in the human nervous system.
    B. Action potentials can spread across longer distances (such as along long axons) than graded potentials do.
    C. Graded potentials can vary in amplitude, but action potentials do not.
    D. Action potentials do exhibit temporal and spatial summation, whereas graded potentials do not.

Questions 11–20: Circle T (true) or F (false). If the statement is false, change the underlined word or phrase so that the statement is correct.

T F 11. Because CNS fibers contain no neurolemma and the neurolemma produces myelin, CNS <u>fibers are all unmyelinated</u>.

T F 12. Neurotransmitter substances are released at <u>synapses and also at neuromuscular junctions</u>.

T F 13. Generally, release of excitatory transmitter by <u>a single presynaptic end bulb</u> is sufficient to develop an action potential in the postsynaptic neuron.

T F 14. <u>Epilepsy</u> is a condition that involves abnormal electrical discharges within neurons of the brain.

T F 15. In the <u>converging</u> circuit, a single presynaptic neuron influences several postsynaptic neurons (or muscle or gland cells) at the same time.

T F 16. Action potentials are measured in <u>milliseconds, which are thousandths of a second</u>.

T F 17. Nerve fibers with a short absolute refractory period can respond to more rapid stimuli than nerve fibers with a long absolute refractory period.

T F 18. The concentration of potassium ions (K^+) is considerably greater inside a resting cell than outside of it.

T F 19. A stimulus that is adequate will temporarily increase permeability of the nerve membrane to Na^+.

T F 20. The brain and spinal nerves are parts of the peripheral nervous system (PNS).

Questions 21–25: Fill-ins. Complete each sentence with the word or phrase that best fits.

______________ 21. The ________ nervous system consists of the sympathetic and parasympathetic divisions.

______________ 22. Application of cold to a painful area can decrease pain in that area because ________.

______________ 23. One-way nerve impulse transmission can be explained on the basis of release of transmitters only from the ________ of neurons.

______________ 24. The neurolemma is found only around nerve fibers of the ________ nervous system so that these fibers can regenerate.

______________ 25. In an axosomatic synapse, an axon's synaptic end bulb transmits nerve impulses to the ________ of a postsynaptic neuron.

ANSWERS TO MASTERY TEST: CHAPTER 12

Arrange

1. A C B
2. A C B

Multiple Choice

3. B
4. C
5. A
6. C
7. A
8. E
9. B
10. D

True–False

11. F. May be myelinated because oligodendrocytes myelinate CNS fibers
12. T
13. F. A number of presynaptic end bulbs
14. T
15. F. Diverging
16. T
17. T
18. T
19. T
20. F. Spinal nerves (as well as some other structures; but not the brain)

Fill-ins

21. Autonomic
22. Cooling of neurons slows down the speed of nerve transmission, for example, of pain impulses
23. End bulbs of axons
24. Peripheral
25. Cell body (soma)

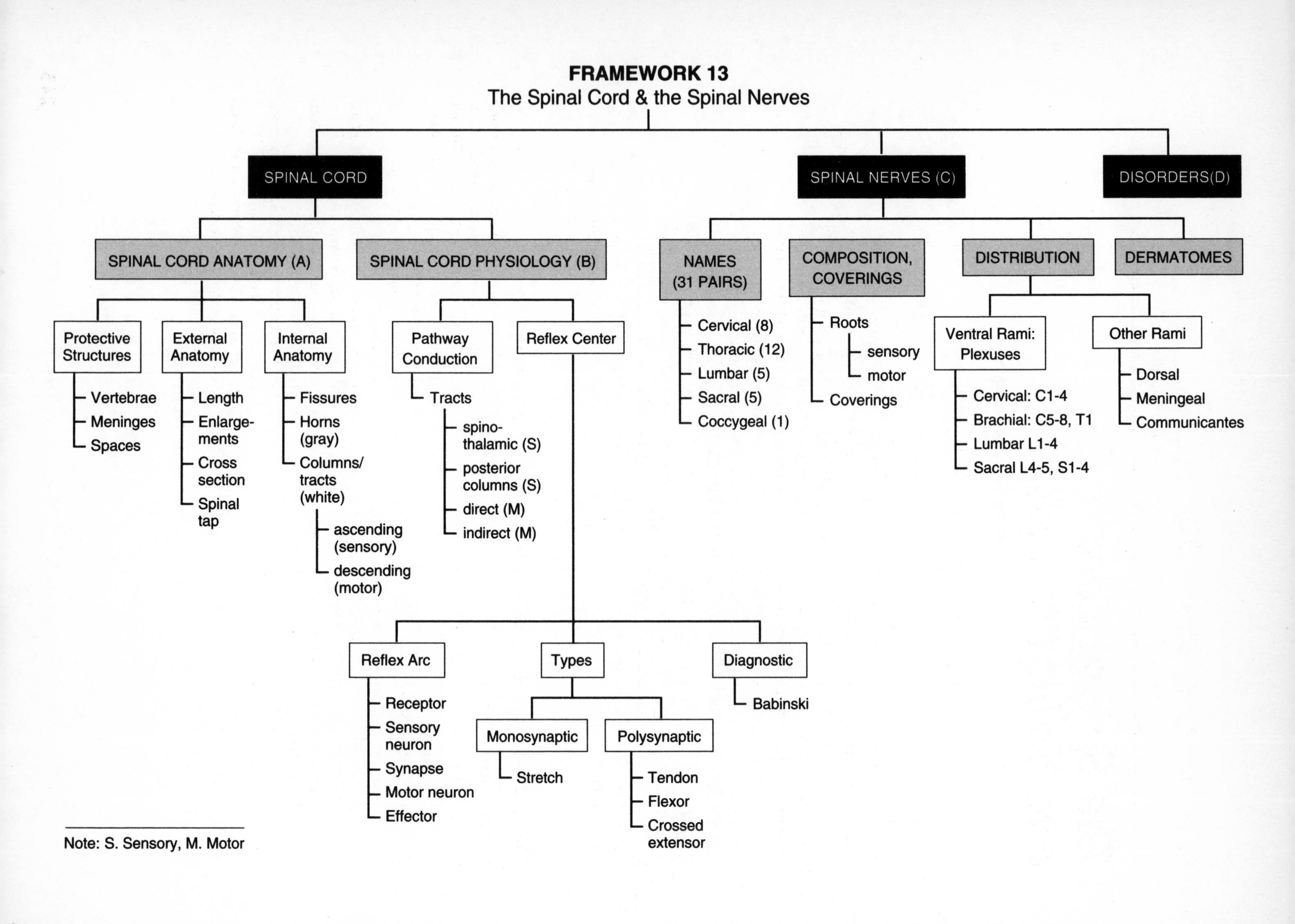

FRAMEWORK 13
The Spinal Cord & the Spinal Nerves
SPINAL CORD
SPINAL NERVES (C)
DISORDERS(D)
SPINAL CORD ANATOMY (A)
SPINAL CORD PHYSIOLOGY (B)
NAMES (31 PAIRS)
COMPOSITION, COVERINGS
DISTRIBUTION
DERMATOMES
Protective Structures
Vertebrae
Meninges
Spaces
External Anatomy
Length
Enlarge-ments
Cross section
Spinal tap
Internal Anatomy
Fissures
Horns (gray)
Columns/ tracts (white)
ascending (sensory)
descending (motor)
Pathway Conduction
Tracts
spino-thalamic (S)
posterior columns (S)
direct (M)
indirect (M)
Reflex Center
Reflex Arc
Receptor
Sensory neuron
Synapse
Motor neuron
Effector
Types
Monosynaptic
Stretch
Polysynaptic
Tendon
Flexor
Crossed extensor
Diagnostic
Babinski
Cervical (8)
Thoracic (12)
Lumbar (5)
Sacral (5)
Coccygeal (1)
Roots
sensory
motor
Coverings
Ventral Rami: Plexuses
Cervical: C1-4
Brachial: C5-8, T1
Lumbar L1-4
Sacral L4-5, S1-4
Other Rami
Dorsal
Meningeal
Communicantes
Note: S. Sensory, M. Motor

C H A P T E R **13**

The Spinal Cord and Spinal Nerves

The spinal cord and spinal nerves serve as the major links in the communication pathways between the brain and all other parts of the body. Nerve impulses are conveyed along routes (or tracts) in the spinal cord, laid out much as train tracks: some head north to regions of the brain, and others carry nerve messages south from the brain toward specific body parts. Spinal nerves branch off from the spinal cord, perhaps like a series of bus lines that pick up passengers (nerve messages) to or from train depots (points along the spinal cord) en route to their final destinations. Organization is critical in this nerve impulse transportation network. Any structural breakdowns—by trauma, disease, or other disorders—can lead to interruption in service with resultant chaos (such as spasticity) or standstill (such as sensory loss or paralysis).

As you begin your study of the spinal cord and spinal nerves, carefully examine the Chapter 13 Topic Outline and Objectives; check off each one as you complete it. To organize your study of the spinal cord and spinal nerves, glance over the Chapter 13 Framework now. Be sure to refer to the Framework frequently and note relationships among key terms in each section.

TOPIC OUTLINE AND OBJECTIVES

A. Spinal cord anatomy

1. Describe the protective structures and gross anatomical features of the spinal cord.

B. Spinal cord physiology

1. Describe the functions of the major sensory and motor tracts of the spinal cord.
2. Describe the functional components of a reflex arc and the ways reflexes maintain homeostasis.

C. Spinal nerves

1. Describe the components, connective tissue coverings, and branching of a spinal nerve.
2. Define plexus, and identify the distribution of nerves of the cervical, brachial, lumbar, and sacral plexuses.
3. Describe the clinical significance of dermatomes.
4. Describe the origin, distribution, and effects of damage to the cervical plexus.
5. Describe the origin, distribution, and effects of damage to the brachial plexus.
6. Describe the origin, distribution, and effects of damage to the lumbar plexus.
7. Describe the origin, distribution, and effects of damage to the sacral plexus.

D. Disorders

WORDBYTES

Now become familiar with the language of this chapter by studying each wordbyte, its meaning, and an example of its use within a term. After you study the entire list, self-check your understanding by writing the meaning of each wordbyte on the line. As you continue through the *Learning Guide*, identify (and fill in) additional terms that contain the same wordbyte.

Wordbyte	Self-check	Meaning	Example(s)
arachn-	______________	spider	*arachn*oid
cauda	______________	tail	*cauda* equina
contra-	______________	opposite	*contra*lateral
conus	______________	cone	*conus* medullaris
dent-	______________	small tooth	*dent*iculated
dura	______________	hard	*dura* mater
endo-	______________	within	*endo*neurium
epi-	______________	over	*epi*neurium
equina	______________	horse	cauda *equina*
inter-	______________	between	*inter*segmental
mater	______________	mother	dura *mater*
para-	______________	abnormal	*pare*sthesia
peri-	______________	around	*peri*neurium
pia	______________	delicate	*pia* mater
soma-	______________	body	*soma*tic
-tome	______________	thin section	derma*tome*

CHECKPOINTS

A. Spinal cord anatomy (pages 420–425)

✓ **A1.** List three general functions of the spinal cord.

✓ **A2.** The spinal cord is part of the *(central? peripheral?)* nervous system.

Figure LG 13.1 Meninges. Color and label as indicated in Checkpoint A3.

✓ **A3.** Refer to Figure LG 13.1 and do the following exercise.

a. Color the meninges to match color code ovals.
b. Label these spaces: *epidural space, subarachnoid space,* and *subdural* space.
c. Write next to each label of a space the contents of that space. Choose from these answers: *cerebrospinal (CSF); interstitial fluid; fat and connective tissue.*

d. Inflammation of meninges is a condition called ______________________.
e. The denticulate ligaments are extensions of *(dura mater? arachnoid? pia mater?)* that

attach laterally to ______________________ along the length of the cord. State functions of these ligaments.

✓ **A4.** Do this activity about your own spinal cord.

a. Identify the location of your own spinal cord. It lies within the vertebral canal, extending from just inferior to the cranium to about the level of your *(waist? sacrum?)*. This level corresponds with about the level of *(L1–L2? L4–L5? S4–S5?)* vertebrae. In other words, the *(spinal cord? vertebral column?)* reaches a lower (more

inferior) level in your body. The spinal cord is about 42 to 45 cm (______ inches) in length.
b. Circle the two regions of your spinal cord that have notable enlargements where nerves exit to your upper and lower limbs:

cervical **thoracic** **lumbar** **sacral** **coccygeal**

✓ **A5.** Match the names of the structures listed in the box with descriptions given.

Ca.	**Cauda equina**	**F.**	**Filum terminale**
Co.	**Conus medullaris**	**S.**	**Spinal segment**

________ a. Tapering inferior end of spinal cord

________ b. Any region of spinal cord from which one pair of spinal nerves arises

________ c. Nonnervous extension of pia mater; anchors cord in place

________ d. "Horse's tail"; extension of spinal nerves in lumbar and sacral regions within subarachnoid space

✓ **A6.** *A clinical correlation*. Answer these questions about meninges and related clinical procedures.

a. State two or more purposes of a spinal tap (or lumbar puncture).

Into which space does the needle enter? *(Epidural? Subdural? Subarachnoid?)*

b. At what level of the vertebral column (not spinal cord) is the needle for a spinal tap

(or lumbar puncture) inserted? Between ____________________ vertebrae. Explain why this location is a relatively safe site for this procedure.

c. Identify level L4 of the vertebral column on yourself by listing two other landmarks

at this level. ____________________ ____________________

d. In some cases anesthetics are introduced into the epidural space, rather than into the subarachnoid space (as in a spinal tap). Explain why an epidural is likely to exert its effects on nerves with slower and more prolonged action than a spinal tap.

State one example of a procedure for which an epidural may be used. ____________

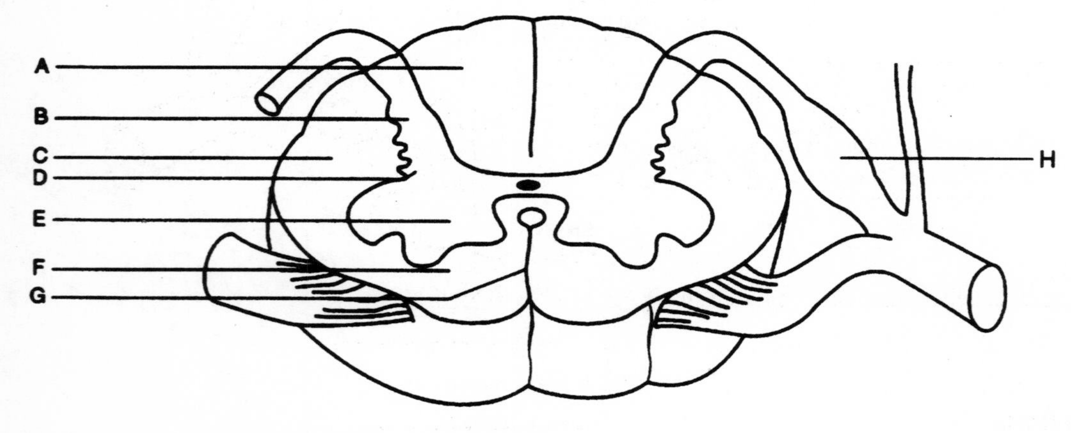

Figure LG 13.2 Outline of the spinal cord, roots, and nerves. Color and label according to Checkpoint A7.

✓ **A7.** Identify the following structures on Figure LG 13.2.

a. Gray matter is found *(within? outside?)* the H-shaped outline, and white matter is located ____________________ that outline. With a lead pencil, shade the gray matter on the right side of the figure only.

b. Now label parts A–G on the left side of the figure.

c. On the right side of the figure, label H.

B. Spinal cord: physiology (pages 425–433)

✓ **B1.** In Checkpoint Al, you listed main functions of the spinal cord. Expand that description in this Checkpoint.

a. One primary function of the spinal cord is to permit ____________________ between nerves in the periphery, such as arms and legs, and the brain by means of the *(tracts? nerves? ganglia?)* located in the white columns of the cord.

b. Another function of the cord is to serve as a ____________________ center by means of spinal nerves. These are attached by _________ roots. The *(anterior? posterior?)* root contains sensory nerve fibers, and the __________________ root contains motor fibers.

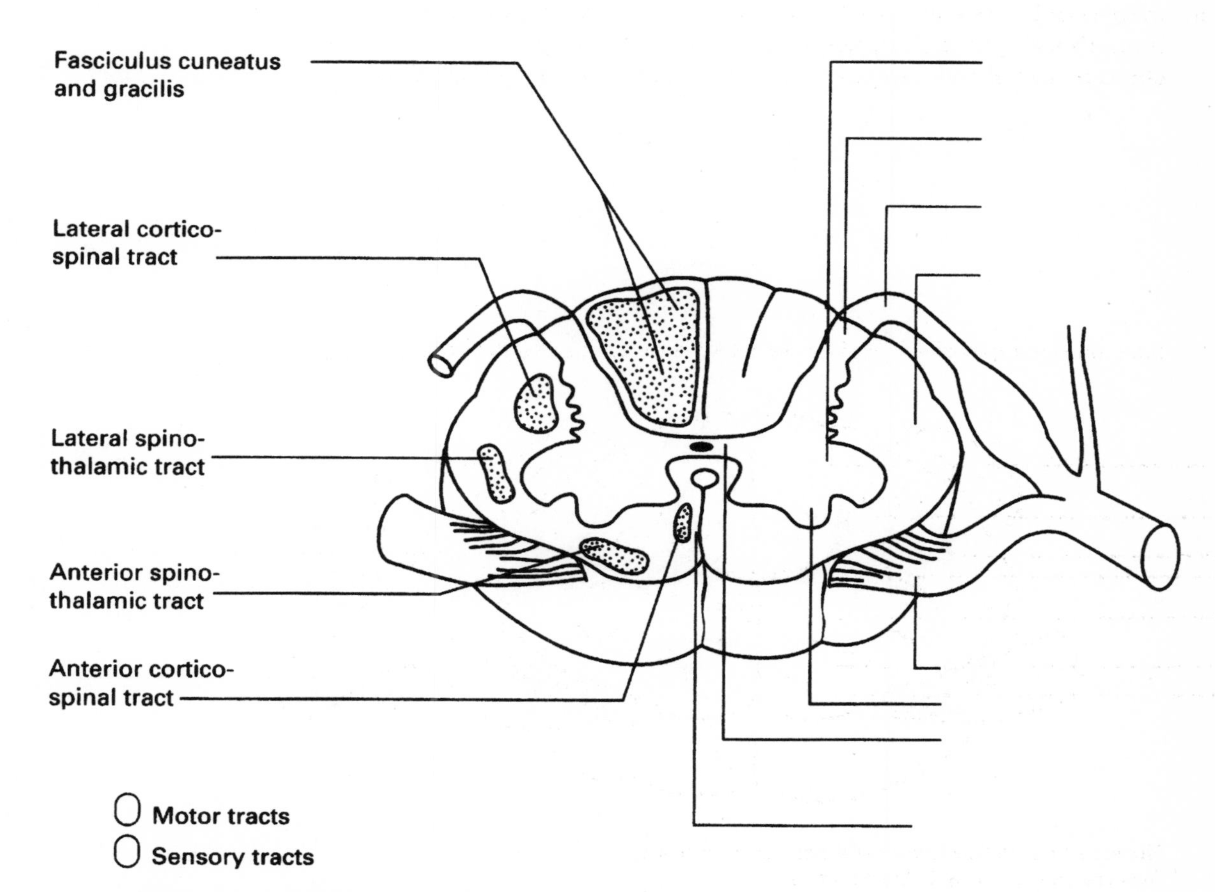

Figure LG 13.3 Major tracts in the spinal cord.
Color and complete as directed in Checkpoints B2, B3, and C1g.

✓ **B2.** Refer to Figure LG 13.3 and do this activity on the conduction function of the spinal cord.

a. Tracts conduct nerve impulses in the *(central? peripheral?)* nervous system. Functionally they are comparable to ____________________ in the peripheral nervous system.

Tracts are located in *(gray horns? white columns?)*. They appear white because they consist of bundles of *(myelinated? unmyelinated?)* nerve fibers.

b. Ascending tracts are all *(sensory? motor?)*, conveying impulses between the spinal cord and the ___________________. All motor tracts in the cord are *(ascending? descending?)*.

c. Color the five tracts on the figure, using color code ovals to demonstrate sensory or motor function of tracts. (*Note:* Tracts are actually present on both sides of the cord but are shown on only one side here.)

d. Write next to the name of each tract its correct functions. Use these answers:

Pain, temperature, itching, tickling, deep pressure, poorly localized touch

Precise voluntary movements

Proprioception, light pressure, vibration, and two-point discriminative touch

e. The name lateral corticospinal indicates that the tract is located in the _______________ white column, that it originates in the *(cerebral cortex? thalamus? spinal cord?)*, and that it ends in the ______________________.

f. The lateral corticospinal tract is *(ascending, sensory? descending, motor?)* and it is a(n) *(direct? indirect?)* tract.

g. The rubrospinal, tectospinal, and vestibulospinal tracts are *(direct? indirect?)* tracts. These are involved with control of: _________.

A. Precise, voluntary movements

B. Automatic movements such as maintenance of muscle tone, posture, and equilibrium

h. Now label all structures with leader lines on the right side of the figure.

B3. *For extra review.* Draw and label the following tracts on Figure LG 13.3. Then color them according to the color code ovals on the figure: *anterior spinocerebellar, posterior spinocerebellar, rubrospinal, tectospinal,* and *vestibulospinal.*

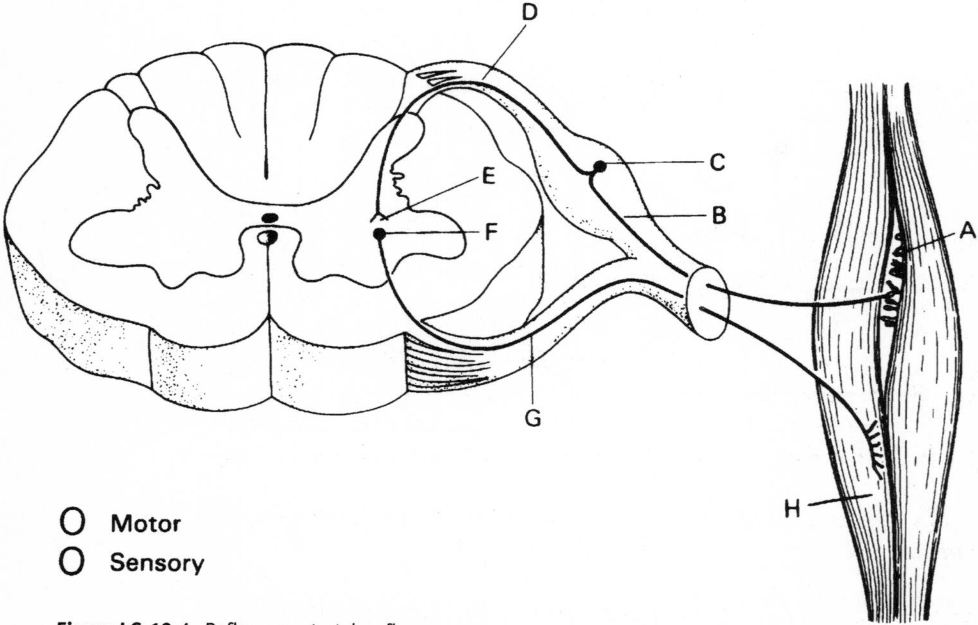

Figure LG 13.4 Reflex arc: stretch reflex.
Color and label as directed in Checkpoints B4 and B5.

✓ **B4.** Checkpoint B1b describes the function of the spinal cord and spinal nerves in reflexes. Color lightly the sensory and motor roots of a spinal nerve on Figure LG 13.4. Select colors according to the color code ovals there.

✓ **B5.** Complete this activity on the function of the spinal cord as a reflex center. Label structures A–H on Figure LG 13.4, using the following terms: *effector, motor neuron axon, motor neuron cell body, receptor, sensory neuron axon, sensory neuron cell body, sensory neuron dendrite,* and *synapse* (integrating center). Note that these structures are lettered in alphabetical order along the conduction pathway of a reflex arc. Add arrows showing the direction of nerve transmission in the arc.

✓ **B6.** Answer these questions about the reflex arc in Figure LG 13.4.

a. How many neurons does this reflex contain? ________ The neuron that conveys the impulse toward the spinal cord is a ____________________ neuron; the one that carries the impulses toward the effector is a ____________________ neuron.

b. This is a *(monosynaptic? polysynaptic?)* reflex arc. The synapse, like all somatic synapses, is located in the *(CNS? PNS?)*.

c. Receptors, in this case located in skeletal muscle, are called ____________________. They are sensitive to changes in ____________________. This type of reflex might therefore be called a ____________________ reflex.

d. Because sensory impulses enter the cord on the same side as motor impulses leave, the reflex is called *(ipsilateral? contralateral? intersegmental?)*.

e. What structure is the effector? ____________________

f. One example of a stretch reflex is the ____________________, in which stretching of the patellar tendon initiates the reflex.

✓ **B7.** Explain how the brain may get the message that a stretch reflex (or other type of reflex) has occurred.

✓ **B8.** Do this exercise describing how tendon reflexes protect tendons.

a. Receptors located in tendons are named *(muscle spindles? tendon organs?)*. They are sensitive to changes in muscle *(length? tension?)*, as when the hamstring muscles are contracted excessively, pulling on tendons.

b. When this occurs, association neurons cause *(excitation? inhibition?)* of this same muscle, so that the hamstring fibers *(contract further? relax?)*.

c. Simultaneously, other association neurons fire impulses that stimulate *(synergistic? antagonistic?)* muscles (such as the quadriceps in this case). These muscles then *(contract? relax?)*.

d. The net effect of such a *(mono? poly?)*-synaptic tendon reflex is that the tendons are *(protected? injured?)*.

✓ **B9.** Contrast stretch, flexor, and crossed extensor reflexes in this learning activity.

a. A flexor reflex *(does? does not?)* involve association neurons, and so it is *(more? less?)* complex than a stretch reflex.

b. A flexor reflex sends impulses to *(one? several?)* muscle(s), whereas a stretch reflex, such as the knee jerk, activates *(one? several?)* muscle(s), such as the quadriceps. A flexor reflex is also known as a ____________________ reflex, and it is ____________________ -lateral.

c. If you simultaneously contract the flexor and extensor (biceps and triceps) muscles of your forearm with equal effort, what action occurs? (Try it.) ____________________
In order for movement to occur, it is necessary for extensors (triceps) to be inhibited while flexors (biceps) are stimulated. The nervous system exhibits such control by a phenomenon known as ____________________ innervation.

d. Reciprocal innervation also occurs in the following instance. Suppose you step on a tack under your right foot. You quickly withdraw that foot by *(flexing? extending?)* your right leg using your hamstring muscles. (Stand up and try it.) What happens to your left leg? You *(flex? extend?)* it. This is an example of reciprocal innervation involving a ____________________ reflex.

e. A crossed extensor reflex is *(ipsilateral? contralateral?)*. It *(may? may not?)* be intersegmental. Therefore, many muscles may be contracted to extend your left thigh and leg to shift weight to your left side and provide balance during the tack episode.

✓ **B10.** Why are deep tendon reflexes (that involve stretching a tendon such as that of the quadriceps femoris muscle in the patellar reflex) particularly helpful diagnostically?

✓ **B11.** *Clinical correlation.* Discuss a commonly used method for assessing health of the nervous system in this Checkpoint.

a. How is the Babinski sign assessed? Demonstrate the assessment technique on a friend.

Note: For (b–d), choose from these answers:

A. Extending the great toe with or without fanning of the other toes (the Babinski sign)

B. Curling toes under (plantar flexion)

b. In children over 18 months and in adults, what is the normal response? *(A? B?)*

c. In babies under 18 months, what is the normal response? *(A? B?)*

d. If corticospinal tracts lack myelination, which response is likely to occur? *(A? B?)*
State two reasons why these tracts may have incomplete myelination.

C. Spinal nerves (pages 433–446)

✓ **C1.** Complete this exercise about spinal nerves.

a. Normally, the human nervous system includes ________ pairs of spinal nerves. Write the number of pairs in each region.

________ Cervical ________ Thoracic ________ Lumbar ________ Sacral ________ Coccygeal

b. Which of these spinal nerves form the cauda equina?

c. Spinal nerves are attached by two roots. The posterior root is *(sensory? motor? mixed?)*,

and the anterior root is ______________________, whereas the spinal nerve is

______________________.

d. Individual nerve fibers are wrapped in a connective tissue covering known as *(endo? epi? peri?)* -neurium. Groups of nerve fibers are held in bundles (fascicles) by

________-neurium. The entire nerve is wrapped with ________-neurium.

e. Spinal nerves branch when they leave the intervertebral foramen. These branches are

called ______________________. Because they are extensions of spinal nerves, rami are *(sensory? motor? mixed?)*.

f. Which ramus is larger? *(Ventral? Dorsal?)* What areas does it supply?

g. What area does the dorsal ramus innervate? Label it on Figure LG 13.3.

h. Name two other branches (rami) and state their functions.

✓ **C2.** Match the plexus names in the box with descriptions. Refer to Figure 13.2, page 422 for help.

B. Brachial	**L. Lumbar**
C. Cervical	**S. Sacral**
I. Intercostal	

________ a. Provides the entire nerve supply for the arm

________ b. Contains origin of phrenic nerve (nerve that supplies diaphragm)

________ c. Forms median, radial, and axillary nerves

________ d. Not a plexus at all, but rather segmentally arranged nerves

________ e. Supplies nerves to scalp, neck, and part of shoulder and chest

________ f. Supplies fibers to the femoral nerve, which innervates the quadriceps, so injury to this plexus would interfere with actions such as lifting the knee (flexing hip)

________ g. Forms the largest nerve in the body (the sciatic), which supplies posterior of thigh and the leg

✓ **C3.** *For extra review.* Match names of nerves in the box with their descriptions. On the line following the description, write the site of origin of the nerve. The first one is done for you.

Axillary	**Musculocutaneous**	**Radial**
Inferior gluteal	**Pudendal**	**Sciatic**

<u>Sciatic</u> a. Consists of two nerves, the tibial and common peroneal; supplies the hamstrings, adductors, and all muscles distal to the knee. <u>L4–S3</u>

________ b. Supplies the deltoid muscle. _______

________ c. Innervates the major flexors of the arm. _______

________ d. Supplies most extensor muscles of the forearm, wrist, and fingers; may be damaged by extensive use of crutches. _______

________ e. Supplies the gluteus maximus muscle. _______

________ f. May be anesthetized in childbirth because it innervates external genitalia and lower part of the vagina. _______

C4. *For additional review* of nerve supply to muscles, complete the table below. Write the name of the plexus with which the nerve is associated, as well as the distribution of that nerve. The first one is done for you. (*Hint:* one answer involves a nerve that is not a spinal nerve, so it is not a part of a plexus.)

Name of Nerve	Plexus	Distribution
a. Lateral pectoral	Brachial	Pectoralis major muscles
b. Medial pectoral		
c. Lesser occipital		
d. Ulnar		
e.		Latissimus dorsi muscle
f. Long thoracic		
g.		Trapezius and sternocleidomastoid muscles (*Hint:* See Table 14.3, page 489 of text.)
h. Femoral		
i.	Lumbar	Cremaster muscle that pulls testes closer to pelvis
j. Obturator		
k.		Gastrocnemius, soleus, tibialis posterior
l.		Tibialis anterior

✓ **C5.** *Clinical thinking.* Trevor's cord is completely transected (severed) just below the C7 spinal nerves. How are the functions listed below likely to be affected? (Remember that nerves that originate below this point would not communicate with the brain and so would lose much of their function.) Explain your reasons in each case. (*Hint:* Refer to Figure 13.2 in the text.)

a. Breathing via diaphragm

b. Movement and sensation of thighs and legs

c. Movement and sensation of the arms

d. Use of muscles of facial expression and muscles that move jaw, tongue, eyeballs

✓ **C6.** *A clinical correlation.* Refer to diagrams in Figure 13.14 and descriptions in Exhibit 3.4 of the text and assume each of those positions. Then select the name of each condition listed in the box that is associated with the nerve injury described below.

CH.	**Clawhand**	**SC.**	**Sciatica**
CV.	**Calcaneovalgus**	**WD.**	**Wristdrop**
ED.	**Erb-Duchenne palsy (waiter's tip)**	**WS.**	**Winged scapula**
FDE.	**Footdrop and equinovarus**		

_________ a. Radial nerve injury leading to inability to extend the wrist and fingers

_________ b. Ulnar nerve palsy with loss of sensation in the little finger

_________ c. Injury to superior roots of the brachial plexus (C5–C6)

_________ d. Injury of the long thoracic nerve supply to serratus anterior muscles

_________ e. Injury to the tibial portion of the sciatic nerve causing "knock-kneed" position

_________ f. Injury to the common peroneal nerve causing plantar flexion and eversion ("bowlegged") position

_________ g. Pain over buttocks and much of the posterior of the lower extremity often due to a herniated L5–S1 disc

C7. Describe the general pattern of dermatomes:

a. In the trunk

b. In the limbs

C8. *A clinical correlation.* Explain how knowledge of dermatomes and related spinal cord segments can be used clinically.

D. Disorders (page 447)

✓ **D1.** Shingles is an infection of the *(central? peripheral?)* nervous system. The causative virus is also the agent of *(chickenpox? measles? cold sores?)*. Following recovery from chickenpox, the virus remains in the body in the *(spinal cord? dorsal root ganglia?)*. At times it is activated and travels along *(sensory? motor?)* neurons, causing *(pain? paralysis?)*.

✓ **D2.** Match names of disorders in the box with definitions below.

N.	**Neuritis**	**Sc.**	**Sciatica**
P.	**Poliomyelitis**	**Sh.**	**Shingles**

________ a. Inflammation of a single nerve

________ b. Also known as infantile paralysis; caused by a virus that may destroy motor cell bodies in the brainstem or in the anterior gray horn of the spinal cord

________ c. Acute inflammation of the nervous system by *Herpes zoster* virus

________ d. Neuritis of a nerve in the posterior of hip and thigh; often due to a slipped disc in the lower lumbar region

ANSWERS TO SELECTED CHECKPOINTS: CHAPTER 13

A1. Processing center for reflexes, integration of afferent and efferent nerve impulses, and conduction pathway for those impulses.

A2. Central (CNS).

Figure LG 13.1A Meninges.

A3. (d) Meningitis. (e) Pia mater, dura mater; anchor the cord and protect it from displacement.

A4. (a) Waist; L1–L2; vertebral column; 16–18. (b) Cervical, lumbar.

A5. (a) Co. (b) S. (c) F. (d) Ca.

A6. (a) To insert anesthetics, antibiotics, chemotherapy or contrast media; to withdraw cerebrospinal fluid (CSF) for diagnostic purposes such as for analysis for blood or microorganisms; subarachnoid. (b) L3–L4 or L4–L5; the spinal cord ends at about L1–L2, so the cord is not likely to be injured at this lower level. (c) The iliac crest and umbilicus (navel) are both at about this level. (d) The anesthetic must penetrate epidural tissues and then all three layers of meninges before reaching nerve fibers, whereas an anesthetic in the subarachnoid space needs to penetrate only the pia mater to reach nerve tissue; childbirth (labor and delivery).

A7 (a) Within, outside; refer to Figure 13.3 of the text. (b) A, posterior white columns; B, posterior gray horn; C, lateral white columns; D, lateral gray horn; E, anterior gray horn; F, anterior white columns; G, anterior median fissure.(c) Dorsal root ganglion.

B1. (a) Conduction (or a pathway), tracts. (b) Reflex; 2; posterior (dorsal), anterior (ventral). See Figure LG 13.3A.

B2. (a) Central; nerves; white columns; myelinated. (b) Sensory, brain; descending. (c–d) See Figure LG 13.3A. (e) Lateral, cerebral cortex, spinal cord (in anterior gray horn). (f) Descending, motor, direct. (g) Indirect; B (h) See Figure LG 13.3A.

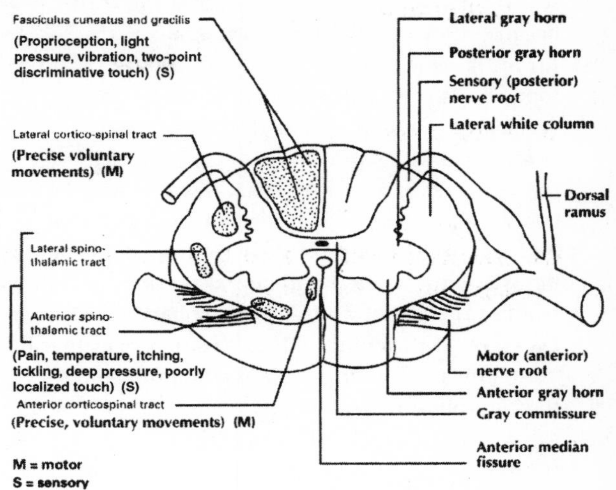

Figure LG 13.3A Major tracts in the spinal cord.

B4. Refer to Figure LG 13.4A; B and D are sensory; G is motor.

B5.

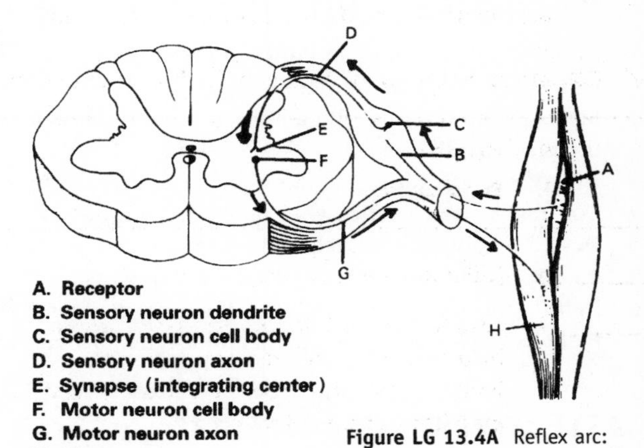

Figure LG 13.4A Reflex arc: stretch reflex.

B6. (a) 2; sensory, motor. (b) Monosynaptic; CNS. (c) Muscle spindles; length (or stretch); stretch. (d) Ipsilateral. (e) Skeletal muscle. (f) Knee jerk (patellar reflex).

B7. Branches from axons of sensory or association neurons travel through tracts to the brain.

B8. (a) Tendon organs; tension. (b) Inhibition, relax. (c) Antagonistic: contract. (d) Poly, protected.

B9. (a) Does, more. (b) Several, one; withdrawal, ipsi. (c) No action; reciprocal. (d) Flexing; extend; crossed extensor. (e) Contralateral; may.

B10. They can readily pinpoint a disorder of a specific spinal nerve or plexus, or portion of the cord, because they do not involve the brain.

B11. (a) Stroking the lateral margin of the sole of the foot. (b) B. (c) A. (d) A; developmental stage (under 18 months) or damage to these tracts in older persons.

C1. (a) 31; 8, 12, 5, 5, 1. (b) Lumbar, sacral, and coccygeal. (c) Sensory, motor, mixed. (d) Endo; peri; epi. (e) Rami; mixed. (f) Ventral: all of the limbs and the ventral and lateral portions of the trunk. (g) Muscles and skin of the back; see Figure LG 13.3A. (h) Meningeal branch supplies primarily vertebrae and meninges; rami communicantes have autonomic functions.

C2. (a) B. (b) C. (c) B. (d) I. (e) C. (f) L. (g) S.

C3. (b) Axillary, C5–C6. (c) Musculocutaneous, C5–C7. (d) Radial, C5–C8, T1. (e) Inferior gluteal, L5–S2. (f) Pudendal, S2–S4.

C5. (a) Not affected because (phrenic) nerve to the diaphragm originates from the cervical plexus (at C3–C5), higher than the transection. So this nerve continues to receive nerve impulses from the brain. (b) Complete loss of sensation and paralysis because lumbar and sacral plexuses originate below the injury and therefore no longer communicate with the brain. (c) Most arm functions are not affected. As shown on Figure 13.13 of the text, the brachial plexus originates from C5 through T1, so most nerves to the arm (those from C5 through C7) still communicate with the brain. (d) Not affected because all are supplied by cranial nerves which originate from the brain.

C6. (a) WD. (b) CH. (c) ED. (d) WS. (e) CV. (f) FDE. (g) S.

D1. Peripheral; chickenpox; dorsal root ganglia; sensory, pain

D2. (a) N. (b) P. (c) Sh. (d) Sc.

MORE CRITICAL THINKING: CHAPTER 13

1. Describe the anatomy and physiology of the meninges, as well as the spaces formed between and surrounding meninges.
2. State several reasons why a lumbar puncture might be done. Give a brief description of the procedure.
3. Describe a reflex arc, including the structure and function of its five components.
4. Contrast a stretch reflex with a flexor (withdrawal) reflex.
5. Describe how reflexes help you to maintain your balance when you quickly pick up your foot in response to stepping on a piece of glass.
6. Define the term plexus and describe what is meant by ventral rami that form a plexus. State examples of two plexuses and the major nerves they form.

MASTERY TEST: CHAPTER 13

Questions 1–4: Arrange the answers in correct sequence.

______ ______ ______ 1. From superficial to deep:
A. Subarachnoid space
B. Epidural space
C. Dura mater

______ ______ ______ 2. From anterior to posterior in the spinal cord:
A. Posterior columns
B. Anterior spinothalamic tract
C. Central canal of the spinal cord

______ ______ ______ ______ 3. The plexuses, from superior to inferior:
A. Lumbar
B. Brachial
C. Cervical
D. Sacral

______ ______ ______ ______ ______ 4. Order of structures in a conduction pathway, from origin to termination:
A. Motor neuron
B. Sensory neuron
C. Integrative center
D. Receptor
E. Effector

Questions 5–11: Circle the letter preceding the one best answer to each question.

5. Herniation (or "slipping") of the disc between L4 and L5 vertebrae is most likely to result in damage to the _______ nerve.
   A. Femoral
   B. Sciatic
   C. Radial
   D. Musculocutaneous
6. All of these tracts are sensory *except:*
   A. Anterior spinothalamic
   B. Lateral spinothalamic
   C. Posterior columns
   D. Lateral corticospinal
   E. Posterior spinocerebellar
7. Choose the *false* statement about the spinal cord.
   A. It has enlargements in the cervical and lumbar areas.
   B. It lies in the vertebral foramen.
   C. It extends from the medulla to the sacrum.
   D. It is surrounded by meninges.
   E. In cross section an H-shaped area of gray matter can be found.
8. All of these structures are composed of white matter *except:*
   A. Posterior root (spinal) ganglia
   B. Lumbar plexus
   C. Sciatic nerve
   D. Ventral ramus of a spinal nerve
9. Which is a *false* statement about the patellar reflex?
   A. It is also called the knee jerk.
   B. It involves a two-neuron, monosynaptic reflex arc.
   C. It results in extension of the leg by contraction of the quadriceps femoris.
   D. It is contralateral.
10. The cauda equina is:
    A. Another name for the cervical plexus
    B. The lumbar and sacral spinal nerve roots extending below the end of the cord and resembling a horse's tail
    C. The inferior extension of the pia mater
    D. A denticulate ligament
    E. A canal running through the center of the spinal cord
11. Choose the one *false* statement.
    A. The Babinski sign *is normal* in a 10-month-old child but *not* in a 10-year-old child.
    B. Femoral nerve injury is more likely to cause loss of function of *hamstring* muscles than of *quadriceps* muscles.
    C. All spinal nerves except C1 emerge from below the vertebra of the same number.
    D. The axillary nerve supplies the deltoid muscle.

Questions 12–20: Circle T (true) or F (false). If the statement is false, change the underlined word or phrase so that the statement is correct.

T F 12. The two main functions of the spinal cord are that it serves as a reflex center and it is the site where sensations are felt.

T F 13. Dorsal roots of spinal nerves are sensory, ventral roots are motor, and spinal nerves are mixed.

T F 14. A tract is a bundle of nerve fibers inside the central nervous system (CNS).

T F 15. Synapses are present in posterior (dorsal) root ganglia.

T F 16. After a person reaches 18 months, the Babinski sign should be negative, as indicated by plantar flexion (curling under of toes and foot).

T F 17. Visceral reflexes are used diagnostically more often than somatic ones because it is easy to stimulate most visceral receptors.

T F 18. Transection (cutting) of the spinal cord at level C6 will result in greater loss of function than transection at level T6.

T F 19. The ventral root of a spinal nerve contains axons and dendrites of both motor and sensory neurons.

T F 20. A lumbar puncture (spinal tap) is usually performed at about the level of vertebrae L1 to L2 because the cord ends between about L3 and L4.

Questions 21–25: Fill-ins. Complete each sentence with the word or short answer that best fits.

____________________ 21. The phrenic nerve innervates the ________.

____________________ 22. An inflammation of the dura mater, arachnoid, and/or pia mater is known as __________.

____________________ 23. Tendon reflexes are ________-synaptic and ________-lateral.

____________________ 24. The layer of the meninges that gets its name from its delicate structure, which is much like a spider's web, is the ________.

____________________ 25. The filum terminale and denticulate ligaments are both composed of __________ mater.

ANSWERS TO MASTERY TEST: CHAPTER 13

Arrange

1. B C A
2. B C A
3. C B A D
4. D B C A E

Multiple Choice

5. B
6. D
7. C
8. A
9. D
10. B
11. B

True–False

12. F. Reflex center, a conduction site
13. T
14. T
15. F. Are not
16. T
17. F. Less often than somatic ones because it is difficult to stimulate most visceral receptors
18. T
19. F. Axons of motor
20. F. L3–L4, L1–L2

Fill-ins

21. Diaphragm
22. Meningitis
23. Poly, ipsi
24. Arachnoid
25. Pia

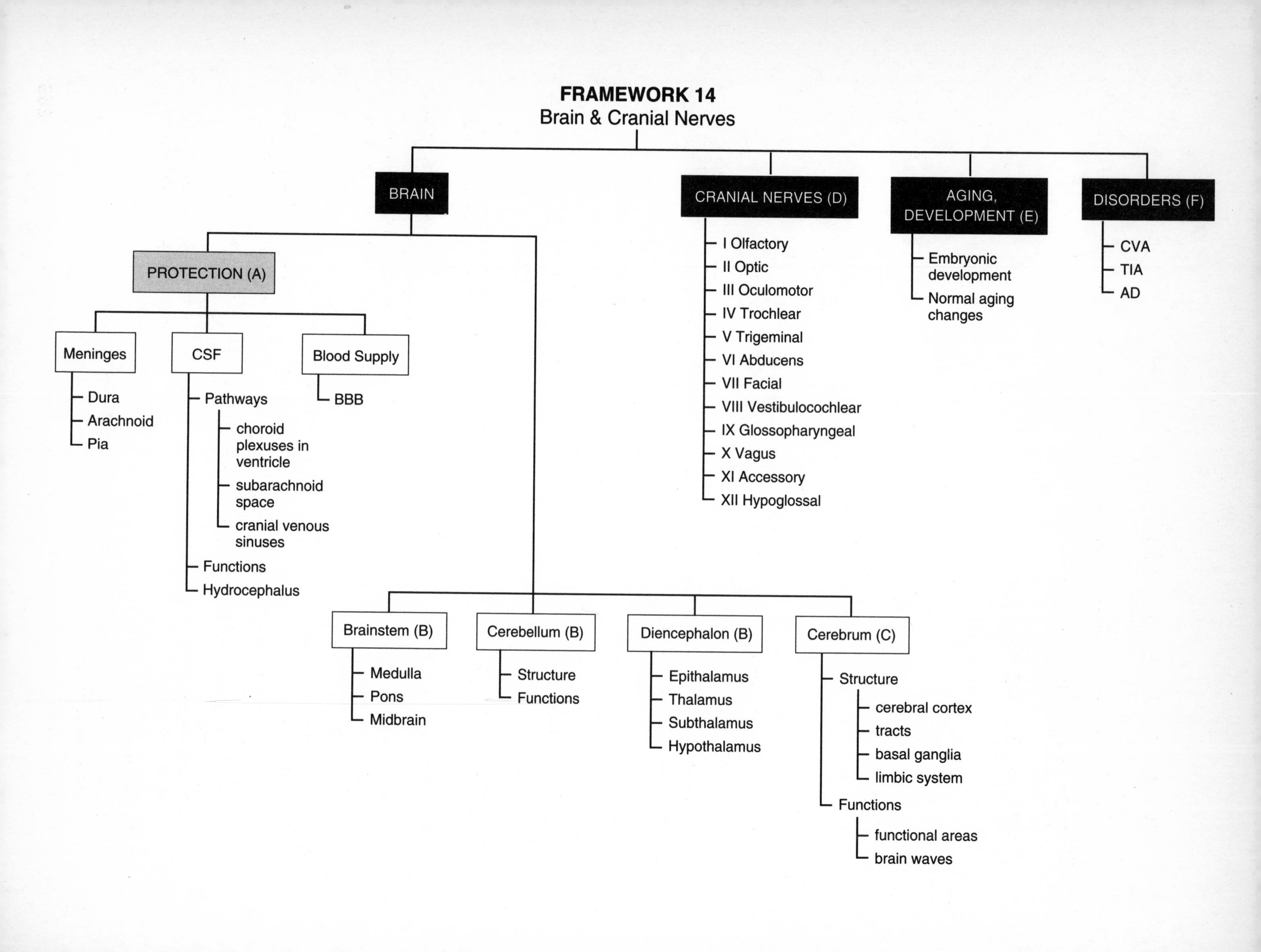

FRAMEWORK 14
Brain & Cranial Nerves
BRAIN
PROTECTION (A)
Meninges
Dura
Arachnoid
Pia
CSF
Pathways
choroid plexuses in ventricle
subarachnoid space
cranial venous sinuses
Functions
Hydrocephalus
Blood Supply
BBB
Brainstem (B)
Medulla
Pons
Midbrain
Cerebellum (B)
Structure
Functions
Diencephalon (B)
Epithalamus
Thalamus
Subthalamus
Hypothalamus
Cerebrum (C)
Structure
cerebral cortex
tracts
basal ganglia
limbic system
Functions
functional areas
brain waves
CRANIAL NERVES (D)
I Olfactory
II Optic
III Oculomotor
IV Trochlear
V Trigeminal
VI Abducens
VII Facial
VIII Vestibulocochlear
IX Glossopharyngeal
X Vagus
XI Accessory
XII Hypoglossal
AGING, DEVELOPMENT (E)
Embryonic development
Normal aging changes
DISORDERS (F)
CVA
TIA
AD

C H A P T E R **14**

The Brain and Cranial Nerves

The brain is the major control center for the global communication network of the body: the nervous system. The brain requires round-the-clock protection and maintenance afforded by bones, meninges, cerebrospinal fluid, a special blood—brain barrier, along with a fail-safe blood supply. This vital control center consists of four major substructures: the brain stem, diencephalon, cerebrum, and cerebellum. Each brain part carries out specific functions and each releases specific chemical neurotransmitters. Twelve pairs of cranial nerves convey information to and from the brain.

Structural defects may occur in construction (brain development) and also with the normal wear and tear that accompanies aging. Just as a complex computer network may experience minor disruptions in service or major shutdowns, disorders within the brain or cranial nerves may lead to minor, temporary changes in nerve functions or profound and fatal outcomes.

As you begin your study of the brain, carefully examine the Chapter 14 Topic Outline and Objectives; check off each one as you complete it. To organize your study of the brain, glance over the Chapter 14 Framework now. Be sure to refer to the Framework frequently and note relationships among key terms in each section.

TOPIC OUTLINE AND OBJECTIVES

A. Overview of brain: organization, protection, and nourishment

1. Identify the major parts of the brain.
2. Describe how the brain is protected.
3. Explain the formation and circulation of cerebrospinal fluid.

B. The brain stem, cerebellum, and diencephalon

1. Describe the structure and functions of the brain stem.
2. Describe the structure and functions of the cerebellum.
3. Describe the components and functions of the diencephalon.

C. The cerebrum

1. Describe the cortex, convolutions, fissures, and sulci of the cerebrum.
2. List and locate the lobes of the cerebrum.
3. Describe the nuclei that comprise the basal ganglia.
4. List the structures and describe the functions of the limbic system.
5. Describe the locations and functions of the sensory, association, and motor areas of the cerebral cortex.

D. Cranial nerves

1. Identify the cranial nerves by name, number, and type, and give the functions of each.

E. Developmental anatomy and aging of the nervous system

1. Describe how the parts of the brain develop.
2. Describe the effects of aging on the nervous system.

F. Disorders

WORDBYTES

Now become familiar with the language of this chapter by studying each wordbyte, its meaning, and an example of its use within a term. After you study the entire list, self-check your understanding by writing the meaning of each wordbyte on the line. As you continue through the *Learning Guide*, identify (and fill in) additional terms that contain the same wordbyte.

Wordbyte	Self-check	Meaning	Example(s)
-algia	____________	pain	neur*algia*, an*alg*esia
callosum	____________	hard	corpus *callosum*
cauda-	____________	tail	*cauda*te
cephalo-	____________	head	hydro*cephal*ic
collicus	____________	little hill	superior *colliculi*
corpus	____________	body	*corpus* callosum
cortex	____________	bark	cerebral *cortex*
decuss-	____________	cross	*decuss*ation
di-	____________	through	*di*encephalon
-ellum	____________	little	cereb*ellum*
epi-	____________	above	*epi*thalmus
enceph-	____________	brain	di*enceph*alon
falx	____________	sickle	*falx* cerebri
glossi-	____________	tongue	hypo*gloss*al
hemi-	____________	half	*hemi*sphere
hypo-	____________	under	*hypo*thalamus
mes-	____________	middle	*mes*encephalon
nigr-	____________	black	substantia *nigr*a
oculo-	____________	eye	*oculo*motor nerve
ophthalm-	____________	eye	*ophthalm*ic
opti-	____________	eye	*opti*c nerve
-phasia	____________	speech	a*phasia*
pineal	____________	pineconelike	*pineal* gland
pons	____________	bridge	*pon*tine nuclei
pros-	____________	before	*pros*encephalon
rhomb-	____________	behind	*rhomb*encephalon
tri-	____________	three	*tri*geminal nerve
trochle-	____________	pulley	*trochle*ar nerve

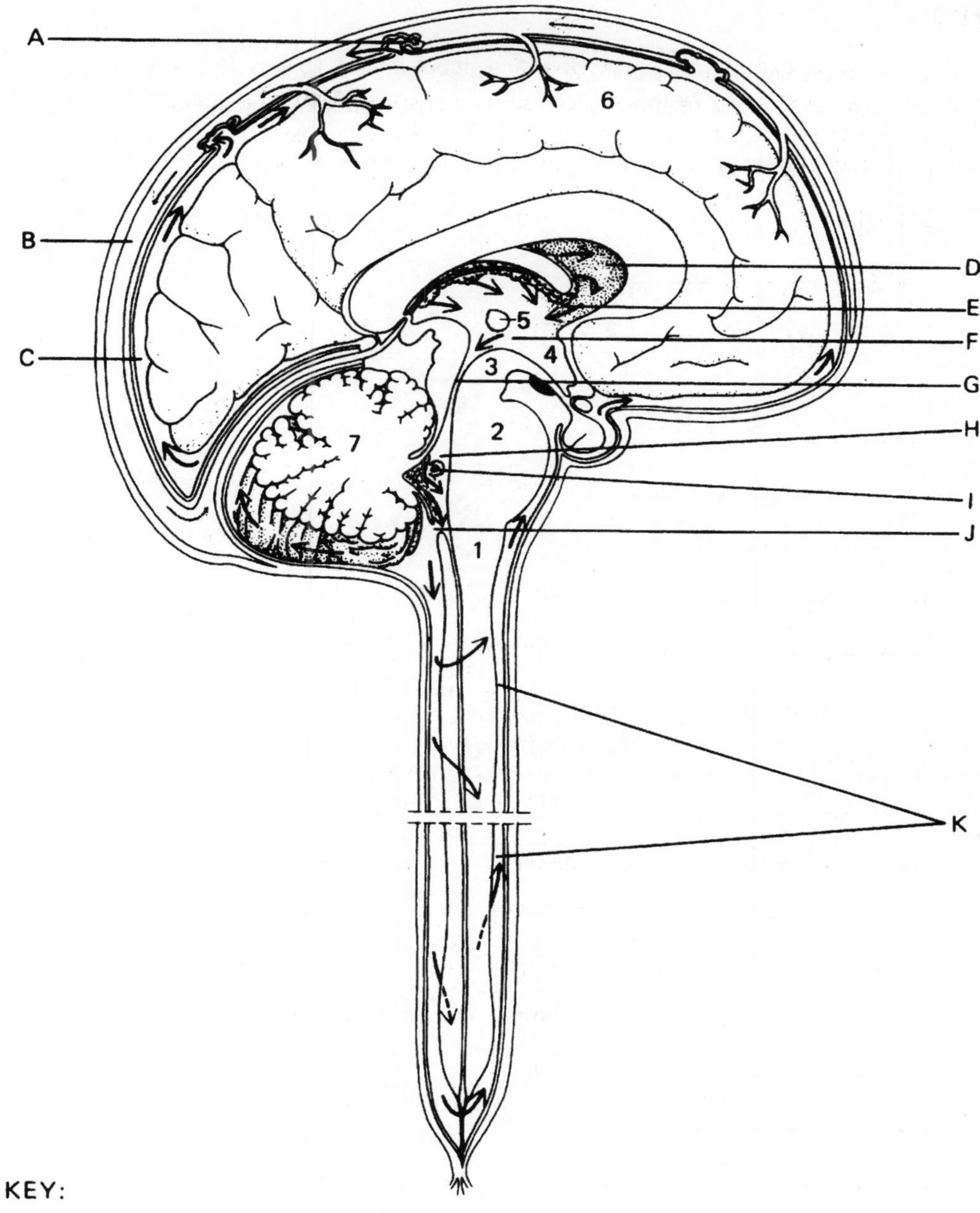

KEY:

1. Medulla oblongata
2. Pons
3. Midbrain
4. Hypothalamus
5. Thalamus
6. Cerebrum
7. Cerebellum

A. Arachnoid villus
B. Cranial venous sinus
C. Subarachnoid space of brain
D. Lateral venticle
E. Interventricular foramen
F. Third ventricle
G. Cerebral aqueduct
H. Fourth ventricle
I. Lateral aperture **(of Luschka)**
J. Median aperture **(of Magendie)**
K. Subarachnoid space of spinal cord

Figure LG 14.1 Brain and meninges seen in sagittal section. Parts of the brain are numbered; refer to Checkpoint A1. Letters indicate pathway of cerebrospinal fluid (CSF); refer to Checkpoint A8.

CHECKPOINTS

A. Overview of the brain: organization, protection, and nourishment (pages 452–458)

✓ **A1.** Identify numbered parts of the brain on Figure LG 14.1. Then complete this exercise.

a. Structures 1–3 are parts of the ______________________.

b. Structures 4 and 5 together form the ______________________.

c. Structure 6 is the largest part of the brain, the ______________________.

d. The second largest part of the brain is structure 7, the ______________________.

A2. Briefly state results of oxygen starvation of the brain for about 4 minutes.

List effects of glucose deprivation of the brain.

✓ **A3.** Describe the blood-brain barrier (BBB) in this exercise.

a. Blood capillaries supplying the brain are *(more? less?)* leaky than most capillaries of the body. This fact is related to a large number of *(gap? tight?)* junctions, as well as an abundance of neuroglia named ______________________ pressed against capillaries.

b. What advantages are provided by this membrane?

c. List two or more substances needed by the brain that do normally cross the BBB.

d. What problems may result from the fact that some chemicals cannot cross this barrier?

✓ **A4.** List three ways in which the brain is protected.

A5. Review the layers of the meninges covering the brain and spinal cord by listing them here. *For extra review,* label the layers on Figure LG 14.1 and review LG Chapter 13, Checkpoint A3.

✓ **A6.** Identify the extensions of dura mater in this Checkpoint. Use answers in the box.

Falx cerebelli	**Falx cerebri**	**Tentorium cerebelli**

a. A tentlike structure that separates the cerebrum from the cerebellum:

_______________________.

b. Separates the two hemispheres of the cerebellum: _______________________.

c. Separates the two hemispheres of the cerebrum: _______________________.

✓ **A7.** Circle all correct answers about CSF.

a. The entire nervous system contains 80–150 mL of CSF. This amount is equal to approximately:
   A. 1 to 2 tablespoons
   B. 1/3 to 2/3 cup
   C. 1 to 2 cups
   D. 1 quart

b. The color of CSF is normally:
   A. Yellow
   B. Clear, colorless
   C. Red
   D. Green

c. Choose the function(s) of CSF.
   A. Serves as a shock absorber for brain and cord
   B. Contains red blood cells
   C. Contains some white blood cells
   D. Contains nutrients such as glucose and proteins, ions, and wastes such as urea

d. Which statement(s) describe its formation?
   A. It is formed by diffusion of substances from blood.
   B. It is formed by filtration and secretion.
   C. It is formed from blood in capillaries called choroid plexuses.
   D. It is formed by ependymal cells that line all four ventricles.

e. Which statement(s) describe its pathway?
   A. It circulates around the brain but not the cord.
   B. It flows inferior to the end of the spinal cord.
   C. It bathes the brain by flowing through the epidural space.
   D. It passes via projections (villi) of the arachnoid into blood vessels (venous sinuses) surrounding the brain.
   E. It is formed initially from blood and finally flows back to blood.

f. An excessive accumulation of CSF within the ventricles is a condition known as:
   A. Hydrarthrosis
   B. Hydrophobia
   C. Hydrocephalus
   D. Hydrocholecystitis

✓ **A8.** To check your understanding of the pathway of cerebrospinal fluid (CSF), list in order the structures through which it passes. Use key letters on Figure LG 14.1. Start at the site of formation of CSF.

B. Brain: brain stem, cerebellum, and diencephalon (pages 458–467)

✓ **B1.** Describe the principal functions of the medulla in this exercise.

a. The medulla serves as a ____________________ pathway for all ascending and descending tracts. Its white matter therefore transmits *(sensory? motor? both sensory and motor?)* impulses.

b. Included among these tracts are the triangular ____________________ tracts which are the principal *(sensory? motor?)* pathways. The main fibers that pass through the pyramids are the *(spinothalamic? corticospinal?)* tracts.

c. Crossing (or ____________________) of fibers occurs in the medulla. This explains why movements of your right hand are initiated by motor neurons that originate in the *(right? left?)* side of your cerebrum. The *(axons? dendrites?)* of these motor neurons decussate in the medulla to proceed down the right lateral

________________-spinal tract.

d. The medulla contains gray areas as well as white matter. Several important nuclei lie in the medulla. Two of these are synapse points for sensory axons that have ascended in the posterior columns of the cord. These two nuclei, named ____________________

and ____________________, transmit impulses for several types of sensations. Name four.

____________________, and ____________________.

e. A hard blow to the base of the skull can be fatal because the medulla is the site of

two vital centers: the ____________________ center, regulating the heart; and the

____________________ center, adjusting the rhythm of breathing.

f. Some input to the medulla arrives by means of cranial nerves; these nerves may serve motor functions also. Which cranial nerves are attached to the medulla?

____

(Functions of these nerves will be discussed later in this chapter.)

✓ **B2.** Summarize important aspects of the pons in this learning activity.

a. The name *pons* means ____________________. It serves as a bridge in two ways.

It contains longitudinally arranged fibers that connect the ____________________

and ____________________ with the upper parts of the brain. It has transverse fibers

that connect the two sides of the ____________________.

b. Cell bodies associated with fibers in cranial nerves numbered ____________________ lie in nuclei in the pons.

c. *(Respiration? Heartbeat? Blood pressure?)* is controlled by the pneumotaxic and apneustic areas of the pons.

✓ **B3.** Describe the midbrain in this exercise.

a. Like the pons and medulla, the midbrain is about 1 inch (________ cm) long.

b. The midbrain is more *(anterior and superior? posterior and inferior?)* compared to the pons and medulla.

c. Cerebral peduncles contain *(axons? cell bodies?)* of *(motor? sensory? both motor and sensory?)* neurons.

d. *(Inferior? Superior?)* colliculi are involved with head movements in response to *(auditory? visual?)* stimuli.

✓ **B4.** You maintain a conscious state, or you wake up from sleep, thanks to the regions of the brain known as the RAS, or R_________________ A_________________ S_________________. These areas are especially sensitive to sensations such as _________________ from ears and _________________ or _________________ from skin. The RAS is part of the reticular formation. Where is the reticular formation located?

B5. Describe the cerebellum

a. Where is it located?

b. Describe its structure. Include these terms in your description: *vermis, lobes, cortex,* and *arbor vitae.*

c. Describe its functions. Use these key terms: *evaluate, coordinate, posture, language.*

✓ **B6.** Identify locations of each of the tracts known as *peduncles* that permit communication between the cerebellum and other brain parts. Fill in lines with the terms in the box.

Inferior	**Middle**	**Superior**

a. Cerebellum → _________________ cerebellar peduncle → Midbrain + thalamus

b. Cerebellum ← _________________ cerebellar peduncle ← Pons ← Cerebral cortex

c. Cerebellum ← _________________ cerebellar peduncle ← Medulla ← Inner ear

✓ **B7.** Describe the *diencephalon* in this Checkpoint.

a. If you look for the diencephalon in most external views of the brain, you will not find it. It is located *(deep within? on the surface of?)* the cerebrum. The diencephalon consists of *(two? four? eight?)* main structures: the thalamus, as well as the ______ -

thalamus, the ______- thalamus, and the ______- thalamus. Which part comprises 80% of the diencephalon? ____________________

b. The epithalamus forms the *(roof? lateral walls? floor?)* of the third ventricle. The epithalamus includes the ____________________ gland and the ____________________ nuclei, as well as the ____________________ plexus of the third ventricle. The pineal gland secretes the hormone named ____________________ and serves as the body's biological ____________________. The habenular nucleus is involved in *(hearing? smell? vision?)*.

c. The subthalamus works in controlling *(sensations? movements?)*.

d. The *(two? four?)* lobes of the thalamus form parts of the *(roof? lateral walls? floor?)* of the third ventricle. A "crossbar" ("intermediate mass") between the two lobes passes through the center of the ____________________ ventricle. (See Figure LG 14.1.)

e. The thalamus is the principal relay station for *(motor? sensory?)* impulses. For example, spinothalamic and lemniscal tracts convey general sensations such as pain, ____________________, ____________________, ____________________, and temperature to the thalamus where they are relayed to the cerebral cortex. Special sense impulses (for vision and hearing) are relayed through the *(geniculate? reticular? ventral posterior?)* nuclei of the thalamus.

f. Although the thalamus plays a primary role in conduction of sensory impulses, it also has other functions. Name three or more.

g. The name hypothalamus indicates that this structure lies *(above? below?)* the thalamus, forming the floor and part of the lateral walls of the ____________________ ventricle.

B8. Expanding on the key words listed below, write a sentence describing major hypothalamic functions.

a. Regulator of visceral activities via control of the ANS

b. Regulating factors to anterior pituitary

c. Feelings (rage or pleasure)

d. Temperature

e. Thirst

f. Reticular formation: arousal and consciousness

✓ **B9.** *For extra review.* Check your understanding of these parts of the brain stem and diencephalon by matching them with the descriptions given below.

H.	**Hypothalamus**	**Mid.**	**Midbrain**	**T.**	**Thalamus**
Med.	**Medulla**	**P.**	**Pons**		

_____ a. It is the principal regulator of visceral activities because it acts as a liaison between the cerebral cortex and autonomic nerves that control viscera.

_____ b. Site of the red nuclei and substantia nigra that help to regulate movement.

_____ c. Cranial nerves III–IV attach to this brain part.

_____ d. Cranial nerves V–VIII attach to this brain part.

_____ e. Cranial nerves VIII–XII attach to this brain part.

_____ f. Feelings of hunger, fullness, and thirst stimulate centers here so that you can respond accordingly.

_____ g. All sensations are relayed through here.

_____ h. Regulation of heart, blood pressure, and respiration occurs by centers located here.

_____ i. It constitutes four-fifths of the diencephalon.

_____ j. It lies under the third ventricle, forming its floor.

_____ k. It forms most of side walls of the third ventricle.

_____ l. Tumor in this region could compress the cerebral aqueduct and cause hydrocephalus.

_____ m. It contains centers for coughing, sneezing, and vomiting.

_____ n. The olivary, gracile, and cuneatus nuclei are located here.

✓ **B10.** *For extra review* of specific locations and functions of nuclei and other masses of gray matter, complete this table. Choose answers from the lists of brain parts and functions. Provide two answers in parts of the table indicated by (2).

Brain part		**Functions**
Cer.	Cerebellum	Emotions, memory
Epi.	Epithalamus	Hearing
Hypo.	Hypothalamus	Movements in general
Med.	Medulla	Movements of head or neck in response to visual stimuli
Mid.	Midbrain	Pupil reflexes
Sub.	Subthalamus	Regulation of anterior pituitary
Thal.	Thalamus	Regulation of posterior pituitary
		Smell
		Taste
		Touch, pressure, vibrations
		Vision

Name of Nucleus	Brain Part	Functions
a. Habenular nucleus	Epi	Smell
b. Subthalamic nucleus		
c. Red nucleus and substantia nigra	(2)	
d. Superior colliculi		
e. Nuclei cuneatus and gracilis		
f. Anterior nucleus		
g. Lateral geniculate nucleus		
h. Medial geniculate nucleus		
i. Ventral posterior nucleus		(2)
j. Mammillary bodies		
k. Median eminence of tuber cinereum		
l. Supraoptic region		

✓ **B11.** Describe *circumventricular organs* (CVOs) in this exercise.

a. Name three organs that are CVOs.

b. CVOs are unique in the brain region in that they *(have? lack?)* the blood–brain barrier. State one advantage and one disadvantage of this structural feature.

C. Brain: cerebrum (pages 467–477)

✓ **Cl.** Complete this exercise about cerebral structure.

a. The outer layer of the cerebrum is called _________________. It is composed of *(white? gray?)* matter. This means that it contains mainly *(cell bodies? tracts?)*.
b. In the margin, draw a line the same length as the thickness of the cerebral cortex. Use a metric ruler. Note how thin the cortex is.
c. The surface of the cerebrum looks much like a view of tightly packed mountains or ridges, called _________________. The parts where the cerebral cortex dips down into valleys are called _________________ (deep valleys) or _________________ (shallow valleys).
d. The cerebrum is divided into halves called _________________. Connecting them is a band of *(white? gray?)* matter called the _________________. Notice this structure in Figure LG 14.1 and in Figures 14.12 and 14.13 in your text.
e. The falx cerebri is composed of *(nerve fibers ? dura mater?)*. Where is it located? _________________ At its superior and inferior margins, the falx is dilated to form channels for venous blood flowing from the brain; these enclosures are called _________________.

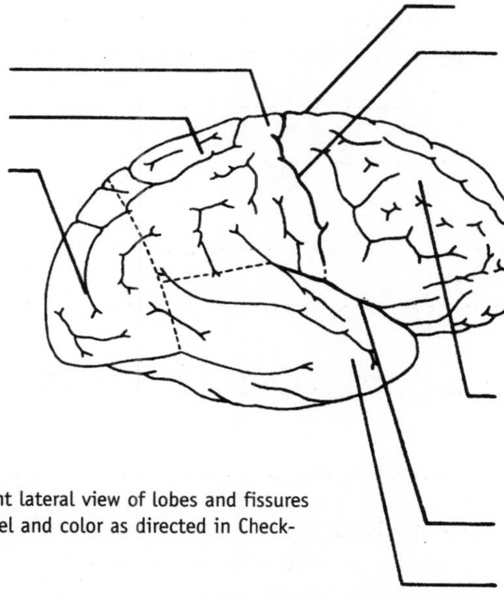

Figure LG 14.2 Right lateral view of lobes and fissures of the cerebrum. Label and color as directed in Checkpoints C2 and 10.

O Motor speech (Broca's) area
O Premotor area
O Primary auditory area
O Primary motor area
O Primary somatosensory (general sensory) area
O Primary visual area
O Auditory association area

✓ **C2.** Label the following structures using leader lines on Figure LG 14.2: *frontal lobe, occipital lobe, parietal lobe, temporal lobe, central sulcus, lateral cerebral sulcus, precentral gyrus, postcentral gyrus.*

✓ **C3.** Match the three types of white matter fibers with these descriptions.

A. Association	C. Commissural	P. Projection

_______ a. The corpus callosum contains these fibers and connects the two cerebral hemispheres.

_______ b. Sensory and motor fibers passing between cerebrum and other parts of the CNS are this type of fiber; the internal capsule is an example.

_______ c. These fibers transmit impulses among different areas of the same hemisphere.

✓ **C4.** Complete this exercise about basal ganglia.

a. Which is more anterior and superior in location? *(Body? Tail?)* of the caudate nucleus. The primary function of this nucleus is brain activity that precedes movements of *(extremities? eyes?)*.

b. Which is more lateral in location? *(Globus pallidus? Putamen?)*. Which structure helps to regulate muscle tone? *(Globus pallidus? Putamen?)*.

c. The "striped body" or *corpus striatum* consists of areas of gray matter, mainly the

_______________________ nucleus and the _______________________located on either side of white matter *(internal capsule)* as seen in text Figure 14.13b.

d. Name a disorder that involves basal ganglia. _______________________ Besides affecting movement, what other functions are attributed to this disorder?

C5. Describe the limbic system in this exercise.

a. This system consists of portions of the cerebrum, thalamus, and hypothalamus. List several component structures.

b. Explain why the limbic system is sometimes called the "visceral" or "emotional" brain.

c. One other function of the limbic system is _______________________. Forgetfulness. such as inability to recall recent events, results partly from impairment of this system.

✓ **C6.** Which condition can be described as visible bruising of the brain with some loss of consciousness? *Concussion? Contusion?*

✓ **C7.** Draw two important generalizations about brain functions in this activity.

a. In general, the anterior of the cerebrum is more involved with *(motor? sensory?)* control, whereas the posterior of the cerebrum is more involved with *(motor? sensory?)* functions. (Take a moment to visualize those activities taking place in the front and back of your own brain.)

b. As a general rule, *(primary? association?)* sensory areas receive sensations and *(primary? association?)* sensory areas are involved with interpretation and memory of sensations. For example, your ability to see the outline of the cerebrum, as in Figure LG 14.2, depends on your *(primary? association?)* visual areas. The fact that you can distinguish this diagram as a cerebrum (not a hand or heart), along with your memory of its structure for your next test, is based on the health of the neurons in your *(primary? association?)* visual areas.

✓ **C8.** Refer to Figure 14.15 in your text as you complete this exercise about sensory areas of the brain.

AAA.	**Auditory association area**	**POA.**	**Primary olfactory area**
CIA.	**Common integrative area**	**PSA.**	**Primary somatosensory area**
PAA.	**Primary auditory area**	**PVA.**	**Primary visual area**
PGA.	**Primary gustatory area**	**VAA.**	**Visual association area**
PMA.	**Primary motor area**		

_______ a. Area 17 located in the occipital lobe

_______ b. Areas 18 and 19 located in the occiptial lobe

_______ c. Areas 1, 2, 3 located in the postcentral gyrus within the parietal lobe

_______ d. Receives impulses for pain, touch, pressure, itch, tickle, hot and cold

_______ e. Receives impulses for smell

_______ f. Areas 41 and 42 in the temporal lobe

_______ g. Area 22 in the temporal lobe

_______ h. Areas 5, 7, 39, and 40 in the parietal lobe that interpret input from different sensory association areas

✓ **C9.** Select answers from the box as you describe regulation of Kelsy's brain as she and her friend Isaac share lunch and study for a test on the nervous system.

AAA.	**Auditory association area**	**PGA.**	**Primary gustatory area**
BSA.	**Broca's speech area**	**PMA.**	**Primary motor area**
FEF.	**Frontal eye field**	**POA.**	**Primary olfactory area**
OFC.	**Orbitofrontal cortex**	**PVA.**	**Primary visual area**
PA.	**Premotor area**	**WA.**	**Wernicke's area**
PAA.	**Primary auditory area**		

_______ a. Kelsey reads a paragraph in her textbook. (2 answers)

_______ b. She begins to pick up a spoon.

_______ c. She uses the spoon to eat her soup. (2 answers)

_______ d. Kelsey tastes the soup and detects the aroma of three spices in the soup. (3 answers)

_______ e. Kelsey speaks to Isaac about brain structure. (3 answers)

_______ f. She hears his response

_______ g. Kelsey interprets his meaning and enjoys the humor of Isaac's response. (2 answers)

✓ **C10.** Color functional areas of the cerebral cortex listed with color code ovals on Figure 14.2.

✓ **C11.** Do this exercise about control of speech.

a. Language areas are usually located in the *(left? right?)* cerebral hemisphere. This is true for most *(persons who are right-handed? persons, regardless of handedness?)*.

b. The term aphasia refers to inability to ____________________. *Fluent aphasia* means inability to *(articulate or form? understand?)* words. *Word blindness* refers to inability to understand *(spoken? written?)* words, whereas *word deafness* is inability to understand ____________________ words.

✓ **C12.** *Critical thinking*. You are helping to care for Laura, a hospital patient with a contusion of the right side of the brain. Which of the following would you be most likely to observe in Laura? Circle letters of correct answers.

A. Her right leg is paralyzed.
B. Her left arm is paralyzed.
C. She cannot speak out loud to you nor write a note to you.
D. She used to sing well but cannot seem to stay on tune now.
E. She has difficulty with concepts involving numbers.
F. Laura appears to have "flat affect": she speaks in a monotone, and appears without emotion.
G. Laura cannot readily differentiate odors

✓ **C13.** Identify the type of brain waves associated with each of the following situations.

A. Alpha	**D. Delta**
B. Beta	**T. Theta**

________ a. Occur in persons experiencing stress and in certain brain disorders

________ b. Lowest frequency brain waves, normally occurring when adult is in deep sleep; presence in awake adult indicates brain damage

________ c. Highest frequency waves, noted during periods of mental activity

________ d. Present when awake but resting, these waves are intermediate in frequency between beta and delta waves

D. Cranial nerves (pages 477–489)

✓ **D1.** Answer these questions about cranial nerves.

a. There are ________ pairs of cranial nerves. They are all attached to the ____________________; they leave the ____________________ via foramina.

b. They are numbered by Roman numerals in the order that they leave the cranium. Which is most anterior? *(I? XII?)* Which is most posterior? *(I? XII?)*

c. All spinal nerves are *(purely sensory? purely motor? mixed?)*. Are all cranial nerves mixed? *(Yes? No?)*

✓ **D2.** Complete the table about the 12 pairs of cranial nerves. (*For extra review* on locations of these nerves, identify their entrance or exit points through the cranial bones in LG Chapter 7, Checkpoint B7.)

Number	Name	Functions
a.	Olfactory	
b.		Vision (not pain or temperature of the eye)
c. III		
d.	Trochlear	
e. V		
f.		Stimulates lateral rectus muscle to abduct eye; proprioception of the lateral rectus
g.	Facial	
h.		Hearing; equilibrium
i. IX		
j.	Vagus	
k. XI		
l.		Supplies muscles of the tongue with motor and sensory fibers

✓ **D3.** *The Big Picture: Looking Ahead and Looking Back.* Refer to figures and exhibits in your text that emphasize structure and function of cranial nerves.

a. Refer to Figure 16.1 and identify the olfactory nerves, which form part of the pathway for *(smell? taste?)*. These are located in mucosa of *(inferior? superior?)* portions of the nose. After passing through the cribriform plate of the *(ethmoid? frontal? sphenoid?)* bone, axons of olfactory nerves synapse with neuron cell bodies located in the olfactory *(bulb? tract?)*. Axons of these neurons then pass through the olfactory *(bulb? tract?)*, which is located at the base of the *(frontal? parietal?)* lobes of the cerebrum.

b. Figure 16.5 shows the optic nerves, which are cranial nerves number *(1? II? VIII?)*. Optic nerves form part of the pathway for *(hearing? taste? vision?)*; each eye has *(one? many?)* optic nerve(s). Notice on Figure 16.9 that optic nerves consist of axons of *(rod or cone? bipolar? ganglion?)* cells within the retina of the eye. These axons may cross in the optic *(chiasma? tract?)* (Figure 16.16) or remain uncrossed before they end in the *(thalamus? hypothalamus?)*. There they synapse with neurons that terminate in the *(occipital? temporal?)* lobe of the cerebral cortex.

c. Cranial nerve VIII is associated with the *(ear? tongue?)* (Figure 16.17). The *(cochlear? vestibular?)* portion of this nerve conveys impulses associated with hearing from the snail-shaped *(cochlea? vestibule?)*. The three semicircular canals contain receptors that sense body position and movement; these neurons send axons to the thalamus via the *(cochlear? vestibular?)* branch of cranial nerve VIII.

d. Cranial nerve VII is the *(facial? trigeminal?)*; it can be seen on Figure 14.5, lying right next to cranial nerve *(II? VIII?)*. In fact, these two nerves both pass through the cranium via the same foramen, the *(internal? external?)* auditory meatus (Figure 16.17). The facial nerve is primarily involved with *(sensation? movement?)* of the face (Exhibit 11.1). It also carries sensations of taste from the *(anterior? posterior?)* two-thirds of the tongue (page 531 of the text).

e. Four cranial nerves transmit impulses as part of the autonomic nervous system (Figure 17.1b). These are cranial nerves numbers ______, ______, ______, and ______.

Which of these nerves stimulates production of tears from lacrimal glands? ________

Which two pairs of nerves stimulate salivary glands? ________ ________

Which pair of nerves innervates organs such as airways, heart, and intestine? ________

f. Which three pairs of cranial nerves control eye movements (Exhibit 11.2)? ________, ________; and ________. Which of these also stimulates the sphincter muscle of the iris to control the size of the pupil (Figure 17.1a and b)? ________

✓ **D4.** Check your understanding of cranial nerves by completing this exercise. Write the name of the correct cranial nerve following the related description.

a. Differs from all other cranial nerves in that it originates from the brainstem and from the spinal cord: ____________________

b. Eighth cranial nerve (VIII): ____________________

c. Is widely distributed into neck, thorax, and abdomen: ____________________

d. Senses toothache, pain under a contact lens, wind on the face: ____________________

e. The largest cranial nerve; has three parts (ophthalmic, maxillary, and mandibular): ____________________

f. Two nerves that contain taste fibers and autonomic fibers to salivary glands: ____________________, ____________________

g. Three cranial nerves that are purely or mainly sensory: ____________________, ____________________, ____________________

✓ **D5.** Write the number of the cranial nerve related to each of the following disorders.

________ a. Bell's palsy

________ b. Inability to shrug shoulders or turn head

________ c. Anosmia

________ d. Strabismus and diplopia (3 answers)

________ e. Blindness

________ f. Trigeminal neuralgia (tic douloureux)

________ g. Paralysis of vocal cords; loss of sensation of many organs; increased heart rate

________ h. Vertigo and nystagmus

________ i. Difficulty swallowing (4 answers)

________ j. Difficulty chewing

________ k. Ptosis of eyelid

E. Developmental anatomy and aging of the nervous system (pages 490–492)

✓ **E1.** Check your understanding of the early development of the nervous system by completing this exercise.

a. The nervous system begins to develop during the third week of gestation when the __________________-derm forms a thickening called the __________________ plate. Soon a longitudinal depression is found in the plate; this is the neural __________________. As neural folds on the sides of the groove grow and meet, they form the neural __________________.

b. Three types of cells form the walls of this tube. Two of these are the marginal layer which forms *(gray? white?)* matter and the mantle layer which becomes *(gray? white?)* matter.

c. The neural crest forms *(peripheral? central?)* nervous system structures such as ganglia as well as spinal and cranial __________________.

d. The anterior portion of the neural plate and tube develops into three enlarged __________________ by the fourth week. These fluid-filled cavities eventually become the __________________ which are filled with __________________ fluid.

e. By week five the primary vesicles have flexed (bent) to form a total of *(five? ten?)* secondary vesicles, described in Checkpoint E2.

✓ **E2.** Fill in the blanks in this table outlining brain development.

a. Primary Vesicles (3)		b. Secondary Vesicles (5)		c. Principal Parts of Brain Formed (7)
Prosencephalon (________-brain)	→	1. Diencephalon	→	1A. __________________
				1B. __________________
				1C. __________________
				1D. __________________
		2. __________________	→	2. Cerebrum
______-encephalon (Midbrain)	→	3. Mesencephalon	→	3A. __________________
______-encephalon (______________)	→	4. ______-encephalon	→	4A. __________________
		5. Met-__________	→	5A. Pons
				5B. __________________

✓ **E3.** Do this exercise describing effects of aging on the nervous system.

a. The total number of nerve cells _______-creases with age. Because conduction velocity *(increases? slows down?)*, the time required for a typical reflex is ________-creased.

b. Parkinson's disease is the most common *(motor? sensory?)* disorder in the elderly.

c. The sense of touch is likely to be *(impaired? heightened?)* in old age, placing the older person at *(higher? lower?)* risk for bums or other skin injury.

F. Disorders, medical terminology (pages 492–493)

F1. CVA refers to ______________________________

CVAs are (*rare? common?*) brain disorders. Describe three causes of CVAs.

F2. Contrast CVA with TIA.

F3. Describe the following aspects of Alzheimer's disease (AD):

a. Progressive changes in behavior

b. Pathological findings of brain tissue

c. Risk factors for AD

✓ **F4.** *For extra review.* Match the name of the disorder with the related description.

Agnosia	**Neuralgia**
Alzheimer's disease (AD)	**Stupor**
Cerebrovascular accident (CVA)	**Transient ischemic attack (TIA)**
Lethargy	

a. Characterized by loss of neurons, formation of beta-amyloid plaques, and neurofibrillary tangles:

____

b. Inability to recognize significance of sensory stimuli, such as tactile or gustatory:

____

c. Temporary cerebral dysfunction caused by interference of blood supply to the brain:

____

d. Most common brain disorder; also called stroke or brain attack: ______________

e. Attack of pain along an entire nerve, for example, tic douloureux: ______________

f. Functional sluggishness: ______________

g. Involves decrease in acetylcholine (ACh), so anticholinesterase (AChE)-inhibitors may help: ______________

ANSWERS TO SELECTED CHECKPOINTS: CHAPTER 14

A1. (a) Brain stem. (b) Diencephalon. (c) Cerebrum. (d) Cerebellum.

A3. (a) Less; tight, astrocytes. (b) The brain is protected from many substances that could harm the brain but are kept from the brain by this barrier. (c) Glucose, oxygen, carbon dioxide, and anesthetics. (d) Certain helpful medications, such as most antibiotics, and chemotherapeutic chemicals cannot cross this barrier.

A4. Skull bones, meninges, cerebrospinal fluid (CSF).

A6. (a) Tentorium cerebelli. (b) Falx cerebelli. (c) Falx cerebri.

A7. (a) B. (b) B. (c) A, C, D. (d) B, C, D. (e) B, D, E. (f) C.

A8. D E F G H I J K C A B.

B1. (a) Conduction; both sensory and motor. (b) Pyramidal, motor; corticospinal. (c) Decussation; left; axons, cortico. (d) Cuneatus, gracilis; pressure, touch, conscious proprioception, and vibrations. (e) Cardiovascular, medullary rhythmicity center. (f) Part of VIII, as well as IX–XII.

B2. (a) Bridge; spinal cord, medulla; cerebellum. (b) V–VII and part of VIII. (c) Respiration.

B3. (a) 2.5. (b) Anterior and superior. (c) Axons, both motor and sensory. (d) Inferior, auditory (or superior, visual).

B4. Reticular activating system; sound, temperature, pain; medulla, pons, and midbrain, as well as portions of the diencephalon (thalamus and hypothalamus) and spinal cord.

B6. (a) Superior. (b) Middle. (c) Inferior. (Note that the names of these tracts are logical, based on locations of the midbrain, pons, and medulla.)

B7. (a) Deep within; four; epi-, sub-, hypo-; thalamus. (b) Roof; pineal, habenular, choroid; melatonin, clock; smell. (c) Movements. (d) Two, lateral walls; third. (e) Sensory; touch, pressure, proprioception; geniculate. (f) Motor control, emotions, and memory. (g) Below, third.

B9. (a) H. (b) Mid. (c) Mid. (d) P. (e) Med. (f) H. (g) T. (h) Med. (i) T. (j) H. (k) T. (1) Mid. (m) Med. (n) Mid.

B10. See Table at right.

B11. (a) Hypothalamus, pineal gland, and pituitary. (b) Lack; advantage: because their capillaries are permeable to most substances, they can monitor homeostatic activities such as fluid balance and hunger: disadvantage: they are permeable to harmful substances, for example, possibly the AIDS virus.

C1. (a) Cerebral cortex; gray; cell bodies. (b) 2 to 4 mm. (c) Gyri; fissures, sulci. (d) Hemispheres; white, corpus callosum. (e) Dura mater; between cerebral hemispheres; superior and inferior sagittal sinuses (see Figure 14.2 in the text).

C2. See Figure LG 14.2A.

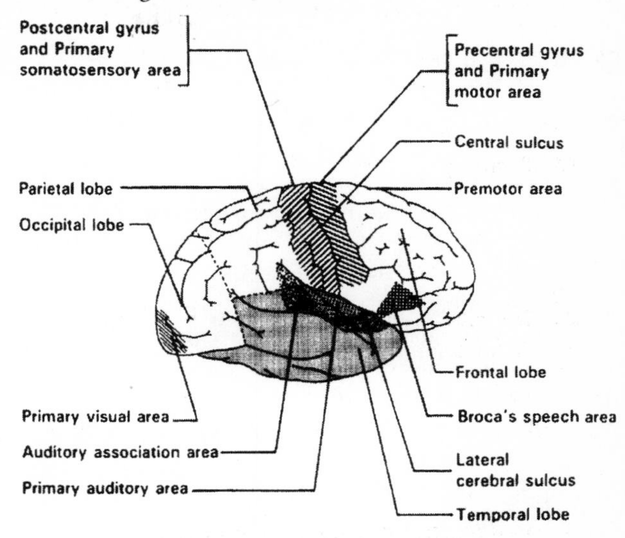

Figure LG 14.2A Right lateral view of lobes and fissures of the cerebrum.

C3. (a) C. (b) P. (c) A.

C4. (a) Body; eyes. (b) Putamen; globus pallidus. (c) Caudate, putamen. (d) Parkinson's disease; memory and emotions (or Huntington's; memory and personality).

C6. Contusion.

C7. (a) Motor, sensory. (b) Primary, association; primary; association.

C8. (a) PVA. (b) VAA. (c) PSA. (d) PSA. (e) POA. (f) PAA. (g) AAA. (h) CIA.

C9. (a) PVA, FEF. (b) PMA. (c) PMA, PA. (d) PGA, POA, OFC. (e) BSA, PMA, and PA. (f) PAA. (g) AAA, WA.

C10. See Figure LG 14.2A.

C11. (a) Left; persons, regardless of handedness. (b) Speak; understand; written, spoken.

Name of Nucleus	Brain Part	Functions
a. Habenular nucleus	Epi	Smell
b. Subthalamic nucleus	**Sub**	**Movements in general**
c. Red nucleus and substantia nigra	(2) **Mid, sub**	**Movements in general**
d. Superior colliculi	**Mid**	(2) **Movement of head or neck in response to visual stimuli; pupil reflex**
e. Nuclei cuneatus and gracilis	**Med**	**Touch, pressure, vibrations**
f. Anterior nucleus	**Thal**	**Emotions, memory**
g. Lateral geniculate nucleus	**Thal**	**Vision**
h. Medial geniculate nucleus	**Thal**	**Hearing**
i. Ventral posterior nucleus	**Thal**	(2) **Taste; touch, pressure, vibrations**
j. Mammillary bodies	**Hypo**	**Smell**
k. Median eminence of tuber cinereum	**Hypo**	**Regulation of anterior pituitary**
l. Supraoptic region	**Hypo**	**Regulation of posterior pituitary**

C12. B D F G.

C13. (a) T. (b) D. (c) B. (d) A.

D1. (a) 12; brain, cranium. (b) I; XII. (c) Mixed; no (the first two are purely sensory).

D2. Refer to answers to LG Chapter 7, Activity B7, page 137.

D3 (a) Smell; superior; ethmoid, bulb; tract, frontal. (b) II; vision, one; ganglion; chiasma, thalamus; occipital. (c) Ear; cochlear, cochlea; vestibular. (d) Facial, VIII; internal; movement; anterior. (e) III, VII, IX, X; VII; VII, IX; X. (f) III, IV, VI; III.

D4. (a) Accessory. (b) Vestibulocochlear. (c) Vagus. (d) Trigeminal. (e) Trigeminal. (f) Facial, glossopharyngeal. (g) Olfactory, optic, vestibulocochlear.

D5. (a) VII. (b) XI. (c) I. (d) III, IV, VI. (e) II. (f) V. (g) X. (h) VIII (vestibular branch). (i) IX-XII. (j) V (mandibular branch). (k) III.

E1. (a) Ecto, neural; groove; tube. (b) White, gray. (c) Peripheral, nerves. (d) Primary vesicles; ventricles of brain, cerebrospinal. (e) Five.

E2.

a. Primary Vesicles (3)	b. Secondary Vesicles (5)	c. Principal Parts of Brain Formed (7)
Prosencephalon (Fore-brain) →	1. Diencephalon →	1A. Epithalamus 1B. Thalamus 1C. Subthalamus 1D. Hypothalamus
	2. Telencephalon →	2. Cerebrum
Mes-encephalon → (Midbrain)	3. Mesencephalon →	3A. Midbrain
Rhomb-encephalon (Hindbrain) →	4. Myel-encephalon →	4A. Medulla oblongata
	5. Met-encephalon →	5A. Pons 5B. Cerebellum

E3. (a) De; slows down, in. (b) Motor. (c) Impaired, higher.

F4. (a) AD. (b) Agnosia. (c) TIA. (d) CVA. (e) Neuralgia. (f) Lethargy. (g) AD.

MORE CRITICAL THINKING: CHAPTER 14

1. Consider what effects are likely to result from the following: (a) blockage of the cerebral aqueduct by a tumor of the midbrain; (b) a severe blow to the medulla; (c) a CVA involving the occipital lobe.
2. Name the structures that form the circumventricular organs (CVOs) and explain why they are so named. Describe the structural characteristic that allows them to monitor chemical changes in the blood supply to the brain.
3. Describe the embryonic development of the brain.
4. Describe the structure and functions of the following: (a) limbic system, (b) diencephalon, (c) reticular formation.
5. Contrast motor and sensory functions of cranial nerves V, VII, and IX.
6. Identify structural changes of the brain that are associated with Alzheimer's disease. Describe roles of the following in these changes: beta-amyloid, and APOE.
7. Contrast a CVA with a TIA.

MASTERY TEST: CHAPTER 14

Questions 1–9: Circle the letter preceding the one best answer to each question.

1. All of these are located in the medulla *except:*
   A. Cuneatus nucleus and gracile nucleus
   B. Cardiac center, which regulates heart
   C. Site of decussation (crossing) of pyramidal tracts
   D. The olive
   E. Origin of cranial nerves V to VIII
2. Which of these is the major function of the postcentral gyrus?
   A. Controls specific groups of muscles, causing their contraction
   B. Receives general sensations from skin, muscles, and viscera
   C. Receives olfactory impulses
   D. Primary visual reception area
   E. Somatosensory association area
3. All of these structures contain cerebrospinal fluid *except:*
   A. Subdural space
   B. Ventricles of the brain
   C. Central canal of the spinal cord
   D. Subarachnoid space
4. Damage to the accessory nerve would result in:
   A. Inability to turn the head or shrug shoulders
   B. Loss of normal speech function
   C. Hearing loss
   D. Changes in heart rate
   E. Anosmia
5. Which statement about the trigeminal nerve is *false?*
   A. It sends motor fibers to the muscles used for chewing.
   B. Pain in this nerve is called trigeminal neuralgia (tic douloureux).
   C. It carries sensory fibers for pain, temperature, and touch from the face, including eyes, lips, and teeth area.
   D. It is the smallest cranial nerve.
6. All of these are functions of the hypothalamus *except:*
   A. Control of body temperature
   B. Release of chemicals (regulating factors) that cause release or inhibition of hormones
   C. Principal relay station for sensory impulses
   D. Regulates viscera via autonomic control
   E. Involved in maintaining sleeping or waking state
7. All of these are located in the midbrain *except:*
   A. Apneustic area
   B. Cerebral peduncles
   C. Cerebral aqueduct
   D. Origin of cranial nerves Ill and IV
   E. Red nucleus
8. All of the following are mixed cranial nerves except cranial nerve number:
   A. II
   B. III
   C. V
   D. VII
   E. IX
9. All of the following contain parts of motor pathways except
   A. Red nucleus
   B. Substantia nigra
   C. Cerebral peduncles
   D. Medial lemniscus

Questions 10–14: Arrange the answers in correct sequence.

______ ______ ______ 10. From superior to inferior:
A. Thalamus
B. Hypothalamus
C. Corpus callosum

______ ______ ______ 11. Order in which impulses are relayed in conduction pathway for vision:
A. Optic nerve
B. Optic tract
C. Optic chiasma

______ ______ ______ ______ 12. Pathway of cerebrospinal fluid, from formation to final destination:
A. Choroid plexus in ventricle
B. Subarachnoid space
C. Cranial venous sinus
D. Arachnoid villi

______ ______ ______ 13. From anterior to posterior:
A. Fourth ventricle
B. Pons and medulla
C. Cerebellum

______ ______ ______ 14. Location of parts of the cerebral cortex from anterior to posterior:
A. Premotor cortex
B. Somatosensory cortex
C. Primary motor cortex
D. Visual cortex

Questions 15–20: Circle T (true) or F (false). If the statement is false, change the underlined word or phrase so that the statement is correct.

T F 15. The thalamus, hypothalamus, and cerebrum are all developed from the <u>forebrain</u> (<u>prosencephalon</u>).

T F 16. The language areas are located in the <u>cerebellar cortex</u>.

T F 17. The limbic system functions in control of <u>emotional aspects of behavior</u>.

T F 18. Mammillary bodies, the infundibulum, as well as preoptic and supraoptic nuclei, are all parts of the <u>thalamus</u>.

T F 19. The corpus callosum contains <u>projection</u> fibers.

T F 20. A cerebrovascular accident (CVA) can be expected to produce <u>more lasting</u> effects than a transient ischemic attack (TIA).

Questions 21–25: Fill-ins. Complete each sentence with the word or phrase that best fits.

_______________ 21. The _______________ is the extension of dura mater that separates the cerebral hemispheres, whereas the _______________ separates the cerebrum from the cerebellum.

_______________ 22. The superior and inferior colliculi, associated with movements of eyeballs and head in response to visual stimuli, are located in the _______________.

_______________ 23. The blood-brain barrier (BBB) is from _______________ of cells of brain capillaries together with neuroglia known as _______________.

_______________ 24. The _______________ controls wakefulness.

_______________ 25. The globus pallidus, putamen, and caudate nucleus are parts of the _______________.

ANSWERS TO MASTERY TEST: CHAPTER 14

Multiple Choice

1. E
2. B
3. A
4. A
5. D
6. C
7. A
8. A
9. D

Arrange

10. C A B
11. A C B
12. A B D C
13. B A C
14. A C B D

True-False

15. T
16. F. Cerebral cortex
17. T
18. F. Hypothalamus
19. F. Commissural
20. T

Fill-ins

21. Falx cerebri; tentorium cerebelli
22. Midbrain
23. Tight junctions, astrocytes
24. Reticular formation or reticular activating system (RAS)
25. Basal ganglia

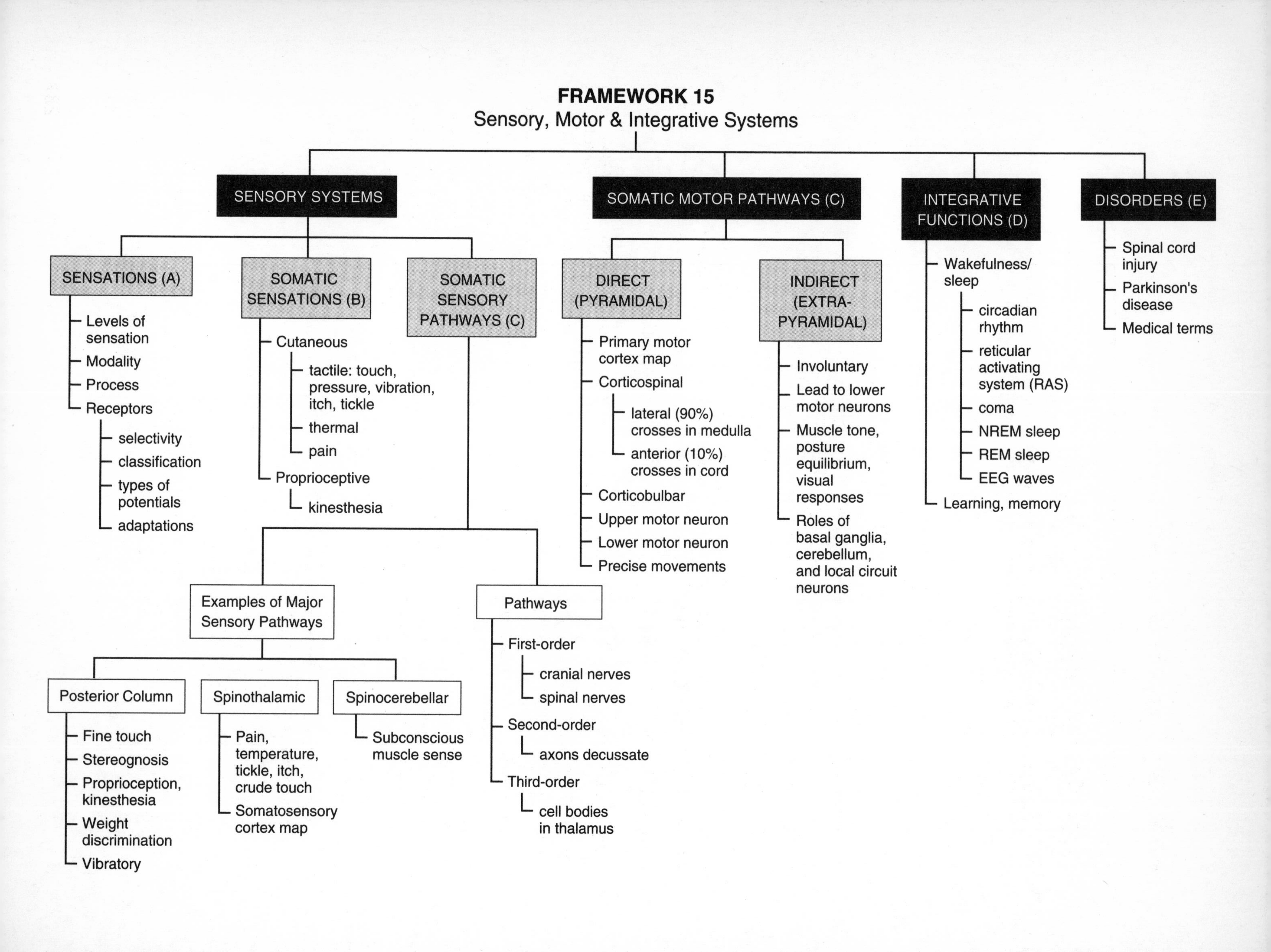

FRAMEWORK 15
Sensory, Motor & Integrative Systems
SENSORY SYSTEMS
SENSATIONS (A)
Levels of sensation
Modality
Process
Receptors
selectivity
classification
types of potentials
adaptations
SOMATIC SENSATIONS (B)
Cutaneous
tactile: touch, pressure, vibration, itch, tickle
thermal
pain
Proprioceptive
kinesthesia
SOMATIC SENSORY PATHWAYS (C)
Examples of Major Sensory Pathways
Posterior Column
Fine touch
Stereognosis
Proprioception, kinesthesia
Weight discrimination
Vibratory
Spinothalamic
Pain, temperature, tickle, itch, crude touch
Somatosensory cortex map
Spinocerebellar
Subconscious muscle sense
Pathways
First-order
cranial nerves
spinal nerves
Second-order
axons decussate
Third-order
cell bodies in thalamus
SOMATIC MOTOR PATHWAYS (C)
DIRECT (PYRAMIDAL)
Primary motor cortex map
Corticospinal
lateral (90%) crosses in medulla
anterior (10%) crosses in cord
Corticobulbar
Upper motor neuron
Lower motor neuron
Precise movements
INDIRECT (EXTRA-PYRAMIDAL)
Involuntary
Lead to lower motor neurons
Muscle tone, posture equilibrium, visual responses
Roles of basal ganglia, cerebellum, and local circuit neurons
INTEGRATIVE FUNCTIONS (D)
Wakefulness/ sleep
circadian rhythm
reticular activating system (RAS)
coma
NREM sleep
REM sleep
EEG waves
Learning, memory
DISORDERS (E)
Spinal cord injury
Parkinson's disease
Medical terms

CHAPTER 15

Sensory, Motor, and Integrative Systems

The last three chapters have introduced the components of the nervous system and demonstrated the arrangement of neurons in the spinal cord, spinal nerves, brain, and cranial nerves. Chapter 15 integrates this information in the study of major nerve pathways including those for (a) general senses (such as touch, pain, and temperature), (b) motor control routes for precise, conscious movements and for unconscious movements, and (c) complex integrative functions such as memory and sleep.

As you begin your study of sensory, motor, and integrative systems, carefully examine the Chapter 15 Topic Outline and Objectives; check off each one as you complete it. To organize your study of this chapter glance over the Chapter 15 Framework now. Be sure to refer to the Framework frequently and note relationships among key terms in each section.

TOPIC OUTLINE AND OBJECTIVES

A. Sensations

1. Define a sensation and discuss the components of sensation.
2. Describe the different ways to classify sensory receptors.

B. Somatic sensations

1. Describe the location and function of the receptors for tactile, thermal, and pain sensations.
2. Identify the receptors for proprioception, and describe their functions.

C. Somatic sensory and motor cortex maps in the cerebral cortex and pathways

1. Describe the location and functions of the primary somatosensory areas and primary motor areas in the cerbral cortex.
2. Describe the neuronal components and functions of the posterior column-medial lemniscus, the anterolateral, and the spinocerebellar pathways.
3. Identify the locations and functions of lower motor neurons.
4. Describe the four neural circuits that provide input to lower motor neurons.
5. Compare the locations and functions of the direct and indirect motor pathways.
6. Explain how the basal ganglia and cerebellum modulate movements.

D. Integrative functions

1. Compare the integrative cerebral functions of wakefulness and sleep, and learning and memory.

E. Disorders

WORDBYTES

Now become familiar with the language of this chapter by studying each wordbyte, its meaning, and an example of its use within a term. After you study the entire list, self-check your understanding by writing the meaning of each wordbyte on the line. As you continue through the *Learning Guide,* identify (and fill in) additional terms that contain the same wordbyte.

Wordbyte	Self-check	Meaning	Example(s)
a-, an-	______________	without	*a*taxia, *an*esthesia
-algesia	______________	pain	an*algesia*
brady-	______________	slow	*brady*kinesia
-ceptor	______________	receiver	proprio*ceptor*
circa-	______________	about	*circa*dian
dia-	______________	a day	circa*dian* rhythm
-esthesia	______________	perception	kin*esthesia*
hemi-	______________	half	*hemi*plegia
in-	______________	not	*in*somnia
-kin-	______________	motion	*kin*esthesia
myo-	______________	muscle	a*myo*tropic
noci-	______________	harmful	*noci*receptors
para-	______________	abnormal, beyond	*par*esthesia
-plegia	______________	paralysis or stroke	hemi*plegia*
propio-	______________	one's own	*proprio*ception
quad-	______________	four	*quad*riplegia
rhiz-	______________	root	*rhiz*otomy
-somnia	______________	sleep	in*somnia*
somat-	______________	of the body	*somat*ic
tact-	______________	touch	*tact*ile
-taxia	______________	coordination	a*taxia*
-tropic	______________	nourishment	amyo*tropic*

CHECKPOINTS

A. Sensation (pages 499–502)

✓ **A1.** Review the three basic functions of the nervous system by listing them in sequence here:

a. Receiving ______________ input

b. ______________; ______________; ______________ information

c. Transmitting ______________ impulses to ______________ or ______________

A2. For a moment, visualize what your life would be like if you were unable to experience any sensations. List the three types of sensations that you believe you would miss most.

_____________________ _____________________ _____________________ Now write a sentence describing how your health and safety might be endangered by lack of ability to perceive sensations.

✓ **A3.** Do this activity about levels of sensation.

a. *(Perception? Sensation?)* is the awareness (either conscious or unconscious) of stimuli, such as the act of seeing a bird, whereas _____________________ is the conscious awareness and interpretation of sensation, such as the recognition of that bird as a bluebird and not a robin.

b. Identify levels of sensation by matching parts of the nervous system listed in the box with descriptions below.

B. Brain stem	**S. Spinal cord**
C. Cerebral cortex	**T. Thalamus**

________ 1. Immediate reflex action is possible without involvement of the brain

________ 2. Involves subconscious motor reactions, such as those that facilitate balance and equilibrium

________ 3. Offers a general (or "crude") awareness of locale and type of sensations, such as awareness of pressure within the leg or that something is being heard; offers no specifics

________ 4. Gives precise information about sensations, such as distinguishing Bach from Baez from the B-52s, as well as memories of hearing a particular musical work

✓ **A4.** The quality that makes the sensation of pain different from the sense of touch, or which distinguishes hearing from vision is called the ___________________ of sensation. Any single sensory neuron carries sensations of *(one modality? many modalities?)*. Different modalities of sensations are distinguished precisely at the level of the *(spinal cord? thalamus? cerebral cortex?)*.

✓ **A5.** Select from answers in the box to correctly categorize different modalities of sensations. Write one answer on each line provided. One answer will not be used; other answers will each be used once.

Awareness of joint positions	**Pain**	**Vibration**
Awareness of movements	**Pressure**	**Vision**
Cold	**Smell**	**Warm**
Equilibrium	**Taste**	
Hearing	**Touch**	

a. Thermal sensations: ________________ ________________

b. Proprioceptive sensations: ________________ ________________

c. Tactile sensations: ________________ ________________ ________________

d. Special senses: ________________ ________________ ________________

________________ ________________

✓ **A6.** Arrange in correct order the components in the pathway of sensation.

IGP.	**Impulse generation and propagation**	**S.**	**Stimulation**
Int.	**Integration**	**T.**	**Transduction**

1. _______ → 2. _______ → 3. _______ → 4. _______
Now match each of these components of sensation with the correct description below.

_______ a. Activation of a sensory neuron

_______ b. Stimulus → graded potential

_______ c. Graded potential → nerve impulse; impulse is conveyed along a first-order neuron into the spinal cord or brainstem up to a higher level of the brain.

_______ d. Nerve impulse → sensation; usually occurs in the cerebral cortex

✓ **A7.** Complete Table 15.1, categorizing the nature of different sensory receptors.

Table LG 15.1 Classification of sensory receptors

Receptor	**Structure**	**Type of Graded Potential**	**Location (-ceptor)**	**Type of Stimulus Detected**
	Free Encapsulated Separate cells	Generator Receptor	Extero- Intero- Proprio-	Mechanoreceptor Thermoreceptor Nociceptor Photoreceptor Chemoreceptor
1. Pressure		Generator		
2. Pain (skin)	Free		Extero-	
3. Pain (stomach)				Nociceptor
4. Vision		Receptor		Photoreceptor
5. Equilibrium			Proprio-	
6. Smell	Separate cells			
7. Cold on skin	Free			Thermoreceptor
8. Changes of O_2 level in blood	(Omit)	(Omit)	Intero-	

A8. Do this exercise on the characteristics of sensations.

a. Arrange these receptors from simplest to most complex structurally:

________ ________ ________

A. Receptors for special senses such as vision or hearing
B. Free nerve endings, such as receptors for pain
C. Receptors for touch or pressure

b. Describe what is meant by the *selectivity* of receptors.

c. Recall getting dressed this morning. The fact that you felt the shirt touching your back just after you put on that shirt, but that the awareness of that touch has

dissipated over time, is known as the characteristic of____________________. Circle the type of receptors that adapt most slowly *(Pain? Pressure? Smell? Touch?)*. Explain how such slow adaptation is advantageous to you.

B. Somatic senses (pages 502–507)

B1. Contrast these terms: *somatic sensations/cutaneous sensations*

✓ **B2.** Do the following exercise about tactile sensations.

a. Name the tactile sensations.

These sensations are sensed by ____________________-receptors. Receptors for these sensations are *(evenly? unevenly?)* distributed throughout the skin of the body.

b. *(Fine? Crude?)* touch refers to ability to determine that something has touched the skin, but its precise location, shape, size, or texture cannot be distinguished.

c. In general, pressure receptors are more *(superficial? deep?)* in location than touch receptors.

d. Sensations of ____________________ result from rapid repetition of sensory impulses from tactile receptors.

✓ **B3.** Match names of receptors with their descriptions. Answers may be used more than once.

C. Corpuscles of touch (Meissner's corpuscles)	**Type I. Type I cutaneous mechanoreceptors (Merkel discs)**
H. Hair root plexuses	**Type II. Type II cutaneous mechanoreceptors (end organs of Ruffini)**
L. Lamellated (Pacinian) corpuscles	
N. Nociceptors	

_______ a. Egg-shaped masses located in dermal papillae, especially in fingertips, palms of hands, and soles of feet

_______ b. Multilayered structures sensitive to pressure and high-frequency vibration

_______ c. Superficial touch receptors (two answers)

_______ d. Slowly adapting touch receptors (three answers)

_______ e. May respond to any type of stimulus if stimulus is strong enough to cause tissue damage

_______ f. Surround hair follicles; can detect breeze blowing against hairs on skin

B4. *For extra review.* Refer to Figure LG 5.1. Identify and label types of receptors on that figure. Add your own drawings of three other types of receptors in appropriate layers of skin. Label those receptors also.

✓ **B5.** For additional review of cutaneous sensations, do this Checkpoint.

a. Explain how the sensation of itch develops.

b. Name three chemicals that may be released from injured tissues that may stimulate pain receptors. _______________, _______________, _______________

c. Fast pain is felt in *(deep? superficial?)* tissues of the body. Circle the qualities that are associated with fast pain.

Aching **Acute** **Burning** **Chronic** **Pricking** **Sharp** **Throbbing**

d. Which type of pain impulses travel along unmyelinated C nerve fibers? *(Fast? Slow?)*

e. Circle the type of pain that arises from receptors in skin.

Deep somatic pain **Superficial somatic pain** **Visceral pain**

✓ **B6.** *The Big Picture: Looking Ahead and a clinical correlation.* Refer to figures in upcoming chapters of your text to help you determine why patients experiencing visceral pain (as during a heart attack or gallbladder attack) may feel pain in locations quite distant from these two organs.

a. Pain impulses that originate from the heart, for example, during a "heart attack" involving the left ventricle (Figure 20.4), enter the spinal cord at the same level as do sensory fibers from skin covering the _______________. This level of the cord is about _______ to _______.

b. Refer to Figure 15.3. Notice that some liver and gallbladder pain is felt in a region quite distant from these organs, that is, in the _______________ region. These organs lie just inferior to the dome-shaped ____________________, which is supplied by the phrenic nerve from the *(cervical? brachial? lumbar?)* plexus. A painful gallbladder can send impulses to the cord via the phrenic nerve, which enters the cord at the

same level (about C _______ or C _______) as nerves from the neck and shoulder. So

gallbladder pain is said to be ____________________ to the neck and shoulder area.

✓ **B7.** *Critical thinking.* Mr. Rummel was in a motorcycle accident that resulted in the severing of his left leg above the knee. Months later he is referred to the university pain clinic for an assessment. He states, "I feel this pain in my left foot, and sometimes my toes itch like crazy, but they're not there to scratch!" How might you interpret his experience?

✓ **B8.** *A clinical correlation.* Match methods of pain relief listed in the box with the following descriptions.

A.	**Anesthetic (Novocaine)**	**O.**	**Opioids such as morphine, codeine, or meperidine (Demerol)**
AAI.	**Aspirin, acetaminophen (Tylenol), ibuprofen (Motrin, Advil)**		

________ a. Medication that alters the quality of pain perception so that pain feels less noxious

________ b. Medication that inhibits formation of chemicals (prostaglandins) that stimulate pain receptors

________ c. Medication that blocks conduction of nerve impulses in first-order neurons

✓ **B9.** Close your eyes for a moment. Reflect on the position of your arms and legs as well as the degree of tension in your muscles and tendons. This awareness of body

position and movements is known as ____________________. Now match names of proprioceptors listed in the box with their descriptions below. Answers may be used more than once.

H.	**Hair cells of the inner ear**	**M.**	**Muscle spindles**
J.	**Joint kinesthetic receptors**	**T.**	**Tendon organs**

________ a. Located within and around articular capsules of synovial joints; help protect joints

________ b. Sensory fibers within most skeletal muscle fibers that help to prevent overstretching of muscles

________ c. Monitor the force of contraction of muscles; help to protect tendons (and their muscles) from excessive tension

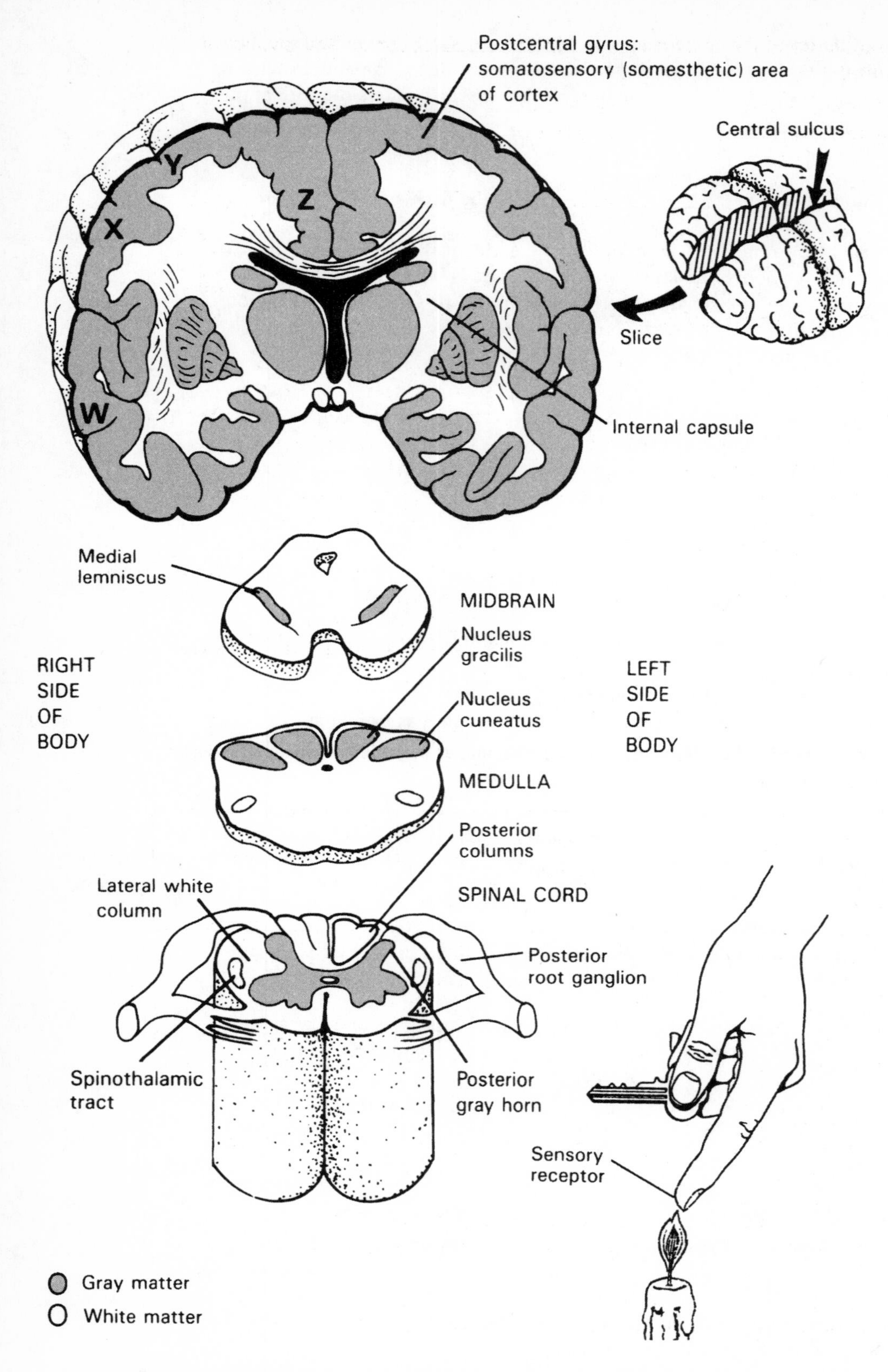

Figure LG 15.1 Diagram of the central nervous system. Refer to Checkpoints C1 and C6.

C. Somatic sensory and motor cortex maps in the cerebral cortex and pathways (pages 508–517)

✓ **C1.** Do this learning activity about sensory reception in the cerebral cortex.

a. Picture the brain as roughly similar in shape to a small, rounded loaf of bread. The crust would be comparable to the cerebral ____________________.

b. Suppose you cut a slice of bread in midloaf that represents the postcentral gyrus. It looks somewhat like the "slice" or section of brain shown at the top of Figure LG 15.1.

Locate the Z on that figure. This portion (next to the ____________________ fissure) receives sensations from the *(face? hand? hip? foot?)*. (Refer to Figure 15.5a, of your text for help.)

c. Suppose a person suffered a loss of blood supply (CVA or "stroke") to the area of the brain marked X in Figure LG 15.1. Loss of sensation in the *(face? hand? leg?)* area would result. Damage to the area marked W would be likely to lead to loss of the sense of *(vision? hearing? taste?)* because the W is located on the temporal lobe.

d. The letters W, X, Y, and Z are marked on the *(right? left?)* side of the brain. (Note that this is an anterior view of this brain section. If a friend stands and faces you, the *right* side of your friend's brain will be in your *left* field of view.) What would be the most profound effects of brain damage to area Y of the postcentral gyrus? Loss of *(sensation? movement?)* to the *(face? hand? foot?)* on the *(right? left?)* side or the body.

e. Certain regions of the body are represented by large areas of the ____________________ ____________________ cortex, indicating that these areas *(are? are not?)* very sensitive. Name two of these areas.

✓ **C2.** Write terms listed in the box after correct definitions below to identify types of sensations you experience.

Crude touch	**Stereognosis**
Kinesthesia	**Fine touch**
Proprioception	

a. Ability to sense touch on the back of your neck but inability to specify precise location of the touch: ____________________

b. Ability to recognize exact location of a sharp pencil touched to your fingertip and ability to distinguish whether one or two sharp pencils are touching nearby points on that fingertip: ____________________

c. Ability to tell—with your eyes closed—the exact position of your arms and legs: ____________________

d. Awareness of precise movements of your fingers as you type a paper: ______________

e. Ability to distinguish a dime from a button based on size, shape, and texture: ____________________

✓ **C3.** Contrast functions of major sensory pathways by selecting answers from the box that match the types of sensations they transmit.

APSC.	**Anterior and posterior spinocerebellar tract**
AST.	**Anterior spinothalamic tract**
LST.	**Lateral spinothalamic tract**
PCML.	**Posterior column-medical lemniscus tract**

_________ a. Pain and temperature

_________ b. Crude touch or pressure

_________ c. Fine two-point discrimination

_________ d. Proprioception, kinesthesia, stereognosis

_________ e. Weight discrimination and vibration

_________ f. Tickle and itch

_________ g. Subconscious proprioceptive impulses to the cerebellum needed for posture, balance, and coordination

✓ **C4.** Sensory pathways involve *(one? two? three?)*-neuron sets that allow nerve impulses from receptors, for example, in skin or muscles, to reach the primary somatosensory area of the cerebral cortex. Identify first-order (I), second-order (II), and third-order (III) neurons by writing I, II, or III next to the related neuron.

_________ a. Crosses to the opposite side of the spinal cord or brainstem and ascends to the thalamus

_________ b. Reaches the spinal cord or brainstem

_________ c. Passes to the postcentral gyrus of the cerebral cortex

✓ **C5.** Write in locations of cell bodies of the three neurons in these two pathways.

Pathway	First-Order	Second-Order	Third-Order
a. Posterior column–medial lemniscus			
b. Anterolateral (spinothalamic)			

✓ **C6.** Check your understanding of pathways by drawing them according to the directions below. Try to draw them from memory; do not refer to the text figures or the table until you finish. Label the *first-, second-,* and *third-order neurons* as *I, II,* and *III.*

a. Draw in *blue* on Figure LG 15.1 the pathway that allows you to accomplish stereognosis so that you can recognize the shape of a key in your left hand. (*Hint:* The correct recognition of the key involves discriminative touch, proprioception, and weight discrimination.)

b. Now draw in *red* the pathway for sensing pain and temperature as the index finger of the left hand comes too close to a flame.

✓ **C7.** Discuss sensory pathways to the cerebellum in this Checkpoint.

a. Circle the two pathways that are the major routes for proprioceptive impulses to the cerebellum.

A. Anterior spinocerebellar tract
B. Posterior spinocerebellar tract
C. Cuneocerebellar
D. Posterior columns

b. Proprioception in the left leg is transmitted by the anterior and posterior spinocerebellar tracts to the cerebellum. These sensory impulses *(are? are not?)* consciously perceived.

c. Circle all functions of the spinocerebellar pathways in this list:
   A. Provide awareness of touch and tickle
   B. Needed for balance and coordination, for example, in walking
   C. Needed to help maintain posture
   D. Provide motor impulses that initiate fine motor movements

d. Syphilis is a sexually transmitted disease (STD) caused by a *(bacterium? fungus? virus?)*. Which stage occurs latest in the life of a syphilis patient? *(Primary? Secondary? Tertiary?)* syphilis. Which of the tracts listed above (in C7a) are most affected by

   tertiary syphilis? ______________________

C8. Review locations of tracts in the spinal cord by referring to LG Chapter 13, Checkpoints B2–B3.

C9. Identify roles of neurons in different sites in the control of movements.

BG. Basal ganglia	**CN. Cerebellar neurons**	**LCN. Local circuit neurons**
LMN. Lower motor neurons	**UMN. Upper motor neurons**	

________ a. Cerebral cortex neurons that are essential for planning, initializing, and directing voluntary movements

________ b. Brainstem neurons that regulate unconscious movements such as muscle tone, posture, and balance

________ c. Known as the final common pathway because these neurons must be activated for movements to occur; located in brain stem or anterior gray horns of spinal cord

________ d. Assist UMNs by monitoring differences between intended movements and those actually performed

________ e. Assist UMNs by initiating and terminating movements, establishing muscle tone, and suppressing unwanted movements

✓ **C10.** Refer to Figure LG 15.2 and do the following exercise about motor control.

a. This "slice" or section of the brain containing the primary motor cortex is located in the *(frontal? parietal?)* lobe.

b. Area P in this figure controls *(sensation? movement?)* of the *(left? right?) (arm? leg? side of the face?)*. Motor neurons in area P would send impulses to the pons to activate

   cranial nerve ____________________ to facial muscles.

c. Describe effects of damage to area R.

✓ **C11.** Show the route of impulses along the principal direct pathway by listing in correct sequence the structures in the box below that comprise the pathway. Refer to Figure 15.8 in the text if you have difficulty with this activity.

_______ _______ _______ _______ _______ _______ _______ _______

A. Anterior gray horn (lower motor neuron)	**E. Lateral corticospinal tract**
B. Midbrain and pons	**F. Medulla, decussation site**
C. Effector (skeletal muscle)	**G. Precentral gyrus (upper motor neuron)**
D. Internal capsule	**H. Ventral root of spinal nerve**

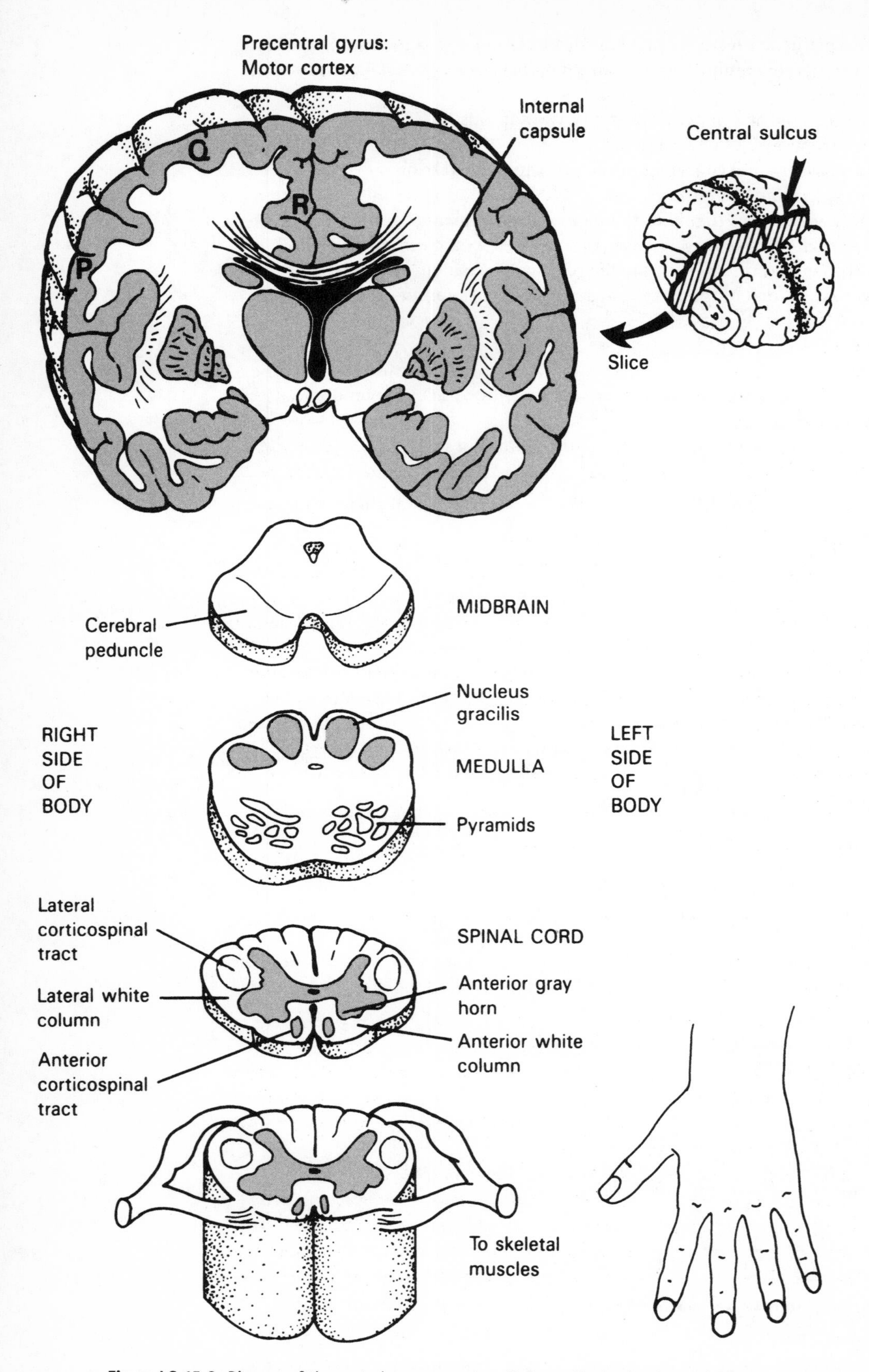

Figure LG 15.2 Diagram of the central nervous system. Refer to Checkpoint C10 and C12.

✓ **C12.** Now draw this pathway on Figure LG 15.2. Show the control of muscles in the left hand by means of neurons in the lateral corticospinal pathway. Label the *upper* and *lower motor neurons.*

✓ **C13.** Complete this exercise on motor pathways.

a. Corticospinal tracts are classified as *(direct? indirect?)* pathways; they control *(precise, voluntary? involuntary?)* movements. Based on their structure, corticospinal tracts are also known as *(pyramidal? extrapyramidal?)* pathways. Circle parts of the cerebrum that contain cell bodies of neurons whose axons form direct motor pathways:

A. Postcentral gyrus B. Precentral gyrus C. Premotor area

b. Axons of about 90% of upper motor neurons (UMNs) pass through *(anterior? lateral?)* corticospinal tracts; these are axons that *(decussated or crossed? did not decussate or cross?)* in the medulla. Lateral corticospinal tracts on the left side of the spinal cord consist of axons that originated in upper motor neurons on the *(right? left?)* side of the motor cortex.

c. These tracts (left lateral corticospinal) then convey impulses to LMNs located in the *(right? left?) (anterior? lateral? posterior?)* gray horns of the spinal cord. In most cases there *(are? are not?)* interneurons between UMNs and LMNs.

d. About ______% of UMN axons pass through ____________________ corticospinal tracts. Because axons of these UMNs *(do? do not?)* decussate in the medulla, anterior corticospinal tracts on the left side of the cord consist of axons that originated on the *(right? left?)* side of the motor cortex. These tracts are therefore *(contra? ipsi?)*-lateral. Most of these *(do? do not?)* eventually decussate at lower levels.

e. Corticobulbar tracts consist of UMNs that terminate in the *(brainstem? anterior gray horn of the cord?)*. Lower motor neurons (LMNs) there pass out through *(cranial? spinal?)* nerves.

f. Destruction of *(upper? lower?)* motor neurons results in spastic paralysis because the brain is not controlling movements. Flaccid paralysis results from damage of *(upper? lower?)* motor neurons.

g. Why are lower motor neurons called the final *common pathway?*

✓ **C14.** *Critical thinking.* Refer to Figure 21.20c as you consider possible effects of cerebrovascular accidents (CVAs or "strokes") involving specific arteries to the brain.

a. Locate the anterior cerebral arteries on the figure. These proceed anteriorly and pass between the two cerebral hemispheres, supplying area R of Figure LG 15.2. Paralysis and/or sensory loss of *(arm and hand? leg and foot? face?)* is most likely to result from interruption of blood flow through these vessels.

b. Locate the middle cerebral arteries. These proceed laterally and then pass medial to the temporal lobe to supply pre- and postcentral gyri. A stroke involving this artery is likely to cause effects on *(sensation only? movement only? sensation and movement?).*

c. Identify the vertebral and basilar arteries and their branches. Strokes involving these vessels *(are? are not?)* likely to cause effects on extrapyramidal pathways. State your rationale.

✓ **C15.** Describe locations of indirect (extrapyramidal) tracts in this exercise:

a. All of these tracts lie *(within? outside of?)* corticospinal tracts. Most of them originate in areas *(within? outside of?)* the cerebral cortex and end in the __________________.

b. Select names of tracts in the box that fit descriptions below. Write one answer on each line.

LRS. Lateral reticulospinal	**TS. Tectospinal**
MRS. Medial reticulospinal	**VS. Vestibulospinal**
RS. Rubrospinal	

________ 1. You tighten your back muscles as you sit up straight in front of your computer.

________ 2. Neurons in the left side of the midbrain convey messages to cranial nerves to direct your head and eyes to the right side of the computer monitor.

________ 3. As you turn your head to the right, muscle tone on your right side is adjusted slightly to help you maintain your balance.

________ 4. Neurons in the left side of the midbrain send messages to the spinal cord and spinal nerves to move your right hand on the keyboard.

C16. Summarize how the basal ganglia and cerebellum, together with associated neurons, help you to carry out coordinated, precise movements, such as those involved in a game of tennis.

✓ **C17.** *A clinical correlation.* Describe conditions that significantly affect movements.

a. *(Huntington's? Parkinson's?)* disease involves changes in basal ganglia associated with decrease in the neurotransmitter *(acetylcholine? dopamine? GABA?)*. Signs of this condition are *(jerky movements and facial twitches? muscle rigidity and tremors?)*.

b. A predominant sign of Huntington's disease (HD) is jerky movements (or "dance") known as __________________ related to loss of neurons that make the inhibitory transmitter named __________________. Signs of this progressive condition are most likely to be first noted at about age *(4 years? 40 years? 70 years?)*. A child of a parent with HD has *(almost no? a 50–50?)* chance of having this condition.

c. Lou Gehrig's disease is a common term for the condition known as ALS or __________________ __________________ __________________. In this condition, muscles are *(overactive and hypertrophied? weak and atrophied?)*. Intellect and sensations *(are? are not?)* typically affected.

d. __________________ literally means lack of coordination and may be due to cerebellar damage. Another sign of cerebellar damage, __________________ tremor, involves shaking during voluntary movements.

D. Integrative functions (pages 517–520)

✓ **D1.** List three types of activities that require complex integration processes by the brain.

✓ **D2.** Complete this exercise about sleep and wakefulness.

a. ____________________ rhythm is a term given to the usual daily pattern of sleep and wakefulness. This rhythm appears to be established by a nucleus within the *(basal ganglia? hypothalamus? thalamus?)*.

b. The RAS (or R____________________ A____________________ S____________________) is a portion of the reticular formation. List three types of sensory input that can stimulate the RAS.

c. Stimulation of the reticular formation leads to *(increased? decreased?)* activity of the cerebral cortex, as indicated by recordings of brain waves. If RAS and cerebral activation are continued, the state of consciousness is maintained. Inactivation or damage

of these areas leads to ____________________ or ____________________. In the state of *(coma? sleep?)*, the person cannot be roused. See page 521 of the text.

d. Name one chemical that can induce sleep: ____________________. This chemical

binds to ______ receptors so neurons of the RAS are *(activated? inhibited?)*. Explain how coffee or tea can keep you awake.

✓ **D3.** Do the following learning activity about sleep.

a. Name two kinds of sleep.

b. In NREM sleep, a person gradually progresses from stage *(1 to 4? 4 to 1?)* into deep sleep. Alpha waves are present in stage(s) *(1? 3 and 4?)*, whereas slow delta waves

characterize stage *(1? 4?)* sleep. Sleep spindles characterize stages __________ and

__________ of NREM sleep.

c. After moving from stage 1 to stage 4 sleep, a person *(moves directly into? ascends to stages 3 and 2 and then moves into?)* REM sleep.

d. Rapid eye movements characterize *(REM? NREM?)* sleep. Most dreaming occurs during *(REM? NREM?)* sleep. EEG readings of REM sleep are more like those in stage *(1? 4?)* of NREM sleep. Muscle tone, skeletal muscle activity, heart rate and blood pressure are all lower during *(REM? NREM?)* sleep.

e. Periods of REM and NREM sleep alternate throughout the night in about ______-minute cycles. During the early part of an 8-hour sleep period, REM periods last about *(5–10? 30? 50?)* minutes; they gradually increase in length until the final REM

period lasts about ______ minutes.

f. With aging, the percentage of REM sleep normally ________-creases.

D4. Define these terms: *learning, memory, plasticity.*

D5. List several parts of the brain that are likely to assist with memory.

✓ **D6.** Do this exercise on memory.

a. Looking up a phone number, dialing it, and then quickly forgetting it is an example of *(short-term? long-term?)* memory.

b. Repeated use of a telephone number can commit it to long-term memory. Such rein-

forcement is known as memory ____________________. *(Most? Very little?)* of the information that comes to conscious attention goes into long-term memory because the brain is selective about what it retains.

c. Some evidence indicates that *(short-term? long-term?)* memory involves electrical and chemical events rather than anatomical changes. Supporting this idea is the fact that chemicals used for anesthesia and shock treatments interfere with *(recent? long-term?)* memory.

D7. Discuss changes in the following anatomical structures or chemicals that may be associated with enhanced long-term memory:

a. Presynaptic terminals, synaptic end bulbs, dendritic branches

b. Glutamate, NMDA glutamate receptors, nitric oxide (NO)

E. Disorders (pages 520–521)

E1. List several causes of spinal cord injury.

✓ **E2.** Match the terms in the box with descriptions below.

H. Hemiplegia	**P. Paraplegia**
M. Monoplegia	**Q. Quadriplegia**

_______ a. Paralysis of one extremity only

_______ b. Paralysis of both legs

_______ c. Paralysis of both arms and both legs

_______ d. Paralysis of the arm, leg, and trunk on one side of the body

✓ **E3.** *A clinical correlation.* Describe changes associated with transection of the spinal cord in this activity.

a. Hemisection of the cord refers to transection of *(half? all?)* of the spinal cord.

b. If the posterior columns (cuneatus and gracile) and lateral corticospinal tracts on the right side of the cord are severed at level T11–T12, symptoms of loss of awareness of muscle sensations (proprioception), loss of touch sensations, and paralysis are likely to occur in the *(right arm? left arm? right leg? left leg?)*. (Circle all that apply.)

c. The period of spinal shock is likely to last for several *(hours to months? years?)*. During this time, the person is likely to experience *(areflexia? hyperreflexia?)*, meaning *(exaggerated? no?)* reflexes. List several signs of spinal shock.

✓ **E4.** Cerebral palsy (CP) is a group of disorders affecting *(movements? sensations?)*. This condition results from brain damage *(in fetal life, at birth, or in infancy? in early adulthood? in later years?)*. CP *(is? is not?)* a progressive disease; it *(is? is not?)* reversible.

✓ **E5.** Complete this Checkpoint about Parkinson's disease (PD).

a. Parkinson's typically affects persons *(in fetal life, at birth, or in infancy? in early adulthood? in later years?)*. PD *(is? is not?)* a progressive disease; it *(is? is not?)* usually hereditary.

b. This condition involves a decrease in the release of the neurotransmitter

_______________. This chemical is normally produced by cell bodies located in

the substantia nigra, which is part of the _______________. Axons lead from

here to the _______________, where dopamine (DA) is released. DA is an *(excitatory? inhibitory?)* transmitter.

c. The basal ganglia neurons produce a neurotransmitter ___________________ (ACh), which is *(excitatory? inhibitory?)* to skeletal muscles. The result is a relative increase in the ratio of ACh to DA in Parkinson's patients. Explain how this imbalance leads to muscle tremor and rigidity.

d. Voluntary movements may be slower than normal; this is the condition of *(brady? tachy?)*-kinesia.

e. Because persons with parkinsonism have a deficit in dopamine, it seems logical to treat the condition by administration of this neurotransmitter. Does this work? *(Yes? No?)* Explain.

f. Why do levodopa and/or anticholinergic medications provide some relief to these patients?

g. How do MAO-inhibitor drugs help PD patients?

ANSWERS TO SELECTED CHECKPOINTS: CHAPTER 15

A1. (a) Sensory. (b) Integrating, associating, and storing. (c) Motor (efferent), muscles or glands.

A3. (a) Sensation, perception. (b1) S; (b2) B; (b3) T; (b4) C.

A4. Modality; one modality; cerebral cortex.

A5. (a) Cold, warm. (b) Awareness of joint positions, awareness of movements. (c) Pressure, touch, vibration. (d) Equilibrium, hearing, smell, taste, vision.

A6. 1. S → 2. T → 3. IGP → 4. Int. (a) S. (b) T. (c) IGP. (d) Int.

A7. See Table LG 15.1A at right.

A8. (a) B C A. (b) Vigorous response of each type of receptor to a specific type of stimulus (such as retina of eye to light)

Table LG 15.1A Classification of sensory receptors

Receptor	Structure	Type of Graded Potential	Location (-ceptor)	Type of Stimulus Detected
	Free Encapsulated Separate cells	Generator Receptor	Extero- Intero- Proprio-	Mechanoreceptor Thermoreceptor Nociceptor Photoreceptor Chemoreceptor
1. Pressure	**Encapsulated**	Generator	**Extero-**	**Mechanoreceptor**
2. Pain (skin)	Free	**Generator**	Extero-	**Nociceptor**
3. Pain (stomach)	**Free**	**Generator**	**Intero-**	Nociceptor
4. Vision	**Separate cells**	Receptor	**Extero-**	Photoreceptor
5. Equilibrium	**Separate cells**	**Receptor**	Proprio-	**Mechanoreceptor**
6. Smell	Separate cells	**Receptor**	**Extero-**	**Chemoreceptor**
7. Cold on skin	Free	Generator	Extero-	Thermoreceptor
8. Changes of O_2 level in blood	(Omit)	(Omit)	Intero-	Chemoreceptor

with little or no response to other types of stimuli (such as lack of retinal response to sound waves). (c) Adaptation; pain; protects you by continuing to provide warning signals of painful stimuli.

B2. (a) Touch, pressure, vibration, as well as itch and tickle; mechano; unevenly. (b) Crude. (c) Deep. (d) Vibration.

B3 (a) C. (b) L. (c) C, Type I. (d) N, Type I, Type II. (e) N. (f) H.

B5. (a) A local inflammation, for example, from a mosquito bite, causes release of chemicals such as bradykinin, which stimulates free nerve endings and lamellated corpuscles. (b) Prostaglandins, kinins, K^+. (c) Superficial; acute, sharp, pricking. (d) Slow. (e) Superficial somatic pain.

B6: (a) Medial aspects of left arm; T1, T4. (b) Shoulder and neck (right side); diaphragm, cervical; 3, 4; referred.

B7. Phantom pain (phantom limb sensation). This results from irritation of nerves in the stump. Pathways from the stump to the cord and brain are still intact and may be stimulated even though the original sensory receptors in his foot are missing. Remember that he feels pain or itching in his brain, not in his leg or foot. Also, the networks of neurons in the brain that previously received sensory impulses may remain active, providing false sensory perceptions.

B8. (a) 0. (b) AA1. (c) A.

B9. Proprioception. (a) J. (b) M. (c) T.

C1. (a) Cortex. (b) Longitudinal, foot. (c) Face; hearing. (d) Right; sensation, hand, left. (Note: some left-handed paralysis also, since some of motor control pathways are in the postcentral gyrus. (e) General sensory or somatosensory, are; lips, face, thumb, and other fingers. (See Figure 15.5a in the text.)

C2. (a) Crude touch. (b) Fine touch. (c) Proprioception. (d) Kinesthesia. (e) Stereognosis.

C3. (a) LST. (b) AST. (c–e) PCML. (f) AST. (g) APSC.

C4. Three. (a) II. (b) I. (c) III.

C5.

Pathway	First-Order	Second-Order	Third-Order
a. Posterior column–medial lemniscus	Posterior root ganglion	Medulla: cuneatus nucleus and gracile nucleus	Thalamus
b. Anterolateral (spinothalamic)	Posterior root ganglion	Posterior gray horn of spinal cord	Thalamus

C6. See Figure LG 15.1A.

C7. (a) A, B. (b) Are not. (c) B, C. (d) Bacterium; tertiary; B and D.

C9. (a) UMN. (b) UMN. (c) LMN. (d) CN. (e) BG.

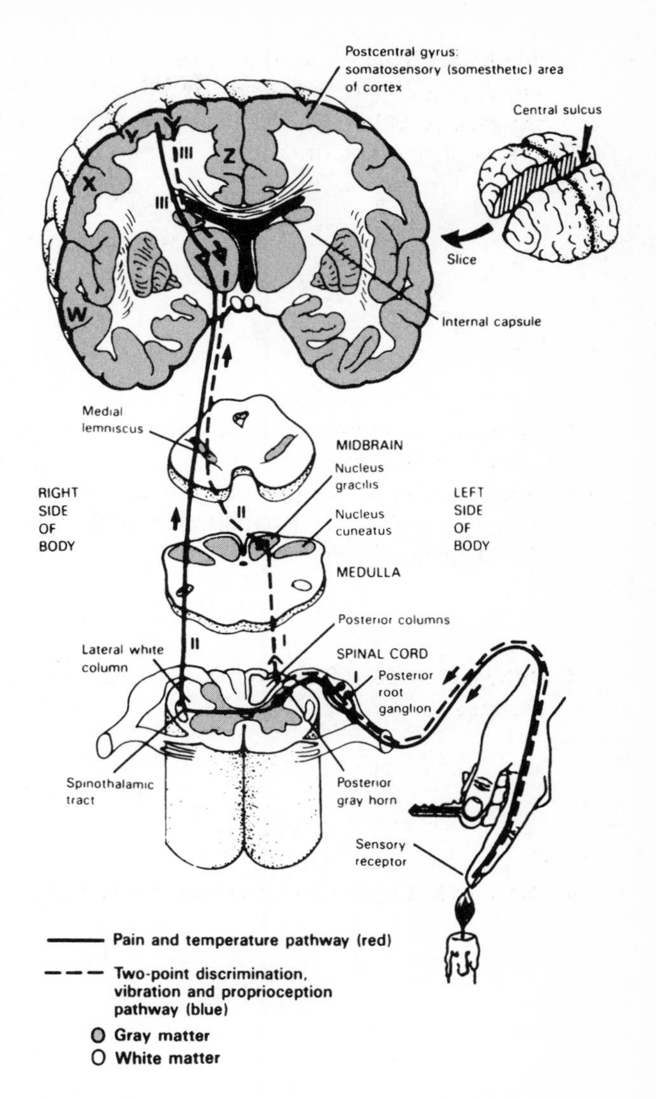

Figure LG 15.1A Diagram of the central nervous system.

C10. (a) Frontal. (b) Movement, left side of the face; VII. (c) Movement of left foot affected (spasticity).

C11. G D B F E A H C.

C12. See Figure 15.2A.

C13. (a) Direct; precise, voluntary; pyramidal; A B and C. (b) Lateral, decussated or crossed; right. (c) Left, anterior; are. (d) 10, anterior; do not, left; ipsi; do. (e) Brainstem; cranial. (f) Upper; lower. (g) Because these neurons, which are activated by input from a variety of sources (within the brain or spinal cord), must function for any movement to occur.

C14. (a) Leg and foot. (b) Sensation and movement. (c) Are, because the brainstem (including reticular formation) and cerebellum are sites of origin of these pathways.

C15. (a) Outside of; outside of, spinal cord. (b1) MRS. (b2) TS. (b3). VS. (b4) RS.

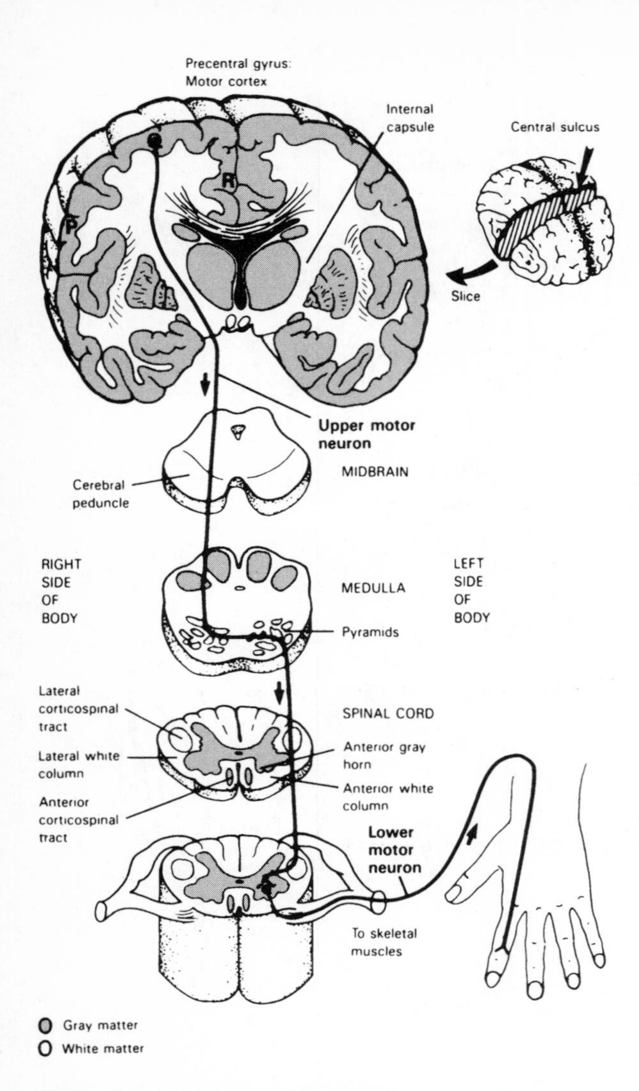

Figure LG 15.2A Diagram of the central nervous system.

C17. (a) Parkinson's, dopamine; muscle rigidity and tremors. (b) Chorea, GABA; 40 years; 50–50. (c) Amyotropic lateral sclerosis; weak and atrophied; are not. (d) Ataxia; intention.

Dl. Wakefulness and sleep, learning and memory, and emotional responses.

D2. (a) Circadian; hypothalamus. (b) Reticular activating system; pain, proprioception, visual or auditory stimuli. (c) Increased, EEG sleep or coma; coma. (d) Adenosine; Al, inhibited; chemicals (caffeine or theophylline) block Al receptors so the RAS remains activated.

D3. (a) Nonrapid eye movement (NREM) and rapid eye movement (REM). (b) 1 to 4; 1; 4; 2, 3. (c) Ascends to stages 3 and 2 and then moves into. (d) REM; REM; 1; REM. (e) 90; 5–10; 50. (f) De.

D6. (a) Short-term. (b) Consolidation; very little. (c) Short-term, recent.

E2. (a) M. (b) P. (c) Q. (d) H.

E3. (a) Half. (b) Right leg (same side as cord is severed and in areas served by nerves inferior to the T11–T12 transection). (c) Hours to months; areflexia, no; decreased heart rate and blood pressure, flaccid paralysis, loss of somatic sensations, and bladder dysfunction.

E4. (a) Movements; in fetal life, at birth, or in infancy; is not; is not.

E5. (a) In later years; is; is not. (b) Dopamine (DA); midbrain; basal ganglia; inhibitory. (c) Acetylcholine, excitatory; lack of inhibitory DA causes a double-negative effect so that extra, unnecessary movements occur (tremor) that may then interfere with normal movements (such as extension at the same time as flexion), resulting in rigidity and masklike facial appearance. (d) Brady. (e) No, DA cannot cross the blood—brain barrier. (f) Levodopa (L-dopa) is a precursor (leads to formation) of DA and anticholinergic medications counteract effects of ACh. (g) MAO (monoamine oxidase) is an enzyme made in the body that normally destroys catecholamines such as norepinephrine or dopamine. Use of MAO inhibitors permits the dopamine to continue acting for longer periods.

MORE CRITICAL THINKING: CHAPTER 15

1. Describe structures and mechanisms that help prevent overstretching of your muscles, tendons, and joints. Be sure to include roles of receptors and describe nerve pathways.
2. Contrast the sensations of proprioception, kinesthesia, and stereognosis. Tell what information these senses provide to you.
3. Contrast each of the following sensory receptors according to the following four characteristics: receptor structure and location, type of graded potential generated, and type of stimulus detected: (a) taste, (b) hearing, (c) light touch, and (d) awareness of muscle tension.
4. Contrast upper motor neurons (UMNs) with lower motor neurons (LMNs) with regard to locations of cell bodies and axons, functions, and effects if damaged.
5. Contrast pyramidal and extrapyramidal pathways according to their specific functions and also locations of the upper and lower motor neurons in each pathway.
6. Discuss whether a CVA affecting a 0.5 cm^2 area of the cerebral cortex or a 0.5 cm^2 area of the internal capsule would be likely to cause greater loss of motor function.
7. Describe the pathological changes and treatment associated with Parkinson's disease.

MASTERY TEST: CHAPTER 15

Questions 1–6: Circle the letter preceding the one best answer to each question.

1. All of these sensations are conveyed by the posterior column—medial lemniscus pathway *except:*
   A. Pain and temperature
   B. Proprioception
   C. Fine touch, two-point discrimination
   D. Vibration
   E. Stereognosis
2. Choose the *false* statement about REM sleep.
   A. Infant sleep consists of a higher percentage of REM sleep than does adult sleep.
   B. Dreaming occurs during this type of sleep.
   C. The eyes move rapidly behind closed lids during REM sleep.
   D. EEG readings are similar to those of stage 4 of NREM sleep.
   E. REM sleep occurs periodically throughout a typical 8-hour sleep period.
3. All of the following are correct statements about the left gracile fasciculus *except:*
   A. It is part of the posterior columns.
   B. It consists of axons whose cell bodies lie in the left dorsal root ganglia.
   C. It transmits impulses along axons that terminate in the medulla.
   D. It transmits nerve impulses that originated in the left upper limb.
4. Choose the *false* statement about pain receptors.
   A. They may be stimulated by any type of stimulus, such as heat, cold, or pressure.
   B. They have a simple structure with no capsule.
   C. They are characterized by a high level of adaptation.
   D. They are found in almost every tissue of the body.
   E. They are important in helping to maintain homeostasis.
5. "Change in sensitivity to a long-lasting stimulus" is a description of the characteristic of sensation called:
   A. Adaptation
   B. Modality
   C. Selectivity
   D. Proprioception
6. All of the following are tactile sensations except:
   A. Touch
   B. Pressure
   C. Itch
   D. Proprioception
   E. Vibration

Questions 7–9: Arrange the answers in correct sequence.

______ ______ ______ ______ 7. In order of occurrence in a sensation:
   A. Transduction of a stimulus to a generator potential
   B. Integration of the impulse into a sensation
   C. Impulse generation and propagation along a nervous pathway
   D. A stimulus

______ ______ ______ ______ 8. Levels of sensation, from those causing simplest, least precise reflexes to those causing most complex and precise responses:
   A. Thalamus
   B. Brain stem
   C. Cerebral cortex
   D. Spinal cord

_____ _____ _____ _____ _____ 9. Pathway for conduction of most of the impulses for voluntary movement of muscles:

A. Anterior gray horn of the spinal cord

B. Precentral gyrus

C. Internal capsule

D. Location where decussation occurs

E. Lateral corticospinal tract

Questions 10–20: Circle T (true) or F (false). If the statement is false, change the underlined word or phrase so that the statement is correct.

T F 10. In general, the left side of the brain controls the right side of the body.

T F 11. The final common pathway consists of upper motor neurons.

T F 12. Damage to the final common pathway is likely to result in flaccid paralysis.

T F 13. The cuneatus fasciculus and the cuneatus nucleus form parts of the pathway for proprioception and most tactile sensations from neck, upper chest, and upper arms.

T F 14. The posterior columns contain second-order neurons in a sensory pathway.

T F 15. Perception is the conscious or unconscious awareness of sensations.

T F 16. The neuron that crosses to the opposite side in sensory pathways is typically the second-order neuron.

T F 17. Damage to the left lateral spinothalamic tract would be most likely to result in loss of awareness of pain and vibration sensations in the left side of the body.

T F 18. Conscious sensations, such as those of sight and touch, can occur only in the cerebral cortex.

T F 19. Pain experienced by an amputee, as if the amputated limb were still there, is an example of phantom pain.

T F 20. Circadian rhythm pertains to the complex feedback circuits involved in producing coordinated movements.

Questions 21–25: Fill-ins. Complete each sentence with the word or phrase that best fits.

_______________ 21. ________ refers to the awareness of muscles, tendons, joints, balance, and equilibrium.

_______________ 22. Inactivation of the ________ results in the state of sleep.

_______________ 23. Rubrospinal, tectospinal, and vestibulospinal tracts are all classified as ________ tracts.

_______________ 24. A CVA affecting the medial portion of the postcentral gyrus of the right hemisphere (area Z in Figure LG 15.1, page 294) is most likely to result in symptoms such as ________.

_______________ 25. Voluntary motor impulses are conveyed from motor cortex to neurons in the spinal cord by ________ pathways.

ANSWERS TO MASTERY TEST: CHAPTER 15

Multiple Choice

1. A
2. D
3. D
4. C
5. A
6. D.

Arrange

7. D A C B
8. D B A C
9. B C D E A

True-False

10. T
11. F. Lower motor
12. T
13. T
14. F. First-order neurons in a sensory
15. F. Conscious awareness and interpretation of meaning
16. T
17. F. Pain and temperature sensations in the right
18. T
19. T
20. F. Events that occur at approximately 24-hour intervals, such as sleeping and waking

Fill-ins

21. Proprioception
22. Reticular activating system (RAS)
23. Extrapyramidal or indirect motor
24. Loss of sensation of left foot
25. Direct (pyramidal or corticospinal) pathways

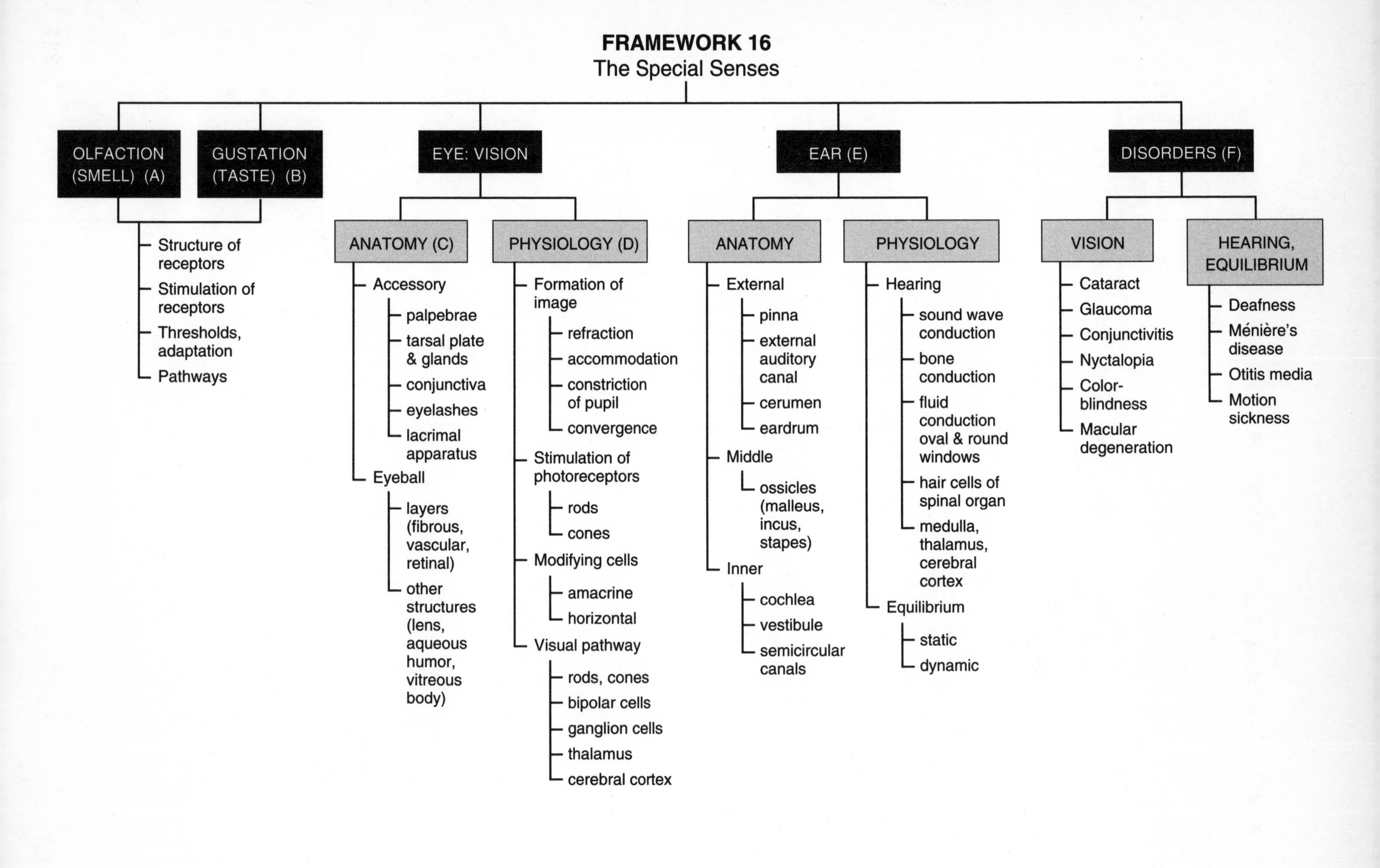

FRAMEWORK 16
The Special Senses
OLFACTION (SMELL) (A)
GUSTATION (TASTE) (B)
Structure of receptors
Stimulation of receptors
Thresholds, adaptation
Pathways
EYE: VISION
ANATOMY (C)
Accessory
palpebrae
tarsal plate & glands
conjunctiva
eyelashes
lacrimal apparatus
Eyeball
layers (fibrous, vascular, retinal)
other structures (lens, aqueous humor, vitreous body)
PHYSIOLOGY (D)
Formation of image
refraction
accommodation
constriction of pupil
convergence
Stimulation of photoreceptors
rods
cones
Modifying cells
amacrine
horizontal
Visual pathway
rods, cones
bipolar cells
ganglion cells
thalamus
cerebral cortex
EAR (E)
ANATOMY
External
pinna
external auditory canal
cerumen
eardrum
Middle
ossicles (malleus, incus, stapes)
Inner
cochlea
vestibule
semicircular canals
PHYSIOLOGY
Hearing
sound wave conduction
bone conduction
fluid conduction oval & round windows
hair cells of spinal organ
medulla, thalamus, cerebral cortex
Equilibrium
static
dynamic
DISORDERS (F)
VISION
Cataract
Glaucoma
Conjunctivitis
Nyctalopia
Color-blindness
Macular degeneration
HEARING, EQUILIBRIUM
Deafness
Ménière's disease
Otitis media
Motion sickness

CHAPTER 16

The Special Senses

All sensory organs contain receptors for increasing sensitivity to the environment. In Chapter 15 you considered the relatively simple receptors and pathways for the general senses of touch, pressure, temperature, pain, and vibration. The special senses include smell, taste, vision, hearing, and equilibrium. Special afferent pathways in many ways resemble general afferent pathways. However, a major point of differentiation is the arrangement of special sense receptors in complex sensory organs, specifically the nose, tongue, eyes, and ears.

As you begin your study of the special senses, carefully examine the Chapter 16 Topic Outline and Objectives; check off each one as you complete it. To organize your study of special senses, glance over the Chapter 16 Framework now. Be sure to refer to the Framework frequently and note relationships among key terms in each section.

TOPIC OUTLINE AND OBJECTIVES

A. Olfaction: sense of smell

1. Describe the olfactory receptors and the neural pathway for olfaction.

B. Gustation: sense of taste

1. Describe the gustatory receptors and the neural pathway for gustation.

C. Vision: anatomy

1. List and describe the accessory structures of the eye and the structural components of the eyeball.
2. Discuss image formation by describing refraction, accommodation, and constriction of the pupil.

D. Vision: physiology

1. Describe the processing of visual signals in the retina and the neural pathway for vision.

E. Hearing and equilibrium

1. Describe the anatomy of the structures in the three main regions of the ear.
2. List the major events in the physiology of hearing.
3. Identify the receptor organs for equilibrium, and describe how they function.

F. Development of the eyes and ears

1. Describe the development of the eyes and ears.

G. Disorders and medical terminology

WORDBYTES

Now become familiar with the language of this chapter by studying each wordbyte, its meaning, and an example of its use within a term. After you study the entire list, self-check your understanding by writing the meaning of each wordbyte on the line. As you continue through the *Learning Guide*, identify (and fill in) additional terms that contain the same wordbyte.

Wordbyte	Self-check	Meaning	Example(s)
aqua-	__________	water	*aque*ous humor
blephar-	__________	eyelid	*blephar*oplasty
chiasm	__________	cross	optic *chiasm*
cochlea	__________	snail-shaped	*cochlea*r duct
fungi-	__________	mushroom	*fungi*form
gust-	__________	taste	*gust*ation
kerat-	__________	cornea	radial *kerat*otomy
-lith	__________	stone	oto*lith*
lute-	__________	yellowish	macula *lute*a
macula	__________	spot	*macula* lutea
med-	__________	middle	otitis *med*ia
ocul-	__________	eye	*oculo*motor nerve
olfact-	__________	smell	*olfact*ory
ophthalm-	__________	eye	*ophthalm*ologist
opsin	__________	related to vision	rho*sopsin*
opt-	__________	eye	*opt*ic nerve, optician
orbit-	__________	eye socket	supra*orbit*al
ossi-	__________	bone	*ossi*cle
ot(o)-	__________	ear	*oto*lith
presby-	__________	old	*presby*opia
rhino-	__________	nose	oto*rhino*laryngology
sclera	__________	hard	*sclera*, oto*scler*osis
tympan-	__________	drum	*tympan*ic membrane
vitr-	__________	glassy	*vitr*eous humor

CHECKPOINTS

A. Olfaction: sense of smell (pages 527–529)

✓ **A1.** Describe olfactory receptors in this exercise.

a. Receptors for smell are located in the *(superior? inferior?)* portion of the nasal cavity.

Receptors consist of ________-polar neurons which are located between

____________________ cells. The distal end of each olfactory receptor cell consists of

dendrites that have cilia known as olfactory ____________________.

b. What is unique about the function of basal stem cells?

c. What is the function of olfactory (Bowman's) glands in the nose?

d. Because smell is a ____________________ sense, receptors in olfactory hair membranes respond to different chemical molecules, leading to a

____________________ and ____________________.

e. Adaptation to smell occurs *(slowly? rapidly?)* at first and then happens at a much slower rate.

✓ **A2.** Complete this exercise about the nerves associated with the nose.

a. Upon stimulation of olfactory hairs, impulses pass to cell bodies and axons; the axons of these olfactory cells form cranial nerves *(I? II? III?)*, the olfactory nerves. These pass from the nasal cavity to the cranium to terminate in the olfactory

____________________ located just inferior to the ____________________ lobes of the cerebrum.

b. Cell bodies in the olfactory bulb then send axons through the olfactory *(nerves? tracts?)* along two pathways. One route extends to the limbic system. What is the effect of projection of sensations of smell to this area?

c. The other olfactory pathway passes to the thalamus and then to the frontal lobes of

the ______________________ cortex. The ______________________ area of the *(left? right?)* hemishphere is particularly important in differentiating odors. (See area

11 of Figure 14.15, page 474 of the text and LG Checkpoint C9d in Chapter 14.)

B. Gustation: sense of taste (pages 529–531)

✓ **B1.** Describe receptors for taste in this exercise.

a. Receptors for taste, or ____________________ sensation, are located in taste buds.

These receptors are protected by a cluster of ____________________ cells, as well as the ____________________ cells that produce the supporting cells. The latter cells ultimately become receptor cells, which have a life span of *(10 days? 10 years?)*.

b. Within each taste bud are about *(5? 50?)* gustatory receptor cells, each with a gustatory hair that projects from the taste bud pore. Stimulation of this hair appears to initiate a ____________________ potential, which results in release of ____________________ from receptor cells. This chemical then activates sensory neurons in cranial nerves (see Checkpoint B3e).

c. Taste buds are located on elevated projections of the tongue called ____________________. The largest of these are located in a V-formation at the back of the tongue. These are ____________________ papillae. ____________________ papillae are mushroom-shaped and are located over the entire surface of the tongue. Pointed ____________________ papillae also cover the entire surface of the tongue, and these *(do? do not?)* contain many taste buds.

✓ **B2.** Do the following exercise about taste.

a. There are only five primary taste sensations: ____________________, ____________________, ____________________, ____________________, and ____________________. The taste buds at the tip of the tongue are most sensitive to ____________________ and ____________________ tastes, whereas the posterior portion is most sensitive to ____________________ taste.

b. The four primary tastes are modified by ____________________ sensations to produce the wide variety of "tastes" experienced. If you have a cold (with a "stuffy nose"), you may not be able to discriminate the usual variety of tastes, largely because of loss of the sense of ____________________.

✓ **B3.** Discuss the physiology of gustation in this exercise.

a. Arrange in correct sequence the events in the sensation of gustation when you eat a salty food.

________ ________ ________ ________

A. Neurotransmitter is released from synaptic vesicles and triggers nerve impulses in sensory neurons of the gustatory pathway.
B. Na^+ from the salty food is dissolved in plasma.
C. Na^+ passes through channels of a gustatory receptor cell, depolarizing that cell.
D. Ca^{2+} channels then open, triggering exocytosis of synaptic vesicles from receptor cells.

b. Sour taste involves flow of _______ ions into specific channels in gustatory receptors which leads to release of neurotransmitter and nerve impulses in taste pathways.

c. Circle the types of tastants that do not open membrane channels but lead to taste by activation of second messengers.

Bitter Salty Sour Sweet Umami

d. The threshold for smell is quite *(high? low?)*, meaning that *(a large? only a small?)* amount of odor must be present in order for smell to occur. Threshold varies for the primary tastes; threshold for *(sour? sweet and salt? bitter?)* is lowest,

while those for ____________________ are highest.

e. Taste impulses are conveyed to the brainstem by cranial nerves ________, ________,

and ________; there, nerve messages travel to lower parts of the brain, including the

____________________, ____________________, and ____________________, and finally stimulate the cerebral cortex.

C. Vision: anatomy (pages 531-541)

Cl. Examine your own eye structure in a mirror. With the help of Figure 16.4 in your text), identify each of the accessory structures of the eye listed below.

Conjunctiva	**Lacrimal caruncle**	**Medial commissure**
Eyebrow	**Lacrimal glands: identify site**	**Palpebrae**
Eyelashes		

✓ **C2.** *For extra review.* Match the structures listed above with the following descriptions. Write the name of the structure on the line provided.

a. Eyelid: ____________________

b. Covers the orbicularis oculis muscle; provides protection: ____________________

c. Short hairs; infection of glands at the base of these hairs is *sty:* ____________________

d. Located in superolateral region of orbit; secrete tears: ____________________

e. Mucous membrane lining the eyelids and covering anterior surface of the eye:

____

f. Point at which the two palpebrae meet laterally: ____________________

g. Reddish elevation at site of medial commissure; produces whitish secretions:

____

✓ **C3.** Do this exercise on accessory eye structures.

a. Lacrimal glands produce lacrimal fluid, also known as ____________________, at the

rate of about ________ mL per day. Identify the location of these glands on yourself.

Arrange these structures in correct order in the pathway of tears:

________ ________ ________ ________

A. Lacrimal sac and nasolacrimal duct	**C. Excretory lacrimal ducts**
B. Lacrimal glands	**D. Lacrimal punctae and canals**

State two or more functions of tears.

b. Each eye is moved by *(3? 4? 6?)* extrinsic eye muscles; these are innervated by cranial nerves ________, ________, and ________. *(For extra review,* see Chapter 11, Checkpoint C3.)

c. The eye sits in the bony socket known as the ____________________ with only the anterior *(one-sixth? one-half?)* of the eyeball exposed.

✓ **C4.** On Figure LG 16.1, color the three layers (or tunics) of the eye using color code ovals. Next to the ovals write the letters of the labeled structures that form each layer. One is done for you. Then label all structures in the eye and check your answers against the key.

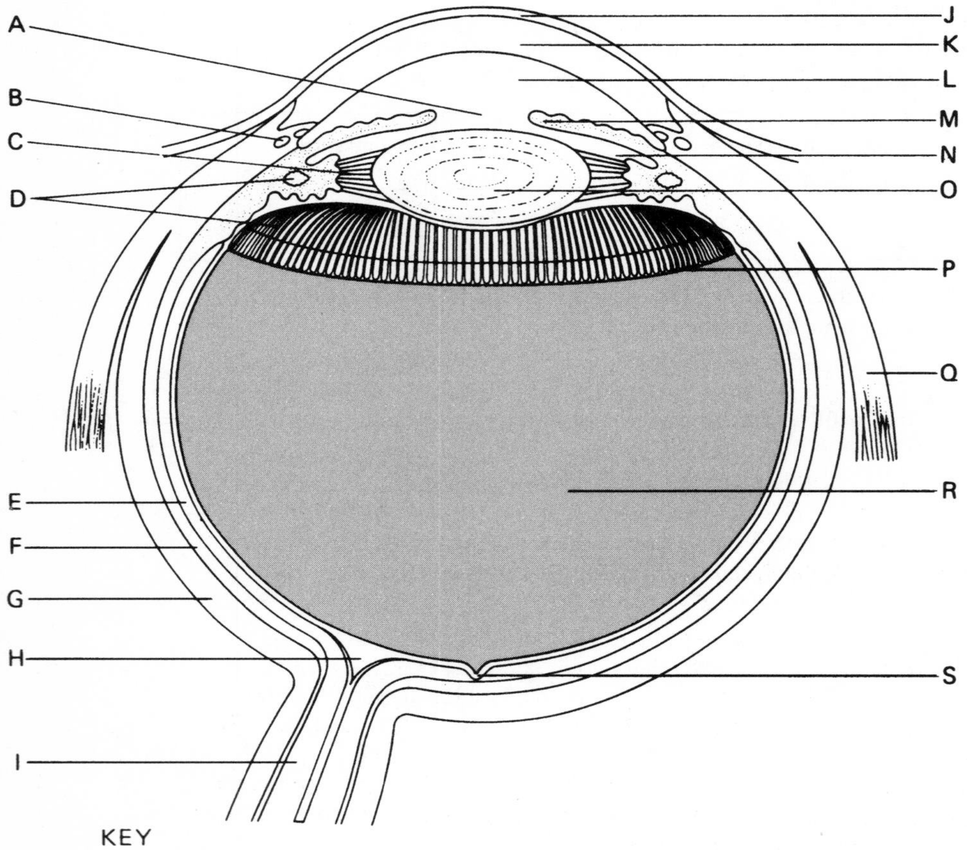

KEY

A. Pupil
B. Scleral venous sinus (Canal of Schlemm)
C. Zonular fibers (suspensory ligaments)
D. Ciliary body
E. Retina
F. Choroid
G. Sclera
H. Optic disk (blind spot)
I. Optic nerve
J. Conjunctiva
K. Cornea
L. Anterior chamber of anterior cavity
M. Iris
N. Posterior chamber of anterior cavity
O. Lens
P. Ora serrata
Q. Extrinsic eye muscle
R. Posterior cavity with vitreous body
S. Central fovea (in macula lutea)

O Fibrous tunic __G__ ____
O Vascular tunic ____ ____ ____
O Nervous tunic ____ ____

Figure LG 16.1 Structure of the eyeball in horizontal section. Color as directed in Checkpoint C4.

✓ **C5.** *Critical thinking.* A cornea that is surgically transplanted *(is? is not?)* likely to be rejected. State your rationale.

✓ **C6.** Complete this exercise about blood vessels of the eye.

a. Retinal blood vessels *(can? cannot?)* be viewed through an ophthalmoscope. Of what advantage is this?

b. The central retinal artery enters the eye at the *(optic disc? central fovea?)*.

C7. Contrast functions of the following eye structures:

a. Circular and radial muscles of the iris

b. Pigmented and neural layers of the retina

✓ **C8.** Do this exercise that describes the retina.

a. After light strikes the back of the retina, nerve impulses pass anteriorly through three zones of neurons. First is the region of rods and cones, or ____________________ zone; next is the ____________________ layer; most anterior is the ____________________ layer.

b. Axons of the ganglion layer converge to form the optic ____________________. This nerve exits from the eye at the point known as the ____________________ disc, or ____________________. The name indicates that no image formation can occur here because no rods or cones are present.

c. What function(s) do *amacrine cells* and *horizontal cells* serve?

d. The point of sharpest vision is known as the area of highest visual __________________. This site is the *(central fovea ? optic disc?)*. It contains only the photoreceptor cells named *(rods? cones?)* and *(no? many?)* bipolar and ganglion cells.

✓ **C9.** Contrast the two types of photoreceptor cells by writing *R* for rods or *C* for cones before the related descriptions.

________ a. About 6 million in each eye; most concentrated in the central fovea of the macula lutea.

________ b. Over 120 million in each eye; located mainly in peripheral regions of the eye.

________ c. Sense color and acute (sharp) vision.

________ d. Used for night vision because they have a low threshold.

✓ **C10.** Describe the lens and the cavities of the eye in this exercise.

a. The lens is composed of *(lipids? proteins?)* known as crystallins arranged in layers, much like an onion. Normally the lens is *(clear? cloudy?)*. A loss of transparency of the lens occurs in the condition called ___________________.

b. The lens divides the eye into anterior and posterior *(chambers? cavities?)*. The posterior cavity is filled with ___________________ body, which has a *(watery? jellylike?)* consistency. This body helps to hold the ___________________ in place. Vitreous body is formed during embryonic life and *(is? is not?)* replaced in later life. Bits of debris in the vitreous body are known as vitreal ___________________.

c. The anterior cavity is subdivided by the ___________________ into two chambers. Which is larger? *(Anterior chamber? Posterior chamber?)* A watery fluid called ___________________ humor is present in the anterior cavity. This fluid is formed by the ___________________ processes, and it is normally replaced about every 90 *(minutes? days?)*.

d. *Critical thinking.* How is the formation and final destination of aqueous fluid similar to that of cerebrospinal fluid (CSF)?

e. State two functions of aqueous humor.

f. Aqueous fluid creates pressure within the eye; in glaucoma this pressure *(increases? decreases?)*. What effects may occur?

✓ **C11.** *For extra review.* Check your understanding by matching eye structures in the box with descriptions below. Use each answer once.

CF.	**Central fovea**	**OD.**	**Optic disc**
Cho.	**Choroid**	**P.**	**Pupil**
CM.	**Ciliary muscle**	**R.**	**Retina**
Cor.	**Cornea**	**S.**	**Sclera**
I.	**Iris**	**SVS.**	**Scleral venous sinus (canal of Schlemm)**
L.	**Lens**		

________ a. "Whites of the eyes"

________ b. Normally clear; if cloudy, it is a cataract

________ c. Blind spot; area in which there are no cones or rods

________ d. Area of sharpest vision; area of densest concentration of cones

________ e. Anteriorly located, avascular coat responsible for about 75% of all focusing done by the eye

________ f. Layer containing neurons

________ g. Brown-black layers prevent reflection of light rays; also nourishes retina because it is vascular

_______ h. A hole; appears black, like a circular doorway leading into a dark room

_______ i. Muscular tissue that regulates the amount of light entering the eye; colored part of the eye

_______ j. Attaches to the lens by means of radially arranged fibers called the zonular fibers

_______ k. Located at the junction of iris and cornea; drains aqueous humor

C12. Define each of these processes involved in *image formation* on the retina.

a. *Refraction* of light

b. *Accommodation* of the lens

c. *Constriction* of the pupil

d. *Convergence* of the eyes

✓ **C13.** Answer these questions about formation of an image on the retina.

a. Bending of light rays so that images focus exactly upon the retina is a process known as _______________. The *(cornea? lens?)* accomplishes most (75%) of the refraction within the eye. Explain why you think this is so.

b. Draw how a letter "e" would look as it is focused on the retina. _______

c. Explain why you see things right side up even though refraction produces inverted images on your retina.

✓ **C14.** Explain how accommodation enables your eyes to focus clearly.

a. A *(concave? convex?)* lens will bend light rays so that they converge and finally intersect. The lens in each of your eyes is *(biconcave? biconvex?)*.

b. When you bring your finger toward your eye, light rays from your finger must converge *(more? less?)* so that they will focus on your retina. Thus for near vision, the anterior surface of your lens must become *(more? less?)* convex. This occurs by a thickening and bulging forward of the lens caused by *(contraction? relaxation?)* of the _______________ muscle. In fact, after long periods of close work (such as reading), you may experience eye strain.

c. What is the *near point of vision* for a young adult? ________ cm (________ inches) For a 60-year-old? ________ cm (________ inches) What factor explains this difference?

✓ **C15.** *A clinical correlation.* Describe errors in refraction and accommodation in this exercise.

a. The normal (or __________________ eye) can refract rays from a distance of ________ meters (________ feet) and form a clear image on the retina.

b. In the nearsighted (or ____________________) eye, images focus *(in front of? behind?)* the retina. This may be due to an eyeball that is too *(long? short?)*. Corrective lenses should be slightly *(concave? convex?)* so that they refract rays less and allow them to focus farther back on the retina.

c. The farsighted person can see *(near? far?)* well but has difficulty seeing *(near? far?)* without the aid of corrective lenses. The farsighted (or ____________________) person requires *(concave? convex?)* lenses. After about age 40, most persons lose the ability to see near objects clearly. This condition is known as ____________________.

✓ **C16.** Choose the correct answers to complete each statement.

a. Both accommodation of the lens and constriction of the pupil involve *(extrinsic? intrinsic?)* muscles, whereas convergence of the eyes involves *(extrinsic? intrinsic?)* muscles.

b. The pupil *(dilates? constricts?)* in the presence of bright light; the pupil *(dilates? constricts?)* in stressful situations when sympathetic nerves stimulate the iris.

D. Vision: physiology (page 541–546)

✓ **D1.** List three principal processes that are necessary for vision to occur.

a.

b.

c.

✓ **D2.** Do this exercise on photoreceptors and photopigments.

a. The basis for the names *rods* and *cones* rests in the shape of the *(inner? outer?)* segments of these cells, that is, the segment closest to the *(choroid and pigmented layer? vitreous body?)*. In which segment are photopigments such as rhodopsin located? *(Inner? Outer?)* In which segment can the nucleus, Golgi complex, and mitochondria be found? *(Inner? Outer?)*

b. How many types of photopigment are found in the human retina? ______ Rhodopsin is the photopigment found in *(rods? cones?)*. How many types of photopigments are found in cones?

c. Photopigments are also known as _______________ pigments and consist of colored *(proteins? lipids?)* in outer segment plasma membranes. In rods, these chemicals pinch off to form stacks of *(bubbles? discs?)*. A new disc is formed every 20–60 *(minutes? days? months?)*; old discs are destroyed by the process of _______________.

d. Any photopigment consists of two parts: *retinal* and *opsin*. What is the function of retinal?

e. Retinal is a vitamin ________ derivative and is formed from carotenoids. List several carotenoid-rich foods.

Explain how such foods may help to prevent *nyctalopia* (night blindness)?

✓ **D3.** Refer to Figure 16.14 in your text, and complete this Checkpoint about response of photopigments to light and darkness.

a. *(Retinal? Opsin?)* appears as a colored, dumbbell-shaped protein embedded in a disc in a rod or cone. Retinal fits snugly into opsin when retinal is in a *(bent? straightened?)* shape. This form of retinal is known as the *(cis? trans?)* form, and it is found when eyes are in *(darkness? light?)*.

b. Conversion of *cis*- to *trans*-retinal is known as _______________. This occurs when retinal *(absorbs light? sits in darkness?)*. Within a minute of exposure to light, the *(cis? trans ?)*-retinal slips out of place in opsin. The separate products (opsin and *trans*retinal) are *(colored? colorless or bleached?)*. This is the first step in the process of conversion of a _______________ (light) into a generator potential; this process is known as *(transduction? integration?)*. As a result, bipolar and then ganglion cells will be activated to begin nerve transmission along the visual pathway (see Checkpoints D6–D10).

c. To make further visual transductions possible, *trans*-retinal must be converted back to *cis*-retinal, a process known as _______________. This process requires *(light? darkness?)* and action of an enzyme known as retinal _______________. Which photopigment regenerates more quickly? *(Rhodopsin? Cone photopigments?)*

✓ **D4.** Explain what happens in your retinas after you walk from a bright lobby into a darkened movie theater.

a. In the lobby the rhodopsin in your rods had been bleached (broken into _______________ retinal and opsin) as the initial step in vision involving rods.

b. Only a period of darkness will permit adequate regeneration of retinal in the *cis* form that can bind to opsin and form _______________. After complete bleaching of rhodopsin, it takes *(5? 1.5?)* minutes to regenerate 50% of rhodopsin and about

__________________ minutes for full regeneration of this photopigment. This process is called *(light? dark?)* adaptation.

c. Also necessary for rhodopsin formation is adequate vitamin A. Explain why a detached retina can result in reduced rhodopsin formation and nyctalopia.

✓ **D5.** Explain how cones help you to see in "living color" by completing this paragraph. Cones contain pigments which *(do? do not?)* require bright light for breakdown (and therefore for a generator potential). Three different pigments are present, sensitive

to three colors: ________________, ________________, and ________________.
A person who is red-green color-blind lacks some of the cones receptive to two colors

(__________________ and __________________) and so cannot distinguish between these colors. This condition occurs more often in *(males? females?)*.

✓ **D6.** In Checkpoint D3 you reviewed the transduction in photoreceptor cells in response to a stimulus (dim light). Refer to Figure LG 16.2 and do this activity describing what happens next in the pathway to vision.

a. First color the large arrows indicating pathways of light and of nerve impulses (visual data processing) and all cells on the figure according to the color code ovals.

b. At point * in the figure, a *(rod? cone?)* synapses with a *(bipolar? ganglion?)* cell. In darkness, an inhibitory neurotransmitter named *(cyclic GMP? glutamate?)* is continually released at this synapse. The effect is to *(de? hyper?)*-polarize bipolar cells so that they *(become activated? stay inactive?)* and no visual information is transmitted to the brain.

c. What causes glutamate to be continually released at this synapse in darkness? Inflow

of *(Ca^{2+}? Na^+? K^+?)*, known as the "__________________ current," causes release of glutamate from synaptic terminals of rods. But what keeps these Na^+ channels open during darkness is a chemical called cyclic *(AMP? GMP?)*. Review these steps by referring to the left side of Figure 16.15 in your text.

d. Now review events (right side of Figure 16.15) that follow a stimulus of dim light exposure to rods. Effects will be roughly *(the same as? opposite to?)* those just described in (b) and (c). An enzyme is activated to break down cGMP. Na^+ channels are therefore *(opened? closed?)*, and the inhibitory glutamate is *(still? no longer?)* released so that bipolar cells (and the rest of the visual pathway) can become activated.

✓ **D7.** Refer again to Figure LG 16.2 and complete this Checkpoint about retinal cells.

a. The figure shows that the retina contains a larger number of *(photoreceptor? bipolar and ganglion?)* cells. In fact, there are between ________ and ________ rods for every one bipolar cell. This arrangement demonstrates *(convergence? divergence?)*.

b. Do cones also exhibit convergence? *(Yes? No?)* State the significance of this fact.

c. What functions do horizontal cells and amacrine cells perform? Label these cells at leader lines on the figure.

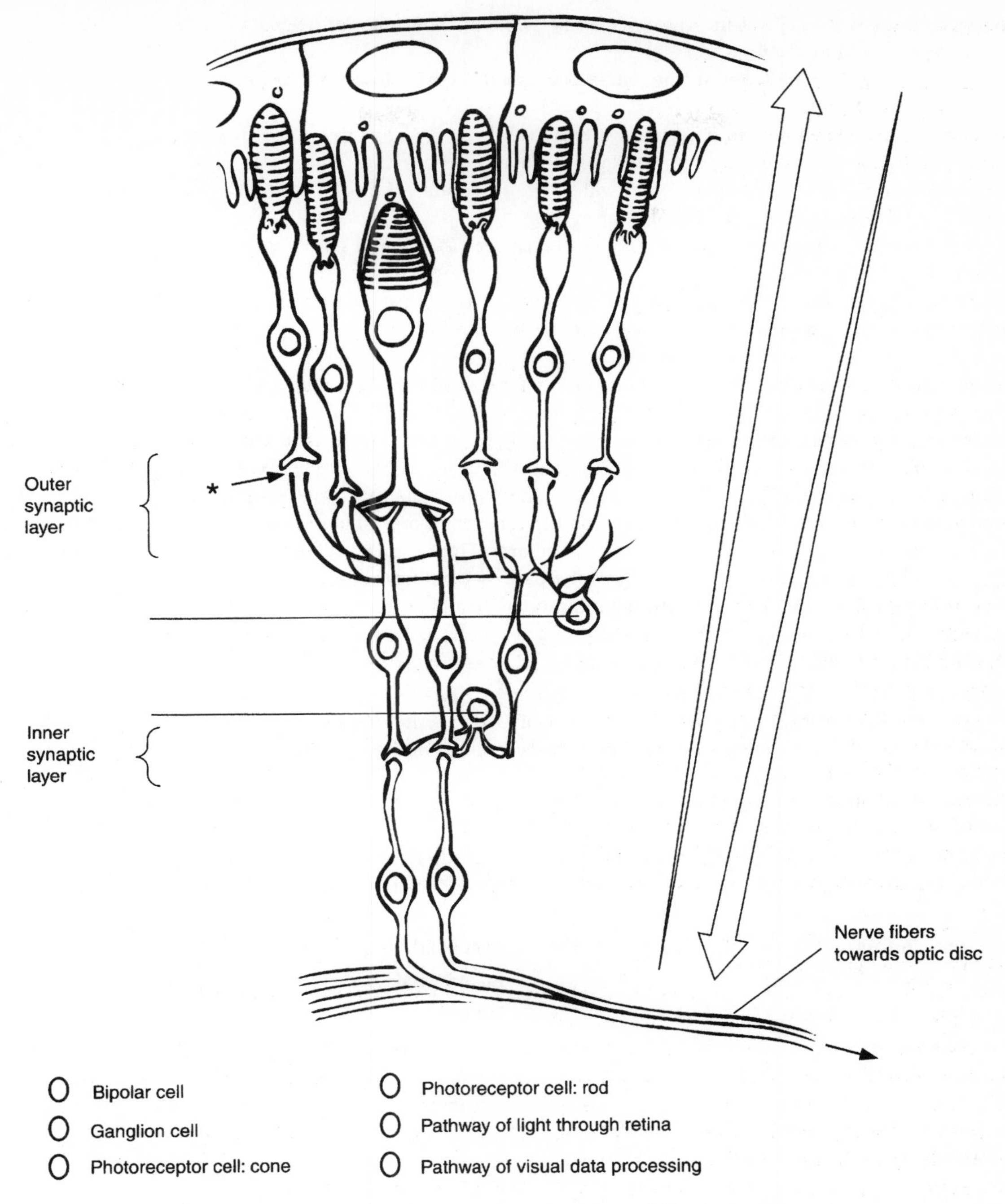

O Bipolar cell
O Ganglion cell
O Photoreceptor cell: cone
O Photoreceptor cell: rod
O Pathway of light through retina
O Pathway of visual data processing

Figure LG 16.2 Microscopic structure of the retina. Color as indicated in Checkpoint D6.

✓ **D8.** Describe the conduction pathway for vision by arranging these structures in sequence. Write the letters in correct order on the lines provided. Also, indicate the four points where synapsing occurs by placing an asterisk (*) between the letters.

B.	**Bipolar cells**	**ON.**	**Optic nerve**
C.	**Cerebral cortex (visual areas)**	**OT.**	**Optic tract**
G.	**Ganglion cells**	**P.**	**Photoreceptor cells (rods, cones)**
OC.	**Optic chiasm**	**T.**	**Thalamus**

______ ______ ______ ______ ______ ______

✓ **D9.** Answer these questions about the visual pathway to the brain. It may be helpful to refer to Figure 16.16 in the text.

a. Hold your left hand up high and to the left so you can still see it. Your hand is in the *(temporal? nasal?)* visual field of your left eye and in the ___________________ visual field of your right eye.

b. Due to refraction, the image of your hand will be projected onto the *(left? right?)* *(upper? lower?)* portion of the retinas of your eyes.

c. All nerve fibers from these areas of your retinas reach the *(left? right?)* side of your thalamus and cerebral cortex.

d. Damage to the right optic tract, right side of thalamus, or right visual cortex would result in loss of sight of the *(left? right?)* visual fields of each eye.

✓ **D10.** *The Big Picture: Looking Back.* Refer to Figure 14.5, page 474 of your text. The optic chiasm is located just anterior to the ___________________ gland. Visual disturbances are likely to signal a tumor in this gland. Pressure of the tumor is likely to be exerted against the more medial fibers within the chiasm. These are fibers that have passed from *(nasal? temporal?)* regions of the retinas and that enter "crossed fiber" regions of optic tracts to reach opposite sides of the brain. (Refer to Figure 16.16 in your text.) As a result of loss of function of nasal retinas, the patient would experience blindness in *(nasal? temporal?)* visual fields, a condition known as "tunnel vision."

E. Hearing and equilibrium (pages 546–557)

✓ **E1.** Refer to Figure LG 16.3 and do the following exercise.

a. Color the three parts of the ear using color code ovals. Then write on lines next to the ovals the letters of structures that are located in each part of the ear. One is done for you.

b. Label each lettered structure using leader lines on the figure.

✓ **E2.** *For extra review.* Select the ear structures in the box that fit the descriptions below. Not all answers will be used.

A.	**Auricle (pinna)**	**OW.**	**Oval window**
AT.	**Auditory (Eustachian) tube**	**RW.**	**Round window**
I.	**Incus**	**S.**	**Stapes**
M.	**Malleus**	**TM.**	**Tympanic membrane**

________ a. Tube used to equalize pressure on either side of tympanic membrane

________ b. Eardrum

________ c. Structure on which stapes exerts piston-like action

________ d. Ossicle adjacent to eardrum

________ e. Anvil-shaped ear bone

________ f. Portion of the external ear shaped like the flared end of a trumpet

✓ **E3.** Check your understanding of structures associated with the middle and inner ears by circling correct answers in this exercise.

a. Which structure permits microorganisms to travel from the throat (nasopharynx) to the middle ear.
   A. External auditory canal   B. Auditory (Eustachian) tube

b. Which is a tiny muscle innervated by the facial nerve; paralysis leads to hyperacusia.
   A. Stapedius   B. Tensor tympani

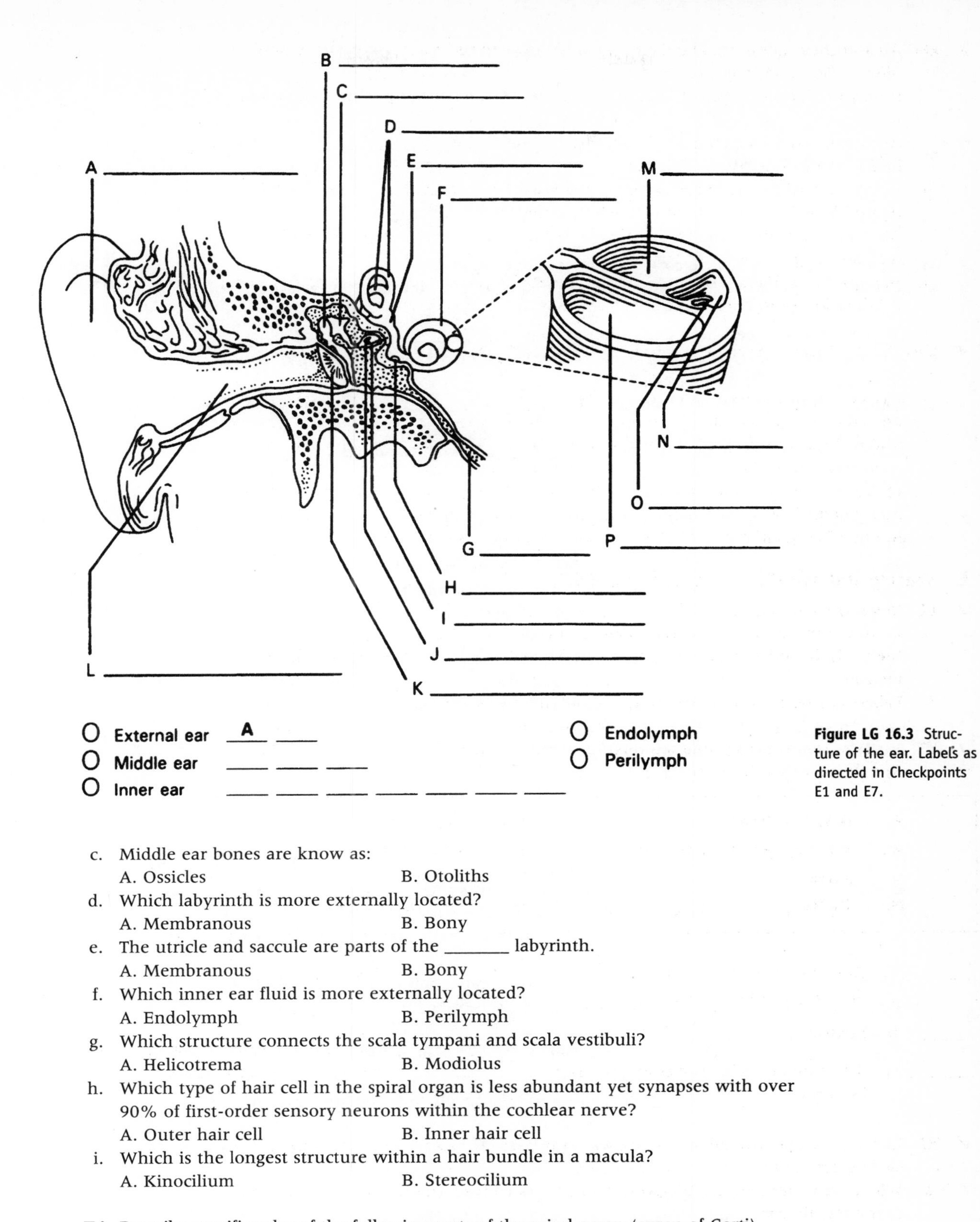

Figure LG 16.3 Structure of the ear. Labels as directed in Checkpoints E1 and E7.

c. Middle ear bones are know as:
   A. Ossicles B. Otoliths
d. Which labyrinth is more externally located?
   A. Membranous B. Bony
e. The utricle and saccule are parts of the _______ labyrinth.
   A. Membranous B. Bony
f. Which inner ear fluid is more externally located?
   A. Endolymph B. Perilymph
g. Which structure connects the scala tympani and scala vestibuli?
   A. Helicotrema B. Modiolus
h. Which type of hair cell in the spiral organ is less abundant yet synapses with over 90% of first-order sensory neurons within the cochlear nerve?
   A. Outer hair cell B. Inner hair cell
i. Which is the longest structure within a hair bundle in a macula?
   A. Kinocilium B. Stereocilium

E4. Describe specific roles of the following parts of the spiral organ (organ of Corti)
a. Tectorial membrane

b. Tip link proteins

c. Transduction channel

✓ **E5.** Fill in the blanks and circle correct answers about sound waves.

a. Sound waves heard by humans range from frequencies of 20 to 20,000 cycles/sec (Hz). Humans can best hear sounds in the range of ________________ Hz.

b. A musical high note has a *(higher? lower?)* frequency than a low note. So frequency is *(directly? indirectly?)* related to pitch.

c. Sound intensity (loudness) is measured in units called ________________.

Normal conversation is at a level of about _______ dB, whereas sounds at _______ dB can cause pain.

✓ **E6.** Summarize events in the process of hearing in this activity. It may help to refer to Figure 16.21 in the text.

a. Sound waves travel through the ______________ and strike the ______________ membrane. Sound waves are magnified by the action of the three ________________ in the middle ear.

b. The ear bone named *(malleus? incus? stapes?)* strikes the *(round? oval?)* window, setting up waves in *(endo-? peri-?)* lymph. This pushes on the floor of the upper scala *(vestibuli? tympani?)*. As a result the cochlear duct is moved, and so is the perilymph in the lower canal, the scala *(vestibuli? tympani?)*. The pressure of the perilymph is finally expended by bulging out the *(round? oval?)* window.

c. As the cochlear duct moves, tiny hair cells embedded in the floor of the duct are stimulated. These hair cells are part of the ________________ organ; its name is based on its spiral arrangement on the ________________ membrane all the way around the 2 3/4 coils of the cochlear duct. *(High? Low?)*-pitched sounds are sensed best toward the apex (helicotrema region) of the cochlea.

d. As spiral organ hair cells are moved by waves in endolymph, stereocilia are bent against the ________________ gelatinous membrane. This movement generates receptor potentials in hair cells that excite nearby sensory neurons of the ________________ branch of cranial nerve _______. The pathway continues to the brainstem, ______________ (relay center), and finally to Brodmann's area of the ______________ lobe of the cerebral cortex.

✓ **E7.** Color endolymph and perilymph and related color code ovals in Figure LG 16.3 on page LG 321. Endolymph contains an unusually *(high? low?)* K^+ level for an extracellular fluid (ECF). State the significance of this fact.

✓ **E8.** Check your understanding of the pathway of fluid conduction in the auditory pathway by placing the following structures in correct sequence. Write the letters on the lines provided.

_______ _______ _______ _______ _______ _______

B.	**Basilar membrane**	**ST.**	**Scala tympani (perilymph)**
C.	**Cochlear duct (endolymph)**	**SV.**	**Scala vestibuli (perilymph)**
O.	**Oval window**	**VM.**	**Vestibular membrane**
RW.	**Round window**		

E9. Explain how hair cells of the spiral organ may be damaged by loud noises.

✓ **E10.** Contrast receptors for hearing and for equilibrium in this summary of the inner ear.

a. Receptors for hearing and equilibrium are all located in the *(middle? inner?)* ear.

All consist of supporting cells and __________________ cells that are covered by a

__________________ membrane.

b. In the spiral organ, which senses __________________, the gelatinous membrane is

called the __________________ membrane. Hair cells move against this membrane as a result of *(sound waves? change in body position?)*.

c. In the macula, located in the *(semicircular canals? vestibule?)*, the gelatinous mem-

brane is embedded with calcium carbonate crystals called __________________. These respond to gravity in such a way that the macula is the main receptor for *(static? dynamic?)* equilibrium. An example of such equilibrium occurs as you are aware of your *(position while lying down? change in position on a careening roller coaster?)*.

d. In the semicircular canals the gelatinous membrane is called the __________________. Its shape is *(flat? like an inverted cup?)*. The cupula is part of the *(crista? saccule?)* located in the ampulla. Change in direction (as in a roller coaster) causes

__________________ to bend hairs in the cupula. Cristae in semicircular canals are therefore receptors primarily for *(static? dynamic?)* equilibrium.

E11. Describe vestibular pathways, including the role of the cerebellum in maintaining equilibrium.

F. Development of the eyes and ears (pages 557–560)

✓ **F1.** Arrange in correct sequence formation of the following structures during development of the eyes.
A. Optic cups (and retina) and lens placodes (and lens vesicles)
B. Optic grooves
C. Optic vesicles

✓ **F2.** Select the structure listed in the box that is most directly formed from each of the tissues described below.

AC. Anterior chamber	**CRA. Central retinal arteries**
NR. Neural portion of the retina	**ON. Optic nerve**
PC. Posterior chamber	**PR. Pigmented portion of the retina**

_______ a. Mesenchyme between iris and lens

_______ b. Mesenchyme between iris and cornea

_______ c. Hyaloid arteries

_______ d. Inner wall of the optic cup

_______ e. Outer layer of the optic cup

_______ f. Axons from neural layer of retina

✓ **F3.** Circle the primary germ layers that contribute to development of ears:

Endoderm Mesoderm Ectoderm

✓ **F4.** Select the part of the ear that is most directly formed from each tissue described below.

E. External ear	**M. Middle ear**	**I. Inner ear**

_______ a. The first pharyngeal pouch

_______ b. The first pharyngeal cleft between first and second pharyngeal pouches

_______ c. Optic placodes which form otic pits and later otic vesicles

G. Disorders, medical terminology (pages 560–561)

✓ **G1.** Match the name of the disorder with the description.

Cat.	**Cataract**	**OM.**	**Otitis media**
Con.	**Conjunctivitis**	**Pr.**	**Presbyopia**
G.	**Glaucoma**	**Pt.**	**Ptosis**
H.	**Hyperacusia**	**Str.**	**Strabismus**
K.	**Keratitis**	**Tin.**	**Tinnitus**
Men.	**Ménière's disease**	**Tra.**	**Trachoma**
MD.	**Macular degeneration**	**V.**	**Vertigo**
Myo.	**Myopia**		

_______ a. Condition requiring corrective lenses to focus distant objects

_______ b. Excessive intraocular pressure may lead to blindness; most common cause of blindness in the United States

_______ c. Earache likely to follow sore throat

_______ d. Pinkeye

_______ e. Abnormally sensitive hearing

_______ f. The leading cause of blindness in persons over age 75

_______ g. The greatest single cause of blindness in the world; caused by the sexually transmitted microorganism, *Chlamydia*

_______ h. Disturbance of the inner ear with excessive endolymph

_______ i. Loss of transparency of the lens

_______ j. Farsightedness due to loss of elasticity of lens, especially after age 40

_______ k. Inflammation of the cornea

_______ l. Sense of spinning or whirling

_______ m. Ringing of the ears

_______ n. Drooping of an eyelid

G2. Discuss causes, signs and symptoms, and treatments of:

a. Macular degeneration

b. Otitis media

G3. Contrast sensorineural deafness with conduction deafness.

ANSWERS TO SELECTED CHECKPOINTS: CHAPTER 16

Al. Superior; bi, supporting; hairs. (b) They continually produce new olfactory receptors that live for only about a month. Because receptor cells are neurons, this "breaks the rule" that mature neurons are never replaced. (c) Produce mucus that acts as a solvent for odoriferous substances. (d) Chemical, generator potential, nerve impulse. (e) Rapidly.

A2. (a) I; bulb, frontal. (b) Tracts; conscious awareness of certain smells, such as putrid odors or fragrance of roses, may lead to emotional responses. (c) Cerebral; orbitofrontal; right.

B1. (a) Gustatory; supporting, basal; 10 days. (b) 50; receptor, neurotransmitter. (c) Papillae; circumvallate; fungiform; filiform, do not.

B2. (a) Sweet, sour, salt, bitter, and umami; sweet and salty, bitter. (b) Olfactory; smell.

B3. (a) B C D A. (b) H^+. (c) Bitter, sweet, umami. (d) Low, only a small; bitter; sweet and salty. (e) VII, IX, X; thalamus, hypothalamus, limbic system.

C2. (a) Palpebra. (b) Eyebrow. (c) Eyelashes. (d) Lacrimal glands. (e) Conjunctiva. (f) Lateral commissure (g) Lacrimal caruncle.

C3. (a) Tears, 1; superolateral to each eyeball: see Figure 16.5b in your text; B C D A; tears protect against infection via the bactericidal enzyme lysozyme and by flushing away irritating substances; tears are signs of emotions. (b) 6; III, IV, VI. (c) Orbit, one-sixth.

C4. Fibrous tunic: G, K; vascular tunic or uvea: D, F, M; nervous tunic: E, S.

C5. Is not. Because the cornea is avascular, antibodies that cause rejection do not circulate there.

C6. (a) Can; health of blood vessels can be readily assessed, for example, in persons with diabetes. (b) Optic disc.

C8. (a) Photoreceptor, bipolar, ganglion. (b) Nerve; optic, blind spot. (c) They modify signals transmitted along optic pathways. (d) Acuity; central fovea; cones, no.

C9. (a) C. (b) R. (c) C. (d) R.

C10. (a) Protein; clear; cataract. (b) Cavities; vitreous, jellylike; retina; is not; floaters. (c) Iris; anterior chamber; aqueous; ciliary, minutes. (d) Both are formed from blood vessels (choroid plexuses) and the fluid finally returns to venous blood. (e) Provides nutrients and oxygen and removes wastes from the cornea and lens; also provides pressure to separate the cornea from lens. (f) Increases; damage to the retina and optic nerve with possible blindness.

C11. (a) S (b) L. (c) OD. (d) CF. (e) Cor. (f) R. (g) Cho. (h) P. (i) I. (j) CM. (k) SVS.

C13. (a) Refraction; cornea; the density of the cornea differs considerably from the air anterior to it. (b) "ə" ("e" inverted 180° and much smaller). (c) Your brain learned early in your life how to "turn" images so that what you see corresponds with locations of objects you touch.

C14. (a) Convex; biconvex. (b) More; more; contraction, ciliary. (c) 10, 4, 80, 31. The lens loses elasticity (presbyopia) with aging.

C15. (a) Emmetropic, 6, 20. (b) Myopic, in front of; long; concave. (c) Far, near; hypermetropic, convex; presbyopia.

C16. (a) Intrinsic, extrinsic. (b) Constricts; dilates.

D1. (a) Formation of an image on the retina. (b) Stimulation of photoreceptors so that a light stimulus is converted into an electrical stimulus (receptor potential and nerve impulse). (c) Transmission of the impulse along neural pathways to the thalamus and visual cortex.

D2. (a) Outer, choroid and pigmented layer; outer; inner. (b) 4; rods; one in each of three types of cones that are sensitive to different colors. (c) Visual, proteins; discs; minutes (one to three per hour), phagocytosis. (d) Retinal is the light-absorbing portion of all photopigments (rhodopsin and the three opsins in cones), so retinal begins the process of transduction of a stimulus (light) into a receptor potential. (e) A; carrots, yellow squash, broccoli, spinach, and liver; these foods lead to production of retinal in rods or cones.

D3. (a) Opsin; bent, *cis*, darkness. (b) Isomerization; absorbs light; *trans*; colorless or bleached; stimulus, transduction. (c) Regeneration; darkness, isomerase; cone photopigments.

D4. (a) *Trans*. (b) Rhodopsin; 5; 30–40; dark. (c) Because the retina has detached from the adjacent pigment epithelium, which is the storage area for vitamin A needed for synthesis of retinal for rhodopsin; difficulty seeing in dim light (nyctalopia) may result.

D5. Do; yellow to red, green, blue; red, green; males.

D6. (a) See Figure LG 16.2A. (b) Rod, bipolar; glutamate; hyper, stay inactive. (c) Na^+, dark; GMP. (d) Opposite to; closed, no longer.

D7. Photoreceptor; 6, 600; convergence. (b) No; one-on-one (cone to bipolar cell) synapsing allows for greater acuity (sharpness) of vision. (c) Enhancement of contrast and differentiation of colors. See Figure LG16.2A

D8. P * B * G ON OC OT * T * C. (Note that some crossing but no synapsing occurs in the optic chiasma.)

D9. (a) Temporal, nasal. (b) Right, lower. (c) Right. (d) Left.

D10. Pituitary; nasal; temporal.

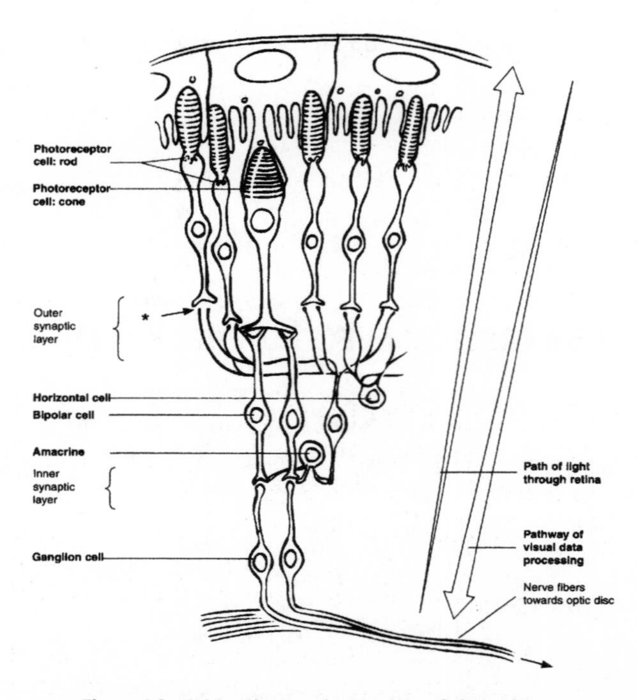

Figure LG 16.2A Microscopic structure of the retina.

El. See Figure LG 16.3A.

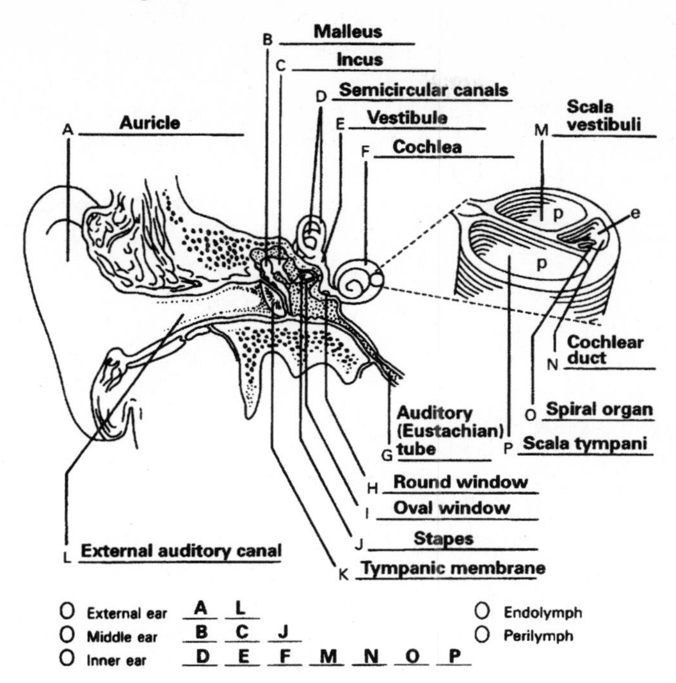

Figure LG 16.3A Diagram of the ear in frontal section

E2. (a) AT. (b) TM. (c) OW. (d) M. (e) I. (f) A.

E3. (a) B. (b) A. (c) A. (d) B. (e) A. (f) B. (g) A. (h) B. (i) A.

E5. (a) 500–5,000. (b) Higher; directly. (c) Decibels (dB); 60, 140.

E6. (a) External auditory canal, tympanic; ossicles. (b) Stapes, oval, peri-; vestibuli; tympani; round. (c) Spiral; basilar; low. (d) Tectorial; cochlear, VIII; thalamus, temporal.

E7. See Figure LG 16.3A. High. Entrance of K^+ into hair cells produces depolarization, thereby initiating nerve impulses in sensory neurons of cochlear nerves.

E8. O SV VM C B ST RW.

E10. (a) Inner; hair, gelatinous. (b) Hearing, tectorial; sound waves. (c) Vestibule, otoliths; static; position while lying down. (d) Cupula; like an inverted cup; crista; endolymph; dynamic.

F1. B C A.

F2. (a) PC. (b) AC. (c) CRA. (d) NR. (e) PR. (f) ON.

F3. All three layers.

F4. (a) M. (b) E. (c) I.

Gl. (a) Myo. (b) G. (c) OM. (d) Con. (e) H. (f) MD. (g) Tra. (h) Men. (i) Cat. (j) Pr. (k) K. (1) V. (m) Tin. (n) Pt.

MORE CRITICAL THINKING: CHAPTER 16

1. Contrast smell and taste according to the following criteria: (a) rate of adaptation, (b) threshold, (c) cranial nerve pathways for each. Also describe how you could demonstrate that much of what most people think of as taste is actually smell.
2. Compare formation, locations, and functions of lacrimal fluid and aqueous fluid.
3. Discuss causes and potential damage to the eye associated with the following conditions: (a) detachment of the retina, (b) glaucoma, (c) cataracts, (d) trachoma.
4. Contrast roles of ions (Na^+, K^+, and Ca^{2+}) in activation of photoreceptor cells of the retina versus hair cells of the cochlea.
5. Explain how cochlear implants function to take the place of cells of the spiral organ (organ of Corti).
6. Explain roles of the outer hair cells in producing otoacoustic emission, and describe how these sounds are used for assessment of hearing.
7. Contrast the anatomy of receptors for static equilibrium with those for dynamic equilibrium.

MASTERY TEST: CHAPTER 16

Questions 1–5: Circle T (true) or F (false). If the statement is false, change the underlined word or phrase so that the statement is correct.

T F 1. Parasympathetic nerves stimulate the circular iris muscle to constrict the pupil, whereas sympathetic nerves cause dilation of the pupil.

T F 2. Convergence and accommodation are both results of contraction of smooth muscle of the eye.

T F 3. Crista, macula, otolith, and spiral organ are all structures located in the inner ear.

T F 4. Both the aqueous body and vitreous humor are replaced constantly throughout your life.

T F 5. The receptor organs for special senses are less complex structurally than those for general senses.

Questions 6–12: Arrange the answers in correct sequence.

______ ______ ______ 6. Layers of the eye, from superficial to deep:
- A. Sclera
- B. Retina
- C. Choroid

______ ______ ______ ______ 7. From anterior to posterior:
- A. Vitreous body
- B. Optic nerve
- C. Cornea
- D. Lens

______ ______ ______ ______ 8. Pathway of aqueous humor, from site of formation to destination:
- A. Anterior chamber
- B. Scleral venous sinus
- C. Ciliary body
- D. Posterior chamber

______ ______ ______ ______ ______ 9. From anterior to posterior:
- A. Anterior chamber
- B. Iris
- C. Lens
- D. Posterior cavity
- E. Posterior chamber

______ ______ ______ ______ ______ 10. Pathway of sound waves and resulting mechanical action:
- A. External auditory canal
- B. Stapes
- C. Malleus and incus
- D. Oval window
- E. Eardrum

______ ______ ______ ______ ______ 11. Pathway of tears, from site of formation to entrance to nose:
- A. Lacrimal gland
- B. Excretory lacrimal duct
- C. Nasolacrimal duct
- D. Surface of conjunctiva
- E. Lacrimal puncta and lacrimal canals

______ ______ ______ ______ ______ 12. Order of impulses along conduction pathway for smell:
- A. Olfactory bulb
- B. Olfactory hairs
- C. Olfactory nerves
- D. Olfactory tract
- E. Primary olfactory area of cortex

Questions 13–20: Circle the letter preceding the one best answer to each question.

13. Which chemical within rods and cones has the function of absorbing light?
    - A. Glutamate
    - B. Opsin
    - C. Retinal
    - D. Magnesium

14. Infections in the throat (pharynx) are most likely to lead to ear infections in the following manner. Bacteria spread through the:
    A. External auditory meatus to the external ear
    B. Auditory (eustachian) tube to the middle ear
    C. Oval window to the inner ear
    D. Round window to the inner ear
15. Choose the *false* statement about rods.
    A. There are more rods than cones in the eye.
    B. Rods are concentrated in the fovea and are less dense around the periphery.
    C. Rods enable you to see in dim (not bright) light.
    D. Rods contain rhodopsin.
    E. No rods are present at the optic disc.
16. Choose the *false* statement about the lens of the eye.
    A. It is biconvex.
    B. It is avascular.
    C. It becomes more rounded (convex) as you look at distant objects.
    D. Its shape is changed by contraction of the ciliary muscle.
    E. Change in the curvature of the lens is called accommodation.
17. Choose the *false* statement about the middle ear.
    A. It contains three ear bones called ossicles.
    B. Infection in the middle ear is called otitis media.
    C. It functions in conduction of sound from the external ear to the inner ear.
    D. The cochlea is located here.
18. Choose the *false* statement about the semicircular canals.
    A. They are located in the inner ear.
    B. They sense acceleration or changes in position.
    C. Nerve impulses begun here are conveyed to the brain by the vestibular branch of cranial nerve VIII.
    D. There are four semicircular canals in each ear.
    E. Each canal has an enlarged portion called an ampulla.
19. Destruction of the left optic tract would result in:
    A. Blindness in the left eye
    B. Loss of left visual field of each eye
    C. Loss of right visual field of each eye
    D. Loss of lateral field of view of each eye ("tunnel vision")
    E. No effects on the eye
20. Mr. Frederick has a detached retina of the lower right portion of one eye. As a result he is unable to see objects in which area in the visual field of that eye?
    A. High in the left
    B. High in the right
    C. Low in the left
    D. Low in the right

Questions 21–25: Fill-ins. Complete each sentence with the word or phrase that best fits.

____________________ 21. In darkness, Na^+ channels are held open by a molecule called ________.

____________________ 22. Name the four processes necessary for formation of an image on the retina: ________ of light rays, ________ of the lens, ________ of the pupil, and ________ of the eyes.

_______________ 23. _______ is the study of the structure, functions, and diseases of the eye.

_______________ 24. The names of three ossicles are _______, _______, and _______.

_______________ 25. Two functions of the ciliary body are _______ and _______.

ANSWERS TO MASTERY TEST: CHAPTER 16

True–False

1. T
2. F. Accommodation, but not convergence, is a result.
3. T
4. F. Aqueous humor, but not vitreous humor, is
5. F. More

Arrange

6. A C B
7. C D A B
8. C D A B
9. A B E C D
10. A E C B D
11. A B D E C
12. B C A D E

Multiple Choice

13. C
14. B
15. B
16. C
17. D
18. D
19. C
20. A

Fill-ins

21. Cyclic GMP
22. Refraction, accommodation, constriction, convergence
23. Ophthalmology
24. Malleus, incus, and stapes
25. Production of aqueous humor and alteration of lens shape for accommodation

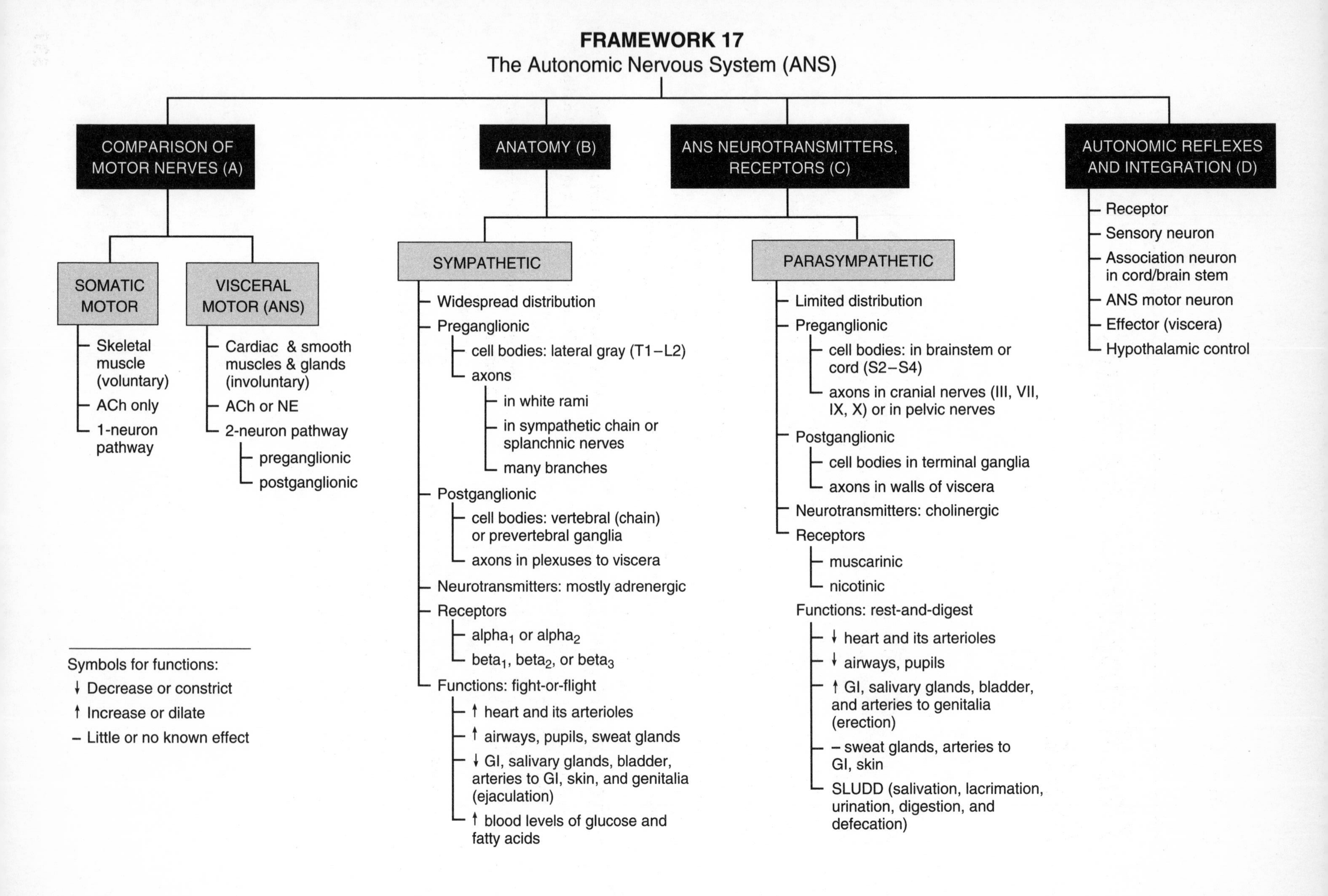
FRAMEWORK 17
The Autonomic Nervous System (ANS)
COMPARISON OF MOTOR NERVES (A)
ANATOMY (B)
ANS NEUROTRANSMITTERS, RECEPTORS (C)
AUTONOMIC REFLEXES AND INTEGRATION (D)
SOMATIC MOTOR
Skeletal muscle (voluntary)
ACh only
1-neuron pathway
VISCERAL MOTOR (ANS)
Cardiac & smooth muscles & glands (involuntary)
ACh or NE
2-neuron pathway
preganglionic
postganglionic
SYMPATHETIC
Widespread distribution
Preganglionic
cell bodies: lateral gray (T1–L2)
axons
in white rami
in sympathetic chain or splanchnic nerves
many branches
Postganglionic
cell bodies: vertebral (chain) or prevertebral ganglia
axons in plexuses to viscera
Neurotransmitters: mostly adrenergic
Receptors
$alpha_1$ or $alpha_2$
$beta_1$, $beta_2$, or $beta_3$
Functions: fight-or-flight
↑ heart and its arterioles
↑ airways, pupils, sweat glands
↓ GI, salivary glands, bladder, arteries to GI, skin, and genitalia (ejaculation)
↑ blood levels of glucose and fatty acids
PARASYMPATHETIC
Limited distribution
Preganglionic
cell bodies: in brainstem or cord (S2–S4)
axons in cranial nerves (III, VII, IX, X) or in pelvic nerves
Postganglionic
cell bodies in terminal ganglia
axons in walls of viscera
Neurotransmitters: cholinergic
Receptors
muscarinic
nicotinic
Functions: rest-and-digest
↓ heart and its arterioles
↓ airways, pupils
↑ GI, salivary glands, bladder, and arteries to genitalia (erection)
– sweat glands, arteries to GI, skin
SLUDD (salivation, lacrimation, urination, digestion, and defecation)
Receptor
Sensory neuron
Association neuron in cord/brain stem
ANS motor neuron
Effector (viscera)
Hypothalamic control
Symbols for functions:
↓ Decrease or constrict
↑ Increase or dilate
– Little or no known effect

C H A P T E R **17**

The Autonomic Nervous System

The autonomic nervous system (ANS) consists of the special branch of the nervous system that exerts unconscious control over viscera. The ANS regulates activities such as heart rate and blood pressure, glandular secretion, and digestion. The two divisions of the ANS—the sympathetic and parasympathetic—carry out a continual balancing act, readying the body for response to stress or facilitating rest and relaxation, according to momentary demands. ANS neurotransmitters hold particular significance clinically because they are often mimicked or inhibited by medications.

As you begin your study of the autonomic nervous system, carefully examine the Chapter 17 Topic Outline and Objectives; check off each one as you complete it. To organize your study of the autonomic nervous system, glance over the Chapter 17 Framework now. Be sure to refer to the Framework frequently and note relationships among key terms in each section.

TOPIC OUTLINE AND OBJECTIVES

A. Comparison of somatic and autonomic nervous systems

1. Compare the structural and functional differences between the somatic and autonomic parts of the nervous system.

B. Anatomy of autonomic motor pathways

1. Describe preganglionic and postganglionic neurons of the autonomic nervous system.
2. Compare the anatomical structure of the sympathetic and parasympathetic divisions of the autonomic nervous system.

C. ANS neurotransmitters and receptors

1. Describe the neurotransmitters and receptors in the autonomic responses.

D. Integration and control of autonomic functions

1. Describe the components of an autonomic reflex.
2. Explain the relationship of the hypothalamus to the ANS.

WORDBYTES

Now become familiar with the language of this chapter by studying each wordbyte, its meaning, and an example of its use within a term. After you study the entire list, self-check your understanding by writing the meaning of each wordbyte on the line. As you continue through the *Learning Guide,* identify (and fill in) additional terms that contain the same wordbyte.

Wordbyte	Self-check	Meaning	Example(s)
auto-	____________	self	*auto*nomic
nomos	____________	law, governing	auto*nomic*
para-	____________	near, beside	*para*vertebral, *para*thyroid
post-	____________	after	*post*ganglionic
pre-	____________	before	*pre*ganglionic
ram(i)-	____________	branch	white *ram*us
splanch-	____________	viscera	*splanch*nic

CHECKPOINTS

A. Comparison of somatic and autonomic nervous systems (pages 566–568)

✓ **A1.** Why is the autonomic nervous system (ANS) so named? Is the ANS entirely independent of higher control centers? Explain.

✓ **A2.** Contrast the somatic and autonomic nervous systems in this table.

	Somatic	**Autonomic**
a. Sensations and movements are mostly *(conscious? unconscious/automatic?)*		
b. Types of tissue innervated by efferent nerves	Skeletal muscle	
c. Efferent neurons are excitatory (E), inhibitory (I), or both (E + I)		

	Somatic	**Autonomic**
d. Examples of sensations (input)		Mostly visceral sensory (such as changes in CO_2 level of blood or stretching or visceral wall)
e. Number of neurons in efferent pathway		
f. Neurotransmitter(s) released by motor neurons	Acetylcholine (ACh)	

✓ **A3.** Name the two divisions of the ANS. ____________________ ____________________

A4. Many visceral organs have dual innervation by the autonomic system.

a. What does dual innervation mean?

✓ b. How do the sympathetic and parasympathetic divisions work in harmony to control viscera?

B. Anatomy of autonomic motor pathways (pages 568–574)

✓ **B1.** Complete this exercise describing structural differences between visceral efferent and somatic efferent pathways.

a. Between the spinal cord and effector, somatic pathways include _________ neuron(s), whereas visceral pathways require _________ neuron(s), known as the pre- ____________________ and ____________________ neurons.

b. Somatic pathways begin at *(all? only certain?)* levels of the cord, whereas visceral routes begin at ____________________ levels of the cord.

✓ **B2.** Contrast preganglionic with postganglionic fibers in this exercise.

a. Consider the sympathetic division first. Its preganglionic cell bodies lie in the lateral gray of segments _________ to _________ of the cord. For this reason, the sympathetic division is also known as the ____________________ outflow.

b. Preganglionic axons *(are? are not?)* myelinated and therefore they appear white. These axons may branch to reach a number of different ____________________ where they synapse. These ganglia contain *(pre-? post-?)* ganglionic neuron cell bodies whose axons proceed to the appropriate organ. Postganglionic axons are *(gray? white?)* because they *(do? do not?)* have a myelin covering.

c. Parasympathetic preganglionic cell bodies are located in two areas. One is the ____________________, from which axons pass out in cranial nerves ________ ________ ________ and ________. A second location is the *(anterior? lateral?)* gray horns of segments ________ through ________ of the cord.

d. Based on location of (pre? post?)-ganglionic neuron cell bodies, the parasympathetic division is also known as the ____________________ outflow. The axons of these neurons extend to *(the same? different?)* autonomic ganglia from those in sympathetic pathways (see Checkpoint B4). There, postganglionic cell bodies send out short *(gray? white?)* axons to innervate viscera.

e. Most viscera *(do? do not?)* receive fibers from both the sympathetic and the parasympathetic divisions. However, the origin of these divisions in the CNS and the pathways taken to reach viscera *(are the same? differ?)*.

✓ **B3.** Contrast these two types of ganglia.

	Posterior Root Ganglia	Autonomic Ganglia
a. Contains neurons that are *(sensory? motor?)*.		
b. Contains neurons in *(somatic? visceral? both?)* pathways.		
c. Synapsing *(does? does not?)* occur here.		

✓ **B4.** Complete the table contrasting types of autonomic ganglia.

	Sympathetic Trunk	Prevertebral	Terminal
a. Sympathetic or parasympathetic		Sympathetic	
b. Alternate name	Paravertebral ganglia or vertebral chain		
c. General location			Close to or in walls of effectors

✓ **B5.** Refer to Figure 17.5 in your text. Trace with your finger along the route of a preganglionic neuron. Now describe the pathway common to *all* sympathetic preganglionic neurons

by listing the structures in correct sequence. ________ ________ ________ ________

A. Ventral root of spinal nerve
B. Sympathetic trunk ganglion
C. Lateral gray of spinal cord (between T1 and L2)
D. White ramus communicans

✓ **B6.** Suppose you could walk along the route of nerve impulses in sympathetic pathways. Start at the sympathetic trunk ganglion. It may be helpful to refer to the "map" provided by Figure 17.5 in your text.

a. What is the shortest possible path you could take to reach a synapse with a postganglionic neuron cell body?

b. You could then ascend and/or descend the sympathetic chain to reach trunk ganglia at other levels. This is important because sympathetic preganglionic cell bodies are

limited in location to ________ to ________ levels of the cord, and yet the

sympathetic trunk extends from ________ to ____________________ levels of the vertebral column. Sympathetic fibers must be able to reach these distant areas to provide the entire body with sympathetic innervation.

c. Now "walk" the sympathetic nerve pathway to sweat glands, blood vessels, or hair muscles in skin or blood vessels of extremities. Again trace this route on Figure 17.5 in the text. Preganglionic neurons synapse with postganglionic cell bodies in sympathetic trunk ganglia (at the level of entry or after ascending or descending). Postgan

glionic fibers then pass through ____________________, which connect to

____________________ and then convey impulses to skin or blood vessels of extremities.

d. Some preganglionic fibers do not synapse as described in (b) or (c) but pass on through trunk ganglia without synapsing there. They course through

____________________ nerves to ____________________ ganglia. Follow their route as

they synapse with postganglionic neurons whose axons form ____________________ en route to viscera.

e. Prevertebral ganglia are located only in the ____________________. In the neck, thorax, and pelvis the only sympathetic ganglia are those of the trunk (see Figure 17.2 in the text). As a result, cardiac nerves, for example, contain only *(preganglionic? postganglionic?)* fibers, as indicated by broken lines in the figure.

f. A given sympathetic preganglionic neuron is likely to have *(few? many?)* branches; these may take any of the paths you have just "walked." Once a branch synapses, it *(does? does not?)* synapse again, because any autonomic pathway consists of just

________ neurons (preganglionic and postganglionic).

✓ **B7.** Test your understanding of these routes by completing the sympathetic pathway shown in Figure LG 17.1. A preganglionic neuron located at level T5 of the cord has

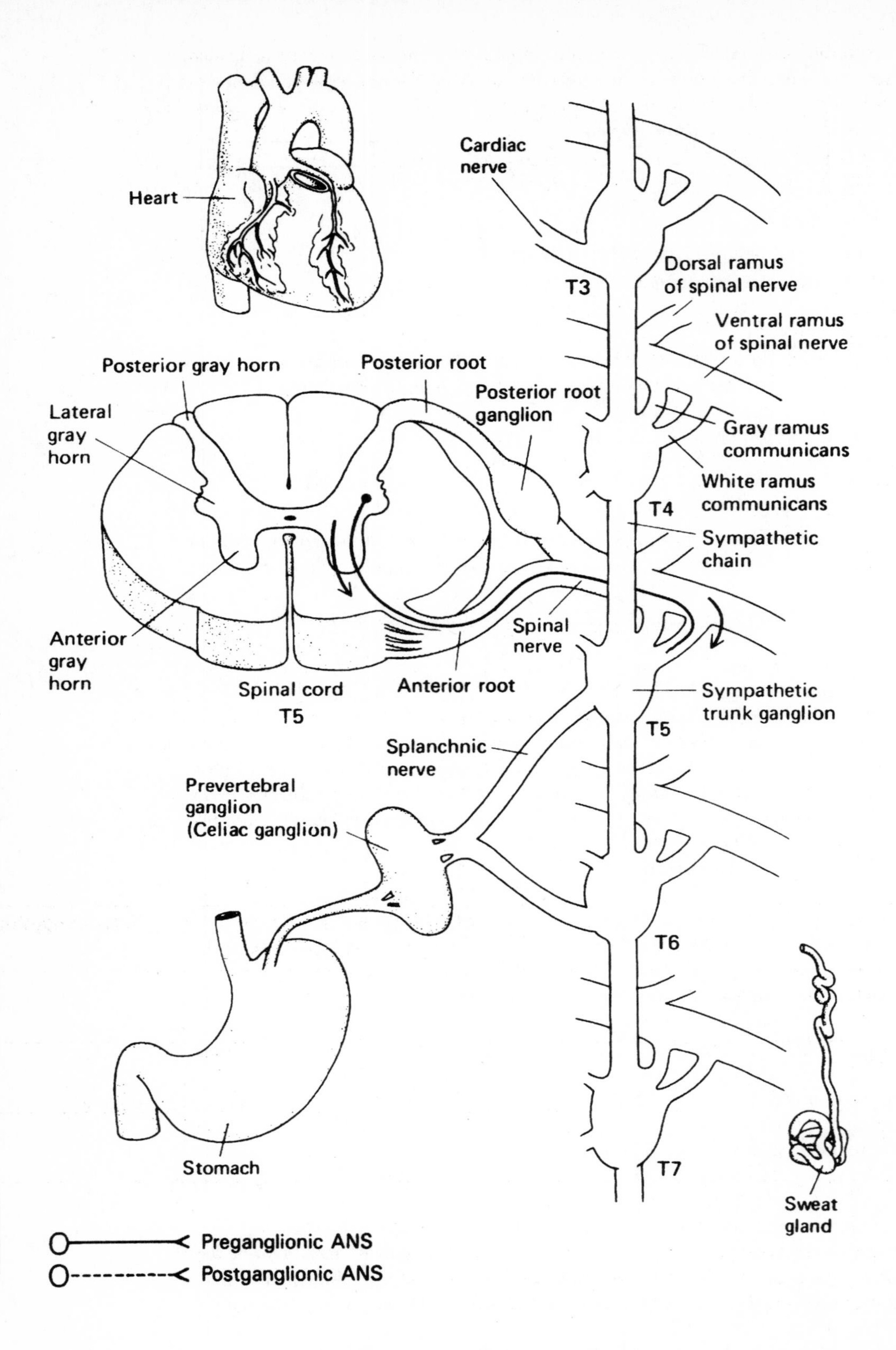

Figure LG 17.1 Typical sympathetic pathway beginning at level T5 of the cord. Complete pathways according to directions in Checkpoint B7.

its axon drawn as far as the white ramus. Finish this pathway by showing how the axon may branch and synapse to eventually innervate these three organs: the heart, a sweat gland in skin of the thoracic region, and the stomach. Draw the preganglionic fibers in solid lines and the postganglionic fibers in broken lines.

✓ **B8.** In Checkpoint B4 you contrasted two types of sympathetic trunk ganglia, namely sympathetic trunk and prevertebral ganglia. For more details on these ganglia, complete this Checkpoint and also B9.

a. The sympathetic trunk ganglia lie in two chains close to vertebrae, ribs, and sacrum. Each chain has *(12–14? 23–26? 31–33?)* ganglia. *(Three? Seven? Eight?)* ganglia lie in the cervical region. The *(superior? middle? inferior?)* cervical ganglion lies next to the second cervical vertebra; its postganglionic fibers innervate *(muscles and glands of the head? the heart?)*.

b. The middle and inferior cervical ganglia lie near the *(third and fourth? sixth and seventh?)* vertebrae. Postganglionic fibers from these ganglia innervate the

______________________.

c. There are *(8–9? 11–12?)* trunk ganglia in the thoracic area; they lie against the *(anterior? posterior?)* of ribs. List six types of viscera innervated by postganglionic fibers that arise from these ganglia.

d. There are _______ pairs of sympathetic trunk ganglia in the lumbar region, _______ in the sacral region and _______ coccygeal ganglion.

✓ **B9.** Do this Checkpoint on autonomic pathways involving splanchnic nerves. These nerves contain *(sympathetic? parasympathetic?)* fibers. Splanchnic nerves are composed of *(pre? post?)*-ganglion fibers that passed through trunk ganglia and *(did? did not?)* synapse there; instead they extend to *(prevertebral? terminal?)* ganglia. Postganglionic cells there pass axons on to viscera located in the __________________. Complete this table about these nerves.

Splanchnic Nerve	Formed from Vertebral Ganglia	Prevertebral Ganglion: Site of Postganglionic Cell Bodies	Viscera Supplied
a. Greater splanchnic n.			
b. Lesser splanchnic n.		Superior mesenteric	
c. Lumbar splanchnic n.	L1–L4		

✓ **B10.** In the past several Checkpoints, you have been focusing on the *(sympathetic? parasympathetic?)* division of the ANS. Now that you are somewhat familiar with the rather complex sympathetic pathways, you will notice that parasympathetic pathways are much simpler. This is due to the fact that there is no parasympathetic chain of ganglia; the only parasympathetic ganglia are __________________ __________________, which are in or close to the organs innervated. Also, parasympathetic preganglionic fibers synapse with *(more? fewer?)* postganglionic neurons. What is the significance of the structural simplicity of the parasympathetic system?

✓ **B11.** Complete the table about the cranial portion of the parasympathetic system.

Cranial Nerve Number	Name	Name of Terminal Ganglion	Structures Innervated
a.			Iris and ciliary muscle of eye
b.		1. Pterygopalatine 2. Submandibular	1. 2.
c.	Glossopharyngeal		
d.		Ganglia in cardiac and pulmonary plexuses and in abdominal plexuses	

✓ **B12.** *For extra review.* Write *S* if the description applies to the sympathetic division of the ANS, *P* if it applies to the parasympathetic division, and *P, S* if it applies to both.

________ a. Also called thoracolumbar outflow

________ b. Has long preganglionic fibers leading to terminal ganglia and very short postganglionic fibers

________ c. Celiac and superior mesenteric ganglia are sites of postganglionic neuron cell bodies

________ d. Sends some preganglionic fibers through cranial nerves

________ e. Has some preganglionic fibers synapsing in vertebral chain (trunk)

________ f. Has more widespread effect in the body, affecting more organs

________ g. Has fibers running in gray rami communicantes to supply sweat glands, hair muscles, and blood vessels

________ h. Has fibers in white rami (connecting spinal nerve with vertebral chain)

________ i. Contains fibers that supply viscera with motor impulses

________ j. The greater and lesser splanchnic nerves contain axons of this division

________ k. The pelvic splanchnic nerves contain axons of this division

✓ **B13.** In Activity B1, we emphasized that autonomic pathways require two neurons. One exception to this rule is the pathway to the ____________________ glands. Because this pair of glands function like modified *(sympathetic? parasympathetic?)* ganglia, no postganglionic neurons lead out to any other organ.

C. ANS neurotransmitters and receptors (pages 575–578)

✓ **C1.** *(All? Most? No?)* viscera receive both sympathetic and parasympathetic innervation, known as ______________ innervation of the ANS. List examples of exceptions here:

a. Sympathetic only: ____

____

b. Parasympathetic only: ____

✓ **C2.** Fill in the blanks to indicate which neurotransmitters are released by sympathetic (S) or parasympathetic (P) neurons or by somatic neurons. (It may help to refer to Figure LG 17.2.) Use these answers:

ACh.	**Acetylcholine (cholinergic)**
NE.	**Norepinephrine = noradrenalin, or possibly epinephrine = adrenaline (adrenergic)**

________ a. All preganglionic neurons (both S and P) release this neurotransmitter

________ b. Most S postganglionic neurons

________ c. A few S postganglionic neurons, namely those to sweat glands

________ d. All P postganglionic neurons

________ e. All somatic neurons

For extra review, color the neurotransmitters on Figure LG 17.2, matching colors you use in the key with those you use in the figure.

✓ **C3.** Do this activity summarizing the effects of autonomic neurotransmitters.

a. Axons that release the transmitter acetylcholine (ACh) are known as *(adrenergic? cholinergic?)*. Those that release norepinephrine (NE, also called noradrenalin) are called ____________________. Most sympathetic postganglionic nerves are said to be *(cholinergic? adrenergic?)*, whereas all parasympathetic nerves are ________________.

b. During stress, the *(sympathetic? parasympathetic?)* division of the ANS prevails, so stress responses are primarily *(adrenergic? cholinergic?)* responses.

c. Another source of NE, as well as epinephrine, is the gland known as the ____________________ ____________________. Therefore, chemicals released by this gland (as in stress) will mimic the action of the *(sympathetic? parasympathetic?)* division of the ANS.

✓ **C4.** Again, refer to Figure LG 17.2 and do this exercise on neurotransmitters and related receptors. Be sure to match the color you use for receptor symbols in the key to the color you use for receptors in the figure. (The symbol S denotes *sympathetic* and P indicates *parasympathetic* in the following statements.)

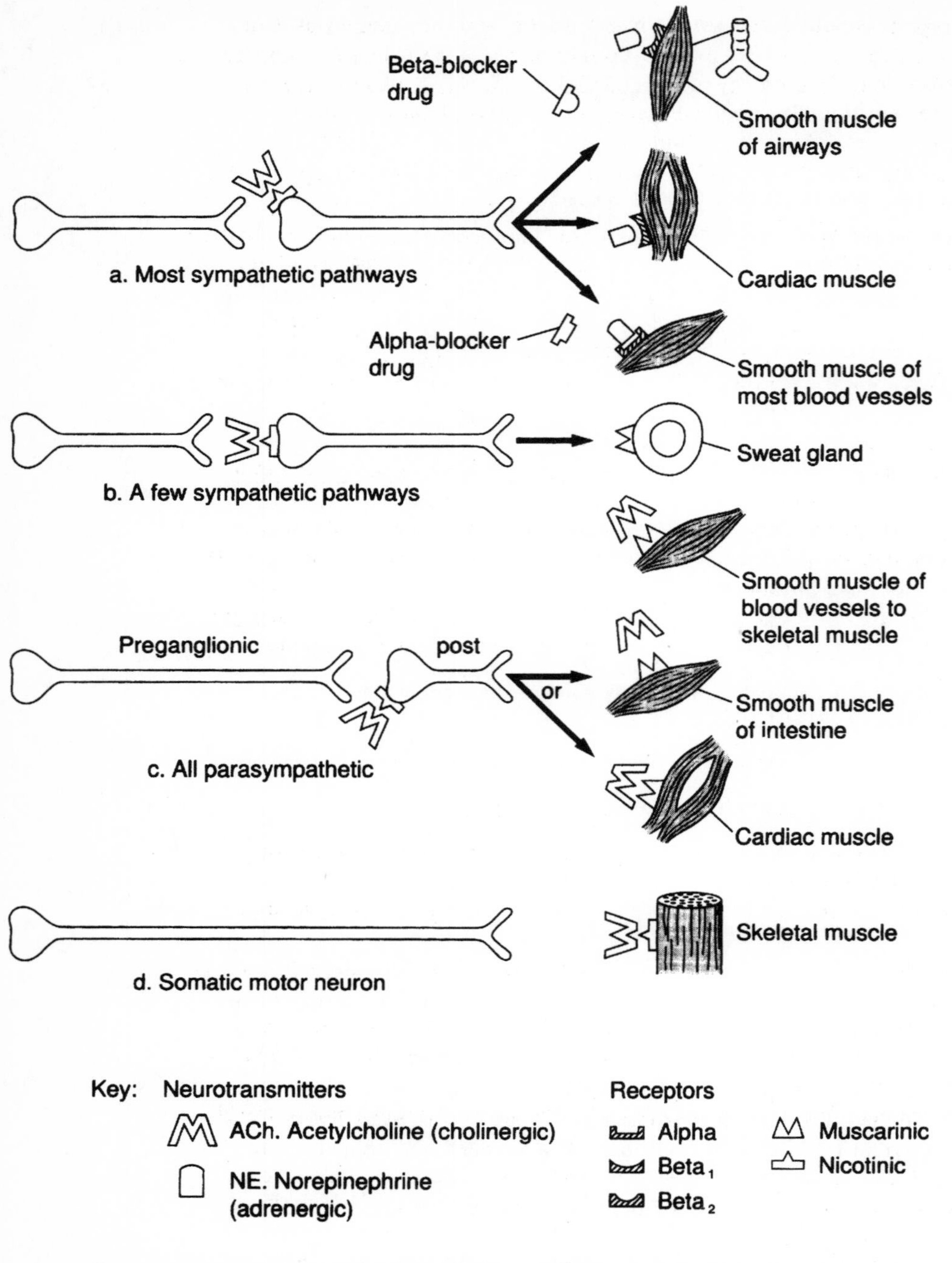

Figure LG 17.2 Diagram of autonomic and somatic neurons, neurotransmitters, and receptors. Note that neurotransmitters and receptors are relatively enlarged. Refer to Checkpoints C2 and C4 to C6.

a. First identify all nicotinic receptors in the figure. Notice that these receptors are found on cell bodies of *(P postganglionic neurons? S postganglionic neurons? autonomic effector cells? somatic effector cells [skeletal muscle]?)*. (Circle all that apply.) In other words, nicotinic receptors are all sites for action of the neurotransmitter *(ACh? NE? either ACh or NE?)*. Activation of nicotinic receptors is *(always? sometimes? never?)* excitatory.

b. On what other types of receptors can ACh act? ____________________. Identify these receptors on the figure. As you do, notice that these receptors are found on all effectors stimulated by *(P? S?)* postganglionic neurons. In addition, muscarinic receptors are found on a few S effectors, such as ____________________. Activation of muscarinic receptors is *(always? sometimes? never?)* excitatory. For example, as P nerves (vagus) release ACh to a muscarinic receptor in cardiac muscle, heart activity *(increases? decreases?)*, but when P nerves (vagus) release ACh to a muscarinic receptor in the wall of the stomach or intestine gastrointestinal activity *(increases? decreases?)*.

c. Explain what accounts for the names nicotinic and muscarinic for ACh receptors. The names are based on the fact that the action of ____________________ on these receptors is mimicked by action of ____________________ on nicotinic receptors and by a poison named ____________________ from mushrooms on muscarinic receptors.

d. *Alpha* and *beta* receptors are found only on effectors innervated by *(P? S?)* nerves; these effectors are stimulated by *(ACh? NE and epinephrine?)*. NE and epinephrine are categorized as ____________________ neurotransmitters.

e. *Critical thinking.* Alex receives word that his brother has been seriously injured in an automobile accident. Among other immediate physical responses, his skin "goes pale." Explain this response. (Refer to Table 17.2 in your text, for help.) Which type of receptor is found in smooth muscle of blood vessels of skin? *(Alpha$_1$? Beta$_1$? Beta$_2$?)* Alpha$_1$ receptors are usually *(excitatory? inhibitory?)*. As sympathetic (the "stress system") nerves release NE, smooth muscle of blood vessels are excited, causing them to *(constrict/narrow? dilate/widen?)*. As a result, blood flows *(into? out of?)* skin, and the skin pales. How may this response help Alex?

✓ **C5.** *A clinical correlation.* Certain medications can mimic the effects of the body's own transmitters. Again refer to Figure LG 17.2 and do this exercise.

a. A beta$_1$-stimulator (or beta-exciter) mimics *(ACh? NE?)*, causing the heart rate and strength of contraction to ________-crease. Name one such drug. (For help, refer to a pharmacology text.) ____________________ Such a drug has a structure that is close enough to the shape of NE that it can "sit on" the beta$_1$ receptor as NE would and activate it. Such a drug is a beta$_1$ *(agonist? antagonist?)*.

b. A beta-blocker (or beta-inhibitor) also has a shape similar to that of NE and so can sit on ____________________ receptors. Name one such drug. ____________________. However, this drug cannot activate the receptor but simply prevents NE from doing so. So a beta-blocker ________-creases heart rate and force of contraction. A beta-blocker is a beta *(agonist? antagonist?)*.

c. Other drugs known as anticholinergics (or ACh-blockers) can take up residence on ACh receptors so that ACh being released from vagal nerve stimulation cannot exert its normal effects. A person taking such a medication may exhibit a(n) *(increased? decreased?)* heart rate.

✓ **C6.** *Critical thinking.* Do this additional exercise on effects of medications on ANS nerves.

a. What types of receptors are located on smooth muscle of bronchi and bronchioles (airways)? *(Alpha? Beta$_1$? Beta$_2$?)* For help refer to Table 17.2 in your text. The effect of the sympathetic stimulation of lungs is to cause *(dilation? constriction?)* of airways, making breathing *(easier? more difficult?)* during stressful times.

b. The effect of NE on the smooth muscle of the stomach, intestines, bladder, and uterus

is also *(contraction? relaxation?)*, so that during stress the body _________-creases activity of these organs and can focus on more vital activities such as heart contractions.

c. From Activity C4e, you may recognize that the sympathetic response during stress causes many blood vessels, such as those in skin, to *(constrict? dilate?)*. Prolonged stress may therefore lead to *(high? low?)* blood pressure. One type of medication used to lower blood pressure is an alpha-*(blocker? stimulator?)* because it tends to dilate these vessels and "pool" blood in nonessential areas like skin, thereby avoiding overloading the heart with blood flow.

✓ **C7.** Do this exercise relating the fate of ANS neurotransmitters and overall effects of the two divisions of the ANS.

a. On the arrows below, write the initials of the enzymes that destroy each transmitter:

ACh ⟶ ~~ACh~~ NE ⟶ ~~NE~~

b. The ANS neurotransmitter that "hangs around" longer (has more lasting effect) is *(ACh? NE?)*. This fact provides one explanation for the longer-lasting effects of *(sympathetic/adrenergic? parasympathetic/cholinergic?)* neurons.

c. State two other rationales for the more lasting and widespread effects of the sympathetic system:

1.

2.

✓ **C8.** Use arrows to show whether parasympathetic (P) or sympathetic (S) fibers stimulate (↑) or inhibit (↓) each of the following activities. Use a dash (—) to indicate that there is no (or virtually no) parasympathetic innervation. The first one is done for you.

a. P __↓__ S __↑__ Dilation of pupil

b. P _______ S _______ Heart rate and blood flow to coronary (heart muscle) blood vessels

c. P _______ S _______ Constriction of blood vessels of skin and abdominal viscera

d. P _______ S _______ Salivation and digestive organ contractions

e. P _______ S _______ Erection or engorgement of genitalia

f. P _______ S _______ Dilation of bronchioles for easier breathing

g. P _______ S _______ Contraction of bladder and relaxation of internal urethral sphincter causing urination

h. P _______ S _______ Contraction of arrector pili of hair follicles causing "goose bumps"

i. P _______ S _______ Contraction of spleen which transfers some of its blood to general circulation, causing increase in blood pressure

j. P _______ S _______ Release of epinephrine and norepinephrine from adrenal medulla

k. P _______ S _______ Secretion of insulin and digestive enzymes from the pancreas

l. P _______ S _______ Increase in blood glucose level by pancreatic secretion of the hormone glucagon and by formation of glucose in the liver

m. P _______ S _______ Coping with stress, fight-or-flight response

✓ **C9.** List the four "E situations" in which the sympathetic nervous system predominates.

C10. List functions of parasympathetic nerves; use the following hints:

a. S L U D D

b. The three "decreases"

D. Integration and control of autonomic functions (page 578–582)

✓ **D1.** Autonomic motor neurons can be stimulated by reflexes initiated by *(somatic sensory neurons? visceral sensory neurons? either somatic or visceral sensory neurons?)*. Visceral sensory receptors are mostly *(extero? intero? proprio?)*-receptors.

✓ **D2.** Arrange in correct sequence structures in the pathway for a painful stimulus at your fingertip to cause an autonomic (visceral) reflex, such as sweating. Write the letters of the structures in order on the lines provided.

_______ _______ _______ _______ _______ _______ _______ _______ _______ _______

A. Association neuron in spinal cord	**F. Nerve fiber in anterior root of spinal nerve**
B. Postganglionic fiber in gray ramus	**G. Nerve fiber in white ramus**
C. Pain receptor in skin	**H. Postganglionic fiber in spinal nerve in brachial plexus**
D. Cell body of postganglionic neuron in trunk ganglion	**I. Sweat gland**
E. Cell body of preganglionic neuron in lateral gray of cord	**J. Somatic sensory neuron**

✓ **D3.** Most visceral sensations *(do? do not?)* reach the cerebral cortex, so most visceral sensations are at *(conscious? subconscious?)* levels.

D4. Explain roles of the hypothalamus in control of autonomic function by answering the following questions.

a. What types of sensory input reach the hypothalamus to evoke autonomic responses?

b. What types of output go to autonomic centers in the brainstem?

c. What types of output go to autonomic centers in the spinal cord?

d. Which parts of the hypothalamus control sympathetic nerves? Which parts control parasympathetic nerves?

D5. Describe how sympathetic nerves are affected by the following conditions. List specific signs and symptoms of each condition.

a. Horner's syndrome

b. Raynaud's disease

c. Autonomic dysreflexia

ANSWERS TO SELECTED CHECKPOINTS: CHAPTER 17

A1. It was thought to be autonomous (self-governing). However, it is regulated by brain centers such as the hypothalamus and brainstem with input from the limbic system and other parts of the cerebrum.

A2.

	Somatic	Autonomic
a. Sensations and movements are mostly *(conscious? unconscious/automatic?)*	**Conscious**	**Unconscious/automatic**
b. Types of tissue innervated by efferent nerves	Skeletal muscle	**Cardiac muscle, smooth muscle, and many glands**
c. Efferent neurons are excitatory (E), inhibitory (I), or both (E + I)	E	E + I
d. Examples of sensations (input)	**Special senses, general somatic senses, or proprioceptors, and possibly visceral sensory, as when pain of appendicitis or menstrual cramping causes a person to curl up (flexion of thighs, legs)**	Mostly visceral sensory (such as changes in CO_2 level of blood or stretching of visceral wall)
e. Number of neurons in efferent pathway	**One (from spinal cord or brain stem to effector)**	**Two (pre- and postganglionic**
f. Neurotransmitter(s) released by motor neurons	Acetylcholine (ACh)	**ACh or norepinephrine (NE)**

A3. Sympathetic and parasympathetic.

A4. (a) Most viscera are innervated by both sympathetic and parasympathetic divisions. (b) One division excites and the other division inhibits the organ's activity.

B1. (a) 1, 2, ganglionic, postganglionic. (b) All, only certain (see Checkpoint B2).

B2. (a) T1, L2; thoracolumbar. (b) Are; ganglia; post-; gray, do not. (c) Brainstem, III, VII, IX, X; lateral, S2, S4. (d) Pre-, craniosacral; different; gray. (e) Do; differ.

B3. (a) Sensory, motor. (b) Both, visceral. (c) Does not, does.

B4.

	Sympathetic Trunk	Prevertebral	Terminal
a. Sympathetic or parasympathetic	**Sympathetic**	Sympathetic	**Parasympathetic**
b. Alternate name	Paravertebral ganglia or vertebral chain	**Collateral or prevertebral (celiac, superior and inferior mesenteric)**	**Intramural (in walls of viscera)**
c. General location	**In vertical chain along both sides of vertebral bodies from base of skull to coccyx**	**In three sites anterior to spinal cord and close to major abdominal arteries**	Close to or in walls of effectors

B5. C A D B.

B6. (a) Immediately synapse in trunk ganglion. (b) T1, L2, C3, sacral. (c) Gray rami, spinal nerves. (d) Splanchnic, prevertebral; plexuses. (e) Abdomen; postganglionic. (f) Many; does not, 2.

B7.

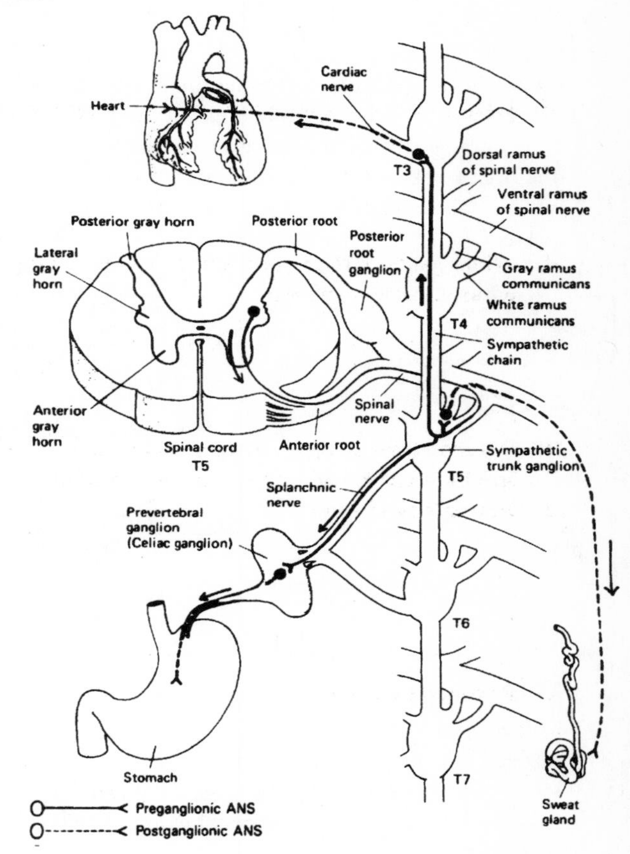

Figure LG 17.1A Typical sympathetic pathway beginning at level T5 of the cord.

B8. (a) 23–26; three; superior, muscles and glands of the head (b) Sixth and seventh; heart. (c) 11–12, posterior; heart, lungs, bronchi, sweat glands, smooth muscle of blood vessels, and hair (arrector pili) muscles. (d) 4–5, 4–5, 1.

B9. Sympathetic; pre, did not; prevertebral; abdomen or pelvis.

Splanchnic Nerve	Formed from Vertebral Ganglia	Prevertebral Ganglion: Site of Postganglionic Cell Bodies	Viscera Supplied
a. Greater splanchnic n.	**T5-T9 (or T10)**	**Celiac**	**Stomach, spleen, liver, kidneys, small intestine**
b. Lesser splanchnic n.	**T10-T11**	Superior mesenteric	**Small intestine, colon**
c. Lumbar splanchnic n.	L1-L4	**Inferior mesenteric**	**Distal colon, rectum; urinary bladder, genitalia**

B10. Sympathetic; terminal ganglia; fewer. Parasympathetic effects are much less widespread than sympathetic, distributed to fewer areas of the body, and characterized by a response of perhaps just one organ, rather than an integrated response in which many organs respond in *sympathy* with one another.

B11.

Cranial Nerve Number	Name	Name of Terminal Ganglion	Structures Innervated
a. III	**Oculomotor**	**Ciliary**	Iris and ciliary muscle of eye
b. VII	**Facial**	1. Pterygopalatine	**1. Nasal mucosa, palate, pharynx, lacrimal glands**
		2. Submandibular	**2. Submandibular and sublingual salivary glands**
c. IX	Glossopharyngeal	**Otic**	**Parotid salivary gland**
d. X	**Vagus**	Ganglia in cardiac and pulmonary plexuses and in abdominal plexuses	**Thoracic and abdominal viscera**

B12. (a) S. (b) P. (c) S. (d) P. (e) S. (f) S. (g) S. (h) S. (i) P, S. (j) S. (k) P.

B13. Adrenal (medulla); sympathetic.

Cl. Most, dual. (a) Sweat glands, arrector pili (hair) muscles, fat cells, kidneys, adrenal glands, spleen, and most blood vessels. (b) None, although lacrimation is primarily parasympathetic. (Refer to Table 17.4 and Figure 17.2 in your text.)

C2. (a) ACh. (b) NE. (c–e) ACh. (Refer to Figure LG 17.2.)

C3. (a) Cholinergic; adrenergic; adrenergic, cholinergic. (b) Sympathetic, adrenergic. (c) Adrenal medulla; sympathetic.

C4. (a) Refer to key for Figure LG 17.2; both P and S postganglionic neurons and somatic effector cells (skeletal muscle); ACh; always. (b) Muscarinic (*Hint:* Take time to notice that in this figure and the muscarinic receptors are ones shaped so that ACh "sits down" in an "M" [for muscarinic] position, whereas nicotinic receptors have ACh attach in the "W" position); P (notice this at top of Table 17.2 in the text; many sweat glands; sometimes; decreases, increases. (c) ACh, nicotine, muscarine. (d) S; NE and epinephrine; catecholamine. (e) Alpha_1; excitatory; constrict/narrow; out of; blood flows into major vessels to increase blood pressure, which may assist in a stress response.

C5. (a) NE, in; isoproterinol (Isoprel) or epinephrine; agonist. (b) Beta; propranolol (Inderal); de; antagonist. (c) Increased (a double negative: the inhibiting vagus is itself inhibited).

C6. (a) Beta_2 dilation, easier. (b) Relaxation, de. (c) Constrict; high; blocker.

C7. (a)

$$\text{ACh} \xrightarrow{\text{AChE}} \cancel{\text{ACh}} \qquad \text{NE} \xrightarrow{\text{COMT or MAO}} \cancel{\text{NE}}$$

(b) NE; sympathetic/adrenergic. (c1) Greater divergence of sympathetic neurons, for example, through the extensive sympathetic trunk chain and through gray rami to all spinal nerves. (c2) Mimicking of sympathetic responses via release of catecholamines (epinephrine and NE) from the adrenal medulla.

C8. (b) P↓, S↑. (c) P—, S↑. (d) P↑, S↓. (e) P↑, S↓. (f) P↓, S↑. (g) P↑, S↓. (h) P—, S↑. (i) P—, S↑. (j) P—, S↑. (k) P↑, S↓. (l) P—, S↑. (m) P↓, S↑.

C9. Exercise, emergency, excitement, and embarrassment.

Dl. Either somatic or visceral afferents; intero.

D2. C J A E F G D B H I.

D3. Do not, subconscious.

MORE CRITICAL THINKING: CHAPTER 17

1. Does the human ANS include more white or gray rami communicantes? Explain.
2. Identify 12 specific organs in your own body that are supplied by autonomic neurons.
3. Explain why the names craniosacral and thoracolumbar are given to the two divisions of the ANS.
4. Contrast locations of autonomic ganglia in sympathetic and parasympathetic divisions of the ANS.
5. Describe a "sympathetic response" that you might exhibit when faced with a threatening situation. Include specific changes likely to occur in your viscera.

MASTERY TEST: CHAPTER 17

Questions 1–6: Circle letters preceding all correct answers to each question.

1. Choose all *true* statements about the vagus nerve.
   A. Its autonomic fibers are sympathetic.
   B. It supplies ANS fibers to viscera in the thorax and abdomen but not in the pelvis.
   C. It causes digestive glands to increase secretions.
   D. Its ANS fibers are mainly preganglionic.
   E. It is a cranial nerve originating from the medulla.
2. Which activities are characteristic of the stress response, or fight-or-flight reaction?
   A. The liver breaks down glycogen to glucose.
   B. The heart rate decreases.
   C. Kidneys increase urine production because blood is shunted to kidneys.
   D. There is increased blood flow to genitalia, causing erect or engorged state.
   E. Hairs stand on end ("goose pimples") due to contraction of arrector pili muscles.
   F. In general, the sympathetic system is active.
3. Which fibers are classified as autonomic?
   A. Any visceral efferent nerve fiber
   B. Any visceral sensory nerve fiber
   C. Nerves to salivary glands and sweat glands
   D. Pain fibers from ulcer in stomach wall
   E. Sympathetic fibers carrying impulses to blood vessels
   F. All nerve fibers within cranial nerves
   G. All parasympathetic nerve fibers within cranial nerves
4. Choose all *true* statements about gray rami.
   A. They contain only sympathetic nerve fibers.
   B. They contain only postganglionic nerve fibers.
   C. They carry impulses from trunk ganglia to spinal nerves.
   D. They are located at all levels of the vertebral column (from cervical to coccyx).
   E. They carry impulses between lateral ganglia and collateral ganglia.
   F. They carry preganglionic neurons from anterior ramus of spinal nerve to trunk ganglion.
5. Which of the following structures contain some sympathetic preganglionic nerve fibers?
   A. Splanchnic nerves
   B. White rami
   C. Sciatic nerve
   D. Cardiac nerves
   E. Ventral roots of spinal nerves
   F. The sympathetic chains
6. Which are structural features of the parasympathetic system?
   A. Ganglia lie close to the CNS and distant from the effector
   B. Forms the craniosacral outflow
   C. Distributed throughout the body, including extremities
   D. Supplies nerves to blood vessels, sweat glands, and adrenal glands
   E. Has some of its nerve fibers passing through lateral (paravertebral) ganglia

Questions 7–12; Circle the letter preceding the one best answer to each question.

7. All of the following axons are cholinergic *except:*
   A. Parasympathetic preganglionic
   B. Parasympathetic postganglionic
   C. Sympathetic preganglionic
   D. Sympathetic postganglionic to sweat glands
   E. Sympathetic postganglionic to heart muscle
8. All of the following are collateral ganglia *except:*
   A. Superior cervical
   B. Prevertebral ganglia

C. Celiac ganglion
D. Superior mesenteric ganglion
E. Inferior mesenteric ganglion

9. Which region of the cord contains no preganglionic cell bodies at all?
   A. Sacral
   B. Lumbar
   C. Cervical
   D. Thoracic
10. Which statement about postganglionic neurons is *false*?
    A. They all lie entirely outside of the CNS.
    B. Their axons are unmyelinated.
    C. They terminate in visceral effectors.
    D. Their cell bodies lie in the lateral gray matter of the cord.
    E. They are very short in the parasympathetic.
11. All of the following answers match types of receptors with correct locations *except:*
    A. Alpha$_1$: blood vessels in skin and in walls of abdominal viscera
    B. Alpha$_1$: salivary gland arterioles
    C. Beta$_1$: smooth muscle fibers in walls of airways
    D. Nicotinic: motor end plates of skeletal muscle fibers
    E. Muscarinic: most sweat glands
12. In which condition are fingers and toes cool and pale due to vasoconstriction of blood vessels there?
    A. Autonomic dysreflexia
    B. Homer's syndrome
    C. Raynaud's disease

Questions 13–20: Circle T (true) or F (false). If the statement is false, change the underlined word or phrase so that the statement is correct.

T F 13. Synapsing occurs in both sympathetic and parasympathetic ganglia.

T F 14. The sciatic, brachial, and femoral nerves all contain sympathetic postganglionic nerve fibers.

T F 15. Under stress conditions the sympathetic system dominates over the parasympathetic.

T F 16. When one side of the body is deprived of its sympathetic nerve supply, as in Horner's syndrome, the following symptoms can be expected: constricted pupil (miosis) and lack of sweating (anhidrosis).

T F 17. Sympathetic cardiac nerves stimulate heart rate, and the vagus slows down heart rate.

T F 18. Viscera do have sensory nerve fibers, and they are included in the autonomic nervous system.

T F 19. All viscera have dual innervation by sympathetic and parasympathetic divisions of the ANS.

T F 20. The sympathetic system has a more widespread effect in the body than the parasympathetic does.

Questions 21–25: Fill-ins. Complete each sentence with the word or phrase that best fits.

________________ 21. The three major types of tissue (effectors) innervated by the ANS nerves are ________.

________________ 22. ________ nerves convey impulses from the sympathetic trunk ganglia to collateral ganglia.

________________ 23. About 80% of the craniosacral outflow (parasympathetic nerves) is located in the ________ nerves.

_________________ 24. Alpha (α) and beta (β) receptors are stimulated by the transmitter

________.

_________________ 25. Drugs that block (inhibit) beta receptors in the heart will cause

_______-crease in heart rate and blood pressure.

ANSWERS TO MASTERY TEST: CHAPTER 17

Multiple Answers

1. B C D E
2. A E F
3. A B C D E G
4. A B C D
5. A B E F
6. B

Multiple Choice

7. E
8. A
9. C
10. D
11. C
12. C

True–False

13. T
14. T
15. T
16. T
17. T
18. T
19. F Most (or some)
20. T

Fill-ins

21. Cardiac muscle, smooth muscle, and glandular epithelium
22. Splanchnic
23. Vagus
24. NE (and also epinephrine)
25. De

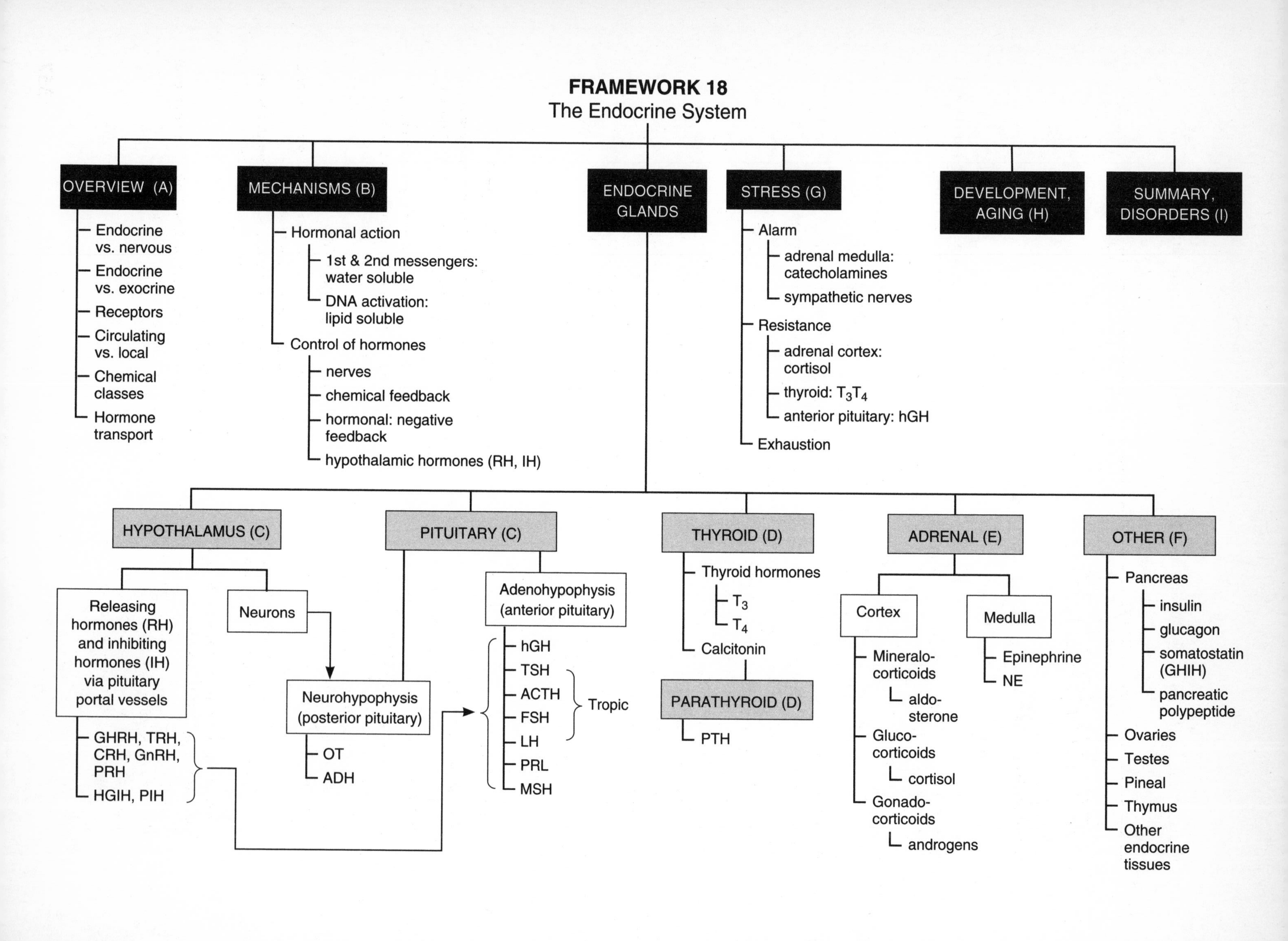

FRAMEWORK 18
The Endocrine System
OVERVIEW (A)
Endocrine vs. nervous
Endocrine vs. exocrine
Receptors
Circulating vs. local
Chemical classes
Hormone transport
MECHANISMS (B)
Hormonal action
1st & 2nd messengers: water soluble
DNA activation: lipid soluble
Control of hormones
nerves
chemical feedback
hormonal: negative feedback
hypothalamic hormones (RH, IH)
ENDOCRINE GLANDS
STRESS (G)
Alarm
adrenal medulla: catecholamines
sympathetic nerves
Resistance
adrenal cortex: cortisol
thyroid: T_3T_4
anterior pituitary: hGH
Exhaustion
DEVELOPMENT, AGING (H)
SUMMARY, DISORDERS (I)
HYPOTHALAMUS (C)
Releasing hormones (RH) and inhibiting hormones (IH) via pituitary portal vessels
GHRH, TRH, CRH, GnRH, PRH
HGIH, PIH
Neurons
PITUITARY (C)
Neurohypophysis (posterior pituitary)
OT
ADH
Adenohypophysis (anterior pituitary)
hGH
TSH
ACTH
FSH
LH
PRL
MSH
Tropic
THYROID (D)
Thyroid hormones
T_3
T_4
Calcitonin
PARATHYROID (D)
PTH
ADRENAL (E)
Cortex
Mineralo-corticoids
aldo-sterone
Gluco-corticoids
cortisol
Gonado-corticoids
androgens
Medulla
Epinephrine
NE
OTHER (F)
Pancreas
insulin
glucagon
somatostatin (GHIH)
pancreatic polypeptide
Ovaries
Testes
Pineal
Thymus
Other endocrine tissues

CHAPTER 18

The Endocrine System

Hormones produced by at least 18 organs exert widespread effects on just about every body tissue. Hormones are released into the bloodstream, which serves as the vehicle for distribution throughout the body. In fact, analysis of blood levels of hormones can provide information about the function of specific endocrine glands. How do hormones "know" which cells to affect? How does the body "know" when to release more or less of a hormone? Mechanisms of action and regulation of hormone levels are discussed in this chapter. The roles of all major hormones and effects of excesses or deficiencies in those hormones are also included. The human body is continually exposed to stressors, such as rapid environmental temperature changes, a piercing sound, or serious viral infection. Adaptations to stressors are vital to survival: however, certain adaptations lead to negative consequences. A discussion of stress and adaptation concludes Chapter 18.

As you begin your study of the endocrine system, carefully examine the Chapter 18 Topic Outline and Objectives: check off each one as you complete it. To organize your study of the endocrine system, glance over the Chapter 18 Framework now. Be sure to refer to the Framework frequently and note relationships among key terms in each section.

TOPIC OUTLINE AND OBJECTIVES

A. Overview: comparison of nervous and endocrine systems; endocrine glands; hormone activity

1. Compare control of body functions by the nervous system and endocrine system.
2. Distinguish between exocrine and endocrine glands.
3. Describe how hormones interact with target-cell receptors.
4. Compare the two chemical classes of hormones based on their solubility.

B. Mechanisms of hormonal action and control of hormone secretion

1. Describe the two general mechanisms of hormonal action.
2. Describe the three types of signals that can control hormone secretion.

C. Hypothalamus and pituitary gland

1. Describe the locations of and relationships between the hypothalamus and the pituitary gland.
2. Describe the location, histology, hormones, and functions of the anterior and posterior pituitary glands.

D. Thyroid and parathyroids

1. Describe the location, histology, hormones, and functions of the thyroid gland.
2. Describe the location, histology, hormones, and functions of the parathyroid glands.

E. Adrenal glands

1. Describe the location, histology, hormones, and functions of the adrenal glands.

F. Pancreas, gonads, pineal gland, thymus gland, and other endocrine cells

1. Describe the location, histology, hormones, and functions of the pancreatic islets.
2. Describe the location, histology, hormones, and functions of the male and female gonads.
3. Describe the location, histology, hormones, and functions of the pineal gland.
4. List the hormones secreted by cells in tissues and organs other than endocrine glands, and describe their functions.

G. The stress response

1. Describe how the body responds to stress.

H. Development and aging of the endocrine system

1. Describe the development of the endocrine glands.
2. Describe the effects of aging on the endocrine system.

I. Summary of hormones and related disorders

WORDBYTES

Now become familiar with the language of this chapter by studying each wordbyte, its meaning, and an example of its use within a term. After you study the entire list, self-check your understanding by writing the meaning of each wordbyte on the line. As you continue through the *Learning Guide*, identify (and fill in) additional terms that contain the same wordbyte.

Wordbyte	Self-check	Meaning	Example(s)
adeno-	______________	gland	*adeno*hypophysis
andro-	______________	man	*andro*gens
crin-	______________	to secrete	endo*crin*e
endo-	______________	within	*endo*crine
exo-	______________	outside	*exo*crine
gen-	______________	to create	diabeto*gen*ic
gonado-	______________	seed	*gonado*tropins
horm-	______________	excite, urge on	*horm*one
insipid-	______________	without taste	diabetes *insipid*us
mellit-	______________	sweet	diabetes *mellit*us
oxy(s)-	______________	swift	*oxy*tocin
para-	______________	around	*para*thyroid, *para*crine
somato-	______________	body	*somato*tropin
thyro-	______________	shield	*thyro*tropin
-tocin	______________	childbirth	oxy*tocin*
-tropin	______________	change, turn	somato*tropin*

A. Overview: comparison of nervous and endocrine systems; endocrine glands; hormone activity (pages 588–590)

✓ **A1.** Compare the ways in which the nervous and endocrine systems exert control over the body. Write *N* (nervous) or *E* (endocrine) next to descriptions that fit each system.

_______ a. Hormones are mediator molecules of this system.

_______ b. Neurotransmitters are mediator molecules of this system.

_______ c. Sends messages to muscles, glands, and neurons only.

_______ d. Sends messages to virtually any part of the body.

_______ e. Effects are generally faster and shorter lived.

_______ f. Norepinephrine is released by cells of this system.

✓ **A2.** Complete the table contrasting exocrine and endocrine glands.

	Secretions transported In	**Examples**
a. Exocrine		
b. Endocrine		

✓ **A3.** Refer to Figure LG 18.1 and do this exercise.

a. Color and label each endocrine gland next to letters A–G.

b. Next to numbers 1–13, label each organ or tissue that contains endocrine cells but is not an endocrine gland exclusively.

✓ **A4.** Write a sentence explaining how a specific hormone "knows" to affect a specific target organ, for example, how thyroid-stimulating hormone (TSH) "knows" to affect the thyroid gland rather than cells of pancreas, ovaries, or muscles.

How does this same principle explain the effectiveness of RU486 (mifepristone)?

✓ **A5.** Describe regulation of hormones in this exercise.

a. When excessive amounts of hormones are present, the number of related receptors is likely to *(decrease? increase?)* so that the overabundant hormone is less effective. This effect is called *(down? up?)*-regulation.

b. Up-regulation makes a target organ *(less? more?)* sensitive to hormones by increase of receptors, for example, when hormone level is *(deficient? excessive?)*.

✓ **A6.** Contrast types of hormones by filling in blanks. Use these answers: *auto, endo, para.*

a. Circulating hormones that exert effects on distant target cells are called

____________________-crines.

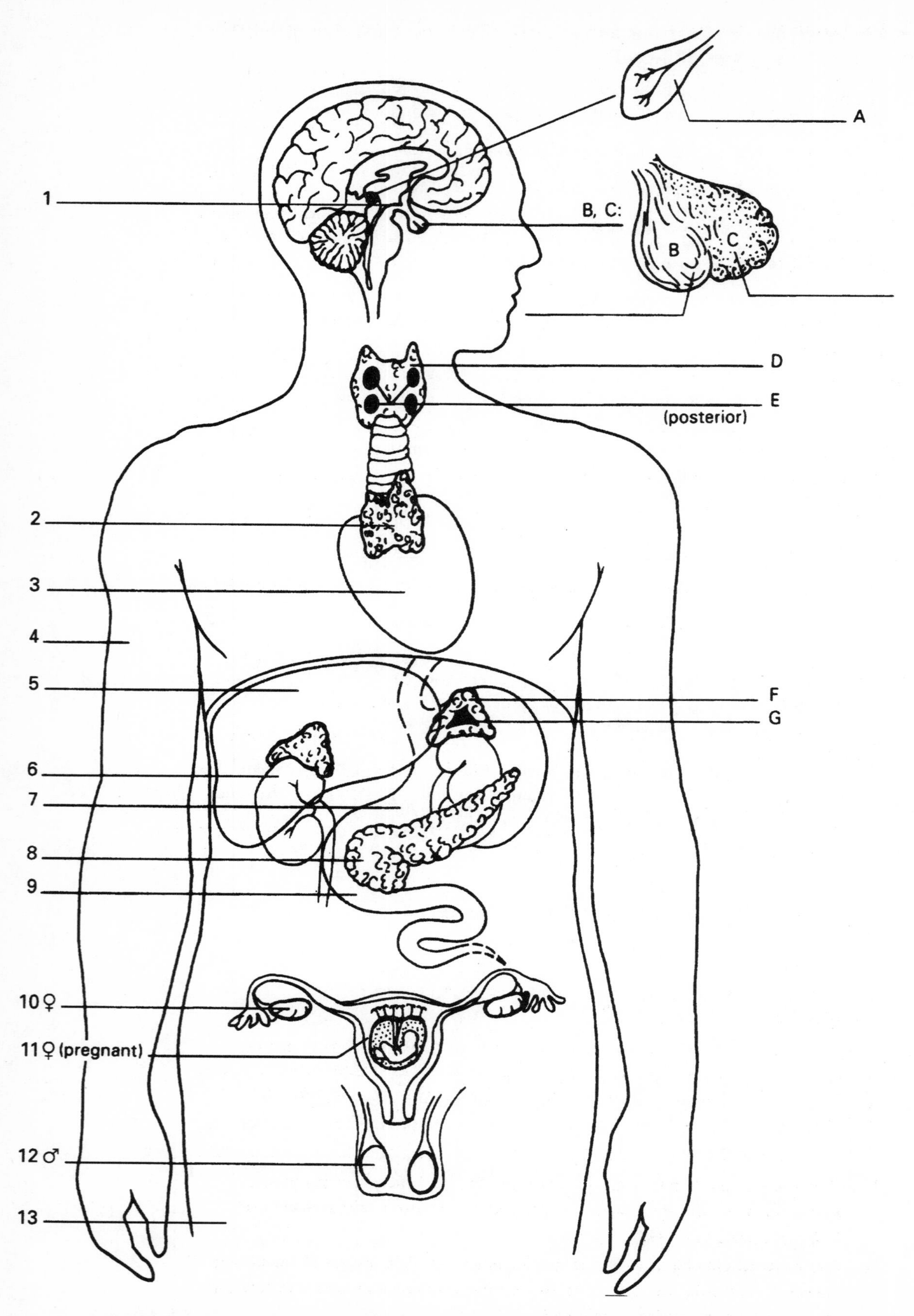

Figure LG 18.1 Diagram of endocrine glands labeled A–G and organs containing endocrine tissue labeled 1–13. Label and color as directed in Checkpoints A3 and C1.

b. Local hormones include ____________________-crines that act upon the same cell that secreted the hormone, as well as ____________________-crines that act on nearby cells.

c. Which of these types of hormones have longest lasting effects? ____________________

d. Nitric oxide (NO) is an example of a ____________________-crines secretion; it is also a neurotransmitter (Chapter 14, Checkpoint).

✓ **A7.** Carefully examine Table 18.2 in your text. Match the following chemical classes with descriptions of hormones listed below. The two short lines following each description will be used in Checkpoints A8 and B1.

A.	**Amines**	**PP.**	**Peptide, protein**
E.	**Eicosanoid**	**S.**	**Steroid**

________ a. Chains of 3 to about 200 amino acids that may form glycoproteins. ______ ______

________ b. Catecholamines, including epinephrine, norepinephrine, and dopamine. ______ ______

________ c. Thyroid hormones (T_3 and T_4). ______ ______

________ d. Derived from 20-carbon fatty acid; examples are prostaglandins and leukotrienes. ______ ______

________ e. Derived from cholesterol; includes hormones from adrenal cortex, ovary, and testis. ______ ______

✓ **A8.** Look back at Checkpoint A7 and write *L* for lipid-soluble or *W* for water-soluble on the first short line following each hormone description in Checkpoint A7.

✓ **A9.** Which category of hormones tend to circulate in the blood attached to transport proteins? *(Lipid-soluble? Water-soluble?)* Are any of these hormones ever not bound to a transport protein?

B. Mechanisms of hormonal action and control of hormone secretion (pages 590–594)

✓ **B1.** Look back at Activity A7. On the second lines following descriptions of hormones (a–e), write letters indicating the expected mechanism of hormonal action. (*Hint:* Remember that each action depends on the chemistry of the hormone.) Use these answers:

I. **This lipid-soluble hormone can pass through the lipid bilayer of the plasma membrane and enter the cell to use *intracellular receptors* that activate DNA to direct synthesis of new proteins.**

P. **This water-soluble hormone cannot pass through the lipid bilayer of the plasma membrane and enter the cell, so it uses *plasma membrane receptors* to activate second messengers such as cyclic AMP.**

✓ **B2.** Study the cyclic AMP mechanism in Figure 18.4 in your text. Then complete this exercise.

a. A hormone such as antidiuretic hormone (ADH) acts as the ____________________ messenger as it carries a message from the ____________________ where it is secreted (in this case the hypothalamus) to the outside of the ____________________ (in this case kidney cells).

b. The hormone then binds to a receptor on the *(inner? outer?)* surface of the plasma membrane, causing activation of __________-proteins.

c. Such activation increases activity of the enzyme ____________________. This enzyme catalyzes conversion of ATP to ____________________, which is known as the ____________________ messenger.

d. Cyclic AMP then activates one or more enzymes known as ____________________ to help transfer *(calcium? phosphate?)* from ATP to a protein, usually an enzyme. The resulting chemical can then set off the target cell's response. In the case of ADH, this is an increase in permeability of kidney cells, with a resulting decrease in ____________________ production.

e. Effects of cyclic AMP are *(short? long?)* lived. This chemical is rapidly degraded by an enzyme named ____________________.

f. A number of hormones are known to act by means of the cyclic AMP mechanism. Included are most of the ____________________-soluble hormones. Name several.

g. Name two or more chemicals besides cyclic AMP that may act as second messengers.

B3. Explain what is meant by amplification of hormone effects.

✓ **B4.** Describe hormonal interactions in this activity.

a. The interaction by which effects of progesterone on the uterus in preparation for pregnancy are enhanced by earlier exposure to estrogen is known as a(n) *(antagonistic? permissive? synergistic?)* effect.

b. Insulin and glucagon exert ____________________ effects on blood glucose levels.

✓ **B5.** Precise regulation of hormone levels is critical to homeostasis. Briefly describe control mechanisms for each hormone listed below. Use these answers.

C. Blood level of a chemical that is controlled by the hormone	
H. Blood level of another hormone	**N. Nerve impulses**

_________ a. Epinephrine

_________ b. Parathyroid hormone

_________ c. Cortisol

✓ **B6.** *(Negative? Positive?)* feedback mechanisms are utilized for most hormonal regulation. For example, a low level of Ca^{2+} circulating through the parathyroid gland causes a(n) *(increase? decrease?)* in release of parathyroid hormone (PTH), which will increase blood level of Ca^{2+}. State one or more example(s) of hormone control by *positive* feedback.

C. Hypothalamus and pituitary gland (pages 594–602)

✓ **Cl.** Complete this exercise about the pituitary gland.

a. The pituitary is also known as the ____________________.

b. Defend or dispute this statement: "The pituitary is the master gland."

c. Where is it located?

d. Seventy-five percent of the gland consists of the *(anterior? posterior?)* lobe, called the ____________________-hypophysis. The posterior lobe (or ____________________-hypophysis) is somewhat smaller.

e. The anterior pituitary is known to secrete *(2? 5? 7?)* different hormones. Write abbreviations for each of these hormones next to the diagram of the anterior pituitary on Figure LG 18.1.

f. Arrange in correct sequence from first to last the vessels that form the blood pathway connecting the hypothalamus to the anterior pituitary. (*Hint:* it may help to refer to Figure 18.5 in your text.)

AHV.	**Anterior hypophyseal veins**
HPV.	**Hypophyseal portal veins**
PPC.	**Primary plexus of capillaries**
SPC.	**Secondary plexus of capillaries**
SHA.	**Superior hypophyseal arteries**

______ ______ ______ ______ ______

Now place an asterisk (*) next to the one answer listed above that consists of vessels located within the anterior pituitary (adenohypophysis) that serve as an exchange site for both hypophysiotropic hormones and anterior pituitary hormones between blood and the anterior pituitary.

g. What advantage is afforded by this special arrangement of blood vessels that link the hypothalamus with the anterior pituitary?

h. Figure 18.5 in your text, shows that the inferior hypophyseal artery, plexus of the infundibular process, and posterior hypophyseal veins form the blood supply to the *(anterior? posterior?)* lobe of the pituitary.

i. List functions of these five types of anterior pituitary cells.

1. Corticotrophs: ______________________________

2. Gonadotrophs: ______________________________

3. Lactotrophs: ______________________________

4. Somatotrophs: ______________________________

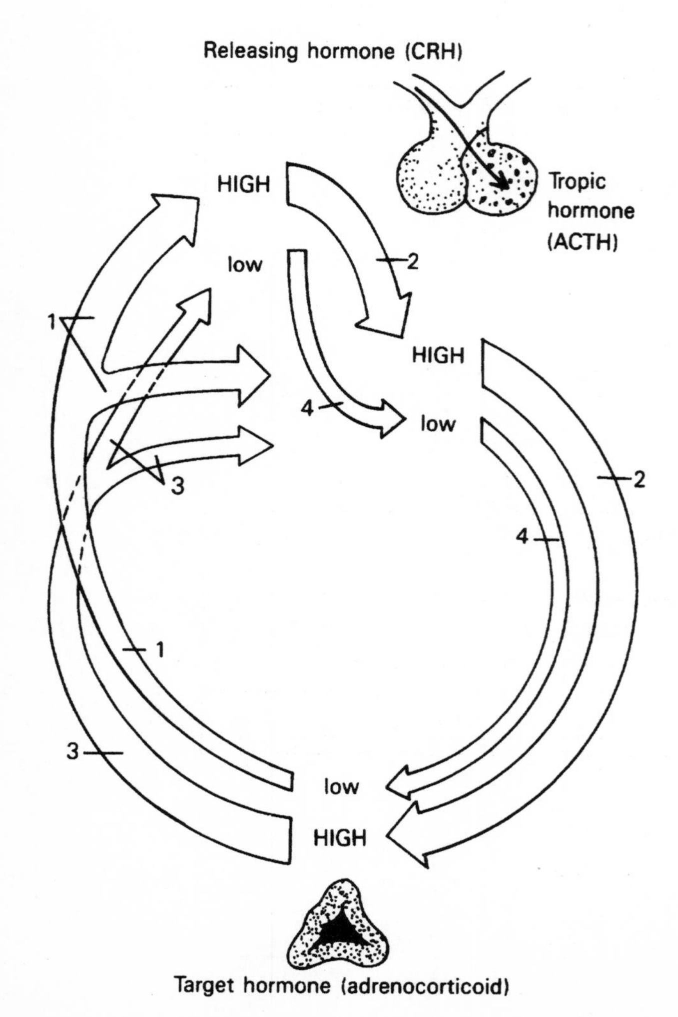

Figure LG 18.2 Control of hormone secretion: example of releasing, tropic, and target hormones. Thickness of arrows indicates amounts of hormones. Color arrows as directed in Checkpoint C2. The branching of arrows 1 and 3 indicates that target hormones affect both the hypothalamus (releasing hormones) and the anterior pituitary (tropic hormones).

O 1 O 3
O 2 O 4

5. Thyrotrophs: ____

j. Define *tropic* hormones or *tropins.*

Which four of the anterior pituitary hormones are tropic hormones?

✓ **C2.** Tropic hormones are involved in feedback mechanisms summarized in steps 1 to 4 in Figure LG 18.2. Trace the pathway of those four steps as you do this exercise.

a. Color the arrow for step 1, in which a low blood level of target hormone, in this example, adrenocorticoid such as cortisone produced by the target gland

_________________, triggers the hypothalamus to secrete a *(high? low?)* level of releasing hormone (CRH). Simultaneously, the low level of target hormones directly stimulates the anterior pituitary to produce a *(high? low?)* level of ACTH.

b. Color the two arrows for step 2. First the high level of releasing hormone (CRH)

passes through blood vessels to the _________________ and stimulates more release of ACTH. A high level of this *(tropic? target?)* hormone in blood passing through the adrenal cortex then stimulates secretion of a *(high? low?)* level of target hormone.

c. Color steps 3 and 4, and note thickness of arrows. Both steps 1 and 3 are *(positive? negative?)* feedback mechanisms, whereas steps 2 and 4 are *(positive? negative?)* feedback mechanisms.

✓ **C3.** Complete this table listing releasing-tropic-target hormone relationships. Fill in the name of a hormone next to each letter and number.

Hypothalamic hormone →	**Anterior Pituitary Hormone** →	**Target Hormone**
a. GHRH (growth hormone-releasing hormone or somatocrinin)		None
b. CRH (corticotropin-releasing hormone)	1. 2.	1. 2. None
c.	TSH (thyrotropin)	1.
d. GnRH (gonadotropic-releasing hormone	1. FSH (follicle stimulating hormone) 2. LH (luteinizing hormone)	1. 2. or testosterone
e.	Prolactin	None

✓ **C4.** Fill in this table naming two hypothalamic inhibiting hormones and their related anterior pituitary hormones. Hypothalamic inhibiting hormones provide additional regulation for the anterior pituitary hormones that *(are? are not?)* tropic hormones. (Notice this in the right column in the table below.)

Hypothalamic hormone →	Anterior Pituitary Hormone →	Target Hormone
a. GHIH (growth hormone-inhibiting hormone or somatostatin)		None
b.	Prolactin	

✓ **C5.** Refer to Figure LG 18.3 as you do this Checkpoint about human growth hormone hGH

a. Effects of hGH are *(direct? indirect?)* because hGH promotes synthesis of

____________________-like growth factors (IGFs). List four sites in the body where IGFs are made in response to hGH.

b. A main function of hGH by means of IGFs is to stimulate growth and maintain size of

____________________ and ____________________. The alternative name for hGH,

____________________, indicates its function: to "change" or "turn" (*trop-*) the body (*soma-*) to growth. Note that somatotropic hormone *(is? is not?)* classified as a true tropic hormone because it *(does? does not?)* stimulate another endocrine gland to produce a hormone.

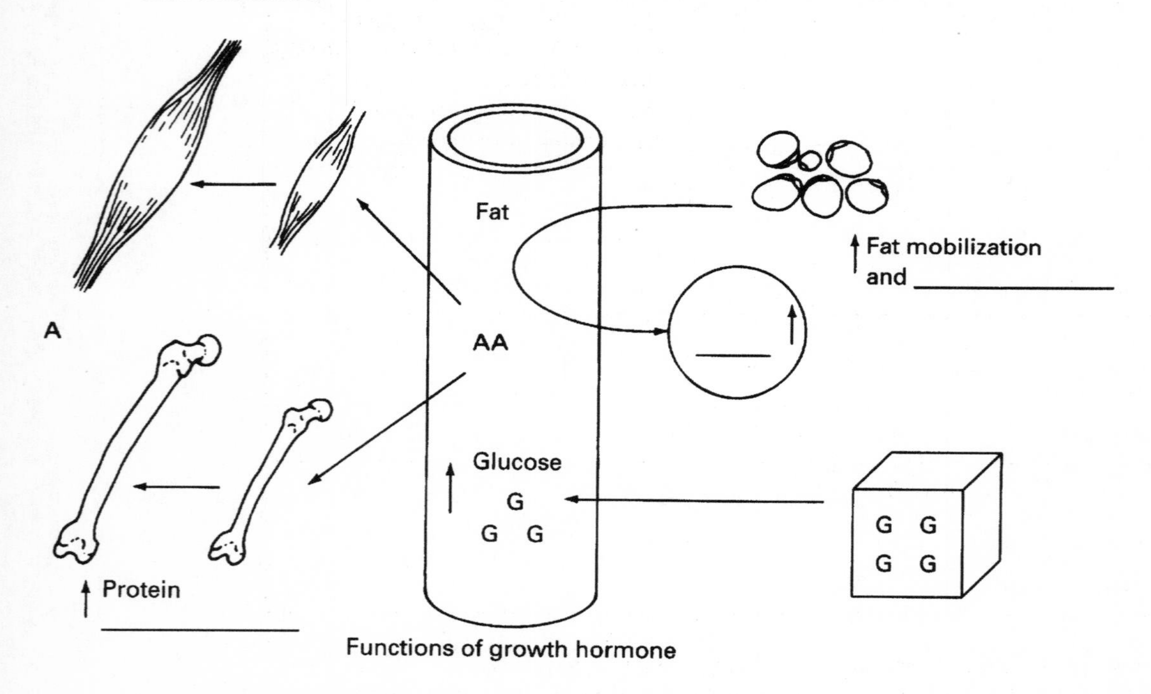

Figure LG 18.3 Functions of growth hormone in regulating metabolism. AA, amino acid; G, glucose. Complete as directed in Checkpoint C5.

c. hGH stimulates growth by *(accelerating? inhibiting?)* entrance of amino acids into

cells where they can be used for __________________ synthesis. Indicate this by filling in the blank on the left side of Figure LG 18.3.

d. hGH promotes release of stored fats and *(synthesis? breakdown?)* of fats, leading to

production of __________________. Show this by filling in the blank on the right side of the figure.

e. Because cells depend on fats as a source of energy, the level of glucose in blood

______-creases and may lead to a condition known as *(hyper? hypo ?)*-glycemia. In this capacity, action of hGH is *(similar? opposite?)* to action of insulin. In fact, because hGH-induced hyperglycemia may overstimulate and "burn out" *(alpha? beta?)* cells

of the pancreas, hGH may be considered __________________-genic.

f. hGH secretion is controlled by two regulating factors: __________________ and

__________________. One stimulus that promotes release of hGH is low blood sugar (for example, during growth when cells require much energy). This condition, *(hvper-? hypo-?)* glycemia, stimulates secretion of *(GHRH? GHIH?)*, which then causes release of *(high? low?)* levels of hGH.

C6. *For extra review.* As you proceed through this chapter, complete the table describing factors that affect production of each of the following hormones. The first one is done for you.

Hormone	How Affected	Factors
a. hGH	Decreased by	HGIH, hGH (by negative feedback of high blood glucose), REM sleep, obesity, increased fatty acids and decreased amino acids in blood; emotional deprivation; low thyroid
b. hGH	Increased by	
c. ACTH	Increased by	
d. ADH	Increased by	

✓ **C7.** Do this exercise about hormonal control of the mammary glands.

a. Milk is produced in mammary glands following stimulation by the hormone

__________________ (______) which is secreted by the *(anterior? posterior?)* pituitary. A different hormone causes ejection of milk from the glands at the time of

the baby's suckling. This hormone, named __________________, is released by the *(anterior? posterior?)* pituitary.

b. Name several hormones that must "prime" breasts for milk production before prolactin and oxytocin can exert their effects.

c. During pregnancy, prolactin levels *(increase? decrease?)* due to increased levels of *(PIH? PRH?)*. During most of a woman's lifetime prolactin levels are low as a result of high levels of *(PIH? PRH?)*. An exception is the time just prior to each menstrual flow when PIH levels *(increase? decrease?)*. Due to this lack of inhibition, prolactin secretion increases, activating breast tissues (and causing tenderness). State two reasons why breasts do not normally produce milk during this premenstrual phase.

✓ **C8.** Complete this Checkpoint on posterior pituitary hormones.

a. Name the two hormones released by the posterior pituitary: ____________________ (OT) and ____________________ (ADH).

b. Although these are released by the posterior pituitary, they are actually synthesized by *(endocrine cells? neurons?)* in the *(anterior pituitary? hypothalamus?)*. These chemicals are then transported by *(fast axonal transport? pituitary portal blood vessels?)* to the posterior pituitary where they are stored.

c. Oxytocin and ADH *(are? are not?)* regulated by tropic-target hormone relationships, as are the four anterior pituitary hormones TSH, _______ _______, and _______. OT and ADH *(are? are not?)* regulated by releasing or inhibiting hormones from the hypothalamus.

d. Functions of oxytocin are better known in human *(females? males?)*. The two target tissues of oxytocin are the uterus, which is stimulated to *(contract? relax?)*, and the breasts, in which milk *(production? letdown?)* is stimulated.

e. On a day when your body becomes dehydrated, for example, by loss of much sweat, your ADH production is likely to _______-crease. ADH causes kidneys to *(retain? eliminate?)* more water, _______-creasing urine output; ADH also _______-creases sweat production. ADH *(dilates? constricts?)* small blood vessels, which allows more blood to be retained in large arteries. Because of this latter function, ADH is also known as ____________________. As a result of all of these mechanisms, ADH _______-creases blood pressure.

f. Effects of diuretic medications are *(similar to? opposite of?)* ADH. In other words, diuretics _______-crease urine production and _______-crease blood volume and blood pressure.

g. Inadequate ADH production, as occurs in diabetes *(mellitus? insipidus?)* results in *(enormous? minute?)* daily volumes of urine. Unlike the urine produced in diabetes mellitus, this urine *(does? does not?)* contain sugar. It is bland or insipid. (Refer to page 626 of your text.)

h. Alcohol *(stimulates? inhibits?)* ADH secretion, which contributes to _______-creased urine production after alcohol intake. For extra review of other factors affecting ADH production, refer to Checkpoint C6d.

C9. Complete parts (1–9) of Table LG 18.1, summarizing names and major functions of the seven hormones made by the anterior pituitary and the two hormones released by the posterior pituitary.

Table 18.1 Summary of major endocrine glands and hormones.

Abbreviation	Name	Source	Principal Functions
1. hGH		Anterior lobe of pituitary	
2.	Thyroid-stimulating hormone		
3. ACTH			
4. FSH			
5. LH			
6. PRL	Prolactin		
7. MSH			Increases skin pigmentation
8.	Oxytocin		
9. ADH			
10. —	Thyroxin		
11.			Decreases blood calcium
12.		Parathyroids	
13. —	Aldosterone		
14. —	Cortisol, cortisone		
15. —	Sex hormones	(4)	
16.	(2)	Adrenal medulla	
17. —	Insulin		
18.		Alpha cells of pancreas	

D. Thyroid and parathyroids (pages 603–608)

✓ **D1.** Describe the histology and hormone secretion of the thyroid gland in this exercise.

a. The thyroid is composed of two types of glandular cells. *(Follicular? Parafollicular?)*

cells produce the two hormones ________________ and ________________.

Parafollicular cells manufacture the hormone ________________.

b. The ion *(Cl^-? I^-? Br^-?)* is highly concentrated in the thyroid because it is an essential component of T_3 and T_4. Each molecule of thyroxine (T_4) contains four of the I^- ions,

whereas T_3 contains ________ ions of I^-. *(T_3? T_4?)* is found in greater abundance in thyroid hormone, but *(T_3? T_4?)* is more potent.

c. Once formed, thyroid hormones *(are released immediately? may be stored for months?)*. T_3 and T_4 travel in the bloodstream bound to plasma *(lipids? proteins?)*.

D2. Briefly describe steps in the formation, storage, and release of thyroid hormones. Be sure to include these terms in your description: *trapping, tyrosines, colloid, peroxidase, T_1 and T_2, pinocytosis, secretion, TBG.*

D3. Release of thyroid hormone is under influence of releasing factor

________________ and tropic hormone ________________. List factors that stimulate TRF and TSH production.

✓ **D4.** Fill in blanks and circle answers about functions of thyroid hormones contrasted to those of growth hormone (hGH).

a. Like hGH, thyroid hormone stimulates protein *(synthesis? breakdown?)*.

b. Unlike hGH, thyroid hormone ________-creases use of glucose for ATP production. Therefore, thyroid hormone *(increases? decreases?)* basal metabolic rate (BMR), releasing heat and *(increasing? decreasing?)* body temperature. This is known as the

________________ effect of thyroid hormones.

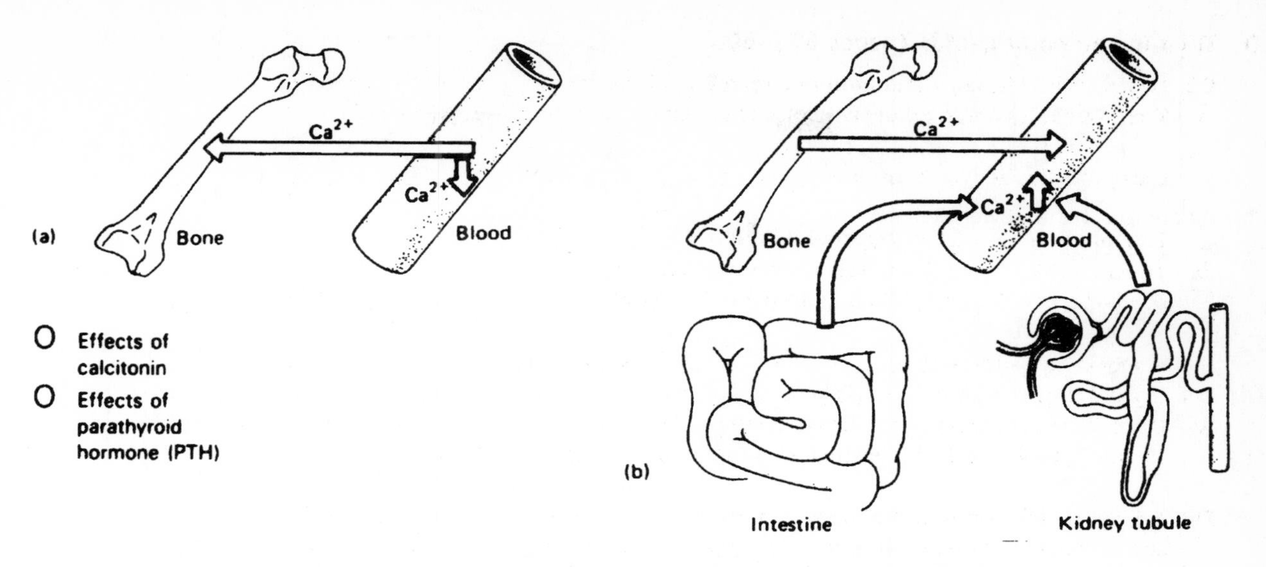

Figure LG 18.4 Hormonal control of calcium ion (Ca^{2+}). (a) Increased calcium storage in bones lowers blood calcium (hypocalcemic effect). (b) Calcium from three sources increases blood calcium level (hypercalcemic effect). Arrows show direction of calcium flow. Color arrows as directed in Checkpoint D5.

c. Thyroid hormone increases *(synthesis? breakdown?)* of fats and excretion of cholesterol in bile, therefore ______-creasing blood cholesterol level.

d. Thyroid hormone *(increases? decreases?)* heart rate and blood pressure and *(increases? decreases?)* nervousness.

e. This hormone is important for development of normal bones and muscles, as well as brain tissue. Therefore, deficiency in thyroid hormone during childhood (a condition called ____________________) *(does? does not?)* lead to retardation as well as small stature.

✓ **D5.** Color arrows on Figure LG 18.4 to show hormonal control of calcium. This figure emphasizes that CT and PTH are *(antagonists? synergists?)* with regard to effects on calcium. One way that PTH increases blood calcium level is by stimulating kidney production of the active form of vitamin _____ (also known as ________________), which facilitates ____ absorption in the intestine. Both CT and PTH act synergistically to *(lower? raise?)* blood phosphate levels.

D6. Complete parts (10–12) of Table LG 18.1, summarizing names and major functions of hormones made by the thyroid and parathyroid glands.

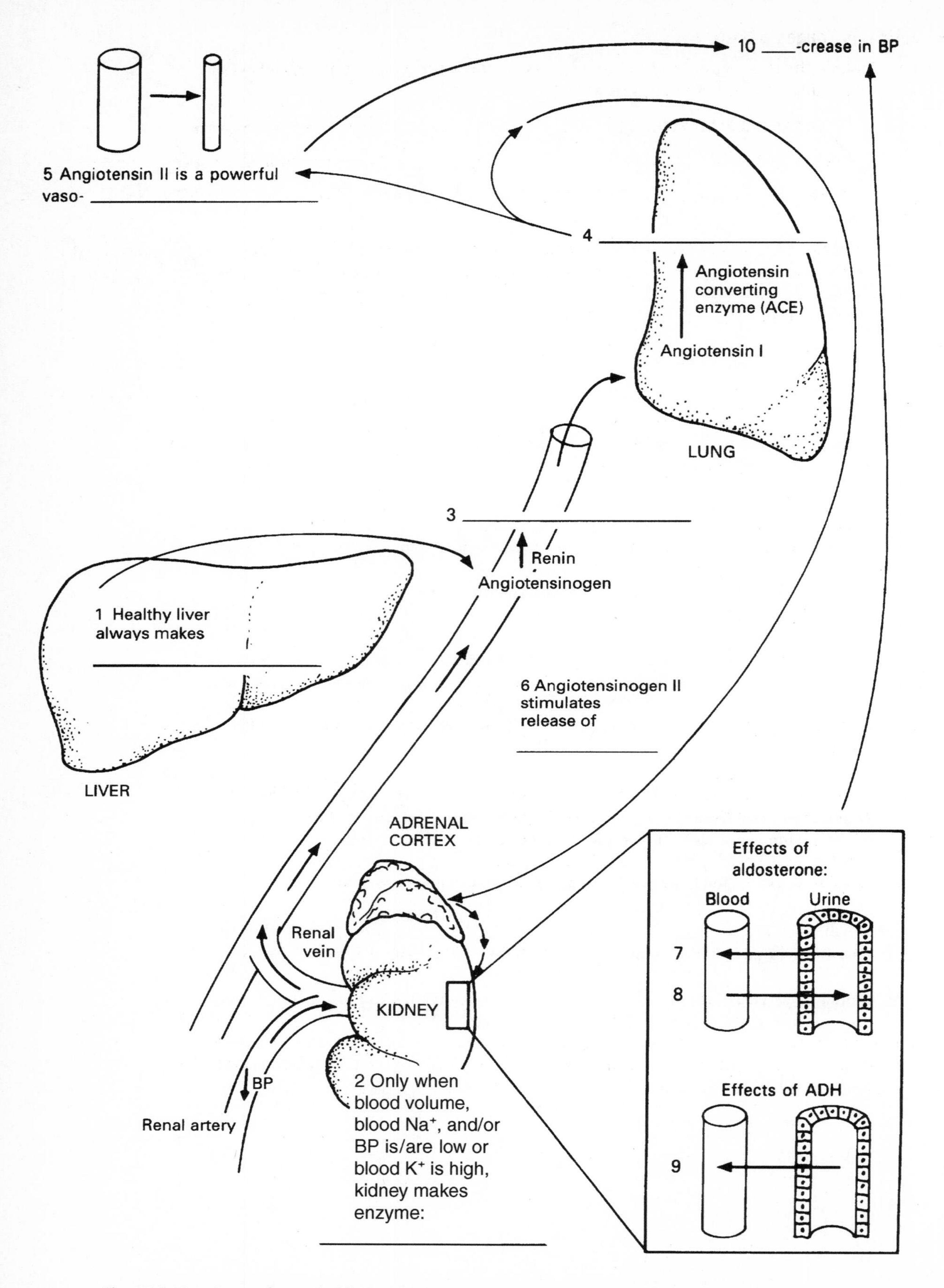

Figure LG 18.5 Hormonal control of fluid, electrolyte balance and blood pressure (BP) by renin, angiotensin, aldosterone, and ADH. Fill in blanks 1–10 as directed in Checkpoint E3.

E. Adrenals (pages 608–614)

✓ **E1.** Contrast the two regions of the adrenal glands in this Checkpoint. Write *C* (cortex) or *M* (medulla) next to related descriptions.

_________ a. More external in location

_________ b. Makes up 80–90% of the adrenal glands

_________ c. Develops from mesoderm

_________ d. Develops from ectoderm

_________ e. Essential for life

E2. In the space below, draw an outline of the adrenal cortex (actual size). Label the *adrenal medulla* and the *adrenal cortex*. Then name the three zones (*zona*) of the adrenal cortex from outermost to innermost. Finally, label your figure with the primary hormones(s) secreted by each zone.

✓ **E3.** Fill in blank lines on Figure LG 18.5 to show mechanisms that control fluid balance and blood pressure.

✓ **E4.** *A clinical correlation.* Determine whether each of the following conditions is likely to *increase* or *decrease* blood pressure.

a. Excessive renin production: ____________________.

b. Use of an angiotensin converting enzyme (A.C.E.)-inhibitor medication:

____________________.

c. Aldosteronism (for example, due to a tumor of the zona glomerulosa):

____________________.

✓ **E5.** Aldosterone is a hormone made by the adrenal *(cortex? medulla?)*. Aldosterone release *(is? is not?)* regulated by adrenocorticotropic hormone (ACTH). State two factors that stimulate aldosterone production.

✓ **E6.** Write the name of the adrenal cortical hormone that is most abundant in each category:

a. Mineralocorticoids: ____________________ c. Androgens: ____________________

b. Glucocorticoids: ____________________

✓ **E7.** In general, glucocorticoids are involved with metabolism and response to stress. Describe their effects in this exercise.

a. These hormones promote *(synthesis? breakdown?)* of muscle proteins with release of amino acids then available for other body functions. In this respect their effects are *(similar? opposite?)* to those of both growth hormone and thyroid hormone.

b. However, glucocorticoids may stimulate conversion of amino acids and lactic acid to glucose, a process known as ____________________. In this way glucocorticoids (like hGH) *(increase? decrease?)* blood glucose.

c. Refer to Figure 18.20 in your text. Note that levels of glucocorticoid (cortisol) and hGH *(increase? decrease?)* during times of stress. By raising blood glucose levels, these hormones make energy available to cells most vital for a stress response. During stress, glucocorticoids also act to *(increase? decrease?)* fluid retention and blood pressure.

d. Cortisol (and the related chemical cortisone) *(enhance? limit?)* the inflammatory process, for example by *(stimulating? inhibiting?)* white blood cells. In other words, these chemicals are *(inflammatory? anti-inflammatory?)* and they ______-crease immune responses.

e. List two or more clinical uses of such glucocorticoids and similar drugs.

Now list side effects of such drugs.

f. *For extra review* of regulation of glucocorticoids, refer to Checkpoints C2, C3, and C6.

✓ **E8.** Discuss congenital adrenal hyperplasia (CAH) in this exercise.

a. CAH is a disorder of the adrenal *(cortex? medulla?)*. In CAH, which hormone is low? *(ACTH? Cortisol? Epinephrine? Testosterone?)*. Which two of these hormone levels are high? ____________________ and ____________________. Explain why ACTH level is altered.

b. What signs or symptoms are likely to be present in adult women with CAH?.

✓ **E9.** Do this exercise on the adrenal medulla.

a. Name the two principal hormones secreted by the adrenal medulla.

____________________ ____________________

Circle the one that accounts for 80% of adrenal medulla secretions.
These hormones are made by *(chromaffin? zona glomerulosa? zona fasciculata?)* cells.

b. In general, effects of these hormones mimic the *(sympathetic? parasympathetic?)* nervous system. List three or more effects.

c. Describe how the adrenal medulla is stimulated to release its hormones.

E10. Complete parts (13–16) of Table LG 18.1, summarizing names and major functions of hormones made by the adrenal glands.

F. Pancreatic islets, ovaries and testes, pineal, thymus, and other hormones and growth factors (page 614–619)

✓ **F1.** The *islets of Langerhans* are located in the ______________________. Do this exercise describing hormones produced there.

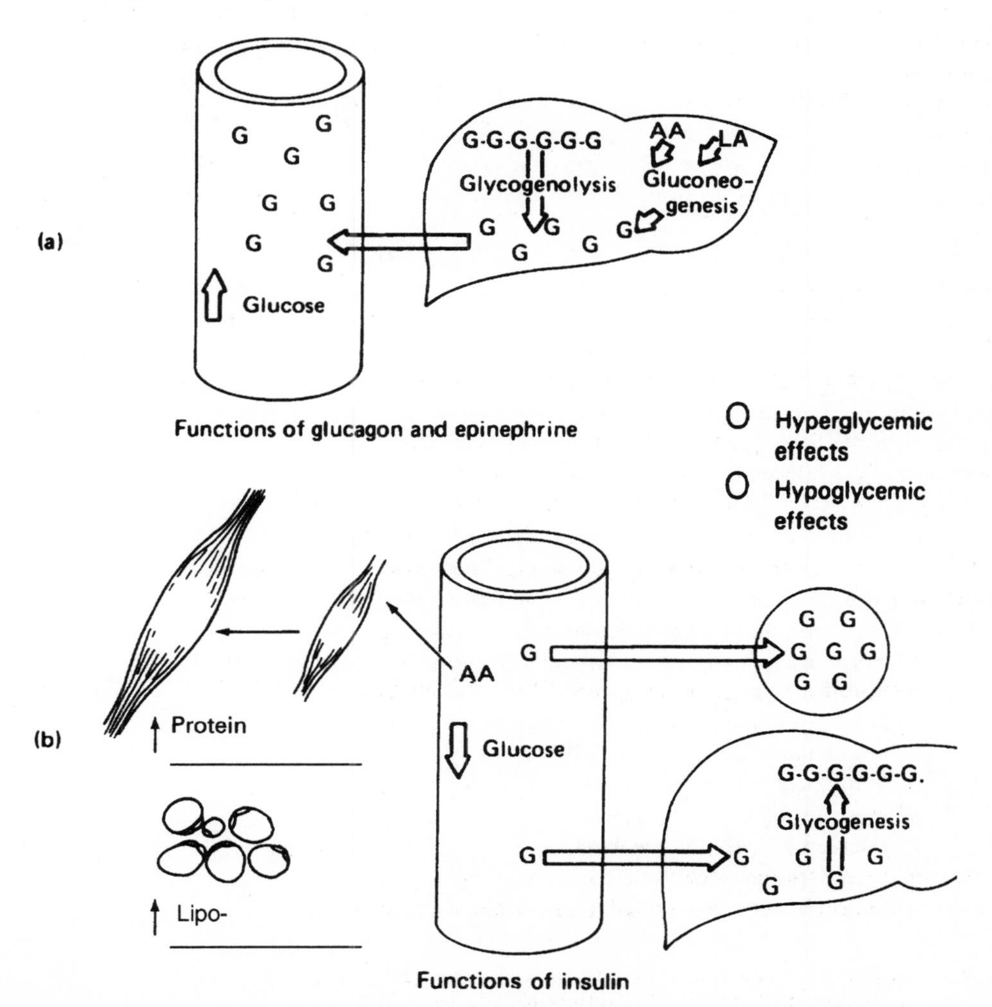

Figure LG 18.6 Hormones that regulate blood glucose level. AA, amino acid; LA, lactic acid; G, glucose; G-G-G-G-G-G, glycogen. Fill in blanks and color according to directions in Checkpoint F1.

a. The name islets of Langerhans suggests that these clusters of ________-crine cells lie amidst a "sea" of exocrine cells within the pancreas. Name the four types of islet cells.

b. Refer to Figure LG 18.6a. Glucagon, produced by *(alpha? beta? delta?)* cells, *(increases? decreases?)* blood sugar in two ways. First, it stimulates the breakdown of ____________________ to glucose, a process known as *(glycogenesis? glycogenolysis? gluconeogenesis?)*. Second, it stimulates the conversion of amino acids and other compounds to glucose. This process is called ____________________. Name one other hormone that stimulates gluconeogenesis. ____________________

c. Is glucagon controlled directly by an anterior pituitary tropic hormone? *(Yes? No?)* In fact, control is by effect of blood ____________________ level directly upon the pancreas. When blood glucose is low, then a *(high? low?)* level of glucagon will be produced. This will raise blood sugar.

d. *For extra review.* Define the terms *glycogen* and *glucagon.*

e. Now look at Figure LG 18.6b. The action of insulin is *(the same as? opposite?)* that of glucagon. In other words, *insulin decreases blood sugar*. (Repeat that statement three times; it is important!) Insulin acts by several mechanisms. Two are shown in the figure. First, insulin *(helps? hinders?)* transport of glucose from blood into cells. Second, it accelerates conversion of glucose to ____________________, the process called ____________________. It also stimulates protein *(breakdown? synthesis?)* and lipo-*(genesis? lysis?)*.

f. Hormones that raise blood glucose levels are called *(hypo? hyper?)*-glycemic hormones. Insulin, the one main hormone that lowers blood glucose level is said to be ____________________-glycemic. To show this, color arrows in Figure 18.6.

✓ **F2.** Contrast effects of hormones by circling effects of cortisol (C), epinephrine (E), glucagon (G), human growth hormone (H), insulin (I), and thyroid hormone (T) where each is noted next to descriptions below. (It may help to refer to Figures LG 18-3 and 18-6. Be sure you fill in all blanks and color as indicated.)

C E G H I   a. Hormone that stimulates glycogenesis (formation of glycogen)
C E G H I   b. Stimulates glycogenesis (breakdown of glycogen to glucose)
C E G H I   c. Stimulates gluconeogenesis (formation of glucose from noncarbohydrates)
C E G H I   d. Stimulates entrance of glucose into cells
C E G H I   e. Increases blood glucose level
C E G H I   f. Decreases blood glucose level
C E G H I T   g. Stimulates protein synthesis
C E G H I T   h. Stimulates lipolysis (breakdown of fatty acids)
C E G H I T   i. Stimulates lipogenesis (formation of fats)

✓ **F3.** Circle *glucagon* or *insulin* to indicate which hormone level is likely to increase by each of the following factors.

a. Low blood glucose level stimulates secretion of *(glucagon? insulin?)* which will raise blood glucose level.

b. During stress, sympathetic nerves stimulate release of *(glucagon? insulin?)*.

c. Parasympathetic nerves and the neurotransmitter ACh stimulate release of *(glucagon? insulin?)* which will lower blood glucose level.

d. GIP released from small intestine after a meal containing glucose stimulates release of *(glucagon? insulin?)*.

F4. Identify types of pancreatic cells that make each of the following hormones. Then write three functions of each hormone.

a. Somatostatin (made by ______ cells of the pancreas):

b. Pancreatic polypeptide (made by ______ cells of the pancreas):

F5. Complete parts (17–18) of Table LG 18.1, summarizing names and major functions of hormones made by the pancreas.

✓ **F6.** Write names of hormones in the box next to related descriptions below. Write one answer on each line.

E.	**Estrogens**	**R.**	**Relaxin**
I.	**Inhibin**	**T.**	**Testosterone**
P.	**Progesterone**		

a. Produced by pregnant women; dilates cervix and increases flexibility of the symphysis pubic during labor and delivery: ______

b. The primary androgen in males: ______

c. The principal female sex hormones: ______ ______

d. Made in ovaries and testes; inhibits FSH in both genders: ______ ______ ______ ______

✓ **F7.** Sex hormones are produced in several locations in the body. List four locations:

F8. Where is the pineal gland (body) located?

Secretion of one pineal hormone, melatonin, increases in *(light? darkness?)*. How may this hormone be related to the form of depression known as seasonal affective disorder (SAD)?

✓ **F9.** What is the general function of the thymus? ______________________ Name four hormones made by this gland.

✓ **F10.** *The Big Picture: Looking Ahead.* Refer to upcoming chapters and answer questions about hormones produced by tissues that not only contain endocrine cells, but that also have other functions. Select answers from the box.

Atrial natriuretic peptide	**Erythropoietin**
CCK, gastrin, and secretin	**Human chorionic gonadotropin (HCG)**
Eicosanoids: thromboxane A_2 and PG	**Vitamin D**

a. Chemical that promotes development of red blood cells (erythropoiesis) (Figure 19.5): ______________________.

b. Chemical that decreases blood pressure by promoting loss of Na^+ and water, and by dilating blood vessels (Table 21.1, page 713) ______________________.

c. Chemicals that constrict blood vessels and therefore increase blood pressure (page 713): ______________________.

d. Hormones that alter secretion of gastric or intestinal secretions and movements (Table 24.4, page 881): ______________________.

e. Hormone made in skin and kidney; aids in absorption of calcium (Table 25.6, page 940): ______________________.

f. Hormone made by the chorion layers of the placenta that stimulates production of estrogens and progesterone to maintain pregnancy (Figure 29.11, page 1077): ______________________.

F11. Write a brief description of functions of each of the following categories of chemicals produced by the body.

a. Leukotrienes (LTs)

b. Prostaglandins (PGs)

c. Growth factors

✓ **F12.** *A clinical correlation.* Aspirin and nonsteroidal anti-inflammatories (NSAIDs) inhibit synthesis of one of the categories of chemicals listed in Checkpoint F11. Which one?

_______________________ Write three therapeutic effects of NSAIDs.

✓ **F13.** Match growth factors listed in the box with descriptions below.

EGF.	**Epidermal growth factor**
PDGF.	**Platelet-derived growth factor**
TAF.	**Tumor angiogenesis factor**

________ a. Produced by normal and tumor cells; stimulates growth of new blood vessels

________ b. Produced in salivary glands; stimulates growth of epithelial cells, fibroblasts, and neurons

________ c. Produced in platelets; appears to play a role in wound healing

G. The stress response (pages 620–622)

G1. Define two categories of stressors and give two examples of *stressors.*

✓ **G2.** Summarize the body's response to stressors in this exercise.

a. The part of the brain that senses the stress and initiates response is the _______________________. It responds by two main mechanisms.

b. First is the *(alarm? resistance?)* reaction involving the adrenal *(medulla? cortex?)* and the _______________________ division of the autonomic nervous system.

c. Second is the _______________________ reaction. Playing key roles in this response are three anterior pituitary hormones: _______________________, _______________________, _______________________, and their associated releasing hormones and target hormones. Therefore, the adrenal *(cortex? medulla?)* hormones are activated.

✓ **G3.** Now describe events that characterize the first (or alarm) stage of response to stress. Complete this exercise.

a. The alarm reaction is sometimes called the _______________________ response.

b. During this stage blood glucose *(increases? decreases?)* by a number of hormonal mechanisms, including those of adrenal medulla hormones _______________________ and _______________________. Glucose must be available for cells to have energy for the stress response.

c. Oxygen must also be available to tissues; the respiratory system *(increases? decreases?)* its activity. Heart rate and blood pressure *(increase? decrease?)*. Blood is

shunted to vital tissues such as the ____________________, ____________________,

and ____________________ and is directed away from reservoir organs such as the

____________________, ____________________, and ____________________.

d. Nonessential activities such as digestion *(increase? decrease?)*. Sweating *(increases? decreases?)* in order to control body temperature and eliminate wastes.

✓ **G4.** Describe hormonal regulation during the resistance stage of the stress response.

a. ATP levels increase as a direct result of actions of two hormones:

____________________ and ____________________.

b. Inflammation is reduced by effects of the hormone ____________________.

c. Describe further effects of each of these hormones during the resistance stage.

1. Glucocorticoids

2. hGH

3. T_3T_4

G5. Describe how the following are related to stress:

a. PTSD

b. Interleukin-I

✓ **G6.** Identify activities of the three stages of response to stress in this Checkpoint. Use answers in the box; they may be used once, more than once, or not at all.

A. Alarm stage	**E.** Exhaustion stage	**R.** Resistance stage

_______ a. Makes large amounts of glucose and oxygen available for a fight-or-flight response.

_______ b. This stage is short-lived.

_______ c. Initiated by sympathetic nerve impulses and adrenal medulla hormones (epinephrine and norepinephrine).

_______ d. Mediated by hypophysiotropic hormones CRH, GHRH, and TRH, which then stimulate anterior pituitary hormones ACTH, hGH, and TSH.

_______ e. Sodium and water retention is promoted by effects of aldosterone in this stage.

_______ f. During this stage, blood chemistry typically returns to nearly normal.

_______ g. Depletion of body reserves and failure of resistance mechanisms occur.

H. Development and aging of the endocrine system (pages 622–623)

✓ **H1.** Match the endocrine gland with the correct description of its embryological origin.

AC.	**Adrenal cortex**	**P.**	**Pancreas**
AM.	**Adrenal medulla**	**PT.**	**Parathyroid**
AP.	**Anterior pituitary**	**T.**	**Thymus**

_______ a. Develops from the foregut area that later becomes part of small intestine

_______ b. Derived from the roof of the mouth; called the hypophyseal (Rathke's) pouch

_______ c. Originates from the neural crest which also produces sympathetic ganglia

_______ d. Derived from tissue from the same region that forms gonads

_______ e. Arise from pharyngeal pouch regions (two answers)

✓ **H2.** Write arrows next to hormone names to indicate whether production of each hormone increases (↑) or decreases (↓) with aging. If the hormone level remains stable in elderly persons, write a dash (—). Note that in most cases tropic hormones (*) tend to increase when their target hormones have decreased.

a. Estrogen (women): ______

b. Testosterone (men): ______

c. FSH and LH (both genders)*: ______

d. T_3/T_4: ______

e. TSH*: ______

f. hGH: ______

g. Adrenocorticoids: ______

h. Insulin: ______

i. PTH: ______

j. Epinephrine and norepinephrine: ______

I. Summary of hormones; disorders and medical terminology (pages 625–627)

✓ **I1.** *For extra review.* Test your understanding of these hormones by writing the name (or abbreviation where possible) of the correct hormone after each of the following descriptions.

a. Stimulates release of growth hormone: ____________________

b. Stimulates testes to produce testosterone: ____________________

c. Promotes protein synthesis of bones and muscles: ____________________

d. Tropic hormone for thyroxin: ____________________

e. Stimulates development of ova in females and sperm in males: ____________________

f. Present in bloodstream of nonpregnant women to prevent lactation:

____

g. Induces sleep: ____________________

h. Stimulates uterine contractions and also breast milk let-down: ____________________

i. Mimic many effects of sympathetic nerves: ____________________

j. Stimulates kidney tubules to produce small volume of concentrated urine:

____

k. Target hormone of ACTH: ____________________

l. Increases blood calcium and decreases blood phosphate: ____________________

m. Decreases blood calcium and phosphate: ____________________

n. Contains iodine as an important component: ____________________

o. Raises blood glucose (three or more answers): ____________________,

____________________, ____________________

p. Lowers blood glucose: ____________________

q. Serves anti-inflammatory functions: ____________________

✓ **I2.** *For extra review.* Match the disorder with the hormonal imbalance.

Ac.	**Acromegaly**	**GD.**	**Graves' disease**
AD.	**Addison's disease**	**M.**	**Myxedema**
C.	**Cretinism**	**PC.**	**Pheochromocytoma**
DI.	**Diabetes insipidus**	**PD.**	**Pituitary dwarfism**
DM.	**Diabetes mellitus**	**T.**	**Tetany**
Go.	**Goiter**		

_________ a. Deficiency of hGH in child; slow bone growth

_________ b. Excess of hGH in adult; enlargement of hands, feet, and jawbones

_________ c. Deficiency of ADH; production of enormous quantities of "insipid" (nonsugary) urine

_________ d. Deficiency of effective insulin; hyperglycemia and glycosuria (sugary urine)

_________ e. Deficiency of thyroxin in child; short stature and mental retardation

_________ f. Deficiency of thyroxin in adult; edematous facial tissues, lethargy

_________ g. Excess of thyroxin; protruding eyes, "nervousness," weight loss

_________ h. Deficiency of thyroxin due to lack of iodine; most common reason for enlarged thyroid gland

_________ i. Result of deficiency of PTH; decreased calcium in blood and fluids around muscles resulting in abnormal muscle contraction

_________ j. Deficiency of adrenocorticoids; increased K^+ and decreased Na^+ resulting in low blood pressure and dehydration

_________ k. Tumor of adrenal medulla causing sympathetic-type responses, such as increased pulse and blood pressure, hyperglycemia, and sweating

✓ **I3.** Hypocalcemia can result from *(hypo? hyper?)*-parathyroid conditions. Decreased blood calcium results in abnormal _________-crease in nerve impulses to muscles, a condition known as _____________________. In other words, there is a(n) *(direct? inverse?)* relationship between blood level of calcium and nerve and muscle activity.

✓ **I4.** Summarize effects of hormones on electrolytes by completing this table. Write *In* for increase and *De* for decrease to indicate the effect of a hormone on blood levels of each ion. Omit portions of the table with —.

Hormone	Ca^{2+}	Mg^{2+}	HPO_4^{2-}	K^+	Na^+
a. PTH	In			—	—
b. Calcitonin		—		—	—
c. Aldosterone	—	—	—		

✓ **I5.** Summarize effects of different hormones on metabolism by completing this table.

	Stimulates Protein Synthesis? Breakdown?	Stimulates Lipolysis?	Hyperglycemic Effect?
a. hGH		Yes? No?	Yes? No?
b. Thyroid hormone		Yes? No?	Yes? No?
c. Cortisol		Yes? No?	Yes? No?
d. Insulin		Yes? No?	Yes? No?

✓ **I6.** Write *A* (for Addison's disease) or *C* (for Cushing's syndrome) next to the manifestations of those adrenal cortex disorders listed below.

_______ a. Abdominal striae

_______ b. Hyperglycemia

_______ c. "Moon face" and "buffalo hump"

_______ d. Decreased resistance to stress and infection

_______ e. Low blood pressure

_______ f. High level of ACTH

✓ **I7.** Contrast type I and type II diabetes mellitus (DM) by writing I or II next to the related descriptions.

_______ a. Also known as maturity-onset DM

_______ b. The more common type of DM

_______ c. Related to insensitivity of body cells to insulin, rather than to absolute insulin deficiency

_______ d. Also known as insulin-dependent diabetes mellitus (IDDM)

_______ e. More likely to lead to serious complications such as ketoacidosis

✓ **I8.** *A clinical correlation.* Mrs. Jefferson has diabetes mellitus. In the absence of sufficient insulin, her cells are deprived of glucose. Explain why she experiences the following symptoms.

a. Hyperglycemia and glucosuria

b. Increased urine production (poly-_______________) and increased water intake

(poly-_______________)

c. Ketoacidosis, a form of _______________-osis

d. Atherosclerosis

ANSWERS TO SELECTED CHECKPOINTS: CHAPTER 18

Al. (a) E. (b) N. (c) N. (d) E. (e) N. (f) N, E.

A2.

	Secretions transported In	Examples
a. Exocrine	Ducts	Sweat, oil, mucus, digestive juices
b. Endocrine	Bloodstream	Hormones

A3. (a) A, pineal; B, posterior pituitary; C, anterior pituitary; D, thyroid; E, parathyroid; F, adrenal cortex; G, adrenal medulla. (b) 1, hypothalamus; 2, thymus; 3, heart; 4, skin; 5, liver; 6, kidney; 7, stomach; 8, pancreas; 9, intestine; 10, ovary; 11, placenta; 12, testis; 13, adipose tissue.

A4. Hormones affect only cells with specific receptors that "fit" the hormone. RU486 blocks receptors for the hormone progesterone and therefore prevents pregnancy because this hormone is needed to establish and maintain a luxuriant uterine lining required for embryonic development.

A5 (a) Decrease; down. (b) More, deficient.

A6. (a) Endo. (b) Auto, para. (c) Endo. (d) Para.

A7 (a) PP. (b) A. (c) A. (d) E. (e) S.

A8. (a) W. (b) W. (c) L (since derived from the non-polar amino acid, tyrosine). (d) W. (e) L.

A9. Lipid-soluble. Yes, about 0.1–10% of lipid-soluble hormones; these are diffusing out of blood so they can bind to receptors and exert their actions.

B1. (a–b) P. (c) I. (d) P. (e) I.

B2. (a) First, endocrine gland, target cell. (b) Outer, G. (c) Adenylate cyclase; cyclic AMP, second. (d) Protein kinases, phosphate; urine. (e) Short; phosphodiesterase. (f) Water; ADH, TSH, ACTH, glucagon, epinephrine, the hypothalamic releasing hormones. (g) Ca^{2+}, cGMP, IP_3, or DAG.

B4. (a) Permissive. (b) Antagonistic.

B5. (a) N. (b) C. (c) H.

B6. Negative; increase. Examples of positive feedback: uterine contractions stimulate more oxytocin, or high estrogen level stimulates high LH level and ovulation.

Cl. (a) Hypophysis. (b) The pituitary regulates many body activities; yet this gland is "ruled" by releasing and inhibiting hormones from the hypothalamus and also by negative feedback mechanisms. (c) In the hypophyseal fossa (sella turcica) just inferior to the hypothalamus. (d) Anterior, adeno; neuro. (e) 7; hGH, PRL, MSH, ACTH, TSH, FSH, LH. (f) SHA PPC HPV SPC* AHV. (g) Hypothalamic hormones reach the pituitary quickly and without dilution because they do not have to traverse systemic circulation. (h) Posterior. (i) 1, Corticotrophs synthesize ACTH, and some make MSH; 2, gonadotrophs make FSH and LH; 3, lactotrophs produce PRL; 4, somatotrophs make hGH; 5, thyrotrophs produce TSH. (j) Hormones that influence other endocrine glands; ACTH, FSH, LH, and TSH.

C2. (a) Adrenal cortex, high; high. (b) Anterior pituitary; tropic, high. (c) Negative, positive.

C3.

Hypothalamic hormone →	Anterior Pituitary Hormone →	Target Hormone
a. GHRH (growth hormone releasing hormone or somatocrinin)	**hGH (human growth hormone)**	None
b. CRH (corticotropin releasing hormone)	1. **ACTH (corticotropin or adrenocorticotropic hormone)** 2. **MSH (melanocyte stimulating hormone)**	1. **Cortisol** 2. None
c. **TRH (thyrotropin releasing hormone)**	TSH (thyrotropin) **or thyroid stimulating hormone**	1. T_3T_4
d. GnRH (gonadotropic releasing hormone	1. FSH (follicle stimulating hormone) 2. LH (luteinizing hormone)	1. **Estrogen** 2. **Estrogen, progesterone** or testosterone
e. **PRH (prolactin releasing hormone)**	Prolactin	None

C4. Are not. (a) hGH (human growth hormone) and also TSH (thyroid-stimulating hormone). (b) PIH (prolactin-inhibiting hormone), which is dopamine; none.

C5. (a) Indirect, insulin; liver, muscle, cartilage, and bone. (b) Bones, skeletal muscles; somatotropin; is not, does not. (c) Accelerating, protein; write *synthesis* on the line at left. (d) Breakdown, ATP; write *breakdown* on the line and *ATP* in the circle. (e) In, hyper; opposite; beta, diabeto. (f) GHRH (somatocrinin), GHIH (somatostatin); hypo, GHRH, high.

C7. (a) Prolactin (PRL), anterior; oxytocin (OT), posterior. (b) Estrogens, progesterone, glucocorticoids, hGH, thyroxine, insulin, and human chorionic somatomammotropin (hCS). (c) Increase, PRH; PIH; decrease; the high PRL level does not last long enough, and mammary glands are not "primed" for milk production as they are during pregnancy.

C8. (a) Oxytocin, antidiuretic hormone. (b) Neurons, hypothalamus; fast axonal transport. (c) Are not, FSH, LH, and ACTH; are not. (d) Females; contract, letdown. (e) In; retain, de, de; constricts; vasopressin; in. (f) Opposite of; in, de. (g) Insipidus, enormous; does not. (h) Inhibits; in.

Dl. (a) Follicular, thyroxine, triiodothyronine: calcitonin. (b) I^-; 3; T_4, T_3. (c) May be stored for months; proteins.

D4. (a) Synthesis. (b) In, increases; increasing; calorigenic. (c) Breakdown, de. (d) Increases, increases. (e) Congenital hypothyroidism; does.

D5. Labels on Figure LG 18.4: (a) Calcitonin (CT). (b) Parathyroid hormone (PT). Antagonists; D, calcitriol or 1, 25-dihydroxy vitamin D_3, calcium, phosphate, and magnesium; lower.

El. (a–c) C. (d) M. (e) C.

E3. 1, The protein angiotensinogen; 2, renin; 3, angiotensin I; 4, angiotensin II; 5, constrictor; 6, aldosterone; 7, Na^+ (and H_2O by osmosis, and Cl^- and HCO_3^- to achieve electrochemical balance); 8, K^+ and H^+; 9, H_2O; 10, In.

E4. (a) Increase (known as *high renin hypertension*). (b) Decrease, such as the antihypertensive captopril (Capoten). (c) Increase.

E5. Cortex; is not; angiotensin II (resulting from the renin-angiotensin-aldosterone pathway) and increased blood level of K^+ (hyperkalemia) since aldosterone helps to lower blood K^+.

E6. (a) Aldosterone. (b) Cortisol. (c) DHEA (dehydroepiandrosterone).

E7. (a) Breakdown; opposite. (b) Gluconeogenesis; increase. (c) Increase; increase. (d) Limit, inhibiting; anti-inflammatory, de. (e) Cortisol for arthritis or severe respiratory infections, nonsteroidal anti-inflammatories (NSAIDS) such as aspirin or ibuprofen for infections or chronic inflammations, and drugs that suppress rejection of transplanted organs. Side effects involve slow wound healing and increased risk of infection.

E8. (a) Cortex; cortisol; ACTH and testosterone; negative feedback: low cortisol stimulates high levels of ACTH. (b) Excessive testosterone levels cause virillism or masculinization with excessive hair and muscle development, deep voice, enlarged clitoris, atrophied breasts.

E9. (a) Epinephrine or adrenaline (80%) and norepinephrine or noradrenaline; chromaffin. (b) Sympathetic: fight-or-flight response such as increase of heart rate, force of heart contraction, and blood pressure; constrict blood vessels and dilate airways; increase blood glucose level and decrease digestion. (c) Direct stimulation by preganglionic sympathetic nerves during stressful periods.

Fl. Pancreas. (a) Endo; alpha, beta, delta, and F. (b) Alpha, increases, glycogen, glycogenolysis; gluconeogenesis; cortisone. (c) No; glucose; high. (d) *Glycogen* is a polysaccharide storage form of glucose; *glucagon* is a hormone that stimulates breakdown of glycogen to glucose. (e) Opposite; helps; glycogen, glycogenesis; synthesis, genesis. (f) Hyper; hypo; glucagon and epinephrine are hyperglycemic; insulin is hypoglycemic.

F2. (a) I. (b) E G. (c) C E G. (d) I. (e) C E G H. (f) I. (g) H I T. (h) C E G H T. (i) I.

F3. (a) Glucagon. (b) Glucagon. (c) Insulin. (d) Insulin.

F6. (a) R. (b) T. (c) E, P. (d) E, P, T, I.

F7. Ovaries, testes, adrenal cortex, and placenta.

F9. Immunity; thymosin, thymic humoral factor, thymic factor, and thymopoietin.

F10. (a) Erythropoietin. (b) Atrial natriuretic peptide. (c) Eicosanoids: thromboxane A and PG F. (d) CCK, gastrin, and secretin. (e) Vitamin D. (f) Human chorionic gonadotropin (HCG).

F12. Prostaglandins (PGs); reduce fever, pain, and inflammation.

F13. (a) TAF. (b) EGF. (c) PDGF.

G2. (a) Hypothalamus. (b) Alarm, medulla, sympathetic. (c) Resistance; ACTH, hGH, TSH; cortex.

G3. (a) Fight-or-flight. (b) Increases; norepinephrine, epinephrine. (c) Increases; increase; skeletal muscles, heart, brain; spleen, GI tract, skin. (d) Decrease; increases.

G4. (a) T_3T_4 and hGH plus other hormones that increase blood levels of glucose, such as cortisol plus epinephrine residual from the alarm stage. (b) Cortisol. (c) See Checkpoint F2 above.

G6. (a–c) A. (d–f) R. (g) E.

H1. (a) P. (b) AP. (c) AM. (d) AC. (e) PT, T.

H2. (a–b) ↓. (c) ↑. (d) ↓. (e) ↑. (f–h) ↓. (i) ↑. (j) —.

I1. (a) GHRH. (b) LH. (c) hGH, thyroid hormone (T_3 T_4), and testosterone; insulin. (d) TSH. (e) FSH. (f) PIH. (g) Melatonin. (h) OT. (i) Epinephrine and norepinephrine. (j) ADH. (k) Adrenocorticoid or glucocorticoid, such as cortisol. (1) PTH. (m) CT. (n) Thyroid hormone (T_3 and T_4). (o) Epinephrine. norepinephrine, glucagon, hGH, cortisol (and indirectly CRH, ACTH, GHRH). (p) Insulin. (q) Adrenocorticoid or glucocorticoid, such as cortisol (and indirectly CRH and ACTH).

I2. (a) PD. (b) Ac. (c) DI. (d) DM. (e) C. (f) M. (g) GD. (h) Go. (i) T. (j) AD. (k) PC.

I3. Hypo, in, tetany; inverse.

I4.

Hormone	Ca^{2+}	Mg^{2+}	HPO_4^{2-}	K^+	Na^+
a. PTH	In	In	De	—	—
b. Calcitonin	De	—	De	—	—
c. Aldosterone	—	—	—	De	In

I5.

	Stimulates Protein Synthesis? Breakdown?	Stimulates Lipolysis?	Hyperglycemic Effect?
a. hGH	Synthesis	Yes	Yes
b. Thyroid hormone	Synthesis	Yes	No (increases use of glucose for ATP production)
c. Cortisol	Breakdown	Yes	Yes
d. Insulin	Synthesis	No	No

I6. (a–d) C. (e-f) A.
I7. (a–c) II. (d-e) I.
I8. (a) Lack of insulin prevents glucose from entering cells, so more is left in blood and spills into urine. (b) Uria, dipsia: glucose in urine draws water (osmotically), and as result of water loss, the patient is thirsty. (c) Acid: because cells have little glucose to use, they turn to excessive fat catabolism which leads to keto-acid formation; also due to lack of effective insulin (which is lipogenic). (d) As fats are transported in extra amounts (for catabolism), some fats are deposited in walls of vessels; fatty (= *athero*) thickening (= *sclerosis*) results.

MORE CRITICAL THINKING: CHAPTER 18

1. Contrast blood supply and hormone production of the anterior and posterior lobes of the pituitary gland.
2. Contrast sources and functions of releasing hormones and tropic hormones.
3. Contrast functions of prolactin and oxytocin in the lactation process.
4. Describe hormonal regulation of blood levels of the following ions: (a) calcium: (b) potassium.
5. Describe the renin-angiotensin-aldosterone (RAA) mechanism for controlling blood pressure.
6. Contrast functions of the following hormones in control of blood pressure: ADH (vasopressin), aldosterone, epinephrine.
7. Discuss advantages and disadvantages of anti-inflammatory effects of glucocorticoids.
8. Discuss the hormone imbalance (whether excessive or deficient) and symptoms of each of the following conditions: acromegaly, Addison's disease, Cushing's syndrome, diabetes insipidus, diabetes mellitus, and myxedema.
9. Describe the roles of feedback systems in the following disorders: (a) a condition in which dietary iodine is decreased, leading to enlargement of the thyroid gland (simple goiter); (b) a condition in which cortisol is decreased (congenital adrenal hyperplasia), leading to enlarged adrenal glands.

MASTERY TEST: CHAPTER 18

Questions 1–2: Arrange the answers in correct sequence.

______ ______ ______ ______ ______ 1. Steps in action of TSH upon a target cell:
A. Adenylate cyclase breaks down ATP.
B. TSH attaches to receptor site.
C. Cyclic AMP is produced.
D. Protein kinase is activated to cause the specific effect mediated by TSH, namely, stimulation of thyroid hormone production.
E. G-protein is activated.

______ ______ ______ ______ 2. Steps in renin-angiotensin mechanism to increase blood pressure:
A. Angiotensin I is converted to angiotensin II.
B. Renin converts angiotensinogen, a plasma protein, to angiotensin I.
C. Angiotensin II causes vasoconstriction and stimulates aldosterone production which causes water conservation and raises blood pressure.
D. Low blood pressure or low blood Na^+ level causes secretion of enzyme renin from cells in kidneys.

Questions 3–10: Circle the letter preceding the one best answer to each question.

3. All of these hormones are synthesized in the hypothalamus *except:*
   A. ADH
   B. GHRH
   C. CT
   D. PIH
   E. Oxytocin

4. Choose the *false* statement about endocrine glands.
   A. They secrete chemicals called hormones.
   B. Their secretions enter extracellular spaces and then pass into blood.
   C. Sweat and sebaceous glands are endocrine glands.
   D. Endocrine glands are ductless.

5. All of the following correctly match a hormonal imbalance with related signs or symptoms *except:*
   A. Deficiency of ADH—excessive urinary output
   B. Excessive aldosterone—low blood pressure
   C. Deficiency of aldosterone—high blood level of potassium and muscle weakness
   D. Excess of parathyroid production—demineralization of bones; high blood level of calcium

6. All of the following hormones are secreted by the anterior pituitary *except:*
   A. ACTH
   B. FSH
   C. Prolactin
   D. Oxytocin
   E. hGH

7. Which one of the following is a function of adrenocorticoid hormones?
   A. Lower blood pressure
   B. Raise blood level of calcium
   C. Convert glucose to amino acids
   D. Lower blood level of sodium
   E. Anti-inflammatory

8. All of these hormones lead to increased blood glucose *except:*
   A. ACTH
   B. Insulin
   C. Glucagon
   D. Growth hormone
   E. Epinephrine

9. Choose the *false* statement about the anterior pituitary.
   A. It secretes at least seven hormones.
   B. It develops from mesoderm.
   C. It secretes tropic hormones.
   D. It is stimulated by releasing factors from the hypothalamus.
   E. It is also known as the adenohypophysis.

10. Choose the *false* statement.
    A. Both ADH and aldosterone tend to lead to retention of water and increase in blood pressure.
    B. PTH activates vitamin D, which enhances calcium absorption.
    C. PTH activates osteoclasts, leading to bone destruction.
    D. Tetany is a sign of hypocalcemia.
    E. ADH and OT pass from hypothalamus to posterior pituitary via pituitary portal veins.

Questions 11–20: Circle T (true) or F (false). If the statement is false, change the underlined word or phrase so that the statement is correct.

T F 11. <u>Polyphagia, polyuria, and hypoglycemia are all</u> symptoms associated with insulin deficiency.

T F 12. <u>Human growth hormone (hGH)</u> carries out its roles by the actions of insulin-like growth factors (IGFs).

T F 13. The hormones of the adrenal medulla mimic the action of <u>parasympathetic</u> nerves.

T F 14. Somatostatin is also known as <u>GHIH</u>, and it is produced by both the <u>hypothalamus</u> and <u>F</u> cells of the <u>pancreatic islets</u>.

T F 15. <u>Epinephrine and cortisol are both</u> adrenocorticoids.

T F 16. Secretion of <u>growth hormone, insulin, and glucagon</u> is controlled (directly or indirectly) by blood glucose level.

T F 17. Most of the feedback mechanisms that regulate hormones are <u>positive (rather than negative)</u> feedback mechanisms.

T F 18. The alarm reaction occurs <u>before</u> the resistance reaction to stress.

T F 19. Most hormones studied so far appear to act by a mechanism involving <u>cyclic AMP</u> rather than by the <u>gene activation</u> mechanism.

T F 20. A high level of thyroxine circulating in the blood will tend to lead to a <u>low</u> level of TRH and TSH.

Questions 21–25: Fill-ins. Complete each sentence with the word or phrase that best fits.

_______________ 21. ________ is a hormone used to induce labor because it stimulates contractions of smooth muscle of the uterus.

_______________ 22. ________ and ________ are two hormones that are classified chemically as catecholamines.

_______________ 23. ________ is a term that refers to the control mechanism by which the number of receptors for a hormone will be decreased when that hormone is present in excessive amounts.

_______________ 24. Type ________ diabetes mellitus is a non-insulin-dependent condition in which insulin level is adequate but cells have decreased sensitivity to insulin.

_______________ 25. Three factors that stimulate release of ADH are ________.

ANSWERS TO MASTERY TEST: CHAPTER 18

Arrange

1. B E A C D
2. D B A C

Multiple Choice

3. C
4. C
5. B
6. D
7. E
8. B
9. B
10. E

True–False

11. F. Polyphagia and polyuria are both or Polyphagia, polyuria, and hyperglycemia are all
12. T
13. F. Sympathetic
14. F. GHIH, hypothalmus, D, pancreatic islets.
15. F. Cortisol, aldosterone, and androgens are all
16. T
17. F. Negative (rather than positive)
18. T
19. T
20. T

Fill-ins

21. Oxytocin (OT or pitocin)
22. Epinephrine or norepinephrine
23. Down-regulation
24. II (maturity onset)
25. Decreased extracellular water, pain, stress, trauma, anxiety, nicotine, morphine, tranquilizers

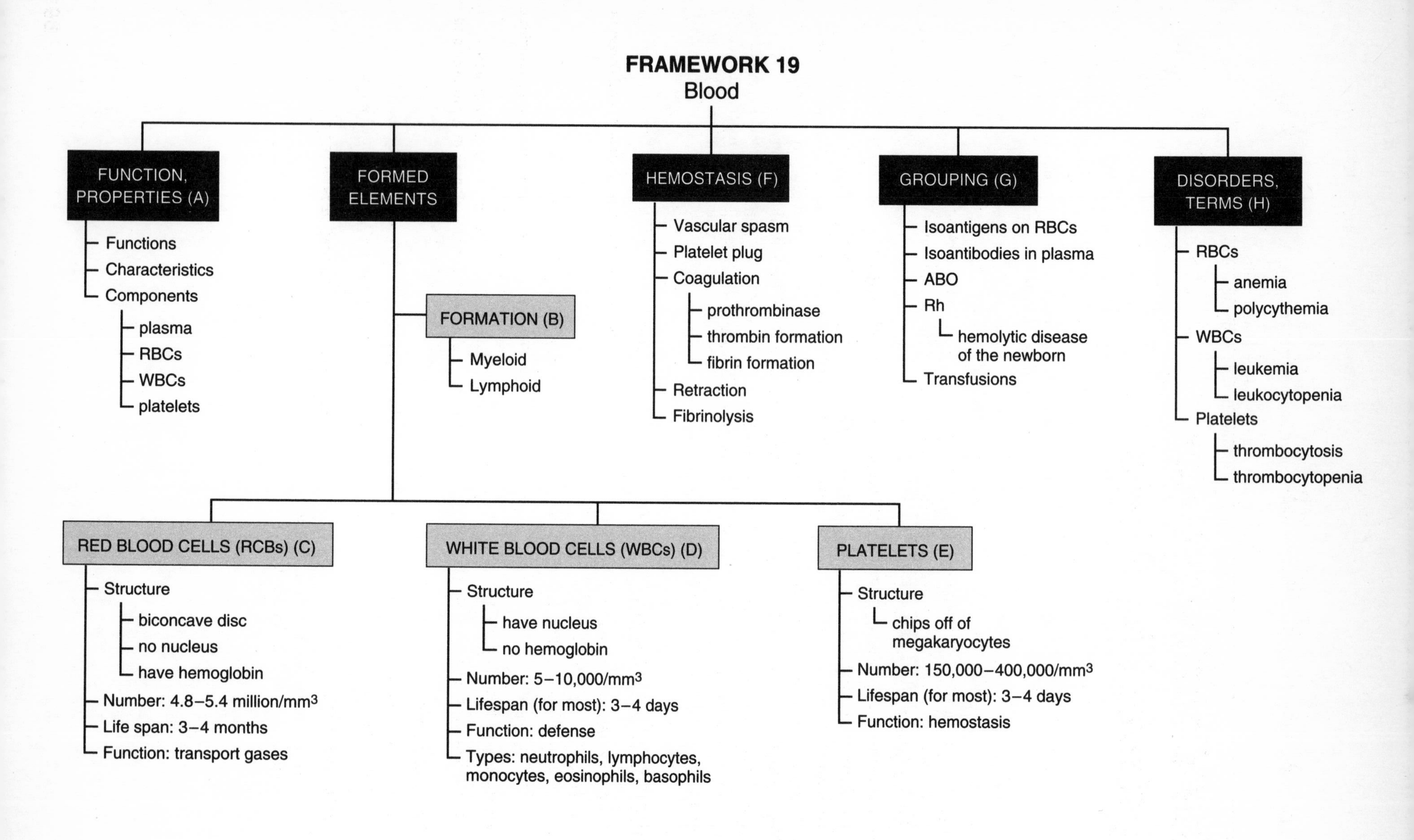

FRAMEWORK 19
Blood
FUNCTION, PROPERTIES (A)
Functions
Characteristics
Components
plasma
RBCs
WBCs
platelets
FORMED ELEMENTS
FORMATION (B)
Myeloid
Lymphoid
HEMOSTASIS (F)
Vascular spasm
Platelet plug
Coagulation
prothrombinase
thrombin formation
fibrin formation
Retraction
Fibrinolysis
GROUPING (G)
Isoantigens on RBCs
Isoantibodies in plasma
ABO
Rh
hemolytic disease of the newborn
Transfusions
DISORDERS, TERMS (H)
RBCs
anemia
polycythemia
WBCs
leukemia
leukocytopenia
Platelets
thrombocytosis
thrombocytopenia
RED BLOOD CELLS (RCBs) (C)
Structure
biconcave disc
no nucleus
have hemoglobin
Number: 4.8–5.4 million/mm3
Life span: 3–4 months
Function: transport gases
WHITE BLOOD CELLS (WBCs) (D)
Structure
have nucleus
no hemoglobin
Number: 5–10,000/mm3
Lifespan (for most): 3–4 days
Function: defense
Types: neutrophils, lymphocytes, monocytes, eosinophils, basophils
PLATELETS (E)
Structure
chips off of megakaryocytes
Number: 150,000–400,000/mm3
Lifespan (for most): 3–4 days
Function: hemostasis

C H A P T E R 19

The Cardiovascular System: The Blood

The major role of the blood is transportation through an intricate system of channels—blood vessels—that reach virtually every part of the body. Blood constantly courses through the vessels, carrying gases, fluids, nutrients, electrolytes, hormones, and wastes to and from body cells. Most of these chemicals "ride" within the fluid portion of blood (plasma). Some, such as oxygen, piggyback on red blood cells. White blood cell "passengers" take short trips in the blood and depart from the vessels at sites calling for defense. Platelets and other clotting factors congregate to plug holes in the network of vessels.

As you begin your study of the blood, carefully examine the Chapter 19 Topic Outline and Objectives; check off each one as you complete it. To organize your study of blood, glance over the Chapter 19 Framework now. Be sure to refer to the Framework frequently and note relationships among key terms in each section.

TOPIC OUTLINE AND OBJECTIVES

A. Functions and properties of blood

1. Describe the functions of blood.
2. Describe the physical characteristics and principal components of blood.

B. Formation of blood cells

1. Explain the origin of blood cells.

C. Red blood cells

1. Describe the structure, functions, life cycle, and production of red blood cells.

D. White blood cells

1. Describe the structure, functions, and production of white blood cells.

E. Platelets

1. Describe the structure, function, and origin of platelets.

F. Hemostasis

1. Describe the three mechanisms that contribute to hemostasis.
2. Identify the stages of blood clotting, and explain the various factors that promote and inhibit blood clotting.

G. Blood groups and blood types

1. Explain the ABO and Rh blood groups.

H. Disorders, medical terminology

WORDBYTES

Now become familiar with the language of this chapter by studying each wordbyte, its meaning, and an example of its use within a term. After you study the entire list, self-check your understanding by writing the meaning of each wordbyte on the line. As you continue through the *Learning Guide*, identify (and fill in) additional terms that contain the same wordbyte.

Wordbyte	Self-check	Meaning	Example(s)
a-, an-	____________	not, without	*an*emia
anti-	____________	against	*anti*body, *anti*coagulant
cardio-	____________	heart	*cardio*vascular
-crit	____________	to separate	hemato*crit*
-cyte	____________	cell	leuko*cyte*
-emia	____________	blood	leuk*emia*
erythro-	____________	red	*erythro*poiesis
ferr-	____________	iron	trans*ferr*in
heme-, hemo-	____________	blood	*hemo*stasis
jaund-	____________	yellow	*jaund*ice
leuko-	____________	white	*leuko*cytosis
mega-	____________	large	*mega*karyocyte
-osis	____________	condition	leukocyt*osis*
-penia	____________	want, lack	leukocyto*penia*
-phil-	____________	love	hemo*phil*iac
phlebo-	____________	vein	*phlebo*tomist
pluri-	____________	several	*pluri*potent
-poiesis	____________	formation	hemo*poiesis*
poly-	____________	many	*poly*cythemia
-rhag-	____________	burst forth	hemor*rhag*e
-stasis	____________	standing still	hemo*stasis*
thromb-	____________	clot	*thromb*us, pro*thromb*in
trans-	____________	across	*trans*ferrin
ven-	____________	vein	*ven*esection

CHECKPOINTS

A. Functions and properties of blood (pages 634–637)

A1. Review relationships among these three body fluids: blood, interstitial fluid, and lymph. (Refer to Figure LG 1.1 of the *Learning Guide*.)

✓ **A2.** Describe functions of blood in this Checkpoint.

a. Blood transports many substances. List six or more.

__________ __________ __________

__________ __________ __________

b. Name the four body systems to which blood delivers wastes or chemicals that lead to wastes.

c. List four aspects of homeostasis regulated by blood. One is done for you.

____pH____ __________

__________ __________

d. Explain how your blood protects you.

✓ **A3.** Circle the answers that correctly describe characteristics of blood.

a. Average temperature:
36°C (96.8°C) 37°C (98.6°C) 38°C (100.4°C)

b. pH:
6.8 7.0–7.1 7.35–7.45

c. Volume of blood in average adult:
1.5–3 liters 4–6 liters 8–10 liters

d. Three hormones that regulate volume and osmotic pressure of blood:
ADH aldosterone ANP epinephrine PTH

A4. Contrast the following procedures: *venipuncture/finger stick.*

✓ **A5.** Refer to Figure 19.1a in your text, and complete this checkpoint. Use these answers.

E.	**Erythrocytes**	**LP.**	**Leukocytes and platelets**	**P.**	**Plasma**

________ a. In centrifuged blood, layers of blood from top to bottom of the tube are:

______ ______ ______

________ b. In centrifuged blood, volume of layers of blood from greatest to least are:

______ ______ ______

________ c. Which layer contains the red blood cells? ______

________ d. Which layer forms the buffy coat? ______

✓ **A6.** Match the names of components of plasma with their descriptions.

A.	**Albumins**	**GAF.**	**Glucose, amino acids, and fats**
E.	**Electrolytes**	**HE.**	**Hormones and enzymes**
F.	**Fibrinogen**	**UC.**	**Urea, creatinine**
G.	**Globulins**	**W.**	**Water**

________ a. Makes up about 92% of plasma

________ b. Regulatory substances carried in blood

________ c. Cations and anions carried in plasma

________ d. Constitutes about 54% of plasma protein

________ e. Made by liver; a protein used in clotting

________ f. Antibody proteins

________ g. Wastes carried to kidneys or sweat glands

________ h. Food substances carried in blood

________ i. Of albumin, fibrinogen, and globulin, found in least abundance in plasma

✓ **A7.** Contrast antigens (Ag) and antibodies (Ab) in this exercise by writing *Ag* or *Ab* next to related descriptions below.

________ a. Viruses or bacteria that enter the body as "invaders"

________ b. Chemicals made by the body to defend against invaders

________ c. Immunoglobulins (Ig's) produced by plasma cells and found in plasma

✓ **A8.** Check your understanding of formed elements of blood in this Checkpoint. Use the answers in the box.

E.	**Erythrocytes**	**L.**	**Leukocytes**	**P.**	**Platelets**

________ a. Not truly cells, but fragments of cells

________ b. Also known as thrombocytes

________ c. Includes neutrophils, lymphocytes, monocytes, eosinophils, and basophils

________ d. Also known as white blood cells

________ e. Makes up over 99% of all formed elements

✓ **A9.** *A clinical correlation.* Answer these questions about hematocrits.

a. A hematocrit consists of the percentage of centrifuged blood that consists of

________________ blood cells.

b. A normal range for hematocrit for adult women is ____________________%; adult

males should have hematocrits in the range of ____________________%. State two reasons for this gender difference in hematocrits.

c. Polycythemia occurs when red blood cell count is *(higher? lower?)* than normal. For example, a hematocrit of *(22? 40? 62?)* would indicate polycythemia. Explain why dehydration may lead to polycythemia.

A10. Describe blood doping in this Checkpoint.

a. This condition is also known as induced *(anemia? polycythemia?)*.

b. Describe the procedure, and discuss the role of Epoetin alfa.

c. State one reason why blood doping is performed.

d. Explain why this practice is dangerous.

B. Formation of blood cells (pages 637–639)

✓ **B1.** Refer to Figure 19.3 in the text and check your knowledge about blood formation.

a. Regulation of *(erythrocyte? leukocyte?)* number occurs by a negative feedback mechanism, whereas alteration of *(erythrocyte? leukocyte?)* number occurs in response to invading microorganisms and antigens.

b. Blood formation is a process known as ____________________. All blood cells arise

from ____ cells.

c. Fill in the blanks in the table to indicate stages in formation of cells.

Stem cell	**→ Progenitor or precursor cell**	**→ Mature cell**
1. Lymphoid stem cell		T lymphocyte
2.	B lymphoblast	
3.	CFU-GM→ myeloblast	
4.	CFU-GM→	Monocyte
5.	CFU-E→ proerythroblast	
6. Myeloid stem cell		Platelet

d. Which of the mature cells above (in Checkpoint B1c) passes through a stage as a reticulocyte immediately before the mature stage? ____________________ (see page 638 of the text). A reticulocyte *(has? lacks?)* a nucleus. Which of the mature cells above (in Checkpoint B1c) later becomes a macrophage? ____________________.

e. Name several structures where blood formation takes place before birth.

f. By about three months before birth, most hemopoiesis takes place in the *(red bone marrow? liver? lymph nodes?)*. Name six or more bones in which red blood cells are formed after birth.

g. Which type of white blood cells complete their development in lymphoid tissues? ____________________ All other types of blood cells develop totally within myeloid tissue, which is another name for ____________________.

h. Erythropoietin is a hormone made mostly in the ____________________. This hormone stimulates production of *(RBCs? WBCs? platelets?)*.

i. Two types of cytokines that stimulate WBC formation are ____________________-____________________ factors (CSFs) and ____________________ (ILs).

✓ **B2.** *A clinical correlation.* Explain how the following chemicals produced via recombinant DNA may be helpful clinically:

a. *EPO* for patients with kidney failure

b. *Granulocyte-macrophage colony-stimulating factor (GM-CSF)* and *granulocyte CSF (G-CSF)* for patients with immune deficiency

c. *TPO*

C. Red blood cells (pages 639–643)

✓ **Cl.** Refer to Figure LG 19.1. Which diagram represents a mature erythrocyte? ________ Note that it *(does? does not?)* contain a nucleus. The chemical named

____________________ accounts for the color of red blood cells (RBCs). Label and color the RBC.

✓ **C2.** Explain the significance of the structure and shape of red blood cells (RBCs) in this Checkpoint.

a. RBCs *(do? do not?)* have a nucleus and mitochondria. State one advantage of this fact.

State one disadvantage.

b. RBCs are normally shaped like ____________________ disks. What advantage does this shape offer?

✓ **C3.** Describe hemoglobin in this exercise.

a. The hemoglobin molecule consists of a central portion, which is the protein *(heme? globin?)* with four polypeptide chains called *(hemes? globins?)*.

b. Each heme contains one ____________________ atom on which a molecule of *(oxygen? carbon dioxide?)* can be transported so that each hemoglobin molecule can carry *(1? 4?)* oxygen molecule(s). Almost all oxygen is transported in this manner. Carbon dioxide has one combining site on hemoglobin; it is the amino acid portion of the *(heme? globin?)*. About *(13? 70? 97?)*% of CO_2 is transported in this manner.

c. Besides 0_2 and $C0_2$, hemoglobin can also bind to the gas *(nitric? nitrous?)* oxide (NO) that is produced by cells lining walls of blood vessels. Describe the effect of NO.

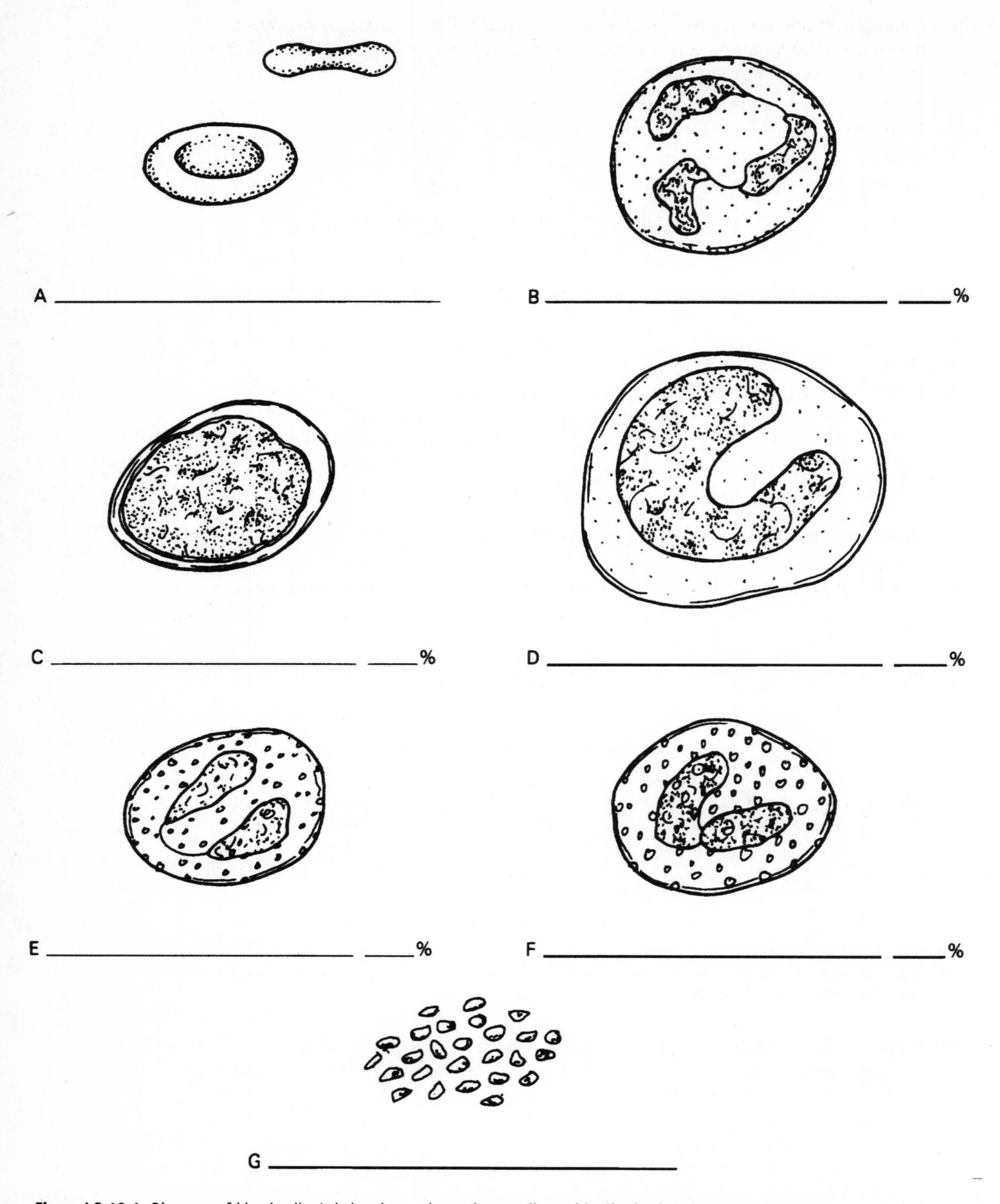

Figure LG 19.1 Diagrams of blood cells. Label, color, and complete as directed in Checkpoints C1, D1, and E2.

✓ **C4.** Fill in blanks in the following description of the RBC life cycle. Select answers from the box. Each answer will be used once except one which will be used three times. It may help to refer to Figure 19.5 in your text.

Amino acids	**Ferritin and hemosiderin**	**Stercobilin**
Bile	**Iron**	**Transferrin**
Bilirubin	**Red bone marrow**	**Urobilin**
Biliverdin	**Spleen, liver, or red bone marrow**	**Urobilinogen**

a. Old RBCs are phagocytosed by macrophages in the ___________________.

b. Hemoglobin is then recycled: globin is broken into ___________________, and hemes are degraded into the metal ___________________ and a noniron portion; see (e) below.

c. Iron is carried through blood attached to the protein ___________________; iron may be stored in muscles attached to the protein ___________________. Once released from muscles, iron is again combined with ___________________. Iron absorbed from foods in the intestine is also attached to ___________________ for transport through blood.

d. Ultimately iron reaches ___________________, where it is used for formation of new ___________________ molecules.

e. What happens to the noniron portion of heme? It is converted to the green pigment named ___________________ and the orange pigment ___________________.

f. Bilirubin has two possible fates. It may pass in blood from bone marrow to liver and be used to form ___________________, or bilirubin may be transformed by intestinal bacteria into ___________________. Urobilinogen may be used to form ___________________ that gives feces its brown color or ___________________ that contributes to the yellow color of urine.

C5. Describe effects of iron overload.

✓ **C6.** After you study Figure 19.6 in your text, explain the roles of the terms in the box in

B_{12}.	**Vitamin B_{12}**	**H.**	**Hypoxia**
E.	**Erythropoietin**	**IF.**	**Intrinsic factor**

________ a. Decrease of oxygen in cells; serves as a signal that erythropoiesis is needed (to help provide more oxygen to tissues)

________ b. Hormone produced by kidneys when they are hypoxic; stimulates erythropoiesis in red bone marrow

________ c. Vitamin necessary for normal hemoglobin formation

________ d. Substance produced by the stomach lining and necessary for normal vitamin B_{12} absorption

red blood cell production.

✓ **C7.** Circle the most normal blood values. (Note that values vary slightly according to age and gender.)

a. Average life of a red blood cell: 4 hours 4 days 4 months 4 years

b. RBC count in a cube this size: ■ (mm^3) 500 5,000 250,000 5 million

c. Number of RBCs produced each second by a healthy human adult:
20 200 2,000 20,000 2 million

d. Number of hemoglobin molecules in one red blood cell:
280 2,800 280,000 280 million

e. Reticulocyte count of person undergoing rapid erythropoiesis, for example, in response to anti-anemic medications:
0.3% 1.0% 3.0%

f. Hemoglobin in adults (g/100 mL blood. See page 647 of the text):
1 8 15 27 41

D. White blood cells (pages 643–645)

✓ **Dl.** Refer to Figure LG 19.1 and do this exercise about leukocytes.

a. Leukocytes *(have? lack?)* hemoglobin, and so these cells are known as *(red blood cells or RBCs? white blood cells or WBCs?).*

b. Label each leukocyte (WBC) on Figure 19.1, and indicate what percentage of the total WBC count is accounted for by each type of WBC. Such a breakdown of white blood cells is known as a ____________________ WBC count.

c. Each white blood cell (WBC) *(has? lacks?)* a nucleus. Which type of WBC has a large kidney-shaped nucleus? ____________________ Which WBC has a nucleus that occupies most of the cell? ____________________. Which is known as a "poly" or PMN? ____________________

d. Which three WBCs are granulocytes? ____________________

e. *For extra review.* Color nucleus, cytoplasm, and granules of all WBCs.

D2. Contrast terms in each pair.

a. *Granular leukocytes/agranular leukocytes*

b. *Polymorphonuclear leukocytes (PMNs)/bands*

c. *Fixed macrophages/wandering macrophages*

✓ **D3.** What are MHC antigens and how do they affect the success rate of transplants?

✓ **D4.** Contrast WBCs with RBCs in this activity.

a. A normal RBC count is about *(700? 5,000–10,000? 250,000? 5 million?)* cells/mm^3, whereas a typical WBC count is _______________ cells/mm^3. In other words, the ratio of RBCs to WBCs is about ________:1.

b. Which cells normally live longer? *(RBCs? WBCs?)* The typical lifespan of a WBC is a few *(days? months? years?)*.

✓ **D5.** Match the answers in the box with descriptions of processes involved in inflammation.

A. Adhesion molecules	**E. Emigration**
C. Chemotaxis	**P. Phagocytosis**

________ a. The process by which WBCs stick to capillary wall cells, slow down, stop, and then squeeze out through the vessel wall

________ b. Chemicals known as selectins made by injured or inflamed capillary cells stick to chemicals on neutrophils, slowing them down

________ c. Attraction of phagocytes to chemicals such as kinins or CSFs in microbes or in inflamed tissue

________ d. Ingestion and disposal of microbes or damaged tissue by neutrophils or macrophages

✓ **D6.** Check your understanding of types of WBCs by matching names of WBCs with descriptions. Answers may be used more than once.

B. Basophils	**L. Lymphocytes**	**N. Neutrophils**
E. Eosinophils	**M. Monocytes**	

_______ a. Constitute the largest percentage of WBCs

_______ b. Produce defensins, lysozymes, and oxidants with antibiotic activity

_______ c. Types of these cells include B cells, T cells, and natural killer (NK) cells

_______ d. They leave the blood, enter tissues, and release heparin, histamine, and serotonin, which are involved in allergic and inflammatory responses

_______ e. Involved in allergic reactions, combat histamines, and provide protection against parasitic worms

_______ f. Form fixed and wandering macrophages that clean up sites of infection

_______ g. Important in phagocytosis (two answers)

_______ h. Classified as agranular leukocytes (two answers)

_______ i. Known as bands in immature state

_______ j. In stained cells, lilac-colored granules are visible

_______ k. In stained cells, red-orange granules are visible

✓ **D7.** Contrast types of lymphocytes in this activity.

a. B lymphocytes (or B cells) are especially effective against *(bacteria and their toxins? cancer or transplanted cells, viruses, fungi, and some bacteria?).*

b. T cells are particularly effective against *(bacteria and their toxins? cancer or transplanted cells, viruses, fungi, and some bacteria?).*

c. Natural killer (NK) cells combat a *(few selected? wide variety?)* of microbes plus some cancer cells.

✓ **D8.** *A clinical correlation.* Mrs. Doud arrives at a health clinic with a suspected acute infection. Complete this Checkpoint about her.

a. During infection, it is likely that Mrs. Doud's leukocyte count will *(in? de?)*-crease. A count of *(4000? 8000? 12,000?)* leukocytes/mm^3 blood is most likely. This condition is known as *(leukocytosis? leukopenia?).*

b. A differential increase in the number of WBCs named ____________________ is most indicative of enhanced phagocytic activity during infection. Neutrophils are most likely to account for *(48? 62? 76?)*% of the total white count in Mrs. Doud's blood. A sign of a chronic infection (such as tuberculosis) is increase in the percentage of cells that can become macrophages; these are ____________________.

✓ **D9.** *A clinical correlation.* Do this exercise on diagnostic clues offered in a differential WBC count. Select the conditions that might he indicated for each specific WBC change listed below. Select answers from the box.

A. Allergic reactions
B. Prolonged, severe illness and/or immunosuppression
C. Lupus, vitamin B_{12} deficiency, radiation
D. Pregnancy, ovulation, stress, and hyperthyroidism

_______ a. Decrease in lymphocyte count

_______ b. Decrease in basophil count

_______ c. Decrease in neutrophil count

_______ d. Increase in basophil or eosinophil count

D10. Briefly describe the procedure of *bone marrow transplant.* Describe the process as well as conditions for which this procedure is used.

E. Platelets (pages 645–647)

✓ **E1.** Circle correct answers related to platelets.

a. Platelets are also known as:

antibodies thrombocytes red blood cells white blood cells

b. Platelets are formed in:

bone marrow tonsils and lymph nodes spleen

c. Platelets are:

entire cells chips off of megakaryocytes

d. A normal range for platelet count is ________ /mm³

5000–1000 4.5–5.5 million 150,000–400,000

e. The primary function of platelets is related to:

O_2 and CO_2 transport blood clotting defense blood typing

✓ **E2.** Label platelets on Figure LG 19.1.

✓ **E3.** List the components of a complete blood count (CBC).

F. Hemostasis (pages 647–651)

✓ **F1.** Hemostasis literally means ________________ (*hemo*)-________________ (*stasis*). List the three basic mechanisms of hemostasis.

a. Vascular ________________

b. ________________ plug formation

c. ________________ (clotting)

✓ **F2.** After reviewing Figure 19.9 in your text, check your understanding of the first two steps of hemostasis in this Checkpoint.

a. When platelets snag on the inner lining of a damaged blood vessel, their shape is altered: they develop a surface that is *(smooth? irregular with extensions?)* that facilitate interaction with other platelets. This phase is known as platelet *(adhesion? plug?)*.

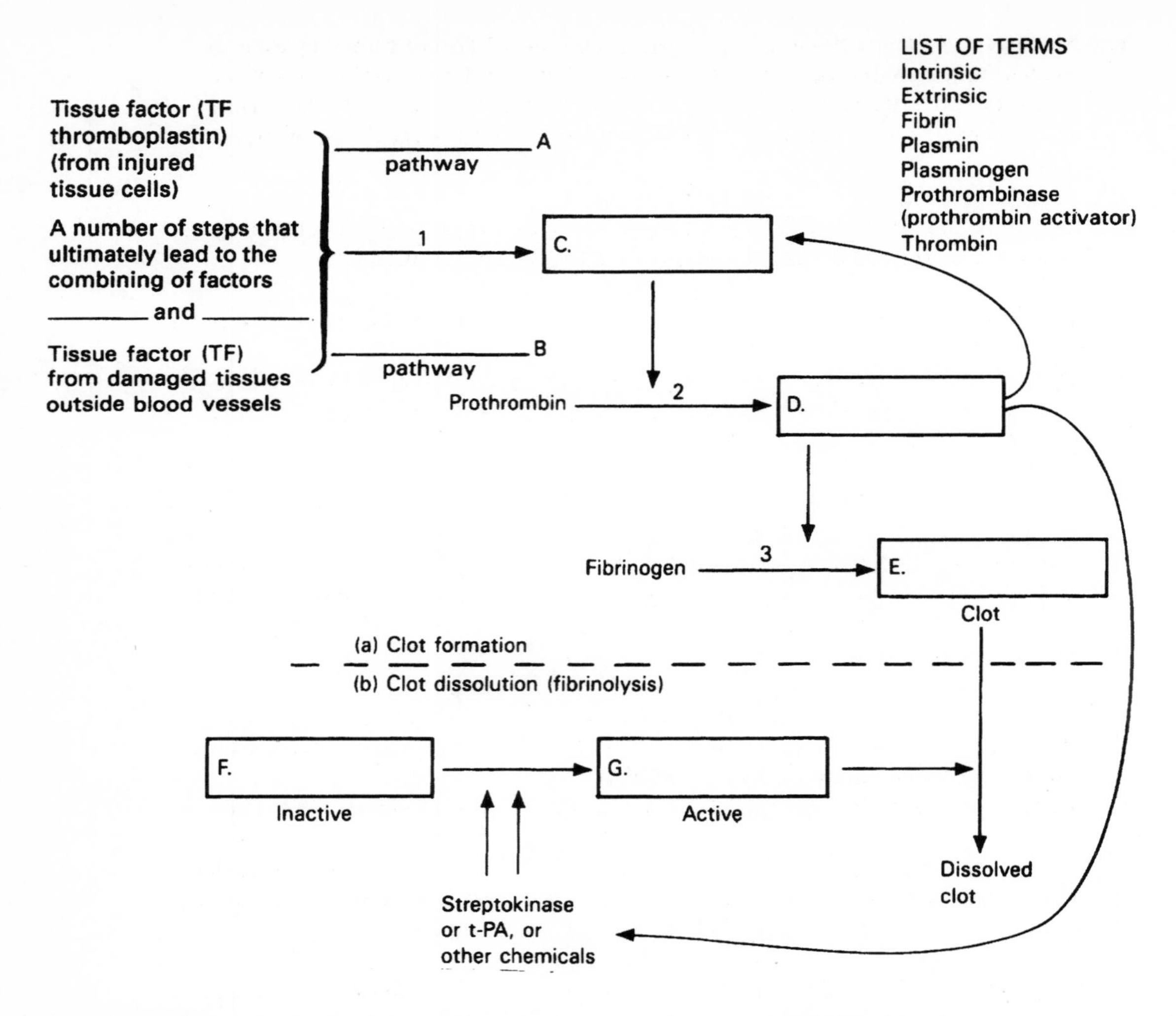

Figure LG 19.2 (a) Summary of three main steps in clot formation. Complete as directed in Checkpoints F5 and F6. (b) Summary of steps in clot dissolution (fibrinolysis). Complete as directed in Checkpoint F7.

b. Platelets then release chemicals. Some of these chemicals *(dilate? constrict?)* vessels. How does this help hemostasis?

Name two of these chemicals. ____________________, ____________________

c. State two roles of the ADP that is released by platelets.

d. The resulting platelet plug is useful in preventing blood loss in *(large? small?)* vessels.

✓ **F3.** The third step in hemostasis, coagulation, requires clotting factors derived from several sources described in this Checkpoint.

a. Most clotting factors are synthesized by *(bone marrow? kidney? liver?)* cells. (Notice this in Table 19.4 of your text.)

b. Others are released by *(red blood cells? platelets?)*.

c. One important factor, called ________________ factor (TF) or thromboplastin, is released from damaged cells. If these cells are within blood or in the lining of blood vessels, then the *(extrinsic? intrinsic?)* pathway to clotting is initiated. If other tissues are traumatized and release tissue factor, then the *(extrinsic? intrinsic?)* pathway is initiated.

✓ **F4.** Summarize the three major stages in the cascade of events in coagulation by filling in blanks A–E in Figure LG 19.2. Choose from the list of terms on the figure.

✓ **F5.** *For extra review.* of events in the process of blood clotting, refer to Figure LG 19.2 and Table 19.4 in your text and do this exercise.

a. Which protein is an insoluble protein? *(Fibrin? Fibrinogen?)* Which of these proteins is always found in blood plasma? *(Fibrin? Fibrinogen?)* What would happen if fibrin were always present in plasma?

b. Fibrinogen is not converted to fibrin unless the enzyme ________________ (D in Figure LG 19.2) is present. What would happen if thrombin were always present in blood?

c. The enzyme required to convert prothrombin to thrombin is ________________. Fill in the two short lines on Figure LG 19.2 with Roman numerals to indicate which two factors combine to form prothrombinase.

d. The figure indicates that thrombin exerts a *(negative? positive?)* feedback effect upon

prothrombinase, and thrombin also activates ________________. As a result, the clot *(enlarges? shrinks?)* in size. Which chemical finally absorbs most of the thrombin

to prevent snowballing of the clot? ________________

e. Notice on Table 19.4 in the text that clotting factors are numbered from 1 to ______;

however there is no Factor ______. Use that table to help you identify the Roman numerals of other clotting chemicals shown in Figure LG 19.2. Write those numerals on the figure.

f. The ion *(Ca^{2+}? K^{+}? Na^{+}?)*, also known as factor ______, plays a critical role in many aspects of coagulation. Write this ion on the figure next to each of the three major steps.

g. How does the cascade of events in blood clotting help to prevent excessive clot formation?

✓ **F6.** Following clot formation, clots undergo two changes. Summarize these.

a. First, clot ________________ occurs, in which the fibrin threads become *(tighter? looser?)*.

b. The lower portion of Figure LG 19.2 shows the next step: clot dissolution or

________________. An inactive enzyme (F on the figure) is activated to (G), which dissolves clots. Label F and G on the figure. F is activated by chemicals produced within the body, for example, tissue plasminogen activator (t-PA), and also by ________________ (D)

✓ **F7.** *A clinical correlation.* Match each chemical with its role in clot formation or dissolution.

H. Heparin	**S, t-PA. Streptokinase and t-PA**
K. Vitamin K	**W. Warfarin (Coumadin)**
PR. Prostacyclin	

_______ a. Required for synthesis of factors II, VII, IX, and X in clot formation. Commercial preparations may be required for newborns who lack the intestinal bacteria that synthesize this chemical and also for persons with malabsorption of fat because this is a fat soluble chemical.

_______ b. Slow-acting anticoagulant that is antagonistic to vitamin K, so decreases synthesis of prothrombin in liver; used to prevent clots in persons with history of clot formation; also the active ingredient in rat poison.

_______ c. Fast-acting anticoagulant that blocks action of thrombin; produced in the body by mast cells and basophils, but also available in pharmacologic preparations.

_______ d. Commercial preparations introduced into coronary arteries can limit clot size and prevent severe heart attacks; classified as fibrinolytic agents because they activate plasminogen (shown on Figure LG 19.2) to dissolve clot.

_______ e. A chemical made by white blood cells and cells lining blood vessels; opposes actions of thromboxane A2.

✓ **F8.** *A clinical correlation.* Match the correct term with the description.

E. Embolus	**T. Thrombus**

_______ a. A blood clot

_______ b. A "clot-on-the-run" dislodged from the site at which it formed (usually a deep vein of the leg); also fat from broken bone, bubble of air, or amniotic fluid traveling through blood, possibly to lung (pulmonary) vessels

✓ **F9.** *Critical thinking.* Address factors related to clotting in this Checkpoint.

a. Mr. B, a chronic alcoholic, needs to have a surgical procedure. Explain why he might be at risk for excessive bleeding.

b. Aspirin is used to *(promote? prevent?)* hemostasis, so it can lower risk of thrombi in cerebral, coronary, and peripheral arteries. Circle the step(s) of hemostasis that it affects. *(Vasoconstriction? Platelet plug formation? Coagulation?)*

G. Blood groups and blood types (pages 651–654)

✓ **G1.** Contrast *agglutination* and *coagulation.*

a. Coagulation (or ____________________) involves coagulation factors found in platelets, plasma, or other tissue fluids. The process *(requires? can occur in absence of?)* red blood cells.

b. Agglutination (or "clumping") of erythrocytes is an antigen-____________________ process that *(does? does not?)* require red blood cells because these are sites of ____________________ used in the agglutination reaction.

✓ **G2.** Do this exercise about factors responsible for blood groups.

a. The surface of RBCs contains chemicals that act as *(antigens? antibodies?)*, also known as iso-________________. Based on these chemicals, blood can be categorized into at least *(2? 4? 24? 100?)* different blood groups. One of these groups, the ABO group, consists of four possible blood types. Name them:

type ______, type ______, type ______, and type ______.

b. Type B blood has *(A? B?)* antigens on RBCs, and *(anti-A? anti-B)* antibodies in blood plasma. In other words, people have antibodies in plasma that *(would? would not?)* react with their own RBC antigens.

G3. Complete the first four columns of the table contrasting blood types within the ABO group.

Blood Group	Percentage of White Population	Percentage of Black Population	Sketch of Blood Showing Correct Antigens and Antibodies	Can Donate Safely to	Can Receive Blood Safely from
a. Type A					A, O
b.	11				
c. Type AB		4	A B B A A B neither anti-A B A nor anti-B		
d.				A, B, AB, O	

✓ **G4.** Complete this exercise about the *Rh system.*

a. The Rh *(+? –?)* type is more common. Rh *(+? –?)* blood has Rh antigens on the surfaces of RBCs.

b. Under normal circumstances plasma of *(Rh+ blood? Rh– blood? both Rh types? neither Rh type?)* contains anti-Rh antibodies.

c. Rh *(+? –?)* persons can develop these antibodies when they are exposed to Rh *(+? –?)* blood, for example, by transfusion.

d. An example of this occurs in fetal-maternal incompatibility when a mother who is Rh *(+? –?)* has a baby who is Rh *(+? –?)* and some of the baby's blood enters the mother's bloodstream. The mother develops anti-Rh antibodies that may cross the placenta in future pregnancies and hemolyze the RBCs of Rh *(+? –?)* babies. Such a

condition is known as ____________________.

e. *A clinical correlation.* Discuss precautions taken with Rh- mothers soon after delivery, miscarriage, or abortion of an Rh+ baby to prevent future problems with Rh incompatibility.

✓ **G5.** Complete this exercise about blood transfusions.

a. Type A blood has *(A? B? both A and B? neither A nor B?)* antigen(s) on RBCs.

b. Type A blood most likely has the *(anti-A? anti-B? both anti-A and anti-B? neither anti-A nor anti-B?)* antibodies in blood plasma. If a person with type A blood receives type B blood, the recipient's *(anti-A? anti-B?)* antibodies will attack the *(A? B?)* antigens on the donor's cells. Are the donor's anti-A antibodies likely to attack the recipient's type A cells and cause a major incompatibility reaction?

c. Type O blood has often been called the universal *(donor? recipient?)* because blood lacks _____________________ of the ABO group. Conversely, Type ____ has been called the universal recipient. Explain why these terms (universal donor and universal recipient) are misleading.

d. Explain what causes hemolysis of RBCs in incompatible blood transfusions.

G6. Now complete the last two columns in the table in Checkpoint G3.

✓ **G7.** Discuss blood typing in this Checkpoint.

a. Blood typing involves checking for the presence of *(antigens? antibodies?)* in blood.

b. A drop of serum from a bottle containing anti-A antibodies is mixed with a sample drop of blood. If the blood clumps (or __________________), then the A antigen *(is? is not?)* present and the blood does contain A antigens. The serum in the bottle used here is comparable to serum from blood that is type *(A? B?)*. Why is serum (rather than plasma) used in this procedure?

c. If similar procedures are performed using anti-B and anti-Rh antisera, and neither of those results demonstrates agglutination, what blood type is in the sample drop of blood? ______

d. If all three drops of blood agglutinate using the above procedure, what blood type is in the sample drop of blood? ______

e. Name another body fluid besides blood that can be tested for a quick and fairly accurate indication of blood type? __________________

G8. Explain how the process of cross-matching differs from blood typing.

H. Disorders, medical terminology (pages 654–655)

✓ **H1.** Match types of anemia with descriptions below.

A.	**Aplastic**	**Hr.**	**Hemorrhagic**	**P.**	**Pernicious**
Hl.	**Hemolytic**	**I.**	**Iron deficiency**	**S.**	**Sickle cell**

________ a. Condition resulting from inadequate diet such as deficiency of iron

________ b. Condition in which intrinsic factor is not produced, so absorption of vitamin B_{12} is inadequate

________ c. Inherited condition in which hemoglobin forms stiff rodlike structures causing erythrocytes to assume sickle shape and rupture, reducing oxygen supply to tissues

________ d. Rupture of red blood cell membranes due to variety of causes, such as parasites, toxins, or antibodies

________ e. Condition due to excessive bleeding, as from wounds, gastric ulcers, heavy menstrual flow

________ f. Inadequate erythropoiesis as a result of destruction or inhibition of red bone marrow

✓ **H2.** Check your understanding of sickle cell anemia in this Checkpoint.

a. Describe the normal shape of red blood cells: ____________________

b. What shape may RBCs assume in the disorder sickle-cell anemia (SCA)?

____________________ Under what conditions are the RBCs likely to do this?

____________________ What are the consequences?

c. Explain why persons with sickle cell anemia are likely to have greater resistance to malaria.

✓ **H3.** Check your understanding of hemophilia in this Checkpoint.

a. Although there are several different forms of hemophilia, all types involve deficiency of ____________________. Types A and B occur primarily in *(males? females?)*.

b. In the most common form of hemophilia, known as hemophilia *(A? B? C?)*, factor ________ is missing.

c. Write three or more signs or symptoms of hemophilia.

H4. Explain why patients with disseminated intravascular clotting (DIC) have both excessive clotting and hemorrhaging occurring simultaneously.

H5. Define and explain the clinical significance of each of these terms.

a. Jaundice

b. Thrombocytopenia

c. Septicemia

d. Cyanosis

ANSWERS TO SELECTED CHECKPOINTS: CHAPTER 19

A2. (a) Oxygen, carbon dioxide and other wastes, nutrients, hormones, enzymes, heat. (b) Respiratory, urinary, integumentary, and digestive. (c) pH, temperature, water (fluids) and dissolved chemicals (such as electrolytes). (d) Contains phagocytic cells, blood clotting proteins, and defense proteins such as antibodies, complement, and interferon.

A3. (a) 38°C (100.4°F). (b) 7.35–7.45. (c) 4–6 liters. (d) ADH, aldosterone, and ANP.

A5. (a) P LP E. (b) P E LP. (c) E. (d) LP.

A6. (a) W. (b) HE. (c) E. (d) A. (e) F. (f) G. (g) UC. (h) GAF. (i) F.

A7. (a) Ag. (b) Ab. (c) Ab.

A8. (a-b) P. (c-d) L. (e) E.

A9. (a) Red. (b) 38-46; 40-54; women lose blood during menstruation, and men produce more red blood cells due to testosterone levels. (c) Higher; 62; loss of water decreases blood plasma and leads to a relative increase in red blood cells.

B1. (a) Erythrocyte, leukocyte. (b) Hemopoiesis (or hematopoiesis); pluripotent stem. (c)

Stem cell	→ Progenitor or Precursor cell	→ Mature cell
1. Lymphoid stem cell	**T lymphoblast**	T lymphocyte
2. **Lymphoid stem cell**	B lymphoblast	**B lymphocyte**
3. **Myeloid stem cell**	CFU-GM → myeloblast	**Neutrophil**
4. **Myeloid stem cell**	CFU-GM → **monoblast**	Monocyte
5. **Myeloid stem cell**	CFU-E → proerythroblast	**Erythrocyte**
6. Myeloid stem cell	**CFU-Meg → megakaryocyte**	Platelet

(d) Red blood cell (or erythrocyte); lacks; monocyte. (e) Yolk sac, liver, spleen, thymus, lymph glands, bone marrow. (f) Red bone marrow; proximal epiphyses of femurs and humeri, flat bones of the skull, sternum, ribs, vertebrae, and hip bones. (g) B and T lymphocytes; bone marrow. (h) Kidneys, RBCs. (i) Colony-stimulating factors (CSFs) and interleukins (ILs).

B2. EPO stimulates RBC production in persons whose failing kidneys no longer produce erythropoietin. (b) GM-CSF and G-CSF stimulate white blood cell formation in persons whose bone marrow function is inadequate, such as transplant, chemotherapy, or AIDS patients. (c) TPO helps to prevent depletion of platelets during chemotherapy.

C1. A; does not; hemoglobin; see Figure 19.4 in your text.

C2. (a) Do not; absence of these organelles leaves more room for O_2 transport; RBCs have a limited life span, partly because the anuclear cell cannot synthesize new components. (b) Biconcave; greater plasma membrane surface area for better diffusion of gas molecules and flexibility for squeezing through small capillaries.

C3. (a) Globin, hemes. (b) Iron (Fe), oxygen, 4. (c) Globin, 13. (d) Nitric; when NO is released from hemoglobin, it dilates blood vessels which improves delivery of blood and oxygen to tissues.

C4. (a) Spleen, liver, or red bone marrow. (b) Amino

acids, iron. (c) Transferrin, ferritin; transferrin; transferrin. (d) Red bone marrow, heme. (e) Biliverdin, bilirubin. (f) Bile; urobilinogen; stercobilin, urobilin.

C6. (a) H. (b) E. (c) B_{12}. (d) IF.

C7. (a) 4 months. (b) 5 million. (c) 2 million. (d) 280 million. (e) 3.0. (f) 15.

Dl. (a) Lack, white blood cells or WBCs. (b) B, neutrophil (60–70); C, lymphocyte (20–25); D, monocyte (3–8); E, eosinophil (2–4); F, basophil (0.5–1.0); differential. (c) Has; monocyte (D); lymphocyte (C); neutrophil (B). (d) B E F. (e) See Figure 19.7 and Table 19.3 in your text.

D3. MHC antigens are proteins on cell surfaces; they are unique for each person. The greater the similarity between major histocompatibility (MHC) antigens of donor and recipient, the less likely is rejection of the transplant.

D4. (a) 5 million, 5,000–10,000; 700. (b) RBCs; days.

D5. (a) E. (b) A. (c) C. (d) P.

D6. (a) N. (b) N. (c) L. (d) B. (e) E. (f) M. (g) M, N. (h) L, M. (i) N. (j) N. (k) E.

D7. (a) Bacteria and their toxins. (b) Cancer or transplanted cells, viruses, fungi, and some bacteria. (c) Wide variety of.

D8. (a) In; 12,000; leukocytosis. (b) Neutrophils (PMNs or polys); 76; monocytes.

D9. (a) B. (b) D. (c) C. (d) A.

El. (a) Thrombocytes. (b) Bone marrow. (c) Chips off of megakaryocytes. (d) 150,000–400,000. (e) Blood clotting.

E2. G.

E3. RBC, WBC, and platelet counts; differential WBC count; hemoglobin and hematocrit (H&H).

F1. Blood standing still (or stopping bleeding). (a) Spasm. (b) Platelet. (C) Coagulation.

F2. (a) Irregular with extensions; adhesion. (b) Constrict; reduces blood flow into injured area; serotonin; thromboxane A2. (c) Activates platelets and makes them stickier, thereby inviting more platelets to the scene. This phenomenon is known as platelet aggregation. (d) Small.

F3. (a) Liver. (b) Platelets. (c) Tissue; intrinsic; extrinsic.

F4. A, extrinsic. B, intrinsic. C, Prothrombinase (prothrombin activator). D, Thrombin. E. Fibrin.

F5. (a) Fibrin; fibrinogen; fibrin clots would form and block vessels inappropriately. (b) Thrombin; fibrinogen would be converted to fibrin excessively, so excessive numbers of clots would form. (c) Prothrombinase; V and X. (d) Positive, platelets; enlarges; fibrin. (e) XIII; VI; I, fibrinogen; II, prothrombin; III, tissue factor (thromboplastin). (f) Ca^{2+}; IV. (g) Active clotting factors (prothrombinase, thrombin, and fibrin) are not normally formed until circumstances call for clot formation.

F6. (a) Retraction, tighter. (b) Fibrinolysis; F, plasminogen; G, plasmin; thrombin.

F7. (a) K. (b) W. (c) H. (d) S, t-PA. (e) PR.

F8. (a) T. (b) E.

F9. (a) Most coagulation factors are normally synthesized in the liver, which is severely damaged by cirrhosis, a liver disease that can be caused by excessive consumption of alcohol. (b) Prevent; all three: vasoconstriction, platelet plug formation, and coagulation.

G1. (a) Clotting; can occur in absence of. (b) Antibody, does, antigens.

G2. (a) Antigens; antigens; 24; A, B, AB, and 0. (b) B, anti-A; would not.

G4. (a) +; +, (b) Neither Rh type. (c) –, +. (d) –, +; +, hemolytic disease of the newborn (HDN) (e) The chemical RhoGAM (anti-Rh antibodies), given soon after delivery, binds to any fetal Rh antigens within the mother, destroying them. Therefore the Rh- mother will not produce anti-Rh antibodies that would attack Rh antigens of future Rh^+ fetuses.

G5. (a) A. (b) Anti-B; anti-B, B. A major reaction is not likely to occur because the donor's antibodies are rapidly diluted in the recipient's plasma, but a mild (minor) reaction may occur because of this incompatibility. (c) Donor, antigens; AB. Blood contains antigens and antibodies of blood groups other than the ABO blood group. (d) The antigen-antibody complex activates complement proteins in plasma. Complement makes RBC membranes leaky so that they burst (hemolyze).

G7. (a) Antigens. (b) Agglutinates, is; B; blood would clot if clotting factors (in plasma) were added. (c) A-negative. (d) AB-positive. (e) Saliva.

H1. (a) I. (b) P. (c) S. (d) H1. (e) Hr. (f) A.

H2. (a) Biconcave discs. (b) Sickled; upon exposure to low oxygen; sickled cells are more rigid so they may lodge in and block small vessels, (c) The gene that causes RBCs to sickle also increases permeability of RBCs to potassium, which then leaves the RBC. The malaria parasite cannot live in RBCs with such low potassium levels.

H3. (a) Some clotting factor; males. (b) A, VIII. (c) Hemorrhaging, including nosebleeds; blood in urine (hematuria); and joint damage and pain.

MORE CRITICAL THINKING: CHAPTER 19

1. Describe mechanisms for assuring homeostasis of red blood cell count.
2. Describe the recycling of red blood cells. Be sure to use all of these terms in your essay: hemoglobin, iron, transferrin, ferritin, bilirubin, bile, urobilinogen, urobilin, and stercobilin.
3. Describe roles of the following types of cells in inflammation (including phagocytosis) and immune responses: neutrophils, monocytes, and lymphocytes.
4. Contrast advantages and disadvantages of blood clotting in the human body. Then describe control mechanisms that provide checks and balances so that clotting does not get out of hand.
5. Explain why athletes might be likely to train in a high-altitude city such as Denver for several weeks immediately before competing in that city.
6. Contrast effects of blood type incidence as you address these questions: (a) Is hemolytic disease of the newborn (HDN) more likely to occur among Asians and Native Americans or among white persons? Explain why. (b) Which cultural group has the greatest incidence of type O blood, and the lowest incidence of the other three types?

MASTERY TEST: CHAPTER 19

Questions 1–3: Arrange the answers in correct sequence.

_____ _____ _____ 1. Events in the coagulation process:
A. Retraction or tightening of fibrin clot
B. Fibrinolysis or clot dissolution by plasma
C. Clot formation

_____ _____ _____ 2. Stages in the clotting process:
A. Formation of prothrombin activator (prothrombinase)
B. Conversion of prothrombin to thrombin
C. Conversion of fibrinogen to fibrin

_____ _____ _____ 3. Arrange from greatest to least in number per mm^3
A. WBCs
B. RBCs
C. Platelets

Questions 4–17: Circle the letter preceding the one best answer to each question.

4. All of the following types of formed elements develop entirely within bone marrow *except:*
   A. Neutrophils
   B. Basophils
   C. Platelets
   D. Erythrocytes
   E. Lymphocytes
5. Which chemicals are forms in which bilirubin is excreted in urine or feces?
   A. Ferritin and bilirubin
   B. Biliverdin and transferrin
   C. Urobilinogen and stercobilin
   D. Interleukins-5 and -7
6. Megakaryocytes are involved in formation of:
   A. Red blood cells
   B. Basophils
   C. Lymphocytes
   D. Platelets
   E. Neutrophils
7. Choose the *false* statement about Norine, who has type O blood.
   A. Norine has neither A nor B antigens on her red blood cells.
   B. She has both anti-A and anti-B antibodies in her plasma.
   C. She is called the universal donor.
   D. She can receive blood safely from both type O and type AB persons.
8. Choose the *false* statement about blood.
   A. Blood is thicker than water.
   B. It normally has a pH of 7.0.
   C. The human body normally contains about 4 to 6 liters of it.
   D. It normally consists of more plasma than cells.
9. Choose the *false* statement about factors related to erythropoiesis.
   A. Erythropoietin stimulates RBC formation.
   B. Oxygen deficiency in tissues serves as a stimulus for erythropoiesis.
   C. Intrinsic factor, which is necessary for RBC formation, is produced in the kidneys.
   D. A reticulocyte count of over 1.5% of circulating RBCs indicates that erythropoiesis is occurring rapidly.

10. Choose the *false* statement about plasma.
    A. It is red in color.
    B. It is composed mainly of water.
    C. Its concentration of protein is greater than that of interstitial fluid.
    D. It contains plasma proteins, primarily albumins.
11. Choose the *false* statement about neutrophils.
    A. They are actively phagocytic.
    B. They are the most abundant type of leukocyte.
    C. Neutrophil count decreases during most infections.
    D. An increase in their number would be a form of leukocytosis.
12. All of the following correctly match parts of blood with principal functions *except:*
    A. RBC: carry oxygen and CO_2
    B. Plasma: carries nutrients, wastes, hormones, enzymes
    C. WBCs: defense
    D. Platelets: determine blood type
13. Which of the following chemicals is an enzyme that converts fibrinogen to fibrin?
    A. Heparin
    B. Thrombin
    C. Prothrombin
    D. Coagulation factor VI
    E. Tissue factor
14. A type of anemia that is due to a deficiency of intrinsic factor production, caused, for example, by alterations in the lining of the stomach, is:
    A. Aplastic
    B. Sickle cell
    C. Hemolytic
    D. Pernicious
15. A type of anemia in which hemoglobin molecules assume rod shapes that alter the shape of RBCs.
    A. Iron-deficiency
    B. Sickle cell
    C. Hemorrhagic
    D. Pernicious
16. A person with a hematocrit of 66 and hemoglobin of 22 is most likely to have the condition named:
    A. Anemia
    B. Polycythemia
    C. Infectious mononucleosis
    D. Leukemia
17. A hematocrit value is normally about ________ the value of the hemoglobin.
    A. The same
    B. One third
    C. Three times
    D. 700 times

Questions 18–20: Circle T (true) or F (false). If the statement is false, change the underlined word or phrase so that the statement is correct.

T F 18. Plasmin is an enzyme that facilitates clot formation.

T F 19. Hemolytic disease of the newborn is most likely to occur with an Rh positive mother and her Rh negative babies.

T F 20. In both black and white populations in the U.S., type O is most common and type AB is least common of the ABO blood groups.

Questions 21–25: Fill-ins. Answer questions or complete sentences with the word or phrase that best fits.

____________________ 21. Name five types of substances transported by blood.

____________________ 22. Name three types of plasma proteins.

____________________ 23. Name the type of leukocyte that is commonly known as a "PMN" or a "poly."

____________________ 24. Write a value for a normal leukocyte count: ________/mm^3.

____________________ 25. All blood cells and platelets are derived from ancestor cells called ________.

ANSWERS TO MASTERY TEST: CHAPTER 19

Arrange

1. C A B
2. A B C
3. B C A

Multiple Choice

4. E
5. C
6. D
7. D
8. B
9. C
10. A
11. C
12. D
13. B
14. D
15. B
16. B
17. C

True–False

18. F. Dissolution or fibrinolysis
19. F. Negative mother and her Rh positive babies
20. T

Fill-ins

21. Oxygen, carbon dioxide, nutrients, wastes, sweat, hormones, enzymes
22. Albumins, globulins, and fibrinogens
23. Neutrophil
24. 5,000–10,000
25. Pluripotent stem cells

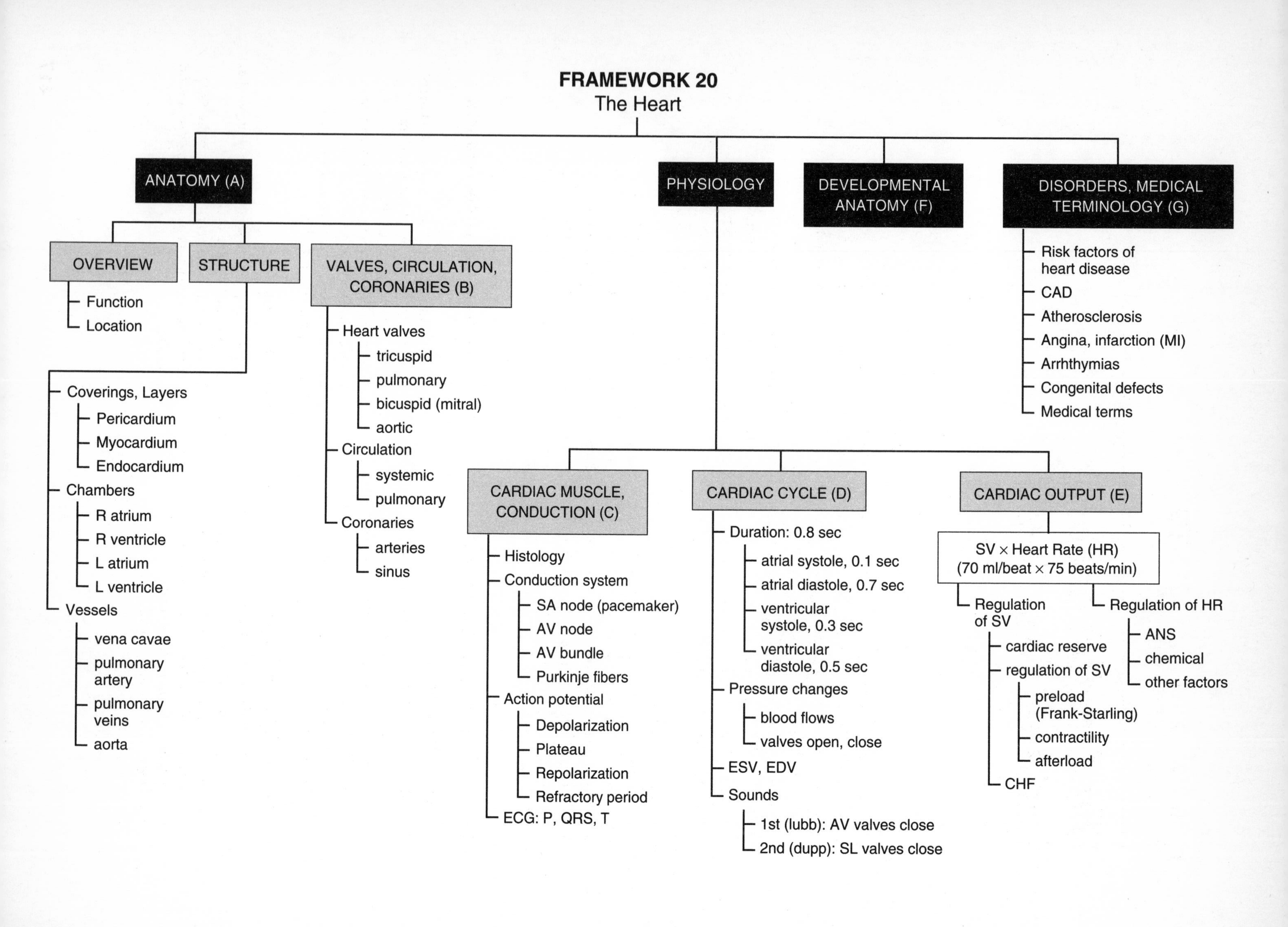

FRAMEWORK 20
The Heart
ANATOMY (A)
PHYSIOLOGY
DEVELOPMENTAL ANATOMY (F)
DISORDERS, MEDICAL TERMINOLOGY (G)
OVERVIEW
STRUCTURE
VALVES, CIRCULATION, CORONARIES (B)
Function
Location
Coverings, Layers
Pericardium
Myocardium
Endocardium
Chambers
R atrium
R ventricle
L atrium
L ventricle
Vessels
vena cavae
pulmonary artery
pulmonary veins
aorta
Heart valves
tricuspid
pulmonary
bicuspid (mitral)
aortic
Circulation
systemic
pulmonary
Coronaries
arteries
sinus
CARDIAC MUSCLE, CONDUCTION (C)
Histology
Conduction system
SA node (pacemaker)
AV node
AV bundle
Purkinje fibers
Action potential
Depolarization
Plateau
Repolarization
Refractory period
ECG: P, QRS, T
CARDIAC CYCLE (D)
Duration: 0.8 sec
atrial systole, 0.1 sec
atrial diastole, 0.7 sec
ventricular systole, 0.3 sec
ventricular diastole, 0.5 sec
Pressure changes
blood flows
valves open, close
ESV, EDV
Sounds
1st (lubb): AV valves close
2nd (dupp): SL valves close
CARDIAC OUTPUT (E)
SV × Heart Rate (HR)
(70 ml/beat × 75 beats/min)
Regulation of SV
cardiac reserve
regulation of SV
preload (Frank-Starling)
contractility
afterload
CHF
Regulation of HR
ANS
chemical
other factors
Risk factors of heart disease
CAD
Atherosclerosis
Angina, infarction (MI)
Arrhythymias
Congenital defects
Medical terms

CHAPTER 20

The Cardiovascular System: The Heart

Make a fist, then open and squeeze tightly again. Repeat this about once a second as you read the rest of this Overview. Envision your heart as a muscular pump about the size of your fist. Unfailingly, your heart exerts pressure on your blood, moving it onward through the vessels to reach all body parts. The heart has one job; it is simply a pump. But this function is critical. Without the force of the heart, blood would come to a standstill, and tissues would be deprived of fluids, nutrients, and other vital chemicals. To serve as an effective pump, the heart requires a rich blood supply to maintain healthy muscular walls, a specialized nerve conduction system to synchronize actions of the heart, and intact valves to direct blood flow correctly. Heart sounds and EGG recordings as well as a variety of more complex diagnostic tools provide clues to the status of the heart.

Is your fist tired yet? The heart ordinarily pumps 24 hours a day without complaint and seldom reminds us of its presence. As you complete this chapter on the heart, keep in mind the value and indispensability of this organ. Start by studying the Chapter 20 Framework and refer to it frequently; note relationships among key terms in each section. As you begin your study of the heart, also carefully examine the Chapter 20 Topic Outline and Objectives; check off each one as you complete it.

TOPIC OUTLINE AND OBJECTIVES

A. Anatomy of the heart

1. Describe the location of the heart, and trace its outline on the surface of the chest.
2. Describe the structure of the pericardium and the heart wall.
3. Discuss the external and internal anatomy of the chambers of the heart.

B. Heart valves and circulation

1. Describe the structure and function of the valves of the heart.
2. Describe the flow of blood through the chambers of the heart and through the systemic and pulmonary circulations.
3. Discuss the coronary circulation.

C. Cardiac muscle and the cardiac conduction system

1. Describe the structural and functional characteristics of cardiac muscle tissue and the conduction system of the heart.
2. Describe how an action potential occurs in cardiac contractile fibers.
3. Describe the electrical events of a normal electrocardiogram (ECG).

D. The cardiac cycle

1. Describe the pressure and volume changes that occur during a cardiac cycle.
2. Relate the timing of heart sounds to the ECG waves and pressure changes during systole and diastole.

E. Cardiac output; exercise and the heart

1. Define cardiac output, and describe the factors that affect it.
2. Explain the relationship between exercise and the heart.

F. Developmental anatomy

1. Describe the development of the heart.

G. Disorders and medical terminology

WORDBYTES

Now become familiar with the language of this chapter by studying each wordbyte, its meaning, and an example of its use within a term. After you study the entire list, self-check your understanding by writing the meaning of each wordbyte on the line. As you continue through the *Learning Guide*, identify (and fill in) additional terms that contain the same wordbyte.

Wordbyte	Self-check	Meaning	Example(s)
angio-	______________________	blood vessel	*angio*gram
ausculta-	______________________	listening	*auscult*ation
cardi-	______________________	heart	*cardi*ologist
coron-	______________________	crown	*coron*ary arteries
-cuspid	______________________	point	tri*cuspid*
ectop-	______________________	displaced	*ectop*ic pacemaker
endo-	______________________	within	*endo*carditis
intercalat-	______________________	inserted between	*intercalat*ed discs
-lunar	______________________	moon	semi*lunar*
myo-	______________________	muscle	*myo*cardial infarction
peri-	______________________	around	*peri*cardium
pulmo-	______________________	lung	*pulmo*nary artery
-sclero-	______________________	hard	arterio*sclero*sis
tri-	______________________	three	*tri*cuspid
vascul-	______________________	small vessel	cardio*vascul*ar

CHECKPOINTS

A. Anatomy of the heart (pages 660–666)

✓ **A1.** Do this exercise about heart functions.

a. What is the primary function of the heart? ____________________

b. Visualize a coffee cup sitting next to a gallon-size milk container. Keep in mind that a gallon holds about ______ quarts or approximately 4 liters. Now take a moment to imagine a number of containers lined up and holding the amount of blood pumped out by the heart.

1. With each heartbeat, one side of the heart pumps a stroke volume equal to about 1/3 cup (70 mL) because 1 cup = about ______ mL.

2. Each minute, one side of the heart pumps enough blood to fill up 1.3 gallons

   (_______ liters).
3. Within the 1440 minutes in a day, a volume of about 1800 gallons (_______ liters) is pumped out of one side of the heart (or a total of twice that amount by both sides of the heart).

c. Reflecting on the work your heart is doing for you, what message would you give to your heart?

✓ **A2.** Closely examine Figure 20.1b in your text. Consider the location, size, and shape of your heart as you do this exercise. Trace its outline on your body.

a. Your heart lies in the ____________________ portion of your thorax, between your two ____________________. About *(one-third? one-half? two-thirds?)* of the mass of your heart lies to the left of the midline of your body.

b. Your heart is about the size and shape of your ____________________.

c. The pointed part of your heart, called the ____________________, coincides with the *(superior? inferior?)* left point, and is located in the ____________________ intercostal space, about _______ cm (_______ inches) from the midline of your body.

d. The inferior surface of the heart lies on the ____________________. The right and left borders of the heart face the ____________________.

e. The anterior surface of the heart faces the ____________________ and the ____________________. Explain how the location of the heart accounts for the effectiveness of cardiopulmonary resuscitation (CPR).

✓ **A3.** Complete this Checkpoint on the structure of the heart wall and its coverings. Use answers in the box.

E.	**Endocardium**	**PC.**	**Pericardial cavity (site of pericardial fluid)**
FP.	**Fibrous pericardium**	**PP.**	**Parietal pericardium**
M.	**Myocardium**	**VP.**	**Visceral pericardium (epicardium)**

a. Arrange in order from most superficial to deepest.

   ________ ________ ________ ________ ________ ________

b. Which layer of the heart is made of muscle that forms the bulk of the weight of the heart? ________

c. In the conditions known as pericarditis and cardiac tamponade, where does fluid accumulate, causing compression of the heart? ________

d. Which layer lines the inside of the heart and forms much of the tissue of valves?

____

For extra review of the pericardium, look again at Checkpoint D2 in Chapter 4, page 78 in the *Learning Guide.*

✓ **A4.** Refer to Figures LG 20.1 and 20.2. Identify all structures with letters A–I by writing labels on lines A–I on Figure LG 20.1. Add leader lines and label structures A–D and G–I on Figure LG 20.2.

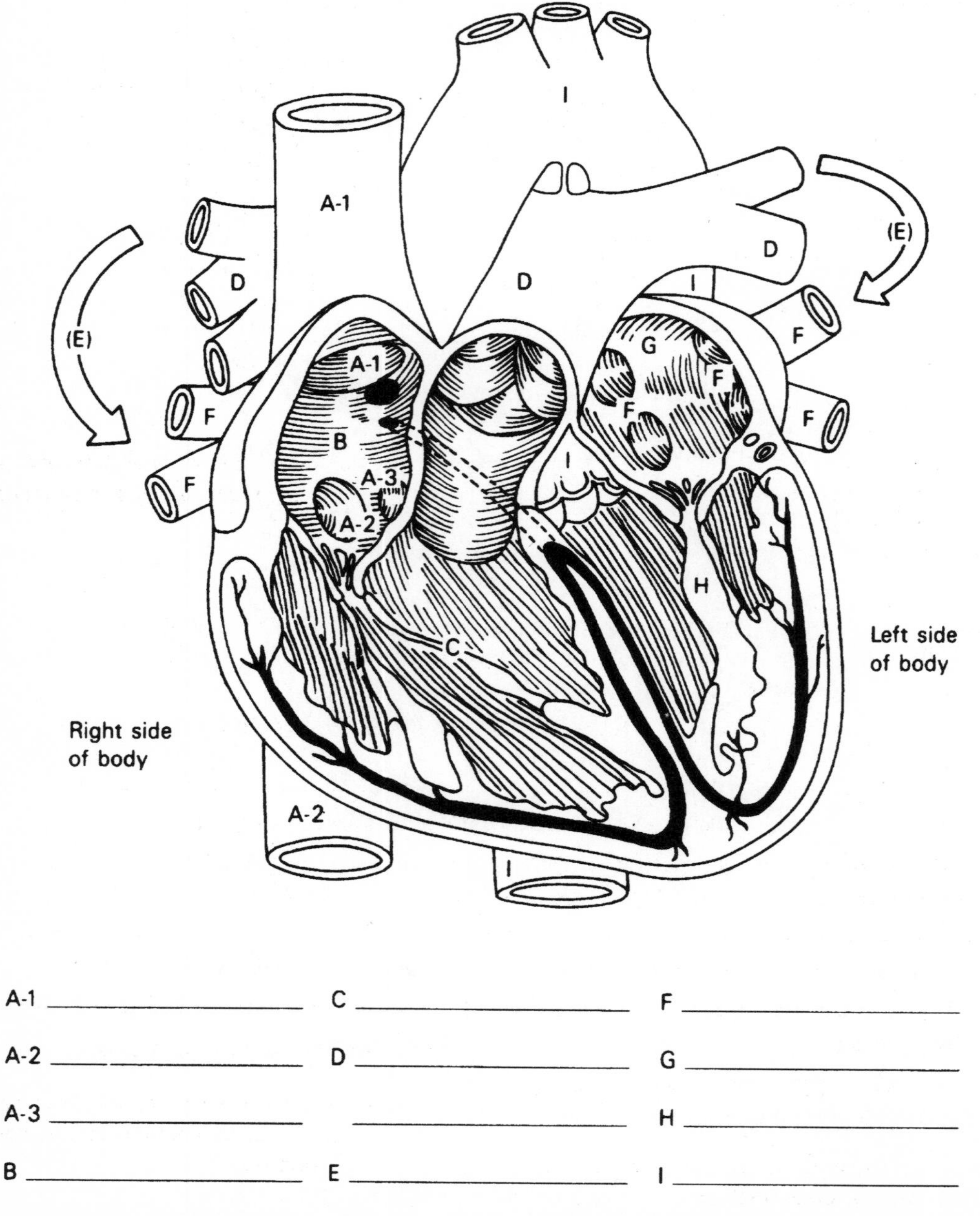

Figure LG 20.1 Diagram of a frontal section of the heart. Letters follow the path of blood through the heart. Label, color, and draw arrows as directed in Checkpoints A4, B2, B5, and C2.

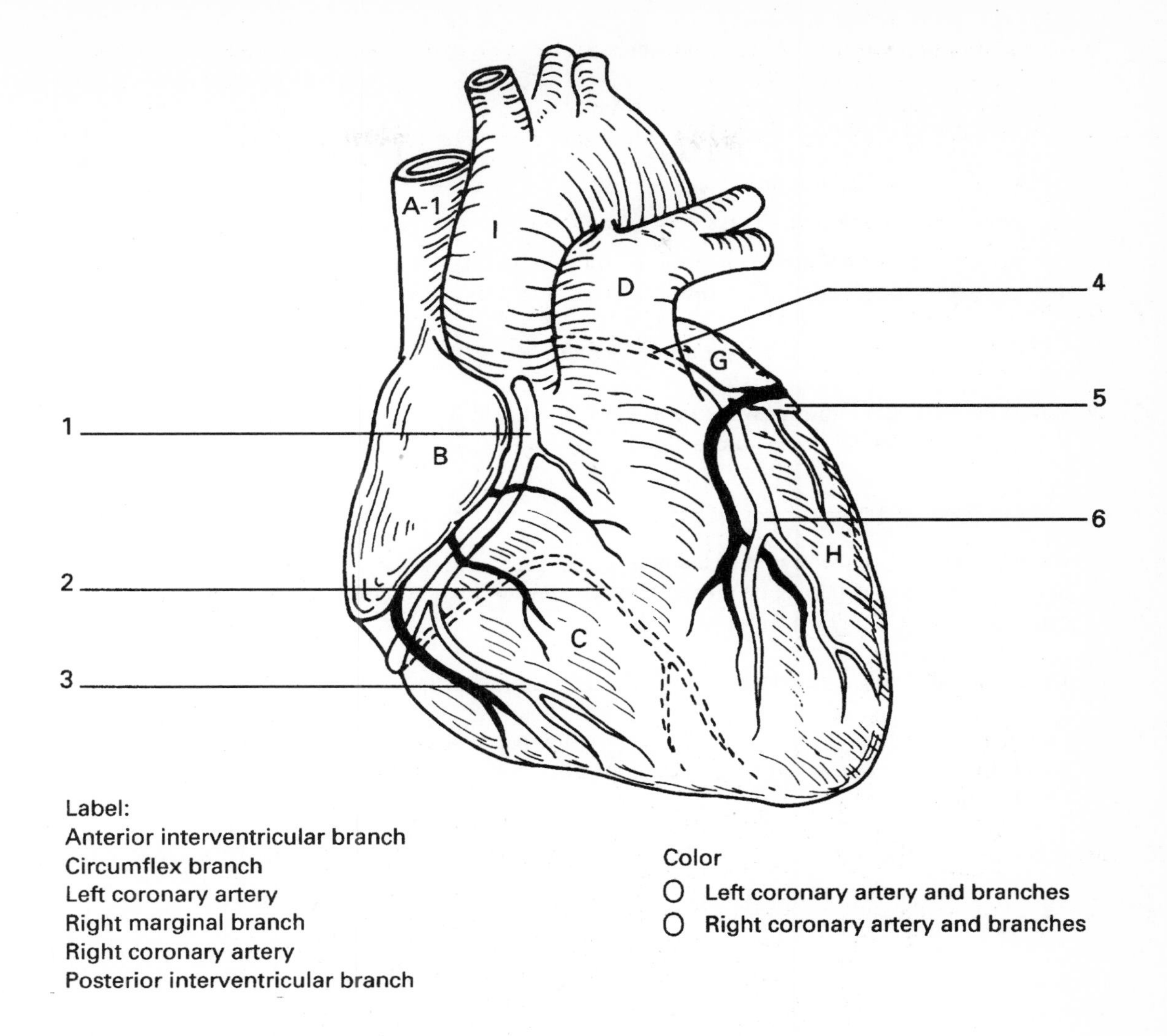

Figure LG 20.2 Anterior of the heart. Label and color as directed in Checkpoints A4, A6, B5, and B9. Cardiac veins are solid black.

✓ **A5.** Check your understanding of heart structure by selecting the terms that fit descriptions below. Not all answers will be used.

A.	**Atria**	**IVS.**	**Interventricular septum**	**PV.**	**Pulmonary vein**
CS.	**Coronary sulcus**	**PA.**	**Pulmonary artery**	**TC.**	**Trabeculae carnae**
CT.	**Chordae tendineae**	**PapM.**	**Papillary muscles**	**V.**	**Ventricles**
IAS.	**Interatrial septum**	**PecM.**	**Pectinate muscles**		

_________ a. Thin-walled chambers that receive blood from veins

_________ b. Blood vessel that carries blood that is rich in oxygen from lungs to left atrium

_________ c. Site of the fossa ovalis, the location of the foramen ovale in fetal life

_________ d. Strong wall separating the two ventricles

_________ e. Sites of most myocardium

_________ f. A groove separating left from right side of the heart; contains fat and coronary blood vessels

_______ g. Irregular ridges and folds of myocardium covered with smooth endocardium

_______ h. Strong tendons that anchor atrioventricular (AV) valves to ventricular muscle preventing eversion of the AV valves

_______ i. Muscles that give a roughened appearance to parts of the atrial walls

_______ j. Nipple-shaped muscles in ventricles that are attached to chordae tendineae

✓ **A6.** On Figure LG 20.2 label the ligamentum arteriosum. In fetal life this is the site of the ductus *(arterious? venosus?)*, which connects the pulmonary *(artery? vein?)* to the

_________________, thereby permitting blood to bypass the lungs.

✓ **A7.** Which ventricle has a thicker wall? *(Right? Left?)* ventricle. The difference is related to the fact that the left ventricle pumps blood to

____ come

whereas the right ventricle pumps blood to _________________. In which chamber of the heart would you imagine greater pressure develops when that chamber contracts? *(Right? Left?)* ventricle.

✓ **A8.** The fibrous skeleton of the heart consists of _________________ tissue that

supports and anchors _________________. Action potentials in atria *(can? cannot?)* be transmitted across this fibrous skeleton. (More about the significance of this fact in Checkpoint C2d).

B. Heart valves and circulation (pages 667–671)

✓ **B1.** The human heart has *(2? 4? 6?)* valves. The primary function of heart valves is to

_________________________________. Check your understanding of heart valves by writing answers from the box next to related descriptions below.

A. Aortic	**P Pulmonary**
B. Bicuspid	**T. Tricuspid**

_______ a. Also called the mitral valve

_______ b. Prevents backflow of blood from right ventricle to right atrium

_______ c. Prevents backflow from pulmonary trunk to right ventricle

_______ d. Prevents backflow of blood into left atrium

_______ e. Have half-moon-shaped leaflets or cusps (two answers)

_______ f. Also called atrioventricular (AV) valves (two answers)

_______ g. Open when ventricular pressure surpasses pressure within the pulmonary artery and aorta (two answers)

✓ **B2.** On Figure LG 20.1, label the four valves that control blood flow through the heart.

✓ **B3.** Contrast valve disorders in this exercise.

a. Stenosis is a failure of a heart valve to *(open? close?)* properly, whereas insufficiency or incompetence is a failure of a heart valve to *(open? close?)* properly.

b. *(Stenosis? Insufficiency or incompetence?)* is a narrowing of a valve by scar formation following an infection or by a congenital defect. Mitral valve prolapse (MVP) is a mild form of valvular *(stenosis? insufficiency or incompetence?)*.

c. In the valve disorder aortic insufficiency, blood flows backwards into the left *(atrium? ventricle)*.

✓ **B4.** *A clinical correlation.* Name the microorganism that causes rheumatic fever (RF).

___________________. What condition is likely to signal the presence of this microbe

and warn of possible effects (sequelae) upon the heart? ___________________.
Which parts of the heart are most likely to be affected?

Critical thinking. as you look at heart physiology, speculate about why these two valves might be more susceptible. *Hint:* Refer to checkpoint A7 above.

✓ **B5.** Refer to Figures LG 20.1 and 20.2. and check your understanding of the pathway of blood through the heart in this Checkpoint.

a. Draw arrows on Figure LG 20.1 to indicate direction of blood flow.
b. Color red the chambers of the heart and vessels that contain highly oxygenated blood; color blue the regions in which blood is low in oxygen and high in carbon dioxide.

✓ **B6.** Refer to Figure LG 20.3 and complete this Checkpoint on blood vessels.

a. Label vessels 1–7. Note that numbers are arranged in the sequence of the pathway of blood flow. Use these labels:

Aorta and other large arteries	**Small artery**
Arteriole	**Small vein**
Capillary	**Venule**
Inferior vena cava	

b. Color vessels according to color code ovals.

B7. On Figure LG 20.3, label the *pulmonary artery, pulmonary capillaries,* and a *pulmonary vein*. All blood vessels in the body other than pulmonary vessels are known as

___________________ vessels because they supply a variety of body systems.

B8. Defend or dispute this statement: "The myocardium receives all of the oxygen and nutrients it needs from blood that is passing through its four chambers."

✓ **B9.** Refer to Figure LG 20.2 and complete this checkpoint.

a. Using the color code ovals on the figure, color the two coronary arteries and their main branches.
b. Label each of the vessels, using the list of labels on the figure.
c. Write another name for the coronary branch known as the "LAD."_______________
d. The left coronary artery (and its branches) supplies blood to all of the chambers of

the heart *except* the ___________________.

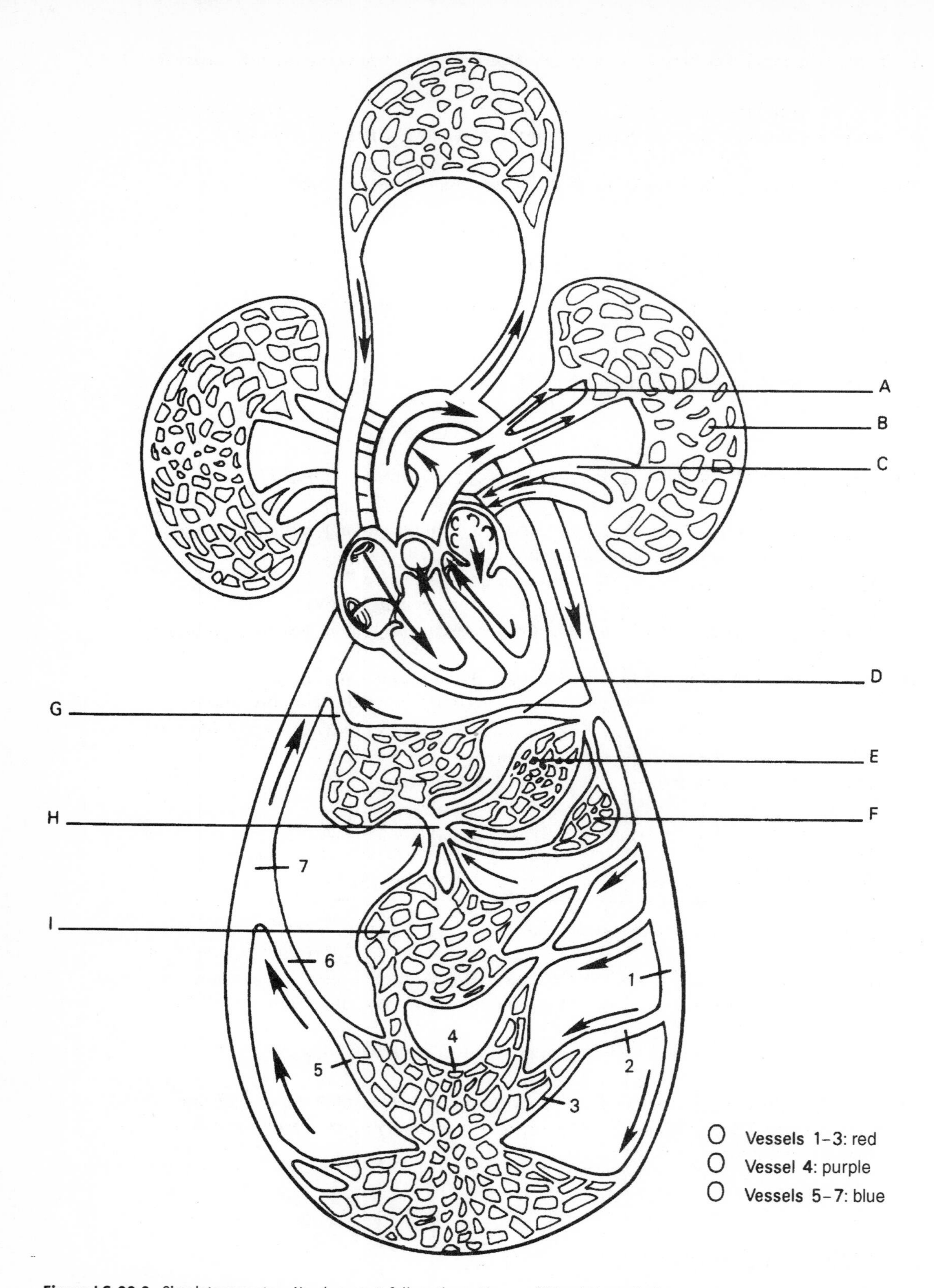

Figure LG 20.3 Circulatory routes. Numbers 1–7 follow the pathway of blood through the systemic blood vessels. Color and label numbered vessels according to directions in Checkpoint B6. Letters A–I refer to Checkpoints B6–B7, and Chapter 21 Checkpoint F1, page LG 466.

✓ **B10.** The coronary sinus functions as *(an artery? a vein?)*. It collects blood that has passed

through coronary arteries and capillaries into ___________________ veins. The coronary sinus finally empties this blood into the *(right? left?) (atrium? ventricle?)*.

✓ **B11.** Define *anastomosis* and explain how this term is relevant to the heart.

B12. Mr. Lasita had a heart attack and now has *reperfusion damage*. Explain.

C. Cardiac muscle and the cardiac conduction system (pages 672–678)

✓ **C1.** Contrast cardiac muscle with skeletal muscle in this exercise.

a. Review these muscle types in the table in Chapter 10, Checkpoint E3, page LG 192.

b. Circle C (cardiac) or S (skeletal) to indicate characteristics of these muscle types.
   1. Contains more mitochondria: C S
   2. Have sarcomeres with actin and myosin arranged in bands: C S
   3. Contains desmosomes and gap junctions that facilitates spread of action potentials to adjacent muscle fibers: C S

✓ **C2.** In this exercise describe how the heart beats regularly and continuously.

a. In embryonic life about _______% of cardiac muscle fibers become *autorhythmic*. State two functions of these specialized cells of the heart.

b. Label the parts of the conduction system on Figure LG 20.1

c. The normal pacemaker of the heart is the *(SA? AV?)* node. Each time the SA node "fires," impulses travel via the conduction system and also by the

___________________ junctions in ___________________ discs of cardiac muscle.

d. What structural feature of the heart makes the AV node and AV bundle necessary for conduction from atria to ventricles?

e. Through which part of the conduction system do impulses pass most slowly? *(SA node? AV node? AV bundle [of His]?)*. Based on the anatomy of this tissue, why does the rate of impulse conduction slow down?

Of what advantage is this slowing? (See text page 436.)

✓ **C3.** Do this exercise about the pacemaker(s) of the heart.

a. On its own, the SA node, located in the *(left? right?)* atrium, normally fires at about ______ times per minute. This rate is *(a little faster than? a little slower than? about the same as?)* a typical resting heart rate. At rest, *(acetylcholine? epinephrine?)* released from *(sympathetic? parasympathetic?)* nerves normally slows the pace of the SA node to modify the heart rate to about ______ beats per minute.

b. How might this "normal pacemaker" be damaged? ________________
If this occurs, then responsibility for setting the pace of the heart may be passed on to the ______ node, which fires at ______ to ______ times per minute.

c. If both the SA and AV nodes fail, then autorhythmic fibers in ________________ may take over with a rate of only ______ to ______ beats per minute. Pacemakers at "other than the normal site" are known as ________________ pacemakers and tend to be *(more? less?)* effective than the SA node in pacing the heart.

✓ **C4.** Describe the physiology of cardiac muscle contraction in this activity.

a. Contractile fibers of the normal heart have a resting membrane potential that is close *(–70? –90?)* mV.

b. Arrange in correct sequence the events in an action potential of cardiac muscle. The blank parentheses are for Activity C4c. One is done for you.

C (RD) → ______ (______) → ______ **or** ______ (______) → ______ (______ → ______ (______)

A. *Slow* Ca^{2+} channels open so Ca^{2+} enters muscle fibers. **P**
B. Combined flow of Na^{+} and Ca^{2+} maintains depolarization for about 250 msec which is about 250X longer than depolarization in skeletal muscle.
C. Na^{+} enters through *fast Na^{+} channels*; voltage rises rapidly to about +20 mV.______
D. A second contraction cannot be triggered during this period, which is longer than for skeletal muscle. ______
E. The presence of Ca^{2+} binding to troponin permits myocardial contraction via sliding of actin filaments next to myosin filaments. ________________
F. K^{+} channels open so K^{+} ions leave the fiber; meanwhile fewer Ca^{2+} ions enter as those channels are closing; voltage returns to resting level. ______

c. Now label events A–F above by filling in parentheses. One is done for you. (For help refer to Figure 20.11 in your text.) Use these answers:

P.	**Plateau**	**Rf.**	**Refractory**
RD.	**Rapid depolarization**	**Rp.**	**Repolarization**

d. State the significance of the prolonged depolarization (250 msec) of cardiac muscle compared to the brief depolarization (1 msec) of skeletal muscle.

e. Explain the significance of the long refractory period of cardiac muscle.

f. *A clinical correlation.* Calcium channel blockers [such as verapamil (Procardia)] tend to *(in? de?)*-crease contraction of the cardiac muscle (and of smooth muscle of coronary arteries). Write one or more condition(s) for which such a medication might be prescribed.

Now name one chemical that enhances flow of Ca^{2+} through these channels to

strengthen contraction of the heart. ____________________

✓ **C5.** Discuss energy supply to cardiac muscle in this exercise.

a. Cardiac muscle depends on *(aerobic? anaerobic?)* respiration which produces *(large? small?)* amounts of ATP.

b. List two structural features of cardiac muscle cells that normally supply this ATP.

c. At rest, *(creatine phosphate? fatty acids? glucose? lacitic acid?)* provide most of the fuel to cardiac muscle. During exercise, *(creatine phosphate? fatty acids? glucose? lactic acid?)* provides more of the fuel.

✓ **C6.** Describe an ECG (or EKG) in this activity.

a. What do the letters ECG (or EKG) stand for?

An ECG is a recording of *(electrical changes associated with impulse conduction? muscle contractions?)* of the heart.

b. On what parts of the body are leads (electrodes) placed for a 12-lead ECG?

c. In Figure LG 20.4c, an ECG tracing using lead *(1? II? III?)* is shown. Label the following parts of that ECG on the figure: *P wave, P-R interval, QRS wave (complex), S-T segment, T wave.*

d. *A clinical correlation.* Answer the following questions with names of parts of an ECG.

1. Indicates atrial depolarization leading to atrial contraction ____________________
2. Prolonged if the AV node is damaged, for example, by rheumatic fever, or in the

   condition described as "bundle (of His) block" ____________________

3. Elevated in acute MI ____________________
4. Elevated in hyperkalemia (high blood level of K^+) ____________________

5. Flatter than normal in coronary artery disease ____________________

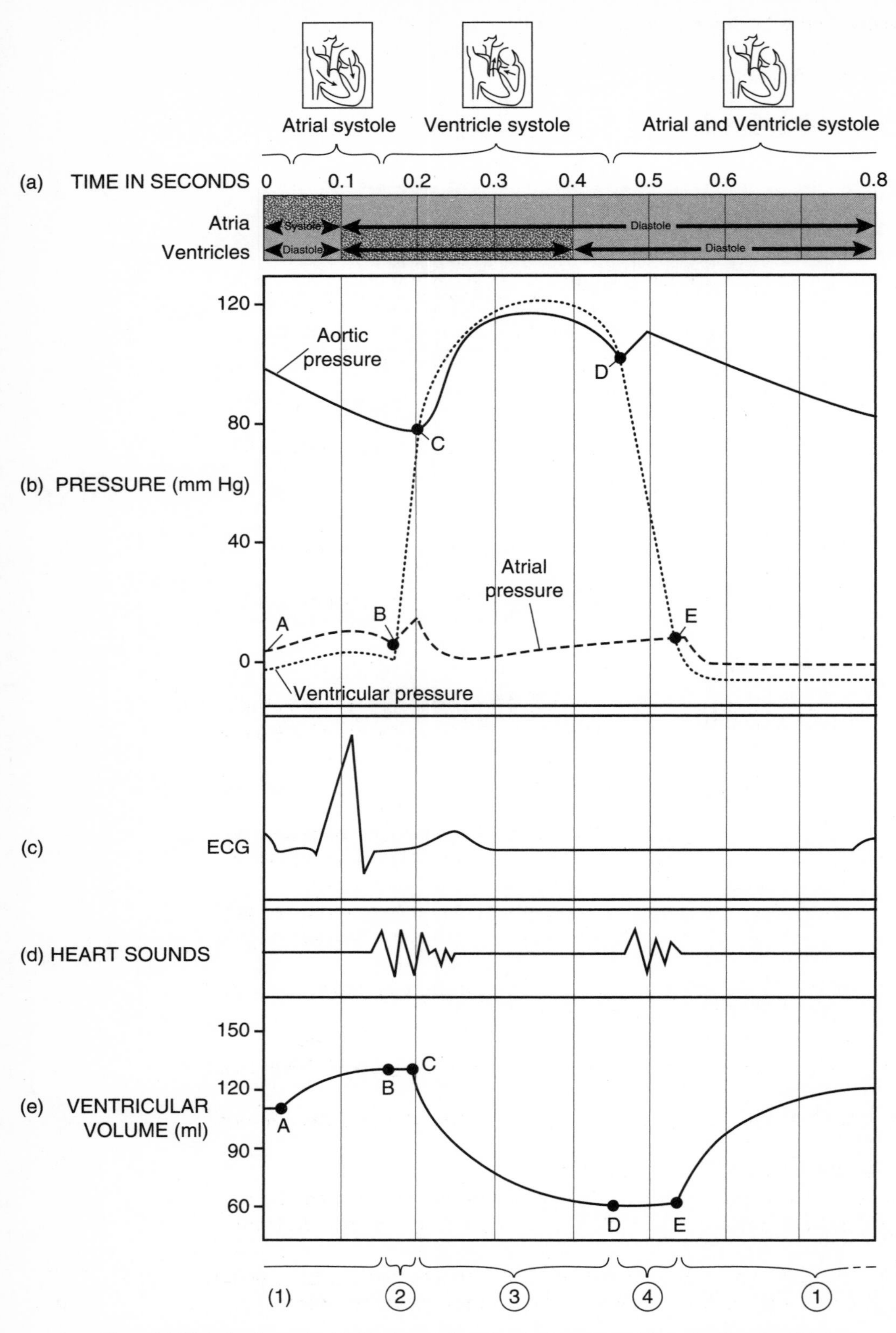

Figure LG 20.4 Cardiac cycle. (a) Systole and diastole of atria and ventricles related to time. (b) Pressures in aorta (intraarterial), atria, and ventricles. Points labeled *A* to *E* refer to Checkpoint D2 and D3. (c) ECG related to cardiac cycle. Label as directed in Checkpoint C6. (d) Heart sounds related to cardiac cycle. Label according to Checkpoint D4. (e) Volume of blood in ventricles. refer to Checkpoints D2j–l4 and D6.

6. Represents ventricular repolarization ____________________

7. Represents ventricular depolarization ____________________

8. Represents atrial repolarization ____________________

D. Cardiac cycle (pages 678–682)

✓ **D1.** Complete the following overview of movement of blood through the heart.

a. Blood moves through the heart as a result of ____________________ and ____________________ of cardiac muscle, as well as ____________________ and ____________________ of valves. Valves open or close due to ____________________ within the heart.

b. Contraction of heart muscle is known as ____________________ whereas relaxation of myocardium is called ____________________.

✓ **D2.** Now refer to Figure LG 20.4 and consider details of the cardiac cycle in this Checkpoint.

a. The duration of one average cycle is _____ sec; *(54? 60? 75?)* complete cardiac cycles (or heart beats) occur per minute (if pulse rate is 75).

b. As the P wave of the ECG occurs, nerve impulses spread across the ____________________. During this initial *(0.1? 0.3?)* sec of the cycle, the atria begin their contraction. As a result, pressure within the atria *(increases? decreases?)*, as shown at point A in Figure LG 20.4b.

c. The ____________________ wave then signals contraction (or ____________________) of the ventricles. Note that although the QRS wave itself happens over a very brief period, the ventricular contraction that follows occurs over a period of *(0.3? 0.5? 0.7?)* sec.

d. Trace a pencil lightly along the curve that shows changes in pressure within the ventricles following the QRS wave. Notice that pressure there *(increases? decreases?)* *(slightly? dramatically?)*. In fact it quickly surpasses atrial pressure (at point ____ in the figure). As a result, AV valves are forced *(open? closed?)*.

e. A brief time later ventricular pressure becomes so great that it even surpasses pressure in the great arteries (pulmonary artery and aorta). This pressure forces blood against the undersurface of the ____________________ valves, *(opening? closing?)* them. This occurs at point _______ on the figure.

f. Continued ventricular systole ejects blood from the heart into the great vessels. In the aorta this typically creates a systolic blood pressure of about *(15? 80? 120?)* mm Hg.

g. What causes ventricular pressure to begin to drop? The cessation of _________-polarization (and contraction) followed by repolarization of these chambers (after the _______ wave) causes the ventricles to go into *(systole? diastole?)*.

h. When the pressure within ventricles drops just below that in the great arteries, blood in these vessels fills the ________________ valves and closes them. This occurs at point ______ in the figure just as ventricular pressure drops precipitously. Note that pressure in the aorta rises briefly causing the ________________ wave after which aortic pressure gradually declines. State the reason for this wave in the aortic pressure curve.

i. When intraventricular pressure becomes lower than that in atria, the force of blood within the atria causes the AV valves to *(open? close?)*. This happens at point ______ in the figure.

j. Note that the AV valves now remains open all the way to point *B* in the next cycle. This permits adequate filling of ventricles before their next contraction. Trace a pencil lightly along the curve on Figure LG 20.4e starting at point *E*. Note that the curve is steepest *(just after E? just before A of next cycle?)*. In other words, the first part of the curve from *E* to *A* of the next cycle (ventricular diastole) is the time when ventricles are filling *(rapidly? slowly?)*. In fact, about ______ percent of all blood that will enter ventricles does so during the first third of ventricular diastole (period *E–A*).

k. The remaining ______ percent enters ventricles between points *A* and *B*, that is, as a result of atrial *(systole? diastole?)*.

l. Summarize changes in ventricular volume by labeling regions 1–4 below Figure LG 20.4e. In part 1, the ventricles are relaxed and they are in a state of *(filling? ejection?)*. In part 2, ventricles are contracting but the semilunar valves have not yet opened. So blood cannot leave the ventricles yet. Thus this is the brief period named isovolumetric *(contraction? relaxation?)*. Once the semilunar valves open, the ventricular volume ______-creases. This period (3) is ventricular ________________. In the short period 4, the ventricles have stopped contracting so pressure is rapidly dropping, but AV valves have not yet opened. So ventricles cannot fill yet. This is the period of ________________. Note that *iso-* means *same*.

✓ **D3.** *For extra review.* Label points *B, C, D,* and *E* on Figure LG 20.4b to indicate which valves open or close at those points.

✓ **D4.** Do this exercise on heart sounds.

a. The first heart sound *(lubb? dupp?)* is produced by turbulence of blood at the *(opening? closing?)* of the ________________ valves. What causes the second sound?

b. Write *first* and *second* next to the parts of Figure LG 20.4d showing each of these sounds.

c. *A clinical correlation.* Using Figure 20.15 in your text, recognize why a stethoscope is placed at several points on the chest in order to best hear different heart sounds. If a stethoscope is placed at the level of about the fifth intercostal space (between ribs 5 and 6) on the left side of the sternum, the *(first? second?)* sound is better heard because turbulence from the *(mitral? aortic semilunar?)* valve is readily detected there. To best hear the second sound (related to SL valve closure), the stethoscope must then be moved in a more *(inferior? superior?)* direction on the chest.

D5. Explain what causes heart murmurs.

✓ **D6.** Complete this exercise about blood volumes during the cardiac cycle.

a. The volume remaining in one ventricle at the end of diastole is known as end-

_______________ volume (EDV). Write EDV next to point C on Figure LG 20.4e. The volume of blood in EDV is about ________ mL.

b. At the end of systole, about ________ mL remains in each ventricle. This volume is known as end- _______________ volume (ESV). ESV is shown at point ________ on Figure LG 20.4e.

c. The amount of blood ejected from each ventricle during systole is known as _______________ volume (SV) and equals about ________ mL. Show the calculation of stroke volume here:

End-diastolic volume (EDV) – End-systolic volume (ESV) = Stroke volume (SV)

________ mL – ________ mL = ________ mL

✓ **D7.** *Critical thinking.* Consider effects of a significant change in heart rate.

a. Recall that a heart beating at about 75 beats/mm has a cardiac cycle of about 0.8 sec duration: ventricles contract for ________ sec and they relax and fill with blood for ________ sec.

b. When the heart rate doubles, for example, to 150 beats/mm, the duration of the cardiac cycle *(doubles? is halved?)*, for example, to 0.4 sec. Which part of the cycle is especially affected? Ventricular *(systole? diastole?)* What are the consequences of this change?

E. Cardiac output; exercise and the heart (pages 683–688)

✓ **E1.** Determine the average cardiac output in a resting adult.

Cardiac output = stroke volume × heart rate

= _______________ mL/stroke × _______________ strokes/min

= _______________ mL/min (_______________ liter/min)

✓ **E2.** At rest Dave has a cardiac output of 5 liters per minute. During a strenuous cross-country run, Dave's maximal cardiac output is 20 liters per minute. Calculate Dave's cardiac reserve.

$$\text{Cardiac reserve} = \frac{\text{maximal cardiac output}}{\text{cardiac output at rest}} =$$

Note that cardiac reserve is usually *(higher? lower?)* in trained athletes than in sedentary persons.

✓ **E3.** The two major factors (shown in Activity El) that control cardiac output are

_____________________ and _____________________.

✓ **E4.** Do this activity about stroke volume (SV).

a. Which statement is true about stroke volume (SV)? (*Hint:* Refer to Checkpoint D6.)

A. SV = ESV – EDV B. SV = EDV – ESV C. $SV = \frac{ESV}{EDV}$ D. $SV = \frac{EDV}{ESV}$

b. List the three factors that control stroke volume (SV):

_____________________ _____________________ _____________________

c. During exercise, skeletal muscles surrounding blood vessels squeeze *(more? less?)* blood back to the heart. This increase in venous return to the heart *(increases? decreases?)* end-diastolic volume (EDV). As a result, muscles of the heart are stretched *(more? less?)* so that preload (= stretching of the heart muscle) *(increases? decreases?)*.

d. Within limits, a stretched muscle contracts with *(greater? less?)* force than a muscle that is only slightly stretched. This is a statement of the _____________________ _____________________ law of the heart. To get an idea of this, blow up a balloon slightly and then let it go. Then blow up a balloon quite full of air (much like the heart when stretched by a large venous return) and let that balloon go. In which stage do the walls of the balloon (heart) compress the air (blood) with greater force?

e. As a result of the Frank-Starling law, during exercise the ventricles of the normal heart contract *(more? less?)* forcefully, and stroke volume *(increases? decreases?)*.

f. State several reasons why venous return might be decreased, leading to reduction in stroke volume and cardiac output.

g. Elevated blood pressure or narrowing of blood vessels tends to _______-crease afterload and therefore to _______-crease stroke volume and cardiac output. Identify two causes of increased afterload.

✓ **E5.** *The Big Picture: Looking Back.* Return to text references given below and answer questions related to cardiac function.

a. In Table 4.4, page 130 cardiac muscle cells appear *(parallel? branching?)*, which is a characteristic *(similar to? different from?)* skeletal muscle. What is the significance of this pattern of cardiac muscle tissue?

b. Refer to Table 17.4, page 580. Sympathetic nerves _____-crease heart rate and force of atrial and ventricular contraction and also *(dilate? constrict?)* coronary vessels.

Therefore sympathetic nerves tend to _________-crease stroke volume and cardiac output.

c. Figure 17.3, page 569, shows that the vagus nerve is a *(sympathetic? parasympathetic?)* nerve. Its effects upon heart muscle, coronary vessels, and cardiac output are *(similar? opposite?)* to those of sympathetic nerves.

d. Table 18.8, page 613, states that epinephrine and norepinephrine have *(sympathomimetic? parasympathomimetic?)* effects. In other words, these hormones, produced by the adrenal *(cortex? medulla?)* have effects similar to those of the sympathetic division of the autonomic nervous system: they prepare the body for a

____________________ response.

✓ **E6.** Summarize factors affecting heart rate (HR), stroke volume (SV), and cardiac output (CO) by completing arrows: ↑ (for increase) or ↓ (for decrease). The first one is done for you.

a. ↑ exercise → ↑ preload
b. Moderate ↑ in preload → | SV
c. ↑ SV (if heart rate is constant) → | CO
d. ↑ in heart contractility → | SV and CO
e. Positively inotropic medication such as digoxin (Lanoxin) → | SV and CO
f. ↑ afterload (for example, due to hypertension) → | SV and CO
g. Hypothermia → | heart rate and CO
h. ↑ heart rate (if SV is constant) → | CO
i. ↑ stimulation of the cardioaccelerator center → | heart rate
j. ↑ sympathetic nerve impulses → | heart strength (and SV) and | heart rate
k. ↑ vagal nerve impulses → | heart rate
l. Hyperthyroidism → | heart rate
m. ↑ Ca^{2+} → | heart strength (and SV) and | heart rate
n. ↓ SV (as in CHF) → | ESV → excessive EDV and preload → | SV and CO

✓ **E7.** *A clinical correlation.* Do this exercise about Ms. Schmidt, who has congestive heart failure (CHF).

a. Her weakened heart becomes overstretched much like a balloon (except much thicker!) that has been expanded 5000 times. Now her heart myofibers are stretched beyond the optimum length according to the Frank-Starling law. As a result, the

force of her heart is ______-creased, and its stroke volume ______-creases (as shown in Activity E6n).

b. Ms. Schmidt has right-sided heart failure, so her blood is likely to back up, distending vessels in *(lungs? systemic regions, such as in neck and ankles?)*; thus

____________________ edema results, with signs such as swollen hands and feet.

c. Write a sign or symptom of left-sided heart failure.

d. An intra-aortic balloon pump (IABP) is especially helpful for *(right? left?)*-sided CHF

because the IABP ______-creases afterload and therefore can ______-crease stroke volume. (See text Table 20.1, page 445.)

e. Describe the procedure of cardiomyoplasty used to assist the heart's pumping capacity.

✓ **E8.** Circle answers that indicate benefits of physical conditioning, such as three to five aerobic workouts each week.

a. ______-crease in maximal cardiac output

b. ______-crease in capillary networks in skeletal muscles

c. ______-crease in resting heart rate

F. Developmental anatomy (pages 688–689)

✓ **F1.** The heart is derived from ____________________-derm. The heart begins to develop during the *(third? fifth? seventh?)* week. Its initial formation consists of two endothelial tubes which unite to form the ____________________ tube.

✓ **F2.** Match the regions of the primitive heart below with the related parts of a mature heart listed below.

A.	**Atria**	**SV.**	**Sinus venosus**
BCTA.	**Bulbus cordis and truncus arteriosus**	**V.**	**Ventricle**

________ a. Superior and inferior vena cava

________ b. Right and left ventricles

________ c. Parts of the fetal heart connected by foramen ovale

________ d. Aorta and pulmonary trunk

G. Disorders and medical terminology (pages 689–692)

G1. List eight risk factors for heart disease. Circle five of those that can be modified by a healthy lifestyle.

a. ______________________

b. ______________________

c. ______________________

d. ______________________

e. ______________________

f. ______________________

g. ______________________

h. ______________________

✓ **G2.** Outline the sequence of changes in atherosclerosis in this learning activity.

a. Atherosclerosis involves damage to the walls of arteries. State several factors that may lead to damage of this lining.

b. The resulting fatty lesion in the arterial lining is known as an atherosclerotic ________________. The rough surface of plaque may snag platelets. These may then cause ________________ formation at the site, possibly leading to unwanted emboli.

c. Chemicals released in the plaque cause thickening of smooth muscle and collagen in the blood vessel wall. What are the ultimate effects of these changes?

✓ **G3.** Match diagnostic techniques and treatments in the box with descriptions below.

C.	**Catheterization**	**CABG.**	**Coronary artery bypass graft**
CA.	**Cardiac angiography**	**PTCA.**	**Percutaneous transluminal coronary angioplasty**

_______ a. Invasive procedure in which a long, slender tube is inserted into an artery or vein under x-ray observation

_______ b. Catheterization with use of contrast dye for visualization of coronary blood vessels or heart chambers

_______ c. Nonsurgical technique to increase blood supply to the heart as arterial plaque is compressed

_______ d. Surgical procedure utilizing a vessel from another part of the body (such as a vein from the thigh) to reroute blood around a blocked portion of a coronary artery

G4. Define *ischemia* and explain how it is related to *angina pectoris.*

✓ **G5.** *A clinical correlation.* Do this exercise about "heart attacks."

a. "Heart attack" is a common name for a myocardial ____________________ (MI).

What does the term *infarction* mean? ____
In what other areas of the body might infarctions also occur? Cerebral infarction:

____________________; pulmonary infarction: ____________________.

b. List two or more immediate causes of an MI.

c. List three treatments for an MI.

✓ **G6.** Do this exercise on arrhythmias.

a. *(All? Not all?)* arrhythmias are serious.

b. Conduction failure across the AV node results in an arrhythmia known as

____________________.

c. Which is more serious? *(Atrial? Ventricular?)* fibrillation. Explain why.

d. Excitation of a part of the heart other than the normal pacemaker (SA node) is known as a(n) __________________ focus. Such excitations may be caused by ingestion of caffeine or by lack of sleep or by more serious problems such as ischemia.

✓ **G7.** Match each congenital heart disease in the box with the related description below.

C.	**Coarctation of aorta.**	**IVSD.**	**Interventricular septal defect**
IASD.	**Interatrial septal defect**	**PDA.**	**Patent ductus arteriosus**

________ a. Connection between aorta and pulmonary artery is retained after birth, allowing backflow of blood to right ventricle.

________ b. Foramen ovale fails to close.

________ c. Septum between ventricles does not develop properly.

________ d. Aorta is abnormally narrowed.

✓ **G8.** Match disorders in the box with descriptions below.

CHF.	**Congestive heart failure**	**PA.**	**Palpitation**
CM.	**Cardiomegaly**	**PT.**	**Paroxysmal tachycardia**
CP.	**Cor pulmonale**		

________ a. Enlarged heart

________ b. Right ventricular hypertrophy resulting from lung disorders that create resistance to blood flow through pulmonary vessels

________ c. Fluttering of the heart

________ d. Sudden period of rapid heart rate

________ e. Inability of the heart to function as an effective pump; blood backs up leading to pulmonary or peripheral edema

ANSWERS TO SELECTED CHECKPOINTS: CHAPTER 20

Al. (a) It is a pump. (b) 4 (one gallon = 4 quarts or 3.86 liters). (b1) 237. (b2) 5 (1.3 gallons = 5 liters). (b3) 7200 (1.3 gallons/mm × 1440 mm/day = 1800 gallons/day = over 7000 liters). (c) Perhaps: "Thanks!" "Great job!" "Amazing!" or "You deserve the best of care, and I intend to see that you get it."

A2. (a) Mediastinal, lungs; two-thirds. (b) Fist. (c) Apex, inferior, fifth, 9, 3.5. (d) Diaphragm; right and left lungs. (e) Sternum and ribs (or rib cartilages); compression of the heart between these bony structures anteriorly and the vertebrae posteriorly forces blood out of the heart.

A3. (a) FP PP PC VP M E. (b) M. (c) PC. (d) E.

A4. See Figure LG 20.1A

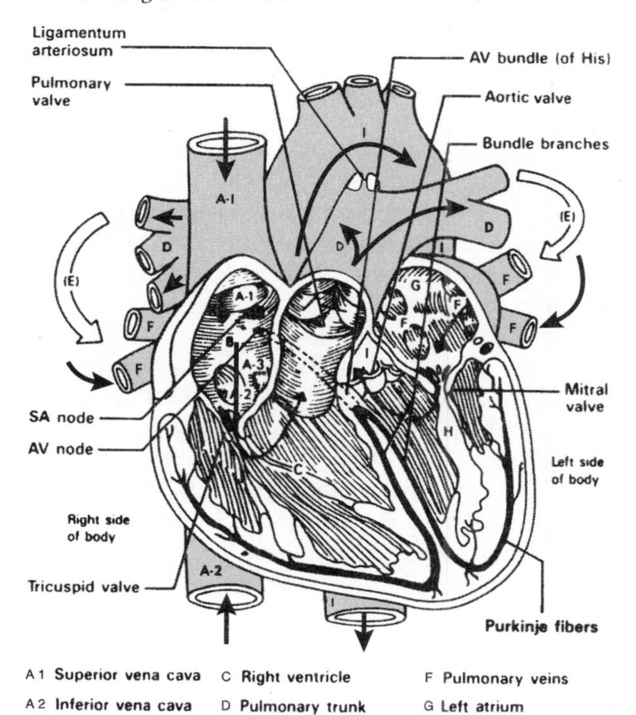

Figure LG 20.1A Diagram of a frontal section of the heart.

A5. (a) A. (b) PV. (c) IAS. (d) IVS. (e) V. (f) CS. (g) TC. (h) CT. (i) PecM. (j) PapM.

A6. See Figure 20.4a in the text; arteriosus, artery, aorta.

A7. Left; all parts of the body except to lungs, only the lungs; left (usually develops pressure 4–5 times greater than that in the right ventricle).

A8. Dense fibrous connective, valves of the heart; cannot.

B1. 4; ensure one-way flow of blood (or prevent back flow [or regurgitation] of blood into the chamber from which blood had just come). (a) B. (b) T. (c) P. (d) B. (e) A, P. (f) B, T. (g) P, A.

B2. See Figure LG 20.1a.

B3. Open, close. (b) Stenosis; insufficiency or incompetence. (c) Ventricle.

B4. Streptococcus; streptococcal sore throat ("strep throat"); valves, particularly the bicuspid (mitral) and the aortic semilunar. These two valves control blood flow through the left side of the heart, which is a higher-pressure system than the right side; therefore, valve damage is more likely, making these valves more vulnerable to effects of RF.

B5. (a) See Figure LG 20.1A. (b) Blue: A–D and first part of E; red: last part of E and F–I.

B6. (a) 1, aorta; 2, small artery; 3, arteriole; 4, capillary; 5, venule; 6, small vein; 7, inferior vena cava. (b) See Figure 21.17 (page 718 of the text.)

B9. (a) Left coronary artery: 4–6; right coronary artery: 1–3. (b) 1, right coronary artery; 2, posterior interventricular branch; 3, marginal branch; 4, left coronary artery; 5, circumflex branch; 6, anterior interventricular branch. (c) Left anterior descending, anterior descending, or anterior interventricular. (d) Right atrium.

B10. Vein; cardiac; right atrium.

B11. Blood supply to a tissue from two or more arteries; coronary artery branches provide many sources of blood to heart to help assure adequate perfusion.

C1. (a) See LG page 196. (b1) C. (b2) C S. (b3) C.

C2. (a) One; pacemaker and impulse conduction. (b) See Figure LG 20.1A. (c) SA; gap, intercalated. (d) The fibrous skeleton (including AV valves) that completely separates atria from ventricles. (e) AV node; fibers have smaller diameter and fewer gap junctions; this 1.0 sec delay affords time for ventricles to fill more completely before they contract.

C3. (a) Right, 100; a little faster than; acetylcholine, parasympathetic, 75. (b) Ischemia as in myocardial infarction (MI); AV, 40, 60. (c) Purkinje fibers, 20, 35; ectopic, less.

C4. (a) – 90. (b, c) C (RD) → A (P) → B or E (both P) → F (Rp) → D (Rf). (d) Myocardium can maintain contractions long enough to pump blood out of the heart. (e) Depolarization (causing contraction) cannot begin until after the refractory period (relaxation, providing time for the heart to fill with blood) is well underway. If the heart contracted all the time, it would never get to fill with blood, so no blood would be in the heart to be pumped out during contractions. (f) De; coronary artery disease (CAD), high blood pressure (hypertension), rapid heart rate (tachycardia); epinephrine.

C5. (a) Aerobic, large. (b) Coronary arteries supply oxygen which is used by the large number of mitochondria in cardiac muscle tissue. (c) Fatty acids; lactic acid.

C6. (a) An electrocardiogram (the recording) or electrocardiograph (the instrument); electrical changes associated with impulse conduction. (b) Arms, legs, and chest. (c) II; refer to Figures 20.12 and 20.14 of the text. (d1) P wave; (d2) P–R interval; (d3) S–T segment; (d4–d6) T wave; (d7) QRS complex; (d8) No wave because masked by QRS complex.

Dl. (a) Contraction, relaxation, opening, closing; pressure changes. (b) Systole, diastole.

D2. (a) 0.8, 75. (b) Atria, 0.1 increases. (c) QRS, systole, 0.3. (d) Increases, dramatically, *B*, closed. (e) Semilunar, opening, *C*. (f) 120. (g) De- T, diastole. (h) Semilunar, *D*, dicrotic; rebound of blood off the closed cups of the aortic (semilunar) valve. (i) Open, *E*. (j) Just after E, rapidly, 75. (k) 25, systole. (l) Filling, contraction, de, ejection, isovolumetric relaxation.

D3. B, AV valves close; C, SL valves open; D, SL valves close; E, AV valves open.

D4. (a) Lubb, closing, AV; turbulence due to closing of SL valves. (b) Refer to Figure 20.15 in the text. (c) First, mitral; superior.

D6. (a) Diastolic; 130. (b) 60; systolic; D or E. (c) Stroke, 70; 130 – 60 = 70.

D7. (a) 0.3, 0.5. (b) Is halved, diastole; because the heart cannot adequately fill with blood during a very brief ventricular diastole, only a very small volume of blood is ejected from the heart during each cardiac cycle.

El. CO = 70 ml/stroke × 75 strokes (beats)/min = 5250 ml/mm = 5.25 liters/min.

E2. Four; 20 liters per min/5 liters per mm = 4; higher.

E3. Stroke volume (SV), heart rate (HR).

E4. (a) B. (b) Preload, contractility, afterload. (c) More; increases; more, increases. (d) Greater; Frank-Starling; the more expanded balloon (heart) exerts greater force on air (blood). (e) More, increases. (f) Lack of exercise, loss of blood (hemorrhage), heart attack (myocardial infarction), or rapid heart rate (shorter diastole: see Checkpoint D7). (g) In, de; vasoconstriction of blood vessels, atherosclerosis.

E5. (a) Branching, different from; allows for more effective spreading of action potentials and contraction of the heart muscle as a unit, rather than as discrete muscle fibers. (b) In, dilate; in. (c) Parasympathetic; opposite. (d) Sympathomimetic; medulla, fight-or-flight.

E6. (a) ↑ exercise → ↑ preload. (b) Moderate ↑ in preload → ↑ SV. (c) ↑ SV (if heart rate is constant) → ↑ CO. (d) ↑ in heart contractility → ↑ SV and CO. (e) Positively inotropic medication such as digoxin (Lanoxin) → ↑ SV and CO. (f) ↑ afterload (for example, due to hypertension) → ↓ SV and CO. (g) Hypothermia → ↓ heart rate and CO. (h) ↑ heart rate (if SV is constant) → ↑ CO. (i) ↑ stimulation of the cardioaccelerator center → ↑ heart rate. (j) ↑ sympathetic nerve impulses → ↑ heart strength (and SV) and ↑ heart rate. (k) ↑ vagal nerve impulses → ↓ heart rate. (1) Hyperthyroidism → ↑ heart rate. (m) ↑ Ca^{2+} → ↑ heart strength (and SV) and ↑ heart rate. (n) ↓ SV (as is CHF) → ↑ ESV → excessive EDV and preload → ↓ SV and CO.

E7. (a) De, de. (b) Systemic regions, such as in neck and ankles; peripheral or systemic. (c) Difficulty breathing or shortness of breath (dyspnea), which is a sign of pulmonary edema. (d) Left, de, in. (e) See text page 445.

E8. (a) In. (b) In. (c) De.

F1. Meso; third; primitive heart.

F2. (a) SV. (b) V. (c) A. (d) BCTA.

G2. (a) Cytomegalovirus, carbon monoxide from smoking, diabetes mellitus, prolonged hypertension, and fatty diet. (b) Plaque; clot (thrombus). (c) Narrowing of the arterial lumen, reducing blood flow and increasing risk of thrombus formation.

G3. (a) C. (b) CA. (c) PTCA. (d) CABG.

G5. Infarction; death of tissue because of lack of blood flow to that tissue; brain, lung. (b) Thrombus (clot), embolus (mobile clot), spasm of coronary artery. (c) Thrombolytic agents (such as t-PA), anticoagulants such as heparin, or surgery such as CABG or coronary angioplasty.

G6. (a) Not all. (b) Heart block. (c) Ventricular; atrial fibrillation reduces effectiveness of the atria by only about 20–30%, but ventricular fibrillation causes the heart to fail as a pump. (d) Ectopic.

G7. (a) PDA. (b) IASD. (c) IVSD. (d) C.

G8. (a) CM. (b) CP. (c) PA. (d) PT. (e) CHF.

MORE CRITICAL THINKING: CHAPTER 20

1. Contrast structure and location of the AV valves with the SL valves.
2. Explain how and why interference in conduction through the AV node, the AV bundle, and bundle branches would alter the ECG and heart function.
3. Describe the structural differences that account for the functional differences between cardiac and skeletal muscle.
4. Relate opening and closing of heart valves to systole and diastole of atria and ventricles.
5. Contrast pulmonary edema with peripheral edema. Define each condition and relate them to different types of congestive heart failure (CHF). Also state two or more prominent signs or symptoms of each.
6. Describe components of a wellness lifestyle that are likely to enhance heart health.

MASTERY TEST: CHAPTER 20

Questions 1–3: Arrange the answers in correct sequence.

______ ______ ______ ______ 1. Pathway of the conduction system of the heart:
- A. AV node
- B. AV bundle and bundle branches
- C. SA node
- D. Purkinje fibers

______ ______ ______ ______ ______ 2. Route of a red blood cell supplying oxygen to myocardium of left atrium and returning to right atrium:
- A. Arteriole, capillary, and venule within myocardium
- B. Branch of great cardiac vein
- C. Coronary sinus leading to right atrium
- D. Left coronary artery
- E. Circumflex artery

______ ______ ______ ______ ______ 3. Route of a red blood cell now in the right atrium:
- A. Left atrium
- B. Left ventricle
- C. Right ventricle
- D. Pulmonary artery
- E. Pulmonary vein

Questions 4–15: Circle the letter preceding the one best answer to each question.

4. All of the following are correctly matched *except:*
   A. Myocardium—heart muscle
   B. Visceral pericardium—epicardium
   C. Endocardium—forms heart valves which are especially affected by rheumatic fever
   D. Pericardial cavity—space between fibrous pericardium and parietal layer of serous pericardium
5. Which of the following factors will tend to decrease heart rate?
   A. Stimulation by cardiac nerves
   B. Release of the transmitter substance norepinephrine in the heart
   C. Activation of neurons in the cardioaccelerator center
   D. Increase of vagal nerve impulses
   E. Increase of sympathetic nerve impulses
6. The average cardiac output for a resting adult is about ______ per minute.
   A. 1 quart
   B. 5 pint
   C. 1.25 gallons (5 liters)
   D. 0.5 liter
   E. 2000 mL
7. When ventricular pressure exceeds atrial pressure, what event occurs first?
   A. AV valves open
   B. AV valves close
   C. Semilunar valves open
   D. Semilunar valves close
8. The second heart sound is due to turbulence of blood flow as a result of what event?
   A. AV valves opening
   B. AV valves closing
   C. Semilunar valves opening
   D. Semilunar valves closing
9. Choose the *false* statement about the circumflex artery.
   A. It is a branch of the left coronary artery.
   B. It provides the major blood supply to the right ventricle.
   C. It lies in a groove between the left atrium and left ventricle.
   D. Damage to this vessel would leave the left atrium with virtually no blood supply.
10. Choose the *false* statement about heart structure.
    A. The heart chamber with the thickest wall is the left ventricle.
    B. The apex of the heart is more superior in location than the base.
    C. The heart has four chambers.
    D. The left ventricle forms the apex and most of the left border of the heart.
11. All of the following are defects involved in Tetralogy of Fallot *except:*
    A. Ventricular septal defect
    B. Right ventricular hypertrophy
    C. Stenosed mitral valve
    D. Aorta emerging from both ventricles
12. Choose the *true* statement.
    A. The T wave is associated with atrial depolarization.
    B. The normal P-R interval is about 0.4 sec.
    C. Myocardial infarction means strengthening of heart muscle.
    D. Myocardial infarction is commonly known as a "heart attack" or a "coronary."
13. Choose the *false* statement.
    A. Pressure within the atria is known as intraarterial pressure.
    B. During most of ventricular diastole the semilunar valves are closed.
    C. During most of ventricular systole the semilunar valves are open.
    D. Diastole is another name for relaxation of heart muscle.
14. Which of the following structures are located in ventricles?
    A. Papillary muscles
    B. Fossa ovalis
    C. Ligamentum arteriosum
    D. Pectinate muscles
15. The heart is composed mostly of:
    A. Epithelium
    B. Muscle
    C. Dense connective tissue

Questions 16–25: Circle T (true) or F (false). If the statement is false, change the underlined word or phrase so that the statement is correct.

T F 16. The blood in the left chambers of the heart contains higher oxygen content than blood in the right chambers.

T F 17. At the point when intraarterial (aortic) pressure surpasses ventricular pressure, semilunar valves open.

T F 18. The pulmonary artery carries blood from the lungs to the left atrium.

T F 19. The normal cardiac cycle does not require direct stimulation by the autonomic nervous system.

T F 20. During about half of the cardiac cycle, atria and ventricles are contracting simultaneously.

Questions 21–25: Fill-ins. Complete each sentence with the word or phrase that best fits.

__________________ 21. Most ventricular filling occurs during atrial ________.

__________________ 22. ________ is a cardiovascular disorder that is the leading cause of death in the U.S.

__________________ 23. An ECG is a recording of ________ of the heart.

__________________ 24. ________ is a term that means abnormality or irregularity of heart rhythm.

__________________ 25. Are atrioventricular and semilunar valves ever open at the same time during the cardiac cycle? ________

ANSWERS TO MASTERY TEST: CHAPTER 20

Arrange

1. C A B D
2. D E A B C
3. C D E A B

Multiple Choice

4. D
5. D
6. C
7. B
8. D
9. B
10. B
11. C
12. D
13. A
14. A
15. B

True–False

16 T
17. F. Close
18. F. Veins carry
19. T
20. F. No part

Fill-ins

21. Diastole
22. Coronary artery disease
23. Electrical changes or currents that precede myocardial contractions
24. Arrhythmia or dysrhythmia
25. No

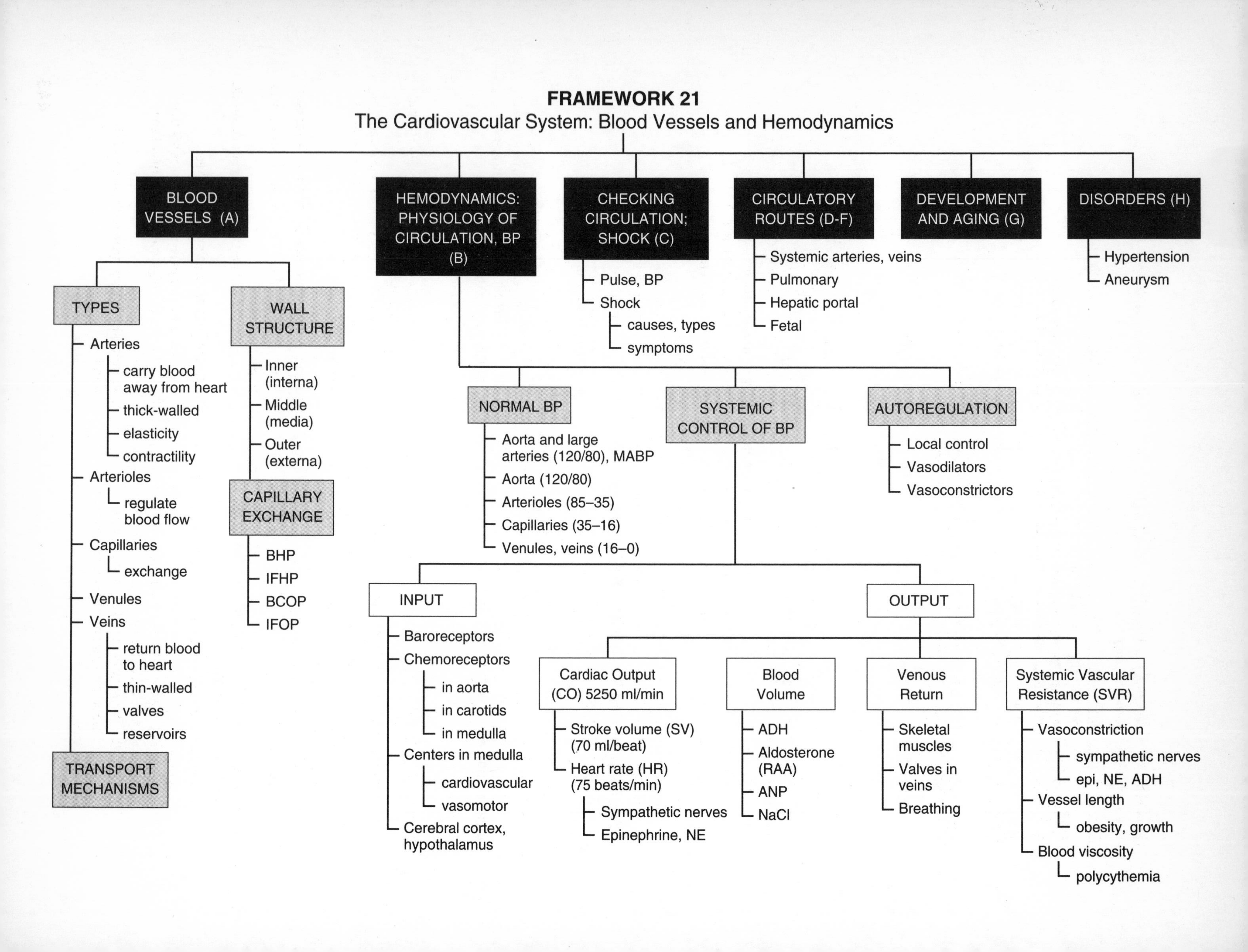

FRAMEWORK 21
The Cardiovascular System: Blood Vessels and Hemodynamics
BLOOD VESSELS (A)
TYPES
Arteries
carry blood away from heart
thick-walled
elasticity
contractility
Arterioles
regulate blood flow
Capillaries
exchange
Venules
Veins
return blood to heart
thin-walled
valves
reservoirs
TRANSPORT MECHANISMS
WALL STRUCTURE
Inner (interna)
Middle (media)
Outer (externa)
CAPILLARY EXCHANGE
BHP
IFHP
BCOP
IFOP
HEMODYNAMICS: PHYSIOLOGY OF CIRCULATION, BP (B)
NORMAL BP
Aorta and large arteries (120/80), MABP
Aorta (120/80)
Arterioles (85–35)
Capillaries (35–16)
Venules, veins (16–0)
SYSTEMIC CONTROL OF BP
INPUT
Baroreceptors
Chemoreceptors
in aorta
in carotids
in medulla
Centers in medulla
cardiovascular
vasomotor
Cerebral cortex, hypothalamus
OUTPUT
Cardiac Output (CO) 5250 ml/min
Stroke volume (SV) (70 ml/beat)
Heart rate (HR) (75 beats/min)
Sympathetic nerves
Epinephrine, NE
Blood Volume
ADH
Aldosterone (RAA)
ANP
NaCl
Venous Return
Skeletal muscles
Valves in veins
Breathing
Systemic Vascular Resistance (SVR)
Vasoconstriction
sympathetic nerves
epi, NE, ADH
Vessel length
obesity, growth
Blood viscosity
polycythemia
AUTOREGULATION
Local control
Vasodilators
Vasoconstrictors
CHECKING CIRCULATION; SHOCK (C)
Pulse, BP
Shock
causes, types
symptoms
CIRCULATORY ROUTES (D-F)
Systemic arteries, veins
Pulmonary
Hepatic portal
Fetal
DEVELOPMENT AND AGING (G)
DISORDERS (H)
Hypertension
Aneurysm

C H A P T E R **21**

The Cardiovascular System: Blood Vessels and Hemodynamics

Blood is pumped by heart action to all regions of the body. The network of vessels is organized much like highways, major roads, side streets, and tiny alleys to provide interconnecting routes through all sections of the body. In city traffic, tension mounts when traffic is heavy or side streets are blocked. Similarly, pressure builds (arterial blood pressure) when volume of blood increases or when constriction of small vessels (arterioles) occurs. In a city, traffic may be backed up for hours, with vehicles unable to deliver passengers to destinations. A weakened heart may lead to similar and disastrous effects, backing up fluid into tissues (edema) and preventing transport of oxygen to tissues (ischemia, shock, and death of tissues).

Careful map study leads to trouble-free travel along the major arteries of a city. A similar examination of the major arteries and veins of the human body facilitates an understanding of the normal path of blood as well as clinical applications, such as the exact placement of a blood pressure cuff over the brachial artery or the expected route of a deep venous thrombus (pulmonary embolism) to the lungs.

As you continue your study of the cardiovascular system, carefully examine the Chapter 21 Topic Outline and Objectives; check off each one as you complete it. To organize your study of blood vessels and hemodynamics, glance over the Chapter 21 Framework now. Be sure to refer to the Framework frequently and note relationships among key terms in each section.

TOPIC OUTLINE AND OBJECTIVES

A. Structure and function of blood vessels

1. Contrast the structure and function of arteries, arterioles, capillaries, venules, and veins.
2. Discuss the pressures that cause movement of fluids between capillaries and interstitial spaces.

B. Hemodynamics: factors affecting circulation and blood pressure

1. Explain the factors that regulate the volume of blood flow.
2. Explain how blood pressure changes throughout the cardiovascular system.
3. Describe the factors that determine mean arterial blood pressure and systemic vascular resistance.
4. Describe the relation between cross-sectional area and velocity of blood flow.
5. Describe how blood pressure is regulated.
6. Define shock, and describe the four types of shock.

C. Checking circulation; shock and homeostasis

1. Define pulse, and define systolic, diastolic, and pulse pressure.
2. Define shock, and describe four types of shock.

D. Circulatory routes: systemic arteries

1. Describe and compare the major routes that blood takes through various regions of the body.
2. Identify the four principal divisions of the aorta.

3. Locate the major arterial branches arising from each divison.
4. Identify the two primary arterial branches of the ascending aorta.
5. Identify the three principal arteries that branch from the arch of the aorta.
6. Identify the visceral and parietal branches of the thoracic aorta.
7. Identify the visceral and parietal branches of the abdominal aorta.
8. Identify the two major branches of the common iliac arteries.

E. Circulatory routes: systemic veins

1. Identify the three systemic veins that return deoxygenated blood to the heart.
2. Identify the three major veins that drain blood from the head.
3. Identify the principal veins that drain the upper limbs.
4. Identify the components of the azygos system of veins.
5. Identify the principal veins that drain the abdomen and the pelvis.
6. Identify the principal superficial and deep veins that drain the lower limbs.

F. Circulatory routes: hepatic portal, pulmonary, and fetal circulation

G. Developmental anatomy of blood vessels and blood; aging of the cardiovascular system

1. Describe the development of blood vessels and blood.
2. Explain the effects of aging on the cardiovascular system.

H. Disorders, medical terminology

WORDBYTES

Now become familiar with the language of this chapter by studying each wordbyte, its meaning, and an example of its use within a term. After you study the entire list, self-check your understanding by writing the meaning of each wordbyte on the line. As you continue through the *Learning Guide*, identify (and fill in) additional terms that contain the same wordbyte.

Wordbyte	Self-check	Meaning	Example(s)
ar-	__________	air	*ar*tery
brady-	__________	slow	*brady*cardia
capillaries	__________	hairlike	*capillaries*
coron-	__________	crown	*coron*aries
fenestr-	__________	window	*fenestr*ated capillaries
hepat-	__________	liver	*hepat*ic artery
med-	__________	middle	tunica *med*ia
phleb-	__________	vein	*phleb*itis
port-	__________	carry	*port*al vein
retro-	__________	backward, behind	*retro*peritoneal
sphygmo-	__________	pulse	*sphygmo*manometer
tachy-	__________	fast	*tachy*cardia

-ter-	__________	to carry	ar*ter*y
tunica	__________	sheathe	*tunica* intima
vaso-	__________	vessel	*vaso*constriction
ven-	__________	vein	*ven*ipuncture

CHECKPOINTS

A. Anatomy of blood vessels and capillary exchange (pages 697–705)

A1. Review types of vessels in the human circulatory system by referring to Figure LG20.3 and Chapter 20, Checkpoint B6, pages LG 423–424.

✓ **A2.** Check your understanding of the structure of blood vessels by matching terms in the box to related descriptions.

L.	**Lumen**	**TM.**	**Tunica media**
TE.	**Tunica externa**	**VV.**	**Vasa vasorum**
TI.	**Tunica intima**		

_______ a. The opening in blood vessels through which blood flows; narrows in atherosclerosis

_______ b. Formed largely of endothelium and basement membrane, it lines lumen; the only layer in capillaries

_______ c. Tiny blood vessels that carry nutrients to walls of blood vessels

_______ d. Composed of muscle and elastic tissue; contraction permits vasoconstriction

_______ e. Strong outer layer composed mainly of elastic and collagenous fibers

✓ **A3.** Match the names of vessels in the box with descriptions below. Answers may be used once, more than once, or not at all.

Aorta and other large arteries	**Medium-sized arteries**
Arterioles	**Small veins**
Capillaries	**Venules**
Inferior vena cava	

a. Known as exchange vessels, they are sites of gas, nutrient, and waste exchange with tissues: __________

b. Known as resistance vessels, they play the primary role in regulating moment-to-moment distribution of blood and in regulating blood pressure: __________

c. Sinusoids of liver are wider, more leaky versions of this type of vessel: __________

d. Reservoirs for about 60% of the volume of blood in the body; vasoconstriction (due to sympathetic impulses) permits redistribution of blood stored here (three answers):

__________ __________ __________

e. Elastic arteries such as carotids or subclavians: __________

f. Flow of blood through here is called microcirculation: __________

g. Muscular arteries such as brachial or radial arteries: __________

A4. Contrast terms in the following pairs:

a. Structure of the tunica media of the following vessels:

1. *conducting arteries/distributing arteries*

2. *artery/vein*

b. Number of capillaries in *muscles and liver/tendons and ligaments*

c. Functions of *thoroughfare channel/true capillary*

d. Structure of *continuous capillary/fenestrated capillary*

✓ **A5.** Refer to Figure 21.4 in your text, and identify structural features of different types of capillaries and sinusoids that permit passage of materials across the vessel wall by four different routes. Select answers from the box.

Direct movement across endothelial membranes	**Intercellular clefts**
Fenestrations	**Pinocytic vesicles**

a. Structural features of very thin endothelial membrane with thin, incomplete, or even absent basement membrane, for example, in sinusoids: ____________________

b. Gaps known as ____________________ between neighboring endothelial cells; may be large, for example, to permit passage of large proteins made in liver cells into the blood in liver sinusoids.

c. Pores known as ____________________ in plasma membranes of endothelial cells, for example, in cells of kidneys, intestinal villi, and choroid plexuses of the brain. The tight seal of the blood–brain barrier is attributed partly to lack of such pores.

d. ____________________ permitting endocytosis and exocytosis for transport of large molecules that cannot cross capillary walls in any other way.

✓ **A6.** Complete this exercise about the structure of veins.

a. Veins have *(thicker? thinner?)* walls than arteries. This structural feature relates to the fact that the pressure in veins is *(more? less?)* than in arteries. The pressure difference is demonstrated when a vein is cut; blood leaves a cut vein in *(rapid spurts? an even flow?)*.

b. Gravity exerts back pressure on blood in veins located inferior to the heart. To counteract this, veins contain ____________________.

c. When valves weaken, veins become enlarged and twisted. This condition is called ____________________. This occurs more often in *(superficial? deep?)* veins. Why?

d. Identify two locations in the gastrointestinal tract where varicosities are relatively common. ____________________ ____________________

e. A vascular (venous) sinus has ____

____

replacing the tunica media and tunica externa. So sinuses have the *(structure but not function? function but not structure?)* of veins. List two places where such sinuses are found.

____________________ ____________________

A7. Define the following terms and explain how they are protective to the body:

a. Anastomosis

b. Collateral circulation

✓ **A8.** Complete this Checkpoint on distribution of blood. It may help to refer to Figure 21.6 in your text.

a. At rest, *(arteries and arterioles? capillaries? venules and veins?)* contain most of the blood in the body. In which vessels is the smallest percentage of blood found in the resting person? *(Arteries and arterioles? Capillaries? Venules and veins?)*

b. *(Sympathetic? Parasympathetic?)* nerve impulses to venules and veins cause vasoconstriction and release of blood from these vessels. Name two locations of venous "reservoirs" that can be activated to release blood when needed.

c. State two examples of circumstances that might activate distribution of reservoir blood. ____________________ ____________________

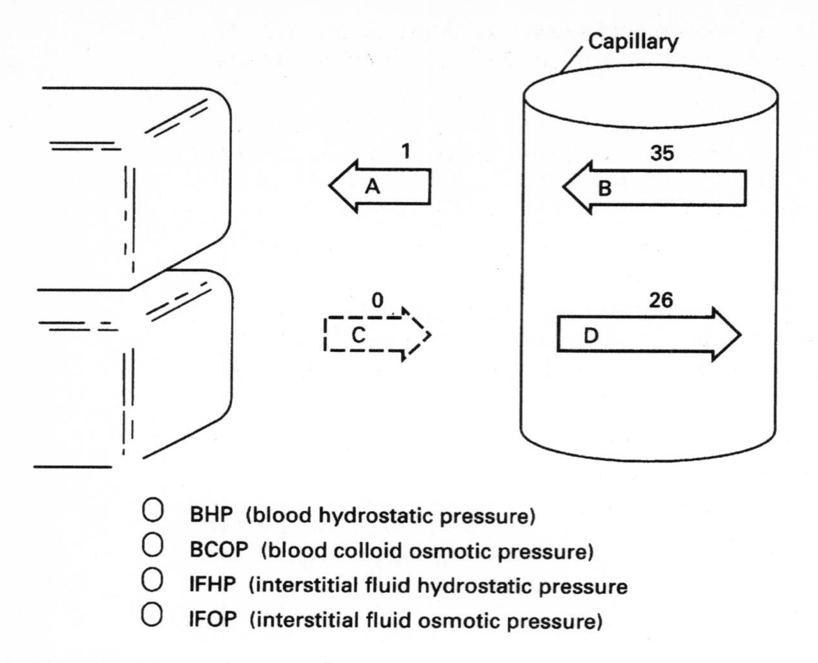

Figure LG 21.1 Practice diagram showing forces controlling capillary exchange and for calculation of NFP at arterial end of capillary. Numbers are pressure values in mm Hg. Color arrows as directed in Checkpoint A10.

✓ **A9.** Identify the types of transport mechanisms involved in capillary exchange by doing this exercise. It may help to refer back to Checkpoint A5. Use these answers:

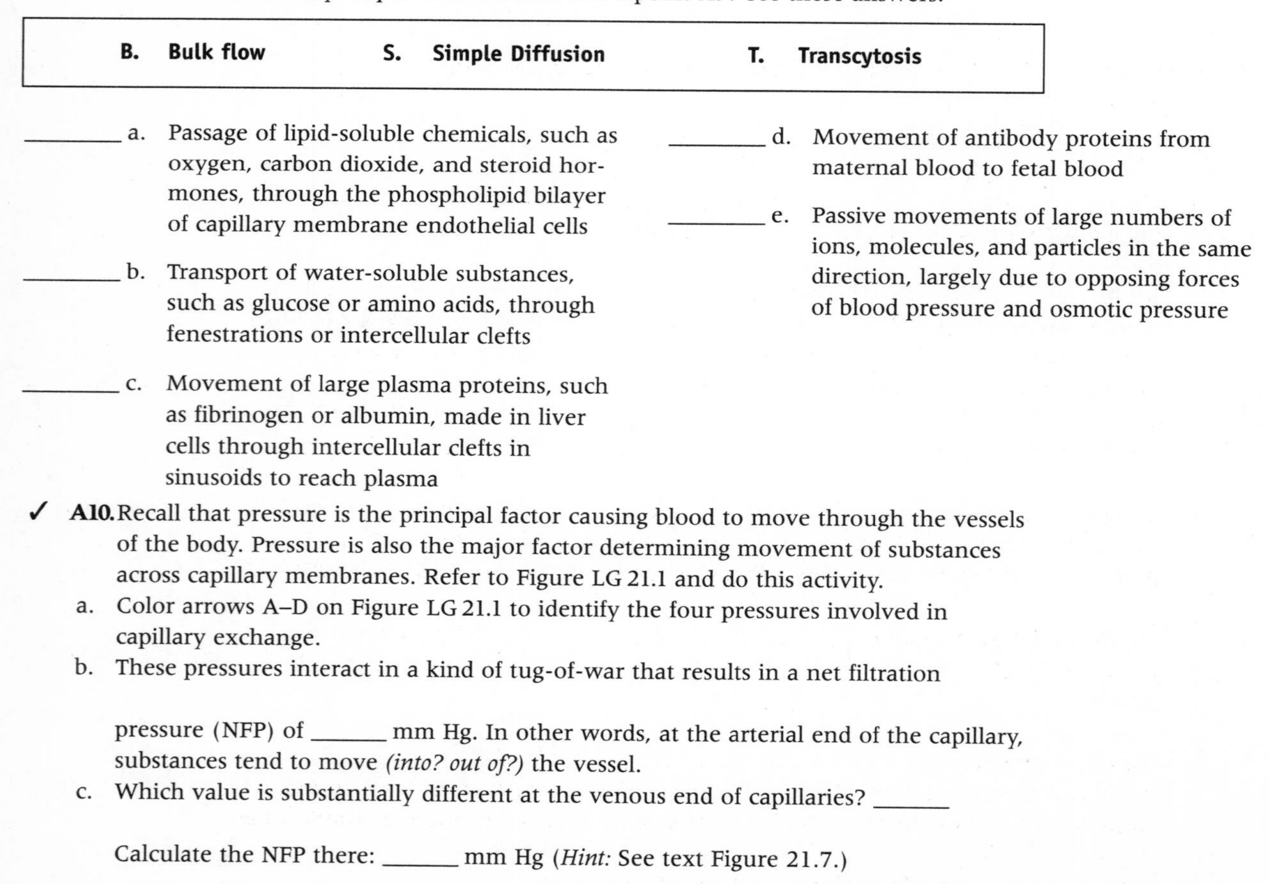

B. Bulk flow	**S. Simple Diffusion**	**T. Transcytosis**

_______ a. Passage of lipid-soluble chemicals, such as oxygen, carbon dioxide, and steroid hormones, through the phospholipid bilayer of capillary membrane endothelial cells

_______ b. Transport of water-soluble substances, such as glucose or amino acids, through fenestrations or intercellular clefts

_______ c. Movement of large plasma proteins, such as fibrinogen or albumin, made in liver cells through intercellular clefts in sinusoids to reach plasma

_______ d. Movement of antibody proteins from maternal blood to fetal blood

_______ e. Passive movements of large numbers of ions, molecules, and particles in the same direction, largely due to opposing forces of blood pressure and osmotic pressure

✓ **A10.** Recall that pressure is the principal factor causing blood to move through the vessels of the body. Pressure is also the major factor determining movement of substances across capillary membranes. Refer to Figure LG 21.1 and do this activity.

a. Color arrows A–D on Figure LG 21.1 to identify the four pressures involved in capillary exchange.

b. These pressures interact in a kind of tug-of-war that results in a net filtration

pressure (NFP) of ______ mm Hg. In other words, at the arterial end of the capillary, substances tend to move *(into? out of?)* the vessel.

c. Which value is substantially different at the venous end of capillaries? ______

Calculate the NFP there: ______ mm Hg (*Hint:* See text Figure 21.7.)

d. The near-equilibrium at the two ends of capillaries is based on ________________'s law of the capillaries. Which system "mops up" the slight amount of fluid that is not drawn back into veins? ________________

e. Excessive amounts of fluid accumulating in interstitial spaces is the condition known as ________________. This may result from high blood pressure in which *(BCOP? BHP?)* is increased, loss of plasma protein in which *(BCOP? BHP?)* is decreased, or *(increase? decrease?)* in capillary permeability.

✓ **A11.** *Critical thinking.* On a random review of patient charts from a hospital, three clients were noted as having edema. Identify the physiological cause that was most likely the cause of each case of edema. Select from these answers:

↑	**BHP.**	**Increased blood hydrostatic pressure**
↑	**PC.**	**Increased permeability of capillaries**
↑	**ECV.**	**Increased extracellular volume**
↓	**PP.**	**Decreased concentration of plasma proteins**

________ a. Paul P, age 72, has a diagnosis of congestive heart failure and experiences what he describes as "spells when I just can't get my breath."

________ b. Tommy R, age 7, received second-degree burns on his left leg, which is swollen and seeping fluids.

________ c. Elaine B, age 41, has kidney failure; she has gained 11 pounds since her last hemodialysis treatment 3 days ago. She reported that she "went overboard eating and drinking at a couple of parties over the weekend." (Two answers)

B. Hemodynamics: factors affecting circulation and blood pressure (pages 705–713)

✓ **B1.** Check your understanding of circulation of blood in this Checkpoint.

a. The typical human body contains about ______ liters of blood. At rest, this same amount of blood is pumped out by the heart each minute, as cardiac ________________ (5 liters/minute).

b. Now name the two factors that determine how rapidly and where this blood is distributed, for example, rapid delivery to brain and muscles, or slow delivery to intestine and skin. (See Checkpoint B3 for more on this topic.)

✓ **B2.** Complete this activity on blood pressure (BP) and mean arterial blood pressure (MABP).

a. In the aorta and brachial artery, blood pressure (BP) is normally about 120 mm Hg immediately following ventricular contraction. This is called *(systolic? diastolic?)* BP. As ventricles relax (or go into ________________), blood is no longer ejected into these arteries. However, the normally *(elastic? rigid?)* walls of these vessels recoil against blood, pressing it onward with a diastolic BP of ______ mm Hg.

b. In a blood pressure (BP) of 120/80, the average of the two pressures (systolic and diastolic) is 100. However the MABP (average or mean blood pressure in arteries) for a BP of 120/80 is actually slightly less than 100; it is about ______.

Reason: during the typical cardiac cycle (0.8 sec), the ventricles are in *(systole? diastole?)* for about two-thirds of the cycle and in systole for only about one-third of the cycle (see Figure LG 20.4, page LG 428). As a result, the MABP is always slightly closer to the value of the *(systolic? diastolic?)* BP.

c. Try calculating MABP for a person with an arterial BP of 120/78. First write the diastolic BP, ______ mm Hg. Then find the difference between systolic BP and diastolic BP a value known as *pulse pressure*: 120 mm Hg – 78 mm Hg = ______ mm Hg. Finally, add one-third of pulse pressure to diastolic BP to arrive at MABP:

$BP_{diastolic} + (1/3 \times \text{pulse pressure}) = MABP$

______ mm Hg + (1/3 × ______ mm Hg) = ______ mm Hg

✓ **B3.** Recall that in Checkpoint B1 you named two factors that determine distribution (or circulation) of cardiac output to the body. Closely examine these two factors in this Checkpoint.

a. One factor is ____________________ (such as blood pressure or MABP). Define blood pressure: the pressure exerted by ____.
Picture the heart pumping harder and faster, much like a dam permitting more water (cardiac output) to pass over it so the river below swells; water (blood) presses forward and laps up against river banks (BP increases). So if cardiac output (CO) increases, BP or MABP are likely to ______-crease. Finish the arrow to show the correct relationship:

↑ CO → | MABP. On the other hand, if CO decreases (as in hemorrhage or shock), then MABP is likely to ______-crease.

b. Now visualize the effects of the second factor (resistance) on distribution of blood. Imagine yourself inside a red blood cell (much like a boat) surging forward at high velocity through the aorta or other large arteries. BP is high there (MABP of 92) as blood flows along freely. But as you enter a small artery or arteriole, your RBC encounters *(more? less?)* resistance to flow as the RBC repeatedly bounces up against and snags the walls of this narrow-diameter vessel. In other words, BP in any given blood vessel is *(directly? inversely?)* related to resistance to flow, while *(directly? inversely?)* related to cardiac output. In fact, as the diameter of a vessel is halved, resistance to flow through that vessel ______-creases *(2? 4? 8? 16?)*-fold.

c. Because CO stays the same within the body at any given time, we can conclude that if resistance to flow is high through a vessel (such as an arteriole or capillary), that BP in that vessel will be *(high? low?)*. Referring to Figure 21.8 in your text, write the normal ranges of BP values for the following types of vessels: arterioles, ______ mm Hg; capillaries, ______ mm Hg; venules and veins, ______ mm Hg; venae cavae, ______ mm Hg.

d. On Figure 21.8 the steepest decline in the pressure curve occurs as blood passes through the *(aorta? arterioles? capillaries?)*. This indicates that the greatest resistance to flow is present in the *(aorta? arterioles? capillaries?)*.

✓ **B4.** Do this exercise on two very important and different relationships between resistance (R) and blood pressure (BP).

a. In Checkpoint B3b, you saw that increase in resistance (R) *within a given blood vessel* causes _______-crease in *BP in that vessel.* Therefore, BP (within a vessel) and R (in that same vessel) are inversely related. Show this by completing the arrow.

$\uparrow R_{\text{within a vessel}} \rightarrow \mid BP_{\text{in that same vessel}}$

For example, the RBC (boat) passing from an artery (wide river) into an arteriole (tributary) meets more resistance to flow in this narrow tributary, thereby _______-creasing BP *within that vessel.* So the increased resistance (R) blood meets as it passes through arterioles causes BP there to decrease from about 85 to 35 mm Hg.

b. However, several factors contribute to changes in *BP in systemic arteries* (usually measured in the brachial artery), for example, increase in BP from a normal _______/_______ to a high BP of 162/98. One of these factors altering systemic BP is resistance. But here the resistance refers to inhibition of flow through many, many small systemic vessels, for example, by sympathetic nerve stimulation of arterioles causing them to narrow. So the term SVR (____________________ vascular ____________________) is used to describe this resistance. Because less blood is engaged in runoff into arterioles, more blood stays in main arteries, thereby increasing systemic BP. Complete arrows to show this:

$\uparrow R_{\text{within systemic arterioles, as in skin or abdomen}} \rightarrow \mid BP_{\text{in systemic arteries}}$

or

$\uparrow SVR \rightarrow \mid BP_{\text{systemic arterial}}$

Consider this analogy: if blood (water) cannot pass from an artery (wide river) into many arterioles (tributaries) because entranceways are blocked by vasoconstriction of arterioles (tributaries are clogged with silt and debris), then blood (water) will remain in the main artery (wide river), _______-creasing BP *in main systemic arteries* (wide rivers).

c. In fact, systemic BP tends to increase if either cardiac output or systemic vascular resistance (SVR) or both of those factors increase. Show this by completing the equation:

$BP_{\text{systemic arterial}} = $ __________ × __________

or

MABP = cardiac output × systemic vascular resistance

d. Describe three factors that may increase SVR leading to increased systemic BP:

1. _______-crease in blood viscosity, for example, due to ____________________

2. _______-crease in total length of blood vessels in the body, for example, due to ____________________

3. ______-crease in diameter (or radius) of blood vessels, for example, by

____________________ of arterioles

e. SVR is also called total ____________________ resistance because the vessels that are targeted to vasoconstrict (decrease radius) and create that resistance are those in

regions of the body (such as skin and ____________________________________) that could be considered more "peripheral," or less crucial, to survival.

✓ **B5.** *A clinical correlation.* Do this exercise about Mr. Tyler, who has been depressed since his wife died two years ago and has gained 50 pounds during that time.

a. Which one of the following factors that increase systemic vascular resistance is most likely to be higher in Mr. Tyler?

A. Blood viscosity
B. Blood vessel length
C. Blood vessel radius

This factor *(does? does not?)* put Mr. Tyler at higher risk for hypertension.

b. Because Mr. Tyler's stress level is often high, his *(sympathetic? parasympathetic?)* nerves are likely to be more active. Which of the three factors (A–C) above is most likely to be affected? *(A? B? C?)* If his arterioles decrease in diameter by 50%, the resistance to

flow increases _______ times. This factor is likely to ______-crease his blood pressure.

B6. Explain how these two factors increase venous return during exercise.

a. Skeletal muscle pump (For help, refer to Chapter 20, Checkpoint E4c, page LG 432.)

b. Respiratory pump

✓ **B7.** Do this exercise about blood flow.

a. Figure 21.11 in your text shows that velocity of blood flow is greatest in *(arteries?*

capillaries? veins?). Flow is slowest in __________________. As a result, ample time is available for exchange between capillary blood and tissues.

b. As blood moves from capillaries into venules and veins, its velocity *(increases? decreases?)*, enhancing venous return.

c. Although each individual capillary has a cross-sectional area which is microscopic, the body contains so many capillaries that their total cross-sectional area is enormous. (For comparison, look at the eraser end on one pencil; you are seeing the cross-sectional area of that one pencil. If you bind together 100 pencils with a strong rubber band and observe their eraser ends as a group, you see that the individual pencils contribute to a relatively large total cross-sectional area.) All of the capillaries

in the body have a total cross-sectional area of about ______ cm^2.

d. *(A direct? An indirect?)* relationship exists between cross-sectional area and velocity of blood flow. This is evident in the aorta also. This vessel has a cross section of only

about ______ cm^2 and has a very *(high? low?)* velocity.

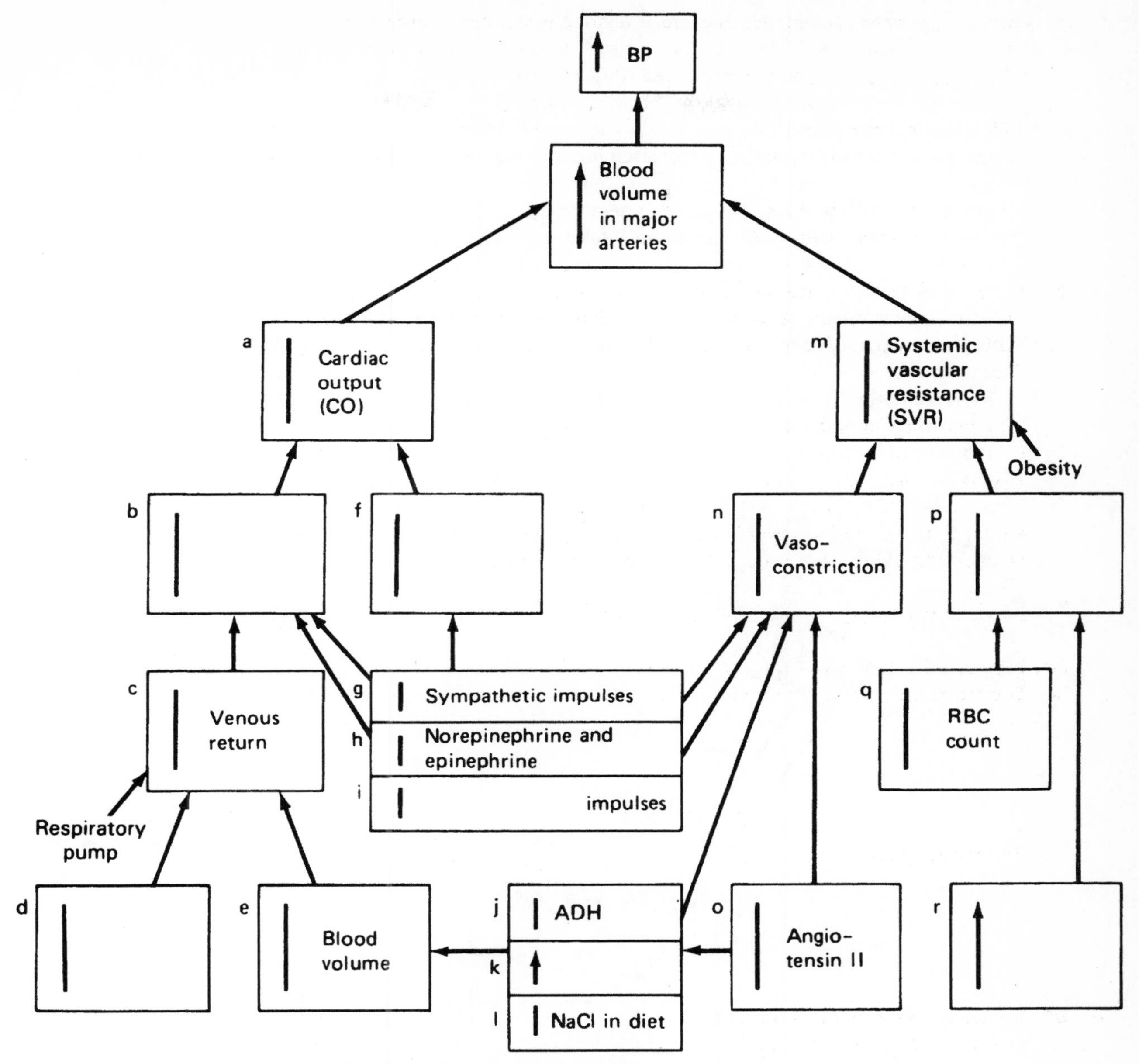

Figure LG 21.2 Summary of factors controlling blood pressure (BP). Arrows indicate increase (↑) or decrease (↓). Complete as directed in Checkpoint B8.

e. *A clinical correlation.* Mr. Fey has an aortic aneurysm. The diameter of his aorta has increased from normal size to about 10 cm. The region of the aorta with the aneurysm therefore has a *(larger? smaller?)* cross-sectional area. One would expect the velocity of Mr. Fey's aortic blood flow to *(in? de?)*-crease. This change may put Mr. Fey at risk for ____________________.

f. On the average, *(1? 5? 10?)* minute(s) is normally required for a blood cell to travel one complete circulatory circuit, for example, heart → lung → heart → foot → heart.

g. Inadequate blood flow to the brain is one cause of fainting, clinically known as ____________________.

✓ **B8.** On Figure LG 21.2 fill in blanks and complete arrows to show details of all of the nervous output, hormonal action, and other factors that increase blood pressure.

✓ **B9.** *A clinical correlation.* Complete this exercise about blood pressure disorders.

a. Hypertension means *(high? low?)* blood pressure. It may be caused by an excess of any of the factors shown in Figure LG 21.2. If directions of arrows in Figure LG 21.2 are reversed, blood pressure can be *(increased? decreased?).*

b. Circle all the factors listed below that tend to decrease blood pressure.
   A. Increase in cardiac output, as by increased heart rate or stroke volume
   B. Increase in (vagal) impulses from cardioinhibitory center
   C. Decrease in blood volume, as following hemorrhage
   D. Increase in blood volume, by excess salt intake and water retention
   E. Increased systemic vascular resistance due to vasoconstriction of arterioles (blood prevented from leaving large arteries and entering small vessels)
   F. Decrease in sympathetic impulses to the smooth muscle of arterioles
   G. Decreased viscosity of blood via loss of blood protein or red blood cells, as by hemorrhage
   H. Use of medications called dilators because they dilate arterioles, especially in areas such as abdomen and skin

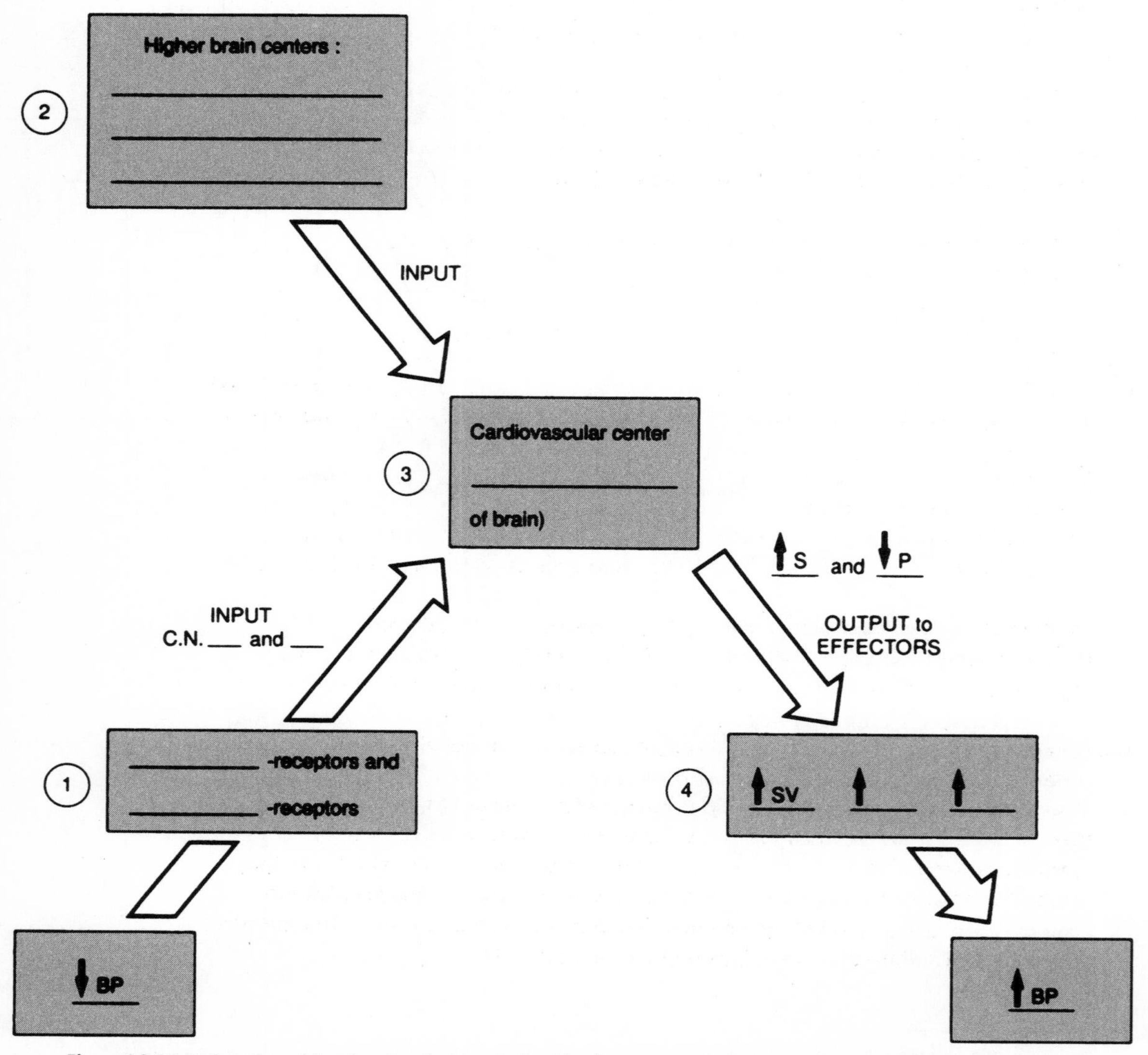

Figure LG 21.3 Overview of input and output regulating blood pressure. Complete as directed in Checkpoint B10.

✓ **B10.** Imagine that your blood pressure (BP) is slightly lower than your body needs right now. Do this exercise to see how your body is likely to become aware of this alteration in homeostasis and respond to increase your blood pressure.

a. Refer to Figure LG 21.3 and fill in box 1 to describe types of receptors that trigger changes in blood pressure. Baroreceptors are sensitive to changes in ________________ ________________, and chemoceptors are activated by chemicals such as ________________. These receptors are located in two locations: ________________ and ________________.

b. Nerve messages from the carotid sinuses travel to the brain via cranial nerves ______, whereas messages from the aorta travel by cranial nerves ______. Write in these cranial nerves (C.N.) on the figure.

c. This *input* reaches the cardiovascular center located within the ________________ of the brain. Fill in box 3 to indicate this region of the brain. This center receives input from other parts of the brain also. Note three of these brain parts in box 2 in the figure.

d. Consider the cardiovascular center as comparable to the control center of an automobile. The center has an accelerator portion (like a gas pedal) and an inhibitor portion (like the ________________ of a car). When baroreceptors report that BP is too low (as if the car is running too slowly), the *(cardiostimulatory? cardioinhibitory?)* center must be activated; in the car analogy, you would *(step on the gas pedal? step on the brakes?)*. At the same time, the cardioinhibitory center must be deactivated; continuing with the car analogy, what response do you make? ________________
What is the outcome for the car/heart?

e. In addition, a vasomotor portion of the cardiovascular center responds to lower blood pressure by *(stimulating? inhibiting?)* smooth muscle of blood vessels, resulting in *(vasoconstriction? vasodilatation?)*. This helps to increase BP by ______-creasing systemic vascular resistance (SVR).

f. Summarize nervous output to effectors from the cardiovascular center by writing in box 4 in Figure LG 21.3 the three factors that are altered. The result of this output is an attempt to restore homeostasis by ______-creasing your blood pressure.

g. Overall, Figure LG 21.3 demonstrates *(positive? negative?)* feedback mechanisms for regulating BP, because a decrease in BP initiates factors to increase BP.

✓ **B11.** *For extra review.* Do the following Checkpoint on additional details on blood pressure regulation.

a. Vasomotor ________________ refers to a moderate level of vasoconstriction normally maintained by the medulla and sympathetic nerves. A sudden loss of sympathetic nerve output, for example, by head trauma or emotional trauma, can reduce this tone and cause BP to *(skyrocket? drop precipitously?)*. This may lead to ________________ (fainting) or even to shock. This can also occur when pressure is exerted on the carotid sinus, for example by a tight collar.

b. Smooth muscle of most small arteries and arterioles contains *(alpha? beta?)* receptors; in response to the neurotransmitter ________________, these vessels vaso-*(constrict? dilate?)*. Sympathetic stimulation of most veins causes them to vaso-*(constrict? dilate?)*, thereby ______-creasing venous return.

c. Vasodilator medications therefore tend to ______-crease BP as blood pools in nonessential areas (such as GI tract or skin). Name two vasodilator chemicals that may be produced by the body: ________________ from cells of the atria, and ________________ from cells lining blood vessels.

d. Baroreceptors involved with the aortic reflex primarily maintain *(cerebral? general systemic?)* BP, whereas receptors that trigger the carotid sinus reflex primarily control ________________ BP.

e. Carotid bodies and aortic bodies are sites of *(baro? chemo?)*-receptors that provide input for control of BP. Low oxygen (known as ________________) or high carbon dioxide (called ________________) will tend to cause *(vasoconstriction? vasodilatation?)* and consequently ______-crease BP. As a result, blood can move more rapidly to the lungs to pick up oxygen and give up carbon dioxide.

f. Write ↑ or ↓ next to each hormone or other chemical listed below to indicate whether the chemical increases or decreases blood pressure. (*Hint*: refer to Figure LG 21.2)

________ 1. ADH

________ 2. Angiotensin II

________ 3. Aldosterone

________ 4. Epinephrine of NE

________ 5. ANP

✓ **B12.** Describe the process of autoregulation in this activity.

a. Autoregulation is a mechanism for regulating blood flow through specific tissue areas. Unlike cardiac and vasomotor mechanisms, autoregulation occurs by *(local? autonomic nervous system?)* control.

b. In other words, when a tissue (such as an active muscle) is hypoxic, that is, has a *(high? low?)* oxygen level, the cells of that tissue release *(vasoconstrictor? vasodilator?)* substances. These may include ________________ acid, built up by active muscle. Other products of metabolism that serve as dilator substance include:

c. Compared to systemic vessels, pulmonary blood vessels respond *(similarly? just the opposite?)* to hypoxic conditions. Arterioles in hypoxic parts of lungs will vaso-

_____________________. How can this mechanism be helpful to the body?

d. Recall from Chapter 19 that the initial step in hemostasis is *(vasodilation? vasospasm?)*. Name two chemicals that produce this effect.

C. Checking circulation; shock and homeostasis (pages 713–717)

C1. Define *pulse*.

C2. In general, where may pulse best be felt? List six places where pulse can readily be palpated. Try to locate pulses at these points on yourself or on a friend.

✓ **C3.** Normal pulse rate is about ______ beats per minute. *Tachycardia* means *(rapid? slow?)*

pulse rate; ____________________ means slow pulse rate. Which condition is more likely to be present in well trained athletes at rest? *(Bradycardia? Tachycardia?)*

✓ **C4.** Describe the method commonly used for taking blood pressure by answering these questions about the procedure.

a. Name the instrument used to check BP. ____________________

b. The cuff is usually placed over the ____________________ artery. How can you tell that this artery is compressed after the bulb of the sphygmomanometer is squeezed?

c. As the cuff is deflated, the first sound heard indicates ____________________ pressure because reduction of pressure from the cuff below systolic pressure permits blood to begin spurting through the artery during diastole.

d. Diastolic pressure is indicated by the point at which the sounds ____________________. At this time blood can once again flow freely through the artery.

e. Sounds heard while taking blood pressure are known as ____________________ sounds.

✓ **C5.** *A clinical correlation.* Mrs. Yesberger has a systolic pressure of 128 mm Hg and diastolic of 78 mm Hg. Write her blood pressure in the form in which BP is usually

expressed: ____________________ Determine her pulse pressure.

Would this be considered a normal pulse pressure? ________
If Mrs. Yesberger were to go into shock, her BP would be likely to ________-crease.

✓ **C6.** Describe cause of shock and compensation for shock in this activity.

a. Shock is said to occur when tissues receive inadequate ____________________ supply to meet their demands for oxygen, nutrients, and waste removal. Shock resulting

from decrease in circulating blood volume is known as ____________________ shock. Write two or more causes of this type of shock.

b. A pulmonary embolism would be most likely to cause ____________________ shock, whereas congestive heart failure or Tetralogy of Fallot would be most likely to lead

to ____________________ shock. Allergic reactions to bee venom may lead to

____________________ shock.

c. Compensatory mechanisms activate *(sympathetic? parasympathetic)* nerves that

_______-crease both heart rate and systemic vascular resistance. These attempts to

increase BP normally work for blood losses up to ______% of total blood volume. One common indicator of this compensatory mechanism is *(warm, pink, and dry? cool, pale,*

and clammy?) skin related to vaso-____________________ and _______-creased sweating.

d. Chemical vasoconstrictors include ____________________, released from the adrenal

medulla; angiotensin II, formed when ____________________ is secreted by kidneys;

and the posterior pituitary hormone, ____________________.

e. Which hormones cause water retention which helps to increase BP in shock?

f. How does hypoxia help to compensate for shock?

✓ **C7.** *A clinical correlation.* Write arrows to indicate signs and symptoms of shock. (↑ = increase; ↓ = decrease)

a. Sweat _______

b. Heart rate (pulse) _______

c. BP _______

d. Blood pH _______

e. Mental status _______

f. Urine volume _______

g. Thirst _______

D. Circulatory routes: systemic arteries (pages 717–736)

✓ **D1.** On Figure LG 20.3 (page LG 424), label the *pulmonary artery, pulmonary capillaries,* and a *pulmonary vein*. Briefly contrast pulmonary and systemic vessels.

Which ventricle pumps with greater force? *(Right? Left?)* Explain.

✓ **D2.** Use Figure LG 21.4 to do this learning activity about systemic arteries.

a. The major artery from which all systemic arteries branch is the ________________. It exits from the chamber of the heart known as the *(right? left?)* ventricle. The aorta can be divided into three portions. They are the ________________ (N1 on the figure), the ________________ (N2), and the ________________ (N3).

b. The first arteries to branch off the aorta are the ________________ arteries. Label these (at O) on the figure. These vessels supply blood to the ________________.

c. Three major arteries branch from the aortic arch. Write the names of these (at K, L, and M) on the figure, beginning with the vessel closest to the heart.

d. Locate the two branches of the brachiocephalic artery. The right subclavian artery (letter ______) supplies blood to ____.
Vessel E is named the *(right? left?)* ________________________________ artery. It supplies the right side of the head and neck.

e. As the subclavian artery continues into the axilla and arm, it bears different names (much as a road does when it passes into new towns). Label G and H. Note that H is the vessel you studied as a common site for measurement of ________________.

f. Label vessels I and J, and then continue the drawing of these vessels into the forearm and hand. Notice that they anastomose in the hand. Vessel *(I? J?)* is often used for checking pulse in the wrist. The deep palmar arch within the hand receives blood mainly from the *(radial? ulnar?)* artery, whereas the superficial palmar arch receives blood mainly from the *(radial? ulnar?)* artery. Both of the arches provide blood to digital arteries that supply blood to ________________.

g. Besides supplying blood to the upper limbs the subclavian arteries, each sends a branch that ascends the neck through foramina in cervical vertebrae. This is the ________________ artery. The right and left vertebral arteries join at the base of the brain to form the ________________ artery. (Refer to D and C on Figure LG 21.5.)

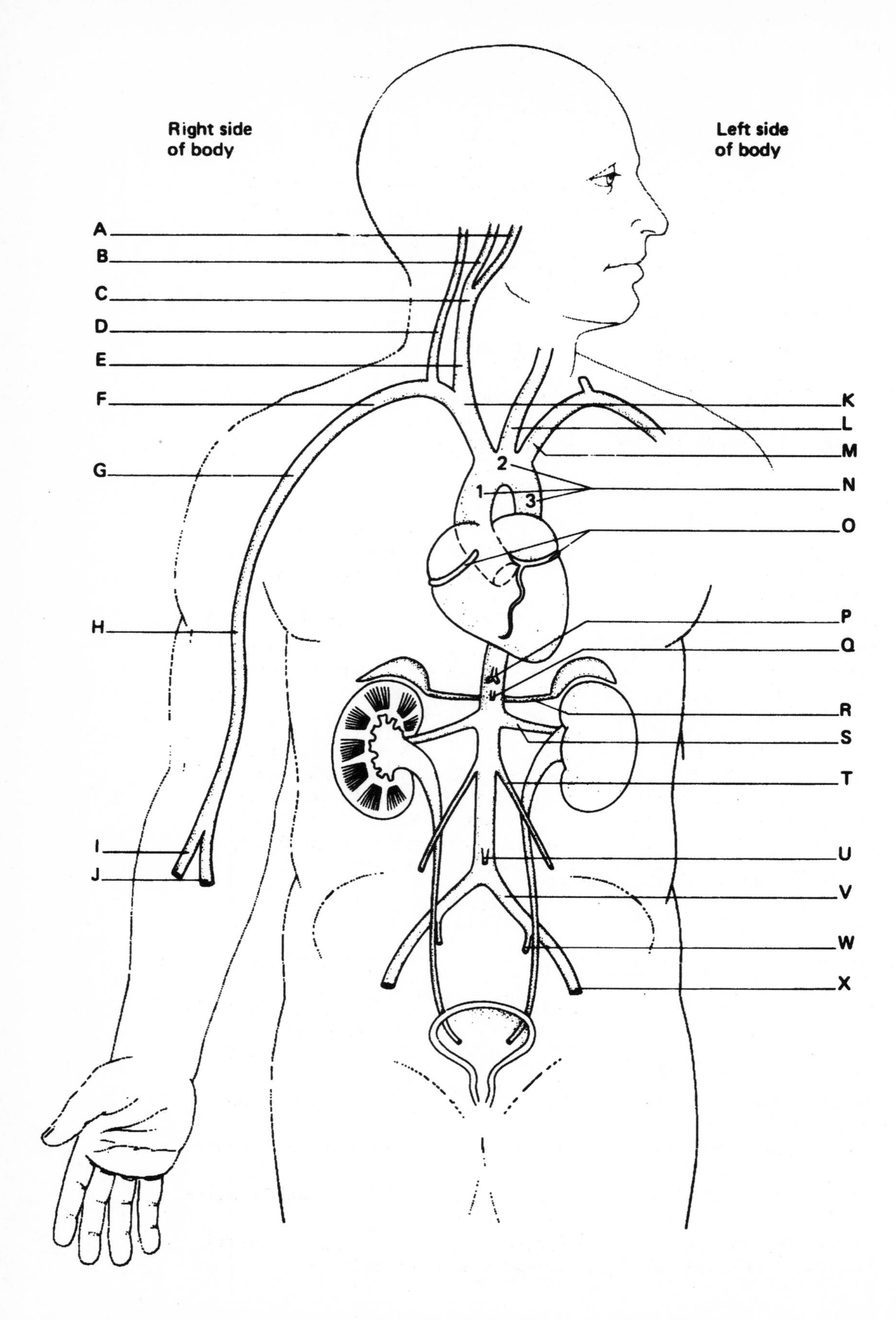

Figure LG 21.4 Major systemic arteries. Identify letter labels as directed in Checkpoints D2, D3, and D7–D9.

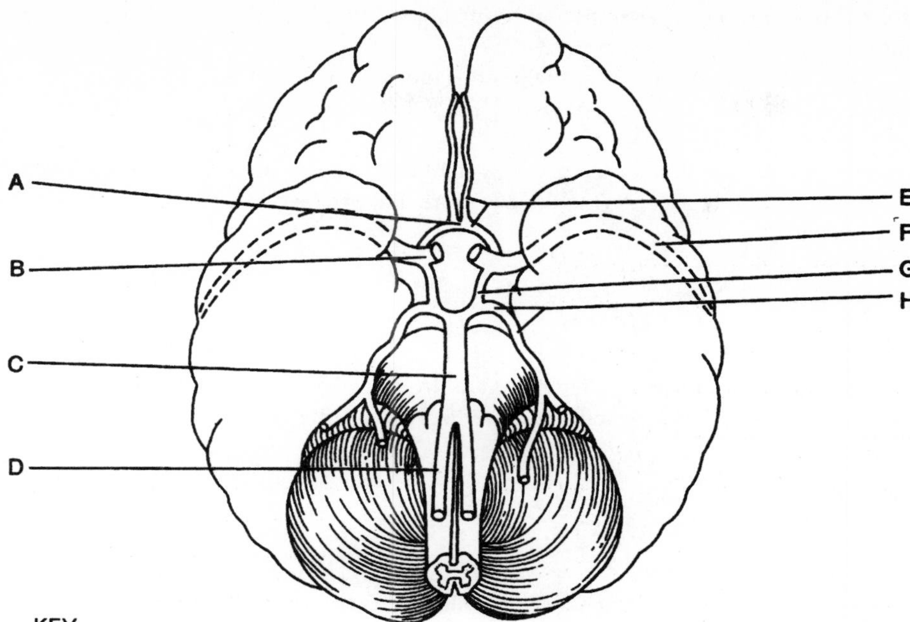

KEY

A. Anterior communicating cerebral artery (1)
B. Internal carotid arteries (cut) (2)
C. Basilar artery (1)
D. Vertebral arteries (2)
E. Anterior cerebral arteries (2)
F. Middle cerebral arteries (2)
G. Posterior communicating cerebral arteries (2)
H. Posterior cerebral arteries (2)

Figure LG 21.5 Arterial blood supply to the brain including vessels of the cerebral arterial circle (of Willis). View is from the undersurface of the brain. Numbers in parentheses indicate whether the arteries are single (1) or paired (2). Refer to Checkpoints D2, D4, and D5.

✓ **D3.** The right and left common carotid arteries ascend the neck in positions considerably *(anterior? posterior?)* to the vertebral arteries. Find a pulse in one of your own carotids. These vessels each bifurcate (divide into two vessels) at about the level of your mandible. Identify the two branches (A and B) on Figure LG 21.4. Which supplies the brain and eye? *(Internal? External?)* What structures are supplied by the external carotids? (See Exhibit 21.3 page 724 in the text, for help.)

✓ **D4.** Refer to Figure LG 21.5 and answer these questions about the vital blood supply to the brain.

a. An anastomosis, called the cerebral arterial circle of ____________________, is located just inferior to the brain. This circle closely surrounds the pituitary gland.

b. The circle is supplied with blood by two pairs of vessels: anteriorly, the *(internal? external?)* carotid arteries and, posteriorly, the ____________________ arteries (joined to form the basilar artery).

c. The circle supplies blood to the brain by three pairs of cerebral vessels, identified on Figure LG 21.5 by letter labels ______ ______, and ______. Vessels A and G complete the circle; they are known as anterior and posterior ____________________ arteries. Now cover the key and correctly label each of the lettered vessels.

✓ **D5.** *Critical thinking.* Relate symptoms to the interruption of normal blood supply to the brain as described in each case. Explain your reasons.

a. A patient's basilar artery is sclerosed. Which part of the brain is most likely to be deprived of normal blood supply? *(Frontal lobes of cerebrum? Occipital lobes of cerebrum, as well as cerebellum and pons?)*

b. Paralysis and loss of some sensation are noted on a patient's right side. Angiography points to sclerosis in a common carotid artery. Which one is more likely to be sclerosed? *(Right? Left?)* Explain.

c. The middle cerebral artery which passes between temporal and frontal lobe (see broken lines on Figure LG 21.5) is blocked. Which sense is more likely to be affected? *(Hearing? Vision?)* Explain.

d. Anterior cerebral arteries pass anteriorly from the circle of Willis and then arch backward along the medial aspect of each cerebral hemisphere. Disruption of blood flow in the right anterior cerebral artery will be most likely to cause effects in the *(right? left?) (hand? foot?)*. *Hint:* The affected brain area would include region Z in Figure LG 15.1 and/or region R in Figure LG 15.2, pages LG 294 and 298. Note that these regions are on the medial aspect of the right hemisphere.

✓ **D6.** Identify which category (listed in the box) fits each artery named below.

PA.	**Parietal abdominal**	**VA.**	**Visceral abdominal**
PT.	**Parietal thoracic**	**VT.**	**Visceral thoracic**

_______ a. Esophageal arteries

_______ b. Lumbar and inferior phrenic arteries

_______ c. Renal arteries

_______ d. Superior phrenic arteries

✓ **D7.** Complete the table describing visceral branches of the abdominal aorta. Identify these on Figure LG 21.4 (vessels P to U). Refer also to Figure LG 24.4, page LG 548.

Artery	Letter	Structures Supplied
a. Celiac (three major branches) 1. 2. 3.		 1. Liver 2. 3.
b.	Q	Small intestine and part of large intestine and pancreas
c.		Adrenal (suprarenal) glands
d. Renal	S	
e.		Testes (or ovaries)
f. Inferior mesenteric		

✓ **D8.** The aorta ends at about level _______ vertebra by dividing into right and left

_____________________ arteries. Label these at V on Figure LG 21.4.

D9. Label the two branches (W and X on Figure LG 21.4) of each common iliac artery. Name structures supplied by each.

a. Internal iliac artery

b. External iliac artery

✓ **D10.** *For extra review.* Suppose that the right femoral artery were completely occluded. Describe an alternate (or collateral) route that would enable blood to pass from the right common iliac artery to the little toe of the right foot.

Right common iliac artery →

D11. *For extra review.* Identify locations of pulses on these three arteries on yourself: femoral, popliteal, dorsalis pedis.

✓ **D12.** *For extra review.* Check your understanding of routes of arterial blood by filling in blanks with names of the two arteries that transport blood directly into arteries or organs listed below. The first one is done for you. Note: L = left; R = right; a. = artery. (*Hint*: refer to text Exhibits 21.1–21.6.)

a. Aorta (descending) → L. renal a. → L. kidney

b. _______________ → _______________ → R. subclavian a.

c. _______________ → _______________ → pancreatic a.

d. _______________ → _______________ → R. gastric a.

e. _______________ → _______________ → R. anterior cerebral a.

f. _______________ → _______________ → circumflex a.

g. _______________ → _______________ → basilar a.

h. _______________ → _______________ → inferior surface of the diaphragm

i. _______________ → _______________ → esophagus

j. _______________ → _______________ → L. side of urinary bladder

E. Circulatory routes: systemic veins (pages 737–750)

✓ **El.** Name the three main vessels that empty venous blood into the right atrium of the

heart. ____________ ____________ ____________

✓ **E2.** Do this exercise about venous return from the head.

a. Venous blood from the brain drains into vessels known as ____________. These are *(structurally? functionally?)* like veins.

b. Name the sinus that lies directly beneath the sagittal suture: ____________

c. Blood from all of the dural venous sinuses eventually drains into the

____________ veins, which descend in the neck. These veins are positioned

close to the ____________ arteries.

d. Where are the jugular arteries and carotid veins located?

✓ **E3.** Imagine yourself as a red blood cell currently located in the *left brachial vein*. Answer these questions.

a. In order for you to flow along in the bloodstream to the *left brachial artery*, which of these must you pass through? *(A brachial capillary? The heart and a lung?)*

b. In order for you to flow from the left brachial vein to the *right arm*, must you pass through the heart? *(Yes? No?)* Both sides of the heart, that is, right and left? *(Yes? No?)* One lung? *(Yes? No?)*

✓ **E4.** Once you are familiar with the pathways of systemic arteries, you know much about the veins that accompany these arteries. Demonstrate this by tracing routes of a drop of blood along the following deep vein pathways.

a. From the thumb side of the left forearm to the right atrium

b. From the medial aspect of the left knee to the right atrium

c. From the right kidney to the right atrium

✓ **E5.** Contrast veins with arteries in this exercise.

a. Veins have *(higher? lower?)* blood pressure than arteries and a *(faster? slower?)* rate of blood flow.

b. Veins are therefore *(more? less?)* numerous, in order to compensate for slower blood flow in each vein. Many of the "extra" veins consist of vessels located just under the skin (in subcutaneous fascia); these are known as _________________ veins.

✓ **E6.** On text Figures 21.26 to 21.28 (pages 743-750), locate the following veins. Identify the location at which each empties into a deep vein.

a. Basilic ___________________________________

b. Median cubital ______________________________

c. Great saphenous ___________________________

d. Small saphenous ___________________________

E7. *A clinical correlation.* Write a short essay describing clinical significance of the great saphenous veins. Be sure to use these three terms: *varicose, intravenous,* and *bypass.*

E8. Veins connecting the superior and inferior venae cavae are named _________________ veins. Identify the azygos system of veins on Figure 21.27 in the text. What parts of the thorax do they drain? Explain how these veins may help to drain blood from the lower part of the body if the inferior vena cava is obstructed.

✓ **E9.** Name the visceral veins that empty into the abdominal portion of the inferior vena cava. Then color them blue on Figure LG 26.1 (page LG 596). Contrast accompanying systemic arteries by coloring them red on that figure. Note on Figure LG 26.1 that the inferior vena cava normally lies on the *(right? left?)* side of the aorta. One way to remember this is that the inferior vena cava is returning blood to the *(right? left?)* side of the heart.

✓ **E10.** *For extra review.* Check your understanding of routes of venous blood by filling in blanks with names of the vein that receives blood directly from the vein(s) listed in each case below. The first one is done for you. Note: L = left; R = right; v. = vein. (*Hint*: refer to text Exhibits 21.8–21.12.)

a. R. internal jugular v. → R. brachiocephalic v.

b. Superior sagittal (cranial venous) sinus → R. transverse sinus → R. sigmoid sinus → ______________

c. R. brachial v. → R. axillary v. → ______________

d. R. and L. brachiocephalic v's → ______________

e. R. deep palmar venous arch → ______________

f. R. superficial palmar venous arch → ______________

g. R. anterior and posterior tibial v's → ______________

h. R. external and internal iliac v's → ______________

i. R. and L common iliac v's → ______________

j. R. ovarian (or testicular) v. → ______________

k. L. ovarian (or testicular) v. → ______________

F. Circulatory routes: hepatic portal, pulmonary, and fetal circulation (pages 751–755)

✓ **F1.** Refer to Figure LG 20.3 (page LG 424) and do the following activity.

a. Label the following vessels: *gastric, splenic,* and *mesenteric vessels.* Note that all of these vessels drain directly into the vessel named *(7, inferior vena cava? H, portal vein to the liver?).* Label that vessel on Figure LG 20.3.

b. Label the vessel that carries oxygenated blood into the liver.

c. In the liver, blood from both the portal vein and proper hepatic artery mixes as it passes through tortuous capillary-like vessels known as ______________. On Figure LG 20.3 label a vessel that collects this blood and transports it to the inferior vena cava.

d. What functions are served by the special hepatic portal circulation?

e. Refer to text Figure 21.29. Blood from most of the wall of the descending and sigmoid colon and the rectum drains into the *(superior? inferior?)* mesenteric vein. Blood from the wall of the ascending colon (as well as the small intestine) drains into the *(superior? inferior?)* mesenteric vein.

✓ **F2.** Contrast pulmonary circulation with systemic circulation in this activity. Write *P* (pulmonary) or *S* (systemic).

a. Which is considered a lower resistance system? List structural features that explain this lower resistance.

b. Which arteries are likely to have a higher blood pressure ______ (BP)? Write typical BP values for these vessels: aorta ______ mm Hg; pulmonary artery ______ mm Hg.

c. Which capillaries normally have a hydrostatic pressure (BP) of about 10 mm Hg? Write one consequence of increased pulmonary capillary BP.

d. Which system has its blood vessels constrict in response to hypoxia? (*Hint:* See checkpoint B12c, page LG 457.) ______ Write one advantage and one disadvantage of this fact.

F3. What aspects of fetal life require the fetus to possess special cardiovascular structures not needed after birth?

✓ **F4.** Complete the Table LG 21.1 below about fetal structures. (Note that the structures are arranged in order of blood flow from fetal aorta back to fetal aorta.)

Table LG 21.1 Six structures unique to fetal circulation.

Fetal Structure	Structures Connected	Function	Fate of Structure After Birth
a.		Carries fetal blood low in oxygen and nutrients and high in wastes	
b. Placenta	(Omit)		
c.	Placenta to liver and ductus venosus		
d.		Branch of umbilical vein, bypasses liver	
e. Foramen ovale			
f.			Becomes ligamentum arteriosum

✓ **F5.** For each of the following pairs of fetal vessels, circle the vessel with higher oxygen content.

a. Umbilical artery/umbilical vein
b. Femoral artery/femoral vein
c. Lower abdominal aorta/thoracic portion of inferior vena cava

G. Developmental anatomy of blood vessels and blood; aging of the cardiovascular system (pages 756–757)

G1. Write a paragraph describing embryonic development of blood vessels. Include the following key terms: *15 to 16 days, mesoderm, mesenchyme, angioblasts, blood island spaces, endothelial, tunica.*

G2. List four sites of blood formation in the embryo and fetus.

✓ **G3.** Draw arrows next to each factor listed below to indicate whether it increases (↑) or decreases (↓) with normal aging.

_______ a. Size and strength of cardiac muscle cells

_______ b. Cardiac output (CO)

_______ c. Maximal heart rate

_______ d. Systolic blood pressure

_______ e. Cerebral blood flow

H. Disorders, medical terminology (pages 758–759)

✓ **H1.** Match types of hypertension listed in the box with descriptions below. One answer will be used twice; one answer will not be used.

Pri.	**Primary**	**Sta-1.**	**Stage 1**
Sec.	**Secondary**	**Sta-3.**	**Stage 3**
		Sta-4.	**Stage 4**

_______ a. Also known as idiopathic (no known cause) hypertension

_______ b. Over 90% of cases of hypertension fit into this category

_______ c. Hypertension that has a known cause

_______ d. BP of 150/94 fits into this category

_______ e. BP of 200/110 fits into this category

✓ **H2.** Match each organ listed in the box with the related disorder that may cause secondary hypertension.

AC.	**Adrenal cortex**	**AM.**	**Adrenal medulla**	**K.**	**Kidney**

_______ a. Excessive production of renin, which catalyzes a step in the formation of angiotensin II, a powerful vasoconstrictor

_______ b. Pheochromocytoma, a tumor that releases large amounts of norepinephrine and epinephrine

_______ c. Release of excessive amounts of aldosterone, which promotes salt and water retention

H3. Describe possible effects of hypertension upon the following body structures:

a. Heart

b. Cerebral blood vessels

c. Kidneys

H4. Explain how each of the following therapeutic measures can help control hypertension.

a. Sodium restriction

b. Cessation of smoking

c. Exercise

d. Stress reduction

e. Vasodilators

f. A.C.E. inhibitors

g. Diuretics

H5. Define *aneurysm* and list four possible causes of aneurysms.

✓ **H6.** *A clinical correlation.* Explain how a deep vein thrombosis (DVT) of the left femoral vein can lead to a pulmonary embolism.

✓ **H7.** Match terms related to blood vessels with descriptions below. Use each answer once.

Angio.	**Angiogenesis**	**Occ.**	**Occlusion**
Art.	**Arteritis**	**Ortho.**	**Orthostatic hypotension**
C.	**Claudication**		

_______ a. Growth of new blood vessels, for example with weight gain, pregnancy, or to increase blood supply to tumor

_______ b. Inflammation of an artery

_______ c. Obstruction of a vessel, such as an artery, due to atherosclerotic plaque

_______ d. Pain upon exercise due to impaired circulation in limbs

_______ e. Significant drop in blood pressure upon standing up

ANSWERS TO SELECTED CHECKPOINTS: CHAPTER 21

A2. (a) L. (b) TI. (c) VV. (d) TM. (e) TE.

A3. (a) Capillaries. (b) Arterioles. (c) Capillaries. (d) Venules, small veins, inferior vena cava (actually all veins and venules). (e) Aorta and other large arteries. (f) Capillaries. (g) Medium-sized arteries.

A5. (a) Direct movement across endothelial membranes. (b) Intercellular clefts. (c) Fenestrations. (d) Pinocytic vesicles.

A6. (a) Thinner; less; an even flow. (b) Valves. (c) Varicose veins; superficial; skeletal muscles around deep veins limit overstretching. (d) Esophagus (esophageal varices) and rectal-anal region (hemorrhoids). (e) Surrounding dense connective tissue (such as the dura mater); function, but not structure; intracranial (such as superior sagittal venous sinus) and coronary sinus.

A8. (a) Venules and veins; capillaries. (b) Sympathetic; veins of skin and of abdominal organs such as liver and spleen. (c) Exercise, hemorrhage.

A9. (a–c) S. (d) T. (e) B.

A10. (a) A, IFOP; B, BHP; C, IFHP; D, BCOP. (b) +10; out of. (c) BHP is about 16 mm Hg; –9. (d) Starling; lymphatic. (e) Edema; BHP, BCOP, increase.

A11. (a) ↑ BHP (pulmonary hypertension due to backlog of blood from a failing left ventricle. (b) ↓ PP (as plasma proteins leaked out of damaged vessels); also ↑ PC. (c) ↑ ECV and ↑ BHP.

B1. (a) 4–6; output. (b) Pressure (such as blood pressure) and resistance to flow.

B2. (a) Systolic; diastole; elastic, 70-80. (b) 93; diastole; diastolic. (c) 78; 120 – 78 = 42; 78 + (1/3 × 42 = 14) = 92.

B3. (a) Pressure; blood on the wall of a blood vessel; in; ↑ CO → ↑ MABP; de. (b) More; inversely, directly; in, 16. (c) Low; 85–35; 35–16; 16–0; close to 0. (d) Arterioles; arterioles.

B4. (a) De; ↑ $R_{\text{within a vessel}}$ → → ↓ $BP_{\text{in that same vessel}}$; de. (b) About 120/80; systemic (vascular) resistance; ↑ $R_{\text{(systemic arterioles skin or abdomen)}}$ → ↑ $BP_{\text{in systemic arteries}}$; or ↑ SVR → ↑ $BP_{\text{systemic arterial}}$; in. (c) $BP_{\text{systemic arteries}}$; = CO × SVR. (dl) In, dehydration or polycythemia. (d2) In, weight gain (or obesity), growth, or pregnancy. (d3) De, vasoconstriction. (e) Peripheral; digestive organs such as liver, spleen, stomach, intestine, as well as kidneys.

B5. (a) B; does. Recall that the body grows an average of about 300 km (200 miles) of new blood vessels for each extra pound of weight. (b) Sympathetic; C; 16; in.

B7. (a) Arteries; capillaries. (b) Increases. (c) 4500–6000. (d) An indirect; 3 to 5, high. (e) Larger; de; clot formation (also shock because pooled blood there does not circulate to tissues). (f) 1. (g) Syncope.

B8. (a) ↑ CO. (b) ↑ Stroke volume. (c) ↑ Venous return. (d) ↑ Exercise. (e) ↑ Blood volume. (f) ↑ Heart rate. (g) ↑ Sympathetic impulses. (h) ↑ Norepinephrine and epinephrine. (i) ↓ Vagal impulses. (j) ↑ ADH. (k) ↑ Aldosterone. (1) ↑ NaCl in diet. (m) ↑ Systemic vascular resistance (SVR). (n) ↑ Vasoconstriction. (o) ↑ Angiotensin II. (p) ↑ Viscosity of blood. (q) ↑ RBC count. (r) ↑ Plasma proteins.

B9. (a) High; decreased. (b) B C F G H.

B10. (a) Box 1: baro, chemo. Blood pressure, H^+, CO_2, and O_2; carotid arteries and aorta. (b) IX, X. (c) Box 3: medulla. Box 2: cerebral cortex, limbic system, and hypothalamus. (d) Brakes; cardiostimulatory, step on the gas pedal; take your foot off the gas; car goes faster/heart beats faster and with greater force of contraction. (e) Stimulating, vasoconstriction; in. (f) ↑ stroke volume (SV), ↑ heart rate (HR), and ↑ systemic vascular resistance (SVR); in. (g) Negative.

B11. (a) Tone; drop precipitously; syncope. (b) Alpha, norepinephrine, constrict; constrict, in. (c) De; atrial natriuretic peptide (ANP), histamine. (d) General systemic, cerebral. (e) Chemo; hypoxia. hypercapnia, vasoconstriction, in. (fl–4) ↑ ; (f5) ↓.

B12. (a) Local. (b) Low, vasodilator; lactic; nitric oxide, K^+, H^+, and adenosine. (c) Just the opposite; constrict; blood is shunted to parts of the lungs that are well oxygenated. (d) Vasospasm; thromboxane A2 and serotonin.

C3. 60–100; rapid, bradycardia; bradycardia.

C4. (a) Sphygmomanometer. (b) Brachial; no pulse is felt or no sound is heard via a stethoscope. (c) Systolic. (d) Change or cease. (e) Korotkoff.

C5. 128/78, 50 (128–78); yes, it is within the normal range; de.

C6. (a) Blood; hypovolemic; loss of blood or other body fluids, or shifting of fluids into interstitial spaces (edema) as with bums. (b) Obstructive, cardiogenic; anaphylactic, neurogenic, or vascular. (c) Sympathetic, in; 10; cool, pale, and clammy, constriction; in. (d) Epinephrine, renin, ADH (vasopressin). (e) Aldosterone and ADH. (f) It causes local vasodilation.

C7. (a–b) ↑ . (c–f) ↓ (g) ↑ .

D1. Refer to Figure 21.17 in your text. The pulmonary artery and its branches carry blood between right ventricle and lungs, whereas pulmonary veins return from the lungs to the left atrium. Systemic arteries transport blood from the left ventricle to all organs of the body, whereas systemic veins return blood to the right atrium. Left (as it must pump blood to a much more extensive set of high resistance vessels).

D2. (a) Aorta; left; ascending aorta, arch of the aorta, descending aorta. (b) Coronary; heart. (c) K, brachiocephalic; L, left common carotid; M, left subclavian. (d) F; in general, the right extremity and right side of thorax, neck, and head; right common carotid. (e) G, axillary; H, brachial; blood pressure. (f) I, radial; J, ulnar; I; radial, ulnar; fingers. (g) Vertebral; basilar.

D3. Anterior; internal; external portions of head (face, tongue, ear, scalp) and neck (throat, thyroid).

D4. (a) Willis. (b) Internal, vertebral. (c) E, F, H; communicating.

D5. (a) Occipital lobes of cerebrum, as well as cerebellum and pons. (b) Left, because this artery supplies the left side of the brain. (Remember that neurons in tracts cross to the opposite side of the body, but arteries do not.) (c) Hearing, because it is perceived in the temporal lobe. (Visual areas are in occipital lobes). (d) Left foot.

D6. (a) VT. (b) PA. (C) VA. (d) PT.

D7.

Artery	Letter	Structures Supplied
a. Celiac (three major branches) 1. **Common hepatic** 2. **Gastric** 3. **Splenic**	**P**	1. Liver, **gallbladder and parts of stomach, duodenum, pancreas** 2. **Stomach and esophagus** 3. **Spleen, pancreas, and stomach**
b. **Superior mesenteric**	Q	Small intestine and part of large intestine and pancreas
c. **Adrenal (suprarenal)**	**R**	Adrenal (suprarenal) glands
d. Renal	S	**Kidneys**
e. **Gonadal (testicular or ovarian)**	**T**	Testes (or ovaries)
f. Inferior mesenteric	**U**	**Parts of colon and rectum**

D8. L4, common iliac.

D10. Right external iliac artery → right lateral circumflex artery (right descending branch) → right anterior tibial artery → right dorsalis pedis artery → to arch and digital arteries in foot.

D12. (b) Aorta (arch) → brachiocephalic a. → (R. subclavian a.) (c) Celiac a. → splenic a. → pancreatic a.) (d) Celiac a. → common hepatic a. → (R. gastric a.) (e) R. common carotid a. → R. internal carotid a. → (R. anterior cerebral a.) (f) Aorta (ascending) → L. coronary a. → (circumflex a.) (g) R. and L. subclavian a.. R. and L. vertebral a.→ (basilar a.) (h) Aorta (descending) → inferior phrenic a. → (inferior surface of the diaphragm.) (i) Aorta (descending) → esophageal a. → (esophagus.) (j) L. common iliac a. → L. internal iliac a. → (L. side of urinary bladder.)

E1. Superior vena cava, inferior vena cava, coronary sinus.

E2. (a) Dural venous sinuses; functionally. (b) Superior sagittal. (c) Internal jugular; common carotid. (d) There are no such vessels.

E3. (a) The heart and a lung. (b) Yes; yes; yes.

E4. (a) Left radial vein → left brachial vein → left axillary vein → left subclavian vein → left brachiocephalic vein → superior vena cava. (Note that there are left and right brachiocephalic veins, but only one artery of that name.) (b) Small veins in medial aspect of left knee → left popliteal vein → left femoral vein → left external iliac vein → left common iliac vein → inferior vena cava. (c) Right renal vein → inferior vena cava.

E5. (a) Lower, slower. (b) More; superficial.

E6. (a) Axillary. (b) Axillary. (c) Femoral. (d) Popliteal.

E9. See key, page 596 of the Guide. Right, right.

E10. (b) R. internal jugular v. (c) R. subclavian v. (d) Superior vena cava. (e) R. radial v. (f) R. ulnar v. (g) Popliteal v. (h) R. common iliac v. (i) Inferior vena cava. (j) Inferior vena cava. (k) R. renal v. (and then inferior vena cava).

F1. (a) E, gastric; F, splenic; I, mesenteric; H, portal vein to the liver. (b) D, hepatic artery. (c) Sinusoids; G, hepatic vein (there are two but only one is shown on Figure LG 20.3. (d) While in liver sinusoids, blood is cleaned, modified, detoxified. Ingested nutrients and other chemicals are metabolized and stored. (e) Inferior; superior.

F2. (a) P; pulmonary arteries have larger diameters and shorter length (a shorter circuit), thinner walls, less elastic tissue. (b) S; 120/80; 25/8 (pulmonary BP is about one-fifth of systemic BP). (c) P; pulmonary edema with shortness of breath and impaired gas exchange. (d) P; an advantage is the diversion of blood to well-ventilated regions. One disadvantage is that chronic vasoconstriction in persons with chronic hypoxia (as in emphysema or chronic bronchitis) may lead to cor pulmonale.

F4. **Table LG 21.1A** Structures in fetal circulation.

Fetal Structure	Structures Connected	Function	Fate of Structure After Birth
a. **Umbilical arteries (2)**	**Fetal internal iliac arteries to placenta**	Carries fetal blood low in oxygen and nutrients and high in wastes	**Medial umbilical ligaments**
b. Placenta	(Omit)	**Site where maternal and fetal blood exchange gases, nutrients, and wastes**	**Delivered as "afterbirth"**
c. **Umbilical vein (1)**	Placenta to liver and ductus venosus	**Carries blood high in oxygen and nutrients, low in wastes**	**Round ligament of the liver (ligamentum teres)**
d. **Ductus venosus**	**Umbilical vein to inferior vena cava**	Branch of umbilical vein, bypasses liver	**Ligamentum venosum**
e. Foramen ovale	**Right and left atria**	**Bypasses lungs**	**Fossa ovalis**
f. **Ductus arteriosus**	**Pulmonary artery to aorta**	**Bypasses lungs**	Becomes ligamentum arteriosum

F5. (a) Umbilical vein. (b) Femoral artery. (c) Thoracic portion of inferior vena cava (because umbilical vein blood enters it via ductus venosus in upper abdomen).

G3. (a) ↓ (b) ↓ . (c) ↓ . (d) ↑ . (e) ↓ .

H1. (a-b) Pri. (c) Sec. (d) Sta-1. (e) Sta-3.

H2. (a) K. (b) AM. (c) AC.

H6. Embolus (a "clot-on-the-run") travels through femoral vein to external and common iliac veins to inferior vena cava to right side of heart. It can lodge in pulmonary arterial branches, blocking further blood flow into the pulmonary vessels.

H7. (a) Angio. (b) Art. (c) Occ. (d) C. (e) Ortho.

MORE CRITICAL THINKING: CHAPTER 21

1. Explain how the structure of the following types of blood vessels is admirably suited to their functions: elastic arteries, muscular arteries, arterioles, metarterioles, and capillaries.
2. Discuss advantages to health offered by collateral circulation. Include the terms anastomoses and end arteries in your discussion.
3. The typical human body contains about 4–6 liters of blood. Describe mechanisms that regulate where this blood goes at any given moment, for example, at rest versus during a period of vigorous exercise. Be sure to include the following terms in your discussion: metarteriole, precapillary sphincter, sympathetic, and vasoconstriction.
4. Describe all mechanisms that will help you to raise your blood pressure in response to a frightening experience. Be sure to include both nervous and hormonal responses.
5. Refer to Figure LG 21.2 and review related sections of Chapter 17 and 18 as you describe the specific mechanism by which each of the following medications or disorders lowers blood pressure: (a) angiotensin converting hormone inhibitor (ACE-I) medications; (b) beta-blocker medications; (c) alpha-blocker medications; (d) diuretics; (e) hemorrhage; and (f) anemia.
6. In most cases, arteries and veins are parallel in both name and locations, such as the brachial artery and brachial vein lying close together in the arm. Identify major arteries and veins that differ in names and locations.

MASTERY TEST: CHAPTER 21

Questions 1–2: Circle the letters preceding all correct answers to each question.

1. In the most direct route from the left leg to the left arm of an adult, blood must pass through all of these structures:
   A. Inferior vena cava
   B. Brachiocephalic artery
   C. Capillaries in lung
   D. Hepatic portal vein
   E. Left subclavian artery
   F. Right ventricle of heart
   G. Left external iliac vein
2. In the most direct route from the fetal right ventricle to the fetal left leg, blood must pass through all of these structures:
   A. Aorta
   B. Umbilical artery
   C. Lung
   D. Ductus arteriosus
   E. Ductus venosus
   F. Left ventricle
   G. Left common iliac artery

Questions 3–9: Circle the letter preceding the one best answer to each question. Note whether the underlined word or phrase makes the statement true or false.

3. Choose the *false* statement.
   A. In order for blood to pass from a vein to an artery, it must pass through chambers of the heart.
   B. In its passage from an artery to a vein a red blood cell must ordinarily travel through a capillary.
   C. The wall of the femoral artery is thicker than the wall of the femoral vein.
   D. Most of the smooth muscle in arteries is in the tunica interna (intima).
4. Choose the *false* statement.
   A. Arteries contain valves, but veins do not.
   B. Decrease in the size of the lumen of a blood vessel by contraction of smooth muscle is called vasoconstriction.
   C. Most capillary exchange occurs by simple diffusion.
   D. Sinusoids are wider and more tortuous (winding) than capillaries.
5. Choose the *false* statement.
   A. Cool, clammy skin is a sign of shock that results from sympathetic stimulation of blood vessels and sweat glands.
   B. Nicotine is a vasodilator that helps control hypertension.
   C. Orthostatic hypotension refers to a sudden, dramatic drop in blood pressure upon standing or sitting up straight.
   D. Most hypertension is primary since it has no identifiable cause.
6. Choose the *false* statement.
   A. The vessels that act as the major regulators of blood pressure are arterioles.
   B. The only blood vessels that carry out exchange of nutrients, oxygen, and wastes are capillaries.
   C. Blood in the umbilical artery is normally more highly oxygenated than blood in the umbilical vein.
   D. At any given moment more than 50% of the blood in the body is in the veins.
7. Choose the *true* statement.
   A. The basilic vein is located lateral to the cephalic vein.
   B. The great saphenous vein runs along the lateral aspect of the leg and thigh.

C. The vein in the body most likely to become varicosed is the femoral.
D. The hemiazygos and accessory hemiazygos veins lie to the left of the azygos.

8. Choose the *true* statement.
   A. A normal blood pressure for the average adult is 160/100.
   B. During exercise, blood pressure will tend to increase.
   C. Sympathetic impulses to the heart and to arterioles tend to decrease blood pressure.
   D. Decreased cardiac output causes increased blood pressure.
9. Choose the *true* statement.
   A. Cranial venous sinuses are composed of the typical three layers found in walls of all veins.
   B. Dural venous sinuses all eventually empty into the external jugular veins.
   C. Internal and external jugular veins do unite to form common jugular veins.
   D. Most parts of the body supplied by the internal carotid artery are ultimately drained by the internal jugular vein.

Questions 10–14: Circle the letter preceding the one best answer to each question.

10. Hepatic portal circulation carries blood from:
    A. Kidneys to liver
    B. Liver to heart
    C. Stomach, intestine, and spleen to liver
    D. Stomach, intestine, and spleen to kidneys
    E. Heart to lungs
    F. Pulmonary artery to aorta
11. Which of the following is a parietal (rather than a visceral) branch of the aorta?
    A. Superior mesenteric artery
    B. Bronchial artery
    C. Celiac artery
    D. Lumbar artery
    E. Esophageal artery
12. All of these vessels are in the leg or foot *except:*
    A. Saphenous vein
    B. Azygos vein
    C. Peroneal artery
    D. Dorsalis pedis artery
    E. Popliteal artery
13. All of these are superficial veins *except:*
    A. Median cubital
    B. Basilic
    C. Cephalic
    D. Great saphenous
    E. Brachial
14. Vasomotion refers to:
    A. Local control of blood flow to active tissues
    B. The Doppler method of checking blood flow
    C. Intermittent contraction of small vessels resulting in discontinuous blood flow through capillary networks
    D. A new aerobic dance sensation

Questions 15–17: Arrange the answers in correct sequence.

______ ______ ______ ______ 15. Route of a drop of blood from the right side of the heart to the left side of the heart:
A. Pulmonary artery
B. Arterioles in lungs
C. Capillaries in lungs
D. Venules and veins in lungs

______ ______ ______ ______ ______ 16. Route of a drop of blood from small intestine to heart:
A. Superior mesenteric vein
B. Hepatic portal vein
C. Small vessels within the liver
D. Hepatic vein
E. Inferior vena cava

______ ______ ______ 17. Layers of blood vessels, from most superficial to deepest:
A. Tunica interna
B. Tunica externa
C. Tunica media

A. Aorta and other arteries	**C. Capillaries**
B. Arterioles	**D. Venules and veins**

Questions 18–20, use the following answers:

______ ______ ______ ______ 18. Blood pressure in vessels, from highest to lowest:

______ ______ ______ ______ 19. Total cross-sectional areas of vessels, from highest to lowest:

______ ______ ______ ______ 20. Velocity of blood in vessels, from highest to lowest:

Questions 21–25: Fill-ins. Complete each sentence or answer the question with the word(s) or phrase that best fits.

________________ 21. During inhalation of air, the diaphragm moves ________-ward, causing a(n) ________-crease in thoracic volume, a(n) ________-crease in thoracic pressure, and ________-crease in abdominal pressure. As a result, venous return to the heart ________-creases.

________________ 22. Name the major factor that creates blood colloid osmotic pressure (BCOP) that prevents excessive flow of fluid from blood to interstitial areas.

________________ 23. Name the two blood vessels that are locations of baroreceptors and chemoreceptors for regulation of blood pressure.

________________ 24. In taking a blood pressure, the cuff is first inflated over an artery. As the cuff is then slowly deflated, the first sounds heard indicate the level of ________ blood pressure.

________________ 25. Most arteries are paired (one on the right side of the body, one on the left). Name five or more vessels that are unpaired.

ANSWERS TO MASTERY TEST: CHAPTER 21

Multiple Answers

1. A C E F G
2. A D G

Multiple Choice

3. D
4. A
5. B
6. C
7. D
8. B
9. D
10. C
11. D
12. B
13. E
14. C

Arrange

15. A B C D
16. A B C D E
17. B C A
18. A B C D
19. C D B A
20. A B D C

Fill-ins

21. Down, in, de, in; in.
22. Plasma protein such as albumin
23. Aorta and carotids
24. Systolic
25. Aorta, anterior communicating cerebral, basilar, brachiocephalic, celiac, common hepatic, splenic, superior and inferior mesenteric, middle sacral, pancreatic, short

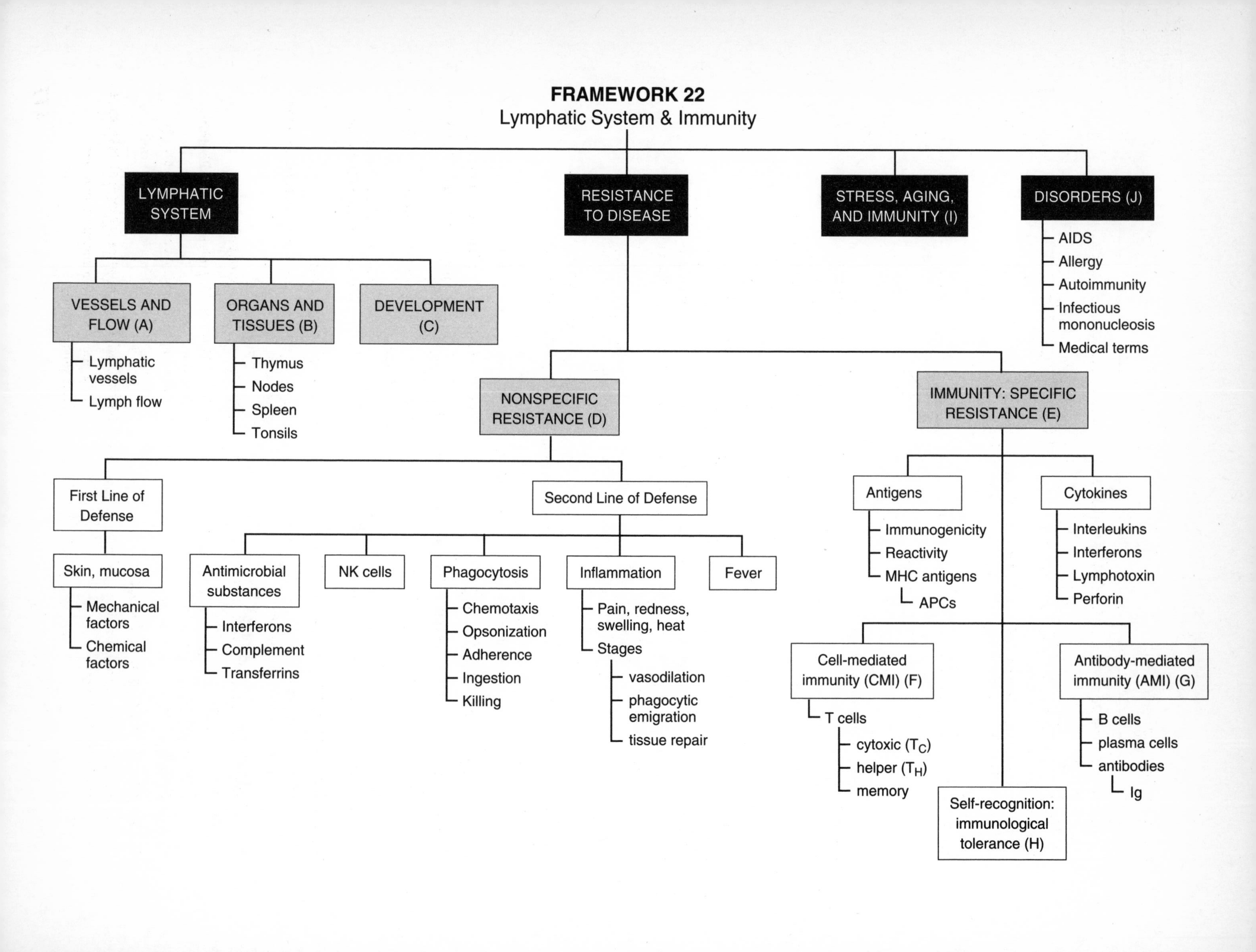

FRAMEWORK 22
Lymphatic System & Immunity
LYMPHATIC SYSTEM
RESISTANCE TO DISEASE
STRESS, AGING, AND IMMUNITY (I)
DISORDERS (J)
AIDS
Allergy
Autoimmunity
Infectious mononucleosis
Medical terms
VESSELS AND FLOW (A)
Lymphatic vessels
Lymph flow
ORGANS AND TISSUES (B)
Thymus
Nodes
Spleen
Tonsils
DEVELOPMENT (C)
NONSPECIFIC RESISTANCE (D)
IMMUNITY: SPECIFIC RESISTANCE (E)
First Line of Defense
Skin, mucosa
Mechanical factors
Chemical factors
Second Line of Defense
Antimicrobial substances
Interferons
Complement
Transferrins
NK cells
Phagocytosis
Chemotaxis
Opsonization
Adherence
Ingestion
Killing
Inflammation
Pain, redness, swelling, heat
Stages
vasodilation
phagocytic emigration
tissue repair
Fever
Antigens
Immunogenicity
Reactivity
MHC antigens
APCs
Cytokines
Interleukins
Interferons
Lymphotoxin
Perforin
Cell-mediated immunity (CMI) (F)
T cells
cytoxic (T_C)
helper (T_H)
memory
Self-recognition: immunological tolerance (H)
Antibody-mediated immunity (AMI) (G)
B cells
plasma cells
antibodies
Ig

CHAPTER 22

The Lymphatic and Immune System and Resistance to Disease

The human body is continually exposed to foreign (nonself) entities: bacteria or viruses, pollens, chemicals in a mosquito bite, or foods or medications that provoke allergic responses. Resistance to specific invaders is normally provided by a healthy immune system made up of antibodies and battalions of T lymphocytes.

Nonspecific resistance is offered by a variety of structures and mechanisms, such as skin, mucus, white blood cells assisting in inflammation, and even a rise in temperature (fever) that limits the survival of invading microbes. The lymphatic system provides sentinels at points of entry (as in tonsils) and at specific sites (lymph nodes) near entry to body cavities (inguinal and axillary regions). The lymphatic system also wards off the detrimental effects of edema by returning proteins and fluids to blood vessels.

As you begin your study of this system, carefully examine the Chapter 22 Topic Outline and Objectives; check off each one as you complete it. To organize your study of the lymphatic system, nonspecific resistance to disease, and immunity, glance over the Chapter 22 Framework now. Be sure to refer to the Framework frequently and note relationships among key terms in each section.

TOPIC OUTLINE AND OBJECTIVES

A. The lymphatic system: vessels and flow

1. Describe the components and major functions of the lymphatic and immune system.
2. Describe the organization of lymphatic vessels.
3. Describe the formation and flow of lymph.

B. The lymphatic system: organs and tissues

1. List and describe the primary and secondary lymphatic organs and tissues.

C. Developmental anatomy of the lymphatic system

1. Describe the development of the lymphatic system.

D. Nonspecific resistance to disease

1. Describe the mechanisms of nonspecific resistance to disease.

E. Specific resistance to disease: immunity

1. Define immunity, and describe how T cells and B cells arise.
2. Explain the relationship between an antigen and an antibody.
3. Compare the functions of cell-mediated immunity and antibody-mediated immunity.

F. Cell-mediated immunity

1. Describe the steps of a cell-mediated immune response.

G. Antibody-mediated immunity

1. Describe the steps in an antibody-mediated immune response.
2. Describe the chemical characteristics and actions of antibodies.

H. Self-recognition and immunological tolerance

1. Describe how self-recognition and self-tolerance develop.

I. Stress, aging and the immune system

1. Describe the effects of stress on the immune system.
2. Describe the effects of aging on the immune system.

J. Disorders, medical terminology

WORDBYTES

Now become familiar with the language of this chapter by studying each wordbyte, its meaning, and an example of its use within a term. After you study the entire list, self-check your understanding by writing the meaning of each wordbyte on the line. As you continue through the *Learning Guide,* identify (and fill in) additional terms that contain the same wordbyte.

Wordbyte	Self-check	Meaning	Example(s)
anti-	______________________	against	*anti*body
auto-	______________________	self	*auto*immune
axilla-	______________________	armpit	*axilla*ry nodes
chyl-	______________________	juice	cisterna *chyl*i
dendr-	______________________	a tree	*dendr*itic cells
-genic	______________________	producing	anti*genic*
hetero-	______________________	other	*hetero*graft
iso-	______________________	same	*iso*graft
lact-	______________________	milk	*lact*eals
lymph-	______________________	clear water	*lymph*atic vessel
meta-	______________________	beyond	*meta*stasis
phago-	______________________	eat	*phago*cytic
-stasis	______________________	to stand	meta*stasis*
xeno-	______________________	foreign	*xeno*graft

CHECKPOINTS

A. The lymphatic system: vessels and flow (pages 765–769)

A1. Summarize the body's defense mechanisms against invading microorganisms and other potentially harmful substances. Be sure to use these terms in your paragraph: *pathogens, resistance, susceptibility.*

✓ **A2.** Write *S* next to specific imm[illegible] responses and *NS* next to nonspecific responses.

_______ a. Barriers to infection, such as skin and mucous membranes

_______ b. Inflammation and phagocytosis

_______ c. Chemicals such as gastric or vaginal secretions, which are both acidic

_______ d. Antigen-antibody responses

_______ e. Destruction of an intruder by a cytotoxic T lymphocyte

_______ f. Antimicrobial substances such as interferons (IFNs) or complement

_______ g. Known as innate defenses, they are present from birth

✓ **A3.** Describe the functions of the lymphatic system in this Checkpoint.

a. Lymphatic vessels drains two types of chemicals from spaces between cells; these are ________________ and ________________. They also transport ________________ and *(water? fat?)*-soluble vitamins from the gastrointestinal (GI) tract to the bloodstream.

b. The defensive functions of the lymphatic system are carried out largely by white blood cells named ______________-cytes. *(B? T?)* lymphocytes destroy foreign substances directly, whereas B lymphocytes lead to the production of ________________.

✓ **A4.** List the components of the lymphatic system.

Check your understanding of locations of lymphatic components by completing this Checkpoint.

a. In subcutaneous tissue, lymph pathways follow *(arterial? venous?)* routes, whereas lymphatic vessels of viscera typically follow ________________ pathways.

b. Lymphatic capillaries are typically present in tissues that *(do? do not?)* have blood capillaries.

c. Explain how lymph differs from interstitial fluid.

A5. Describe relationships between terms in each pair.

a. *Lymphatic capillary epithelium/one-way valve*

b. *Anchoring filaments/edema*

c. *Lacteal/chyle*

✓ **A6.** Answer the following questions about the largest lymphatics.

a. In general, the thoracic duct drains *(three-fourths? one-half? one-fourth?)* of the body's lymph. This vessel starts in the lumbar region as a dilation known as the

__________________. This cisterna chyli receives lymph from which areas of the body?

b. In the neck the thoracic duct receives lymph from three trunks. Name these.

c. One-fourth of the lymph of the body drains into the ____________________ duct. The regions of the body from which this vessel collects lymph are:

d. The thoracic duct empties into the *(right? left?)* ____________________ *(artery? vein?)* close to its junction with the internal jugular vein. The right lymphatic duct empties into *(the same? a different?)* vein on the right side.

e. In summary, lymph fluid starts out originally in plasma or cells and circulates as

____________________ fluid. Then as lymph it undergoes extensive cleaning as it passes through nodes. Finally, fluid and other substances in lymph are returned to

____________________ in veins of the neck.

✓ **A7.** *Critical thinking*. Explain why a person with a broken left clavicle might exhibit these two symptoms:

a. Edema, especially of the left arm and both legs

b. Excess fat in feces (steatorrhea)

A8. Describe the general pathway of lymph circulation. *For extra review.* Refer to Chapter 1, Checkpoint C6 and Figure LG 1.1, page 8 in the *Learning Guide.*

✓ **A9.** The lymphatic system does not have a separate heart for pumping lymph. Describe two principal factors that are responsible for return of lymph from the entire body to major blood vessels in the neck. (Note that these same factors also facilitate venous return.)

B. The lymphatic system: organs and tissues (pages 770–774)

✓ **B1.** Write *P* next to descriptions of primary lymphatic organs and *S* next to descriptions of secondary lymphatic organs and tissues.

_______ a. Origin of B cells and pre-T cells

_______ b. Sites where most immune responses occur

_______ c. Includes lymph nodes, spleen, and lymphatic nodules

_______ d. Includes red bone marrow and thymus

✓ **B2.** Describe the thymus gland in this exercise.

a. It is located in the *(neck? mediastinum?)*. Its position is just posterior to the _________________ bone. Its size is relatively *(large? small?)* in childhood, ______-creasing in size after age 10 to 12 years.

b. Like lymph nodes, the thymus contains a medulla surrounded by a _______________. Within the cortex are densely packed _______________-cytes. The thymus is the site of maturation of *(B? T?)* cells *(only in childhood? throughout life?)*, a process facilitated by _______________ cells and also by thymic hormones produced by _______________ cells.

c. The thymic medulla consists mainly of *(immature? more mature?)* T cells, along with several other types of cells. Some epithelial cells are located in thymic (Hassell's) _______________. Describe the function of these cells.

✓ **B3.** Complete this checkpoint on lymph node structure.

a. Use terms in the box to label structures A-E on Figure LG 22.1.

ALV.	**Afferent lymphatic vessel**	**MS.**	**Medullary sinus**
ELV.	**Efferent lymphatic vessel**	**TS.**	**Trabecular sinus**
GC.	**Germinal center**		

b. Now write one answer from the box above on each line next to related descriptions.

1. Vessels for entrance of lymph into node: ______
2. Vessel for exit of lymph out of node: ______
3. Region composed of antigen-presenting cells, B cells that become either plasma cells or memory cells, and macrophages: ______

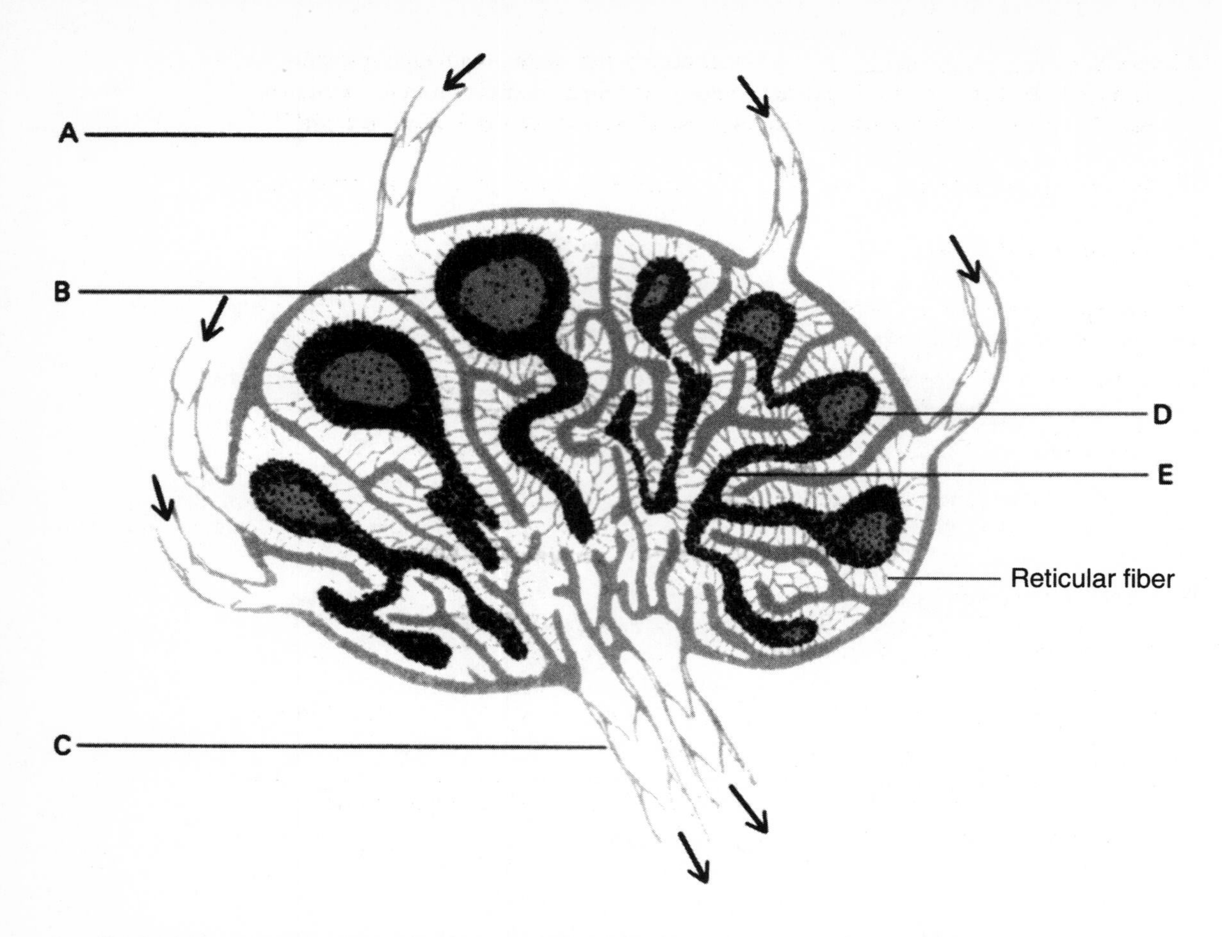

Figure LG 22.1 Structure of a lymph node. Arrows indicate the direction of lymph flow. Label as directed on Checkpoint B3.

4. Developed from a primary lymphatic nodule in response to recognition of an antigen: ______

5. Example of a channel for flow and filtration of lymph fluid: ______ ______

6. Regions of outer cortex: ______ ______

B4. Describe how lymph is "processed" as it circulates through lymph nodes. How does this function of lymph nodes explain the fact that nodes ("glands") become enlarged and tender during infection?

✓ **B5.** Spread, or ________________, of cancer may occur via the lymphatic system. Lymph nodes that have become cancerous tend to be enlarged, firm, and *(tender? non-tender?)*. Cancerous nodes also tend to be *(mobile? fixed via "tentacle-like extensions"?)*.

✓ **B6.** The term stroma refers to *(connective tissue scaffolding? functional cells?)* of an organ such as a lymph node, spleen, or liver. The term parenchyma refers to ________________.

✓ **B7.** The spleen is located in the *(upper right? upper left? lower right? lower left?)* quadrant of the abdomen, immediately inferior to the ________________. List three functions of:

a. The white pulp ________________________________

b. The red pulp ________________________________
Immune functions *(are? are not?)* likely to decline post-splenectomy.

✓ **B8.** Complete this exercise about lymphatic nodules.

a. The acronym MALT refers to M________________ A________________ L________________ T________________. Name four systems that contain MALT:

________________ ________________ ________________ ________________

b. State advantages of the following locations of aggregations of lymphatic tissue:
1. Peyer's patches

2. Tonsils and adenoids

c. Which tonsils are most commonly removed in a tonsillectomy? ________________

Which ones are called adenoids? ________________

C. Developmental anatomy of the lymphatic system (pages 774–775)

✓ **C1.** The lymphatic system derives from ________________-derm, beginning about the *(fifth? seventh? tenth?)* week of gestation.

✓ **C2.** Lymphatic vessels arise from lymph sacs that form from early *(arteries? veins?)*. Name the embryonic lymph sac that develops vessels to each of the following regions of the body.

a. Pelvic area and lower limbs: ________________ lymph sac

b. Upper limbs, thorax, head, and neck: ________________ lymph sac. The left jugular lymph sac eventually develops into the upper portion of the main collecting lymphatic vessel, the ________________ duct.

D. Nonspecific resistance to disease (pages 775-779)

Dl. Briefly contrast the two types of resistance to disease: specific vs. nonspecific. (For help, refer back to Checkpoint A2.)

✓ **D2.** *A clinical correlation.* Match the following conditions or factors with the nonspecific resistance that is lost in each case.

Nonspecific Resistance Lost

_______ a. Cleansing of oral mucosa

_______ b. Cleansing of vaginal mucosa

_______ c. Flushing of microbes from urinary tract

_______ d. Lacrimal fluid with lysozyme

_______ e. Respiratory mucosa with cilia

_______ f. Closely packed layers of keratinized cells

_______ g. HCl production; pH 1 or 2

_______ h. Sebum production

Condition/Factor

A. Long-term smoking
B. Excessive dryness of skin
C. Use of anticholinergic medications that decrease salivation
D. Vaginal atrophy related to normal aging
E. Dry eyes related to decreased production of tears, as in aging
F. Decreased urinary output, as in prostate enlargement
G. Skin wound
H. Partial gastrectomy (removal of part of the stomach)

✓ **D3.** *For extra review.* Describe other nonspecific defenses more fully in this exercise.

a. Interferons (IFNs) are proteins produced by cells such as _______________,

_______________, and _______________ that are infected with viruses.

IFNs then stimulate *(infected? uninfected?)* cells to produce _______________ proteins.

b. List the three types of interferons.

c. State two locations of complement proteins. _______________ and _______________
Explain why these proteins are called complement. (More about complement in Checkpoint G6.)

d. NK cells, or _______________ _______________ cells, are a type of *(lymphocyte? neutrophil?)*. About *(5-10? 60-70?)* percent of lymphocytes are NK cells. These cells attack a *(selected few? wide variety?)* of infectious microbes. One mechanism utilized by NK cells involves their release of chemicals known as

_______________ that perforate plasma membranes of invading microbes.

e. The primary type of white blood cells that are involved in phagocytosis is the

_______________. Other leukocytes, known as _______________, travel to

infection sites and develop into highly phagocytic _______________ cells. Name five organs that contain fixed macrophages.

✓ **D4.** Match each of the steps in phagocytosis with the related description. Note that descriptions are listed in chronological sequence.

A. Adherence	**I. Ingestion**
C. Chemotaxis	**K. Killing**

________ a. Phagocytic cells "sniff' the delectable fragrance of invading microbes and especially savor bacteria with the tasty *complement* coating. "Ah, dinner soon!"

________ b. Phagocytes trap the invaders, possibly wedging them against a blood vessel or a clot, and then attach to the surface of the invader.

________ c. Phagocytic tentacles called pseudopods surround the tasty morsels forming a culinary treat, the *phagosome*. "Slurp!"

________ d. Lysosomal enzymes, oxidants, and digestive enzymes pounce on the microbial meal, a has-been within a short time. *Residual bodies*, like the pits of prunes, are discarded, and the phagocytic face smiles in contentment.

D5. Do this exercise on inflammation.

a. Defend or dispute this statement: "Inflammation is a process that can be both helpful and harmful."

b. List the four cardinal signs or symptoms of inflammation.

________________ ________________

________________ ________________

c. Briefly describe the three basic stages of inflammation.

1.

2.

3.

✓ **D6.** *For extra review.* Select the number that best fits each description of a phase of inflammation. Use these answers:

1. Vasodilation and increased capillary permeability
2. Phagocytic migration
3. Repair

________ a. Brings defensive substances to the injured site and helps remove toxic wastes

________ b. Histamine, prostaglandins, kinins, leukotrienes, and complement enhance this process

________ c. Causes redness, warmth, and swelling of inflammation

________ d. Localizes and traps invading organisms

________ e. Occurs within an hour after initiation of inflammation; involves emigration, chemotaxis, and leukocytosis

________ f. White blood cell and debris formation that may lead to abscess or ulcer

✓ **D7.** Identify the chemicals that fit the descriptions below.

C.	**Complement**	**LT.**	**Leukotrienes**
H.	**Histamine**	**PG.**	**Prostaglandins**
I.	**Interferon**	**T.**	**Transferrins**
K.	**Kinins**		

________ a. Produced by damaged or inflamed tissues; aspirin and ibuprofen neutralize these pain-inducing chemicals

________ b. Produced by mast cells and basophils (two answers)

________ c. Made by some leukocytes and fibroblasts; stimulates cells near an invading virus to produce antiviral proteins that interfere with survival of the virus; effects are enhanced by fever

________ d. Induces vasodilation, permeability, chemotaxis, and irritation of nerve endings (pain)

________ e. Neutrophils are attracted by these chemicals present in the inflamed area (two answers)

________ f. Proteins that inhibit bacterial growth by depriving microbes of iron

D8. Define these terms and relate each one to inflammation or to poor circulation:

a. Pus

b. Ulcer

c. Abscess

✓ **D9.** Describe how interleukins may be involved in fever production.

E. Specific resistance to disease: immunity (pages 780–784)

✓ **El.** Contrast specific and nonspecific resistance in this exercise.

a. *(Nonspecific? Specific?)* resistance is also known as *immunity* and is studied in the branch of science known as ____________________.

b. Substances that are recognized as foreign and also provoke an immune response are known as *(antigens? antibodies?)*. Immune responses involve memory which causes a(n) *(slower and milder? even more rapid and vigorous?)* response to previously encountered antigens.

c. *For extra review,* look back at Checkpoint A2.

✓ **E2.** Complete this exercise about cells involved with immune responses.

a. Name the two categories of cells that carry out immune responses.

____________________ and ____________________ Where do lymphocytes originate?

b. Some immature lymphocytes migrate to the thymus and become *(B? T?)* cells. Here T cells develop immunocompetence, meaning that these cells have the ability to

____.

Some T cells become $CD4^+$ cells and others become ________ cells, based on the type of ____ in their plasma membranes.

c. Where do B cells mature into immune cells? ____________________

✓ **E3.** Contrast two types of immunity by writing AMI before descriptions of *antibody-mediated immunity* and CMI before those describing *cell-mediated immunity.*

________ a. Especially effective against microbes that enter cells, such as viruses and parasites

________ b. Especially effective against bacteria present in extracellular fluids

________ c. Involves plasma cells (derived from B cells) that produce antibodies

________ d. Utilizes killer T cells (derived from $CD4^+$ T cells) that directly attack the antigen

________ e. Facilitated by helper T ($CD4^+$) cells

✓ **E4.** Do this exercise on antigens.

a. An antigen is defined as "any chemical substance which, when introduced into the body, ____."

In general, antigens are *(parts of the body? foreign substances?)*.

b. Complete the definitions of the two properties of antigens.

1. Immunogenicity: ability to __________________ specific antibodies or specific T cells

2. Reactivity: ability to ____________________ specific antibodies or specific T cells

c. An antigen with both of these characteristics is called a(n) ____________________.

d. A partial antigen is known as a ______________________. It displays *(immunogenicity? reactivity?)* but not ______________________. For example, for the hapten penicillin to evoke an immune response (immunogenicity), it must form a complete antigen by combining with a ______________________. Persons who have this particular protein are said to be ______________________ to penicillin.

e. Describe the chemical nature of antigens.

f. Explain why plastics used for valves or joints are not likely to initiate an allergic response and be rejected.

g. Can an entire microbe serve as an antigen? *(Yes? No?)* List the parts of microbes that may be antigenic.

h. If you are allergic to pollen in the spring or fall or to certain foods, the pollen or foods serve as *(antigens? antibodies?)* to you.

i. Antibodies or specific T cells form against *(the entire? only a specific region of the?)* antigen. This region is known as the ______________________. Most antigens have *(only one? only two? a number of?)* antigenic determinant sites.

j. List the final destinations of antigens.

E5. Explain how human cells have the ability to recognize, bind to, and then evoke an immune response against over a billion different antigenic determinants. Describe two aspects.

a. Genetic recombination

b. Somatic mutations

✓ **E6.** Describe roles of major histocompatibility complex (MHC) antigens in this exercise.

a. Describe the "good news" (how they help you) and the "bad news" (how they may be harmful) about MHC antigens.

b. For what purpose is histocompatibility testing performed?

c. Contrast two classes of MHC antigens according to types of cells incorporating the MHC into plasma membrane.

MHC-I

MHC-II

d. Which of the following do T cells normally ignore?

A. MHC with peptide fragment from foreign protein

B. MHC with peptide fragment from self-protein

e. If the body does fail to recognize its own tissues as "self," but instead considers them antigenic (foreign), the resulting condition is known as an ________________ disorder.

✓ **E7.** Describe the processing and presenting of antigens in this activity.

a. For an immune response to occur, either _______ or _______ cells must recognize the presence of a foreign antigen. *(B? T?)* cells can recognize antigens located in extracellular fluid (ECF) and not attached to any cells. However, for *(B? T?)* cells to recognize antigens, the antigens must have been processed by a cell, and then they must be presented in association with MHC-I or MHC-II self-antigens (as described in Checkpoint E6c).

b. Define exogenous antigens.

Cells that present these antigenic proteins are called APCs, or a___________________

p___________________ c___________________. These cells are generally found *(in lymph nodes? at sites where antigens are likely to enter the body, such as skin or mucous membranes?)*. Name three types of APCs.

c. Place in correct sequence the six boxed codes that describe the steps APCs or other body cells take to initiate an immune response to intruder antigens. Use the lines provided.

Bind to MHC–II.	**Bind peptide fragment to MHC–II**
Exo/insert.	**Exocytosis and insertion of antigen fragment-MHC–II into APC plasma membrane.**
Fusion.	**Fusion of vesicles of peptide fragments with MHC–II**
Partial dig.	**Partial digestion of antigen into peptide fragments**
Phago/endo.	**Phagocytosis or endocytosis of antigen**
Present to T.	**Present antigen to T cell in lymphatic tissue**

1. ___________________ → 2. ___________________ → 3. ___________________

4. ___________________ → 5. ___________________ → 6. ___________________

d. Although only selected cells can present exogenous antigens, most cells of the body can present _______-genous antigens. State an example of endogenous antigens.

✓ **E8.** Match the name of each cytokine with the description that best fits.

Gamma-IFN.	**Gamma interferon**	**LT.**	**Lymphotoxin**
IL–1.	**Interleukin-1**	**MMIF.**	**Macrophage migration inhibiting factor**
IL–2.	**Interleukin-2**		
IL–4.	**Interleukin-4**	**P.**	**Perforin**
IL–5.	**Interleukin–5**	**TNF.**	**Tumor necrosis factor**

________ a. Causes plasma cells to secrete antibodies, specifically, IgA's

________ b. Made by activated helper T cells; stimulates B cells, leading to IgE production

________ c. Known as T cell growth factor, it is made by helper T cells; needed for almost all immune responses; causes proliferation of cytotoxic T cells (as well as B cells) and activates NK cells

________ d. Formerly called macrophage activating factor (MAF), stimulates phagocytosis by macrophages and neutrophils; enhances AMI and CMI responses

________ e. Prevents macrophages from leaving the site of infection

________ f. Destroys target cells by perforating their cell membranes

________ g. Kills cells by fragmentation of their DNA

________ h. Therapeutic for persons with multiple sclerosis (MS)

E9. *A clinical correlation*. List one or more disorders for which each of these cytokines is used as a treatment:

a. Alpha interferon (Intron A)

b. Beta-interferon (Betaseron)

c. Interleukin-2 (IL-2)

F. Cell-mediated immunity (pages 785-787)

F1. Describe the stages of cell-mediated immunity (CMI) by completing the following outline that provides key terms. (*Suggestion*: Refer to Figure 22.14 in the text.)

a. Antigen recognition (car-starting analogy) and activation

1. T cell receptors (TCRs)

2. Costimulators

3. What is the meaning of *anergy*?

b. Proliferation and differentiation of T cells in secondary lymphatic organs and tissues.

1. Activated T cells

2. Clones of three different T cells (See also Checkpoint F2)

c. Elimination of the intruder by chemicals from T_c cells (*Hint*: refer to Table 22.2 and page 787 in the text.)
1. Perforin

2. Lymphotoxin

3. Gamma-IFN

✓ **F2.** Match T cells to descriptions of their roles. (*Suggestion*: Refer to Checkpoints E7 and E8 above.)

M. **Memory T cells**	**T_C.** **Cytotoxic T cells**	**T_H.** **Helper T cells**

_______ a. Recognize antigen fragments associated with MHC–I molecules

_______ b. Recognize antigen fragments associated with MHC–II molecules

_______ c. Called $CD8^+$ cells because developed from cells with CD8 protein; to become cytolytic, must be costimulated by IL–2 or other cytokines from T_H cells

_______ d. Called $CD4^+$ cells because developed from cells with CD4 protein; activated by APCs and costimulated by IL–1 and IL–2; produce IL–2, IL–4, and IL–5

_______ e. Programmed to recognize the original invader; can initiate dramatic responses to reappearance of the intruder

F3. Define *immunological surveillance.*

Name the three types of cells responsible for immunological surveillance.

This immune function is most helpful in eliminating *(all cancer cells? only cancer cells due to cancer-causing viruses?)*. Explain why transplant patients taking immunosuppressants are at risk for virus-associated cancers.

G. Antibody-mediated immunity (pages 788–792)

G1. Describe antibody-mediated immunity (AMI) in this activity.

a. Where do B cells perform their functions? *(At sites where invaders enter the body? In lymphatic tissue?)*

b. Now describe the steps in AMI, using these key terms or phrases as a guide.

1. Activation of B cells, BCR

2. Follicular dendritic cells, APC's

3. Antigen-MHC–II

4. Costimulation by IL's

5. Plasma cells

6. Four or five days

7. Memory B cells

8. Antibodies with identical structure to original BCRs

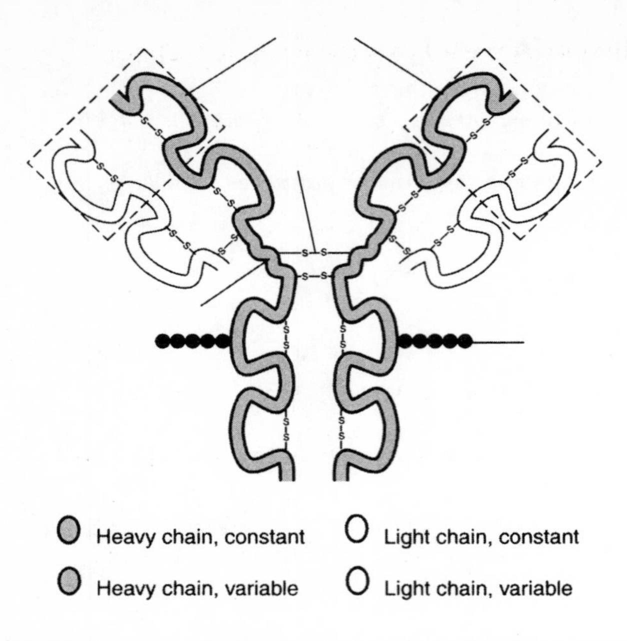

Figure LG 22.2 Diagram of an antibody molecule. Color and label as directed in Checkpoint G2.

✓ **G2.** Refer to Figure LG 22.2 and check your understanding of antibody structure in this Checkpoint.

a. Chemically, all antibodies are composed of glycoproteins named ________________. They consist of *(2? 4? 8?)* polypeptide chains. The two heavy chains contain about ________ amino acids; light chains contain about ________ amino acids. Crosshatch the *heavy chains* on the figure.

b. *(Constant? Variable?)* portions are diverse for different antibodies because these serve as binding sites for different antigens. *(Constant? Variable?)* portions *(differ? are identical?)* for all antibodies within the same class. Color constant and variable portions of both chains as indicated by color code ovals on the figure. Also label antibody *hinge region, antigen binding sites, disulfide bonds,* and *carbohydrates.*

✓ **G3.** Because glycoproteins forming antibodies are involved in immunity, they are called immunoglobulins, abbreviated ________. The different classes of Ig's are distinguishable by the *(constant? variable?)* portions of the antibody structure. Write the name of the related class of Ig next to each description.

________ a. The first antibodies to be secreted after initial exposure to antigen, they are short lived; they destroy invading microbes by agglutination and lysis.

________ b. The only type of Ig to cross the placenta, it provides specific resistance of newborns. Also significantly enhances phagocytosis and neutralizing of toxins in persons of all ages because it is the most abundant type of antibody.

________ c. Found in secretions such as mucus, saliva, tears, and breast milk; protects against oral, respiratory, and vaginal infections.

________ d. Located on mast and basophil cells and involved in allergic reactions, for example, to certain foods, pollen, or bee venom.

________ e. Antigen receptors on B cells; activate B cells to produce antibodies

G4. List five mechanisms that antibodies may use to attack and inactivate antigens.

G5. *A clinical correlation.* Define monoclonal antibodies (MAb's). These antibodies are produced by ____________________ cells. Explain how a *hybridoma* cell is produced.

Describe use of monoclonal antibodies clinically.

✓ **G6.** Complete this Checkpoint on the role of complement in immunity. (Review Checkpoint D3c above.)

a. The complement system consists of more than 30 different *(proteins? lipids?)*. Name several of these.

b. The *(classical? alternative?)* pathway is triggered by binding of antibodies to antigens such as bacteria. The alternative pathway begins its cascade of reactions by interaction of *(antibodies? polysaccarides?)* on invading antigens with complement factors.

c. The three overall effects of complement are:

1. *(Constriction? dilation?)* of blood vessels which ______-creases movement of WBCs into the infected area.
2. Coating of microbes, a process known as ____________________, which promotes phagocytosis.
3. Formation of the membrane attack complex (MAC) which leads to ____________________ of the invading microbes.

✓ **G7.** Do this exercise on immunizations.

a. Once a specific antigen has initiated an immune response, either by infection or by a(n) *(initial? booster dose?)* immunization, the person produces some long-lived B and T cells called ____________________.

b. The level of antibodies rises *(slowly? rapidly?)* after the initial immunization. This level measured in serum is known as the antibody ____________________. Which antibody titer rises first? *(IgA? IgG? IgM?)* Which rises next? ________

c. Upon subsequent exposure to the same antigen, such as during another infection or by a ____________________ dose of vaccine, memory cells provide a *(more? less?)* intense response. Antibodies produced by this secondary response, also known as ____________________ ____________________, have a *(higher? lower?)* affinity for the antigen compared to antibodies produced in the primary response.

✓ **G8.** *A clinical challenge.* Identify types of immunity described below by selecting from answers:

AAAI.	**Artificially acquired active immunity**
AAPI.	**Artificially acquired passive immunity**
NAAI.	**Naturally acquired active immunity**
NAPI.	**Naturally acquired passive immunity**

________ a. As 31-year-old Bud tore apart an old shed, a rusty nail entered his left hand. Bud, who reported that he had not had a tetanus shot since he was "about 14," received a shot of tetanus immunoglobulins at the emergency room.

________ b. Amy has provided her 3-month-old baby Steffy with temporary immunity by the antibodies that crossed over the placenta during Amy's pregnancy and also by antibodies in milk from breast feeding.

________ c. Kim took her baby Jamie to the clinic for Jamie's regularly scheduled MMR (measles-mumps-rubella) immunization.

H. Self-recognition and immunological tolerance (pages 793–795)

✓ **H1.** The ability of your T cells to avoid reacting with your own body proteins is known as immunological ____________________, B cells *(also? do not?)* display this tolerance. Loss of such tolerance results in ____________________ disorders.

H2. Describe each of the processes listed below, which are steps in development of immunological tolerance involving T cells.

a. Positive selection

b. Deletion

c. Anergy

Which of the above steps *(a? b? c?)* appears to be the main mechanism that B cells use to prevent responses to self-proteins? ______

H3. "The ability to recognize tumor antigens as nonself" is a definition of the characteristic of immunological *(surveillance? tolerance?)*. Describe how each of the following forms of therapy may help cancer patients whose own immunological surveillance is failing.

a. LAK cells

b. TILs

I. Stress, aging and the immune system (page 795)

I1. Define *psychoneuroimmunology (PNI)* and state an example of how PNI concepts can maximize your resistance against disease.

✓ **I2.** Complete arrows to indicate immune changes that usually accompany aging.

a. Rate of production of antibodies in response to an infection: |

b. Risk of autoimmune conditions by production of antibodies against self: |

c. Response to vaccines: |

J. Disorders, medical terminology (pages 797–800)

✓ **J1.** Answer these questions about AIDS.

a. AIDS refers to a________________ i________________-d________________ s________________.

b. The causative agent for AIDS has been identified as the h___________________

i___________________ v___________________. What is the major effect of this virus?

c. List the groups of persons that have been most infected in the United States.

d. In the U.S. most AIDS patients are *(female? male?)*. Name the three groups of people

with greatest *increase* in incidence of HIV in the U.S. ___________________

___________________ ___________________. Worldwide, _____ million people are

infected with HIV. Globally, HIV is largely transmitted by *(homosexual? heterosexual?)* contacts and by infected blood.

e. Like other viruses, the HIV virus *(does? does not?)* depend on host cells for replication.

The HIV virus is a ___________________-virus because its genetic code is carried in *(DNA? RNA?)*. What role does reverse transcriptase play in a retrovirus?

Name AIDS drugs that work by inhibiting this enzyme.

f. The outer covering over the HIV viral RNA contains glycoproteins known as *(CD4? GP120?)*; these serve as "docking proteins." These chemical protrusions permit the HIV virus to attach (or "dock") on *(CD4? GP120?)* receptors on human immune cells. Name three types of cells that may be infected by the HIV virus.

___________________ ___________________ ___________________ Which chemical is part of the covering over the HIV virus that helps to fuse with the lipid layer of host cells? *(CCR5? CXCR4? GP41?)*

g. Once inside human cells, the HIV virus causes the human cells to become "factories" that produce perhaps *(a thousand? a million? billions of?)* HIV viruses per day. The T4 cells themselves *(are? are not?)* destroyed in the process.

h. The person is "HIV-positive" when *(T cells drop to under 200/mm^3? anti-HIV antibodies are identified in the blood stream?)* The person is diagnosed as having AIDS when *(T cells drop to under 200/mm3? anti-HIV antibodies are identified in the blood stream?)*, or when opportunistic infections occur. Define this type of infection.

i. List five or more early signs or symptoms of HIV infection.

j. What is *HAART therapy?*

Besides reverse transcriptase inhibitors (RTIs), this therapy (also known as

___________________ therapy) includes drugs known as PIs (or

___________________ inhibitors). How do PIs help to control HIV?

k. Besides measuring HIV antibodies in blood and T4 ($CD4^+$) count, how else can the severity of HIV/AIDS be evaluated?

l. Circle the four fluids through which the HIV virus has been found to be transmitted in sufficient quantities to be infective.

Blood	Breast milk	Mosquito venom	Saliva
Semen	Tears	Vaginal fluids	

✓ **J2.** Match the four types of allergic responses in the box with the correct descriptions below

I. Type I	**II. Type II**	**III. Type III**	**IV. Type IV**

________ a. Known as *cell-mediated* reactions, or delayed-type-hypersensitivity reactions, these lead to responses such as those that occur after exposure to poison ivy or after tuberculin testing (Mantoux test)

________ b. Antigen-antibody *immune complexes* trapped in tissues activate complement and lead to inflammation, as in glomerulonephritis or rheumatoid arthritis (RA)

________ c. Caused by IgG or IgM antibodies directed against RBCs or other cells, as in, for example, response to an incompatible transfusion; called a *cytotoxic* reaction

________ d. Involves IgEs produced by mast cells and basophils; effects may be localized (such as swelling of the lips) or systemic, such as acute *anaphylaxis*

✓ **J3.** Match the condition in the box with the description below.

A.	**Autoimmune disease**	**NHL.**	**Non-Hodgkin's Lymphoma**
HD.	**Hodgkin's disease**	**SPL.**	**Splenomegaly**
IM.	**Infectious mononucleosis**	**SCID.**	**Severe combined immunodeficiency disease**
LA.	**Lymphadenopathy**		

_________ a. Contagious infection of young women especially; B lymphocytes become infected by EBV virus and then enlarge to resemble monocytes; classic signs are sore throat, enlarged and tender lymph nodes, and fever

_________ b. Enlarged, tender lymph glands

_________ c. Enlarged spleen

_________ d. Multiple sclerosis (MS), rheumatoid arthritis (RA), insulin-dependent diabetes mellitus, and systemic lupus erythematosus (SLE) are examples.

_________ e. Of HD and NHL, the malignancy with the poorer prognosis

_________ f. Rare, inherited condition invloving lack of both T cells and B cells

✓ **J4.** Match types of transplants in the box with descriptions below.

ALLO.	**Allograph**	**AUTO.**	**Autograft**	**XENO.**	**Xenograft**

_________ a. Between animals of different species

_________ b. Between different parts of the same person, for example, healthy skin to regions of burned skin

_________ c. Between individuals of same species, but of different genetic backgrounds, such as a mother's kidney donated to her daughter or a blood transfusion

ANSWERS TO SELECTED CHECKPOINTS: CHAPTER 22

A2. (a-c) NS. (d-e) S. (f-g) NS.

A3. (a) Interstitial fluid and proteins; lipids (or fats), fat. (b) Lympho; T, antibodies.

A4. Lymph, lymph vessels, and lymph organs such as tonsils, spleen, and lymph nodes. (a) Venous, arterial. (b) Do. (c) Although similar chemically, lymph is present within lymph vessels, whereas interstitial fluid is located between cells.

A6. (a) Three-fourths; cisterna chyli; digestive and other abdominal organs and both lower limbs. (b) Left jugular, left subclavian, and left bronchomediastinal. (c) Right lymphatic; right upper limb and right side of thorax, neck, and head. (d) Left subclavian vein; the same. (e) Interstitial; plasma.

A7. Fracture of this bone might block the thoracic duct, which enters veins close to the left clavicle. Lymph therefore backs up with these results: (a) Extra fluid remains in tissue spaces normally drained by vessels leading to the thoracic duct. (b) Lymph capillaries in the intestine normally absorb fat from foods. Slow lymph flow can decrease such absorption leaving fat in digestive wastes.

A9. Skeletal muscle contraction squeezing lymphatics which contain valves that direct flow of lymph. Respiratory movements. (*Hint*: see Mastery Test Chapter 21, question #21.)

B1. (a) P. (b) S. (c) S. (d) P.

B2. (a) Mediastinum; sternum; large, de. (b) Cortex; lympho; T, throughout life, dendritic, specialized epithelial. (c) More mature; corpuscles; possible site of T cell death.

B3. (a) A, ALV; B, TS; C, ELV; D, GC; E, MS. (b1) ALV. (b2) ELV. (b3) GC. (b4) GC. (b5) TS and MS. (b6) TS and GC.

B5. Metastasis; nontender; fixed via "tentacle-like extensions" (hence the name *cancer*, meaning *crablike*).

B6. Connective tissue scaffolding; functional cells.

B7. Upper left, diaphragm. (a) Immune functions of B and T cells and phagocytosis of pathogens in blood. (b) Phagocytosis of defective or aging RBCs and platelets, storage of platelets, and hemopoiesis in fetal life. Are.

B8. (a) Mucosa-associated lymphoid tissue; digestive, respiratory, urinary, and reproductive. (b1) Located in the wall of the small intestine, this tissue can protect against microbes found in food or present in excessive numbers in the GI tract; (b2) Because tonsils and adenoids surround the pharynx (throat), they protect against inhaled or ingested microbes. (c) Palatine; pharyngeals.

C1. Meso, fifth.

C2. Veins. (a) Posterior. (b) Jugular; thoracic.

D2. (a) C. (b) D. (c) F. (d) E. (e) A. (f) G. (g) H. (h) B.

D3. (a) Lymphocytes, macrophages, fibroblasts; uninfected, antiviral. (b) Alpha, beta, and gamma. (c) In blood plasma and on plasma membranes; when activated, they "complement" or enhance other defense mechanisms. (d) Natural killer, lymphocyte; 5–10; wide variety; perforins. (e) Neutrophil; monocytes, macrophages; skin, liver, lungs, brain, and spleen.

D4. (a) C. (b) A. (c) I. (d) K.

D6. (a–d) 1. (e) 2. (f) 3.

D7. (a) PG. (b) H, LT. (c) I. (d) K. (e) K, C (and PG somewhat). (f) T.

D9. Bacterial toxins may trigger release of interleukin (IL-1) which resets the body's thermostat (hypothalamus).

E1. (a) Specific, immunology. (b) Antigens; even more rapid and vigorous.

E2. (a) B cells (or B lymphocytes), T cells (or T lymphocytes); from stem cells in bone marrow. (b) T; perform immune functions against specific antigens if properly stimulated; $CD8^+$, antigen-receptor protein. (c) Bone marrow.

E3. (a) CMI. (b) AMI. (c) AMI. (d) CMI. (e) AMI and CMI

E4. (a) Is recognized as foreign and provokes an immune response; foreign substances. (b1) Stimulate production of; (b2) React with (and potentially be destroyed by). (c) Complete antigen. (d) Hapten; reactivity, immunogenicity; body protein; allergic. (e) Large, complex molecules, such as proteins, nucleic acids, lipoproteins, glycoproteins, or complex polysaccharides. (f) They are made of simple, repeating subunits that are not likely to be antigenic. (g) Yes; flagella, capsules, cell walls, as well as toxins made by bacteria. (h) Antigens. (i) Only a specific region of the; antigenic determinant (epitope); a number of. (j) Spleen, lymph nodes, or MALT.

E6. (a) Good news: they help T cells to recognize foreign invaders because antigenic proteins must be processed and presented in association with MHC antigens before T cells can recognize the antigenic proteins; bad news: MHC antigens present transplanted tissue as "foreign" to the body, causing it to be rejected. (b) Organ transplant such as kidney or liver or tissue transplant (such as bone marrow). (c) MHC–I molecules are in plasma membranes of all body cells except in RBCs; MHC–II antigens are present on APCs, thymic cells, and T cells activated by previous exposure to the antigen. (d) B (basis of the principle of self-tolerance). (e) Autoimmune.

E7. (a) B or T: B; T. (b) Antigens from intruders outside of body cells, such as bacteria, pollen, foods, cat hair; antigen presenting cells; at sites where antigens are likely to enter the body, such as skin or mucous membranes; macrophages, B cells, and dendritic cells (including Langerhans cells in skin). (c) 1. Phago/endo. 2. Partial dig. 3. Fusion. 4. Bind to MHC-II. 5. Exo/insert. 6. Present to T. (d) Endo; viral proteins made after a virus infects a cell and takes over cell metabolism or abnormal proteins made by cancer cells.

E8. (a) IL-5 (b) IL–4. (c) IL–2. (d) Gamma-IFN. (e) MMIF. (f) P. (g) LT. (h) IL–1, TNF.

F2. (a) T_C. (b) T_H. (c) T_C. (d) T_H. (e) M.

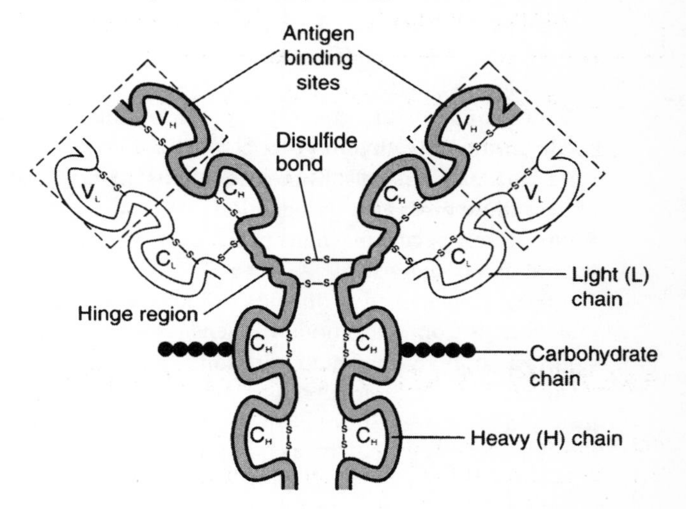

Figure LG 22.2A Diagram of an antibody molecule.

G2. (a) Globulins; 4; 450, 220. See Figure LG 22.2A. (b) Variable; constant, are identical. See Figure LG 22.2A.

G3. Ig's; constant. (a) M. (b) G. (c) A. (d) E. (e) D.

G6. (a) Proteins; proteins C1–C9 and proteins B, D, and P (properdin). (b) Classical; polysaccharides. (c1) Dilation, in. (c2) Opsonization. (c3) Cytolysis (or bursting of cells).

G7. (a) Initial, memory cells. (b) Slowly; titer; IgM; IgG. (c) Booster, more; immunological memory, higher.

G8. (a) AAPI. (b) NAPI. (c) AAAI.

H1. Tolerance: also; autoimmune.

I2. (a) ↓. (b) ↑. (c) ↓.

J1. (a) Acquired immunodeficiency syndrome. (b) Human immunodeficiency virus: by destroying the immune system, the disease places the patient at high risk for opportunistic infections (normally harmless), such as PCP, and for cancers such as KS or invasive cervical carcinoma. (c) Gay men who have not used safe sexual practices, IV drug users, hemophiliacs who received contaminated blood products before 1985 (when testing blood products become available), and heterosexual partners of HIV-infected persons. (d) Male; people of color, women, and teenagers; 35–40, heterosexual. (e) Does; retro, RNA; directs DNA synthesis in host cells modeled on the HIV RNA code; AZT, ddl, ddc, and d4T. (f) GP120; CD4; $CD4^+$ (helper T cells), macrophages, and dendritic cells; GP41. (g) Billions of; are. (h) Anti-HIV antibodies are identified in the blood stream; T cells drop to under 200/μl; infections that do not occur in persons with normal immunity. (i) Initially, flu-like illness such as sore throat, swollen lymph glands, fatigue and night sweats; later, diarrhea, weight loss, serious respiratory infections, lymphoma, neuropathy, and dementia. (j) Highly active antiretroviral therapy; triple, protease; they inhibit an enzyme (protease) needed to assemble the protein coat around newly-formed HIV viruses. (k) Measurement of viral load by amount of RNA in plasma. (l) Blood, semen, vaginal fluids, and breast milk.

J2. (a) IV. (b) III. (c) II. (d) I.

J3. (a) IM. (b) LA. (c) SPL. (d) A. (e) NHL. (f) SCID

J4. (a) XENO. (b) AUTO. (c) ALLO.

MORE CRITICAL THINKING: CHAPTER 22

1. Contrast lymphatic flow with blood circulation. Incorporate into your discussion the differences between (a) blood and lymph, (b) forces controlling flow of blood and of lymph, (c) lymph capillaries and blood capillaries, and (d) major lymph vessels and major blood vessels.
2. List the four cardinal signs or symptoms of inflammation. Then describe the processes that occur during inflammation that account for these signs or symptoms. Be sure to include roles of these chemicals: histamine, kinins, prostaglandins (PGs), and complement. Also discuss these processes: vasodilation, emigration, and tissue repair.
3. Contrast nonspecific mechanisms for resistance known as the first line of defense and the second line of defense.
4. Contrast antigens and antibodies. Include several examples of each. Further distinguish endogenous from exogenous antigens, as well as complete and partial antigens.
5. Contrast antibody-mediated immunity (AMI) and cell-mediated immunity (CMI).
6. Contrast each of the following cytokines by describing types of cells producing the chemicals and also functions of each cytokine: interferons (alpha, beta, and gamma IFNs), interleukins (IL-I, IL–2, IL–4, IL–5), lymphotoxin, and perforin.
7. Differentiate the five classes of immunoglobulins.
8. Contrast helper T ($CD4^+$) cells with cytotoxic T ($CD8^+$) cells.

MASTERY TEST: CHAPTER 22

Questions 1–2: Arrange the answers in correct sequence.

_____ _____ _____ _____ _____ 1. Flow of lymph through a lymph node:
A. Afferent lymphatic vessel
B. Medullary sinus
C. Cortical sinus
D. Efferent lymphatic vessel

_____ _____ _____ _____ _____ 2. Activities in humoral immunity, in chronological order:
A. B cells develop in bone marrow or other part of the body.
B. B cells differentiate and divide (clone), forming plasma cells.
C. B cells migrate to lymphoid tissue.
D. B cells are activated by specific antigen that is presented.
E. Antibodies are released and are specific against the antigen that activated the B cell.

Questions 3–16: Circle the letter preceding the one best answer to each question. Take particular note of underlined terms in selected questions.

3. Both the thoracic duct and the right lymphatic duct empty directly into:
   A. Axillary lymph nodes
   B. Superior vena cava
   C. Cisterna chyli
   D. Subclavian arteries
   E. Subclavian veins
4. Which lymphatic trunk drains lymph primarily from the left upper limb?
   A. Left jugular trunk
   B. Left bronchomediastinal trunk
   C. Left subclavian trunk
   D. Left lumbar trunk
5. All of these are examples of nonspecific defenses *except:*
   A. Antigens and antibodies
   B. Saliva
   C. Complement
   D. Interferon
   E. Skin
   F. Phagocytes
6. All of the following correctly match lymphocytes with their functions *except:*
   A. Helper T cells: stimulate B cells to divide and differentiate
   B. Cytotoxic ($CD8^+$) T cells: produce cytokines that attract and activate macrophages and lymphotoxins that directly destroy antigens
   C. Natural killer (NK) cells: suppress action of cytotoxic T cells and B cells
   D. B cells: become plasma cells that secrete antibodies
7. All of the following match types of allergic reactions with correct descriptions *except:*
   A. Type I: anaphylactic shock, for example, from exposure to iodine or bee venom
   B. Type II: transfusion reaction in which recipient's IgG or IgM antibodies attack donor's red blood cells
   C. Type III: Involves Ag–Ab-complement complexes that cause inflammations, as in rheumatoid arthritis (RA)
   D. Type IV: cell-mediated reactions that involve immediate responses by B cells, as in the TB skin test
8. Choose the one *false* statement about T cells.
   A. T_C cells are best activated by antigens associated with <u>both MHC–I and MHC–II</u> molecules.
   B. T_H cells are best activated by antigens associated with <u>MCH–II</u> molecules.
   C. For T_C cells to become cytolytic, <u>they need costimulation by IL–2</u>.
   D. Perforin is a chemical released by <u>T_H cells</u>.
9. Choose the *false* statement about lymphatic vessels.
   A. Lymph capillaries are <u>more</u> permeable than blood capillaries.
   B. Lymphatics have <u>thinner</u> walls than veins.
   C. Like arteries, lymphatics contain <u>no</u> valves.
   D. Lymph vessels are <u>blind-ended</u>.
10. Choose the *false* statement about nonspecific defenses.
    A. Complement functions protectively by <u>being converted into histamine</u>.
    B. Histamine <u>increases</u> permeability of capillaries so that leukocytes can more readily reach the infection site.

C. Complement and properdin are both proteins found in blood plasma.
D. Opsonization enhances phagocytosis.

11. Choose the *false* statement about T cells.
 A. Some are called memory cells.
 B. They are called T cells because they are processed in the thymus.
 C. They are involved primarily in antibody-mediated immunity (AMI).
 D. Like B cells, they originate from stem cells in bone marrow.
12. Choose the *false* statement about lymph nodes.
 A. Lymphocytes are produced here.
 B. Lymph nodes are distributed evenly through out the body, with equal numbers in all tissues.
 C. Lymph may pass through several lymph nodes in a number of regions before returning to blood.
 D. Lymph nodes are shaped roughly like kidney (or lima) beans.
13. Choose the *false* statement about lymphatic organs.
 A. The palatine tonsils are the ones most often removed in a tonsillectomy.
 B. The spleen is the largest lymphatic organ in the body.
 C. The thymus reaches its maximum size at age 40.
 D. The spleen is located in the upper left quadrant of the abdomen.
14. Choose the *false* statement.
 A. Skeletal muscle contraction aids lymph flow.
 B. Skin is normally more effective than mucous membranes in preventing entrance of microbes into the body.
 C. Interferon is produced by viruses.
 D. An allergen is an antigen, not an antibody.
15. Choose the one *true* statement about AIDS.
 A. It appears to be caused by a virus that carries its genetic code in DNA.
 B. Persons at highest risk are men and women who are homosexual.
 C. Persons with AIDS experience a decline in helper T cells.
 D. AIDS is transmitted primarily through semen and saliva.
16. All of the following are cytokines known to be secreted by helper T cells except:
 A. Gamma interferon
 B. Interleukin–1
 C. Interleukin–2
 D. Interleukin–4

Questions 17–20: Circle T (true) or F (false). If the statement is false, change the underlined word or phrase so that the statement is correct.

T F 17. Antibodies are usually composed of one light and one heavy polypeptide chain.

T F 18. T_C cells are especially active against slowly growing bacterial diseases, some viruses, cancer cells associated with viral infections, and transplanted cells.

T F 19. Immunogenicity means the ability of an antigen to react with a specific antibody or T cell.

T F 20. A person with autoimmune disease produces fewer than normal antibodies.

Questions 21–25: Fill-ins. Complete each sentence with the word or phrase that best fits.

____________________ 21. _______ lymphocytes provide antibody-mediated immunity, and _______ lymphocytes and macrophages offer cellular (cell-mediated) immunity.

____________________ 22. Helper T cells are also known as _______ cells.

____________________ 23. List the four fundamental signs or symptoms of inflammation.

____________________ 24. The class of immunoglobins most associated with allergy are Ig _______.

____________________ 25. Three or more examples of mechanical factors that provide nonspecific resistance to disease are _______.

ANSWERS TO MASTERY TEST: CHAPTER 22

Arrange

1. A C B D
2. A C D B E

Multiple Choice

3. E
4. C
5. A
6. C
7. D
8. D
9. C
10. A
11. C
12. B
13. C
14. C
15. C
16. B

True–False

17. F. Two light and two heavy (chains)
18. T
19. F. Stimulate formation of
20. F. Antibodies against the individual's own tissues

Fill-Ins

21. B, T
22. T_H cells, T4 cells, and $CD4^+$ cells.
23. Redness, warmth, pain, and swelling
24. E
25. Skin, mucosa, cilia, epiglottis; also flushing by tears, saliva, and urine

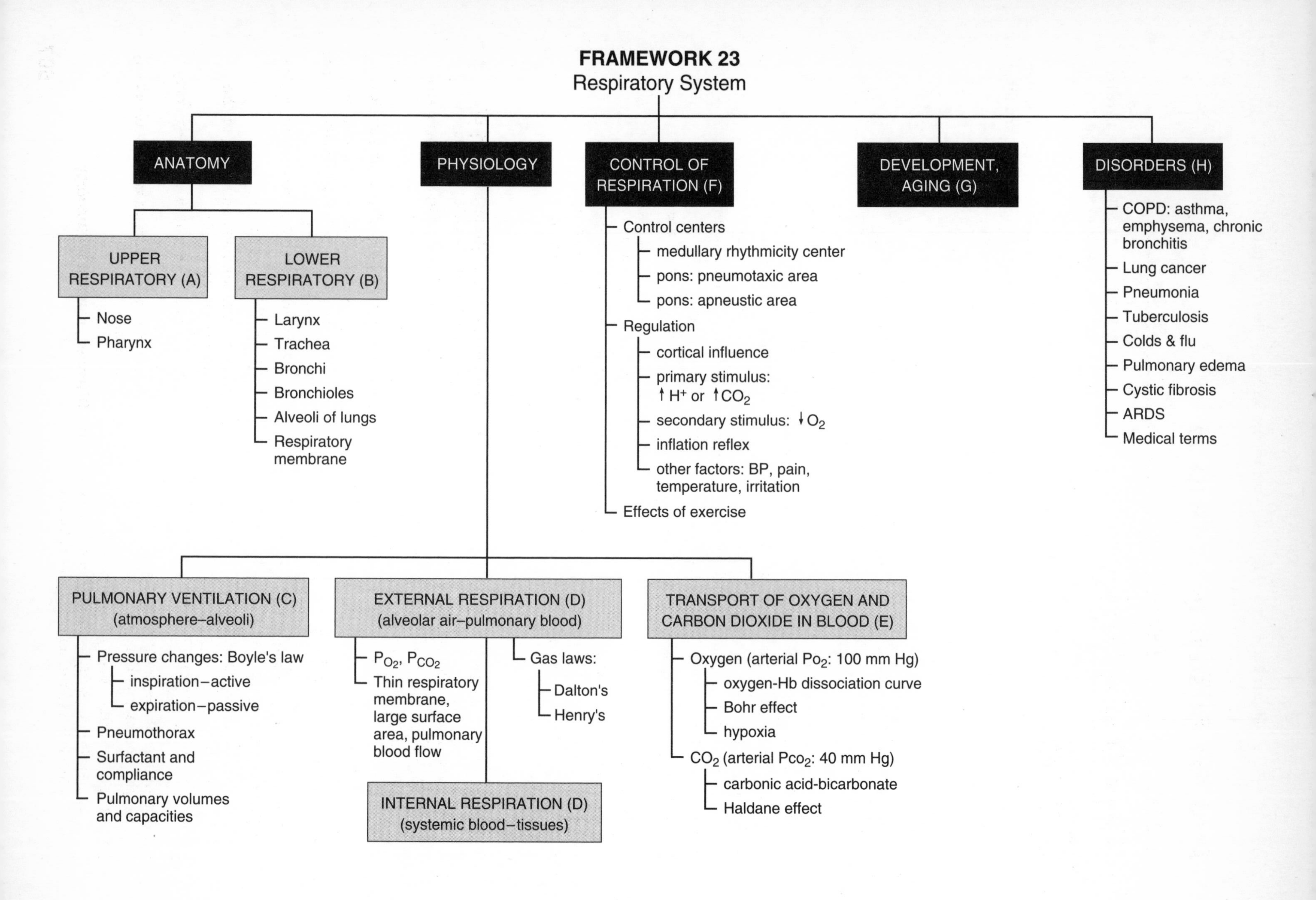

FRAMEWORK 23
Respiratory System
ANATOMY
UPPER RESPIRATORY (A)
Nose
Pharynx
LOWER RESPIRATORY (B)
Larynx
Trachea
Bronchi
Bronchioles
Alveoli of lungs
Respiratory membrane
PHYSIOLOGY
PULMONARY VENTILATION (C) (atmosphere–alveoli)
Pressure changes: Boyle's law
inspiration–active
expiration–passive
Pneumothorax
Surfactant and compliance
Pulmonary volumes and capacities
EXTERNAL RESPIRATION (D) (alveolar air–pulmonary blood)
P_{O_2}, P_{CO_2}
Thin respiratory membrane, large surface area, pulmonary blood flow
Gas laws:
Dalton's
Henry's
INTERNAL RESPIRATION (D) (systemic blood–tissues)
TRANSPORT OF OXYGEN AND CARBON DIOXIDE IN BLOOD (E)
Oxygen (arterial P_{O_2}: 100 mm Hg)
oxygen-Hb dissociation curve
Bohr effect
hypoxia
CO_2 (arterial P_{CO_2}: 40 mm Hg)
carbonic acid-bicarbonate
Haldane effect
CONTROL OF RESPIRATION (F)
Control centers
medullary rhythmicity center
pons: pneumotaxic area
pons: apneustic area
Regulation
cortical influence
primary stimulus: $\uparrow H^+$ or $\uparrow CO_2$
secondary stimulus: $\downarrow O_2$
inflation reflex
other factors: BP, pain, temperature, irritation
Effects of exercise
DEVELOPMENT, AGING (G)
DISORDERS (H)
COPD: asthma, emphysema, chronic bronchitis
Lung cancer
Pneumonia
Tuberculosis
Colds & flu
Pulmonary edema
Cystic fibrosis
ARDS
Medical terms

CHAPTER 23

The Respiratory System

Oxygen is available all around us. But to get an oxygen molecule to a muscle cell in the stomach or a neuron of the brain requires coordinated efforts of the respiratory system with its transport adjunct, the cardiovascular system. Oxygen traverses a system of ever-narrowing and diverging airways to reach millions of air sacs (alveoli). Each alveolus is surrounded by a meshwork of pulmonary capillaries, much like a balloon encased in a nylon stocking. Here oxygen changes places with the carbon dioxide wastes in blood, and oxygen travels through the blood to reach the distant stomach or brain cells. Interference via weakened respiratory muscles (as in normal aging) or extremely narrowed airways (as in asthma or bronchitis) can limit oxygen delivery to lungs. Decreased diffusion of gases from alveoli to blood (as in emphysema, pneumonia, or pulmonary edema) can profoundly affect total body function and potentially lead to death.

Take a couple of deep breaths and, as you begin your study of the respiratory system, carefully examine the Chapter 23 Topic Outline and Objectives. Check off each one as you complete it. To further organize your study of the respiratory system, glance over the Chapter 23 Framework. Be sure to refer to the Framework frequently and note relationships among key terms in each section.

TOPIC OUTLINE AND OBJECTIVES

A. Anatomy: upper respiratory system

1. Describe the anatomy and histology of the nose, pharynx, larynx, trachea, bronchi, and lungs.
2. Identify the functions of each respiratory system structure.

B. Anatomy: lower respiratory system

C. Physiology of respiration: pulmonary ventilation; lung volumes and capacities

1. Describe the events that cause inspiration and expiration.
2. Define the various lung volumes and capacities.

D. Physiology of respiration: exchange of oxygen and carbon dioxide

1. Explain Dalton's law and Henry's law.
2. Describe the exchange of oxygen and carbon dioxide in external and internal respiration.

E. Physiology of respiration: transport of oxygen and carbon dioxide in blood

1. Describe how blood transports oxygen and carbon dioxide.

F. Control of respiration; effects of exercise

1. Explain how the nervous system controls breathing and list the factors that alter the rate and depth of breathing.
2. Describe the effects of exercise on the respiratory system.

G. Developmental anatomy and aging of the respiratory system

1. Describe the development of the respiratory system.
2. Describe the effects of aging on the respiratory system.

H. Disorders, medical terminology

WORDBYTES

Now become familiar with the language of this chapter by studying each wordbyte, its meaning, and an example of its use within a term. After you study the entire list, self-check your understanding by writing the meaning of each wordbyte on the line. As you continue through the *Learning Guide*, identify (and fill in) additional terms that contain the same wordbyte.

Wordbyte	Self-check	Meaning	Example(s)
atel-	____________	incomplete	*atel*ectasis
-baros	____________	pressure	hyper*baric* chamber
-centesis	____________	puncture	thoraco*centesis*
cricoid	____________	ringlike	*cricoid* cartilage
dia-	____________	through	*dia*phragm
dys-	____________	bad, difficult	*dys*pnea
-ectasis	____________	dilation	atel*ectasis*
epi-	____________	over	*epi*glottis
ex-	____________	out	*ex*piration
hyper-	____________	over	*hyper*baric
in-	____________	in	*in*spiration
intra-	____________	within	*intra*pulmonic
meatus	____________	opening, passageway	inferior *meatus*
meter	____________	measure	spiro*meter*
-plasty	____________	to mold or to shape	rhino*plasty*
-pnea	____________	breathe	a*pnea*
pneumo-	____________	air	*pneumo*thorax
pulmo-	____________	lung	*pulmo*nary
rhin-	____________	nose	*rhin*orrhea
spir-	____________	breathe	in*spir*ation, *spir*ometer

CHECKPOINTS

A. Anatomy: upper respiratory system (pages 806–810)

A1. Explain how the respiratory and cardiovascular systems work together to accomplish gaseous exchange among the atmosphere, blood, and cells.

✓ **A2.** Identify parts of the respiratory system in this activity.

a. Arrange respiratory structures listed in the box in correct sequence from first to last in the pathway of inspiration.

ADA.	**Alveolar ducts and alveoli**	**MN.**	**Mouth and nose**
BB.	**Bronchi and bronchioles**	**P.**	**Pharynx**
L.	**Larynx**	**T.**	**Trachea**

_______ → _______ → _______ → _______ → _______ → _______

b. Write asterisks (*) next to structures forming the upper respiratory system.

c. Most of the structures listed in the box above are parts of the *(conducting? respiratory?)* portion of the respiratory system.

✓ **A3.** Match each term related to the nose with the description that best fits.

External nares	**Paranasal sinuses and meatuses**
Internal nares (choanae)	**Rhinoplasty**
Nasal cartilages	**Vestibule**

a. Mucous membrane-lined spaces and passageways: ____________________

b. Nostrils: ____________________

c. Anterior portion of nasal cavity: ____________________

d. Openings from posterior of nose into nasopharynx: ____________________

e. Repair of the nose: ____________________

f. Known as septal, alar, and lateral nasal, these provide a structural framework of the nose: ____________________

✓ **A4.** Name the structures of the nose that are designed to carry out each of the following functions.

a. Warm, moisten, and filter air

b. Sense smell

c. Assist in speech

✓ **A5.** Write *LP* (laryngopharynx), *NP* (nasopharynx), or *OP* (oropharynx) to indicate the location of each of the following structures.

_____ a. Adenoids

_____ b. Palatine tonsils

_____ c. Lingual tonsils

_____ d. Opening (fauces) from oral cavity

_____ e. Opening into larynx and esophagus

A6. On Figure LG 23.1, cover the key and identify all structures associated with the nose, palate, and pharynx. Color structures with color code ovals.

B. Anatomy: lower respiratory system (pages 810–820)

✓ **B1.** Match parts of the larynx with the descriptions that best fit.

A.	**Arytenoid cartilages**	**RG.**	**Rima glottidis**
C.	**Cricoid cartilage**	**T.**	**Thyroid cartilage**
E.	**Epiglottis**		

_____ a. Also known as the Adam's apple; longer in males than in females

_____ b. Includes a leaflike portion that covers the airway during swallowing

_____ c. A pair of triangular-shaped cartilages attached to muscles that control vocal cords

_____ d. Forms the inferior portion of the larynx; a complete ring

_____ e. Space between the vocal cords through which air passes during ventilation

B2. Explain how the larynx normally prevents food from entering the trachea. (*Hint:* Notice the arrow on Figure LG 23.1.)

B3. Tell how the larynx produces sound. Explain how pitch is controlled and what causes male pitch usually to be lower than female pitch. Describe roles played by other structures in producing recognizable speech.

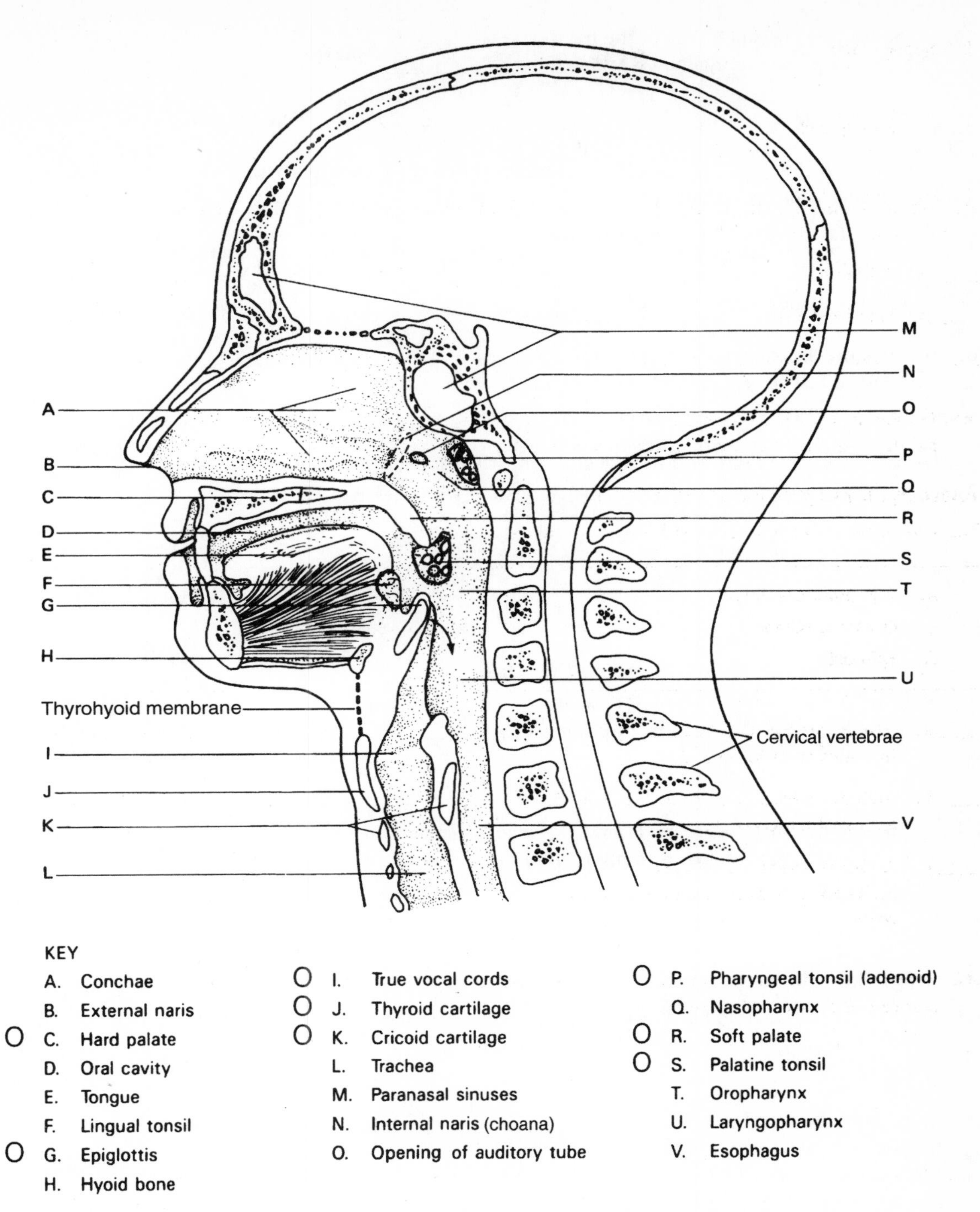

KEY

- A. Conchae
- B. External naris
- O C. Hard palate
- D. Oral cavity
- E. Tongue
- F. Lingual tonsil
- O G. Epiglottis
- H. Hyoid bone
- O I. True vocal cords
- O J. Thyroid cartilage
- O K. Cricoid cartilage
- L. Trachea
- M. Paranasal sinuses
- N. Internal naris (choana)
- O. Opening of auditory tube
- O P. Pharyngeal tonsil (adenoid)
- Q. Nasopharynx
- O R. Soft palate
- O S. Palatine tonsil
- T. Oropharynx
- U. Laryngopharynx
- V. Esophagus

Figure LG 23.1 Sagittal section of the right side of the head with the nasal septum removed. Color and identify labeled structures as directed in Checkpoint A6. Arrow refers to Checkpoint B2.

✓ **B4.** Complete this Checkpoint about the trachea.

a. The trachea is commonly known as the ____________________. It is located *(anterior? posterior?)* to the esophagus. About ________ cm (________ inches) long, the trachea terminates at the ____________________, where the trachea leads to a Y-shaped intersection with the primary bronchi.

b. The tracheal wall is lined with a ____________________ membrane and strengthened by 16–18 C-shaped rings composed of ____________________.

✓ **B5.** Describe the functional advantages provided by each of these parts of the trachea:

a. Pseudostratified ciliated epithelium lining the mucosa

b. C-shaped cartilage rings with the trachealis muscle completing the "ring" posteriorly

c. Sensitivity of the mucosa at the carina

✓ **B6.** *A clinical correlation.* Maggie has aspirated a small piece of candy. Dr. Lennon expects to find it in the right bronchus rather than in the left. Why?

✓ **B7.** Bronchioles *(do? do not?)* have cartilage rings. How is this fact significant during an asthma attack?

B8. Refer to Figure LG 23.2 and do this activity. Use Figure 23.8 (page 815 of your text) for help.

a. Identify tracheobronchial tree structures J–N, and color each structure indicated with a color code oval.

b. Color the two layers of the pleura and the diaphragm according to color code ovals. Also, cover the key and identify all parts of the diagram from A to I.

✓ **B9.** Write the correct term next to each description.

a. Inflammation of the pleura: ____________________

b. Excess fluid in the pleural space: ____________________

c. Procedure to remove fluid from the pleural space: ____________________

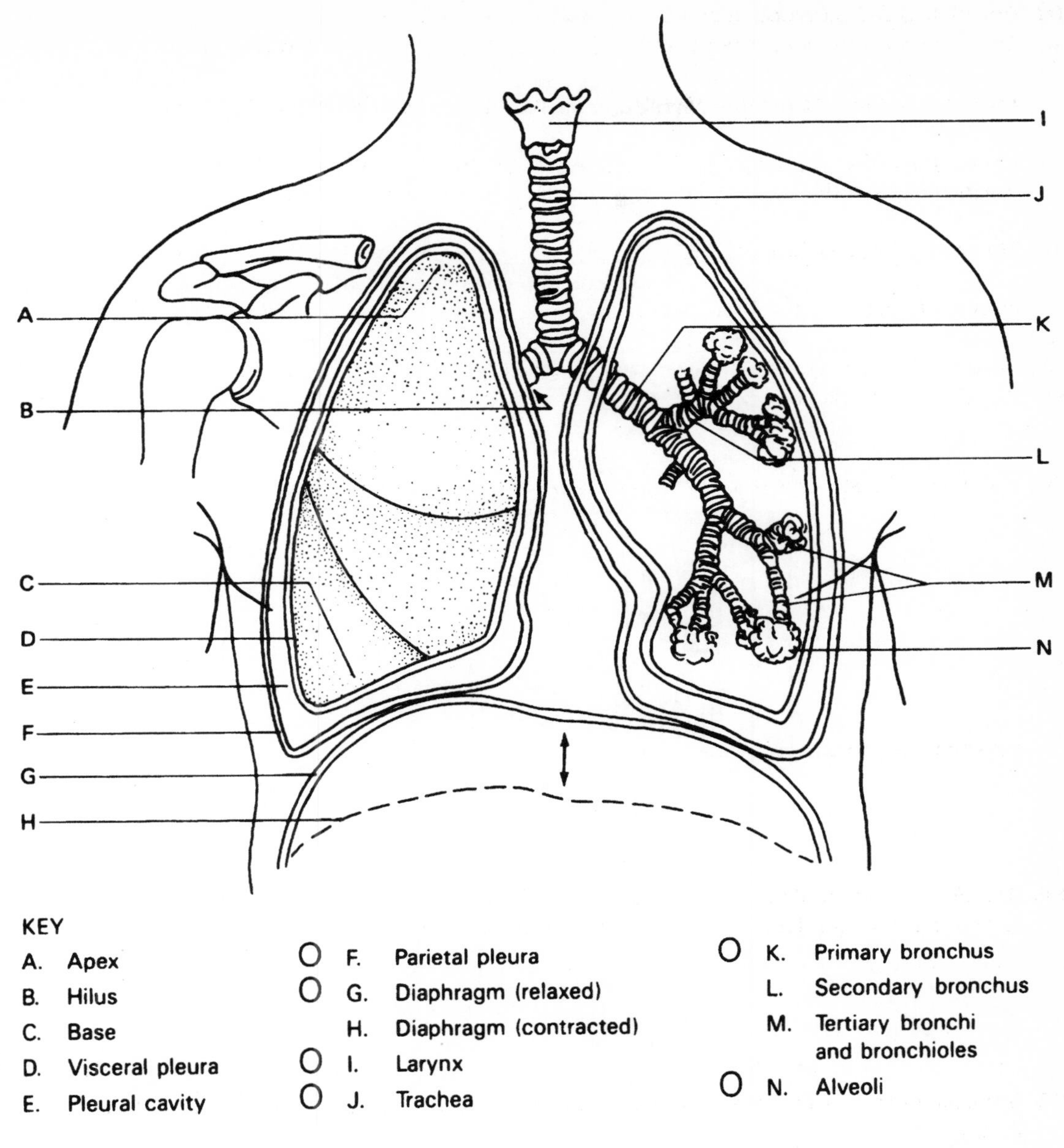

KEY

A. Apex
B. Hilus
C. Base
O D. Visceral pleura
E. Pleural cavity
O F. Parietal pleura
O G. Diaphragm (relaxed)
H. Diaphragm (contracted)
O I. Larynx
O J. Trachea
O K. Primary bronchus
L. Secondary bronchus
M. Tertiary bronchi and bronchioles
O N. Alveoli

Figure LG 23.2 Diagram of lungs with pleural coverings and bronchial tree. Color and identify labeled structures as directed in Checkpoints B8 and B10.

✓ **B10.** Answer these questions about the lungs. (As you do the exercise, locate the parts of the lung on Figure LG 23.2.)

a. The broad, inferior portion of the lung that sits on the diaphragm is called the ____________________. The upper narrow apex of each lung extends just superior to the ____________________. The costal surfaces lie against the ____________________.

b. Along the mediastinal surface is the ____________________ where the root of the lung is located. This root consists of ____.

c. Answer these questions with *right* or *left*.
In which lung is the cardiac notch located? ____________________

Which lung has just two lobes and so only two lobar bronchi? ____________________

Which lung has a horizontal fissure? ____________________

Write *S* (superior), *M* (middle), and *I* (inferior) on the three lobes of the right lung.

✓ **B11.** Describe the structures of a lobule in this exercise.

a. Arrange in order the structures through which air passes as it is about to enter a lobule en route to alveoli. ________ ________ ________

b. In order for air to pass from alveoli to blood in pulmonary capillaries, it must pass through the ____________________ ____________________ (respiratory) membrane. Identify structures in this pathway: A to E on Figure LG 23.3.

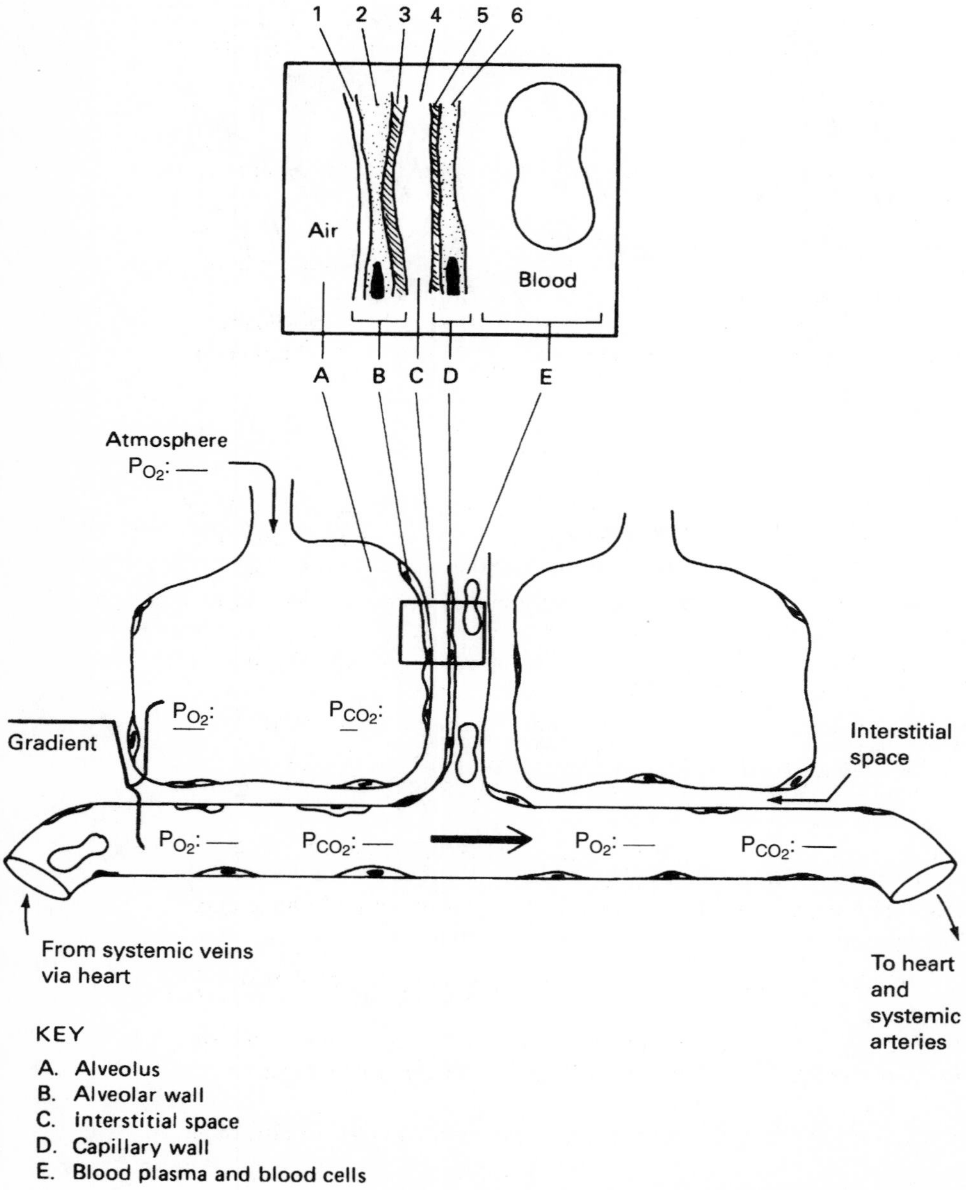

KEY

A. Alveolus
B. Alveolar wall
C. interstitial space
D. Capillary wall
E. Blood plasma and blood cells

Figure LG 23.3 Diagram of alveoli and pulmonary capillary. Insert enlarges respiratory membrane. Numbers and letters refer to Checkpoint B11. Complete partial pressures (in mm Hg) as directed in Checkpoint D4.

A. Alveolar ducts	**R. Respiratory bronchiole**	**T. Tertiary bronchi**

c. *For extra review.* Now label layers 1–6 of the respiratory membrane in the insert in Figure LG 23.3.

d. *A clinical correlation.* Normally the respiratory membrane is extremely *(thick? thin?)*. Suppose pulmonary capillary blood pressure rises dramatically, pushing extra fluid out of blood. The presence of excess fluid in interstitial areas and ultimately in alveoli

is known as pulmonary ________________. Diffusion occurs *(more? less?)* readily through the fluid-filled membrane.

e. The alveolar epithelial layer of the respiratory membrane contains three types of cells. Forming most of the layer are flat *(squamous? septal? macrophage?)* (Type I alveolar) cells which are ideal for diffusion of gases. Septal (Type II alveolar) cells produce

an important phospholipid and lipoprotein substance called ________________, which reduces tendency for collapse of alveoli due to surface tension. Macrophage

cells help to remove ________________ from lungs.

✓ **B12.** *The Big Picture: Looking Back.* Return to earlier chapters in your text and Guide as you contrast the dual blood supply to the lungs.

a. Which vessels supply oxygen-rich blood to the lungs? *(Bronchial? Pulmonary?)* arteries. These vessels carry blood directly from the *(aorta? right ventricle?)*. Review these on Figure 21.21 and Exhibit 21.4 (pages 727–728) in your text.

b. Which veins return blood carrying CO_2 from lung tissue toward the heart (Figure 21.27 and Exhibit 21.10 (pages 744–745) in your text, and Chapter LG 21, Checkpoint E8, page LG 465)?

________________ or ________________ veins.

c. Refer to Chapter 20 in your *Learning Guide,* Checkpoints B8 and C2 (pages LG 423 and 425). Which vessel contains more highly oxygenated blood? Pulmonary *(artery? vein?)*.

d. Return to Chapter 21, Checkpoint F2, page LG 467, and answer these questions.

1. Which system of vessels has higher blood pressure? *(Pulmonary? Systemic?)* One reason for this difference in blood pressure is that the extensive (lengthy) network of arterioles and capillaries in the *(pulmonary? systemic?)* circulation provides greater resistance to blood flow.

2. Which vessels respond to hypoxia by vasoconstricting so that blood bypasses hypoxic regions and more efficiently circulates to oxygen-rich regions? *(Pulmonary? Systemic vessels, for example, those in skeletal muscle?)* This phenomenon

is known as ________________-________________ coupling.

C. Physiology of respiration: pulmonary ventilation; lung volumes and capacities (pages 821–827)

✓ **C1.** Refer to Figure LG 23.4 and visualize the process of respiration in yourself as you do this exercise.

a. Label the first phase of respiration at (1) on the figure. This process involves transfer of air from your environment to *(alveoli? pulmonary capillaries?)* in your lungs.

Inflow of air rich in *(O_2? CO_2?)* is known as ______-spiration, whereas outflow of air

concentrated in *(O_2? CO_2?)* is known as ______-spiration.

b. Phase 2 of respiration involves exchange of gases between ________________ in alveoli and ________________ in pulmonary capillaries. This process is known as *(external? internal?)* respiration. Label phase 2 on the figure.

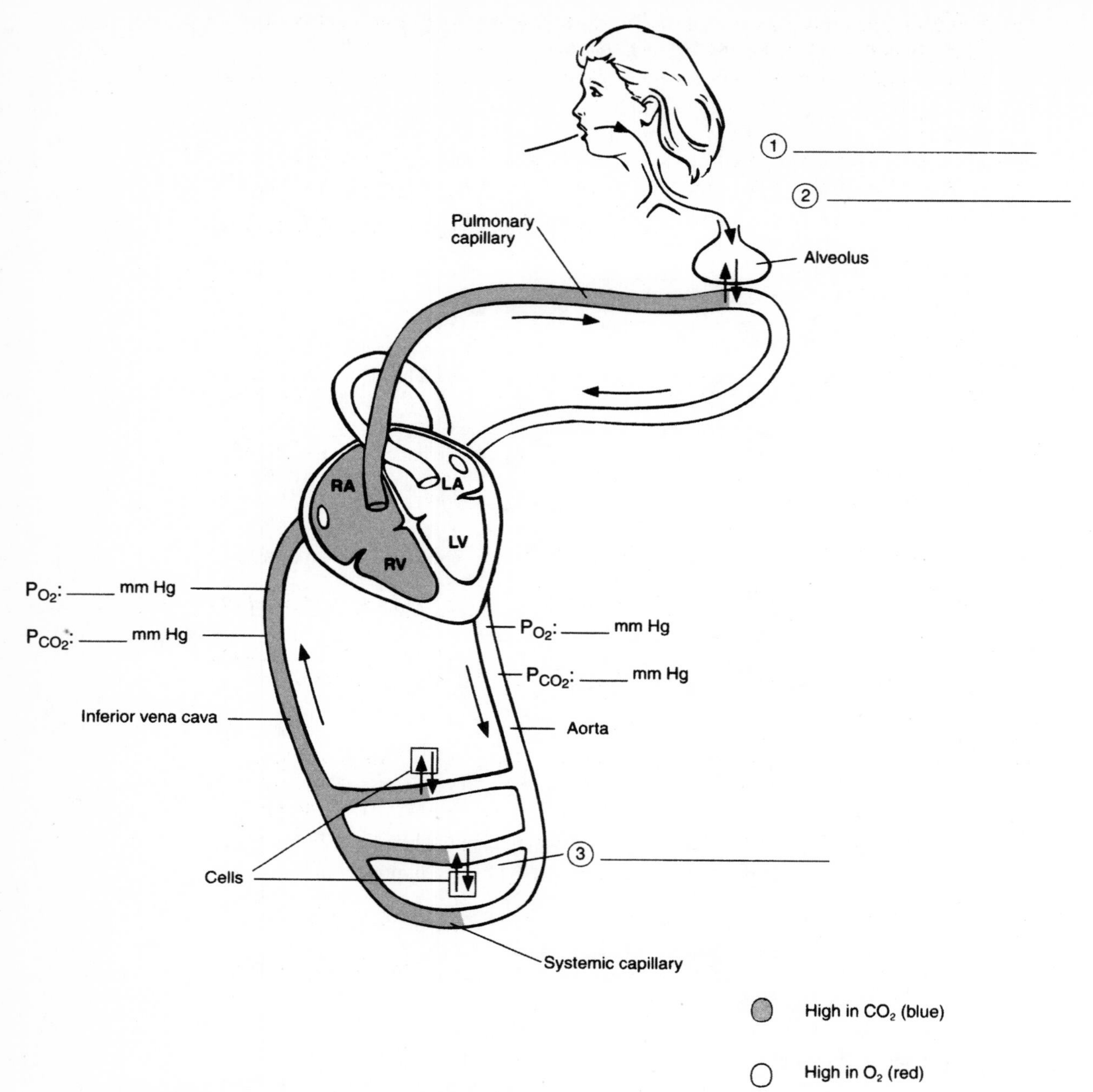

Figure LG 23.4 Phases of respiration and circulatory routes.
Numbers refer to Checkpoint C1. Fill in short lines according to Checkpoint D4.

c. After oxygenated blood returns to your heart via your pulmonary *(artery? veins?)*, blood is pumped to your tissues where gas exchange occurs between ________________ and ________________. This process is known as *(external? internal?)* respiration. Label phase 3 on the figure.

d. Blood high in CO_2, then returns to the heart where it passes directly into the pulmonary *(artery? vein?)* to reach the lungs. Here diffusion of CO_2 from pulmonary capillaries to alveoli occurs; this is another effect of *(external? internal?)* respiration (phase ________). Finally, you exhale this air, a process that is part of *(external respiration? pulmonary ventilation?)* (phase ________).

e. Which process takes place within tissue cells, utilizing O_2 and producing CO_2 as a waste product? *(Cellular? Internal?)* respiration.

f. *For extra review,* color all blood vessels on Figure LG 23.4 according to color code ovals on the figure.

✓ **C2.** Refer to Figure 23.15 in the text and describe the process of ventilation in this exercise.

a. In diagram (a), just before the start of inspiration, pressure within the lungs (called ________________________) is ________ mm Hg. This is *(more than? less than? the same as?)* atmospheric pressure.

b. At the same time, pressure within the pleural cavity (called ________________________ pressure) is ________ mm Hg. This is *(more than? less than? the same as?)* alveolar and atmospheric pressure at sea level.

c. The first step in inspiration occurs as the muscles in the floor and walls of the thorax contract. These are the ________________________ and ________________________ muscles. Note in diagram (b) (and also in Figure LG 23.2) that the size of the thorax *(increases? decreases?)*. Because the two layers of pleura tend to adhere to one another, the lungs will *(increase? decrease?)* in size also.

d. Increase in volume of a closed space such as the pleural cavity causes the pressure there to *(increase? decrease?)* to ________ mm Hg. Because the lungs also increase in size (due to pleural cohesion), alveolar pressure also *(increases? decreases?)* to ________ mm Hg. This inverse relationship between volume and pressure is a statement of ________________________'s law.

e. A pressure gradient is now established. Air flows from high pressure area *(alveoli? atmosphere?)* to low pressure area *(alveoli? atmosphere?)*. So air flows *(into? out of?)* lungs. Thus *(inspiration? expiration?)* occurs. By the end of inspiration, sufficient air will have moved into the lungs to make pressure there equal to atmospheric pressure, that is, ________ mmHg.

f. Inspiration is a(n) *(active? passive?)* process, whereas expiration is normally a(n) *(active? passive?)* process resulting from ________________________ recoil. Identify two factors that contribute to this recoil.

During expiration, the diaphragm moves *(inf? sup?)*-eriorly and the sternum moves *(ant? post?)*-eriorly.

g. During labored breathing, accessory muscles help to force air out. Name two of these sets of muscles.

✓ **C3.** Answer these questions about compliance.

a. Imagine trying to blow up a new balloon. Initially, the balloon resists your efforts. Compliance is the ability of a substance to yield elastically to a force; in this case it is the ease with which a balloon can be inflated. So a new balloon has a *(high? low?)* level of compliance, whereas a balloon that has been inflated many times has *(high? low?)* compliance.

b. Similarly, alveoli that inflate easily have *(high? low?)* compliance. The presence of a coating called ____________________ lining the inside of alveoli prevents alveolar walls from sticking together during ventilation and so *(increases? decreases?)* compliance.

c. Surfactant is made by type *(I? II? III?)* alveolar cells of the lungs, which produce this chemical especially during the final weeks before birth. A premature infant may lack adequate production of surfactant; this condition is known as ____________________.

d. Collapse of all or part of a lung is known as ____________________ (See page 846 of the text.) Such collapse may result from lack of surfactant or from other factors. Name several of these factors.

✓ **C4.** Complete the arrows to indicate whether resistance to air flow increases (↑) or decreases (↓) in each case.

a. Inhalation, causing dilation of bronchi and bronchioles: |

b. Increase in sympathetic nerve impulses to relax smooth muscle of airways: |

c. COPD such as bronchial asthma or emphysema, with obstruction or collapse or airways: |

C5. Refer to Table 23.1 (page 826) in your text. Read carefully the descriptions of modified respiratory movements such as sobbing, yawning, sighing, or laughing, while demonstrating those actions yourself.

C6. Contrast these types of breathing patterns:

a. Tachypnea/dyspnea (page 846 in the text)

b. Costal breathing/diaphragmatic breathing

C7. Each minute the average adult takes ________ breaths (respirations). Check your own respiratory rate and write it here: ________ breaths per minute.

✓ **C8.** Of the total amount of air that enters the lungs with each breath, about *(99%? 70%? 30%? 5%?)* actually enters the alveoli. The remaining amount of air is much like the last portion of a crowd trying to rush into a store: it does not succeed in entering the alveoli during an inspiration but just reaches airways and then is quickly ushered out during the next expiration. Such air is known as anatomic ________________ and constitutes about ________ mL of a typical breath.

✓ **C9.** Maureen, who weighs 150 pounds, is breathing at the rate of 15 breaths/minute. Her tidal volume is 480 mL/breath. Her minute ventilation is ________ mL. Her alveolar ventilation rate is likely to be about ________ mL/mm.

✓ **C10.** Match the lung volumes and capacities with the descriptions given. You may find it helpful to refer to Figure 23.17 in the text.

ERV.	**Expiratory reserve volume**	**RV.**	**Residual Volume**
$FEV_{1.0}$	**Forced expiratory volume in one second**	**TLC.**	**Total lung capacity**
FRC.	**Functional residual capacity**	**VC.**	**Vital capacity**
IRV.	**Inspiratory reserve volume**	**V_T.**	**Tidal volume**

________ a. The amount of air taken in with each inspiration during normal breathing is called ________.

________ b. At the end of a normal expiration the volume of air left in the lungs is called ________. Emphysemics who have lost elastic recoil of their lungs cannot exhale adequately, so this volume will be abnormally large.

________ c. Forced exhalation can remove some of the air in FRC. The maximum volume of air that can be expired beyond normal expiration is called ________. This volume will be small in emphysema patients.

________ d. Even after the most strenuous expiratory effort, some air still remains in the lungs; this amount, which cannot be removed voluntarily, is called ________.

________ e. The volume of air that represents a person's maximum breathing ability is called ________. This is the sum of ERV, V_T and IRV.

________ f. Adding RV to VC gives ________.

________ g. The excess air a person can take in after a normal inhalation is called ________.

________ h. The amount of the vital capacity that can be forced out in 1 second is called ________.

✓ **C11.** Indicate normal values for each of the following.

a. V_T = ________ ml (________ liter)

b. TLC = ________ ml (________ liter)

c. VC = ________ ml (________ liter)

d. FRC = ________ ml (________ liter)

e. RV = ________ ml (________ liter)

C12. State the medical and legal significance of *minimal volume.*

D. Physiology of respiration: exchange of oxygen and carbon dioxide (pages 827–831)

✓ **D1.** Check your understanding of gas laws by matching the correct law with the condition that it explains.

B. Boyle's law	D. Dalton's law	H. Henry's law

________ a. If a patient breathes air highly concentrated in oxygen (as in a hyperbaric chamber), a higher percentage of oxygen will dissolve in the blood and tissues. This is the principle underlying the use of a hyperbaric chamber.

________ b. The total atmospheric pressure (760 mm Hg) is due mostly to pressure caused by nitrogen, partly to P_{O_2} and slightly to P_{CO_2}.

________ c. Under high P_{N_2} more nitrogen dissolves in blood. The *bends* occurs when pressure decreases and nitrogen forms bubbles in tissue as it comes out of solution.

________ d. As the size of the thorax increases, the pressure within it decreases.

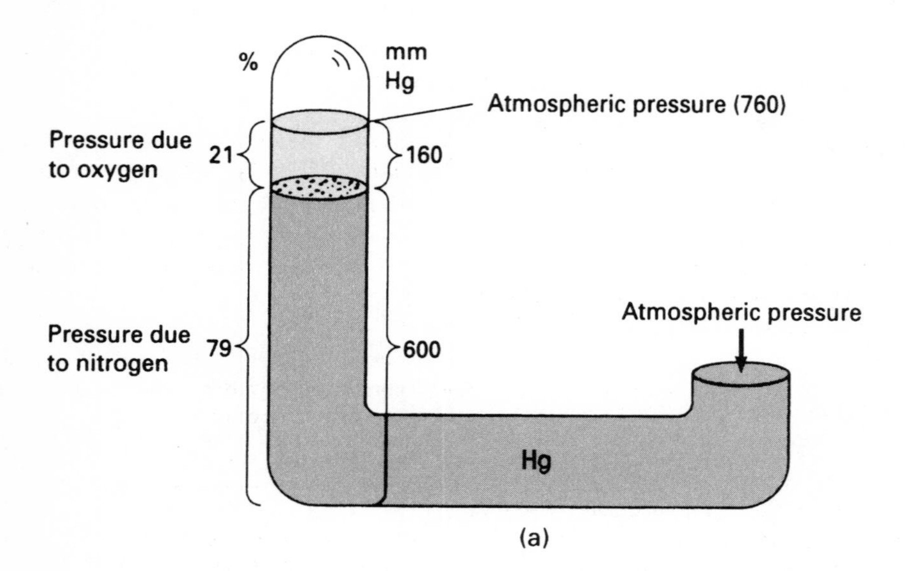

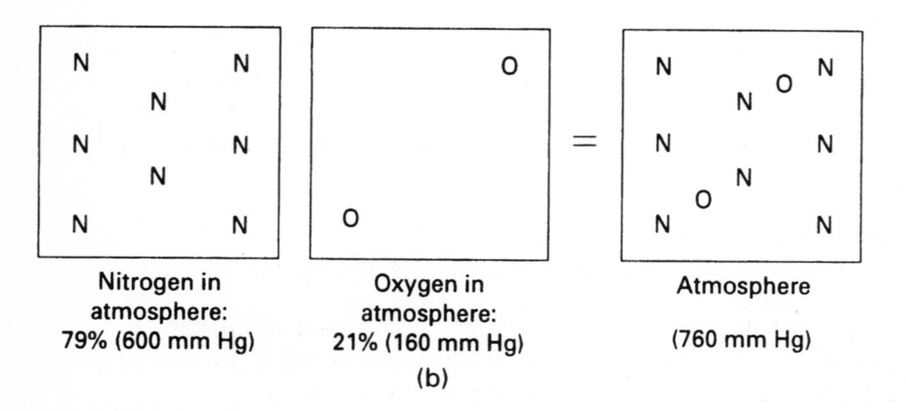

Figure LG 23.5 Atmospheric pressure. (a) Effect of gas molecules on column of mercury (Hg). (b) Partial pressures contributing to total atmospheric pressure. Color according to directions in Checkpoint D2.

✓ **D2.** Refer to Figure LG 23.5 and do this exercise about Dalton's law.

a. The atmosphere contains enough gaseous molecules to exert pressure upon a column of mercury to make it rise about _______ mm. The atmosphere is said to have a pressure of 760 mm Hg.

b. Air is about _______% nitrogen and _______% oxygen. To show this, circle the nitrogen molecules in green and oxygen molecules in red in Figure LG 23.5b. Note that carbon dioxide is not represented in the figure because only about _______% of air is CO_2.

c. Dalton's law explains that of the total 760 mm of atmospheric pressure, a certain amount is due to each type of gas. Determine the portion of the total pressure due to nitrogen molecules:

$79\% \times 760$ mm Hg = _______ mm Hg

This is the partial pressure of nitrogen, or _______.

d. Calculate the partial pressure of oxygen.

e. In Figure LG 23.5a, color the part of the Hg column due to N_2 pressure (up to the 600 mm Hg mark) *green*. Then color the portion of the Hg column due to 0_2 pressure *red*.

✓ **D3.** Complete this clinical correlation activity.

a. Although most of the atmospheric air inspired is *(CO_2? N_2? O_2?), (most? little?)* of this gas normally dissolves in blood plasma because of the low _______________ of this gas in plasma. But in compressed air, *(more? less?)* N_2 gas dissolves in plasma. Deep sea divers can avoid having excessive release of bubbles of this gas in their tissues (the condition known as _______________ sickness) by ascending from the depths *(rapidly? slowly?)*.

b. Hyperbaric oxygenation causes *(more? less?)* O_2 to dissolve in Mr. Costa's blood since he is exposed to *(0.3-0.4? 3-4?)* atmosphere(s) of pressure. Explain why this procedure kills anaerobic bacteria.

c. As Jenny performs CPR on her neighbor, Mr. Chenin, he receives air that is about *(21%? 16%? 13.6%?)* O_2 and about _______% CO_2. Explain the significance of these gas values.

✓ **D4.** Answer these questions about external and internal respiration.

a. A primary factor in the diffusion of gas across a membrane is the difference in concentration of the gas (reflected by _______________ pressures) on the two sides of the membrane. On the left side of Figure LG 23.3 write values for pO_2 (in mm Hg) in each of the following areas. (Refer to text Figure 23.18 for help.)

Atmospheric air (Recall this value from Figure LG 23.5.)

Alveolar air (Note that this value is lower than that for atmospheric pO_2 because some alveolar O_2 enters pulmonary blood and because of the presence of water vapor.)

b. Calculate the P_{O_2} difference (gradient) between alveolar air and blood entering lungs and write that number on the figure.

_______ mm Hg – _______ mm Hg = _______ mm Hg

c. Three other factors that increase exchange of gases between alveoli and blood are: *(large? small?)* surface area of lungs; *(thick? thin?)* respiratory membrane; and *(increased? decreased?)* blood flow (perfusion) through lungs (as in exercise).

d. By the time blood leaves the lungs to return to heart and systemic arteries, its P_{O_2} is normally *(greater than? about the same as? less than?)* P_{O_2} of alveoli. Write the correct value on Figure LG 23.3.

e. Now fill in all three pCO_2 values on Figure LG 23.3.

f. *A clinical correlation.* Joyce, age 62, has had blood drawn from her radial artery to determine her arterial blood gases. Her P_{O_2} is 56 and her P_{CO_2} is 48. Are these typical values for a healthy adult? _______

g. Fill in blood gas values (P_{O_2} and P_{CO_2}) on the short lines on Figure LG 23.4, page LG 516.

✓ **D5.** With increasing altitude the air is "thinner," that is, gas molecules are farther apart, so atmospheric pressure is lower. Atop a 25,000-foot mountain, this pressure is only 282 mm Hg. Oxygen still accounts for 21% of the pressure. What is P_{O_2} at that level?

From this calculation you can see limitations of life (or modifications that must be made) at high altitudes. If atmospheric P_{O_2} is 59.2, neither alveolar nor blood P_{O_2} could surpass that level.

✓ **D6.** *A clinical correlation.* Match the clinical cases listed below with the factors (in box) that can decrease diffusion across the respiratory *(alveolar-capillary or a-c)* membrane. On each line that follows, write a rationale for this decrease. The first one is done for you.

DD.	**Diffusion distance**	**PP.**	**Partial pressure of gases**
SA.	**Surface area for gas exchange**	**SMW.**	**Solubility and molecular weight of gases**

__**PP**__ a. Morphine has been prescribed for Ms. Iudica to alleviate her pain. Rationale: **Morphine slows the respiratory rate, so decreases alveolar *p*O_2.**

_______ b. Mr. Schmidt has interstitial viral pneumonia, with excessive accumulation of fluid in interstitial spaces. Rationale: _______________________.

_______ c. Mrs. McLaughlin was diagnosed with emphysema nine years ago. Rationale: _______________________.

_______ d. Dr. Wu has a left-sided congestive heart failure (CHF) with pulmonary edema. Arterial blood gas (ABG) analysis indicates her P_{O_2} is 54 mm Hg and P_{CO_2} is 43 mm Hg. Rationale: _______________________.

E. Physiology of respiration: transport of oxygen and carbon dioxide in blood (pages 831–835)

✓ **E1.** Answer these questions about oxygen transport. Refer to Figure 23.19 in your text.

a. One hundred mL of blood contains about ______ mL of oxygen. Of this, about 19.7 mL is carried as ____________________. Only a small amount of oxygen is carried in the dissolved state because oxygen has a *(high? low?)* solubility in blood or water.

b. Oxygen is attached to the ____________________ atoms in hemoglobin. The chemical formula for oxyhemoglobin is ___________. When hemoglobin carries all of the oxygen it can hold, it is said to be fully ____________________. High P_{O_2} in alveoli will tend to *(increase? decrease?)* oxygen saturation of hemoglobin.

c. Refer to Figure 23.20 in your text. Note that arterial blood, with a P_{O_2} of about 100–105 mm Hg, has its hemoglobin ______% saturated with oxygen. (This may be expressed as S_{O_2} = ______%.)

d. List four factors that will enhance the dissociation of oxygen from hemoglobin so that oxygen can enter tissues. (*Hint:* Think of conditions within active muscle tissue.)

e. *For extra review.* Demonstrate the effect of temperature on the oxygen-hemoglobin dissociation curve. Draw a vertical line on Figure 23.22 of the text at P_{O_2} = 40 (the value for venous blood). Compare the percentage saturation of hemoglobin (SO_2) at these two body temperatures:

37°C (98.6°F) ______% S_{O_2} 43°C (109.4°F) ______% S_{O_2}

In other words, at a higher body temperature, for example in high fever, *(more? less?)* oxygen will be attached to hemoglobin, while *(more? less?)* oxygen will enter tissues to fuel metabolism. This is known as a "shift to the *(left? right?)*" of the oxygen-hemoglobin dissociation curve.

f. By the time blood enters veins to return to the heart, its oxygen saturation of hemoglobin (S_{O_2}) is about ______%. Note on Figure 23.20 of your text that although p_{O_2} drops from 100 in arterial blood to 40 in venous blood, oxygen saturation drops *(more? less?)* dramatically, that is, from 98% to 75%. Of what significance is this?

g. Fetal hemoglobin (Hb-F) carries 30% *(more? less?)* oxygen than adult hemoglobin (Hb-A). State the significance of this fact.

✓ **E2.** Carbon monoxide has about _______ times the affinity that oxygen has for hemoglobin. State the significance of this fact.

✓ **E3.** Complete this Checkpoint about transport of carbon dioxide.

a. Write the percentage of CO_2 normally carried in each of these forms: _______% is present in bicarbonate ion (HCO_3^-); _______% is bound to the globin portion of hemoglobin; _______% is dissolved in plasma.

b. Carbon dioxide (CO_2) produced by cells of your body diffuses into red blood cells (RBCs) and combines with water to form ____________________.

c. Carbonic acid tends to dissociate into two products. One is H^+ which binds to ____________________. The other product is ____________________ (bicarbonate), which is carried in *(RBCs? plasma?)* in exchange for a *(K^+? Cl^-?)* ion that shifts into the RBC.

d. Now write the entire sequence of reactions described in (b) and (c). Be sure to include the enzyme that catalyzes the first reaction. Notice that the reactions show that increase in CO_2 in the body (such as when respiratory function is inadequate) will tend to cause a buildup of acid (H^+) in the body. *(Repeat this sentece several times as it is significant!)*

CO_2 + _______ ——————→ (enzyme) _______ → H^+ + ____________________

H^+ ↓ Binds To Hb

____________________ ↓ Shifts to plasma in exchange for Cl^-

e. The lower the amount of $Hb\text{-}O_2$ (oxyhemoglobin) in red blood cells, the *(higher? lower?)* the amount of CO_2 that hemoglobin can carry. This principle, known as the *(Bohr? Haldane?)* effect, means that hypoxic tissues (such as during exercise) can get rid of their CO_2 faster.

E4. Note that as the red blood cells reach lung capillaries, the same reactions you just studied occur, but in reverse. Study Figure 23.24b in the text carefully. Then list the major steps that occur in the lungs so that CO_2 can be exhaled.

F. Control of respiration; effects of exercise (pages 835–841)

✓ **F1.** Complete the table about respiratory control areas. Indicate whether the area is located in the medulla (M) or pons (P).

Name	M/P	Function
a.	M	Controls rhythm; consists of inspiratory and expiratory areas
b. Pneumotaxic		
c.		Prolongs inspiration and inhibits expiration

✓ **F2.** Answer these questions about respiratory control.

a. The main chemical change that stimulates respiration is increase in blood level of ________, which is directly related to *(decrease in pO_2? increase in pCO_2?)* of blood. (See Checkpoint E3d above.)

b. Cells most sensitive to changes in blood CO_2 are located in the *(medulla? pons? aorta and carotid arteries?)*. These are known as *(central? peripheral?)* chemoreceptors.

c. An increase in arterial blood P_{CO_2} is called ____________________. Write an arterial P_{CO_2} value that is hypercapnic. ________ mm Hg *(Even slight? Only severe?)* hypercapnia will stimulate the respiratory system, leading to *(hyper? hypo?)*-ventilation.

d. State two locations of chemoreceptors sensitive to changes in P_{O_2}.
____________________ ____________________. These are known as *(central? peripheral?)* chemoreceptors. *(Even slight? Only large?)* decreases in P_{O_2} level of blood will stimulate these chemoreceptors and lead to hyperventilation. Give an example of a P_{O_2} low enough to evoke such a response. ________ mm Hg. This condition is *(hypoxia? hypercapnia?)*

e. Increase in body temperature (as in fever), as well as stretching of the anal sphincter, will cause ________-crease in the respiratory rate.

f. Take a deep breath. Imagine the ____________________ receptors in your airways being stimulated. These will cause *(excitation? inhibition?)* of the inspiratory and apneustic areas, resulting in expiration. This reflex, known as the __________________ reflex, prevents overinflation of the lungs.

g. The term ____________________ means a brief cessation of breathing.

✓ **F3.** *A clinical correlation.* Identify types of hypoxia likely to be present in each situation. Use these answers: *anemic, histotoxic, hypoxic, ischemic.*

a. Cyanide poisoning: ____________________ hypoxia

b. Decreased circulation, as in circulatory shock: ____________________ hypoxia

c. Low hematocrit due to hemorrhage or iron-poor diet: ____________________ hypoxia

d. Chronic bronchitis or pneumonia: ____________________ hypoxia

✓ **F4.** Indicate whether the following factors will cause apnea (*A*), an increase in respiratory rate (*In*) or a decrease (*De*) in respiratory rate.

________ a. Fever

________ b. Prolonged somatic pain

________ c. Sudden, severe pain

________ d. Sudden plunge into cold water

________ e. Sudden drop in blood pressure

________ f. Activation of the limbic system during emotional anxiety

________ g. Proprioceptive sensations that accompany exercise

✓ **F5.** Identify physiological adjustments made during exercise in this Checkpoint.

a. During exercise, blood flow to the lungs (known as pulmonary ____________________) will *(increase? decrease?)*. The rate of diffusion of oxygen from alveolar air into

pulmonary blood may increase ____________________-fold. This rate is known as

oxygen ____________________ ____________________.

b. At the onset of moderate exercise, ventilation abruptly increases primarily in *(rate? depth?)*, providing more oxygen to active muscles. This change in ventilation is due primarily to *(neural? chemical or physical?)* signals that alert the medulla.

c. The gradual increase in ventilation that follows is due to a slight increase in

(pO_2? P_{CO_2} ?), a slight decrease in *(P_{O_2}? P_{CO_2}?)*, and a slight ______-crease in temperature that accompany exercise.

F6. Write an essay in which you describe effects of smoking on respiratory efficiency. Include these terms in your essay: *terminal bronchioles, hemoglobin, mucus, cilia, and elastic fibers.*

G. Developmental anatomy and aging of the respiratory system (pages 841–842)

✓ **G1.** Describe development of the respiratory system in this exercise.

a. The laryngotracheal diverticulum is derived from ________________-derm. List structures formed from this bud.

b. Identify portions of the respiratory system derived from mesoderm.

✓ **G2.** Describe possible effects of these age-related changes.

a. Chest wall becomes more rigid as bones and cartilage lose flexibility.

b. Decreased macrophage and ciliary action of lining of respiratory tract.

H. Disorders, medical terminology (pages 844–846)

✓ **H1.** Describe COPD in this Checkpoint.

a. COPD stands for C________________ O________________ P________________ D________________. In most cases, COPD *(is? is not?)* preventable. Name the two principal types of COPD: ________________ and ________________

b. In emphysema, alveolar walls are broken down so that lungs contain *(more? fewer?)* alveoli that are *(larger? smaller?)* in size. As a result of alveolar destruction, the surface area of the respiratory membrane ______-creases, and the amount of gas diffusion ______-creases also.

c. This breakdown of alveolar walls is related to effects of smoking and decrease in an enzyme known as alpha-1-____________. Since this enzyme normally inhibits proteases, then in the absence of the enzyme, proteases are *(more? less?)* active. One protease, known as ________________, is then free to destroy elastic fibers in alveolar walls so that alveoli remain inflated. Consequently, *(more? less?)* air is trapped in lungs, leading to the classic "________________ chested" appearance of an emphysema patient.

d. Chronic bronchitis involves inflammation of bronchi with *(deficient? excessive?)* mucus production and edematous mucosa lining airways. As a result, airways are *(dilated? narrowed?)*.

e. Asthma (sometimes classified as a COPD) involved changes in mucus and mucosa much like chronic bronchitis. In addition, smooth muscle of airways undergoes broncho-*(dilation? spasm?)*. List several triggers for asthma attacks.

✓ **H2.** Lung cancer is the number *(1? 3? 10?)* cause of cancer death in the United States.

At the time of diagnosis, distant metastases are present in about ______% of patients.

Lung cancer is ______ to ______ times more common in smokers than in nonsmokers. Secondhand smoke *(is? is not?)* known to lead to lung cancer deaths. The most common form of lung cancer, known as __________________ carcinoma, occurs in the *(alveoli? airways?)*, leading to blockage of airways, and inability to get air. (The American Lung Association's motto is, "If you can't breathe, nothing else __________________.")

✓ **H3.** Match the condition with the correct description.

A.	**Asphyxia**	**PE.**	**Pulmonary edema**
CF.	**Cystic fibrosis**	**PN.**	**Pneumonia**
COR.	**Coryza**	**SIDS.**	**Sudden infant death syndrome**
D.	**Dyspnea**	**TA.**	**Tachypnea**
E.	**Emphysema**	**TB.**	**Tuberculosis**
H.	**Hemoptysis**		

________ a. Permanent inflation of lungs due to loss of elasticity; rupture and merging of alveoli, followed by their replacement by fibrous tissue

________ b. Inherited disease in which ducts become obstructed with thick mucus; affects airways as well as digestive and reproductive organs

________ c. Acute infection or inflammation of alveoli, which fill with fluid

________ d. Oxygen starvation

________ e. Caused by a species of *Mycobacterium*; lung tissue is destroyed; incidence is higher in persons with AIDS

________ f. Difficult or painful breathing; shortness of breath

________ g. Rapid breathing rate

________ h. Common cold; caused by rhinoviruses, and directly related to stress

________ i. Abnormal accumulation of fluid in alveoli, for example, due to left ventricular failure

________ j. A condition that may be prevented by having newborns sleep on their backs ("back to sleep")

________ k. Blood in sputum

H4. Explain how the abdominal thrust (Heimlich maneuver) helps to remove food that might otherwise cause death by choking.

ANSWERS TO SELECTED CHECKPOINTS: CHAPTER 23

A2. (a-b) MN* → P* → L → T → BB → ADA. (c) Conducting.

A3. (a) Paranasal sinuses and meatuses. (b) External nares. (c) Vestibule. (d) Internal nares (choanae). (e) Rhinoplasty. (f) Nasal cartilages.

A4. (a) Mucosa lining nose, septum, conchae, meati, and sinuses; lacrimal drainage; coarse hairs in vestibule. (b) Olfactory region lies superior to superior nasal conchae. (c) Sounds resonate in nose and paranasal sinuses.

A5. (a) NP. (b–d) OP. (e) LP.

B1. (a) T. (b) E. (c) A. (d) C. (d) RG.

B4. (a) Windpipe; anterior; 12 (5), carina. (b) Mucous, hyaline cartilage.

B5. (a) Protects against dust. (b) The open part of the cartilage ring faces posteriorly, permitting slight expansion of the esophagus during swallowing; the trachealis muscle provides a barrier preventing food from entering the trachea. (c) Promotes cough reflex, which helps deter food or foreign objects from moving from trachea into bronchi.

B6. The right bronchus is more vertical and slightly wider than the left.

B7. Do not; muscle spasms can collapse airways.

B9. (a) Pleurisy or pleuritis. (b) Pleural effusions. (c) Thoracentesis.

B10. (a) Base; clavicle; ribs. (b) Hilus; bronchi, pulmonary vessels, and nerves. (c) Left; left; right; refer to Figure 23.10, page 817 of the text.

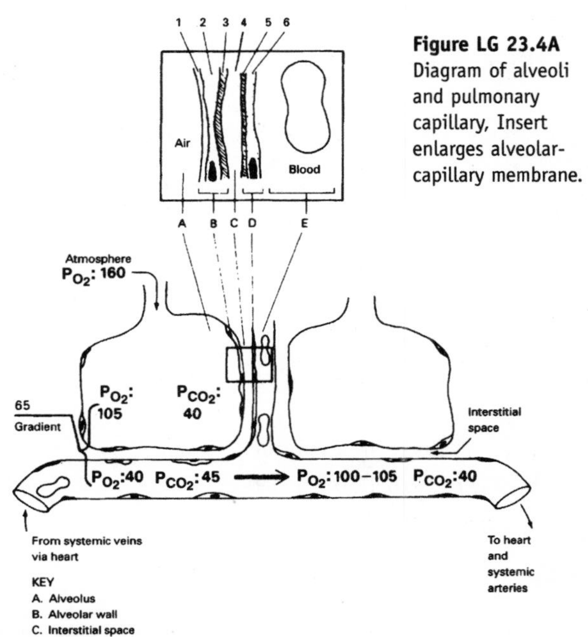

Figure LG 23.4A Diagram of alveoli and pulmonary capillary, Insert enlarges alveolar-capillary membrane.

B11. (a) T R A. (b) Alveolar-capillary. See KEY to Figure LG 23.3A. (c) 1, Surfactant; 2, alveolar epithelium; 3, epithelial basement membrane; 4, interstitial space; 5, capillary basement membrane; 6, capillary endothelium. (d) Thin; edema; less. (e) Squamous; surfactant; debris.

B12. (a) Bronchial; aorta. (b) Azygos, accessory hemiazygos. (c) Vein. (dl) Systemic; systemic; (d2) pulmonary; ventilation-perfusion.

C1. (a) Label (1) pulmonary ventilation; alveoli; O_2, in, CO_2, ex. (b) Air, blood; label (2) external respiration. (c) Veins, blood in systemic capillaries, tissue cells; label (3) internal respiration. (d) Artery, external, 2; pulmonary ventilation, 1. (e) Cellular. (f) Refer to Figure 21.17, p. 718 in your text.

C2. (a) Alveolar or intrapulmonic, 760; the same as. (b) Intrapleural, 756; less than. (c) Diaphragm, external intercostal; increases; increase. (d) Decrease, 754; decreases, 758; Boyle. (e) Atmosphere, alveoli; into; inspiration; 760. (f) Active, passive, elastic; recoil of elastic fibers stretched during inspiration and inward pull of surface tension; sup, post. (g) Abdominal muscles and internal intercostals.

C3. (a) Low, high. (b) High; surfactant, increases. (c) II, respiratory distress syndrome (RDS). (d) Atelectasis; conditions creating external pressure on lungs, such as air in the pleural cavity (pneumothorax) or pleural effusions, or conditions that inhibit reinflation of lungs, such as bronchitis.

C4. (a) ↓ (b) ↓ (c) ↑ .

C8. 70% (350/500); dead space, 150 (or weight of the person).

C9. 7,200 mL/min (= 15 breaths/min × 480 mL/breath); 4950 (= 15 breaths/min × 330 mL/breath).

C10. (a) V_T. (b) FRC. (c) ERV. (d) RV. (e) VC. (f) TLC. (g) IRV. (h) $FEV_{1.0}$.

C11. (a) 500 (0.5). (b) 6,000 (6). (c) 4,800 (4.8). (d) 2,400 (2.4). (e) 1,200 (1.2).

Dl. (a) H. (b) D. (c) H. (d) B.

D2. (a) 760. (b) 79 (or 78.6), 21 (or 20.9); 04. (c) 600 (or 597.4); P_{N_2}. (d) 21 (or 20.9)% × 760 mm Hg = about 160 (158.8) mm Hg = P_{O_2}.

D3. (a) N_2, little, solubility; more; decompression, slowly. (b) More; anaerobes cannot live in the presence of oxygen; see page 828 of the text. (c) 16%; the air Jenny exhales into Mr. Chenin is lower in O_2 than atmospheric air (21%); however it can still meet Mr. Chenin's needs for O_2. The level of CO_2 administered in Jenny's exhaled air (4.5%) is higher than that in atmospheric air (0.04%), but the extra CO_2 may stimulate Mr. Chenin's respiratory centers (See Checkpoint F2a.)

D4. (a) Partial; see Figure LG 23.3A. (b) 105 – 40 = 65. (c) Large, thin, increased. (d) About the same as; 100. (e) See Figure LG 23.3A. (f) No. Typical radial arterial values are same as for alveoli or blood leaving lungs: P_{O_2} = 100, P_{CO_2} = 40. Joyce's values indicate inadequate gas exchange. (g) Labels for Figure LG 23.4; inferior vena cava, P_{O_2} = 40 mm Hg and P_{CO_2} = 45 mm Hg; aorta, P_{O_2} = 100 mm Hg and P_{CO_2} = 40 mm Hg.

D5. 282 × 21% = 59.2 mm Hg.

D6. (b) DD; the excess fluid increases the thickness (diffusion distance) of the a-c membrane. (c) SA; emphysema destroys alveolar walls. (d) SMW. Because the a-c membrane is considerably more permeable to CO_2, she can exhale CO_2; so her P_{CO_2} is only slightly elevated above a normal of 40. Because O_2 exhibits lower solubility in the a–c membrane, her P_{O_2} (54) is considerably lower than normal (about 100). Therefore, her hypoxia is much more severe than her hypercapnia.

E1. (a) 20; oxyhemoglobin; low. (b) Iron; HbO_2, saturated; increase. (c) 98; 98. (d) Increase in temperature, P_{CO_2}, acidity, and BPG. (e) 67; 40; less, more; right. (f) 75; less. In the event that respiration is temporarily halted, even venous blood has much oxygen attached to hemoglobin and available to tissues. (g) More; this partially compensates for the inefficient fetal circulation and the low oxygen saturation of maternal blood.

E2. 200; oxygen-carrying capacity of hemoglobin is drastically reduced in carbon monoxide poisoning.

E3. (a) 78, 13, 0. (b) H_2CO_3 (carbonic acid). (c) Hemoglobin (as H•Hb); HCO_3^-, plasma, Cl^-. (d) CO_2 + $H_2O \xrightarrow{\text{carbonic anhydrase}} H_2CO_3$ (carbonic acid) → H+ + HCO_3^- (bicarbonate). (e) Higher; Haldane.

F1.

Name	M/P	Function
a. **Medullary rhythmicity**	M	Controls rhythm; consists of inspiratory and expiratory areas
b. Pneumotaxic	P	**Limits inspiration and facilitates expiration**
c. **Apneustic**	P	Prolongs inspiration and inhibits expiration

F2. (a) H^+, increase in P_{CO_2}, (b) Medulla; central. (c) Hypercapnia; any value higher than 40; even slight, hyper. (d) Aortic and carotid bodies; peripheral; only large; usually below 60; hypoxia. (e) In. (f) Stretch; inhibition; inflation (Hering-Breuer). (g) Apnea.

F3. (a) Histotoxic. (b) Ischemic. (c) Anemic. (d) Hypoxic.

F4. (a–b) In. (c–d) A. (e–g) In.

F5. (a) Perfusion, increase; three; diffusing capacity. (b) Depth; neural. (c) P_{CO_2}, P_{O_2}, in.

G1. (a) Endo; lining of larynx, trachea, bronchial tree, and alveoli. (b) Smooth muscle, cartilage, and other connective tissues of airways.

G2. (a) Decreased vital capacity, and so decreased P_{O_2}, (b) Risk of pneumonia.

H1. (a) Chronic obstructive pulmonary disease; is (by avoiding smoking); emphysema and chronic bronchitis. (b) Fewer, larger; de, de. (c) Antitrypsin; more; elastase; more; barrel. (d) Excessive; narrowed. (e) Spasm; allergens such as pollen, food, emotional upset, or exercise (exercise-induced asthma).

H2. 1; 55; 10–30; is; bronchogenic, airways; matters.

H3. (a) E. (b) CF. (c) PN. (d) A. (e) TB. (f) D. (g) TA. (h) COR. (i) PE. (j) SIDS. (k) H.

MORE CRITICAL THINKING: CHAPTER 23

1. Contrast location, structure, and function of each of these parts of the respiratory system: vestibule/conchae, nasopharynx/oropharynx, epiglottis/thyroid cartilage, primary bronchi/bronchioles, parietal pleura/visceral pleura.
2. Write job descriptions for the three types of cells that form alveoli.
3. Contrast Boyle's, Henry's, and Dalton's laws, and relate each law to your respiratory function.
4. All lung "capacities" consist of combinations of lung "volumes." Identify which volumes contribute to: vital capacity, total lung capacity, and functional residual capacity. Speculate as to why functional residual capacity is so named.
5. Ruth has just had arterial blood gases (ABGs) drawn. Results indicate that her arterial pO_2 is 64 mm Hg and her P_{O_2} is 52 mm Hg. Determine if each of these values is high, normal, or low?
6. Contrast the Bohr effect with the Haldane effect on the gas-carrying capacity of hemoglobin.
7. Define and state one or more reasons for performing the following procedures: rhinoplasty, tracheostomy, intubation, bronchography, bronchoscopy, thoracentesis, hyperbaric oxygenation, and Heimlich (abdominal thrust) maneuver.

MASTERY TEST: CHAPTER 23

Questions 1–4: Arrange the answers in correct sequence.

_____ _____ _____ 1. From first to last, the steps involved in inspiration:
A. Diaphragm and intercostal muscles contract.
B. Thoracic cavity and lungs increase in size.
C. Alveolar pressure decreases to 758 mm Hg.

_____ _____ _____ 2. From most superficial to deepest:
A. Parietal pleura
B. Visceral pleura
C. Pleural cavity

_____ _____ _____ _____ _____ 3. From superior to inferior:
A. Bronchioles
B. Bronchi
C. Larynx
D. Pharynx
E. Trachea

_____ _____ _____ _____ _____ _____ 4. Pathway of inspired air:
A. Alveolar ducts
B. Bronchioles
C. Lobar bronchi
D. Primary bronchi
E. Segmental bronchi
F. Alveoli

Questions 5–9: Circle the letter preceding the one best answer to each question.

5. All of the following terms are matched correctly with descriptions *except:*
A. Internal nares: choanae
B. External nares: nostrils
C. Posterior of nose: vestibule
D. Pharyngeal tonsils: adenoids

6. Which of these values (in mm Hg) would be normal for P_{O_2} of blood in the femoral artery?
A. 40 D. 160
B. 45 E. 760
C. 100 F. 0

7. Pressure and volume in a closed space are inversely related, as described by ________ law.
A. Boyle's D. Henry's
B. Starling's E. Bohr's
C. Dalton's

8. Choose the correct formula for carbonic acid:
A. HCO_3^- D. HO_3C_2
B. H_3CO_2 E. H_2C_3O
C. H_2CO_3

9. A procedure in which an incision is made in the trachea and a tube inserted into the trachea is known as a(n):
A. Tracheostomy
B. Bronchogram
C. Intubation
D. Pneumothorax

Questions 10–20: Circle T (true) or F (false). If the statement is false, change the underlined word or phrase so that the statement is correct.

T F 10. In the chloride shift Cl^- moves into red blood cells in exchange for H^+.

T F 11. When chemoreceptors sense increase in PCO_2 or increase in acidity of blood H^+ respiratory rate will normally be stimulated.

T F 12. Both increased temperature and increased acid content tend to cause oxygen to bind more tightly to hemoglobin.

T F 13. Under normal circumstances intrapleural pressure is always less than atmospheric.

T F 14. The pneumotaxic and apneustic areas controlling respiration are located in the pons.

T F 15. Fetal hemoglobin has a lower affinity for oxygen than maternal hemoglobin does.

T F 16. Most CO_2, is carried in the blood in the form of bicarbonate.

T F 17. The alveolar wall does contain macrophages that remove debris from the area.

T F 18. Intrapulmonic pressure means the same thing as intrapleural pressure.

T F 19. Inspiratory reserve volume is normally larger than expiratory reserve volume.

T F 20. The P_{O_2} and P_{CO_2} of blood leaving the lungs are normally about the same as P_{O_2} and P_{CO_2} of alveolar air.

Questions 21–25: Fill-ins. Complete each sentence with the word or phrase that best fits.

____________________ 21. The process of exchange of gases between alveolar air and blood in pulmonary capillaries is known as ________.

____________________ 22. Take a normal breath and then let it out. The amount of air left in your lungs is the capacity called ________ and it usually measures about ________ mL.

____________________ 23. The Bohr effect states that when more H^+ ions are bound to hemoglobin, less ________ can be carried by hemoglobin.

____________________ 24. The epiglottis, thyroid, and cricoid cartilages are all parts of the ________.

____________________ 25. ________ is a chemical that lowers surface tension and therefore increases inflatability (compliance) of lungs.

ANSWERS TO MASTERY TEST: CHAPTER 23

Arrange

1. A B C
2. A C B
3. D C E B A
4. D C E B A F

Multiple Choice

5. C
6. C
7. A
8. C
9. A

True–False

10. F. HCO_3^-
11. T
12. F. Dissociate from
13. T
14. T
15. F. Higher
16. T
17. T
18. F. Alveolar
19. T
20. T

Fill-ins

21. External respiration (or pulmonary respiration, or diffusion)
22. Functional residual capacity (FRC), 2400
23. Oxygen
24. Larynx
25. Surfactant

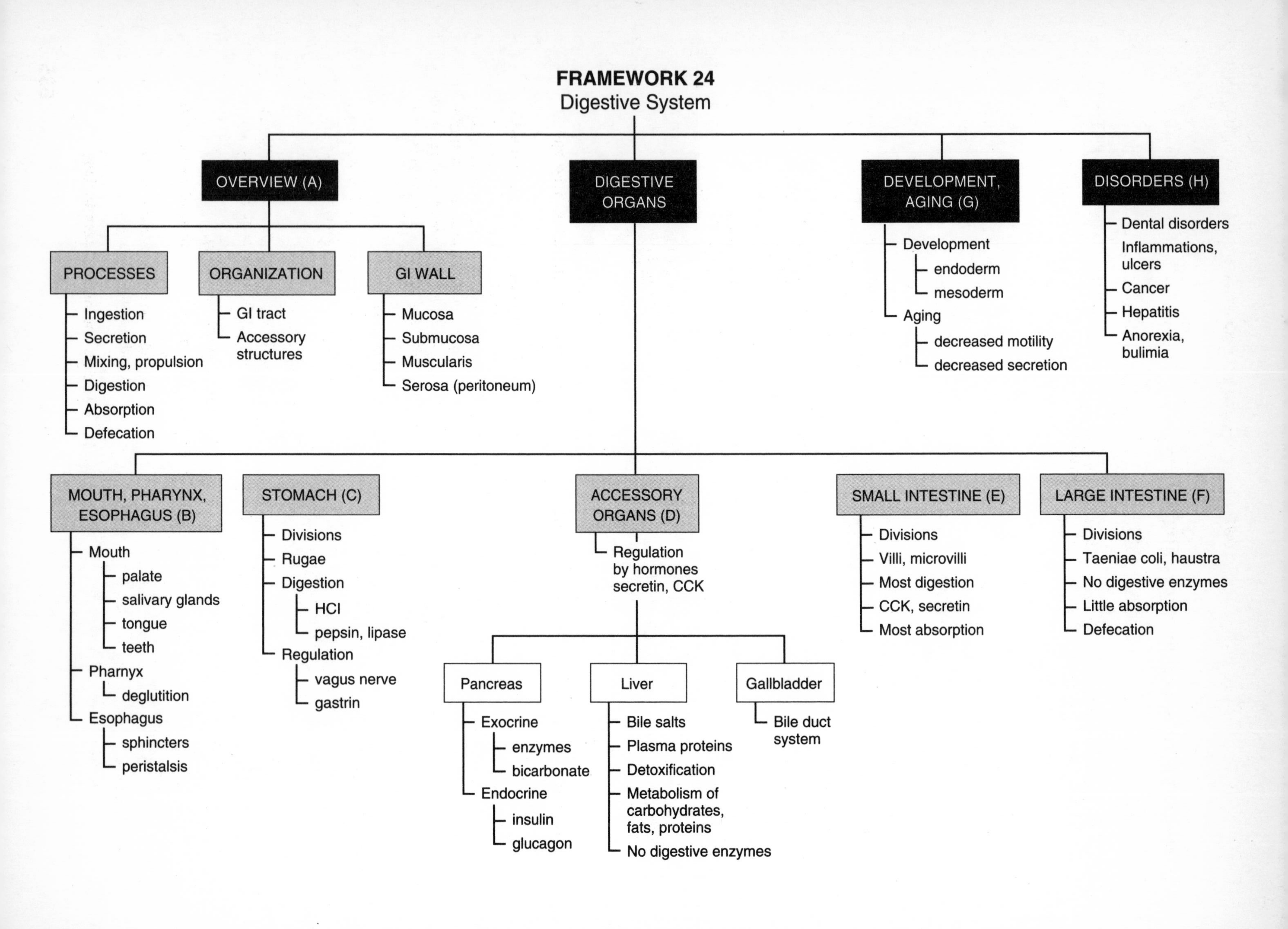

FRAMEWORK 24
Digestive System
OVERVIEW (A)
DIGESTIVE ORGANS
DEVELOPMENT, AGING (G)
DISORDERS (H)
PROCESSES
Ingestion
Secretion
Mixing, propulsion
Digestion
Absorption
Defecation
ORGANIZATION
GI tract
Accessory structures
GI WALL
Mucosa
Submucosa
Muscularis
Serosa (peritoneum)
Development
endoderm
mesoderm
Aging
decreased motility
decreased secretion
Dental disorders
Inflammations, ulcers
Cancer
Hepatitis
Anorexia, bulimia
MOUTH, PHARYNX, ESOPHAGUS (B)
Mouth
palate
salivary glands
tongue
teeth
Pharnyx
deglutition
Esophagus
sphincters
peristalsis
STOMACH (C)
Divisions
Rugae
Digestion
HCl
pepsin, lipase
Regulation
vagus nerve
gastrin
ACCESSORY ORGANS (D)
Regulation by hormones secretin, CCK
Pancreas
Exocrine
enzymes
bicarbonate
Endocrine
insulin
glucagon
Liver
Bile salts
Plasma proteins
Detoxification
Metabolism of carbohydrates, fats, proteins
No digestive enzymes
Gallbladder
Bile duct system
SMALL INTESTINE (E)
Divisions
Villi, microvilli
Most digestion
CCK, secretin
Most absorption
LARGE INTESTINE (F)
Divisions
Taeniae coli, haustra
No digestive enzymes
Little absorption
Defecation

CHAPTER 24

The Digestive System

Food comes in very large pieces (like whole oranges, stalks of broccoli, and slices of bread) that must fit into very small spaces in the human body (like liver cells or brain cells). The digestive system makes such change possible. Foods are minced and enzymatically degraded so that absorption into blood en route to cells becomes a reality. The gastrointestinal (GI) tract provides a passageway complete with mucous glands to help food slide along, muscles that propel food forward, muscles (sphincters and valves) that regulate flow, and secretions that break apart foods or modify pH to suit local enzymes. Accessory structures such as teeth, tongue, pancreas, and liver lie outside the GI tract; yet each contributes to the mechanical and chemical dismantling of food from bite-sized to right-sized pieces. The ultimate fate of the absorbed products of digestion is the story line of metabolism (in Chapter 25).

As you begin your study of the digestive system, carefully examine the Chapter 24 Topic Outline and Objectives. Check off each one as you complete it. You will also find food for thought in the Chapter 24 Framework. Be sure to refer to the Framework frequently and note relationships among key terms in each section.

TOPIC OUTLINE AND OBJECTIVES

A. Overview

1. Describe the organs of the digestive system.
2. Describe the basic processes of the digestive system.
3. Describe the layers that form the wall of the gastrointestinal tract.
4. Describe the peritoneum and its folds.

B. Mouth, pharynx, and esophagus

1. Identify the locations of the salivary glands, and describe the functions of their secretions.
2. Describe the structure and functions of the tongue.
3. Identify the parts of a typical tooth, and compare deciduous and permanent dentitions.
4. Describe the location and function of the pharynx.
5. Describe the location, anatomy, histology, and functions of the esophagus.

C. Stomach

1. Describe the location, anatomy, histology, and functions of the stomach.

D. Accessory organs: pancreas, liver, and gallbladder

1. Describe the location, anatomy, histology, and function of the pancreas.
2. Describe the location, anatomy, histology, and functions of the liver and gallbladder.
3. Describe the site of secretion and actions of gastrin, secretin, and cholecystokinin.

E. Small intestine

1. Describe the location, anatomy, histology, and functions of the small intestine.

F. Large intestine

1. Describe the location, anatomy, histology, and functions of the large intestine.

G. Development and aging

1. Describe the development of the digestive system.
2. Describe the effects of aging on the digestive system.

H. Disorders and medical terminology

WORDBYTES

Now become familiar with the language of this chapter by studying each wordbyte, its meaning, and an example of its use within a term. After you study the entire list, self-check your understanding by writing the meaning of each wordbyte on the line. As you continue through the *Learning Guide*, identify (and fill in) additional terms that contain the same wordbyte.

Wordbyte	Self-check	Meaning	Example(s)
amyl-	______________	starch	*amyl*ase
-ase	______________	enzyme	malt*ase*
bucca-	______________	cheeks	*bucca*l cavity
caec-, cec-	______________	blind	*cecum*
chole-	______________	bile, gall	*chole*static jaundice
cholecyst-	______________	gall bladder	*cholecyst*ectomy
cyst	______________	bladder	*cyst*itis
chym-	______________	juice	*chym*otrypsin
dent-	______________	tooth	*dent*ures
-ectomy	______________	removal of	append*ectomy*
entero-	______________	intestine	*entero*kinase
gastro-	______________	stomach	*gastr*in
gingiv-	______________	gums	gingivitis
hepat-	______________	liver	*hepato*cytes
ileo-	______________	ileum	*ileo*cecal valve
jejun-	______________	empty	*jejuno*ileostomy
lacteal	______________	milky	*lacteals*
lith-	______________	stone	chole*lith*iasis
odont-	______________	tooth	peri*odont*al ligament
or-	______________	mouth	*or*al
ortho-	______________	straight	*ortho*dontist
-ose	______________	sugar	lact*ose*
retro-	______________	behind	*retro*peritoneal
-rrhea	______________	to flow	dia*rrhea*
stoma-	______________	mouth, opening	colo*stomy*
teniae-	______________	flat	*taen*ia coli
vermi-	______________	worm	*vermi*form appendix

CHECKPOINTS

A. Overview (pages 852–857)

A1. Explain why food is vital to life. Give three specific examples of uses of foods in the body.

✓ **A2.** Cover the key and identify all digestive organs on Figure LG 24.1. Then color the structures with color code ovals; these five structures are all *(parts of the gastrointestinal tract? accessory structures?).*

✓ **A3.** List the six basic activities of the digestive system.

_______________ _______________ _______________

_______________ _______________ _______________

✓ **A4.** _______________ digestion occurs by action of enzymes (such as those in saliva) and intestinal secretion, whereas _______________ digestion involves action of the teeth and muscles of the stomach and intestinal wall.

✓ **A5.** Color layers of the GI wall indicated by color code ovals on Figure LG 24.2.

✓ **A6.** Now match the names of layers of the GI wall with the correct description.

Muc.	**Mucosa**	**Ser.**	**Serosa**
Mus.	**Muscularis**	**Sub.**	**Submucosa**

_______ a. Also known as the peritoneum, it forms mesentery and omentum

_______ b. Consists of epithelium, lamina propria, and muscularis mucosae

_______ c. Connective tissue containing glands, nerves, blood and lymph vessels

_______ d. Consists of an inner circular layer and an outer longitudinal layer

A7. Describe the tissues listed below by completing this table. Identify the layer in which each tissue is located, type of tissue in each case, and functions of these tissues.

	Layer of GI Wall	Type of Tissue	Function(s)
a. Enteroendocrine tissue			Secretes hormones
b. MALT tissue		Lymphatic	
c. Submucosal plexus (of Meissner)	Submucosa and mucosa		
d. Myenteric plexus (of Auerbach)		ANS nerve tissue	

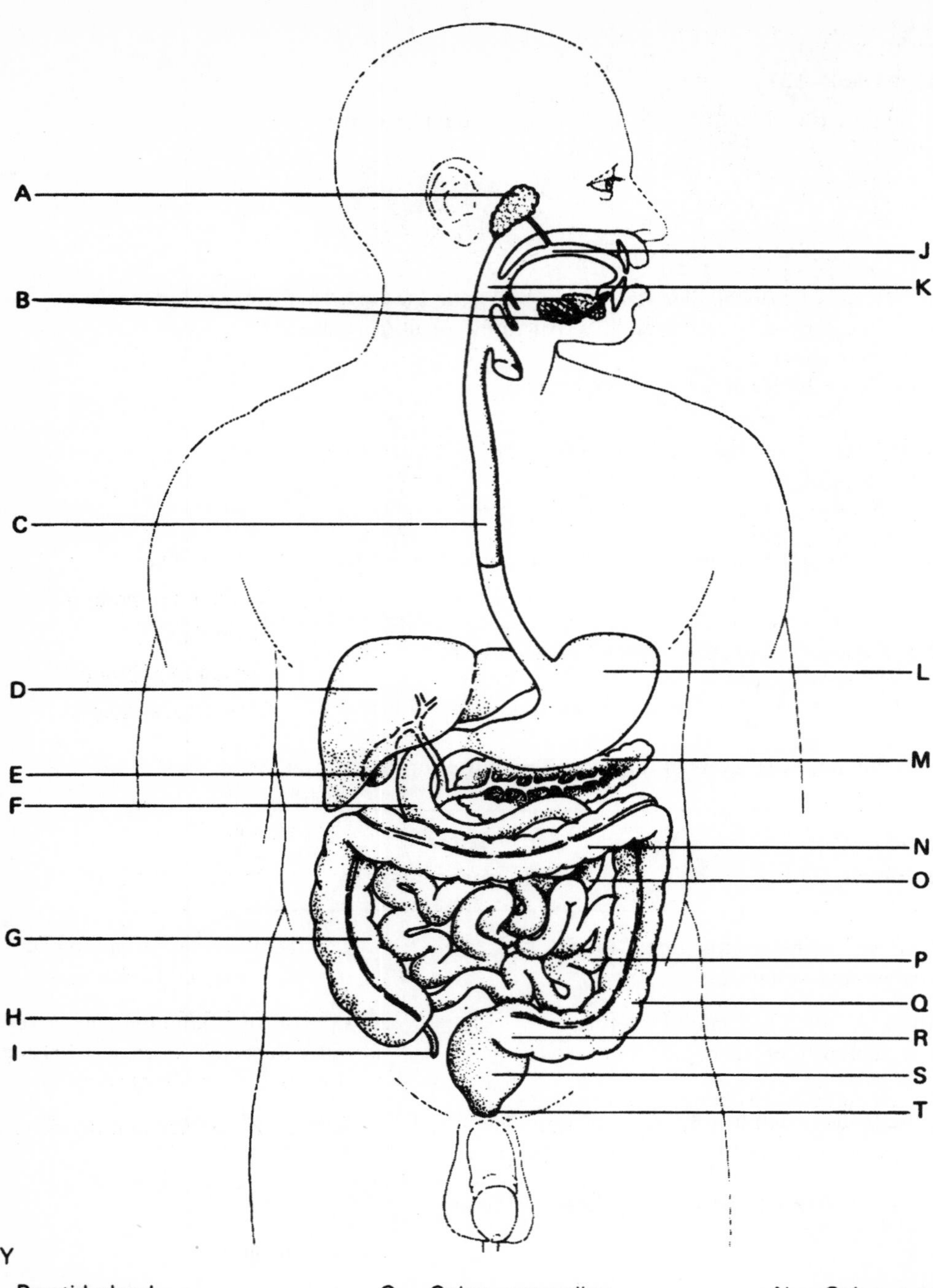

KEY

- A. Parotid gland
- B. Sublingual and submandibular glands
- C. Esophagus
- D. Liver
- E. Gallbladder
- F. Duodenum
- G. Colon, ascending
- H. Cecum
- I. Vermiform appendix
- J. Mouth
- K. Pharynx
- L. Stomach
- M. Pancreas
- N. Colon, transverse
- O. Jejunum
- P. Ileum
- Q. Colon, descending
- R. Colon, sigmoid
- S. Rectum
- T. Anus

Figure LG 24.1 Organs of the digestive system. Color, draw arrows, and identify labeled structures as directed in Checkpoints A2, B2 and F3.

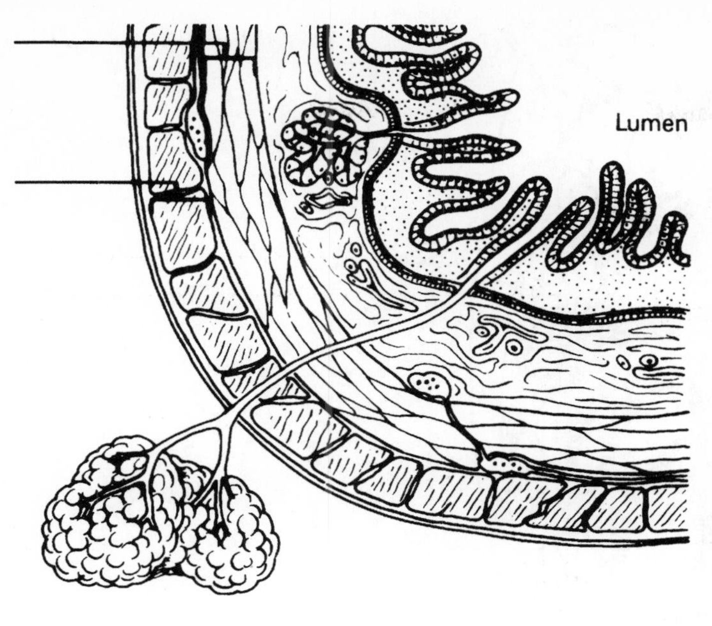

KEY
O Mucous membrane
O Muscularis
O Serous membrane
O Submucosa layer

Figure LG 24.2 Gastrointestinal (GI) tract and related gland seen in cross section. Refer to Checkpoint A5 and color layers of the wall according to color code ovals. Label muscularis layers.

✓ **A8.** Match the names of these peritoneal extensions with the correct descriptions.

F.	**Falciform ligament**	**M.**	**Mesentery**
G.	**Greater omentum**	**Meso.**	**Mesocolon**
L.	**Lesser omentum**		

_________ a. Attaches liver to anterior abdominal wall

_________ b. Binds intestine to posterior abdominal wall; provides route for blood and lymph vessels and nerves to reach small intestine

_________ c. Binds part of large intestine to posterior abdominal wall

_________ d. "Fatty apron"; covers and helps prevent infection in small intestine

_________ e. Suspends stomach and duodenum from liver

A9. Define the following terms.

a. Enteric nervous system (ENS)

b. Peritonitis

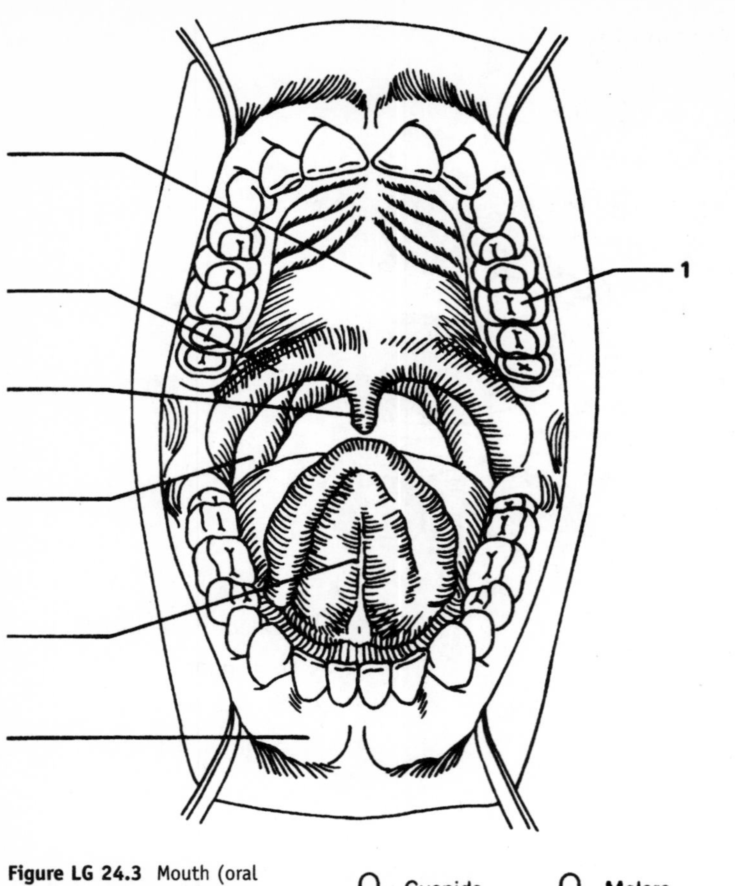

Figure LG 24.3 Mouth (oral cavity). Complete as directed in Checkpoints B1 and B7.

O Cuspids
O Incisors
O Molars
O Premolars (bicuspids)

B. Mouth, pharynx, and esophagus (pages 857–866)

✓ **B1.** Refer to Figure LG 24.3 and label the following structures.

Hard palate	Palatine tonsils	Uvula of soft palate
Lingual frenulum	Palatoglossal arch	Vestibule

B2. Identify the three salivary glands on Figure LG 24.1 and visualize their locations on yourself.

✓ **B3.** Complete this exercise about salivary glands.

a. Which glands are largest? *(Parotid? Sublingual? Submandibular?)*

Which are inflamed by the viral infection known as mumps? ____________________

b. Which secrete the thickest secretion due to presence of much mucus?

____

c. About 1 to 1 1/2 *(tablespoons? cups? liters?)* of saliva are secreted daily.

d. State three or more functions of saliva.

e. The pH of the mouth is appropriate for action of salivary amylase. This is about pH *(2? 6.5? 9?)*.

f. List five or more chemicals that are in saliva.

g. List four or more factors that can stimulate increase in saliva production.

h. *(Sympathetic? Parasympathetic?)* nerves cause dryness of the mouth, whereas stimulation of saliva production occurs by *(sympathetic? parasympathetic?)* nerves, specifically

in cranial nerves ____________________ and ____________________.

B4. Contrast terms in each pair:

a. *Intrinsic muscles of the tongue/extrinsic muscles of the tongue*

b. *Fungiform papillae/filiform papillae*

c. *Lingual frenulum/ankyloglossia*

✓ **B5.** Arrange the following in correct sequence:

______ ______ ______ a. From most superficial to deepest:

A. Root	**B. Neck**	**C. Crown**

______ ______ ______ b. From most superficial to deepest within a tooth:

A. Enamel	**B. Dentin**	**C. Pulp cavity**

______ ______ ______ c. From hardest to softest:

A. Enamel	**B. Dentin**	**C. Pulp cavity**

✓ **B6.** List two functions of the periodontal ligament.

✓ **B7.** Refer to Figure LG 24.3 and do this exercise on dentitions.

a. How many teeth (total) are in a complete set of deciduous (baby) teeth?

_______ How many are in a complete set of permanent teeth? _______

b. Color the teeth in the lower right of Figure LG 24.3 as indicated by color code ovals.

c. Number the teeth (1–8) on the upper right of the diagram according to order of eruption. One (the first one to erupt) is done for you.

d. Which teeth are most likely to become impacted? _____________________

✓ **B8.** Check your understanding of digestion by completing this Checkpoint.

a. *(Deglutition? Mastication?)* is a term that means chewing. Food is formed into a soft mass known as a *(bolus? chyme?)*.

b. Enzymes are used for *(chemical? mechanical?)* digestion of food. The main enzyme in saliva is salivary *(amylase? lipase?)*. This enzyme converts _____________________ into smaller carbohydrates. Most starch *(is? is not?)* broken down by the time food leaves the mouth. What inactivates amylase after about an hour in the stomach?

c. Where is lingual lipase made? *(Salivary glands? Tongue? Tonsils?)* This enzyme starts the breakdown of _____________________.

✓ **B9.** The term *deglutition* means _____________________. Match each stage of deglutition with the correct description.

E. Esophageal	**P. Pharyngeal**	**V. Voluntary**

_______ a. Soft palate and epiglottis close off respiratory passageways.

_______ b. Tongue pushes food back into oropharynx.

_______ c. Peristaltic contractions push bolus from pharynx to stomach.

✓ **B10.** Do this activity on the esophagus.

a. The esophagus is about _______ cm (_______ in.) long. It ends inferiorly in the *(stomach? duodenum?)*.

b. The muscularis of the superior one-third of the esophagus consists of *(smooth? striated?)* muscle, whereas the inferior one-third consists of *(smooth? striated?)* muscle similar to that found in the stomach.

c. Failure of the lower esophageal sphincter to close results in the stomach contents entering the esophagus and distending it and in irritation of the esophageal lining by the *(acidic? basic?)* contents of the stomach. The resulting condition is known as _____________________. Pain associated with either of these conditions may be confused with pain originating in the _____________________. Write four or more suggestions for controlling GERD.

✓ **B11.** *The Big Picture: Looking Back.* Refer to figures and exhibits in earlier text chapters that relate to digestive activity in the mouth.

a. Exhibit 11.4 and Figure 11.7 (pages 326–327) demonstrate the genioglossus, hyoglossus, and styloglossus. These are *(extrinsic? intrinsic?)* muscles of the tongue and are innervated by cranial nerve *(V? VII? XII?)*.

b. Table 17.4 (page 580) demonstrates that increased secretion of saliva is a function of the *(sympathetic? parasympathetic?)* division of the autonomic nervous system.

c. Table 14.2 (page 485) describes cranial nerves. Which ones carry sensory nerves for taste as well as the autonomic nerves stimulating salivary glands? Cranial nerves

________ and ________. Pain in an upper tooth is transmitted by the *(mandibular?*

maxillary? ophthalmic?) branch of the trigeminal nerve (cranial nerve ________).

B12. Summarize digestion in the mouth, pharynx, and esophagus by completing parts *a* and *b* of Table LG 24.1 (next page).

C. Stomach (pages 866–874)

✓ **C1.** Refer to Figure LG 24.1 and complete this Checkpoint.

a. Arrange these regions of the stomach according to the pathway of food from first to last.

Body	Fundus	Cardia	Pylorus

____________________ → ____________________ →

____________________ → ____________________

b. Match the terms in the box with the correct description.

GC. Greater curvature	LC. Lesser curvature

________ 1. More lateral and inferior in location

________ 2. Attached to lesser omentum

________ 3. Attached to greater omentum

c. In which condition does the pyloric sphincter fail to relax normally? *(Pyloric stenosis? Pylorospasm?)* Write one sign or symptom of this condition.

✓ **C2.** Complete this table about gastric secretions.

Name of Cell	Type of Secretion	Function of Secretion
a. Chief (zymogenic)		
b. Mucous		
c.	HCl	
d.	Gastrin	

Table LG 24.1 Summary of digestion.

Digestive Organs	Carbohydrate	Protein
a. Mouth, salivary glands 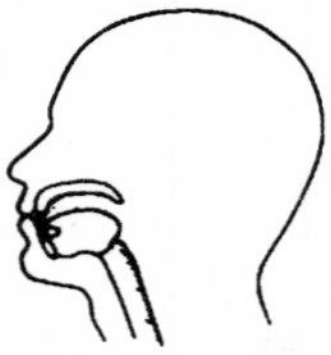	Salivary amylase: digests starch to maltose	
b. Pharynx, esophagus 		
c. Stomach 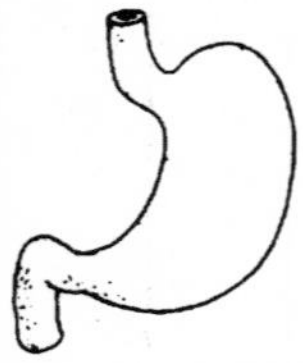		
d. Pancreas 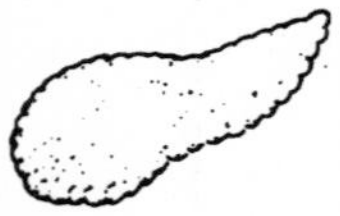		
e. Intestinal juices 		
f. Liver 	No enzymes for digestion of carbohydrates	
g. Large intestine 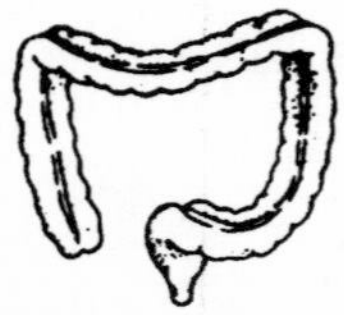	No enzymes for digestion of carbohydrates	No enzymes for digestion of proteins

Table LG 24.1 Summary of digestion. *(continued)*

Lipid	**Mechanical**	**Other Functions**
Lingual lipase		
	Deglutition, peristalsis	
		1. Secretes intrinsic factor 2. Produces hormone stomach gastrin
Pancreatic lipase: digests about 80% of fats		
No enzymes for digestion of lipids		

✓ **C3.** How does the structure of the stomach wall differ from that in other parts of the GI tract?

a. Mucosa

b. Muscularis

✓ **C4.** Answer these questions about chemical digestion in the stomach.

a. Pepsin is most active at very *(acid? alkaline?)* pH.

b. State two factors that enable the stomach to digest protein without digesting its own cells (which are composed largely of protein).

c. If mucus fails to protect the gastric lining, the condition known as

_____________________ may result.

d. Another enzyme produced by the stomach is _____________________, which digests

_____________________. In adults it is quite *(effective? ineffective?)*. Why?

e. Describe the "alkaline tide" that is likely to occur after a relatively large meal. (*Hint*: refer to Figure 24.13.)

✓ **C5.** Answer these questions about control of gastric secretion.

a. Name the three phases of gastric secretion: _____________________,

_____________________, and _____________________. Which of these causes gastric

secretion to begin when you smell or taste food? _____________________

b. Gastric glands are stimulated mainly by the _____________________ nerves. Their fibers are *(sympathetic? parasympathetic?)*. Two stimuli that cause vagal impulses to

stimulate gastric activity are _____________________ and _____________________.
Three emotions that inhibit production of gastric secretions are

_____________________, _____________________, and _____________________.

c. Distention of the stomach triggers the ________________ phase. Two effects of this stimulus are *(increase? decrease?)* in parasympathetic impulses and release of the hormone ________________. What else triggers release of gastrin?

Gastrin travels through the blood to all parts of the body; its target cells are gastric glands and gastric smooth muscle, which are *(stimulated? inhibited?)*, and the pyloric and ileocecal sphincters, which are *(contracted? relaxed?)*.

d. *A clinical correlation.* Both acetylcholine, released by vagal nerves, and gastrin *(stimulate? inhibit?)* HCl production. Their effect is *(greater? less?)* in the presence of histamine, which is made by ________________ cells in the stomach wall. Describe the effect of H_2 blockers such as cimetidine (Tagamet) on the stomach (See page 899 of the text.).

e. When food (chyme) reaches the intestine, nerves initiate the ________________ reflex, which *(stimulates? inhibits?)* further gastric secretion. Two hormones released by the intestine also inhibit gastric secretion as well as gastric motility, so the effect of these hormones is to *(stimulate? delay?)* gastric emptying. Name these two hormones:

________________ and ________________.

✓ **C6.** Food stays in the stomach for about ________________ hours. Which food type leaves the stomach most quickly? ________________ Which type stays in the stomach longest? ________________ Forcible expulsion of stomach contents through the mouth is known as vomiting or ________________. This process is controlled by the vomiting center located in the ________________. Prolonged vomiting *(can be? is never?)* serious. Explain why.

C7. Complete part *c* of Table LG 24.1, describing the role of the stomach in digestion.

D. Accessory organs: pancreas, liver, gallbladder (pages 874–881)

Dl. Refer to Figure LG 24.4 and color all of the structures with color code ovals according to colors indicated.

✓ **D2.** Complete these statements about the pancreas.

a. The pancreas lies posterior to the ________________.

b. The pancreas is shaped roughly like a fish, with its head in the curve of the ________________ and its tail nudging up next to the ________________.

c. The pancreas contains two kinds of glands. Ninety-nine percent of its cells produce *(endocrine? exocrine?)* secretions. One type of these secretions is an *(acid? alkaline?)* fluid to neutralize the chyme entering the small intestine from the stomach.

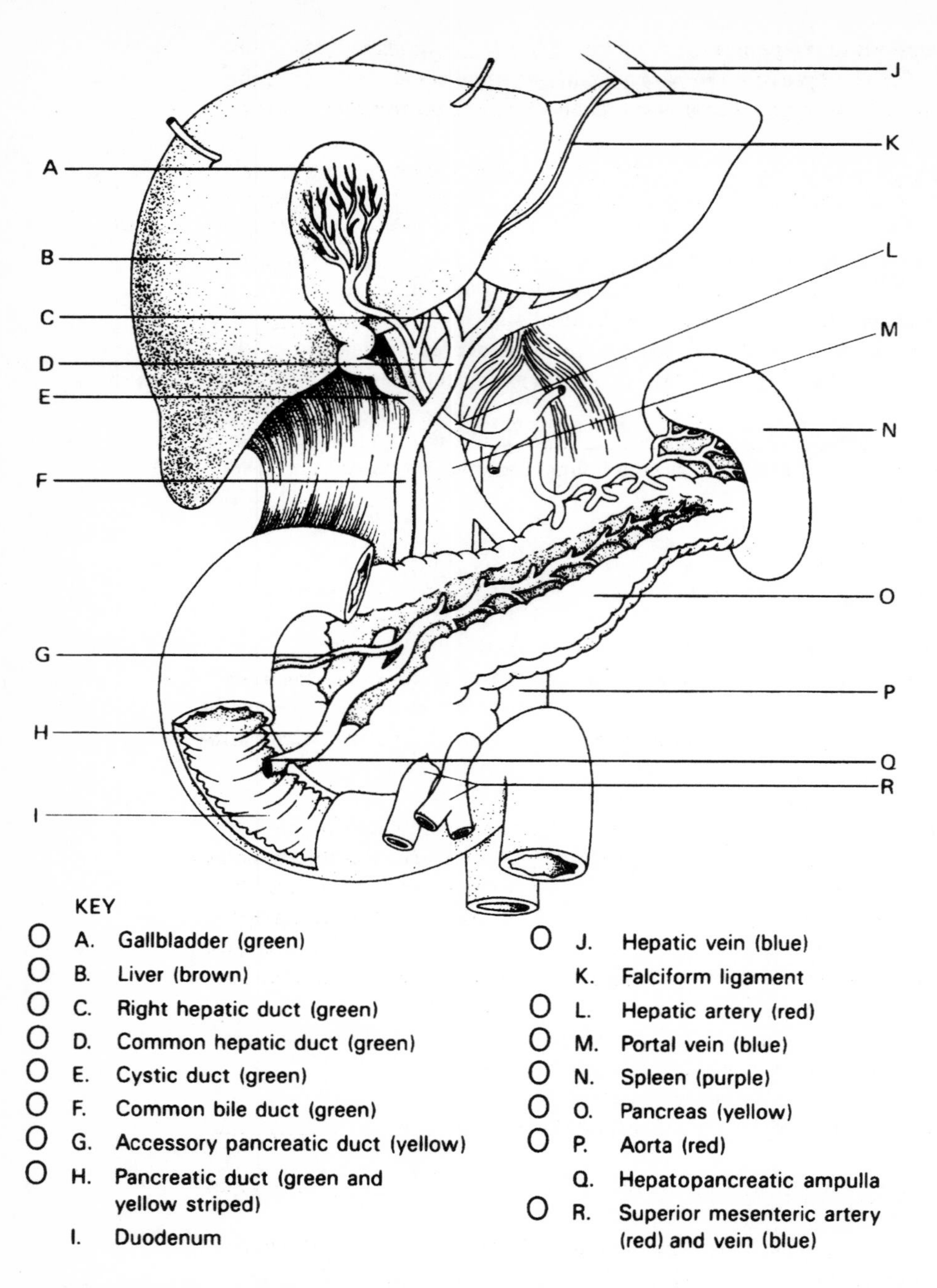

Figure LG 24.4 Liver, gallbladder, pancreas, and duodenum, with associated blood vessels and ducts. (Stomach has been removed). Refer to Checkpoints D1, D2 and D6; color and draw arrows as directed.

d. One enzyme in the pancreatic secretions is trypsin; it digests *(fats? carbohydrates? proteins?)*. Trypsin is formed initially in the inactive form (trypsinogen) and is activated by *(HCl? NaHCO3? enterokinase?)*. Trypsin itself serves as an activator in the formation of the protease named ________________ and the peptidase named ________________.

e. Most of the amylase and lipase produced in the body are secreted by the pancreas. Describe the functions of these two enzymes.

f. All exocrine secretions of the pancreas empty into ducts (_______ and _______ on Figure LG 24.4). These empty into the _______________.

g. *A clinical correlation.* In most persons the pancreatic duct also receives bile flowing through the _______________ (F on the figure). State one possible complication that may occur if a gallstone blocks the pancreatic duct (as at point H on the figure).

h. The endocrine portions of the pancreas are known as the _______________ _______________. List hormones they secrete. _______________, _______________, _______________, _______________. Typical of all hormones, these pass into *(ducts? blood vessels?)*, specifically into vessels that empty into the _______________ vein.

D3. Fill in part *d* of Table LG 24.1, describing the role of the pancreas in digestion.

✓ **D4.** Describe regulation of the pancreas in this exercise.

a. The pancreas is stimulated by *(sympathetic? parasympathetic?)* nerves, specifically the _______________ nerves, and also by hormones produced by cells located in the wall of the _______________.

b. What is the chemical nature of chyme that stimulates release of CCK? Chyme that *(is acidic? contains partially digested fats and proteins?)*. CCK then activates the pancreas to secrete fluid rich in *(HCO_3^-? digestive enzymes such as trypsin, amylase, and lipase?)*.

c. What is the chemical nature of chyme that stimulates release of secretion? Chyme that *(is acidic? contains partially digested fats and proteins?)*. Secretin then activates the pancreas to secrete fluid rich in *(HCO_3^-? digestive enzymes such as trypsin, amylase, and lipase?)*.

✓ **D5.** Answer these questions about the liver.

a. This organ weighs about _______ kg (_______ pounds). It lies in the _______________ quadrant of the abdomen. Of its two main lobes, the _______________ is the largest.

b. The _______________ ligament separates the right and left lobes. In the edge of this ligament is the ligamentum teres (round ligament of the liver), which is the obliterated _______________ vein.

c. Blood enters the liver via vessels named _______________ and _______________. In liver lobules blood mixes in channels called _______________ before leaving the liver via vessels named _______________.

✓ **D6.** Complete Figure LG 24.4 as directed.

a. Identify the pathway of bile from liver to intestine by following the structures you colored green in that figure: C D F H. The general direction of bile is from *(superior to inferior? inferior to superior?)*.

b. Now draw arrows to show the direction of blood through the liver. In order to reach the inferior vena cava, blood must flow *(superiorly? inferiorly?)* through the liver.

c. Write letters of the three structures that form the *portal triad* on the undersurface of the liver. ______ ______ ______

✓ **D7.** While blood is in liver sinusoids, hepatic cells have ample opportunity to act on this blood, modify it, and add new substances to it. Check your understanding of important functions of the liver in this Checkpoint.

a. Bile secreted from the liver is composed largely of the pigment named ____________ which is a breakdown product of __________________ cells. Excessive amounts of this pigment give skin a yellowish color, a condition known as __________________.
Two functions of bile are __________________ of fats and __________________ of fats (and fat-soluble vitamins). One of the breakdown products of __________________ gives feces the normal brown color.

b. Name several plasma proteins synthesized by the liver.

c. Identify two or more types of chemicals that are detoxified (metabolized) by the liver.

d. List three types of cells phagocytosed by the liver.

e. The liver can convert excess glucose to ________________ and ________________ and also reverse those processes. (More on this in Chapter 25.)

f. The liver stores the four fat-soluble vitamins named ________, ________, ________, and ________. It also stores a vitamin necessary for erythropoiesis, namely vitamin ________. The liver also cooperates with kidneys and skin to activate vitamin ________.

g. Note that the liver *(does? does not?)* secrete digestive enzymes.

D8. Complete part *f* of Table LG 24.1

✓ **D9.** Check your understanding of accessory structures of the digestive system by naming the organ that best fits each description below. Select answers in the box.

G. Gallbladder	**L. Liver**	**P. Pancreas**

________ a. Has both exocrine and endocrine functions

________ b. Its primary functions are to store and concentrate bile

________ c. Under influence of CCK, it contracts and ejects bile into the cystic duct

________ d. Responds to CCK by secreting digestive enzymes

________ e. Bilirubin, absorbed from worn-out red blood cells, is secreted into bile here

________ f. Contains phagocytic cells known as stellate reticuloendothelial (Kupffer's) cells

________ g. Site of carbohydrate, fat, and protein metabolism

________ h. Removed in a cholecystectomy

✓ **D10.** Complete this activity about the "big three" hormones of the gut by selecting correct answers describing functions of each hormone. Complete arrows as ↑ if the hormone increases, stimulates, or promotes a certain activity; complete the arrow as ↓ if the hormone decreases, inhibits, or slows an activity.

a. Gastrin: | secretion of gastric juices, including HCl and pepsinogen; | gastric motility; | contraction of the lower esophageal sphincter; and | contraction of the pyloric sphincter. In summary, gastrin | digestive activity of the stomach.

b. CCK: | secretion of pancreatic juices rich in enzymes; | ejection of bile from gallbladder into bile duct system; | opening of hepatopancreatic ampulla (sphincter of Oddi) so that bile can flow into duodenum; | gastric emptying (by | contraction of pyloric sphincter and | satiety); | growth of pancreas; and | effects of secretin. The overall effect of CCK is to | digestive activity in the small intestine.

c. Secretin: | secretion of bicarbonate (in pancreatic juice and bile); | secretion of gastric juice; | effects of CCK; | growth of pancreas. The overall effect of CCK is to | digestive activity in the small intestine.

d. In summary, gastrin tends to | gastric activity, and CCK and secretin tend to | pancreas, liver, and gallbladder activity.

E. Small intestine (pages 881–891)

✓ **E1.** Describe the structure of the intestine in this Checkpoint.

a. The average diameter of the small intestine is:
A. 2.5 cm (1 inch) B. 5.0 cm (2 inches)

b. Its average length in a living person is about:
A. 3 m (10 feet) B. 6 m (20 feet)

c. Name its three main parts (in sequence from first to last):

_______________ → _______________ → _______________

d. Which part of the small intestine is shortest in length? _______________

e. Which part connects to the cecum of the large intestine? _______________

✓ **E2.** Match the correct term related to the intestine to the description that fits.

ALF.	**Aggregated lymphatic follicles (Peyer's patches)**	**MALT.**	**Mucosa-associated lymphoid tissue**
BB.	**Brush border**	**MV.**	**Microvilli**
CF.	**Circular folds**	**PC.**	**Paneth cells**
DG.	**Duodenal (Brunner's) glands**	**SC.**	**S cells and CCK cells**
GC.	**Goblet cells**	**V.**	**Villi**
L.	**Lacteal**		

_______ a. Fingerlike projections up to 1 mm high that give the intestinal lining a velvety appearance and increase absorptive surface

_______ b. Microscopic fingerlike projections (200 million per mm^2) of plasma membrane

_______ c. Mucosal ridges (about 10 mm high) that increase intestinal surface area and cause chyme to move in a spiral pattern

_______ d. Clusters of lymphatic follicles located primarily in the wall of the ileum

_______ e. Submucosal glands that secrete protective alkaline fluids

_______ f. Lymphatic nodules within the mucous membrane of the small intestine

_______ g. Fuzzy line marking the tips of microvilli; site of intestinal digestive enzymes

_______ h. Lymphatic capillary found within each villus

_______ i. Cells at the base of villi that control level of microbes

_______ j. By carrying out phagocytosis and secretion of lysozyme cells that secrete mucus

E3. Name eight brush border enzymes.

These enzymes function primarily *(in the intestinal lumen? at the brush border?)*. Describe the action of one of these enzymes, *α-dextrinase*. (*Hint:* see page 886.)

✓ **E4.** Contrast types of movements in the small intestine by writing *MMC* for migrating mobility complex and *S* for segmentation next to related descriptions.

________ a. Peristalsis; "pushes" chyme along

________ b. Mixing (not pushing) movements as chyme sloshed back and forth

________ c. Movements that occur about 8 to 12 times a minute

________ d. Duration of a typical movement is 1.5–2 hours

________ e. Movements that are strongest in parts of the small intestine with large volumes

✓ **E5.** Write the main steps in the digestion of each of the three major food types.

a. *Carbohydrates*
Polysaccharides → ________________________________ → monosaccharides

b. *Proteins*
Proteins → ________________________ → ________________________

c. *Lipids*
Neutral fats → emulsified fats → ____________________ ____________________

✓ **E6.** *A clinical correlation.* Persons with lactose intolerance lack the enzyme

____________________, which is necessary for digestion of lactose found in foods

such as ________________. Write a rationale for the usual symptoms of this condition.

E7. Complete the description of functions of intestinal juices by filling in part *e* of Table LG 24.1. Then review digestion of each of the three major food groups by reading your columns vertically.

✓ **E8.** *For extra review* of roles of GI organs in digestion, identify which chemicals in the list in the box are made by each of the following organs. Write one answer on each line provided.

Alb.	**Albumin**	**IF.**	**Intrinsic factor**
Amy.	**Amylase**	**Ins.**	**Insulin**
B.	**Bile**	**Lip.**	**Lipase**
CCK.	**Cholecystokinin**	**Muc.**	**Mucus**
Chol.	**Cholesterol**	**MLSD.**	**Maltase, lipase, sucrase, a-dextrinase**
DA.	**Dipeptidases, aminopeptidases**		
E.	**Enterokinase**	**NP.**	**Nucleosidases, phosphatases**
FP.	**Fibrinogen and prothrombin**	**P.**	**Pepsinogen**
Gas.	**Gastrin**	**RD.**	**Ribonuclease, deoxyribonuclease**
Glu.	**Glucagon**	**S.**	**Secretin**
Gly.	**Glycogen**	**TCP.**	**Trypsinogen, chymotrypsinogen, procarboxypeptidase**
HCl.	**Hydrochloric acid**		

a. Salivary glands and tongue: _______ _______ _______

b. Pharynx and esophagus: _______

c. Stomach: _______ _______ _______ _______ _______ _______

d. Pancreas: _______ _______ _______ _______ _______ _______

e. Small intestine: _______ _______ _______ _______ _______ _______ _______

f. Liver: _______ _______ _______ _______ _______

g. Large intestine: _______

✓ **E9.** *For extra review* of secretions involved with digestion, match names of secretions listed for Checkpoint E8 with descriptions below. Blank lines follow some descriptions. On these indicate whether the secretion is classified as an enzyme (E), hormone (H), or neither of these (N). The first one is done for you.

__CCK__ a. Causes contraction of gallbladder and relaxation of the sphincter of the hepatopancreatic ampulla so that bile enters duodenum __H__

________ b. Stimulates production of alkaline pancreatic fluids and bile ________

________ c. Stimulates pancreas to produce secretions rich in enzymes such as lipase and amylase ________

________ d. Promote growth and maintenance of pancreas; enhance effects of each other (two answers)

________ e. Increases gastric activity (secretion, motility, and growth) ________

________ f. Inhibit stomach motility and/or secretions (two answers)

________ g. Precursors to protein-digesting enzymes (two answers)

________ h. Activates pepsinogen ________

________ i. The active form of this enzyme starts protein digestion in the GI tract

________ j. Activates the inactive precursor to form trypsin

________ k. Plasma proteins (two answers)

________ l. A storage form of carbohydrate

________ m. Most effective of this type of fat-digesting enzyme is produced by the pancreas

________ n. Emulsifies fats before they can be digested effectively ________

________ o. Starch-digesting enzymes secreted by salivary glands and pancreas

________ p. Intestinal enzymes that complete carbohydrate breakdown, resulting in simple sugars

________ q. Digest DNA and RNA ________

________ r. Brush border enzymes that carry out digestion on the surface of villi (three answers)

________ s. Complete digestion of proteins into amino acids

✓ **E10.** Describe the absorption of end products of digestion in this exercise. (*Hint:* refer to text Figure 24.25.)

a. Almost all absorption takes place in the *(large? small?)* intestine. Absorption is a two-step process. Products of digestion must first be absorbed into *(blood or lymph capillaries? epithelial cells lining the intestine?)* and then be transported into *(blood or lymph capillaries? epithelial cells lining the intestine?)*.

b. Glucose and some amino acids move across the brush border into epithelial cells lining the intestine by *(primary'? secondary?)* active transport coupled with active transport of ________. Fructose enters epithelial cells by *(diffusion? facilitated diffusion?)*. All simple sugars then cross the basolateral surface of epithelium to enter capillaries by the process of ____________________.

c. Simple sugars, amino acids, and short-chain fatty acids are absorbed into *(blood? lymph?)* capillaries located in ____________________ in the intestinal wall. These vessels lead to the ____________________ vein and then to the ____________________ for storage or metabolism.

d. Long-chain fatty acids and monoglycerides first combine with ____________________ salts to form *(micelles? chylomicrons?)*. This enables fatty acids and monoglycerides to enter epithelial cells in the intestinal lining and soon enter lacteals leading to the *(portal vein? thoracic duct?)*. Most bile salts are ultimately *(eliminated in feces? recycled to the liver?)*. This cycle is known as ____________________ circulation.

e. Aggregates of fats coated with ____________________ are known as chylomicrons. After traveling through lymph and blood, chylomicrons reach hepatocytes in the ____________________ or are stored in ____________________ where triglycerides are reformed.

f. About ________ liters (or ________ quarts) of fluids are ingested or secreted into the GI tract each day. Of this fluid, about ________ liters come from ingested food and about ________ liters derive from GI secretions. All but about 1 liter of the 9 liters is reabsorbed by *(facilitated diffusion? osmosis?)* into blood capillaries in the walls of the *(small? large?)* intestine. Several hundred milliliters of fluid are also reabsorbed each day into the *(small? large?)* intestine.

g. Normally about ________ mL of fluid exits each day in feces. When inadequate water reabsorption occurs, as in *(constipation? diarrhea?)*, then ____________________ such as Na^+ and Cl^- are also lost also.

✓ **E11.** *The Big Picture: Looking Ahead.* Refer to Table 25.6 on page 940 in the text, and answer these questions.

a. Which vitamins are absorbed with the help of bile? *(Water-soluble? Fat-soluble?)* Circle the fat-soluble vitamins: A B_{12} C D E K

b. Inadequate bile production or obstruction of bile pathways may lead to signs or symptoms related to deficiencies of fat-soluble vitamins. List several.

E12. Explain the following about alcohol absorption.

a. Why it is a good idea to eat some fatty foods with alcohol.

b. Whether women or men are more likely to develop higher blood alcohol level after consumption of the same amount of alcohol. Explain.

E13. Complete part *e* of Table LG 24.1.

F. Large intestine (pages 891–896)

✓ **F1.** The large intestine is so named based on its *(diameter? length?)* compared to that of the small intestine. The total length of the large intestine is about _______ m (______ ft). More than 90% of its length consists of the part known as the ____________________.

✓ **F2.** Arrange the parts of the large intestine listed in the box in correct sequence in the pathway of wastes

AnC.	**Anal canal**	**R.**	**Rectum**
AsC.	**Ascending colon**	**SC.**	**Sigmoid colon**
C.	**Cecum**	**SF.**	**Splenic flexure**
DC.	**Descending colon**	**TC.**	**Transverse colon**
HF.	**Hepatic flexure**		

_______ → _______ → _______ → _______ → _______ →

_______ → _______ → _______ → _______

F3. Identify the regions of the large intestine in Figure LG 24.1. Draw arrows to indicate direction of movement of intestinal contents.

✓ **F4.** Contrast different portions of the GI tract by identifying structures or functions associated with each. Use these answers:

LI. Large intestine	**SI.** Small intestine	**Sto.** Stomach

_______ a. Has thickened bands of longitudinal muscle known as teniae coli

_______ b. Pouches known as haustra give this structure its puckered appearance

_______ c. Its fat-filled peritoneal attachments are known as epiploic appendages

_______ d. Bacteria here decompose bilirubin to stercobilin, which gives feces its brown color

_______ e. Has rugae

_______ f. Has villi and microvilli

_______ g. (Vermiform) appendix is attached to the cecum here

_______ h. Diverticuli are outpouchings where the muscularis here has weakened

_______ i. Ileocecal valve is located here (two answers)

F5. Contrast terms in each pair:

a. *Gastroileal reflex/gastrocolic reflex*

b. *Haustral churning/mass peristalsis*

c. *Diarrhea/constipation*

✓ **F6.** Complete this activity describing chemicals associated with digestive wastes (feces).

a. Flatus (gas) in the colon due to the gases ____________________,

____________________, ____________________; these are products of bacterial fermentation of *(carbohydrates? proteins? fats?)*.

b. Odors due to the chemicals ____________________ and ____________________; these are products of bacterial breakdown of *(carbohydrates? proteins and amino acids? fats?)*.

c. Brown color due to the chemical ____________________, a product of bacterial

decomposition of ____________________.

d. Hidden (or ____________________) blood can be detected in feces; its presence can be

used in diagnosing ____________________.

✓ **F7.** Feces are formed by the time chyme has remained in the large intestine for about

_______ hours. List the chemical components of feces.

F8. Describe the process of defecation. Include these terms: *stretch receptors, rectal muscles, sphincters, diaphragm,* and *abdominal muscles.*

✓ **F9.** Describe helpful effects of dietary fiber in this activity.

a. *(Soluble? Insoluble?)* fibers tend to speed up passage of digestive wastes. Potential health benefits include lowering risk for a number of disorders, including

_________________ (hard, dry stool) and colon cancer. Circle the two best sources of insoluble fiber in the following list.

apples broccoli oats and prunes vegetable skins wheat bran

b. Write one health benefit of soluble fiber.

c. Write three sources of soluble fiber from the list above.

F10. Complete part *g* of Table 24.1, describing the role of the large intestine in digestion.

G. Development and aging (page 897)

✓ **G1.** Indicate which germ layer, endoderm (*E*) or mesoderm (*M*), gives rise to each of these structures.

_______ a. Epithelial lining and digestive glands of the GI tract

_______ b. Liver, gallbladder, and pancreas

_______ c. Muscularis layer and connective tissue of submucosa

✓ **G2.** Identify which GI structures will form from each of the following embryonic structures.

a. Stomodeum _________________

b. Proctodeum _________________

c. Foregut

d. Hindgut

✓ **G3.** *A clinical correlation.* List physical changes with aging that may lead to a decreased desire to eat among the elderly population.

a. Related to the upper GI tract (to stomach)

b. Related to the lower GI tract (stomach and beyond)

H. Disorders, medical terminology (pages 899–901)

H1. Explain how tooth decay occurs. Include the roles of bacteria, dextran, plaque, and acid. List the most effective known measures for preventing dental caries.

H2. Briefly describe these disorders, stating possible causes of each.

a. Periodontal disease

b. Peptic ulcer

✓ **H3.** Match the terms with the descriptions.

AG.	**Ankyloglossia**	**Dys.**	**Dysphagia**
AN.	**Anorexia nervosa**	**E.**	**Eructation**
B.	**Bulimia**	**F.**	**Flatus**
Cho.	**Cholecystitis**	**Hem.**	**Hemorrhoids**
Colit.	**Colitis**	**Hep.**	**Hepatitis**
Colos.	**Colostomy**	**Htb.**	**Heartburn**
Con.	**Constipation**	**P.**	**Peptic ulcer**
Dia.	**Diarrhea**		

________ a. Incision of the colon, creating artificial anus

________ b. Inflammation of the liver

________ c. Inflammation of the colon

________ d. Burning sensation in region of esophagus and stomach; probably due to gastric contents in lower esophagus

________ e. Frequent defecation of liquid feces

________ f. Inflammation of the gallbladder

________ g. Infrequent or difficult defecation

________ h. Craterlike lesion in the GI tract due to acidic gastric juices, *Helicobacter pylori* bacteria, and/or drugs such as NSAIDs

________ i. Excess air (gas) in stomach or intestine, usually expelled through anus

________ j. Binge-purge syndrome

________ k. Loss of appetite and self-imposed starvation

________ l. Difficulty in swallowing

________ m. Belching

________ n. Condition of being "tongue-tied"

ANSWERS TO SELECTED CHECKPOINTS: CHAPTER 24

A2. Accessory structures.

A3. Ingestion, secretion, digestion, absorption, defecation

A4. Chemical, mechanical.

A5.

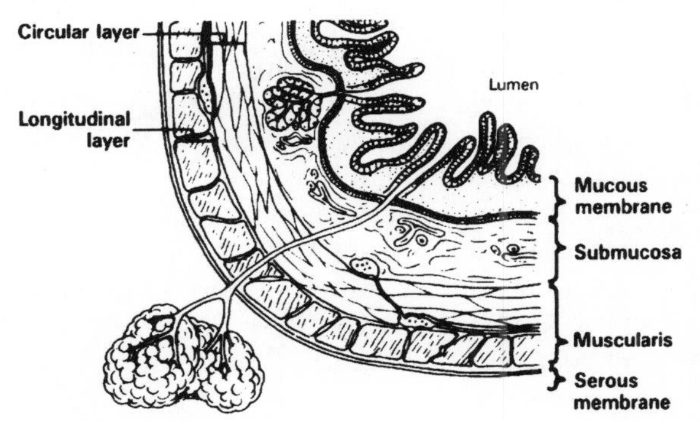

Figure LG 24.2A Gastrointestinal (GI) tract and related gland seen in cross section.

A6. (a) Ser. (b) Muc. (c) Sub. (d) Mus.

A8. (a) F. (b) M. (c) Meso. (d) G. (e) L.

B1.

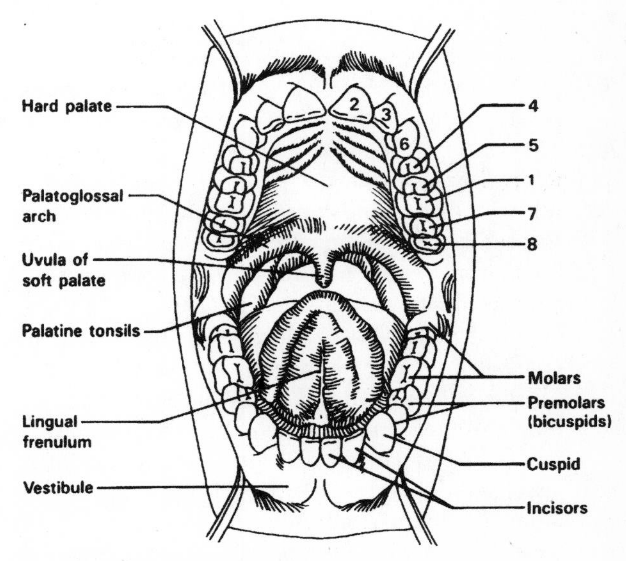

Figure LG 24.3A Mouth (oral cavity).

B3. (a) Parotid; parotid. (b) Sublingual. (c) Liters. (d) Dissolving medium for foods, lubrication, waste removal medium, and source of antibodies, lysozyme. (e) 6.5. (f) Water, Na^+, K^+, Cl^-, HCO_3^-, HPO_4^{2-}, urea, uric acid, lysozyme, salivary amylase, and IgA. (g) Smell, sight, touch, or memory of food; nausea; swallowing of irritating food; nerve impulses. (h) Sympathetic, parasympathetic VII and IX.

B5. (a) C B A. (b) A B C. (c) A B C.

B6. Anchors tooth to bone and acts as a shock absorber.

B7. (a) 20; 32. (b–c) See Figure LG 24.3A. (d) Third molars or "wisdom teeth."

B8. (a) Mastication; bolus. (b) Chemical; amylase; starch; is not; acidic pH of the stomach. (c) Tongue; triglycerides (or fats).

B9. Swallowing. (a) P. (b) V. (c) E.

B10. (a) 25 (10); stomach. (b) Striated, smooth. (c) Acidic, gastroesophageal reflux disease (GERD); heart (heartburn); avoid smoking, alcohol, certain foods, large meals, and lying down just after meals; take H2-blocker medications.

B11. (a) Extrinsic, XII. (b) Parasympathetic. (c) VII, IX; maxillary, V.

C1. (a) Cardia → fundus → body → pylorus. (b1) GC. (b2) LC. (b3) GC. (c) Pylorospasm; vomiting.

C2.

Name of Cell	Type of Secretion	Function of Secretion
a. Chief (zymogenic)	Pepsinogen Gastric lipase	Precursor of pepsin Splits triglycerides
b. Mucous	Mucus	Protects gastric lining from acid and pepsin
c. Parietal	HCl Intrinsic factor	Activates pepsinogen to pepsin; inhibits gastrin secretion; stimulates secretion of secretin and CCK; kills microbes in food. Facilitates absorption of vitamin B_{12}
d. Enteroendocrine (G) cells	Gastrin	Stimulates secretion of HCl and pepsinogen; contracts lower esophageal sphincter; increases gastric motility and relaxes pyloric and ileocecal sphincters

C3. (a) Arranged in rugae when the stomach is empty. (b) It has three, rather than two, layers of smooth muscle; the extra one is an oblique layer located inside the circular layer.

C4. (a) Acid. (b) Pepsin is released in the inactive state (pepsinogen); alkaline mucus protects the stomach lining from pepsin. (c) Ulcer. (d) Gastric lipase, emulsified fats, as in butter; ineffective; its optimum pH is 5 or 6, and most fats have yet to be emulsified (by bile from liver). (e) Many H^+ and Cl^- ions are secreted from parietal cells into the lumen of the stomach to help digest a large meal. This occurs as parietal cells form carbonic acid that dissociates into H^+ and HCO_3^-. The H^+s are pumped into the stomach (as just described); but the HCO_3^- moves out of parietal cells (in exchange for the Cl^- lost from those cells) and enters (is retained in) blood. The pH or blood and urine may then increase (indicating alkalinity).

C5. (a) Cephalic, gastric, intestinal; cephalic. (b) Vagus; parasympathetic; sight, smell, or thought of food and presence of food in the stomach; anger, fear, anxiety. (c) Gastric; increase, gastrin; presence of partially digested proteins, alcohol, and caffeine in the stomach; stimulated, relaxed. (d) Stimulate; greater, mast; H_2 blockers decrease HCl production which could otherwise lead to "acid stomach" or peptic ulcer. (e) Enterogastric, inhibits; delay; secretin, CCK.

C6. Two to four; carbohydrates; fats (triglycerides); emesis; medulla; can be; loss of HCl can lead to alkalosis (increase in blood pH).

D2. (a) Stomach. (b) Duodenum, spleen. (c) Exocrine; alkaline. (d) Proteins; enterokinase; chymotrypsin (and elastase), carboxypeptidase. (e) Amylase digests carbohydrates, including starch; lipase digests lipids. (f) G, H; duodenum. (g) Common bile duct; pancreatic proteases such as trypsin, chymotrypsin, and carboxypeptidase may digest tissue proteins of the pancreas itself. (h) Pancreatic islets (of Langerhans); insulin, glucagon, somatostatin, and pancreatic polypeptide; blood vessels, portal.

D4. (a) Parasympathetic, vagus, small intestine. (b) Contains partially digested fats and proteins; digestive enzymes such as trypsin, amylase, and lipase. (c) Is acidic; HCO_3^-.

D5. (a) 1.4, 3.0; upper right; right. (b) Falciform; umbilical. (c) Hepatic artery, portal vein; sinusoids. hepatic veins.

D6. (a) Superior to inferior. (b) Draw arrows from L and M toward B and then J; superiorly. (c) F, L, and M

D7. (a) Bilirubin, red blood; jaundice; emulsification, absorption; stercobilin. (b) Clotting proteins (prothrombin and fibrinogen), albumin, and globulins. (c) Products of protein digestion, such as ammonia; medications such as sulfa drugs or penicillin; and steroid hormones. (d) Worn-out RBCs, WBCs, and some bacteria. (e) Glycogen, fat. (f) A, D, E, K: B_{12}, D. (g) Does not.

D9. (a) P. (b–c) G. (d) P. (e–g) L. (h) G.

D10. (a) All ↑ except ↓ contraction of pyloric sphincter. (b) All ↑ except ↓ gastric emptying. (c) All ↑ except, ↓ gastric juice. (d.) ↑; ↑.

E1. (a) A. (b) A. (c) Duodenum → jejunum → ileum. (d) Duodenum. (e) Ileum.

E2. (a) V. (b) MV. (c) CF. (d) ALF. (e) DG. (f) MALT. (g) BB. (h) L. (i) PC. (j) GC.

E4. (a) MMC. (b) S. (c) S. (d) MMC. (e) S.

E5. (a) Shorter-chain polysaccharides such as a-dextrins (5–10 glucoses), trisaccharides (such as maltotriose), or disaccharides (such as maltose). (b) Shorter-chain polypeptides → dipeptides and amino acids. (c) Fatty acids, monoglycerides, and glycerol.

E6. Lactase, milk and other dairy products; gas (flatus) and bloating because bacteria ferment the undigested lactose present in the GI tract.

E8. (a) Amy, Lip, Muc. (b) M. (c) Gas, HCl, IF, Lip, Muc, P. (d) Amy, Glu, Ins, Lip, RD, TCP. (e) CCK, DA, E, Muc, MLSD, NP, S. (f) Alb, B, Chol, FP, Gly. (g) Muc.

E9. (a) CCK, H. (b) S, H. (b) CCK, H. (d) CCK and S. (e) Gas, H. (f) CCK and S. (g) P, TCP. (h) HCl, N. (i) P. (j) E. (k) Alb and FP. (l) Gly. (m) Lip. (n) B, N. (o) Amy. (p) MLSD. (q) NP. (r) MLSD, DA, and NP. (s) DA.

E10. (a) Small; epithelial cells lining the intestine, blood or lymph capillaries. (b) Secondary, Na^+; facilitated diffusion; facilitated diffusion. (c) Blood, villi; portal, liver. (d) Bile, micelles; thoracic duct; recycled to the liver; enterohepatic. (e) Protein; liver, adipose. (f) 9.3 (9.8); 2, 7; osmosis, small; large. (g) 100–200; diarrhea, electrolytes.

E11. (a) Fat-soluble: A D E K. (b) Examples: A, night blindness and dry skin; D, rickets or osteomalacia due to decreased calcium absorption; K, excessive bleeding.

F1. Diameter: 1.5 (5.0); colon.

F2. C → AsC → HF → TC → SF → DC → SC → R → AnC.

F4. (a–d) LI. (e) Sto. (f) SI. (g–h) LI. (i) SI, LI.

F6. (a) Methane, carbon dioxide, hydrogen; carbohydrates. (b) Indole, skatole; proteins and amino acids. (c) Stercobilin, bilirubin. (d) Occult; colorectal cancer.

F7. 3–10: water, salts, epithelial cells, bacteria and products of their decomposition, as well as undigested foods.

F9. (a) Insoluble; constipation; vegetable skins and wheat bran. (b) May lower blood cholesterol level. (c) Apples, broccoli, oats and prunes.

G1. (a) E. (b) E. (c) M.

G2. (a) Oral cavity. (b) Anus. (c) Pharynx, esophagus, stomach, and part of the duodenum. (d) Remainder of duodenum through most of the transverse colon.

G3. (a) Decreased taste sensations, gum inflammation (pyorrhea) leading to loss of teeth and loose-fitting dentures, and difficulty swallowing (dysphagia). (b) Decreased muscle tone and neuromuscular feedback may lead to constipation.

H3. (a) Colos. (b) Hep. (c) Colit. (d) Htb. (e) Dia. (f) Cho. (g) Con. (h) P. (i) F. (j) B. (k) AN. (l) Dys. (m) E. (n) AG.

MORE CRITICAL THINKING: CHAPTER 24

1. Contrast the superior and inferior portions of the esophagus, as well as the stomach, and small and large intestines with regard to (a) types of epithelium in the mucosa, and (b) types of muscle in the muscularis.
2. Place in correct sequence each of these structures in the pathway of food through the GI tract. Then briefly describe each structure: anal canal, ascending colon, cecum, duodenum, esophagus, hepatic flexure, ileocecal sphincter, oropharynx, pyloric sphincter, sigmoid colon.
3. Contrast functions of chemicals within each grouping: amylase/trypsin/lipase; gastrin/CCK/secretin.
4. Describe nine functions of the liver. Determine which ones are essential to survival.
5. Identify with which digestive organs the following structures are associated: dentin, lacteal, frenulum linguae, uvula, pyloric region, ileum, brush border, villi, cecum, haustra, and sinusoids.
6. Identify with which digestive organs the following conditions are associated: periodontal membrane disease, pyorrhea, cholecystitis, cirrhosis, hepatitis A-E, diverticulosis, appendicitis, and achalasia.

MASTERY TEST: CHAPTER 24

Questions 1–10: Circle the letter preceding the one best answer to each question.

1. Which of these organs is not part of the GI tract but is an accessory organ?
   A. Mouth
   B. Pancreas
   C. Stomach
   D. Small intestine
   E. Esophagus
2. The main function of salivary and pancreatic amylase is to:
   A. Lubricate foods
   B. Help absorb fats
   C. Digest polysaccharides to smaller carbohydrates
   D. Digest disaccharides to monosaccharides
   E. Digest polypeptides to amino acids
3. Which enzyme is most effective at pH 1 or 2?
   A. Gastric lipase
   B. Maltase
   C. Pepsin
   D. Salivary amylase
   E. Pancreatic amylase
4. Choose the *true* statement about fats.
   A. They are digested mostly in the stomach.
   B. They are the type of food that stays in the stomach the shortest length of time.
   C. They stimulate release of gastrin.
   D. They are emulsified and absorbed with the help of bile.
5. Which of the following is a hormone, not an enzyme?
   A. Gastric lipase
   B. Gastrin
   C. Pepsin
   D. Trypsin
6. Which of the following is under only nervous (not hormonal) control?
   A. Salivation
   B. Gastric secretion
   C. Intestinal secretion
   D. Pancreatic secretion
7. All of the following are enzymes involved in protein digestion *except:*
   A. Amylase
   B. Trypsin
   C. Carboxypeptidase
   D. Pepsin
   E. Chymotrypsin
8. All of the following chemicals are produced by the walls of the small intestine *except:*
   A. Lactase
   B. Secretin
   C. CCK
   D. Trypsin
   E. Brush border enzymes
9. Migrating motility complex (MCC) is a form of ______ in the small intestine.
   A. Peristalsis
   B. Segmentation
   C. Haustral churning
   D. Mastication
10. Choose the *false* statement about layers of the wall of the GI tract.
    A. Most large blood and lymph vessels are located in the submucosa.
    B. The mesenteric plexus is part of the muscularis layer.
    C. Most glandular tissue is located in the layer known as the mucosa.
    D. The mucosa layer forms the peritoneum.

Questions 11–15: Arrange the answers in correct sequence.

______ ______ ______ 11. Arrange from greatest to least in number in a complete set of adult dentition:
A. Molars
B. Premolars
C. Cuspids (canines)

______ ______ ______ ______ 12. GI tract wall, from deepest to most superficial:
A. Mucosa
B. Muscularis
C. Serosa
D. Submucosa

______ ______ ______ ______ ______ 13. Pathway of chyme:
A. Ileum
B. Jejunum
C. Cecum
D. Duodenum
E. Pylorus

______ ______ ______ ______ ______ 14. Pathway of bile:
A. Bile canaliculi
B. Common bile duct
C. Common hepatic duct
D. Right and left hepatic ducts
E. Hepatopancreatic ampulla and duodenum

______ ______ ______ ______ ______ 15. Pathway of wastes:
A. Ascending colon
B. Transverse colon
C. Sigmoid colon
D. Descending colon
E. Rectum

Questions 16–20: Circle T (true) or F (false). If the statement is false, change the underlined word or phrase so that the statement is correct.

T F 16. In general, the sympathetic nervous system stimulates salivation and secretions of the gastric and intestinal glands.

T F 17. The principal chemical activity of the stomach is to begin digestion of protein.

T F 18. The esophagus produces no digestive enzymes or mucus.

T F 19. Cirrhosis and hepatitis are diseases of the liver.

T F 20. The greater omentum connects the stomach to the liver.

Questions 21–25: Fill-ins. Complete each sentence with the word or phrase that best fits.

____________ 21. Stomach motility is _______-creased by CCK and _______-creased by stomach gastrin and parasympathetic nerves.

____________ 22. Mumps involves inflammation of the ________ salivary glands.

____________ 23. Most absorption takes place in the ________, although some substances, such as ________, are absorbed in the stomach.

____________ 24. The clinical specialty that deals with diseases of the stomach and intestines is ________.

____________ 25. Epithelial lining of the GI tract, as well as the liver and pancreas, are derived from ________-derm.

ANSWERS TO MASTERY TEST: CHAPTER 24

Multiple Choice

1. B
2. C
3. C
4. D
5. B
6. A
7. A
8. D
9. A
10. D

Arrange

11. A B C
12. A D B C
13. E D B A C
14. A D C B E
15. A B D C E

True–False

16. F. Inhibits
17. T
18. F. No digestive enzymes, but it does produce mucus
19. T
20. F. Connects the stomach and duodenum to the transverse colon and drapes over the coils of the small intestine.

Fill-ins

21. De, in
22. Parotid
23. Small intestine; alcohol
24. Gastroenterology
25. Endo

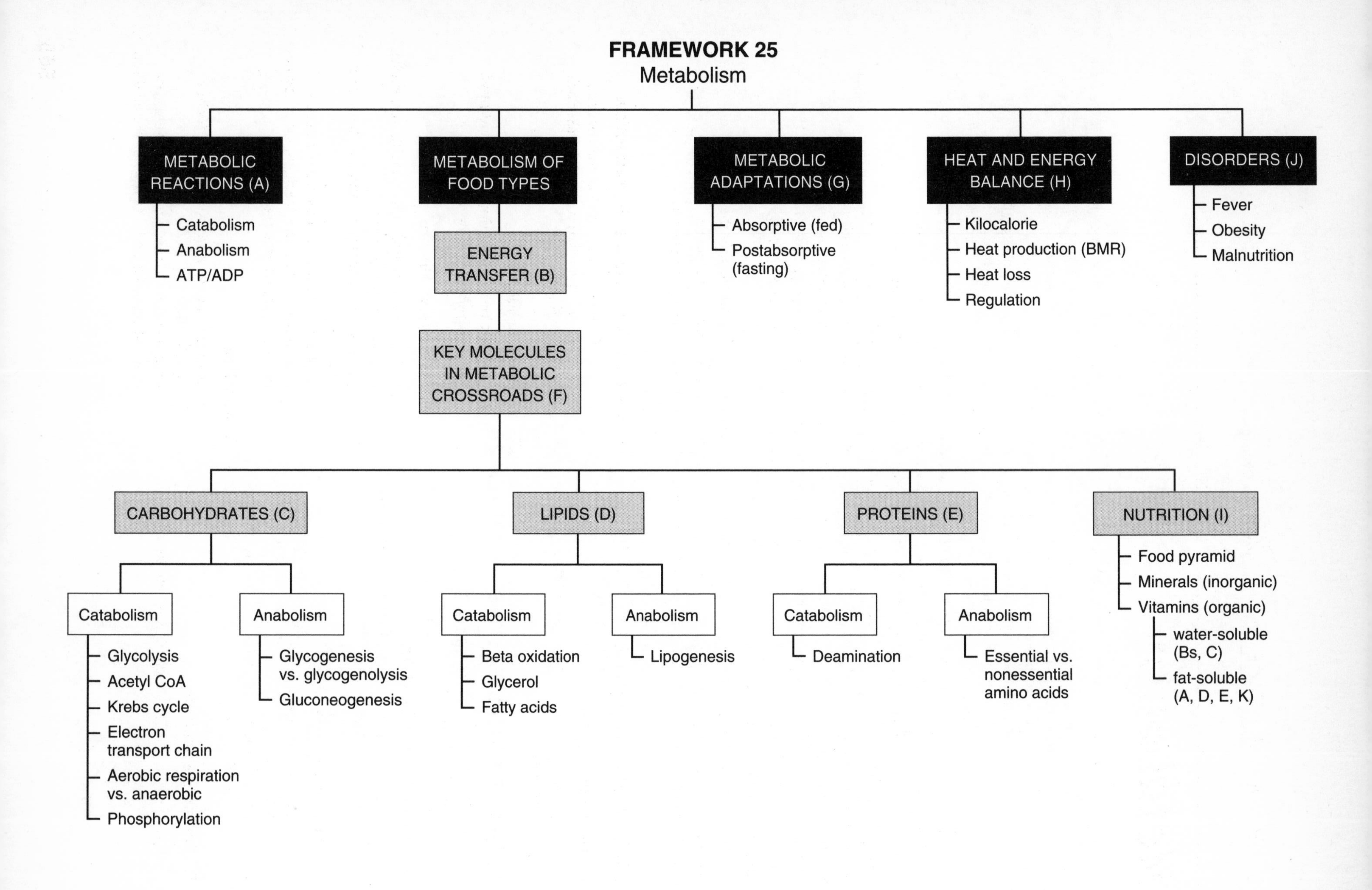
FRAMEWORK 25
Metabolism
METABOLIC REACTIONS (A)
Catabolism
Anabolism
ATP/ADP
METABOLISM OF FOOD TYPES
ENERGY TRANSFER (B)
KEY MOLECULES IN METABOLIC CROSSROADS (F)
CARBOHYDRATES (C)
Catabolism
Glycolysis
Acetyl CoA
Krebs cycle
Electron transport chain
Aerobic respiration vs. anaerobic
Phosphorylation
Anabolism
Glycogenesis vs. glycogenolysis
Gluconeogenesis
LIPIDS (D)
Catabolism
Beta oxidation
Glycerol
Fatty acids
Anabolism
Lipogenesis
PROTEINS (E)
Catabolism
Deamination
Anabolism
Essential vs. nonessential amino acids
NUTRITION (I)
Food pyramid
Minerals (inorganic)
Vitamins (organic)
water-soluble (Bs, C)
fat-soluble (A, D, E, K)
METABOLIC ADAPTATIONS (G)
Absorptive (fed)
Postabsorptive (fasting)
HEAT AND ENERGY BALANCE (H)
Kilocalorie
Heat production (BMR)
Heat loss
Regulation
DISORDERS (J)
Fever
Obesity
Malnutrition

CHAPTER 25

Metabolism

Ingested foods, once digested, absorbed, and delivered, are used by body cells. This array of biochemical reactions within cells is known as metabolism. Products of digestion may be reassembled, as in formation of human protein from the amino acids of meat or beans, or in synthesis of fats from ingested oils. These reactions are examples of anabolism. The flip side of metabolic currency is catabolism, as in the breakdown of complex carbohydrates stored in liver or muscle to provide simple sugars for quick energy. Catabolism releases energy that maintains body heat and provides ATP to fuel activity. Vitamins and minerals play key roles in metabolism—for example, in enzyme synthesis and function.

The Chapter 25 Framework provides an organizational preview of the metabolism of the major food groups. Study the key terms and concepts it presents. As you begin this chapter, carefully examine the Chapter 25 Topic Outline and check off each objective as you complete it.

TOPIC OUTLINE AND OBJECTIVES

A. Metabolic reactions

1. Explain the role of ATP in anabolism and catabolism.

B. Energy transfer

1. Describe oxidation-reduction reactions.
2. Explain the role of ATP in metabolism.

C. Carbohydrate metabolism

1. Describe the fate, metabolism, and functions of carbohydrates.

D. Lipid metabolism

1. Describe the lipoproteins that transport lipids in the blood.
2. Describe the fate, metabolism, and functions of lipids.

E. Protein metabolism

1. Describe the fate, metabolism, and functions of proteins.

F. Summary; key molecules in metabolism

1. Identify the key molecules in metabolism, and describe the reactions and the products they may form.

G. Metabolic adaptations

1. Compare metabolism during the absorptive and postabsorptive states.

H. Heat and energy balance

1. Define basal metabolic rate (BMR), and explain several factors that affect it.
2. Describe the factors that influence body heat production.
3. Explain how normal body temperature is maintained by negative feedback loops involving the hypothalamic thermostat.

I. Nutrition

1. Describe how to select foods to maintain a healthy diet.
2. Compare the sources, functions, and importance of minerals and vitamins in metabolism.

WORDBYTES

Now become familiar with the language of this chapter by studying each wordbyte, its meaning, and an example of its use within a term. After you study the entire list, self-check your understanding by writing the meaning of each wordbyte on the line. As you continue through the *Learning Guide*, identify (and fill in) additional terms that contain the same wordbyte.

Wordbyte	Self-check	Meaning	Example(s)
ana-	______________________	upward	*ana*bolism
calor-	______________________	heat	*calor*ie
cata-	______________________	downward	*cata*bolism
de-	______________________	remove, from	*de*amination
-gen	______________________	to form	gluconeo*gen*esis
glyco-	______________________	sugar	*glyco*lytic
-lysis	______________________	breakdown	hydro*lysis*
neo-	______________________	new	glucon*eo*genesis

CHECKPOINTS

A. Metabolic reactions (pages 907–908)

✓ **A1.** Write brief descriptions of the three possible "fates" of foods absorbed by the GI tract.

a. ____

b. ____

c. ____

✓ **A2.** Complete this table comparing catabolism with anabolism.

Process	Definition	Releases or Uses Energy	Examples
a. Catabolism			
b. Anabolism			

A3. ATP stands for a________________ t________________ p________________.
Explain how it functions as the "energy currency" (or "money machine cash") of the cell. (For help, refer to Chapter 2, Checkpoint E17, page 35 of the *Learning Guide*.)

✓ **A4.** *For extra review,* refer to LG Chapter 10, Checkpoint D1c, page LG 188, and notice the difference between the structure of ATP and ADP. ADP has its *(1? 2? 3?)* phosphate "trailers" attached to the adenosine "truck" by one low-energy and one high-energy bond. ATP has *(1? 2? 3?)* phosphate "trailers" attached to the adenosine "truck" by one low-energy and *(1? 2? 3?)* high-energy bond(s). Therefore, *(ATP? ADP?)* contains more energy that can be utilized for cell activities.

✓ **A5.** About *(1–2%? 10–20%? 40%? 65–70%?)* of the energy released in catabolism is available for cellular activities. The rest of the energy is converted to ________________.

B. Energy transfer (pages 908–909)

✓ **B1.** Complete this overview of catabolism (resulting in energy production).

a. Organic nutrients such as glucose are rich in hydrogen; energy is contained within the C — H bonds. In catabolism, much of the energy in glucose is ultimately released and stored in the high-energy molecule named ________________. Write the chemical formulas to show this overall catabolic conversion:

____________ → ________________ + ______________ + energy stored in ATP.
Glucose *carbon dioxide* *water*

Note that the inorganic compound carbon dioxide is energy-*(rich?; poor?)* because it lacks C — H bonds.

b. The reaction shown is an oversimplification of the entire catabolic process, which actually entails many steps. Each pair of hydrogen atoms (2H) removed from glucose consists of two components: one hydrogen ion or proton lacking orbiting electrons, expressed as *(H^+? H^-?)*, and a hydrogen nucleus with two orbiting electrons, known as a *(hydride? hydroxyl?)* ion, and expressed as *(H^+? H^-?)*.

c. Hydrogens removed from glucose must first be trapped by H-carrying coenzymes, such as ________, ________, or ________; all of these coenzymes are derivatives of a vitamin named ________________. Ultimately the Hs combine with oxygen to form ________________. In other words, glucose is dehydrogenated (loses H) while oxygen is reduced (gains H).

d. Catabolism involves a complex series of stepwise reactions in which hydrogens are transferred. Because oxygen is the final molecule that serves as an oxidizer, catabolism is often described as biological *(reduction? oxidation?)*.

e. Along the way, energy from the original C–H bonds of glucose (and present in the electrons of H) is trapped in ________________. This process is known as ________________ because it involves the addition of a phosphate to ADP. A more complete expression of the total catabolism of glucose (shown briefly in (a) above) is included below. Write the correct labels under each chemical: *oxidized, oxidizer, reduced, reducer:*

Glucose + O_2 + ADP → CO_2 + H_2O + ATP
(__________) (__________) (__________) (__________)

✓ **B2.** Check your understanding of oxidation-reduction reactions and generation of ATP involved in metabolism by circling the correct answer in each case.

a. Which molecule is more reduced and therefore contains more chemical potential energy (can release more energy when catabolized)?
A. Lactic acid ($C_3H_6O_3$)
B. Pyruvic acid ($C_3H_4O_3$)

b. Which is more oxidized and therefore now contains less chemical potential energy since it has given up Hs to some coenzyme?
   A. Lactic acid ($C_3H_6O_3$)
   B. Pyruvic acid ($C_3H_4O_3$)
c. Which is more likely to happen when a molecule within the human body is oxidized?
   A. The molecule gives up Hs
   B. The molecule gains Hs
d. Which coenzyme is more reduced, indicating that it has received Hs when some other molecule was oxidized?
   A. NAD^+
   B. $NADH + H^+$
e. Which molecule contains more chemical potential energy (can release more energy when catabolized)?
   A. ATP
   B. ADP
f. Which is an example of oxidative phosphorylation?
   A. Creatine-P + ADP $\rightarrow$ creatine + ATP
   B. Generation of ATP by energy released by the electron transfer chain
g. Which is the substrate in the following example of substrate-level phosphorylation? Creatine-P + ADP $\rightarrow$ creatine + ATP
   A. ATP
   B. Creatine

C. Carbohydrate metabolism (pages 909–920)

✓ **C1.** Answer these questions about carbohydrate metabolism.

a. The story of carbohydrate metabolism is really the story of ____________________ metabolism because this is the most common carbohydrate (and in fact the most common energy source) in the human diet.
b. What other carbohydrates besides glucose are commonly ingested? How are these converted to glucose?

c. Just after a meal, the level of glucose in the blood *(increases? decreases?)*. Cells use some of this glucose; by ____________________ glucose, they release energy.
d. List three or more mechanisms by which excess glucose is used or discarded.

e. Increased glucose level of blood is known as ____________________. This can occur after a meal containing concentrated carbohydrate or in the absence of the hormone ____________________ because this hormone *(facilitates? inhibits?)* entrance of glucose into cells by increasing insertion GluT4 which is a ____________________ transporter into plasma membranes of most cells. Lack of functional insulin occurs in the condition known as ____________________.
f. Name two types of cells that do not require insulin for glucose uptake so that glucose entry is always possible (provided blood glucose level is adequate).

g. As soon as glucose enters cells, glucose combines with ____________________ to form glucose-6-phosphate. This process is known as ____________________ and is catalyzed by enzymes called ____________________. What advantage does this process provide?

✓ **C2.** Describe the process of glycolysis in this exercise.

a. Glucose is a ________-carbon molecule (which also has hydrogens and oxygens in it). During glycolysis each glucose molecule is converted to two ________-carbon molcules named ____________________. *(A lot? A little?)* energy is released from glucose during glycolysis. This process requires *(one? many?)* step(s) and *(does? does not?)* require oxygen.

b. The initial steps of glycolysis also involve ____________________-ation of glucose, along with conversion of this carbohydrate to another six-carbon sugar, ____________________. These preliminary steps in glycolysis are catalyzed by the enzyme ____________________, called the key regulator of glycolysis. This enzyme is most active when the levels of *(ADP? ATP?)* are high and *(ADP? ATP?)* are low (indicating the need for generation of ATP by glycolysis). When ATP levels are already high, then cellular glucose is instead used for *(ana? cata?)*-bolism of *(glycogen? $CO_2 + H_2O$?)*.

c. The phosphorylated fructose is broken down into two *(2? 3? 6?)*-carbon molecules named glucose 3-phosphate (________) and dihydroxyacetone phosphate. The G 3–P molecules are then *(oxidized? reduced?)*, giving up their hydrogens to *(FAD? NAD?)*.

d. The end result of glycolysis is formation of two ____________________ acids and a net gain of *(2? 3? 4?)* ATP molecules.

e. The fate of pyruvic acid depends on whether sufficient ____________________ is available. If it is, pyruvic acid undergoes chemical change in phases 2 and 3 of glucose catabolism (see below). Those processes *(do? do not?)* require oxygen; that is, they are *(aerobic? anaerobic?)*.

f. If the respiratory rate cannot keep pace with glycolysis, insufficient oxygen is available to break down pyruvic acid. In the absence of sufficient oxygen, pyruvic acid is temporarily converted to ____________________. This is likely to occur during active ____________________.

g. Several mechanisms prevent the accumulation of an amount of lactic acid that might be harmful. The liver changes some back to ____________________. Excess P_{CO_2} caused by exercise *(stimulates? inhibits?)* respiratory rate, so more oxygen is available for breakdown of the pyruvic acid.

✓ **C3.** Review glucose catabolism up to this point and summarize the remaining phases by doing this exercise. Refer to Figure LG 25.1. A six-carbon compound named (a) ____________________ is converted to two molecules of (b) ____________________

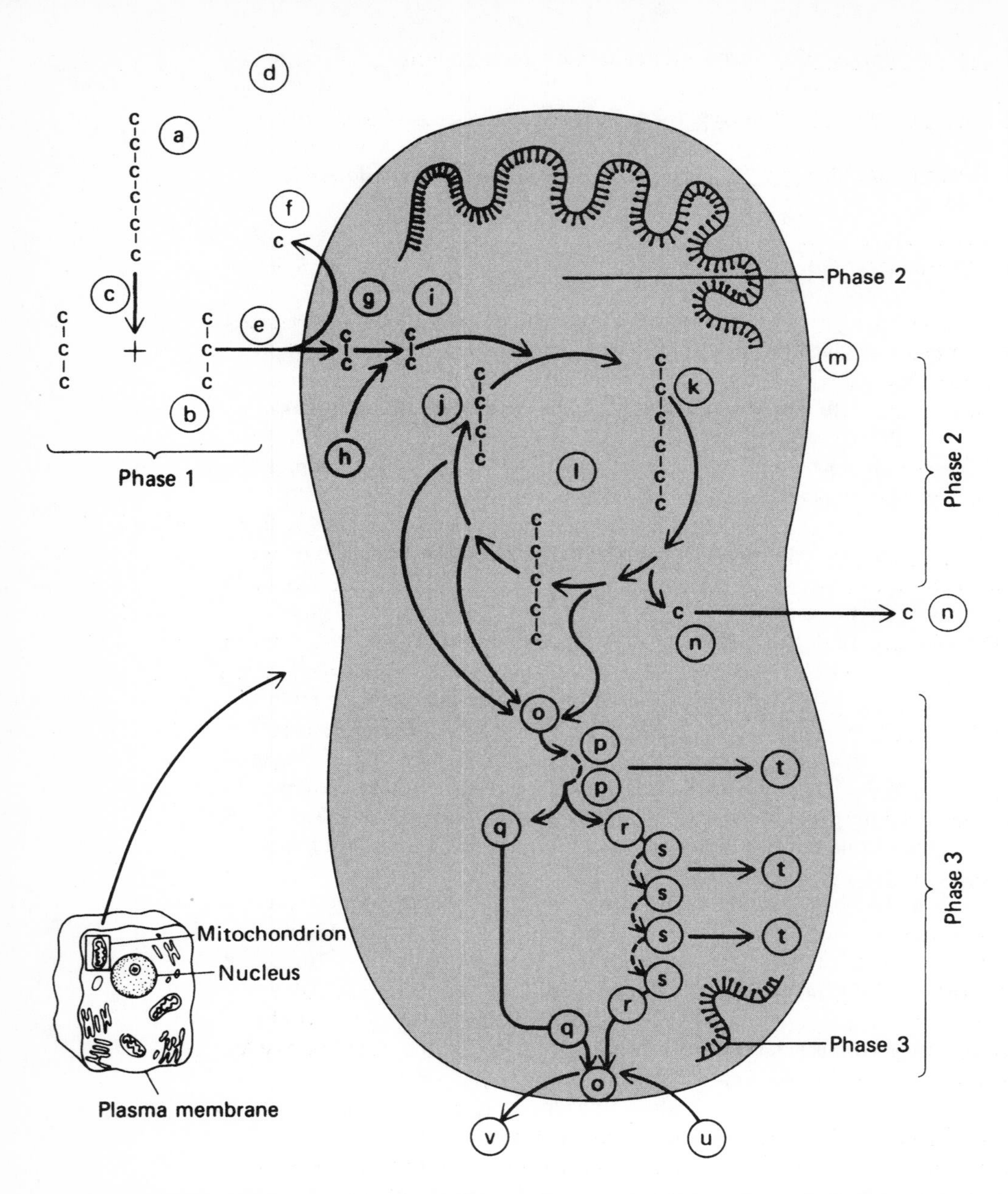

Figure LG 25.1 Summary of glucose catabolism, described in three phases. Phase 1 (glycolysis) occurs in the cytosol. Phase 2 (Krebs cycle) takes place in the matrix of mitochondria. Phase 3 (electron transport) takes place on inner mitochondrial membrane. Letters refer to Checkpoint C3.

via a number of steps, together called (c) ____________________, occurring in the

(d) ____________________ of the cell.

In order for pyruvic acid to be further broken down, it must undergo a

(e) ____________________ step. This involves removal of the one-carbon molecule,

(f) ____________________. The remaining two-carbon (g) ____________________

group is attached to a carrier called (h) ____________________, forming

(i) ____________________. (This compound hooks on to a four-carbon compound called (j) ____________________. The result is a six-carbon molecule of

(k) ____________________. The name of this compound is given to the cycle of

reactions called the (1) ____________________ cycle, occurring in the

(m) ____________________.
Two main types of reactions occur in the Krebs cycle. One is the decarboxylation

reaction, in which (n) ________ molecules are removed, and hydrogens removed at these steps combine with coenzymes FAD or NAD. Notice these locations on Figure 25.7 of your text. (What happens to the CO_2 molecules that are removed?) As a result, the six-carbon citric acid is eventually shortened to regenerate oxaloacetic acid, giving the cyclic nature to this process.

The other major type of reaction (oxidation–reduction) involves removal of

(o) ____________________ atoms during oxidation of compounds such as isocitric and a-ketoglutaric acids. (See Figure 25.7 in your text.) These hydrogens are carried off

by two enzymes named (p) ____________________ and "trapped" for use in phase 3. In phase 3, hydrogen atoms on coenzymes NAD and FAD (and later coenzyme Q) are

ionized to (q) ________ and (r) ________. Electrons are shuttled along a chain of

(s) ____________________. (Refer to Checkpoint C4.) During electron transport, energy in these electrons (derived from hydrogen atoms in ingested foods) is

"tapped" and ultimately stored in (t) ________.

Finally, the electrons, depleted of some of their energy, are reunited with hydrogen

ions and with (u) ____________________ to form (v) ____________________.
Notice why your body requires oxygen—to keep drawing hydrogen atoms off of nutrients (in biological oxidation) so that energy from them can he stored in ATP for use in cellular activities.

✓ **C4.** Now describe details of the electron transport chain in this activity.

a. The first step in this chain is transfer of high-energy electrons from NADH + H^+ to the coenzyme *(FMN? Q?)*. As a result this carrier is *(oxidized? reduced?)* to

____________________. From which vitamin is this carrier derived? ________

b. As electrons are passed on down the chain, $FMNH_2$ is *(oxidized? reduced?)* back to

____________________.

c. Which mineral is the portion of cytochromes on which electrons are ferried?

________ Cytochromes are positioned toward the *(beginning? end?)* of the transport chain.

d. Circle two other minerals that are known to be involved in the electron transport chain and are hence necessary in the diet.
Calcium   Copper   Iodine   Sulfur   Zinc

e. The last carrier in the chain is *(coenzyme Q? cytochrome a_3?)*. It passes electrons to

____________________, the final oxidizing molecule. This is a key step in the process of *(aerobic? anaerobic?)* cellular respiration.

f. Where are the electron carriers located? Circle the correct answer.

MM.	**Mitochondrial matrix**
IMM.	**Inner mitochondrial membrane (cristae)**
OMM.	**Outer mitochondrial membrane**

g. The carriers are grouped into *(two? three? four?)* complexes. Each complex acts to pump *(protons = H^+? electrons = e^-?)* from the *(matrix? space between inner and outer mitochondrial membranes?)* into the ____________________.

h. As a result, a gradient of H^+ is set up across the *(inner? outer?)* mitochondrial membrane, resulting in potential energy known as the ____________________ force. How do these protons (H^+) travel back into the mitochondrial matrix?

i. The channels through which H^+ diffuses contain the enzyme known as ATP ____________________. Because the formation of the resulting ATP is driven by an energy source that consists of a gradient and subsequent flow of a *chemical* (specifically ________), this mechanism is known as ____________________ generation of ATP.

✓ **C5.** Summarize aerobic respiration in this activity.

a. Complete the chemical equation showing the overall reaction for aerobic respiration.

______ + 6 ______ + 36–38 ADPs + 36–38 P → 6 ______ + 6 ______ + ______ ATPs

Glucose, Oxygen, Carbon dioxide, Water

b. Fill in the correct number of high-energy molecules (ATPs or GTPs) yielded from processes in aerobic respiration.

1. Substrate-level phosphorylation via oxidation of glucose to pyruvic acid in glycolysis: ______ ATPs.
2. Oxidative phosphorylation in electron transport chain (ETS) of six reduced NADs (NADH + $2H^+$) resulting from glycolysis that takes place under aerobic conditions: ______ ATPs.
3. Oxidative phosphorylation in electron transport chain (ETS) of two reduced NADs (NADH + $2H^+$) yielded from transition step (formation of acetyl CoA): ______ ATPs.
4. Oxidative phosphorylation in electron transport chain (ETS) of six reduced NADs (NADH + 2H+) yielded from the Krebs cycle: ______ ATPs.
5. Oxidative phosphorylation in electron transport chain (ETS) of two reduced FADS yielded from the Krebs cycle: ______ ATPs. (Notice that each $FADH_2$ yields fewer ATPs than NADH + $2H^+$ because $FADH_2$ enters the ETS at a lower level.)
6. Substrate-level phosphorylation via oxidation via succinyl CoA to succinic acid in the Krebs cycle: ______ GTPs.

7. Grand total of ATPs or GTPs resulting from complete catabolism of glucose by glycolysis, Krebs cycle, electron transport, and resulting phosphorylation of ADP

   to ATP (or GDP to GTP): ______ ATPs and/or GTPs

✓ **C6.** Most of the 36 to 38 ATPs generated from the total oxidation of hydrogens derive from *(glycolysis? the Krebs cycle?)*. Of those generated by the Krebs cycle, most are formed from the oxidation of hydrogens carried by *(NAD? FAD?)*. About what percentage of the energy originally in glucose is stored in ATP after glucose is completely catabolized? *(10%? 40%? 60%? 99%?)*

✓ **C7.** Complete the exercise about the reaction shown below.

$$\text{Glucose} \underset{(2)}{\overset{(1)}{\rightleftharpoons}} \text{Glycogen}$$

a. You learned earlier that excess glucose may be stored as glycogen. In other words,

   glycogen consists of large branching chains of ____________________. Name the process of glycogen formation by labeling (1) in the chemical reaction.

b. Between meals, when glucose is needed, glycogen can be broken down again to release glucose. Label (2) above with the name of this process. [Note that a number of steps are actually involved in both processes (1) and (2).]

c. Which of these two reactions is anabolic? ____________________

d. Where is most (75%) of glycogen in the body stored'? ____________________

e. Identify a hormone that stimulates reaction (1). Write its name on the upper arrow in the reaction above. As more glucose is stored in the form of glycogen, the blood

   level of glucose ______-creases. Now write below the lower arrow the names of two hormones that stimulate glycogenolysis.

f. In order for glycogenolysis to occur, a ____________________ group must be added to glucose as it breaks away from glycogen. The enzyme catalyzing this reaction is

   known as ____________________. Does the reverse reaction [shown in (2)] require phosphorylation also? *(Yes? No?)*

C8. Define *gluconeogenesis* and briefly discuss how it is related to other metabolic reactions.

Name three hormones that stimulate gluconeogenesis.

✓ **C9.** Circle the processes at left and the hormones at right that lead to increased blood glucose level (hyperglycemia).

Processes	**Hormones**
Glycogenesis	Insulin
Glycogenolysis	Glucagon
Glycolysis	Epinephrine
Gluconeogenesis	Cortisol
	Thyroid hormone
	Growth hormone

For extra review. Refer to Figures LG 18.3 and LG 18.6, pages 366 and 375 of the *Learning Guide.*

D. Lipid metabolism (pages 920–923)

✓ **D1.** Most lipids are *(polar? nonpolar?)*, and therefore are hydro-*(philic? phobic?)*. Check your understanding of mechanisms that make lipids water-soluble in blood by completing this Checkpoint. Match answers from the box with related descriptions below.

A.	**Apoproteins**	**LDLs.**	**Low-density lipoproteins**
C.	**Chylomicrons**	**VLDLs.**	**Very low-density lipoproteins**
HDLs.	**High-density lipoproteins**		

_______ a. Designated by letters A, B, C, D, and E, these proteins surround an inner core of lipids within lipoproteins. These chemicals can serve as "docking" molecules that bind to receptors on body cells.

_______ b. Named so because these molecules are composed mainly (close to 50%) of protein which has high density; prevent accumulation of cholesterol in blood, so called the "good" cholesterol.

_______ c. Named so because these molecules are composed of small amounts (25%) of protein but large amounts (50%) of cholesterol (which has low density). These carry 75% of total blood cholesterol, and are called the "bad" cholesterol because excessive numbers of these form atherosclerotic plaques in arteries.

_______ d. Made largely (50%) of triglycerides, their numbers increase with a high-fat diet, leading ultimately to high blood levels of LDLs.

_______ e. Formed almost exclusively (85%) of triglycerides derived from dietary fats, they travel through lymphatic pathways in the GI tract, and then into blood to reach adipose or muscle tissue.

✓ **D2.** *Critical thinking.* Complete this exercise on cholesterol in diet and health.

a. Many foods bear the label "no cholesterol." Does a diet of such foods assure a low blood cholesterol?

b. Total cholesterol level should be below _______ mg/deciliter, and the level of LDLs (which make up most of total cholesterol) should be less than _______ mg/dL. "Total cholesterol to HDL" ratio should be *(less than? greater than?)* 3:1. Two ways to improve (lower) this ratio are to ______-crease total cholesterol level and to ______-crease HDL level. Regular aerobic exercise can achieve both of those goals.

c. Complete the table below as you evaluate and comment on each client's cholesterol profile. (Note that total cholesterol includes VLDLs also, so that LDLs + HDLs do not add up to total cholesterol.)

Client	Total cholesterol*	LDL*	HDL*	Cholesterol: HDL ratio
1. Dave Comment:	296	226	40	296/40 = 7.4:1
2. Juan Comment:	198	138	45	198/45 = ______
3. Denise Comment:	202	160	38	______
4. Jerry Comment:	222	140	74	______

* mg/dL blood

✓ **D3.** About 98% of all of the energy reserves stored in the body are in the form of *(fat? glycogen?)*. About *(10%? 25%? 50%?)* of the stored triglycerides (fats) in the body are located in subcutaneous tissues. Is the fat stored in these areas likely to be "the same fat" that was located there two years ago? *(Yes? No?)* Explain.

✓ **D4.** Do this exercise about fat metabolism.

a. The initial step in catabolism of triglycerides is their breakdown into

______ and ______, a process called ______-lysis

and catalyzed by enzymes called ______-ases.

b. Glycerol can then be converted into ______ (G 3-P); this molecule can then enter glycolytic pathways to increase ATP production. If the cell does not

need to generate ATP, then G 3-P can be converted into ______.

This anabolic step is an example of gluco-______.

c. Recall that fatty acids are long chains of carbons with attached hydrogens and a few oxygens. Two-carbon pieces are "snipped off" of fatty acids by a process called

______________, occurring in the ______________.

d. These two-carbon pieces (______________) attach to coenzyme ______.

The resulting molecules, named ______________ may enter the Krebs cycle. Complete catabolism of one 16-carbon fatty acid can yield a net of *(36–38? 129?)* ATPs.

e. When large numbers of acetyl CoAs form. they tend to pair chemically:

Acetic acid	+	acetic acid	→ ______________
(2C)		(2C)	(4C)

The presence of excessive amounts of these acids in blood *(raises? lowers?)* blood pH. Because these acids are called "keto" acids, this condition is known as

______________.

f. Slight alterations of acetoacetic acid leads to formation of β-hydroxybutyric acid

and ______________. Collectively, these three chemicals are known as

______________ bodies. Formation of ketone bodies is known as

______________; it takes place in the ______________.

g. Circle the cells that can use acetoacetic acid for generation of ATP:

Brain cells	Hepatic cells
Cardiac muscle cells	Cells of the renal cortex

h. Explain how the blood level of acetoacetic acid is normally maintained at a low level.

i. Ketogenesis occurs when cells are forced to resort to fat catabolism. State two or more reasons why cells might carry out excessive fat catabolism leading to ketogenesis and ketosis.

j. *A clinical correlation.* Explain why a person whose diabetes is out of control might have sweet breath.

E. Protein metabolism (pages 923–925)

✓ **E1.** Which of the three major nutrients (carbohydrates, lipids, and proteins) fulfills each function?

a. Most direct source of energy; stored in body least: ____________________

b. Constitutes almost all of energy reserves, but also used for body-building:

____

c. Usually used least for fuel; used most for body structure and for regulation:

____

✓ **E2.** Throughout your study of systems of the body, you have learned about a variety of roles of proteins. List at least six functions of proteins in the body, using one or two words for each function. Include some structural and some regulatory roles.

____________________ ____________________ ____________________

____________________ ____________________ ____________________

✓ **E3.** List names of hormones that enhance transport of amino acids into cells or otherwise promote protein synthesis (anabolism).

✓ **E4.** Refer to Figure LG 25.2 and describe the uses of protein.

Two sources of proteins are shown: (a) ________________ and (b) ________________.

Amino acids from these sources may undergo a process known as (c) ________________ to form new body proteins. If other energy sources are used up, amino acids may undergo catabolism. The first step is (d) ____________________, in which an amino group (NH_2) is removed and converted (in the liver) to (e) ____________________, a component of (f) ____________________ which exits in urine. The remaining portion of the amino acid may enter (g) ____________________ pathways at a number of points (Figure 25.15 in your text). In this way, proteins can lead to formation of (h) ____________________.

E5. Contrast essential amino acids with nonessential amino acids.

How many amino acids are considered essential to humans? ________

✓ **E6.** Phenylketonuria (PKU) is a metabolic disorder in which the amino acid ____________________ cannot be converted into tyrosine. A screening test for this disorder *(is? is not?)* available. The sweetener ____________________ (Nutrasweet) contains phenylalamine, so children with PKU should not consume this.

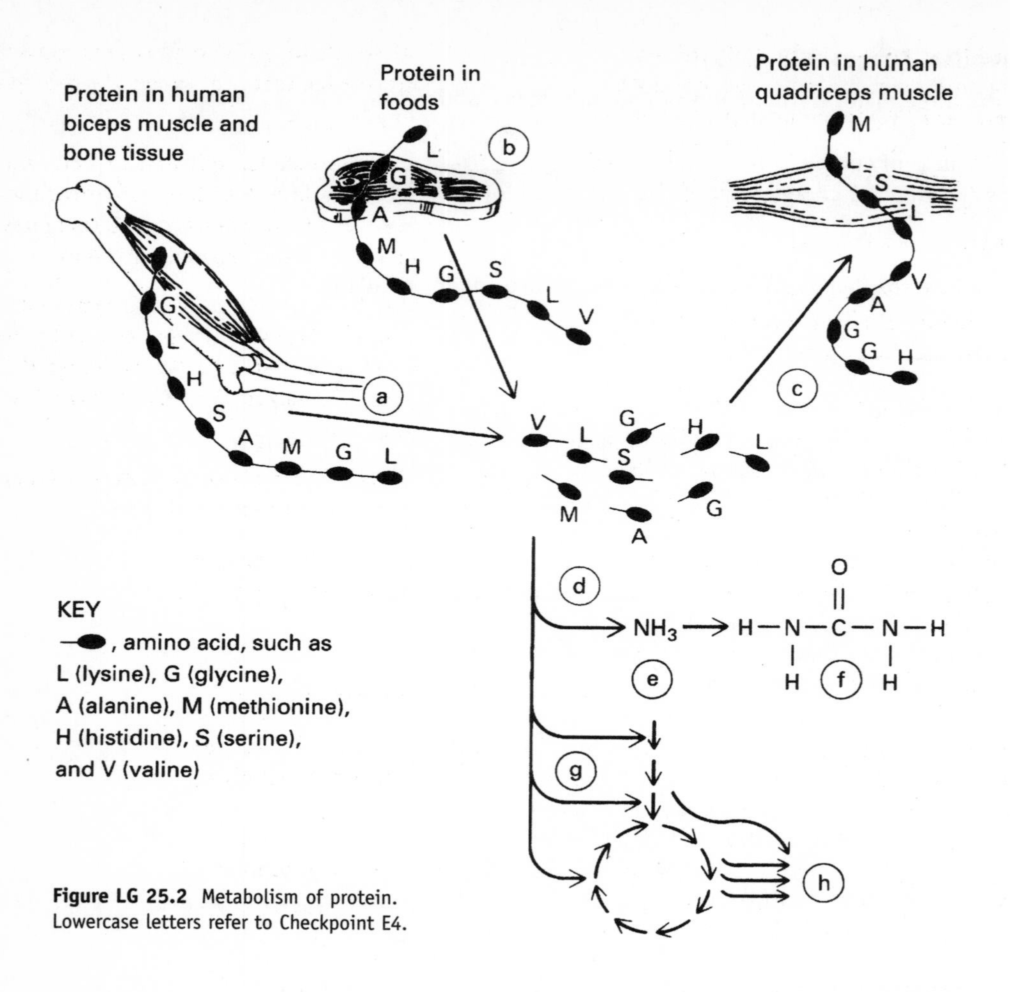

Figure LG 25.2 Metabolism of protein. Lowercase letters refer to Checkpoint E4.

F. Key molecules at metabolic cross roads (pages 925–927)

✓ **F1.** Key molecules in metabolism provide "crossroads" among the three different categories of nutrients: carbohydrates, ________________, and ________________. These key molecules provide pathways that permit you to put on weight (fat) from eating excessive carbohydrates or even proteins, and allow you to be energized (generate ____________________) by eating any of the three food types. Refer to Figure LG 25.16 in the text and identify roles of each of these key molecules (in the box) in metabolism.

Acetyl CoA.	**Acetyl coenzyme A**	**PA.**	**Pyruvic acid**
G 6-P.	**Glucose 6-phosphate**		

________ a. Two carbon molecule formed from pyruvic acid when oxygen is available; leads into Krebs cycle and electron transfer system

________ b. Entrance molecule for fatty acids and ketone bodies into Krebs cycle; also the key molecule in conversion of excessive carbohydrates to fat

_______ c. Formed from beta-oxidation of fatty acids during fat catabolism; excessive amounts of these (as in diabetics lacking adequate insulin or persons who are starving or dieting excessively) can lead to ketone body formation and potentially lethal ketoacidosis (see page 923 of the text)

_______ d. Three carbon molecule that feeds into aerobic metabolism if oxygen is available, otherwise can form lactic acid

_______ e. Similar to amino acid alanine except for one amino group; molecule by which amino acids (from protein in foods) can provide energy (via catabolic pathways)

_______ f. Molecule into which glucose is converted (reversible reaction) when glucose enters any body cell

_______ g. Can form oxaloacetic acid to feed into the Krebs cycle or lead into gluconeogenesis, resulting in synthesis of carbohydrates (glucose) from proteins (alanine)

_______ h. Molecule by which glycogen can enter into pathways to be used as fuel by liver or muscle cells, as in "carbo loading" before a marathon race (see page 919 of the text)

_______ i. Precursor for ribose portion of RNA and deoxyribose portion of DNA

✓ **F2.** *For extra review.* Focus on interconversions of carbohydrates and fats in this Checkpoint.

a. Fill in the blanks in the pathway of conversion of excessive carbohydrates to body fat. Use the same answers as for Checkpoint F1.

Glycogen → _______________ → _______________ → _______________ → fatty acid portion of fats

b. Mammals, including humans, *(can? cannot?)* convert acetyl CoA into pyruvic acid. As a result, fatty acids *(can? cannot?)* be used to form glucose. Gluconeogenesis involving fats can occur, however, by the following pathway:

The _______________ portion of triglycerides → glyceraldehyde 3-phosphate

(G 3-P) → _______________.

G. Metabolic adaptations (pages 927–931)

✓ **G1.** Contrast these two metabolic states by writing *A* if the description refers to the absorptive state and *P* if it refers to the postabsorptive state.

_______ a. Period when the body is fasting and is challenged to maintain an adequate blood glucose level

_______ b. State during which glucose (stored in glycogen in liver and muscle) is released

_______ c. State during which most systems (excluding nervous system) switch over to use of fatty acids as energy sources

_______ d. Time when the body is absorbing nutrients from the GI tract

_______ e. State during which the principal concern is formation of stores of glucose (as glycogen) and fat

_______ f. State when insulin is released under stimulation from glucose-dependent insulinotropic peptide (GIP); insulin then facilitates transport of glucose and amino acids into cells

_______ g. Period dominated by anti-insulin hormones, such as glucagon and epinephrine

✓ **G2.** Do this exercise about maintenance of glucose level between meals.

a. The normal blood glucose level in the postabsorptive state is about *(30–50? 70–110? 150–170? 600–700?)* mg glucose/100 mL blood.

b. Now list four sources of glucose that may he called upon during the postabsorptive state.

____

____

____

____

c. Can fatty acids serve as a source of glucose? *(Yes? No?)* Explain.

d. Fats can be used to generate ATP by being broken down into __________________, which can then enter the Krebs cycle. Therefore, fatty acids are said to provide a glucose-*(sparing? utilizing?)* mechanism between meals.

✓ **G3.** Identify the hormones (in the box) that stimulate each metabolic process listed below. Write one answer on each line. (Refer to Tables 25.3 and 25.4 in your text for help.)

Cor.	**Cortisol**	**IGF.**	**Insulinlike growth factors**
Epi.	**Epinephrine**	**Ins.**	**Insulin**
Glu.	**Glucagon**	**Thy.**	**Thyroid hormone (T3/T4)**

a. Glycogenolysis: ______ ______

b. Glycogenesis: ______

c. Gluconeogenesis: ______ ______

d. Protein synthesis: ______ ______ ______

e. Lipogenesis: ______

f. Lipolysis: ______ ______ ______ ______

g. Protein breakdown: ______

✓ **G4.** *For extra review,* identify which processes (a-g) in Checkpoint G3 are *anabolic* and which are *catabolic.*
Anabolic (*Hint:* terms may include "-genesis" or *creation*) ______ ______ ______ ______

Catabolic (*Hint:* terms may include "-lysis" or *breakdown*) ______ ______ ______

✓ **G5.** *For extra review,* identify the hormones (in the box for Checkpoint G3) that are

hyperglycemic, that is, tend to increase blood glucose. ______ ______ ______ ______
These four hormones *(mimic insulin? are anti-insulin?)* in action, and they are more active during the *(absorptive? postabsorptive?)* state because the body requires more glucose then. During this period, *(sympathetic? parasympathetic?)* nerves stimulate

release of the hormone __________________ from the adrenal medulla.

✓ **G6.** *A clinical correlation.* Describe metabolic changes during fasting or starvation.

a. Contrast these terms: fasting/starving.

b. State the major factor that determines the life span possible without food.

c. Explain what accounts for the "wasting" appearance of people who are starving.

d. Explain why extremely high blood levels of ketone bodies (which provide up to 2/3 of the energy to the brain after many weeks of starvation) may be dangerous and even lethal for example, leading to coma and death.

H. Heat and energy balance (pages 932–936)

✓ **H1.** Contrast *calorie* and *kilocalorie*.

If you ingest, digest, absorb, and metabolize a slice of bread, the energy released from the bread equals about 80 to 100 *(cal? kcal?)*.

H2. Define basal metabolic rate (BMR).

✓ **H3.** Catabolism of foods *(uses? releases?)* energy. Most of the heat produced by the body comes from catabolism by *(oxidation? reduction?)* of nutrients.

✓ **H4.** Discuss BMR in this exercise.

a. BMR may be determined by measuring the amount of ____________________ consumed within a period of time, because oxygen is necessary for the metabolism of foods. Normally, for each liter of oxygen consumed, the body metabolizes enough food to release about ________ kcal.

b. Suppose you consume 15 liters of oxygen in an hour. Your body would release ________ kcal of heat in that hour. At this metabolic rate, you would need to consume ________ kcal per day to meet your metabolic needs.

c. A BMR of 20% below the standard value is likely to be due to *(hyper? hypo?)*-thyroid.

✓ **H5.** Do this activity on regulation of body temperature.

a. Which temperature is normally higher? *(Core? Shell?)* Explain how an increase in core temperature, for example, above 112–114°F can be fatal.

b. Circle the factor in each pair that is likely to lead to a higher metabolic rate.

1. Age: 12 years old 52 years old
2. Body temperature: 98.6°F 103.6°F
3. State of activity: Running 2 miles Typing at computer
4. Epinephrine or thyroid hormone: Increased Decreased
5. Specific dynamic action related to ingestion of: Protein Carbohydrate

H6. Write the percentage of heat loss by each of the following routes (at room temperature). Then write an example of each of these types of heat loss. One is done for you.

a. Radiation _______ ____

b. Evaporation _______ ____

c. Convection __**15**__ **Cooling by draft while taking a shower**

d. Conduction _______ ____

✓ **H7.** During an hour of active exercise Bill produces 1 liter of sweat. The amount of heat loss (cooling) that accompanies this much evaporation is _______ Cal. On a humid day, *(more? less?)* of Bill's sweat occurs, so Bill would be cooled *(more? less?)* on such a day.

✓ **H8.** The body's "thermostat" is located in the preoptic area of the *(eyes? hypothalamus? medulla? skin?)*. If the body temperature needs to be raised, nerve impulses from the preoptic area are sent to the heat-*(losing? promoting?)* center, which is primarily *(parasympathetic? sympathetic?)*. An example of such a response is vaso-*(constriction? dilation?)* of blood vessels in skin, resulting in heat *(loss or release? conservation?)*.

H9. Explain why elderly persons are at greater risk for hypothermia.

Write seven or more signs or symptoms of hypothermia.

✓ **H10.** Complete this Checkpoint about regulation of food intake.

a. The site within the brain that is the location of feeding and satiety centers is the *(medulla? hypothalamus?)*. When the satiety center is active, an individual will feel *(hungry? satiated or full?)*.

b. Circle all of the answers that will keep the feeding center active and enhance the desire to eat.
   ↓ Glucose in blood
   ↓ Amino acids in blood
   ↓ Fats entering intestine (with related ↓ in CCK release)
   ↓ Environmental temperature
   ↓ Stretching of stomach

c. Describe the functions of *leptin.*

I. Nutrition (pages 936–941)

I1. List three primary functions of nutrients.

✓ **I2.** List the six principal classes of nutrients.

Circle the nutrient needed in greatest quantities

✓ **I3.** Fill in the number of calories needed each day by typical persons in each category.

________ a. Active women, teen girls, and most men ________ c. Elderly adults

________ b. Active men and teen boys ________ d. Children

✓ **I4.** Fill in the percentage of calories needed each day in each category of nutrients.

a. Carbohydrates: ________%

b. Fats: no more than ________% with no more than ________% of total calories in the form of saturated fats

c. Proteins: ________%

✓ **I5.** *Critical thinking.* Keep in mind that carbohydrates and proteins each provide about 4 Cal/g and fats provide about 9 Cal/g. Considering the information in Checkpoint I4, determine the number of grams of each of the three major nutrient groups appropriate for a healthy diet of 2000 Cal/day. One is done for you.

a. Carbohydrates

b. Fats

c. Proteins **2000 Cal × (0.12–0.15) = 240–300 Cal/day. At 4 Cal/g =** ***60–75g/day.***

I6. Write seven guidelines for healthy eating.

✓ **I7.** Using the Food Pyramid (page 937 in the text), identify the number of servings suggested per day for each food group listed.

________ a. Bread, cereal, rice, pasta

________ b. Vegetables

________ c. Fruits

________ d. Meat, fish, poultry, dry beans, eggs, nuts

________ e. Milk, yogurt, and cheese

________ f. Fats, oils, sweets

✓ **I8.** Define *minerals.*

Minerals make up about ______% of body weight and are concentrated in the

__________________.

✓ **I9.** Study Table 25.5 in your text. Then check your understanding of minerals by doing this matching exercise

Ca.	**Calcium**	**Fe.**	**Iron**	**Na.**	**Sodium**
Cl.	**Chlorine**	**I.**	**Iodine**	**P.**	**Phosphorus**
Co.	**Cobalt**	**K.**	**Potassium**	**S.**	**Sulfur**
F.	**Fluorine**	**Mg.**	**Magnesium**		

________ a. Main anion in extracellular fluid, part of HCl in stomach; component of table salt

________ b. Involved in generation of nerve impulse, helps to regulate osmosis, acts in buffer systems

________ c. Most abundant mineral in the body. found mostly in bones and teeth; necessary for normal muscle contraction and for blood clotting

________ d. Important component of hemoglobin and cytochromes

________ e. Main cation inside of cells; used in nerve transmission

________ f. Essential component of thyroxin

________ g. Constituent of vitamin B_{12}, so necessary for red blood cell formation

________ h. Important component of amino acids, vitamins, and hormones

________ i. Improves tooth structure

________ j. Found mostly in bones and teeth; important in buffer system; component of ATP, DNA and RNA

✓ **I10.** Circle the correct answer. The primary purpose of vitamins is:

A. Synthesis of body structures
B. Regulation of body activities

✓ **I11.** Vitamins are *(organic? inorganic?)*. Most vitamins *(can? cannot?)* be synthesized in the body. In general, what are the functions of vitamins?

✓ **I12.** Contrast the two principal groups of vitamins, and list the main vitamins in each group.

a. Fat-soluble ______________________

b. Water-soluble ______________________

I13. Defend or dispute this statement: "Most persons who eat a balanced diet do need to take vitamin or mineral supplements."

✓ **I14.** Select the vitamin that fits each description.

A	B_1	B_2	B_{12}	C	D	E	K

________ a. This serves as a coenzyme that is essential for blood clotting, so it is called the anti-hemorrhagic vitamin; synthesized by intestinal bacteria.

________ b. Its formation depends upon sunlight on skin and also on kidney and liver activation; necessary for calcium absorption.

________ c. Riboflavin is another name for it; a component of FAD; necessary for normal integrity of skin, mucosa, and eye.

________ d. This vitamin acts as an important coenzyme in carbohydrate metabolism; deficiency leads to beriberi.

________ e. Formed from carotene, it is necessary for normal bones and teeth; it prevents night blindness.

________ f. This substance is also called ascorbic acid; deficiency causes anemia, poor wound healing, and scurvy.

________ g. This coenzyme, the only B vitamin not found in vegetables, is necessary for normal erythropoiesis; absorption from GI tract depends on intrinsic factor.

________ h. Also known as tocopherol, it is necessary for normal red blood cell membranes: deficiency is associated with sterility in some animals.

J. Disorders (pages 942–943)

✓ **J1.** Arrange in order the events believed to occur in the production of fever of 39.4°C (103°F) and recovery from this state. ______ ______ ______ ______ ______ ______

A. 39.4°C (103°F) temperature is reached and maintained.
B. Prostaglandins (PGs) cause the hypothalamic "thermostat" to be reset from normal 37.0°C (98.6°F) to a higher temperature such as 39°C (103°F).
C. A "chill" occurs as skin feels cool (due to vasoconstriction) and shivering occurs—attempts to raise body temperature to the 39.4°C (103°F) designated by the hypothalamus.

D. Source of pyrogens is removed (for example, bacteria are killed by antibiotics), lowering hypothalamic "thermostat" to 37.5°C (98.6°F).
E. "Crisis" occurs. Obeying hypothalamic orders, the body shifts to heat loss mechanisms such as sweating and vasodilation, returning body temperature to normal.
F. Infectious organisms cause phagocytes to release interleukin-l, which then causes hypothalamic release of PGs.

J2. In what ways is a fever beneficial?

J3. Define obesity, and state several causes of obesity.

J4. Contrast *heatstroke* wih *heat exhaustion.*

✓ **J5.** *A clinical correlation.* Which condition involves deficiency in both protein and calories? *(Kwashiorkor? Marasmus?)* Explain why an individual with kwashiorkor may have a large abdomen even though the person's diet is inadequate.

ANSWERS TO SELECTED CHECKPOINTS: CHAPTER 25

Al. (a) Immediate energy supply. (b) Building blocks. (c) Storage for future use for (a) or (b).

A2.

Process	Definition	Releases or Uses Energy	Examples
a. Catabolism	Breakdown of complex organic compounds into smaller ones	Energy released as heat or stored in ATP	Glycolysis, Krebs cycle, glycogenolysis
b. Anabolism	Synthesis of complex organic molecules from smaller ones	Uses energy (ATP)	Synthesis of protein, fats, or glycogen

A4. 2; 3, 2; ATP.

A5. 40%; heat.

B1. (a) ATP; $C_6H_{12}O_6 \rightarrow CO_2 + H_2O$ + energy stored in ATP: poor. (b) H^+, hydride, H^-. (c) NAD^+, $NADP^+$, FAD; niacin; H_2O. (d) Oxidation. (e) ATP; phosphorylation;

Glucose + O_2 + ADP →
(reducer) (oxidizer)

CO_2 + H_2O + ATP
(oxidized) (reduced)

B2. (a) A (b) B. (c) A. (d) B. (e) A. (f) B. (g) B.

Cl. (a) Glucose. (b) Starch, sucrose, lactose; by enzymes located mostly in liver cells. (c) Increases, oxidizing. (d) Stored as glycogen, converted to fat or protein and stored, excreted in urine. (e) Hyperglycemia; insulin, facilitates, glucose; diabetes mellitus. (f) Neurons and liver cells. (g) Phosphate; phosphorylation, kinases: cells can utilize the trapped glucose since phosphorylated glucose cannot pass out of most cells.

C2. (a) 6; 3, pyruvic acid; a little; many, does not. (b) Phosphoryl, fructose; phosphofructokinase; ADP, ATP; ana, glycogen. (c) 3, G 3-P; oxidized, NAD. (d) Pyruvic, 2. (e) Oxygen; do, aerobic. (f) Lactic acid; exercise. (g) Pyruvic acid; stimulates.

C3. (a) Glucose. (b) Pyruvic acid. (c) Glycolysis. (d) Cytosol. (e) Transition. (f) CO_2. (g) Acetyl. (h) Coenzyme A. (i) Acetyl coenzyme A. (j) Oxaloacetic acid. (k) Citric acid. (1) Citric acid (or Krebs). (m) Matrix of mitochondria. (n) CO_2, which are exhaled. (o) Hydrogen. (p) NAD and FAD. (q) H^+. (r) Electrons (e^-) (or H^-). (s) Cytochromes or electron transfer system. (t) ATP. (u) Oxygen. (v) Water.

C4. (a) FMN: reduced; $FMNH_2$; B_2 (riboflavin). (b) Oxidized, FMN. (c) Iron (Fe); end. (d) Copper and sulfur. (e) Cytochrome a_3; oxygen; aerobic. (f) IMM. (g) Three; protons = H^+, matrix, space between inner and outer mitochondrial membranes. (h) Inner, proton motive; across special channels in the inner mitochondrial membrane. (i) Synthetase; H^+, chemiosmotic.

C5. (a) $C_6H_{12}O_6$ + $6O_2$: + 36 or 38 ADPs
Glucose Oxygen

+ 36 or 38 (P) → $6CO_2$ + $6H_2O$ +
Carbon Water

36–38 ATPs. (b1) 2; (b2) 4–6. (b3) 6. (b4) 18. (b5) 4. (b6) 2. (b7) 36–38.

C6. The Krebs cycle; NAD; 40.

C7. (a) Glucose; 1: glycogenesis. (b) 2: Glycogenolysis. (c) Glycogenesis. (d) Muscles. (e) Insulin; de; glucagon and epinephrine. (f) Phosphate; phosphorylase; yes.

C9. Processes: glycogenolysis and gluconeogenesis; hormones: all except insulin.

D1. Nonpolar, phobic. (a) A. (b) HDLs. (c) LDLs. (d) VLDLs. (e) C.

D2. (a) No. Cholesterol comes not only from diet, but also from synthesis of cholesterol, for example, from saturated fatty acids (as in palm oil, coconut oil, or partially hydrogenated margarine), within the liver. (b) 200, 130; less than; de, in. (c1) Dave's seriously high levels of cholesterol, LDLs, and his ratio of 7.4 put him at almost four times the risk of a heart attack as Juan with a cholesterol of 198. (c2) Juan's total cholesterol is good, but his ratio of 198/45 = 4.5 puts him at increased risk. (c3) Denise's LDL level is high; raising her HDLs would improve her 202/38 ratio of 5.3. (c4) Jerry, at age 74, still runs six miles most days, so keeps his HDLs high (normal range is 39–96). His is the best ratio: 222/72 = 3.0, even though his total cholesterol is borderline-high. (He's working on cutting back on saturated fats.)

D3. Fat: 50%: no. it is continually catabolized and resynthesized.

D4. (a) Fatty acids. glycerol, lipo, lip. (b) Glyceraldehyde 3-phosphate; glucose; neogenesis. (c) Beta oxidation, matrix of mitochondria. (d) Acetic acid, A; acetyl CoA (or acetyl coenzyme A); 129. (e) Acetoacetic acid; lowers; ketoacidosis. (f) Acetone; ketone; ketogenesis, liver. (g) Cardiac muscle cells, cells of the renal cortex, and (during starvation) brain cells. (h) Acetoacetic acid is normally catabolized for ATP production as fast as the liver generates it. (i) Examples of reasons: starvation, fasting diet, lack of glucose in cells due to lack of insulin (diabetes mellitus), excess growth hormone (GH) which stimulates fat catabolism. (j) As acetone (formed in ketogenesis) passes through blood in pulmonary vessels. some is exhaled and detected by its sweet aroma.

E1. (a) Carbohydrates. Exception: cardiac muscle and cortex of the kidneys utilize acetoacetic acid from fatty acids preferentially over glucose. (b) Lipids. (c) Proteins.

E2. See Exhibit 2.6 in your text.
E3. Insulin (as well as insulinlike growth factors, IGFs), thyroid hormones (T3 and T4), estrogen, and testosterone.
E4. (a) Worn-out cells as in bone or muscle. (b) Ingested foods. (c) Anabolism. (d) Deamination. (e) Ammonia. (f) Urea. (g) Glycolytic or Krebs cycle. (h) $CO_2 + H_2O$ + ATP.
E6. Phenylalanine; is; aspartame.
Fl. Proteins and lipids; ATP. (a–c) Acetyl-CoA. (d-e). PA. (f-i) G 6-P.
F2. (a) G 6-P, PA, acetyl CoA. (b) Cannot; cannot; glycerol, glucose.
G1. (a–c) P. (d–f) A. (g) P.
G2. (a) 70–110. (b) Liver glycogen (4-hour supply); muscle glycogen and lactic acid during exercise; glycerol from fats; as a last resource, amino acids from tissue proteins. (c) No. See Checkpoint F2b on page 579. (d) Acetic acids (and then acetyl CoA); sparing.
G3. (a) Glu Epi. (b) Ins. (c) Cor Glu. (d) IGF Ins Thy. (e) Ins. (f) Cor Epi IGF Thy. (g) Cor.
G4. Anabolic: b–e. Catabolic: a, f, and g.
G5. Cor Glu Epi Thy (or all except Ins and IGF); are anti-insulin; postabsorptive; sympathetic, epinephrine.
G6. (a) Fasting is short term (a few hours or days) food deprivation; starvation refers to long periods of up to months without food. (b) The amount of adipose tissue (and also availability of water to prevent dehydration). (c) When dietary glucose is not available, amino acids from muscles and fatty acids from adipose stores are used for ATP production. (d) Glucose (normally the primary fuel for the nervous system: See page 929 of the text) is not available, and fatty acids cannot pass the blood-brain barrier. Ketone bodies can be used as energy sources, but high levels lead to ketoacidosis (a form of metabolic acidosis) which depresses the central nervous system.
H1. (a) A kilocalorie is the amount of heat required to raise 1000 grams (= 1000 mL = 1 liter) of water 1°C; a kilocalorie (kcal) = 1000 calories (cal); kcal.
H3. Releases; oxidation.
H4. (a) Oxygen; 4.8. (b) 72 (= 4.8 kcal/liter oxygen × 15 liters/hr); 1728 (= 72 kcal/hr × 24 hrs/day. (c) Hypo.
H5. (a) Core; hyperthermia denatures (changes chemical configurations of) proteins. (b1) 12 years old; (b2) 103.6°F; (b3) Running 2 miles; (b4) Increased; (B5) Protein.
H7. 580 (= 0.58 cal/ml water × 1000 ml/liter); less, less.
H8. Hypothalamus; promoting, sympathetic; constriction, conservation.
H10. (a) Hypothalamus; satiated or full. (b) All answers. (c) Acts on the hypothalmus to inhibit the desire to eat; also stimulates energy expenditure.
I2. Carbohydrates, proteins, lipids, minerals, vitamins, and water; water.
I3. (a) 2200. (b) 2800. (c) 1600. (d) 2200.
I4. (a) 50–60. (b) 30, 10. (c) 12–15.
I5. (a) 2000 Cal × (0.50–0.60) = 1000–1200 Cal/day. At 4 Cal/g = 250–300 g/day. (b) 2000 Cal × (0.30) = less than 600 cal/day. At 9 Cal/g = 600/9 = 67 g/day of total fat: a maximum of 10% should be saturated fat = 22 g/day.
I7. (a) 6–11. (b) 3–5. (c) 2–4. (d) 2–3. (e) 2–3. (f) Sparingly.
I8. Inorganic substances; 4, skeleton.
I9. (a) Cl. (b) Na. (c) Ca. (d) Fe. (e) K. (f) I. (g) Co. (h) S. (i) F. (j) P.
I10. B.
I11. Organic; cannot; most serve as coenzymes, maintaining growth and metabolism.
I12. (a) A D E K. (b) B complex and C.
I14. (a) K. (b) D. (c) B_2. (d) B_1. (e) A. (f) C. (g) B_{12}. (h) E.
J1. F B C A D E.
J5. Marasmus; protein deficiency (due to lack of essential amino acids) decreases plasma proteins. Blood has less osmotic pressure, so fluids exit from blood. Movement of these into the abdomen (ascites) increases the size of the abdomen. Fatty infiltration of the liver also adds to the abdominal girth.

MORE CRITICAL THINKING: CHAPTER 25

1. Contrast anabolism and catabolism. Give one example of each process in metabolism of carbohydrates, lipids, and proteins.
2. Explain the significance of B-complex vitamins in normal metabolism. Include roles of niacin (B_1), which forms coenzymes NAD^+ and $NADP^+$, riboflavin (B_2), which forms coenzymes FAD and FMN, and pantothenic acid, which forms coenzyme A.
3. Describe interconversions of carbohydrates, lipids, and proteins, including roles of the three key molecules, acetyl CoA, pyruvic acid, and G 6-P.
4. Explain why red blood cells derive all of their ATP from glycolysis. (*Hint:* Which organelles do they lack?)

5. When you wake up at 7 AM and have not eaten since dinner last night, which hormones are helping you to maintain an adequate blood level of glucose? How do they accomplish this?
6. Lil is outside on a snowy day without a warm coat. Her skin is pale and chilled and she begins to shiver. Explain how these responses are attempts of her body to maintain homeostasis of temperature. What other responses might raise her body temperature?
7. Describe guidelines for healthy eating and discuss roles foods may play in disorders such as cancer and diabetes

MASTERY TEST: CHAPTER 25

Questions 1–12: Circle the letter preceding the one best answer to each question.

1. Which of these processes is anabolic?
   A. Pyruvic acid → CO_2 + H_2O + ATP
   B. Glucose → pyruvic acid + ATP
   C. Protein synthesis
   D. Digestion of starch to maltose
   E. Glycogenolysis
2. Which of the following can be represented by the equation: Glucose → pyruvic acids + small amount ATP?
   A. Glycolysis
   B. Transition step between glycolysis and Krebs cycle
   C. Krebs cycle
   D. Electron transport system
3. Which hormone is said to be hypoglycemic because it tends to lower blood sugar?
   A. Glucagon
   B. Glucocorticoids
   C. Growth hormone
   D. Insulin
   E. Epinephrine
4. The complete oxidation of glucose yields all of these products *except:*
   A. ATP
   B. Oxygen
   C. Carbon dioxide
   D. Water
5. All of these processes occur exclusively or primarily in liver cells except one, which occurs in virtually all body cells. This one is:
   A. Gluconeogenesis
   B. Beta oxidation of fats
   C. Ketogenesis
   D. Deamination and conversion of ammonia to urea
   E. Krebs cycle
6. The process of forming glucose from amino acids or glycerol is called:
   A. Gluconeogenesis
   B. Glycogenesis
   C. Ketogenesis
   D. Deamination
   E. Glycogenolysis
7. All of the following are parts of fat absorption and fat metabolism *except:*
   A. Lipogenesis
   B. Beta oxidation
   C. Chylomicron formation
   D. Ketogenesis
   E. Glycogenesis
8. Which of the following vitamins is water soluble? A, C, D, E, K?
9. At room temperature most body heat is lost by:
   A. Evaporation
   B. Convection
   C. Conduction
   D. Radiation
10. Choose the *false* statement about vitamins.
    A. They are organic compounds.
    B. They regulate physiological processes.
    C. Most are synthesized by the body.
    D. Many act as parts of enzymes or coenzymes.
11. Choose the *false* statement about temperature regulation.
    A. Some aspects of fever are beneficial.
    B. Fever is believed to be due to a "resetting of the body's thermostat."
    C. Vasoconstriction of blood vessels in skin will tend to conserve heat.
    D. Heat-producing mechanisms which occur when you are in a cold environment are primarily parasympathetic.
12. Which of the following processes involves a cytochrome chain?
    A. Glycolysis
    B. Transition step between glycolysis and Krebs cycle
    C. Krebs cycle
    D. Electron transport system

Questions 13–14. Arrange the answers in correct sequence.

_____ _____ _____ _____ _____ _____ 13. Steps in complete oxidation of glucose:
A. Glycolysis takes place.
B. Pyruvic acid is converted to acetyl CoA.
C. Hydrogens are picked up by NAD and FAD; hydrogens ionize.
D. Krebs cycle releases CO_2 and hydrogens.
E. Oxygen combines with hydrogen ions and electrons to form water.
F. Electrons are transported along cytochromes, and energy from electrons is stored in ATP.

_____ _____ _____ 14. Using the Food Pyramid as a guide, arrange these food groups according to number of servings suggested per day, from greatest to least:
A. Meat, fish, poultry, dry beans, eggs, and nuts
B. Vegetables
C. Bread, cereal, rice, pasta

Questions 15–20: Circle T (true) or F (false). If the statement is false, change the underlined word or phrase so that the statement is correct.

T F 15. Complexes of carrier molecules utilized in the electron transport system are located in the mitochondrial matrix.

T F 16. Catabolism of carbohydrates involves oxidation, which is a process of addition of hydrogens.

T F 17. Nonessential amino acids are those that are not used in the synthesis of human protein.

T F 18. The complete oxidation of glucose to CO_2, and H_2O yields 4 ATPs.

T F 19. Anabolic reactions are synthetic reactions that release energy.

T F 20. Carbohydrates, proteins, and vitamins provide energy and serve as building materials.

Questions 21–25: Fill-ins. Complete each sentence with the word or phrase that best fits.

_______________ 21. Catabolism of each gram of carbohydrate or protein results in release of about ________ kcal, whereas each gram of fat leads to about ________ kcal.

_______________ 22. Pantothenic acid is a vitamin used to form the coenzyme named ________.

_______________ 23. ________ is the mineral that is most common in extracellular fluid (ECF); it is also important in osmosis, buffer systems, and nerve impulse conduction.

_______________ 24. Aspirin and acetaminophen (Tylenol) reduce fever by inhibiting synthesis of ________ so that the body's "thermostat" located in the ________ is reset to a lower temperature.

_______________ 25. In uncontrolled diabetes mellitus, the person may go into a state of acidosis due to the production of ________ acids resulting from excessive breakdown of ________.

ANSWERS TO MASTERY TEST: CHAPTER 25

Multiple Choice

1. C
2. A
3. D
4. B
5. E
6. A
7. E
8. C
9. D
10. C
11. D
12. D

Arrange

13. A B D C F E
14. C B A

True–False

15. F. Mitochondrial inner (cristae) membrane
16. F. Oxidation, removal of hydrogens
17. F. Can be synthesized by the body
18. F. 36 to 38
19. F. Synthetic reactions that require energy
20. F. Carbohydrates and proteins but not vitamins

Fill-Ins

21. 4, 9
22. Coenzyme A
23. Sodium (Na^+)
24. Prostaglandins (PGs), hypothalamus
25. Acetoacetic (keto-), fats

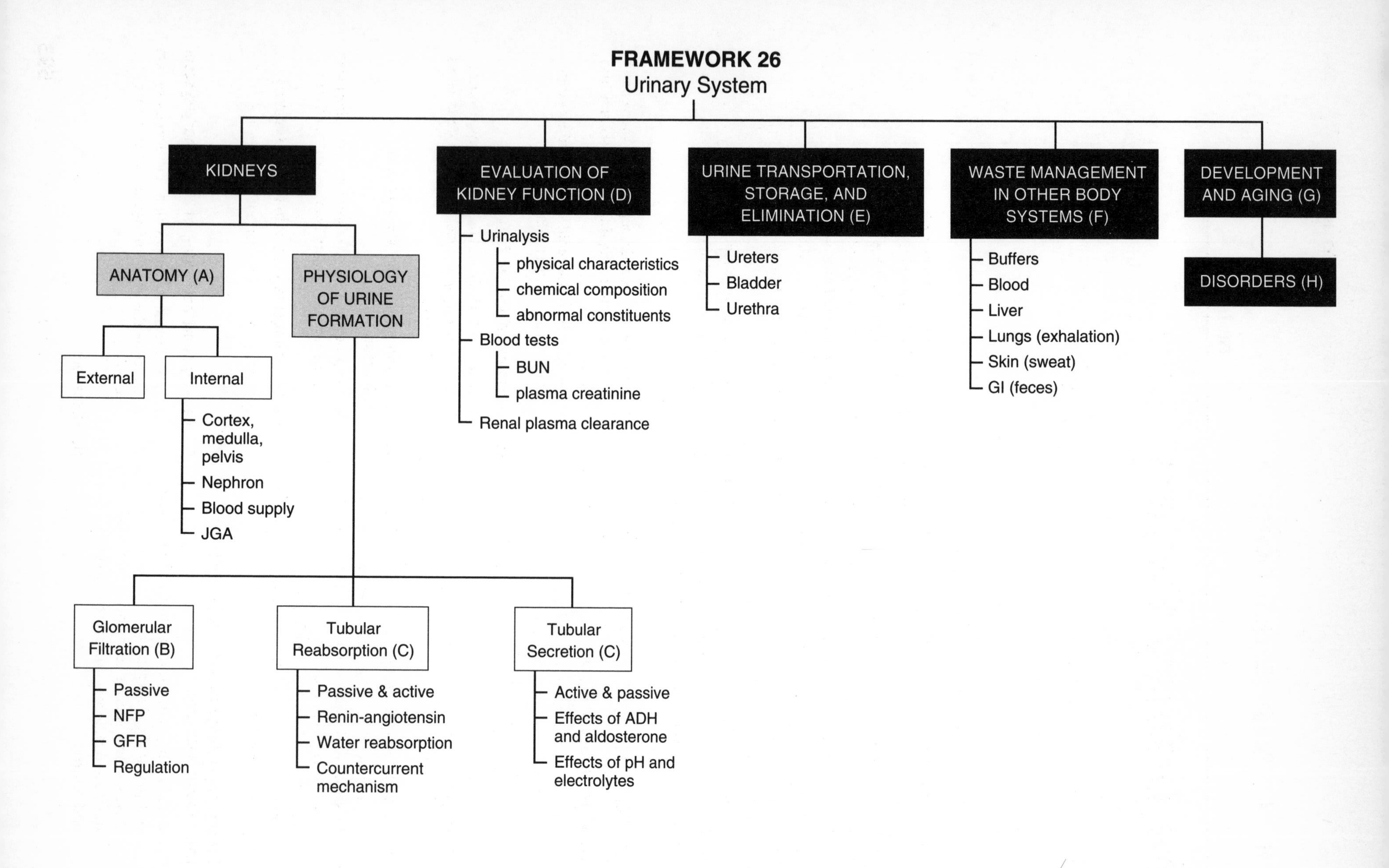

FRAMEWORK 26
Urinary System
KIDNEYS
ANATOMY (A)
External
Internal
Cortex, medulla, pelvis
Nephron
Blood supply
JGA
PHYSIOLOGY OF URINE FORMATION
Glomerular Filtration (B)
Passive
NFP
GFR
Regulation
Tubular Reabsorption (C)
Passive & active
Renin-angiotensin
Water reabsorption
Countercurrent mechanism
Tubular Secretion (C)
Active & passive
Effects of ADH and aldosterone
Effects of pH and electrolytes
EVALUATION OF KIDNEY FUNCTION (D)
Urinalysis
physical characteristics
chemical composition
abnormal constituents
Blood tests
BUN
plasma creatinine
Renal plasma clearance
URINE TRANSPORTATION, STORAGE, AND ELIMINATION (E)
Ureters
Bladder
Urethra
WASTE MANAGEMENT IN OTHER BODY SYSTEMS (F)
Buffers
Blood
Liver
Lungs (exhalation)
Skin (sweat)
GI (feces)
DEVELOPMENT AND AGING (G)
DISORDERS (H)

C H A P T E R **26**

The Urinary System

Wastes like urea or hydrogen ions (H^+ are products of metabolism that must be removed or they will lead to deleterious effects. The urinary system works with lungs, skin, and the GI tract to eliminate wastes. Kidneys are designed to filter blood and selectively determine which chemicals stay in blood or exit in the urine. The process of urine formation in the kidney requires an intricate balance of factors such as blood pressure and hormones (ADH and aldosterone). Examination of urine (urinalysis) can provide information about the status of the blood and possible kidney malfunction. A healthy urinary system requires an exit route for urine through ureters, bladder, and urethra.

Start your study of the urinary system with a look at the Chapter 26 Framework and familiarize yourself with the key concepts and terms. Examine the Chapter 26 Topic Outline and Objectives. Check off each one as you complete it.

TOPIC OUTLINE AND OBJECTIVES

A. Overview; kidney anatomy

1. List the functions of the kidneys.
2. Describe the external and internal gross anatomical features of the kidneys.
3. Trace the path of blood flow through the kidneys.
4. Describe the structure of renal corpuscles and renal tubules.

B. Kidney physiology: glomerular filtration

1. Identify the three basic tasks accomplished by nephrons and collecting ducts, and indicate where each task occurs.
2. Describe the filtration membrane.
3. Discuss the pressures that promote and oppose glomerular filtration.

C. Kidney physiology: tubular reabsorption, and tubular secretion

1. Describe the routes and mechanisms of tubular reabsorption and secretion.
2. Describe how specific segments of the renal tubule and collecting duct reabsorb water and solutes.
3. Describe how specific segments of the renal tubule and collecting duct secrete solutes into urine.
4. Describe how the renal tubule produces dilute and concentrated urine.

D. Evaluation of kidney function

1. Define urinalysis and describe its importance.
2. Define renal plasma clearance and describe its importance.

E. Urine transportation, storage, and elimination

1. Describe the anatomy, histology, and physiology of the ureters, urinary bladder, and urethra.

F. Waste management in other body systems

1. Describe the ways that body wastes are handled.

G. Development and aging of the urinary system

1. Describe the development of the urinary system.
2. Describe the effects of aging on the urinary system.

H. Disorders

WORDBYTES

Now study the following parts of words that may help you better understand terminology in this chapter.

Wordbyte	Self-check	Meaning	Example(s)
af-	______________	toward	*af*ferent
angi-	______________	blood vessel	mes*angi*al cells
anti-	______________	against	*anti*diuretic
azot-	______________	nitrogen-containing	*azot*emia
calyx	______________	cup	major *calyx*
cyst-	______________	bladder	*cyst*ostomy
densa	______________	dense	macula *densa*
ef-	______________	out	*ef*ferrent
-ferent	______________	carry	af*ferent*
-genic	______________	producing	myo*genic*
glomus	______________	ball	*glom*erular
insipid	______________	without taste	diabetes *insipid*us
juxta-	______________	next to	*juxta*glomerular
macula	______________	spot	*macula* densa
mes-	______________	middle	*mes*angial cells
mictur-	______________	urination	*micturi*tion reflex
myo-	______________	muscle	*myo*genic
nephro-	______________	kidney	*nephro*toxic drugs
podo-	______________	foot	*podo*cyte
recta	______________	straight	vasa *recta*
ren-	______________	kidney	*ren*al vein
retro	______________	behind	*retro*peritoneal
-ulus	______________	small	glomer*ulus*
urin-	______________	urinary	*urin*alysis

CHECKPOINTS

A. Overview; kidney anatomy (pages 949–959)

✓ **A1.** Select terms in the box that best label descriptions of different kidney functions designed to maintain homeostatic balance of blood and other body fluids.

pH	**pH**	**P.**	**Pressure (BP)**
E.	**Electrolyte concentration**	**V.**	**Volume**
G.	**Glucose**	**W.**	**Wastes**

_________ a. Regulation of Ca^+, Cl^-, K^+ and Na^+

_________ b. Mechanisms utilizing renin and changes in resistance of renal blood vessels

_________ c. Mechanisms utilizing ADH and aldosterone

_________ d. Alterations in elimination or conservation of H^+ or HCO_3^-

_________ e. Deamination of the amino acid glutamine which facilitates gluconeogenesis

_________ f. Removal of ammonia, urea, uric acid, and creatinine

✓ **A2.** Identify the organs that make up the urinary system on Figure LG 26.1 and answer the following questions about them.

a. The kidneys are located at about *(waist? hip?)* level, between _________ and _________ vertebrae. Each kidney is about _________ cm (_________ in.) long. Visualize kidney location and size on yourself.

b. The kidneys are in an extreme *(anterior? posterior?)* position in the abdomen. They are described as ____________________ because they are posterior to the peritoneum.

c. Identify the parts of the internal structure of the kidney on Figure LG 26.1. Then color all structures as indicated by color code ovals.

d. Define *nephroptosis*. Identify a population at increased risk for this condition, and describe problems that may result from nephroptosis.

✓ **A3.** Select terms that match each description related to kidneys. Choose from these terms. Not all terms will be used.

Adipose capsule	**Renal calyx**	**Renal fascia**
Nephrology	**Renal capsule**	**Renal papilla**
Parenchyma	**Renal columns**	**Renal pyramids**

____________________ a. Cuplike structure that drains urine from papillary ducts and empties in renal pelvis

____________________ b. Outermost layer of connective tissue that anchors kidneys to abdominal wall

____________________ c. Study of kidneys

____________________ d. Functional part of organ, such as kidney

____________________ e. Makes up most of renal medulla

____________________ f. Portion of renal cortex located between renal pyramids

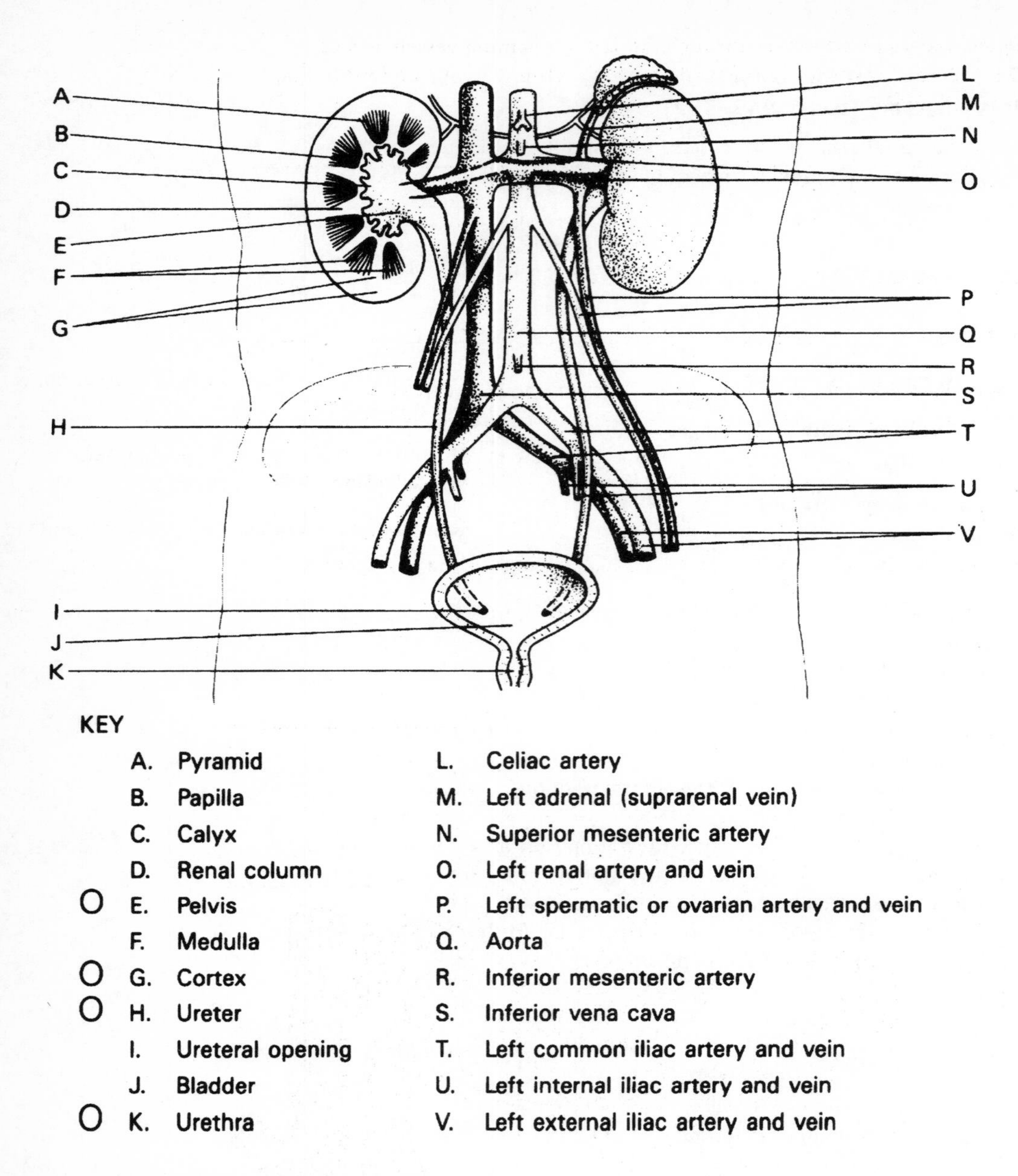

Figure LG 26.1 (Left side of diagram) Organs of the urinary system.
Structures A to G are parts of the kidney. Identify and color as directed in Checkpoint A2.
(Right side of diagram) Blood vessels of the abdomen. Color as directed in Chapter 21, Checkpoint E9, page 465 of the Learning Guide.

✓ **A4.** About what percentage of cardiac output passes through the kidneys each minute?

________% This amounts to about ________ mL/min.

✓ **A5.** Follow the pathway that blood takes through kidneys by naming vessels 1–8 on Figure LG 26.2. Some arrows are shown; add more to reinforce your understanding of the pathway of blood.

1. ____________________ 2. ____________________ 3. ____________________

4. ____________________ 5. ____________________ 6a. ____________________

6b. ____________________ 7. ____________________

Which of these vessels extends most deeply into the renal medulla? ______________

✓ **A6.** Explain what is unique about the blood supply of kidneys in this activity.

a. In virtually all tissues of the body, blood in capillaries flows into vessels named

____________________. However, in kidneys, blood in glomerular capillaries flows

directly into ____________________ ____________________.

b. Which vessel has a larger diameter? *(Afferent? Efferent?)* arteriole. State one effect of this difference.

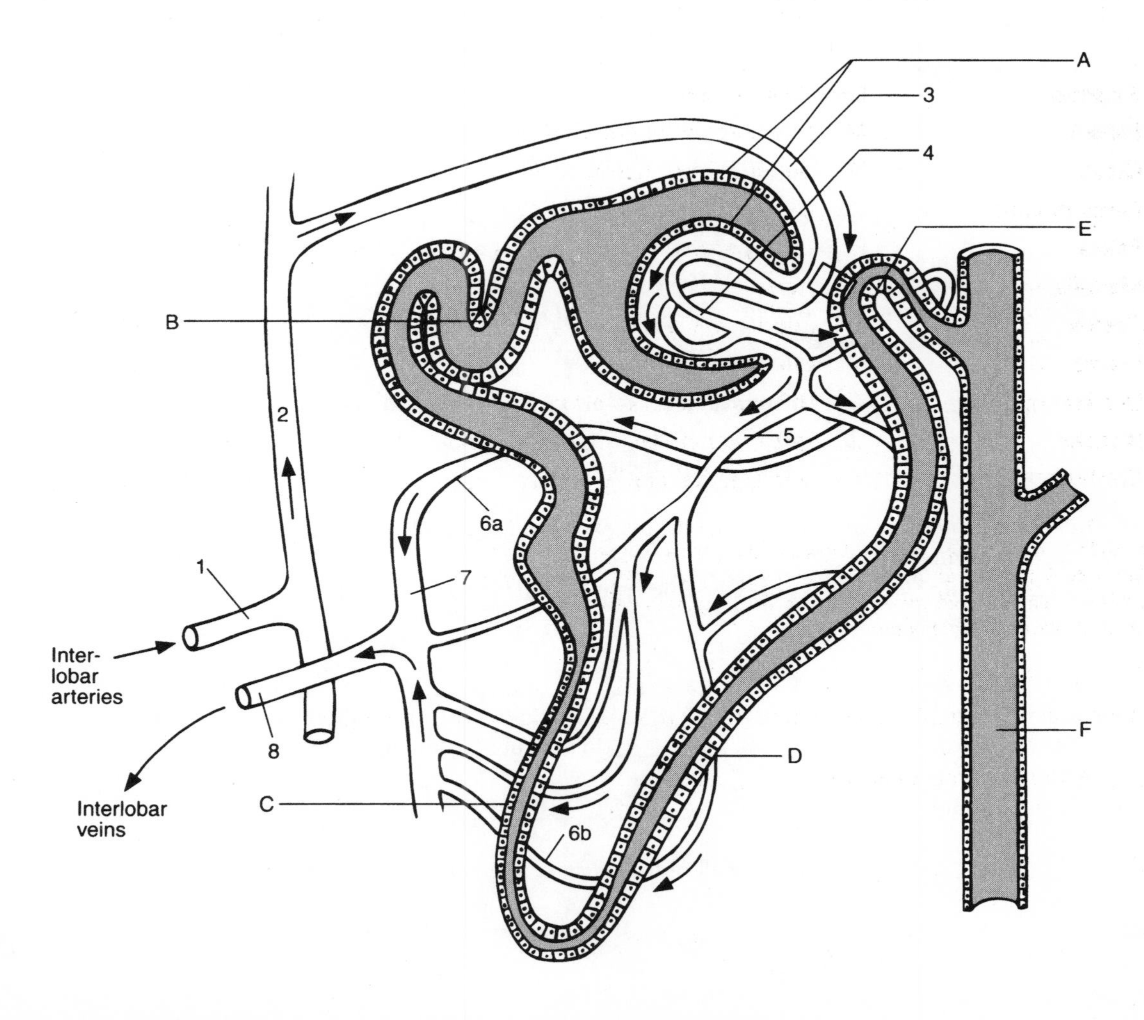

Figure LG 26.2 Diagram of a nephron. Letters refer to Checkpoints A7 and A9. Numbers refer to Checkpoints A5 and A7. The boxed area is used in Checkpoint A10.

✓ **A7.** Do this exercise about the functional unit of the kidney, a nephron.

a. On Figure LG 26.2, label parts A–E of the nephron, with letters arranged according to the flow of urine. Several structure E's empty into a single structure F, known as a

_____________________ _____________________.

b. Which one of those structures forms part of the renal corpuscle? _______________
The other part of the renal corpuscle is a cluster of capillaries known as a

_________________ and numbered *(4? 6b?)* in the figure.

✓ **A8.** Contrast two types of nephrons by writing *C* for cortical nephrons and *JM* for juxtamedullary nephrons.

_______ a. Also known as short-looped nephrons because these lie only in the renal cortex

_______ b. Constitute about 15–20% of nephrons

_______ c. Have a thin (squamous epithelial) layer in the initial portion of the ascending limb of loop of Henle

_______ d. Permit significant variations in concentration of urine

✓ **A9.** Do this exercise on microscopic structure of the nephron.

a. As water and other substances are "cleaned" (filtered) out of glomerular blood, they

collect in the _________________, which lies between _________________ and parietal layers of the capsule. The *(parietal? visceral?)* layer of the glomerular (Bowman's) capsule forms the outermost boundary of the renal capsule.

b. Identify the part of the renal tubule on the figure that consists of cuboidal epithelium with a brush border of many microvilli that increase the absorptive surface area.

_________________ Which part consists of simple squamous epithelium?

____

c. In which regions are *principal cells* located? *(B and C? E and F?)* These cells are sensitive to two hormones. Name them. _________________ and

_________________. The role of intercalated cells—also located in E and F—is to

regulate _________________. See page 970 of the text.

✓ **A10.** Describe the juxtaglomerular apparatus (JGA) in this exercise. Label JGA cells in the boxed structure in Figure LG 26.2

a. The JGA consists of *(two? four?)* types of cells. One type is located in the final portion of the *(ascending? descending?)*

limb of the loop of Henle. These cells, known as _______________ _______________

cells, monitor concentration of _________________ in the tubular lumen.

b. The second type of cell is a _________________ _________________

_________________ cell located in the wall of the *(afferent arteriole? glomerulus?)*.

These cells are known as _________________ cells; they secrete the chemical

known as _________________, which regulates _________________.

A11. Mrs. Amundson has undergone a right *nephrectomy*. How is her remaining kidney likely to compensate?

B. Kidney physiology: glomerular filtration (pages 959–964)

✓ **B1.** Do this exercise on urine formation.

a. List the three steps of urine formation.

____________________ ____________________ ____________________

b. Which of these steps involve passage of substances out of blood and into the "forming urine"? ____________________ ____________________ Which step involves selective return of valuable chemicals from "forming urine" back into blood?

____

c. What percentage of plasma entering nephrons is filtered out of glomerular blood and passed into the capsular space? About *(1%? 15–20%? 48–60%? 99%?)*. This percentage is known as the filtration ____________________. The actual amount of glomerular filtrate is about *(180? 60? 1–2?)* liters/day in males, and about ______ liters/day in females. In fact, about ______ times the total volume of blood plasma is filtered out of blood in the course of a day. So blood is "cleaned" by kidneys many times each day.

d. Of this filtrate, about ____________________ liters/day of fluid are reabsorbed in the "second step" of urine formation. This is about _______% of the glomerular filtrate. So the total amount of urine eliminated from kidneys daily is usually about _______ liters (or quarts).

✓ **B2.** Complete this Checkpoint on the filtration membrane.

a. Walls of glomerular capillaries are *(more? less?)* "leaky" than other capillaries of the body. State a functional advantage of this feature.

b. The three portions of the sandwichlike membrane consists of:

1. Glomerular capillary ____________________ cells with large pores called *(filtration slits? fenestrations?)* between these cells. The size of these pores prevents ____________________ and ____________________ from passing out of blood.
2. An acellular layer known as the ____________________ ____________________
3. Modified simple squamous epithelial cells known as *(pedicels? pococytes?)* that are part of the *(visceral? parietal?)* layer of the glomerular (Bowman's) capsule. Tiny "feet" called *(pedicels? pococytes?)* wrap around glomerular capillaries leaving spaces known as called *(filtration slits? fenestrations?)*. These spaces are small enough to *(permit? prohibit?)* filtration of most plasma proteins such as albumin out of blood.

c. The structure of the filtration membrane allows only certain substances to *pass freely* out of blood and into filtrate ("forming urine"). Circle those substances that do pass uninhibited into filtrate:

Albumin (and other large proteins)	**Glucose**	**Small proteins**
	Na^+ and other ions	**Urea**
Amino acids	**Platelets**	**Water**
Ammonia	**Red blood cells**	**White blood cells**

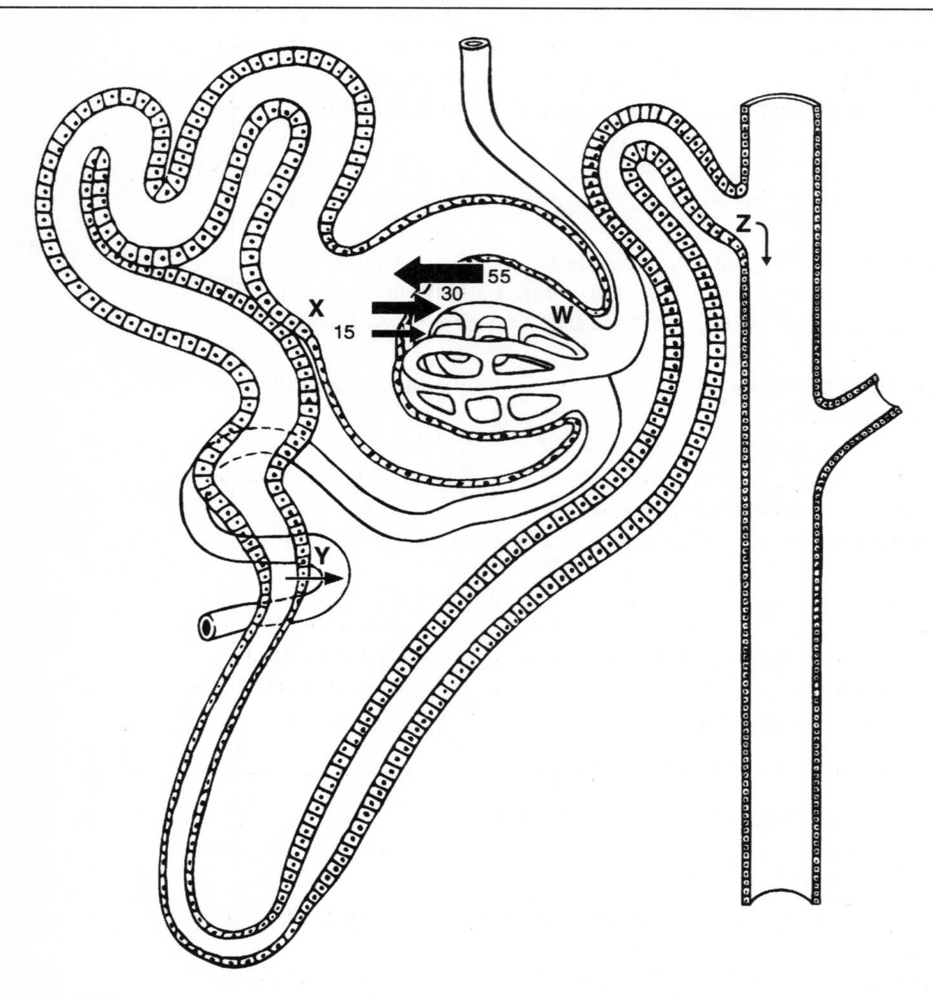

Figure LG 26.3 Diagram of a nephron with renal tubule abbreviated. Numbers refer to Checkpoint B4; letters refer to Checkpoints B3, C1, and C2.

✓ **B3.** Refer to Figure LG 26.3. Now describe the first step in urine production in this checkpoint. Glomerular filtration is a process of *(pushing? pulling?)* fluids and solutes out of

____________________ and into the fluid known as ____________________. The direction of glomerular filtration is shown by the thick arrow from W to X on the figure.

b. *(Most substances? Only a few types of substances?)* are forced out of blood in the process of glomerular filtration. Refer to Figure LG 26.3, area W (plasma in glomeruli) and area X (filtrate just beyond the capsule). During filtration all solutes are freely filtered

from blood (area W) to filtrate (area X) except ____________________. Why is so little protein filtered?

c. Blood pressure (or glomerular blood hydrostatic pressure or GBHP) in glomerular

capillaries is about _______ mm Hg. This value is *(higher? lower?)* than that in other capillaries of the body. This extra pressure is accounted for by the fact that the diameter of the efferent arteriole is *(larger? smaller?)* than that of the afferent arteriole. Picture three garden hoses connected to each other, representing the afferent arteriole, glomerular capillaries, and efferent arteriole. The third one (efferent arteriole) is extremely narrow; it creates such resistance that pressure builds up in the first two, fluids will be forced out of pores in the middle (glomerular) hose.

d. Describe three structural features of nephrons that enhance their filtration capacity.

e. Mesangial cells help to regulate glomerular filtration. Their name indicates their

location which is ____________________. When these cells *(contract? relax?)*, the

surface area of glomerular capillaries increases, leading to a(n) ________-crease in glomerular filtration rate (GFR).

✓ **B4.** Glomerular blood hydrostatic pressure (GBHP) is not the only force determining the amount of filtration occurring in glomeruli.

a. Name two other forces (shown on Figure 26.9 in your textbook).

____________________ ____________________

b. Normal values for these three forces are given on Figure LG 26.3. Write the formula for calculating net filtration pressure (NFP); draw an arrow beneath each term to show the direction of the force in Figure LG 26.3. Then calculate a normal NFP using the values given.

NFP = glomerular blood hydrostatic pressure – ________ – ________

Normal NFP = ________ mm Hg – ________ mm Hg – ________ mmHg

= ________ mmHg

Note that blood (hydrostatic) pressure pushes *(out of? into?)* glomerular blood and is largely counteracted by the other two forces.

c. Damage to glomerular capillaries caused by kidney disease allows protein to be lost

from blood into urine, a condition known as ____________________-uria. As a result,

BCOP will ________-crease and NFP will ________-crease. In addition, loss of plasma proteins causes increase of fluid in interstitial areas of the tissues, the condition

known as ____________________.

d. *For extra review* of calculation of NFP, look back at Chapter 21, Checkpoint A10 on LG page 448.

✓ **B5.** *A clinical correlation.* Show the effects of alterations of these pressures in a pathological situation. Determine NFP of the patient whose pressure values are shown below.

Glomerular blood hydrostatic pressure = 44
Blood colloid osmotic pressure = 30
Capsular filtrate hydrostatic pressure = 15

NFP = ____________________

Note which value(s) is(are) abnormal and suggest causes.

✓ **B6.** Describe GFR in this exercise.

a. To what do the letters GFR refer?

g ____________________ f ____________________ r ____________________

b. Fill in the blanks with normal values for GFR:

a. Females: _______ mL/min × 1440 min/day* = 151,200 mL/day = _______ liters/day**

b. Males: _______ mL/min × 1440 min/day = _______ liters/day

*1440 min/day = 60 minutes/hour × 24 hours/day

**rounded to 150 liters/day, on page 960 in your text

c. *A clinical correlation.* Describe several causes of altered GFR. GFR is likely to

decrease if blood pressure in kidneys _______-creases, for example, during stress

when sympathetic nerves decrease the diameter of the _______-ferent arteriole. Imagine that the first of the three garden hoses in Checkpoint B3c is narrowed, so *(more? less?)* blood flows into the glomerulus (middle hose). Either enlarged prostate gland or

kidney ____________________ lodged in the ureter back up urine into renal tubules of kidneys, opposing glomerular blood pressure, and this will decrease GFR also.

As indicated in Checkpoint B4c, GFR is likely to _______-crease in the presence of edema caused by proteinuria.

d. What consequences may be expected to accompany an abnormally low GFR?

✓ **B7.** List three categories of mechanisms that control GFR and ultimately help to regulate blood pressure in the body:

____________________ ____________________ ____________________

One autoregulation mechanism involves vasoconstriction of afferent arterioles

(leading to _______-creased GFR) in response to stretching of these arterioles due to renal blood flow accompanying high blood pressure. This means of protecting kidneys from overwhelming blood flow is called the *(myogenic? tubuloglomerular feedback?)* mechanism.

✓ **B8.** *A clinical correlation.* Mrs. K's blood pressure has dropped slightly. Explain how another renal autoregulation mechanism functions to ensure adequate renal blood flow.

a. When Mrs. K's systemic blood pressure decreases, her NFP and GFR are likely to *(increase? decrease?)*, causing filtrate to move more *(rapidly? slowly?)* through the renal tubules. As a result, *(more? less?)* fluid and salts are resorbed in the proximal convoluted tubules and loop of Henle.

b. Consequently, the amount of fluid slowly passing the macula densa is *(high? low?)* in volume and concentration of salts. This condition causes JGA cells to produce *(more? less?)* nitric oxide (NO), a vasodilator. As afferent arterioles then dilate, *(more? less?)* blood flows into glomerular capillaries, causing NFP and GFR to *(increase? decrease?)*.

c. This restoration of homeostasis by renal autoregulation occurs by a *(positive? negative?)* feedback mechanism. The name autoregulation indicates that neither the autonomic nervous system nor any chemical made outside the kidneys is involved. The control *(is? is not?)* intrinsic to kidneys.

✓ **B9.** During stress the *(sympathetic? parasympathetic?)* nerves to kidneys tend to cause vasoconstriction of the *(afferent? efferent?)* arteriole. (Look back at Checkpoint B6c.)

The resulting _______-crease in GFR leads to _______-creased urine output. This

helps to _______-crease blood volume and blood pressure, providing blood to brain, heart, and skeletal muscle for "fight-or-flight" responses.

✓ **B10.** Contrast hormonal effects on GFR in this Checkpoint.

a. Angiotensin II _______-creases GFR by *vaso-(constricting? dilating?)* both afferent and efferent arterioles.

b. ANP, secreted by the ________________ when blood volume there _______-creases,

relaxes glomerular ________________ cells, thereby ______-creasing the surface

area of glomerular capillaries. ANP therefore _______-creases GFR and urinary

output. This negative feedback mechanism ultimately _______-creases blood volume returning to the heart.

C. Kidney physiology: tubular reabsorption, and tubular secretion (pages 965–977)

✓ **C1.** Refer to Figure LG 26.3 and do this exercise about formation of urine.

a. We have already seen that glomerular filtration results in movement of substances from *(blood to filtrate/forming urine? filtrate/forming urine to blood?)*, as shown by the arrow between W and X. If this were the only step in urine formation, all the substances in the filtrate would leave the body in urine. Note from Table 26.3 in your

text that the body would produce _______ liters of urine each day, and it would contain many valuable substances. Obviously some of these "good" substances must be drawn back into blood and saved.

b. Recall that blood in most capillaries of the body flows into vessels named

___________________, but blood in glomerular capillaries flows into

___________________ and then into ___________________. This unique arrangement permits blood to recapture some of the substances indiscriminately pushed out during filtration. This occurs during the second step of urine formation, called

___________________. This process moves substances in the *(same? opposite?)* direction as glomerular filtration, as shown by the direction of the arrow at area Y of Figure LG 26.3. The *(proximal? distal?)* convoluted tubule is the site of most reabsorption.

✓ **C2.** Refer to Table 26.3 on page 965 in your text, and notice the two columns on the far right. Write "Y" above REABSORBED: RETURNED TO BLOOD and "Z" above URINE EXCRETED. Then refer to related areas Y and Z on Figure LG 26.3. Discuss the results of tubular reabsorption in this learning activity. (*Hint:* cross reference with Figure 26.20, page 976 in the text also.)

a. Which solutes are 100% reabsorbed into area Y so that virtually none remains in urine (area Z)?

b. About what percentage of water that is filtered out of blood is reabsorbed back into

blood? _______% Identify two ions that are about 99% reabsorbed. _______ _______

c. Which ion is 100% reabsorbed but still shows up in urine because some of this ion is

secreted (step 3 in urine formation) from blood to urine? _______

d. *(Urea? Uric acid?)* is about 90% reabsorbed, whereas ___________________ is about 45% reabsorbed.

e. Identify the solute that is filtered from blood and not reabsorbed.

___________________ State the clinical significance of this fact.

f. Tubular reabsorption is a *(nonselective? discriminating?)* process that enables the body to save valuable nutrients, ions, and water.

✓ **C3.** Refer to Figure 26.11 (page 966) and do this exercise on reabsorption routes.

a. Reabsorption occurs by two possible routes: *between* renal tubule cells, the method

known as ________-cellular reabsorption, or *across* tubule cells, the process called

________-cellular reabsorption.

b. Identify the apical and basolateral portions of renal tubule cells on Figure 26.11. The apical portion of these cells faces the *(lumen containing "forming urine"? peritubular capillaries?)*, whereas the bases of cells lie closer to the *(lumen containing "forming urine"? peritubular capillaries?)*. Now arrange in correct sequence the steps of

reabsorption of Na^+ in the proximal convoluted tubules: _____ _____ _____

A. Sodium pumps actively expel Na^+ from basolateral portions of tubule cells, a process utilizing about large amounts of ATP (See Checkpoint C4).

B. Na^+ passively diffuses from negatively charged fluid within the lumen into apical portions of tubular cells.

C. Na^+ diffuses into pericapillary blood.

Reabsorption of Na^+ is extremely important since more Na^+ is filtered from glomeruli

than any other substance except ____________________, and Na^+ reabsorption is closely linked to *(facultative? obligatory?)* water reabsorption—which normally

accounts for about _______% of water reabsorption.

C4. *For extra review.* Describe reabsorption mechanisms in this exercise.

a. About _______% of total resting metabolic energy is used for reabsorption of Na^+ in renal tubules by *(primary? secondary?)* active transport mechanisms. Explain how active transport of Na^+ also promotes osmosis of water and diffusion of substances such as K^+, Cl^-, HCO_3^-, and urea.

b. Reabsorption of glucose, amino acids, and lactic acid occurs almost completely in the *(proximal? distal?)* convoluted tubule, and the process utilizes Na^+ *(symporters? antiporters?)*. This mechanism is an example of *(primary? secondary?)* active transport because it is the sodium pump, not the symporter, that uses ATP. If blood concentra-

tion of a substance such as glucose is abnormally high, the renal ____________________ is surpassed. In other words, the blood level of glucose is greater than the T_m,

(or ____________________ ____________________).

✓ **C5.** *A clinical correlation.* Eileen and Marlene are both "spilling sugar" in urine; that is, they have ________________-uria. Discuss possible causes of glycosuria in each case, considering their blood sugar levels.

a. Eileen: glucose of 340 mg/100 mL plasma

b. Margaret: glucose of 80 mg/100 mL plasma

✓ **C6.** Identify the part of the renal tubule that best fits the description. Use these answers. (*Hint:* Refer to text Figure 26.20, page 976.)

A.	**Ascending limb of the loop of Henle**	**DCT.**	**Distal convoluted tubule**
CD.	**Collecting duct**	**PCT.**	**Proximal convoluted tubule**
D.	**Descending limb of the loop of Henle**		

________ a. This portion of the renal tubule with a prominent brush border permits tubular reabsorption of most filtered water and almost 100% of filtered nutrients (such as glucose).

________ b. This is the first site where osmosis of water is not necessarily coupled to reabsorption of other filtrates.

________ c. Little or no water is reabsorbed here because these cells are impermeable to water.

________ d. Variable amounts of Na^+ and water may be reabsorbed here under the influence of aldosterone and ADH. (two answers)

________ e. Absorption of almost all HCO_3^- and at least half of Na^+, K^+, Cl^-, and urea occurs here.

✓ **C7.** The third step in urine production is ____________________. It involves movement of substances from *(blood to urine? urine to blood?)*. In other words, tubular secretion is movement of substances in *(the same? opposite?)* direction as movement occurring in filtration. List four or more substances that are secreted by the process of tubular secretion.

✓ **C8.** Explain how tubular secretion helps to control pH. Refer to figures in the text and do this exercise.

a. Jenny s respiratory rate is slow, and her arterial pH is 7.33. Her blood therefore contains high levels of ________. Her blood is therefore somewhat *(acidic? alkaline?)*. Explain why.

b. Jenny's kidneys can assist in restoring normal pH (Figure 26.13a, b). The presence of acid (H^+) stimulates the PCT cells to eliminate H^+. In order for this to happen, another cation present in the tubular lumen must exchange places with H^+. What ion enters kidney tubule cells? ________ It joins HCO_3^- to form ____________________. This process requires *(antiporters? symporters?)*.

✓ **C9.** Do this exercise on regulation of blood level of potassium ion (K^+).

a. What percentage of filtered K^+ is normally reabsorbed in the PCT, loop of Henle, and DCT? (*Hint:* Refer to Table 26.3 in the text) ______%

b. Explain why kidneys might need to secrete variable amounts of K^+ from blood to urine.

c. When kidney reabsorption of the cation Na^+ increases, secretion of K^+ (also a cation) ______-creases. Aldosterone *(enhances? inhibits?)* such processes, causing lowered blood K^+ levels. On the other hand, aldosterone deficiency may lead to *(low? high?)* blood levels of K^+, and even cardiac arrest.

✓ **C10.** Do this exercise on hormonal regulation of urine volume and composition. (See also Checkpoint B10.)

a. Aldosterone and ADH are hormones that act on ____________________ cells located in *(PCT and loop of Henle? last part of DCT as well as collecting ducts?)*. Aldosterone is made by the *(adrenal medulla? adrenal cortex? hypothalamus?)*. This hormone stimulates production of a protein that helps pump *(Na^+? K^+?)* and also water from urine to blood. So aldosterone tends to *(in? de?)*-crease blood pressure.

b. ADH causes principal cells to become *(more? less?)* permeable to water by insertion of a water channel protein known as ____________________ into plasma membrane of these cells. As a result, reabsorption of water ______-creases, leading to ______-crease of blood pressure.

c. ADH is made by the __________________ and released from the *(anterior? posterior?)* pituitary. It is regulated by a *(positive? negative?)* feedback mechanism because

decreased body fluid leads to _____-crease of ADH release and retention of body fluid.

d. Diabetes insipidus involves _____-crease of ADH and _____-crease of urinary output.

✓ **C11.** Now describe chemicals involved in other mechanisms that regulate GFR by completing this checkpoint. Select the chemical that matches each description. Use the following list of answers. (It may help to refer to Figure LG 18.5, page 372 in the *Learning Guide.*)

ACE.	**Angiotensin converting enzyme**	**Ang II.**	**Angiotensin II**
ADH.	**Antidiuretic hormone**	**ANP.**	**Atrial natriuretic peptide**
Ald.	**Aldosterone**	**Epi.**	**Epinephrine**
Ang I.	**Angiotensin I**	**R.**	**Renin**

_______ a. A vasoconstrictor that also stimulates thirst and release of aldosterone and ADH

_______ b. Produced by adrenal cortex, it causes collecting ducts to reabsorb more Na^+ and water

_______ c. Released from the posterior pituitary, this hormone also increases blood volume

_______ d. This hormone increases GFR; it opposes actions of renin, aldosterone, and ADH and suppresses secretion of those chemicals

_______ e. Released under sympathetic stimulation; leads to arteriolar vasoconstriction and decreased GFR

_______ f. Of all chemicals in the above list, the one that tends to decrease blood pressure

✓ **C12.** State an example of a situation in which your body needs to produce dilute urine in order to maintain homeostasis.

Now describe how your body can accomplish this by doing the following learning activity.

a. To form dilute urine, kidney tubules must reabsorb *(more? fewer?)* solutes than usual. Two factors facilitate this. One is permeability of the ascending limb to Na^+, K^+, and Cl^- *(and also? but not?)* to water. Thus solutes enter interstitium and then capillaries, but water stays in urine.

b. The second requirement for dilute urine is a *(high? low?)* level of ADH. The function

of ADH is to _____-crease permeability of distal and collecting tubules to water. Less ADH forces more water to stay in urine, leading to a *(hyper? hypo?)*-osmotic urine.

✓ **C13.** State two reasons why it might be advantageous for the body to produce a concentrated urine.

Now refer to Figure LG 26.4 and describe mechanisms for concentrating urine in this exercise.

a. To help you clearly differentiate microscopic parts of the kidney, first color yellow the renal tubules (containing urine that is forming), color red the capillaries (containing blood), and color green all remaining areas (the interstitium). (Recall that kidneys are composed of over one million nephrons. The interstitium is all of the matrix between nephrons.) Also draw arrows to show route of "forming urine" within the renal tubule.

b. Now find [1] on the figure. This marks the endpoint of the figure where concentrated urine exits from the nephron. Such urine is known as *(hypo? hyper?)*-tonic (or ____________________-osmotic) urine. To produce a small amount of urine that is highly concentrated in wastes, the rate of reabsorption must _______-crease. Therefore more fluid from the urine forming within the lumen of the collecting duct will pass back into blood. (See [3] on the figure.)

c. This leads to the key question of how to increase tubular reabsorption. The answer lies in establishment of a solute gradient with the highest concentration in the *(outer? inner?)* medullary interstitium (IMI), around the inner portions of the long-loop juxtamedullary tubules. Solute concentration in this region may reach a maximum of *(300? 600? 1200?)* mOsm, as compared to only _______ mOsm in the cortical and outer medullary interstitium. (Note these values on the figure.)

d. Because the final segments of the collecting ducts reside in the inner medullary interstitium (IMI) region, urine in these ducts is exposed to this significantly hypertonic environment, shown at [2] on the figure. Much like a desert (or a dry sponge or blotter) surrounding a small stream of fluid, this "dry" (hypertonic) interstitium acts to draw water out of the urine. Ultimately, the fluid in the interstitium will pass into ____________________ blood ([3] on the figure).

e. So how can this hypertonic interstitium be developed? Two mechanisms are involved. The first depends on the relative ____________________ of different portions of the renal tubule to water and to several solutes. The second is the ____________________ mechanism (discussed in Checkpoint C14).

f. Consider the first mechanism. On Figure LG 26.4, find the thick portion of the tubule (next to the [4]). This is part of the *(ascending? descending?)* tubule. The thick *(squamous? columnar?)* epithelium here causes this region to be *(permeable? impermeable?)* to water in the absence of ADH. However, ions such as Na^+, ________ and ________ can be moved (by _______-porters) from the tubule into interstitium and then to blood here. Show this by writing names of these chemicals in circles on the figure. (Note that water is forced to stay inside the ascending limb.)

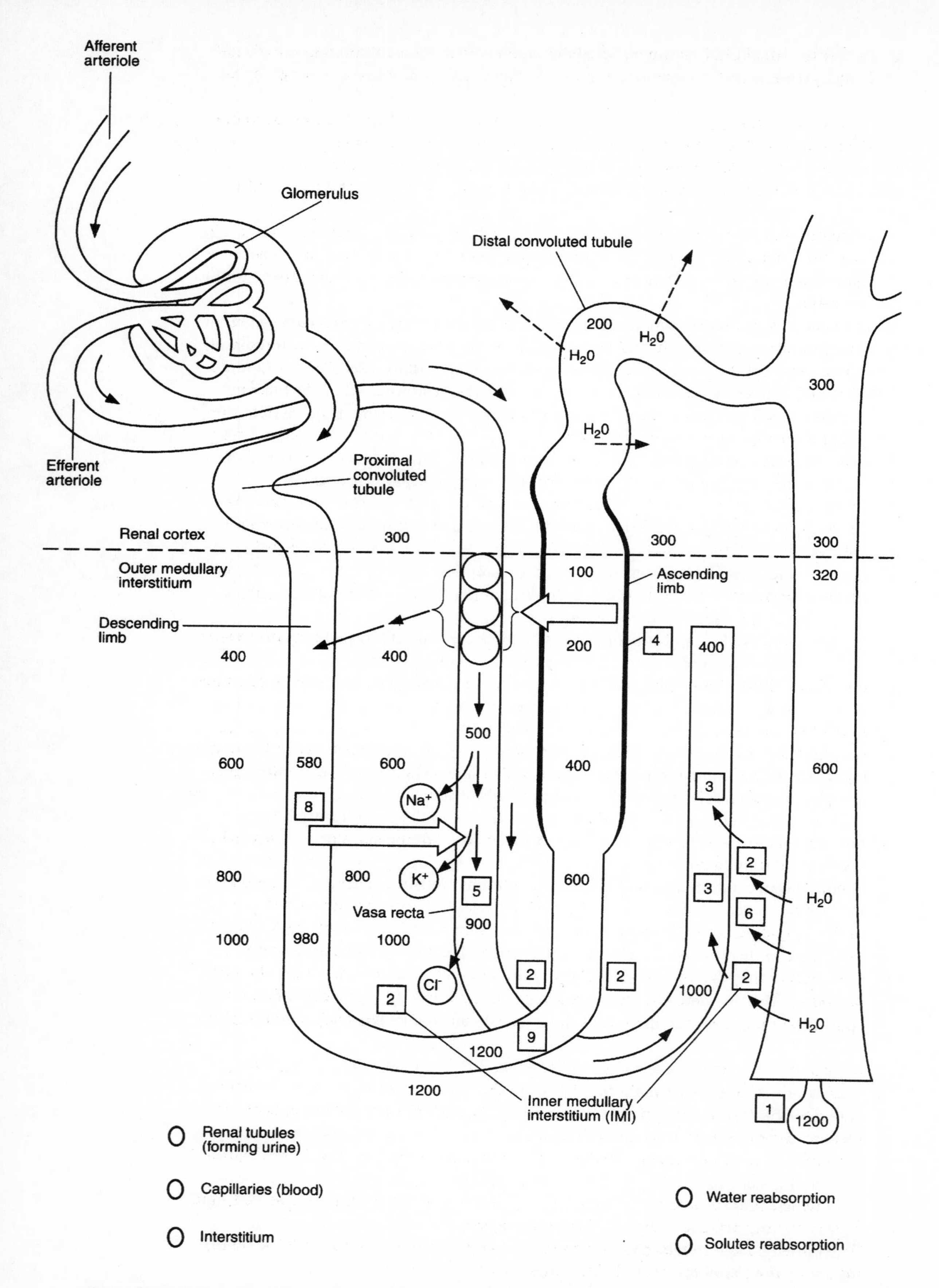

Figure LG 26.4 The concentration of urine in the nephron. Numbers refer to checkpoint C13. Color as directed in Checkpoints C13 and C14.

g. As a result, the fluid in the outer medullary and cortical interstitium becomes *(more? less?)* concentrated (or hypertonic). Many of these ions then move into long capillaries known as vasa ____________________ ([5] on the figure), carrying ions deep into the inner medullary interstitium (IMI) ([2]) and increasing hypertonicity.

h. As the "desert" (IMI) exerts an osmotic (pulling) force, water is drawn out of urine into IMI and then into capillary blood ([2] and [3] on the figure). Of course, this mechanism depends on the hormone ________, which permits normally impermeable collecting tubule cells to develop water pores (aquaporin-2) and therefore to become more water permeable. By itself this mechanism is effective in increasing urine concentration.

i. But urine concentration can be enhanced even further by the role played by urea. Concentration of urea within these tubules is increased as water moves out of the tubule lumen ([2] on the figure). Because this final portion of the collecting tubule is *(permeable? impermeable?)* to urea, this solute can diffuse out into the IMI, further contributing to the "desert-like" hypertonicity. Show this by writing "urea" near the [6]. The urea factor increases renal capability for water reabsorption and concentration of urine even more.

j. In summary, we have discussed a urine-concentrating mechanism based on differences of specific regions of renal tubule in permeability to water and urea. The resulting concentration gradient consists of a relatively hypotonic renal cortex or outer medullary region (300 mOsm) with an inner medullary interstitium (IMI) highly concentrated in the waste ____________________, as well as ions ________ ________, and ________. This IMI "desert" (or blotter) is ready to pull (osmose) water out of urine. But this can happen only when the hormone ________ is released as the signal that the body needs to "save fluid." ADH then causes ____________________ cells in collecting ducts to become more permeable, allowing water to leave them and move into IMI and then into capillary blood. The final outcome (at [1] on the figure) is a *(large? small?)* volume of *(dilute? concentrated?)* urine.

✓ **C14.** Refer to Figure LG 26.4 and check your understanding of the countercurrent mechanism in this checkpoint.

a. The term *countercurrent* derives from the fact that the "forming urine" in the descending limbs of renal tubules runs parallel and in the *(same? opposite?)* direction(s) to that in the ascending limb. This is also true of blood flow in the comparable regions of the vasa recta. Draw arrows in the vasa recta to show direction of blood flow there.

b. This mechanism, like that discussed in Checkpoint C13 leads to production of *(dilute? concentrated?)* urine. The countercurrent mechanism *(also? does not?)* depend(s) on the concentration gradient in the interstitium. The *(ascending? descending?)* limb is highly permeable to water but not to solutes, whereas the ____________________ limb exhibits just the reverse situation: it is permeable to *(water? solutes?)* *(and? but not to?)* water. Show this on Figure LG 26.4 by coloring large, hollow arrows to indicate correct locations of water or solute reabsorption.

c. A consequence of movement of water out of the descending limb ([8] in the figure) is increased concentration of solutes (________ mOsm) in the renal tubule as that fluid rounds the bend at the "hairpin" ([9] in the figure).

d. Explain why urine becomes more and more dilute, finally falling to *(100? 300? 600?)* mOsm by the top of the ascending tubule.

e. But as urine flows into the DCT and collecting tubules, water will be pulled out (into the IMI "desert") at times when ADH is present. So in the presence of this hormone, it is possible to produce highly concentrated urine (_______ mOsm) that is *(two? four? ten?)* times the osmolality of blood plasma and glomerular filtrate (_______ mOsm).

f. The next time you finish a vigorous exercise session, pause for a moment to reflect upon the amazing intricacy of these mechanisms that allow you to limit your fluid loss in urine because you have just eliminated so much fluid from your ____________________ glands!

✓ **C15.** Do this exercise on actions of diuretics.

a. A chemical that mimics the action of ADH is called a(n) *(antidiuretic? diuretic?)*. Such a chemical _______-creases urinary output.

b. A medication that increases urinary output is called a(n) ____________________. Removal of extra body fluid in this way is likely to _______-crease blood pressure.

c. Most diuretics cause loss of *(K^+? Na^+?)* from blood plasma; therefore patients must be sure to ingest adequate food or a supplement containing this ion.

✓ **C16.** *For extra review,* match the names of diuretics with descriptions below.

Thi.	**Thiazides (Diurel)**
Alc.	**Alcohol**
Caf.	**Caffeine (as in coffee, tea, or cola)**
Fur.	**Furosemide (Lasix)**
Spir.	**Spironolactone (aldactone)**

_______ a. Unlike most diuretics, this diuretic is potassium-sparing because it inhibits action of aldosterone; therefore clients do not normally need to take K^+ supplements

_______ b. Inhibits secretion of ADH

_______ c. Promote NaCl (and water) loss from distal convoluted tubule (DCT)

_______ d. Naturally occurring diuretic that inhibits Na^+ (and water) reabsorption

_______ e. The most potent diuretic drugs; known as loop diuretics

C17. *A clinical correlation.* Mrs. Pressman is taking furosemide (Lasix) to lower her blood pressure. Answer the following questions.

a. Why is this diuretic called a *loop diuretic*?

b. How does the action of this diuretic lower blood pressure?

c. Mrs. Pressman's nurse reminds her to eat a banana or orange each day. Why?

d. Some time later Mrs. Pressman is given a prescription for spironolactone (Aldactone). Speculate as to why her doctor changed her prescription.

D. Evaluation of kidney function (pages 977–979)

✓ **D1.** Review factors that influence urine volume in this checkpoint.

a. *Blood pressure.* When blood pressure drops, ________________ cells of the kidney release ________________, which ultimately catalyzes formation of ________________. This substance *(increases? decreases?)* blood pressure in several ways: directly, because it serves as a *(vasoconstrictor? vasodilator?)*, and also stimulates retention of the chemicals _______ and _______; and indirectly, because it stimulates release of ________________ which causes retention of the same chemicals. As a result, blood volume and blood pressure will _______-crease.

b. *Blood concentration.* The hypothalamic hormone _______ helps *(conserve? eliminate?)* fluid. When body fluids are hypertonic, ADH is released, causing *(increased? decreased?)* tubular reabsorption of water back into blood.

c. *Temperature.* On hot days skin gives up *(large? small?)* volumes of fluid via sweat glands. As blood becomes more concentrated, the posterior pituitary responds by releasing *(more? less?)* ADH, so urine volume *(increases? decreases?)*.

d. *Diuretic medications* tend to ______-crease urinary output and ______-crease blood pressure. For extra review, return to Checkpoint C15.

D2. Describe the following characteristics of urine. Suggest causes of variations from the normal in each case.

a. Volume: ______ mL/day ________________________________

b. Color: ______________ ________________________________

c. Turbidity: ______________ ________________________________

d. Odor: ______________ ________________________________

e. pH: ______ ________________________________

f. Specific gravity: ______ ________________________________

D3. The following substances are not normally found in urine. Explain what the presence of each might indicate. (*Hint:* Refer to Table 26.6, page 978.)

a. Albumin

b. Bilirubin

c. Casts

D4. Contrast the following pairs of conditions. Describe what is present in urine in each case and possible causes.

a. *Glucosuria/acetonuria*

b. *Hematuria/pyuria*

c. *Vaginitis due to Candida albicans/vaginitis due to Trichomonas vaginalis*

✓ **D5.** *A clinical correlation.* Of the following factors circle the ones that are likely to increase above normal in a patient with kidney failure. State the rationale for your answer.

BUN Plasma creatinine GFR Renal plasma clearance of glucose

✓ **D6.** For a chemical to work effectively in renal clearance tests for evaluation of renal function, that chemical *(must? must not?)* be readily filtered and *(must? must not?)* be readily absorbed. Name a chemical often used for testing renal clearance

_______________.

D7. Compare and contrast *hemodialysis* and *continuous ambulatory peritoneal dialysis* (CAPD) as methods of cleansing the blood.

E. Urine transportation, storage, and elimination (pages 979–981)

✓ **E1.** Arrange in correct sequence the structures through which urine flows once it is formed in collecting ducts:

MajC.	**Major calyx**	**Ureter.**	**Ureter**
MinC.	**Minor calyx**	**Ureth.**	**Urethra**
PapD.	**Papillary duct**	**UrinB.**	**Urinary Bladder**
RenP.	**Renal pelvis**		

Collecting duct → _______________ → _______________ →

_______________ → _______________ → _______________ →

_______________ → _______________ → outside the body

✓ **E2.** Which organs in the list above (Checkpoint E1) are singular in the human body; in other words, only one of that organ is normally present in each person?

____

✓ **E3.** Refer to Figure LG 26.1, page LG 596, and complete the following exercise about the ureters.

a. Ureters connect _______________ to _______________. Ureters are about

_______ cm (_______ in.) long. What causes urine to flow through these tubes?

b. What prevents urine from moving (retrograde) back into the ureters during contraction of the urinary bladder?

c. Kidneys and ureters are located in the *(anterior? posterior?)* of the abdominopelvic cavity. They are said to be ________________-peritoneal.

d. Ureters enter the urinary bladder at two corners of the *(detrusor? trigone?)* muscle. The third corner marks the opening of the ________________.

e. Both the ureters and the urinary bladder are lined with ________________ epithelium. What advantage does this type of epithelium offer in these organs?

✓ **E4.** In the micturition reflex, *(sympathetic? parasympathetic?)* nerves stimulate the ________________ muscle of the urinary bladder and cause relaxation of the internal sphincter. What stimulus initiates the micturition reflex?

✓ **E5.** Refer to Figures LG 28.2 and 28.3 (pages 652 and 655 in your *Guide*), and complete the following exercise about the bladder and the urethra.

a. The urinary bladder is located in the *(abdomen? true pelvis?)*. Two sphincters lie just inferior to it. The *(internal? external?)* sphincter is under voluntary control.

b. Urine leaves the bladder through the ________________. In females the length of this tube is ______ cm (______ in.); in males it is about ______ cm (______ in.). Write in correct sequence (according to the pathway of urine) the three portions of the male urethra:

________________ → ________________ → ________________

In addition to urine, what other fluid passes through the male urethra?

____

✓ **E6.** Ureters, urinary bladder, and urethra are lined with ________________ membrane. What is the clinical significance of that fact?

E7. Contrast *urinary incontinence* and *retention*. List possible causes of each.

F. Waste management in other body systems (pages 981–982)

F1. Complete this list of body structures and tissues, besides urinary organs, that perform excretory functions. Then list the substances they eliminate or detoxify.

	Structure or tissue	Substances eliminated or made less toxic
1.	______________	Excess H^+ is bound ______________
2.	______________	______________
3.	Liver ______________	______________
4.	______________	CO_2, heat, and some water eliminated ______________
5.	Sweat glands ______________	______________
6.	GI tract (via anus) ______________	______________

✓ **F2.** Which of these "waste management" systems acts most like garbage trucks or sewer lines to transport waste to kidneys for removal there? ______________

G. Development and aging of the urinary system (page 982)

✓ **G1.** Do this exercise about urinary system development.

a. Kidneys form from ______________-derm, beginning at about the ______________ week of gestation.

b. Which develops first? *(Pro? Meso? Meta?)*-nephros? Which extends most superiorly in location? ________-nephros. Which one ultimately forms the kidney? ________-nephros.

c. The urinary bladder develops from the original ______________, which is derived from ______________-derm.

✓ **G2.** GFR is likely to ______-crease by about 50% between age 40 and 70. This change is a function of:

a. ______-creased renal blood flow (related to decreased cardiac output)

b. ______-creased kidney mass.

✓ **G3.** Two other common problems among older adults are inability to control voiding (______________) and increased frequency of ______________ (UTIs). Explain why UTIs may occur more often in elderly males.

Nocturia may also occur more among the aged population. Define *nocturia*.

H. Disorders, medical terminology (pages 985–986)

✓ **H1.** Kidney stones are known as renal ________________. Lithotripsy is a *(surgical? nonsurgical?)* technique for destroying kidney stones by high-frequency

________________ waves. Suggest two or more health practices that may help you to avoid formation of kidney stones.

✓ **H2.** List three or more signs or symptoms of urinary tract infections (UTIs).

H3. Discuss and contrast two types of renal failure in this learning activity.

a. *(Acute? chronic?)* kidney failure is an abrupt cessation (or almost) of kidney function. List several causes of acute renal failure (ARF).

b. *(Acute? chronic?)* renal failure is progressive and irreversible. Describe changes that occur during the three stages of CRF.

✓ **H4.** Match each term in the box with descriptions below.

A.	**Azotemia**	**E.**	**Enuresis**
C.	**Cystitis**	**GN.**	**Glomerulonephritis**
		PCD.	**Polycystic disease**

_______ a. Inherited disorder in which kidneys (and other organs such as liver or pancreas) develop multiple fluid-filled cavities

_______ b. Inflammation of the urinary bladder

_______ c. Bed wetting

_______ d. Inflammation of the kidney involving glomeruli; may follow strep infection

_______ e. Presence of nitrogen-containing chemicals such as creatinine or urea in blood

ANSWERS TO SELECTED CHECKPOINTS: CHAPTER 26

A1. (a) E. (b) P. (c) V. (d) pH. (e) G. (f) W.

A2. (a) Waist, T12, L2; 10-12 (4-5). (b) Posterior; retroperitoneal. (c) See KEY to Figure LG 26.1. (d) Drooping kidney which is most likely to occur in extremely thin persons. This condition can lead to kinked ureter with diminished urine flow, formation of kidney stones, and pain.

A3. (a) Calyx. (b) Renal fascia. (c) Nephrology. (d) Parenchyma. (e) Renal pyramids. (f) Renal columns.

A4. 20-25; 1200.

A5. 1, Arcuate artery; 2, interlobular artery; 3, afferent arteriole; 4, glomerulus; 5, efferent arteriole; 6a, peritubular capillaries; 6b, vasa recta; 7, interlobular vein; 8, arcuate vein. Vasa recta.

A6. (a) Venules; efferent arterioles. (b) Afferent; increases glomerular blood pressure. (See Checkpoint B3c.)

A7. (a) A, glomerular (Bowman's) capsule; B, proximal convoluted tubule; C, descending limb of the loop of Henle; D, ascending limb of the loop of Henle; E, distal convoluted tubule; collecting duct (F). (b) A, glomerular (Bowman's) capsule; glomerulus, 4.

A8. (a) C. (b–d) JM.

A9. (a) Capsular (Bowman's) space, visceral; parietal. (b) B (proximal convoluted tubule), C (descending limb of the loop of Henle), and the first part of D (ascending limb of the loop of Henle) of juxtaglomerular nephrons. (c) E and F; ADH and aldosterone; blood and urine pH.

A10. (a) Two; ascending; macula densa, NaCl. (b) Modified smooth muscle, afferent arteriole; juxtaglomerular (JG); renin, blood pressure. (See text Figure 26.6.)

B1. (a) Glomerular filtration, tubular reabsorption, tubular secretion. (b) Glomerular filtration and tubular secretion; tubular reabsorption. (c) 15–20%; fraction; 180; 150; 65. (d) 178–179 in males and 148–149 in females; 99; 1–2 (= 180 – 178 or 179).

B2. (a) More; blood passing through glomerular vessels in kidneys can be cleaned out. (b1) Endothelial, fenestrations, blood cells and platelets; (b2) Basal lamina; (b3) Podocytes, visceral; pedicels, filtration slits; prohibit. (c) Amino acids, ammonia, glucose, Na^+ and other ions, small proteins, urea, and water.

B3. (a) Pushing, blood, filtrate. (b) Most substances; protein molecules. Many proteins are too large to pass through healthy filtration membranes, and the basement membrane restricts passage of proteins. (c) 55; higher; smaller. (Refer to Checkpoint A6.) (d) Glomerular capillaries are long, thin, and porous and have high blood pressure. (e) Between afferent and efferent arterioles; relax, in.

B4. (a) Blood colloid osmotic pressure (BCOP) and capsular hydrostatic pressure (CHP). (b) NFP = GBHP – CHP – BCOP. NFP = 55 – 15 – 30; 10; out of. (d) Protein; de, in; edema.

B5. NFP = –1 mm Hg = 44 – 30 – 15. The NFP of –1 indicates that no net filtration would occur; no urine (anuria) would result. In fact some fluids would pass from filtrate back into blood due to the low glomerular blood pressure possibly related to hemorrhage.

B6. (a) Glomerular filtration rate. (b) 105, 151.2 (or 150); 125, 180. (c) De, af; less; stones; in. (d) Additional water and waste products (such as H^+ and urea) normally eliminated in urine are retained in blood, possibly resulting in hypertension and acidosis.

B7. Autoregulation, neural and hormonal mechanisms; de; myogenic.

B8. (a) Decrease, slowly; more. (b) Low; more; more, increase. (c) Negative; is.

B9. Sympathetic, afferent; de, de, in.

B10. (a) De, constricting. (b) Atria of the heart, in, mesangial, in; in; de.

C1. (a) Blood to filtrate/forming urine: 180. (b) Venules; efferent arterioles, capillaries: tubular reabsorption; opposite; proximal.

C2. (a) Glucose and bicarbonate (HCO_3^-). (b) 99; Na^+ and Cl^-. (c) K^+. (d) Uric acid, urea. (e) Creatinine; creatinine clearance is often used as a measure of glomerular filtration rate (GFR). One hundred percent of the creatinine (a breakdown product of muscle protein) filtered or "cleared" out of blood will show up in urine because 0% is reabsorbed. (f) Discriminating.

C3. (a) Para, trans. (b) Lumen containing "forming urine"; peritubular capillaries; B A C; water; obligatory, 90.

C5. Glycos. (a) Eileen's hyperglycemia could be caused by ingestion of excessive amounts of carbohydrates or by hormonal imbalance, such as deficiency in effective insulin (diabetes mellitus) or excess growth hormone. Eileen's glucose level of 340 mg/100 ml plasma exceeds the renal threshold of about 200 mg/100 ml plasma because kidneys have a limited transport maximum (T_m) for glucose. (b) Although Margaret's blood glucose is in the normal (70–110 mg/per 100 ml) range, she may have a rare kidney disorder with a reduced transport maximum (T_m) for glucose.

C6. (a) PCT. (b) D. (c) A. (d) DCT and CD. (e) PCT.

C7. Tubular secretion; blood to urine; the same; K^+, H^+, ammonium (NH_4^+), creatinine, penicillin, and para-aminohippuric acid.

C8. (a) CO_2; acidic; increasing levels of CO_2 tend to cause increase of H^+: $CO_2 + H_2O \rightarrow H_2CO_3 \rightarrow H^+ + HCO_3^-$. (Review content on transport of CO_2 in Chapter 23.) (b) Na^+; $NaHCO_3^-$ (sodium bicarbonate); antiporters.

C9. (a) 100. (b) Dietary K^+ deficit or loss of K^+ accompanying tissue trauma or use of certain medications (such as most diuretics) leads to need for increased plasma K^+ levels via reabsorption. Dietary K^+ excesses (such as some salt substitutes) require extra elimination (secretion) of K^+. (c) In; enhances; high.

C10. (a) Principal, last part of DCT as well as collecting ducts; adrenal cortex; Na^+ in. (b) More, aquaporin-2 in, in. (c) Hypothalamus, posterior; negative, in. (d) De, in.

C11. (a) Ang II. (b) Ald. (c) ADH. (d) ANP. (e) Epi. (f) ANP.

C12. During periods of high water intake or decreased sweating. (a) More; but not to. (b) Low; in; hypo.

C13. Water retention by kidneys raises blood pressure as plasma volume is increased and also helps prevent or reverses dehydration, for example, when drinking water is unavailable or when extra water loss occurs via sweating or fever. (a) Refer to Figure LG 26.4A. (b) Hyper, hyper; in. (c) Inner; 1200, 300. (d) Capillary. (e) Permeability; countercurrent. (f) Ascending; columnar, impermeable; K^+, Cl^-; sym; see Figure LG 26.4A. (g) More; recta. (h) ADH. (i) Permeable; see Figure LG 26.4A. (j) Urea, Na^+, K^+, Cl^-; ADH; principal; small, concentrated.

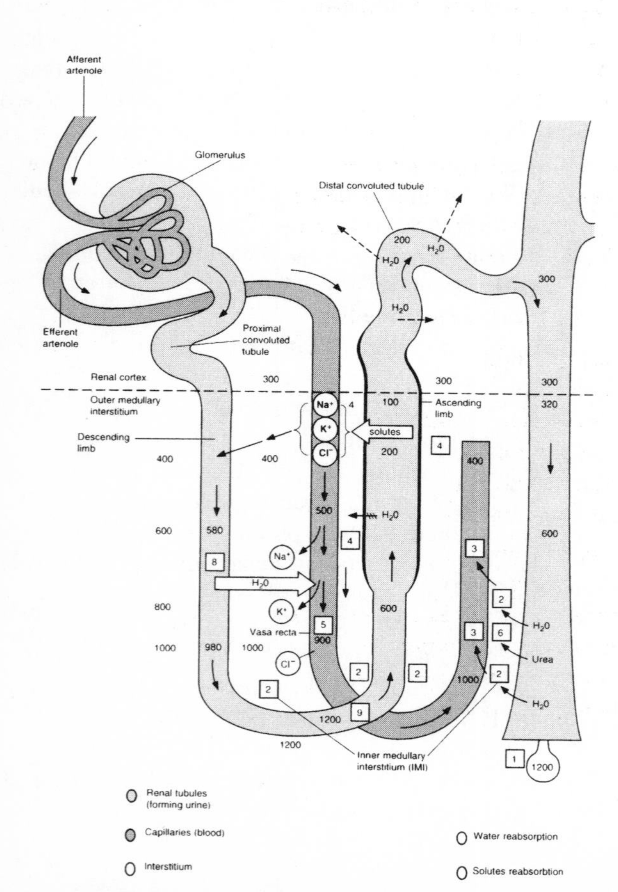

Figure LG 26.4A The concentration of urine in the nephron.

C14. (a) Opposite; see Figure LG 26.4A. (b) Concentrated; also; descending, ascending, solutes, but not to; see Figure LG 26.4A. (c) 1200. (d) 100; see Checkpoint C13f. (e) 1200, four, 300. (f) Sweat.

C15. (a) Antidiuretic; de. (b) Diuretic (or "water pill"); de. (c) K^+.

C16. (a) Spir. (b) Alc. (c) Thi. (d) Caf. (e) Fur.

D1. (a) Juxtaglomerular, renin, angiotensin II; increases, vasoconstrictor, Na^+, H_2O; aldosterone; in. (b) ADH, conserve; increased. (c) Large; more, decreases. (d) In, de.

D5. BUN and plasma creatinine both increase in blood because less urea and creatinine are filtered from blood. Because GFR decreases, kidneys "clear out" less glucose from blood (renal plasma clearance of glucose decreases), resulting in hyperglycemia.

D6. Must, must not; creatinine.

E1. PapD → MinC → MajC → RenP → Ureter → UrinB → Ureth.

E2. Urin B, Ureth.

E3. (a) Kidneys, urinary bladder; 25–30 (10–12); gravity, hydrostatic pressure, and peristaltic contractions. (b) No anatomical valve is present, but filling of the urinary bladder compresses ureteral openings, thus forcing urine through the open urethral sphincter. (c) Posterior; retro. (d) Trigone; urethra. (e) Transitional; the epithelium stretches and becomes thinner as these organs fill with urine.

E4. Parasympathetic, detrusor; stretching the walls of the bladder by more than 200–400 mL (about 0.8–1.6 cups) of urine.

E5. (a) True pelvis; external. (b) Uréthra; 4 (1.5); 15–20 (6–8); prostatic → membranous → spongy; semen.

E6. Mucous; microbes can spread infection from the exterior of the body (at urethral orifice) along mucosa to kidneys.

F2. #2 blood in the list on page 982.

G1. (a) Meso, third. (b) Pro; pro; meta. (c) Cloaca (urogenital sinus portion), endo.

G2. De. (a) De. (b) De.

G3. Incontinence of bladder; urinary tract infections; increased incidence of prostatic hypertrophy leads to urinary retention; excessive urination at night.

H1. Calculi; nonsurgical, shock; drink plenty of water (8–10 glasses/day) to keep urine dilute, avoid excessive calcium, and discuss with a dietitian and/or urologist a diet that minimizes extremes of urine pH.

H2. Burning with urination or painful urination (dysuria); frequency of urination due to irritation of the urinary tract; if the infection spreads up the ureters to kidneys (located at waist level), low back pain may result.

H4. (a) PCD. (b) C. (c) E. (d) GN. (e) A.

MORE CRITICAL THINKING: CHAPTER 26

1. Discuss the significance of the structure and arrangement of blood vessels within a nephron.
2. Describe how blood pressure is increased by mechanisms initiated by renin and aided by angiotensin I and II, aldosterone, and ADH.
3. Tamika, an accident victim, was admitted to the hospital 3 hours ago. Her chart indicates that she had been hemorrhaging at the scene of the accident. Tamika's nurse monitors her urinary output and notes that it is currently 12 mL/hr. (Normal output is 30–60 mL/hr.) Explain why a person in such severe stress is likely to have decreased urinary output (oliguria).
4. Describe several mechanisms that allow you to: (a) assure adequate blood flow into kidneys; (b) prevent loss of glucose and amino acids in your urine; and (c) produce more concentrated urine (conserving water) when you produce large amounts of sweat.
5. Who is more likely to be at higher risk for urinary tract infections (UTIs)? Explain why in each case. (a) 30-year-old women or men. (b) 30-year-old men or 70-year-old men.
6. Discuss signs and symptoms that are likely to be present in a person who has chronic renal failure (CRF).

MASTERY TEST: CHAPTER 26

Questions 1–6: Arrange the answers in correct sequence.

_____ _____ _____ 1. From superior to inferior:
A. Ureter
B. Bladder
C. Urethra

_____ _____ _____ _____ 2. From most superficial to deepest:
A. Renal capsule
B. Renal medulla
C. Renal cortex
D. Renal pelvis

_____ _____ _____ _____ _____ 3. Pathway of glomerular filtrate:
A. Ascending limb of loop of Henle
B. Descending limb of loop of Henle
C. Collecting tubule
D. Distal convoluted tubule
E. Proximal convoluted tubule

_____ _____ _____ _____ _____ 4. Pathway of blood:
A. Arcuate arteries
B. Interlobular arteries
C. Renal arteries
D. Afferent arteriole
E. Interlobar arteries

_____ _____ _____ _____ _____ 5. Pathway of blood:
A. Afferent arteriole
B. Peritubular arteries and vasa recta
C. Glomerular capillaries
D. Venules and veins
E. Efferent arteriole

_____ _____ _____ _____ _____ _____ 6. Order of events to restore low blood pressure to normal:
A. This substance acts as a vasoconstrictor and stimulates aldosterone.
B. Renin is released and splits off part of angiotensinogen to form angiotensin I.
C. Angiotensin I is converted into angiotensin II.
D. Decrease in blood pressure or sympathetic nerves stimulate juxtaglomerular cells.
E. Under influence of this hormone Na^+ and H_2O reabsorption occurs.
F. Increased blood volume raises blood pressure.

Questions 7–12: Circle the letter preceding the one best answer to each question.

7. Which of these is a normal constituent of urine?
A. Albumin
B. Urea
C. Glucose
D. Casts
E. Acetone

8. Which is a normal function of the bladder?
A. Oliguria
B. Nephrosis
C. Calculi formation
D. Micturition

9. Which parts of the nephron are composed of simple squamous epithelium?
A. Ascending and descending limbs of loop of Henle
B. Glomerular capsule (parietal layer) and descending limb of loop of Henle
C. Proximal and distal convoluted tubules
D. Distal convoluted and collecting tubules

10. Choose the one *false* statement.
A. An increase in glomerular blood hydrostatic pressure (GBHP) causes a decrease in net filtration pressure (NFP).
B. Glomerular capillaries have a higher blood pressure than other capillaries of the body.
C. Blood in glomerular capillaries flows into arterioles, not into venules.
D. Vasa recta pass blood from the efferent arteriole toward venules and veins.

11. Choose the one *true* statement about nephron structure and function.
A. Afferent arterioles offer more resistance to flow than do efferent arterioles.
B. An efferent arteriole normally has a larger diameter than an afferent arteriole.
C. The juxtaglomerular apparatus (JGA) consists of cells of the proximal convoluted tubule (PCT) and the afferent arteriole.
D. Most water reabsorption normally takes place across the proximal convoluted tubule (PCT).

12. Which one of the hormones listed here causes increased urinary output and natriuresis?
A. ANP
B. ADH
C. Aldosterone
D. Angiotensin II

Questions 13–20: Circle T (true) or F (false). If the statement is false, change the underlined word or phrase so that the statement is correct.

T F 13. Antidiuretic hormone (ADH) decreases permeability of distal and collecting tubules to water, thus increasing urine volume.

T F 14. A ureter is a tube that carries urine from the bladder to the outside of the body.

T F 15. Diabetes insipidus is a condition caused by a defect in production of insulin or lack of sensitivity to this hormone by principal cells in collecting ducts.

T F 16. A person taking diuretics is likely to urinate less than a person taking no medications.

T F 17. Blood colloid osmotic pressure (BCOP) is a force that tends to push substances from blood into filtrate.

T F 18. The thick portion of the ascending limb of the loop of the nephron is permeable to Na^+ and Cl^- but not to water.

T F 19. The countercurrent mechanism permits production of small volumes of concentrated urine.

T F 20. In infants 2 years old and under, urinary retention is normal.

Questions 21–25: Fill-ins. Complete each sentence with the word or phrase that best fits.

_______________ 21. The glomerular capsule and its enclosed glomerular capillaries constitute a ________.

_______________ 22. A ________ consists of a renal corpuscle and a renal tubule.

_______________ 23. When 90% of nephron function is lost, a person is said to be in ________.

_______________ 24. Sympathetic impulses cause greater constriction of ________ arterioles with resultant ______-crease in GFR and ______-crease in urinary output.

_______________ 25. A person with chronic renal failure is likely to be in acidosis because kidneys fail to excrete ________, may be anemic because kidneys do not produce ________, and may have symptoms of hypocalcemia because kidneys do not activate ________.

ANSWERS TO MASTERY TEST: CHAPTER 26

Arrange

1. A B C
2. A C B D
3. E B A D C
4. C E A B D
5. A C E B D
6. D B C A E F

Multiple Choice

7. B
8. D
9. B
10. A
11. D
12. A

True–False

13. F. Increases; decreasing
14. F. A kidney to the bladder
15. F. ADH
16. F. More
17. F. Pull substances from filtrate into blood
18. T
19. T
20. F. Incontinence

Fill-ins

21. Renal corpuscle
22. Nephron
23. End-stage renal failure
24. Afferent, de, de
25. H^+, erythropoietic factor, vitamin D

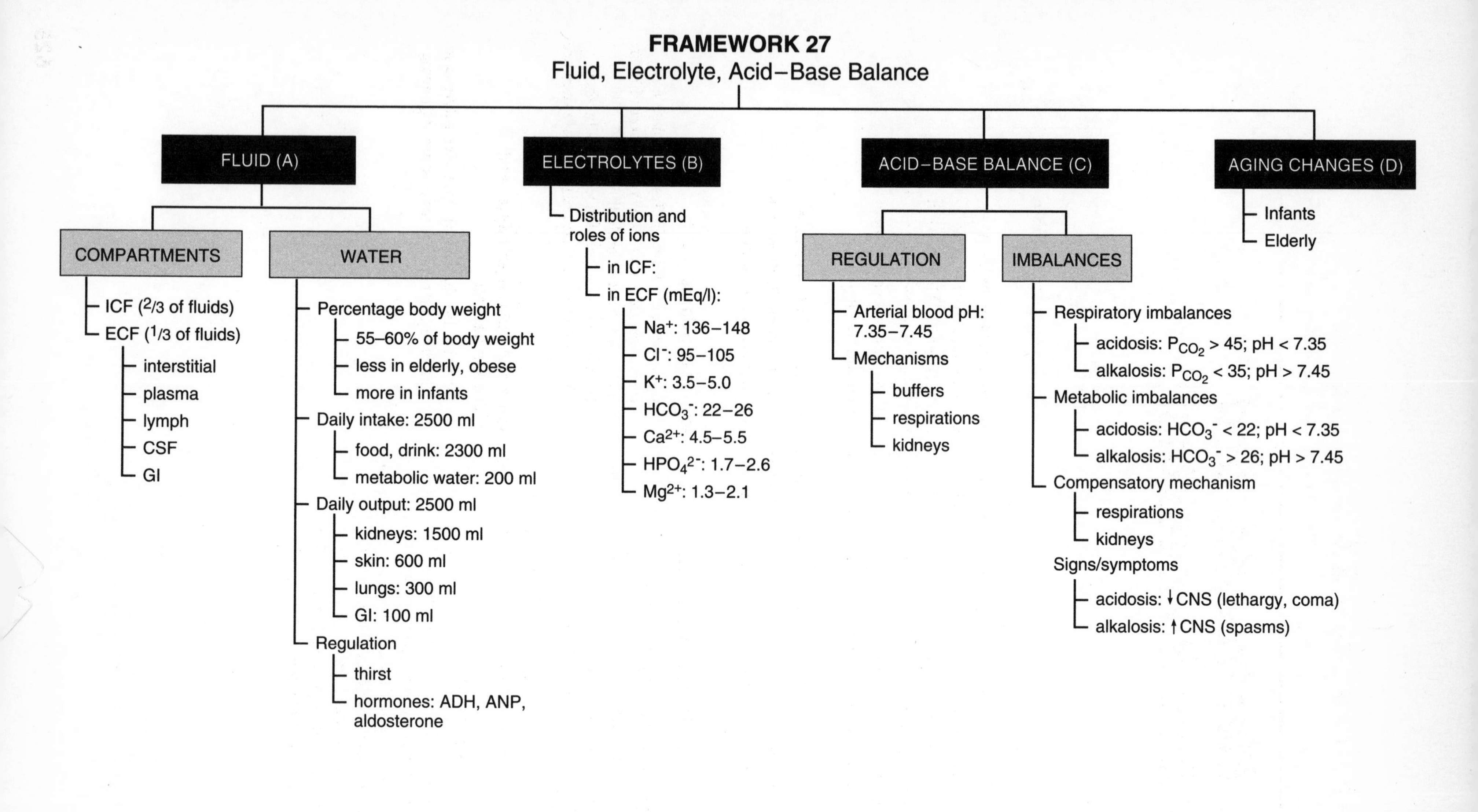
FRAMEWORK 27
Fluid, Electrolyte, Acid–Base Balance
FLUID (A)
COMPARTMENTS
ICF (2/3 of fluids)
ECF (1/3 of fluids)
interstitial
plasma
lymph
CSF
GI
WATER
Percentage body weight
55–60% of body weight
less in elderly, obese
more in infants
Daily intake: 2500 ml
food, drink: 2300 ml
metabolic water: 200 ml
Daily output: 2500 ml
kidneys: 1500 ml
skin: 600 ml
lungs: 300 ml
GI: 100 ml
Regulation
thirst
hormones: ADH, ANP, aldosterone
ELECTROLYTES (B)
Distribution and roles of ions
in ICF:
in ECF (mEq/l):
Na^{+}: 136–148
Cl^{-}: 95–105
K^{+}: 3.5–5.0
HCO_3^{-}: 22–26
Ca^{2+}: 4.5–5.5
HPO_4^{2-}: 1.7–2.6
Mg^{2+}: 1.3–2.1
ACID–BASE BALANCE (C)
REGULATION
Arterial blood pH: 7.35–7.45
Mechanisms
buffers
respirations
kidneys
IMBALANCES
Respiratory imbalances
acidosis: $P_{CO_2} > 45$; pH < 7.35
alkalosis: $P_{CO_2} < 35$; pH > 7.45
Metabolic imbalances
acidosis: $HCO_3^{-} < 22$; pH < 7.35
alkalosis: $HCO_3^{-} > 26$; pH > 7.45
Compensatory mechanism
respirations
kidneys
Signs/symptoms
acidosis: ↓CNS (lethargy, coma)
alkalosis: ↑CNS (spasms)
AGING CHANGES (D)
Infants
Elderly

CHAPTER 27

Fluid, Electrolyte, and Acid–Base Homeostasis

For the maintenance of homeostasis, levels of fluid, ions or electrolytes, acids and bases must be kept within acceptable limits. Without careful regulation of pH, for example, nerves will become overactive (in alkalosis) or underfunction, leading to coma (in acidosis). Alterations in blood levels of electrolytes can be lethal: excess K^+ can cause cardiac arrest, and decrease in blood Ca^{2+} can result in tetany of the diaphragm and respiratory failure. Systems you studied in Chapter 23 (respiratory) and Chapter 26 (urinary), along with buffers in the blood, normally achieve fluid, electrolyte, and acid–base balance.

As you begin your study of fluids, electrolytes, and acid–base balance, carefully examine the Chapter 27 Topic Outline and Objectives; check off each one as you complete it. Be sure to refer to the Framework frequently and note relationships among key terms in each section.

TOPIC OUTLINE AND OBJECTIVES

A. Fluid compartments and fluid balance

1. Compare the locations of intracellular fluid (ICF) and extracellular fluid (ECF), and describe the various fluid compartments of the body.
2. Describe the sources of water and solute gain and loss, and explain how each is regulated.

B. Electrolytes in body fluids

1. Compare the electrolyte composition of the three fluid compartments: plasma, interstitial fluid, and intracellular fluid.
2. Discuss the functions of sodium, chloride, potassium, bicarbonate, calcium, phosphate, and magnesium ions, and explain how their concentrations are regulated.

C. Acid–base balance

1. Compare the roles of buffers, exhalation of carbon dioxide, and kidney excretion of H^+ in maintaining pH of body fluids.
2. Define acid–base imbalances, describe their effects on the body, and explain how they are treated.

D. Aging and fluid, electrolyte, and acid–base balance

1. Describe the changes in fluid, electrolyte, and acid–base balance that may occur with aging.

WORDBYTES

Now become familiar with the language of this chapter by studying each wordbyte, its meaning, and an example of its use within a term. After you study the entire list, self-check your understanding by writing the meaning of each wordbyte on the line. As you continue through the *Learning Guide,* identify (and fill in) additional terms that contain the same wordbyte.

Wordbyte	Self-check	Meaning	Example(s)
extra-	____________________	outside	*extra*cellular
hyper-	____________________	above	*hyper*ventilation
hypo-	____________________	below	*hypo*tonic
inter-	____________________	between	*inter*stitial
intra-	____________________	within	*intra*cellular
natri-	____________________	sodium	hypo*natr*emia
-osis	____________________	condition of	alkal*osis*

CHECKPOINTS

A. Fluid compartments and fluid balance (pages 992–997)

✓ **A1.** Do this exercise about body fluid by answering questions about 40-year-old Alex, who is 193 cm (6′ 4″) tall and weighs 100 kg (220 pounds). (*Hint:* Refer to Figure 27.1.)

a. Fill in the expected weight in each category:

1. _______ kg of solids (such as fats, protein, and calcium)

2. _______ kg of fluid within cells

3. _______ kg of fluid in interstitial fluid

4. _______ kg of fluid in blood plasma

b. The total amount of fluid in Alex's body amounts to about _______ kg, whereas the amount of his extracellular fluid is about _______ kg.

✓ **A2.** About 67% of body fluids are *(inside of? outside of?)* cells. Check your understanding of fluid compartments by circling all fluids that are considered *extracellular fluids (ECF):*

Blood plasma
Aqueous humor and vitreous humor of the eyes
Endolymph and perilymph of ears
Fluid within liver cells
Fluid immediately surrounding liver cells
Lymph
Pericardial fluid
Synovial fluid
Cerebrospinal fluid (CSF)

For extra review, refer to Figure LG 1.1, page 8 of the *Learning Guide.*

✓ **A3.** *A clinical correlation.* Is a 45-year-old woman who has a water content of 55% likely to be in fluid balance? Explain.

✓ **A4.** Circle the person in each pair who is more likely to have a higher proportion of water content.

a. 37-year-old woman/67-year-old woman

b. Female/male

c. Lean person/obese person

✓ **A5.** Daily intake and output (I and O) of fluid usually both equal _______ mL (_______ liters). Of the intake, about _______ mL are ingested in food and drink, and _______ mL is produced by catabolism of foods (discussed in Chapter 25).

✓ **A6.** Now list four systems of the body that eliminate water. Indicate the amounts lost by each system on an average day. One is done for you.

a. Integument (skin): 600 mL/day

b. ____________________ : _______ mL/day

c. ____________________ : _______ mL/day

d. ____________________ : _______ mL/day

From this exercise, it is clear that the largest route for elimination of fluid is normally via: (circle answer)

exhalation from lungs   feces   sweat   urine

On a day when you exercise vigorously, how would you expect routes of fluid output to change?

✓ **A7.** Describe mechanisms for replacement of fluid that result from reduction of the body's fluid volume.

a. Loss of fluid volume is a condition known as *(de? re? over?)*-hydration. This condition stimulates thirst in three ways. One is ______-creased blood volume (and blood pressure), which stimulates release of ____________________ from kidneys and leads to production of ____________________ II. This chemical plays several roles in increasing blood pressure, one of which is to *(stimulate? inhibit?)* thirst receptors located in the ____________________.

b. Identify the other two factors that stimulate thirst receptors during dehydration:

______-crease in blood osmotic pressure, as well as ______-creased saliva production.

c. List three causes of dehydration.

In these cases, is it advisable to begin fluid replacement before or after onset of thirst? *(Before? After?)*

✓ **A8.** Do this exercise on relationships between water and solutes.

a. Define osmolarity of body fluids.

b. Write two categories of changes in the body that can increase blood osmolarity.

c. Identify the hormones that alter blood levels of NaCl and water:

1. Aldosterone promotes *(excretion? retention?)* of Na^+ and Cl^- by kidneys.
2. Angiotension II ultimately has the *(same? opposite?)* effect on blood levels of NaCl compared to aldosterone. Explain why.

   Angiotensin is formed when kidneys produce the chemical named ______________ in response to several factors such as (see Figure LG 18.5, page LG 372):

3. The name ANP indicates that this hormone promotes *(excretion? retention?)* of Na^+ and Cl^- by kidneys since the "N" in ANP stands for ________________ (or loss of Na^+ in urine). ANP is released from the ____________________ when they are stretched by a large volume of blood. Natriuresis will then lead to loss of both NaCl and ____________________, returning blood volume to normal.
4. The name ADH indicates that this hormone promotes *(promotes? opposes?)* diuresis which is defined as a *(large? small?)* volume of urine. A high level of ADH therefore causes a *(high? low?)* urine output. ADH causes insertion of water-channel proteins called ________________ into cells in collecting tubules of kidneys, causing a(n) ______-crease in permeability of those cells to water. Excessive amounts of ADH can lead to so much water retention that blood osmolarity (and Na^+ concentration) ______-creases significantly. (See Checkpoint A10).

✓ **A9.** Circle the factors that increase fluid output:

A. ADH
B. Aldosterone
C. Diarrhea
D. Increase in blood pressure (BP)
E. Fever
F. Hyperventilation
G. Atrial natriuretic peptide (ANP)

✓ **A10.** Do this exercise on movement of water between compartments.

a. List several examples of causes of low blood levels of Na^+ (hyponatremia).

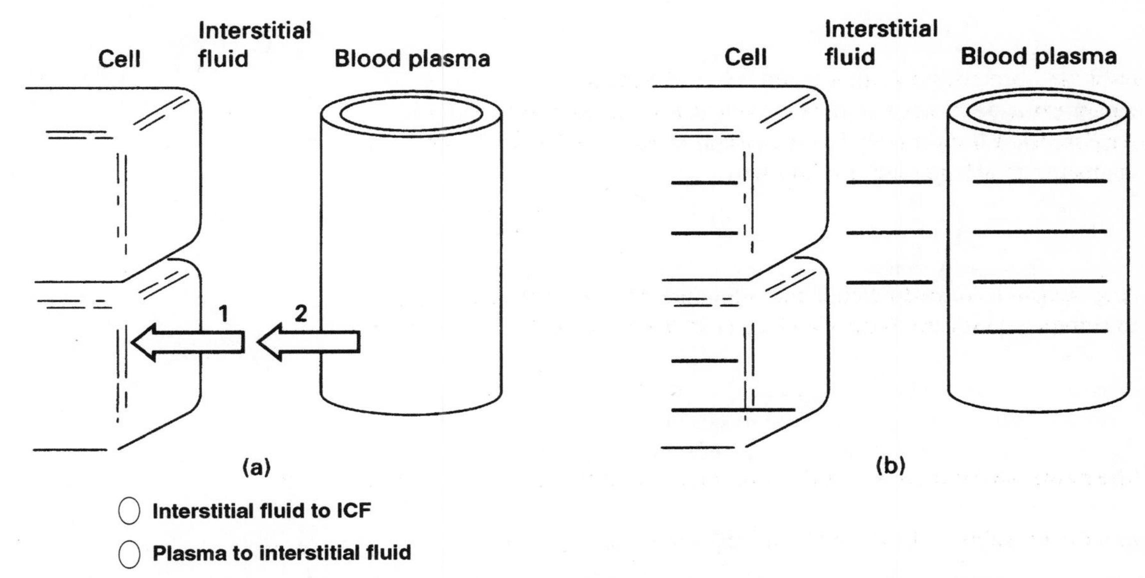

◯ **Interstitial fluid to ICF**

◯ **Plasma to interstitial fluid**

Figure LG 27.1 Diagrams of fluid compartments (a) Direction of fluid shift resulting from Na^+ deficit. Color according to Checkpoint A10. (b) Write in electrolytes according to Checkpoint B5.

b. Show on Figure LG 27.1a the shift of fluid resulting from loss of Na^+. Color arrows 1 and 2 according to color code ovals. Note that this figure also shows effects likely to occur in the large intestine if a *(hypertonic? hypotonic or tap water?)* enema is administered.

✓ **A11.** In this section on regulation of water balance, you have seen that electrolyte levels of the body *(do? do not?)* affect fluid levels, and fluid levels *(do? do not?)* affect electrolyte levels. Explain.

B. Electrolytes in body fluids (pages 997–1000)

B1. List four general functions of electrolytes.

____

____

____

____

B2. Define each term: *mole, mmole/liter, mEq/liter.*

✓ **B3.** Circle all of the answers that are classified as electrolytes:

A. Proteins in solutions
B. Anions
C. Glucose
D. K^+ and Na^+
E. H^+

✓ **B4.** Circle the electrolytes that are anions.

Ca^{2+}	HCO_3^- (bicarbonate)	Lactate	Cl^-
HPO_4^{2-} (phosphate)	Na^+	H^+	K^+

✓ **B5.** On Figure LG 27.1b write the chemical symbols for electrolytes that are found in greatest concentrations in each of the three compartments. Write one symbol on each line provided. Use the following list of symbols. (*Hint:* Refer to Figure 27.6 in the text.)

Cl^-.	**Chloride**	**Mg^{2+}.**	**Magnesium**
HCO_3^-.	**Bicarbonate**	**Na^+.**	**Sodium**
HPO_4^{2-}.	**Phosphate**	**Protein$^-$.**	**Protein**
K^+.	**Potassium**		

✓ **B6.** Analyze the information in Figure LG 27.1b by answering these questions.

a. Circle the two compartments that are most similar in electrolyte concentration.
cell   blood plasma   interstitial fluid (IF)

b. List two major differences in electrolyte content of these two similar compartments.

c. Most body protein is located in which compartment? ____________________

d. Protein anions in intracellular fluid are more likely to be bound to the cation *(Na^+? K^+?).*

B7. Check your understanding of the functions and disorders related to major electrolytes by completing Table LG 27.1 (next page).

Table LG 27.1 Summary of major blood electrolytes

Electrolyte (Name and Symbol)	Normal Range (SerumLevel), in mEq/liter	Principal Functions	Signs and Symptoms of Imbalances: Hyper-	Hypo-
a.	136–148			
b.				Hypochloremia: muscle spasms, alkalosis, ↓ respirations
c.		Most abundant cation in ICF; helps maintain fluid volume of cells; functions in nerve transmission		
d.			Hypercalcemia: weakness, lethargy, confusion, stupor, and coma	
e.	1.7–2.6	Helps form bones, teeth; important components of DNA, RNA, ATP, and buffer system		
f. Magnesium (Mg^{2+})				

✓ **B8.** *A clinical correlation.* Norine is having her serum electrolytes analyzed. Complete this exercise about her results.

a. Her Na^+ level is 128. Norine's value indicates that she is likely to be in a state of *(hyper? hypo?)*-natremia. Write two or more signs or symptoms of this electrolyte imbalance.

b. A diagnosis of hypochloremia would indicate that Norine's blood level of ________ ion is lower than normal. A normal range for serum chloride (Cl^-) level is ________ mEq/liter. One cause of low chloride is excessive ____________________.

c. A normal range for K^+ in serum is 3.5–5.0 mEq/liter. This range is considerably *(higher? lower?)* than that for Na^+. This is reasonable since K^+ is the main cation in *(ECF? ICF?)*, yet lab analysis of serum electrolytes examines K^+ in *(ECF? ICF?)*.

d. Norine's potassium (K^+) level is 5.6 mEq/liter, indicating that she is in a state of hyper-____________________. If Norine's serum K^+ continues to increase, for example to 8.0 mEq/l, her most serious risk, possibly leading to death, is likely to be

____________________.

e. A normal range for Mg^{2+} level of blood is 1.3–2.1 mEq/liter. Norine's electrolyte report indicates a value of 1.1 mEq/liter for Mg^{2+}. This electrolyte imbalance is

known as ____________________. List three or more factors related to nutritional state that could lead to this condition.

f. Norine's Ca^{2+} level is 3.9 mEq/]iter. A normal range is 4.5–5.5 mEq/liter. This

electrolyte imbalance is known as ____________________. This condition is often associated with *(decreased? elevated?)* phosphate levels, because blood levels of calcium and phosphate are *(directly? inversely?)* related. Low blood levels of both Ca^{2+} and Mg^{2+} cause overstimulation of the central nervous system and muscles. List two symptoms of these electrolyte imbalances.

✓ **B9.** *The Big Picture: Looking Back* and *A clinical correlation*. Refer to figures in earlier chapters of the *Learning Guide* or the text and complete this Checkpoint about regulation and imbalances of electrolytes.

a. On Figure LG 3.2, page LG 45, you identified the major anions and cations inside of cells and in fluids surrounding cells. When significant tissue trauma occurs, injured cells are likely to release their contents, leading to increased blood levels of the major ICF cation *(K^+? Na^+?)*. This condition is know as hyper-*(calc? kal? natr?)*-emia.

b. In Figure LG 10.1, page LG 181, you identified sarcoplasmic reticulum (SR) in muscle tissue. Which electrolyte is sequestered inside of SR and released into sarcoplasm surrounding thick and thin myofilaments only when this ion is needed to trigger muscle contraction? *(Ca^{2+}? Cl^-? K^+ Na^+?)*

c. On Figure LG 12.2, page LG 230, you demonstrated the major anions and cations in and around nerve cells. Which major ECF cation is permitted to pass through voltage-gated channels in neuron plasma membranes to initiate an action potential? *(K^+? Na^+?)*

d. On Figure LG 12.3, page LG 232, you identified that the voltage of a resting

membrane potential is typically _______ mV. The major ICF cation in neurons is *(K^+? Na^+?)*. In hypokalemia, nerve cells (as well as body fluids) would be deficient in K^+ and so would be more likely to have a resting membrane potential of *(–65? –80?)*. This means a greater stimulus would be necessary to raise the membrane potential to

threshold of about _______ mV. In hypokalemia, nerves and muscles are therefore likely to be *(hyper? hypo ?)*-active. Write two or more signs or symptoms of hypokalemia related to nerves or muscles.

e. Figure LG 18.4, page LG 371, shows that *(calcitonin? PTH?)* enhances movement of calcium into bones; this occurs by stimulation of osteo-*(blasts? clasts?)* in bone. PTH

______-creases calcium in blood by drawing calcium from several sources. Name

three. __________________ __________________ __________________ In fact, excessive PTH can lead to *(hyper? hypo?)*-calcemia.

f. Figure LG 18.5, page LG 372, emphasizes the effect of aldosterone on blood levels of several electrolytes. Aldosterone tends to increase blood levels of *(H^+? K^+? Na^+ ?)* and decrease blood levels of *(H^+? K^+? Na^+ ?)*. An overproduction of aldosterone (as in primary aldosteronism) is likely to lead to *(hypo? hyper?)*-natremia and *(hypo? hyper?)*-kalemia.

g. The same figure points out that ADH ______-creases reabsorption of water from urine into blood. The more water reabsorbed, the *(more? less?)* concentrated Na^+ will be in blood. What effect would overproduction of ADH (as in the syndrome of inappropriate ADH production [SIADH]) have upon blood level of Na^+? Leads to *(hypo? hyper?)*-natremia. What effect would deficient production of ADH (as in diabetes insipidus) have upon blood level of Na^+? Urine output would greatly

______-crease, leading to dehydration and *(hypo? hyper?)*-natremia.

h. Refer to text Figure 23.24 (page 836) and complete the chemical formulas showing formation of bicarbonate.

$CO_2 + H_2O \rightarrow$ ____________________ (carbonic acid) $\rightarrow$ ____________________ (bicarbonate) $+ H^+$

In fact, this equation indicates how most (______%) of CO_2 in blood is transported as bicarbonate. Bicarbonate also serves as a buffer for *(acids? bases?)*. An arterial blood level of bicarbonate of 32 mEq/ml is *(high? normal? low?)*. Excessive blood levels of

bicarbonate can be reduced as bicarbonate is excreted by ____________________.

✓ **B10.** *A clinical correlation.* Edema is an abnormal increase of *(intracellular fluid? interstitial fluid? plasma?)*. Describe three or more factors that may lead to edema.

C. Acid–base balance (pages 1001–1006)

✓ **C1.** Complete this Checkpoint about acid–base regulation.

a. The pH of systemic arterial blood should be maintained between ______ and ______.

b. Name the three major mechanisms that work together to maintain acid–base balance.

____________________ ____________________ ____________________

c. List four important buffer systems in the body.

____________________ ____________________

____________________ ____________________

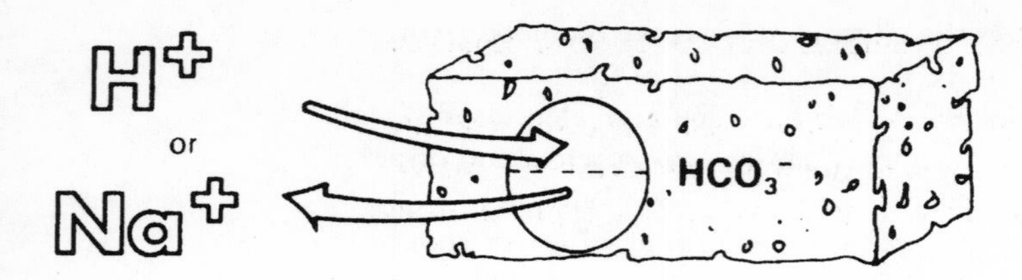

a. H^+ and Na^+ : only one of you can join HCO_3^- at any one time.

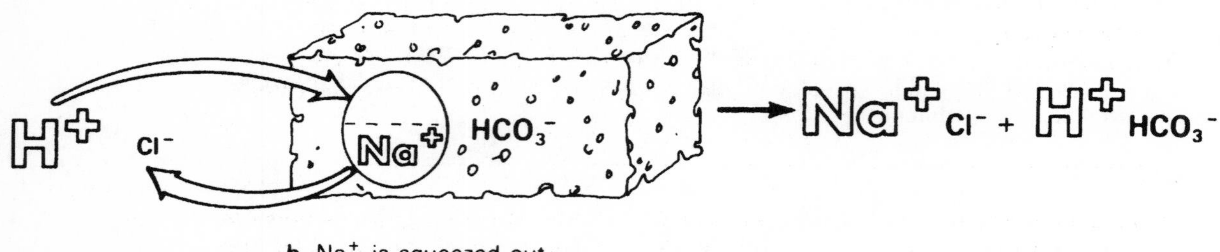

b. Na^+ is squeezed out as H^+ is drawn in.

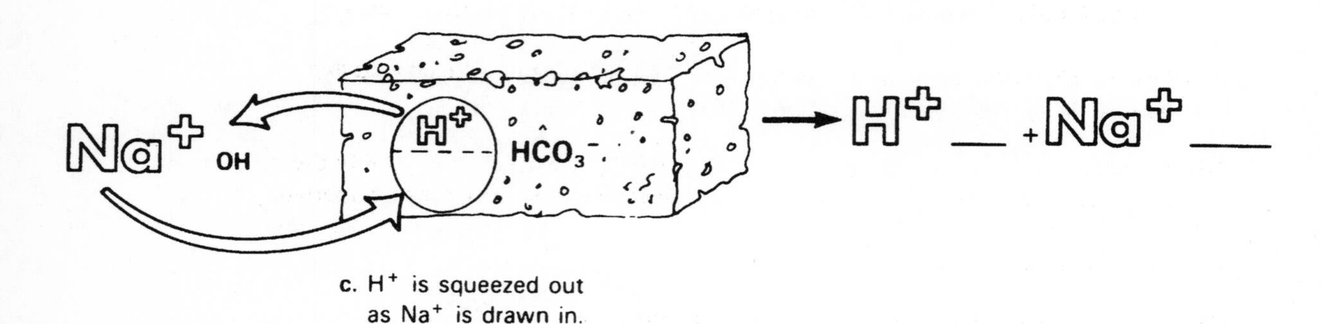

c. H^+ is squeezed out as Na^+ is drawn in.

O H^+ and arrows showing movement of H^+

O Na^+ and arrows showing movement of Na^+

Figure LG 27.2 Buffering action of carbonic acid—sodium bicarbonate system. Complete as directed in Checkpoint C2.

✓ **C2.** Refer to Figure LG 27.2 and complete this exercise about buffers.

a. In Figure LG 27.2, you can see why buffers are sometimes called "chemical sponges." They consist of an anion (in this case bicarbonate, or HCO_3^-) that is combined with a cation, which in this buffer system may be either H^+ or Na^+. When H^+ is drawn to the HCO_3^- "sponge," the weak acid ____________________ is formed. When Na^+ attaches to HCO_3^-, the weak base ____________________ is present. As you color H^+ and Na^+, imagine these cations continually trading places as they are "drawn in or squeezed out of the sponge."

b. Buffers are chemicals that help the body to cope with *(strong? weak?)* acids or bases that are easily ionized and cause harm to the body. When a strong acid, such as hydrochloric acid (HCl), is added to body fluids, buffering occurs, as shown in Figure LG 27.2b. The easily ionized hydrogen ion (H^+) from HCl is "absorbed by the sponge" as Na^+ is released from the buffer/sponge to combine with Cl^-. The two products,

____________________ and ____________________, are weak (less easily ionized).

c. Now complete Figure LG 27.2c by writing the names of products resulting from buffering of a strong base, NaOH. Color H^+ (and its arrow) and Na^+ (and its arrow) and fill in products to show this exchange reaction.

d. When the body continues to take in or produce excess strong acid, the concentration

of the ____________________ member of this buffer pair will decrease as it is used up in the attempt to maintain homeostasis of pH.

e. Though buffer systems provide rapid response to acid–base imbalance, they are limited because one member of the buffer pair can be used up, and they can convert only strong acids or bases to weak ones; they cannot eliminate them. Two systems of the body that can actually eliminate acidic or basic substances are

____________________ and ____________________.

✓ **C3.** *For extra review.* Check your understanding of the carbonic acid-bicarbonate buffer in this Checkpoint. When a person actively exercises, *(much? little?)* CO_2 forms, leading to H_2CO_3 production. Consequently, *(much? little?)* H^+ will be added to body tissues. In order to buffer this excess H^+ (to avoid drop in pH and damage to tissues), the *(H_2CO_3? $NaHCO_3$?)* component of the buffer system is called upon. The excess H^+

then replaces the ________ ion of $NaHCO_3$, in a sense "tying up" the potentially harmful H^+.

✓ **C4.** Phosphates are more concentrated in *(ECF? ICF?)*. (For help, see Figure 27.6 in the text.) Name one type of cell in which the phosphate buffer system is most important.

____

C5. Write the reaction showing how phosphate buffers strong acid.

✓ **C6.** Which is the most abundant buffer system in the body? ____________________ Circle the part of the amino acid shown below that buffers acid. Draw a square around the part of the amino acid that buffers base.

$$NH_2-\overset{\displaystyle R}{\underset{\displaystyle H}{\overset{|}{\underset{|}{C}}}}-COOH$$

C7. Explain how the hemoglobin buffer system buffers carbonic acid, so that an acid even weaker than carbonic acid (that is, hemoglobin) is formed.
For extra review. See Chapter 23, Checkpoint E3, page 524 in the *Learning Guide* and Figure 23.24 in the text.

✓ **C8.** Complete this Checkpoint about respiration and kidneys related to pH.

a. Hyperventilation will tend to *(raise? lower?)* blood pH because, as the person exhales CO_2, less CO_2 is available for formation of ________________ acid and free hydrogen ion.

b. A slight decrease in blood pH will tend to *(stimulate? inhibit?)* the respiratory center and then will *(increase? decrease?)* respirations.

c. The kidneys regulate acid–base balance by altering their tubular secretion of ________ or elimination of ________ ion. Name the two parts of the renal tubule that are most involved with these processes: ________________ and ________________.

✓ **C9.** An acid–base imbalance caused by abnormal alteration of the respiratory system is classified as *(metabolic? respiratory?)* acidosis or alkalosis. Any other cause of acid–base imbalance, such as urinary, hormonal, or digestive tract disorder, is identified as *(metabolic? respiratory?)* acidosis or alkalosis.

✓ **C10.** *Critical thinking.* Indicate which of the four categories of acid–base imbalances listed in the box is most likely to occur as a result of each condition described below.

MAcid.	**Metabolic acidosis**	**RAcid.**	**Respiratory acidosis**
MAlk.	**Metabolic alkalosis**	**RAlk.**	**Respiratory alkalosis**

________ a. Decreased blood level of CO_2, as a result of hyperventilation

________ b. Decreased respiratory rate in a patient taking overdose of morphine

________ c. Excessive intake of antacids

________ d. Prolonged vomiting of stomach contents (which are high in HCl content)

________ e. Ketosis in uncontrolled diabetes mellitus

________ f. Decreased respiratory minute volume in a patient with emphysema or fractured rib

________ g. Excessive loss of bicarbonate from the body, as in prolonged diarrhea or renal dysfunction

✓ **C11.** Refer to the table below and do this exercise on diagnosis of acid–base imbalances.

Patient	**Arterial pH**	**P_{CO_2} (mm Hg)**	**HCO_3^- (m Eq/liter)**
Normal	acidosis 7.35–7.45 alkalosis	alkalosis 35–45 acidosis	acidosis 22–26 alkalosis
Patient W	← 7.29	52* →	30** →
Patient X	7.50 →	← 28*	23**
Patient Y	← 7.31	← 32**	← 19*
Patient Z	7.44	53** →	35* →

Arrows indicate which values have increased above normal (→) or decreased below normal (←). Absence of an arrow means the value is normal. * indicates *primary value* associated with the initial acid–base problem. ** indicates *secondary value* which may be due to compensatory mechanisms for acid–base imbalances.

a. First refer to normal values for arterial blood pH, P_{CO_2}, and HCO_3^-. Acidosis is indicated by arterial blood pH less than *(7.35? 7.45?)*. Keep in mind that acidity and pH are *(directly? inversely?)* related; that is, as acidity of blood increases, its pH ______-creases. Which patients have blood pH that indicate acidosis? Patients *(W? X? Y? Z?)*

b. Look at the blood chemistry values for patient W, a 61-year-old man with diagnoses of emphysema and pneumonia. The table indicates that a P_{CO_2} value greater than *(35? 45?)* may be an indicator of acidosis. Patients with chronic respiratory problems retain CO_2, as indicated by this patient's P_{CO_2} of 52, which is *(higher? lower?)* than normal. The arrow drawn to the right of the 52 indicates that such high CO_2 levels have led this patient into acidosis.

c. Patient X is a 27-year-old woman who has just arrived in the emergency room. She is hyperventilating and in a state of high anxiety after an accident that has critically injured her child. Hyperventilation tends to *(increase? decrease?)* blood P_{CO_2}, as shown by her value of ________ mm Hg. The arrow next to the 28 points left to show the effect (alkalosis) of this lowered blood P_{CO_2}.

d. Both patients W and X have acid–base imbalances caused by changes in respiration. Therefore patient W's imbalance is classified as *(respiratory? metabolic?)* acidosis, and patient X's imbalance is known as ____________________ ____________________.

e. In respiratory acid–base imbalances, the value that is altered is always *(P_{CO_2}? HCO_3^-?)*. For clarity, call this the *primary* value associated with respiratory acid–base imbalances; this value is marked by * in the table. The alteration is very logical; for example, increased P_{CO_2} is associated with ____________________-osis because CO_2 leads to production of carbonic acid, and decreased P_{CO_2} indicates ____________________-osis because less carbonic acid is made.

f. In order to maintain homeostasis, the body will try to compensate for acid–base imbalances by altering the *secondary* (or "other") value *in the same direction* as the primary value, so that the two values are more matched or in balance. The secondary value is marked with ** in the table. Where compensation is occurring, you will see the arrow for the secondary value (**) going in the same direction as the primary value (*). For respiratory acidosis, in which P_{CO_2} (the primary value) ______-creases, we can expect compensatory mechanisms to ______-crease the secondary value (HCO_3^-), possibly to a value greater than normal, in an attempt to make pH more alkaline. It is even possible that a person in respiratory acidosis could move into metabolic alkalosis due to overcompensation, for example, by reduced kidney excretion of sodium bicarbonate ($NaHCO_3$).

g. Does it appear that patient W has done some compensating for his respiratory acidosis? *(Yes? No?)* This is indicated by his HCO_3^- value, which is *(within? outside?)* the normal range. This patient has *(fully? partially?)* compensated respiratory acidosis because his pH is still abnormal.

h. If patient W had his HCO_3^- value within the normal range, this would indicate *(full? partial? no?)* compensation. If patient W's values were pH of 7.39, P_{CO_2} of 52 mm Hg, and HCO_3^- of 30, this return to the normal pH range would indicate *(fully? partially? un-?)* compensated respiratory acidosis.

i. Patient X's HCO_3^- value (23 mEq) is *(high? normal? low?)*, indicating that she has *(fully? partially? un-?)* compensated respiratory alkalosis. Although she may eventually excrete HCO_3^- from her ____________________ to match the decreased P_{CO_2} level, this has not yet occurred because her anxiety and hyperventilation have been relatively short-term.

j. Patient Y is in diabetic ketoacidosis, is lethargic, and is hyperventilating. The lowered HCO_3^- level indicates that the patient is using bicarbonate to buffer ketoacids resulting from uncontrolled diabetes. The primary value (*) affected is *(P_{CO_2}? HCO_3^-?)*, indicating *(respiratory? metabolic?)* acid–base imbalance. What factor indicates partial compensation?

k. Patients with ketoacidosis or other metabolic acidosis (such as from excessive aspirin intake) may even move into ________________ alkalosis by compensatory hyperventilation. In that case, the P_{CO_2} would continue to drop, and pH would return to normal and go right on to a value greater than *(7.35? 7.45?)*.

l. Patient Z has a history of peptic ulcer and had been vomiting for days preadmission. The patient states that she had taken a "bunch of antacids" for her ulcer. Does her pH indicate a pH imbalance? *(Yes? No?)* Are her other two values normal? *(Yes? No?)* What type of acid–base imbalances have normal pH but abnormal P_{CO_2} and HCO_3^-? *(Fully? Partially? Un-?)* compensated imbalances.

m. Patient Z's ulcer, vomiting (loss of gastric HCl), and ingestion of antacids, as well as the level of her pH (7.44) so close to the alkaline end of the normal range, all suggest metabolic ________________-osis. Her compensatory mechanisms apparently include *(hypo? hyper?)*-ventilation, with retention of CO_2 which has helped to *(raise? lower?)* her pH back into the normal range.

n. Complete this Checkpoint by writing complete diagnoses (including degree of compensation) for each patient in the table. Remember these key points (stated in Checkpoint C11e and j): P_{CO_2} is the factor altered in *respiratory* acid-base imbalances, whereas HCO_3^- is altered in *metabolic* acid-base imbalances.

✓ **C12.** Describe effects of acid-bases imbalances on the central nervous system (CNS).

a. Acidosis has the effect of *(stimulating? depressing?)* the (CNS). Write one or more signs of acidosis.

b. Alkalosis is more likely to cause *(spasms? lethargy?)* because this acid-base imbalance *(stimulates? depresses?)* the CNS.

D. Aging and fluid, electrolyte, and acid–base balance (page 1006)

✓ **D1.** Complete this tablet contrasting fluids and electrolytes in infants and adults. Select correct answers from those given in *italics*.

Factor (answers)	Infant	Adult
a. Percentage of total body mass that is water: *(55–60% 70–90%)*		
b. Metabolic rate (with resulting metabolic wastes and "turnover" of water) *(Higher? Lower?)*	Higher	
c. Ability of kidneys to concentrate urine (retaining water) *(Greater? Less?)*		
d. Body surface area (which permits loss of fluid by evaporation) *(Greater? Less?)*		
e. Respiration rate (which permits loss of fluid in expiration) *(Greater? Less?)*		
f. Concentration of K^+ (if higher, leads towards acidosis) *(Higher? Lower?)*		

D2. *Critical thinking.* Older adults may experience increased risk for changes in fluid, electrolytes, and acid–base as a result of normal aging changes. Write possible causes and signs or symptoms associated with each imbalance. Check your answers against Table 27.2 in the text.

Imbalance	Possible causes	Signs or symptoms
a. Hypokalemia		
b. Hyponatremia		
c. Hypernatremia		
d. Acidosis (respiratory or metabolic)		

ANSWERS TO SELECTED CHECKPOINTS: CHAPTER 27

A1. (a1) 40; (a2) 40; (a3) 16: (a4) 4. (b) 60, 20.

A2. Inside of; all answers except fluid within liver cells.

A3. The percentage is within the normal range; however, more information is needed because the fluid must be distributed correctly among the three major compartments: cells (ICF) and ECF in plasma and interstitial fluid. For example, excessive distribution of fluid to interstitial space occurs in the fluid imbalance known as edema.

A4. (a) 37-year-old woman. (b) Male. (c) Lean person.

A5. 2500 (2.5); 2300, 200.

A6. (Any order) (a) Kidney, 1500. (b) Lungs, 300. (c) GI, 100. Urine. With exercise, output of fluid increases greatly via sweat glands and somewhat via respirations. To compensate, urine output decreases, and intake of fluid increases.

A7. (a) De: de, renin, angiotensin; stimulate, hypothalamus. (b) In. de. (c) Sweating, vomiting, diarrhea; before.

A8. (a) The degree of concentration or osmotic pressure of body fluids. (b) Loss of fluid (dehydration) or gain of electrolytes (such as by a high intake of salty foods). (c1) Retention. (c2) Same, because angiotensin II promotes release of aldosterone from the adrenal cortex; renin; dehydration or low renal artery blood pressure, low blood Na^+ level or high blood K^+ level. (c3) Excretion, natriuretic; atria; water. (c4) Opposes, large; low; aquaporin-2, in; de.

A9. C–G.

A10. (a) Excessive loss of Na^+ (as by sweating or diarrhea) followed by drinking large volumes of pure water, decreased intake of Na^+; (see causes of hyponatremia in Table 27.2 of the text.) (b) 1, Shift of fluid leading to overhydration or water intoxication of brain cells, possibly leading to convulsions and coma; 2, shift of fluid (from plasma to replace IF that entered cells), possibly leading to hypovolemic shock. Hypotonic or tap water.

A11. Do, especially Na^+ (the major cation in ECF), which "holds" (retains) water, do. Note (in Checkpoint A8c4) that excessive water retention dilutes blood levels of Na^+, possibly leading to hyponatremia.

B3. All except C (glucose).

B4. Cl^-, HCO_3^- (bicarbonate), HPO_4^{2-} (phosphate), lactate.

B5.

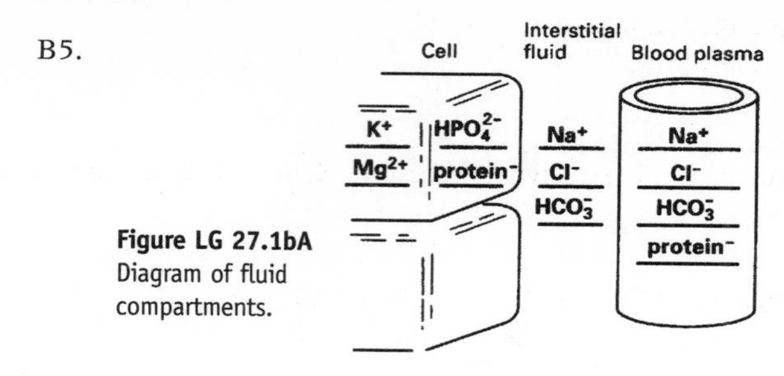

Figure LG 27.1bA Diagram of fluid compartments.

B6. (a) Blood plasma and interstitial fluid. (b) Plasma contains more plasma protein (which accounts for its high colloidal osmotic pressure), and slightly less sodium and less chloride than IF. (c) Intracellular fluid (ICF). (d) K^+.

B8. (a) Normal serum Na^+ ranges from 136 to 148; hypo; examples: headache, low blood pressure, tachycardia, and weakness, possibly leading to confusion, stupor, and coma if her serum Na^+ continues to drop. (b) Cl^-; 95–105; emesis or use of certain diuretics. (c) Lower; ICF, ECF. (d) Kalemia; ventricular fibrillation resulting in cardiac arrest. (e) Hypomagnesemia; malnutrition, malabsorption, diarrhea, alcoholism, or excessive lactation. (f) Hypocalcemia; elevated, inversely; tetany, convulsions.

B9. (a) K^+; kal. (b) Ca^{2+}. (c) Na^+. (d) –70; K^+; –80; –55; hypo; flaccid paralysis, fatigue, changes in cardiac muscle and GI muscle function. (e) Calcitonin, blasts; in; bones, intestine, kidneys; hyper. (f) Increases Na^+ and decreases H^+ and K^+, hyper, hypo. (g) In; less; hypo; in, hyper. (h) H_2CO_3, HCO_3^-; 79; acids; high; kidneys.

B10. Interstitial fluid; hypertension (which increases filtration at arteriolar ends of capillaries), excessive NaCl intake (which causes fluid retention), congestive heart failure (which backs up blood into venules and capillary beds), or blockage of lymph flow (for example, by cancer metastasis to lymph nodes).

C1. (a) 7.35–7.45. (b) Buffers, respirations, and kidney excretion. (c) Carbonic acid-bicarbonate, phosphate, hemoglobin-oxyhemoglobin, protein.

C2. (a) Carbonic acid, or H_2CO_3; sodium bicarbonate, or $NaHCO_3$. (b) Strong; the salt sodium chloride (NaCl), the weak acid H_2CO_3 ($H \bullet HCO_3$). (c) Water (H_2O or $H \bullet OH$) and sodium bicarbonate ($NaHCO_3$). (d) $NaHCO_3$ or weak base. (e) Respiratory (lungs), urinary (kidneys).

C3. Much; much; $NaHCO_3$; Na^+.
C4. ICF; kidney cells.
C6. Protein;

$$\text{Buffers acid}-\underset{}{\boxed{NH_2}}-\underset{\displaystyle H}{\overset{\displaystyle R}{\overset{|}{\underset{|}{C}}}}-\boxed{COOH}-\text{Buffers base}$$

C8. (a) Raise, carbonic. (b) Stimulate, increase. (c) H^+, HCO_3^-; PCT and collecting ducts.
C9. Respiratory; metabolic.
C10. (a) RAlk. (b) RAcid. (c) MAlk. (d) MAlk. (e) MAcid. (f) RAcid. (g) MAcid.
C11. (a) 7.35; inversely; de: W and Y. (b) 45 (except when pCO_2 is high due to hypoventilation as compensation for metabolic alkalosis: see Patient Z); higher. (c) Decrease. 28. (d) Respiratory, respiratory alkalosis. (e) P_{CO_2}: acid. alkal. (f) In, in. (g) Yes: outside; partially. (h) No: fully. (i) Normal, un: kidneys. (j) HCO_3^-, metabolic; P_{CO_2} is decreased also related to hyperventilation. (k) Respiratory; 7.45. (1) No; no; fully. (m) Alkal; hypo. lower. (n) Patient W: respiratory acidosis. partially compensated; X: respiratory alkalosis, uncompensated: Y: metabolic acidosis, partially compensated; Z: metabolic alkalosis, fully compensated.
C12. Depressing; lethargy, stupor, or coma. (b) Spasms, stimulates.
D1. (a) 70–90%, 55–60%. (b) (Higher), lower. (c) Less, greater. (d–e) Greater; less. (f) Higher; lower.

MORE CRITICAL THINKING: CHAPTER 27

1. Contrast the chemical composition of intracellular fluid with that of plasma and interstitial fluid.
2. Explain the importance of maintaining fluid balance among the three major fluid compartments of the body.
3. Explain why changes in fluids result in changes in concentration of electrolytes in interstitial and intracellular fluids. Be sure to state which two electrolytes normally exert the greatest effect on this balance.
4. Describe effects of ADH, ANP, and aldosterone on fluid and electrolyte balance.
5. Describe how you might be able to distinguish two persons with acid–base imbalances: one in acidosis and one in alkalosis.
6. Contrast causes of respiratory acidosis and metabolic acidosis. State several examples of causes.

MASTERY TEST: CHAPTER 27

Questions 1–13: Circle T (true) or F (false). If the statement is false, change the underlined word or phrase so that the statement is correct.

T F 1. Hyperventilation tends to raise pH.
T F 2. Chloride is the major intracellular anion.
T F 3. Under normal circumstances fluid intake each day is greater than fluid output.
T F 4. During edema there is increased movement of fluid out of plasma and into interstitial fluid.
T F 5. Parathyroid hormone causes an increase in the blood level of calcium.
T F 6. When aldosterone is in high concentrations, sodium is conserved (in blood) and potassium is excreted (in urine).
T F 7. Elderly persons are at lower risk compared to young adults for acidosis and sodium imbalances.
T F 8. About two-thirds of body fluids are found in intracellular fluid (ICF) and one-third of body fluids are found in extracellular fluid (ECF).
T F 9. The body of an infant contains a higher percentage of water than the body of an adult.
T F 10. Starch, glucose, HCO_3^- Na^+, and K^+ are all electrolytes.
T F 11. Protein is the most abundant buffer system in the body.
T F 12. Sodium plays a much greater role than magnesium in osmotic balance of ECF because more mEq/liter of sodium than magnesium are in ECF.
T F 13. The pH of arterial blood is higher than that of venous blood because of a lower level of CO_2 and H^+.

Questions 14–20: Circle the letter preceding the one best answer to each question.

14. Three days ago Ms. Weldon fractured several ribs in an auto accident, and her breathing has been shallow since then. Her arterial blood pH is 7.29, P_{CO_2} is 50 mm Hg, and HCO_3^- is 30 mEq/liter.

    Her acid-base imbalance is most likely to be:
    A. Respiratory acidosis
    B. Respiratory alkalosis
    C. Metabolic acidosis
    D. Metabolic alkalosis
15. Ms. Weldon's acid–base imbalance (question 14) appears to be:
    A. Fully compensated
    B. Partially compensated
    C. Uncompensated
16. Mr. Wong is in cardiac care post-myocardial infarction. His arterial blood pH is 7.31, P_{CO_2} is 39 mm Hg, and HCO_3^- is 20 mEq/liter. His acid–base imbalance is most likely to be:
    A. Metabolic acidosis, fully compensated
    B. Metabolic acidosis, partially compensated
    C. Metabolic acidosis, uncompensated
17. Metabolic alkalosis imbalances are indicated by a(n) ________ in arterial blood pH and altered ________ level.
    A. Increase; P_{CO_2}
    B. Decrease; P_{CO_2}
    C. Increase; HCO_3^-
    D. Decrease; HCO_3^-
18. Which term refers to a lower than normal blood level of sodium?
    A. Hyperkalemia
    B. Hypokalemia
    C. Hypernatremia
    D. Hyponatremia
    E. Hypercalcemia
19. Which two electrolytes are in greatest concentration in ICF?
    A. Na^+ and Cl^-
    B. Na^+ and K^+
    C. K^+ and Cl^-
    D. K^+ and HPO_4^{2-}
    E. Na^+ and HPO_4^{2-}
    F. Ca^{2+} and Mg^{2+}
20. All of the following factors will tend to increase fluid output except:
    A. Fever
    B. Vomiting and diarrhea
    C. Hyperventilation
    D. Increased glomerular filtration rate
    E. Decreased blood pressure

Questions 21–25: Fill-ins. Complete each sentence with the word or phrase that best fits.

________________ 21. List four factors that increase production of aldosterone.

________________ 22. On an average day, the greatest volume of fluid output is via the ________ system.

________________ 23. To compensate for acidosis, kidneys will increase secretion of ________, and in alkalosis kidneys will eliminate more ________.

________________ 24. Normally blood serum contains about ________ mEq/liter of Na+ and ________ mEq/liter of K+.

________________ 25. The three main mechanisms used by the body to compensate for acid–base imbalances are ________________, ________________, and ________________.

ANSWERS TO MASTERY TEST: CHAPTER 27

True–False

1. T
2. F. Extracellular anion
3. F. Equals
4. T
5. T
6. T
7. F. Higher
8. T
9. T
10. F. Not all. (Starch and glucose are nonelectrolytes.)
11. T
12. T
13. T

Multiple Choice

14. A
15. B
16. C
17. C
18. D
19. D
20. E

Fill-ins

21. Decreased blood volume or cardiac output, decrease of Na^+ or increase of K^+ in ECF, or physical stress.
22. Urinary
23. H^+, HCO_3^-.
24. 136 to 148; 3.5 to 5.0
25. Buffers, respirations, and renal output of H^+ or HCO_3^-.

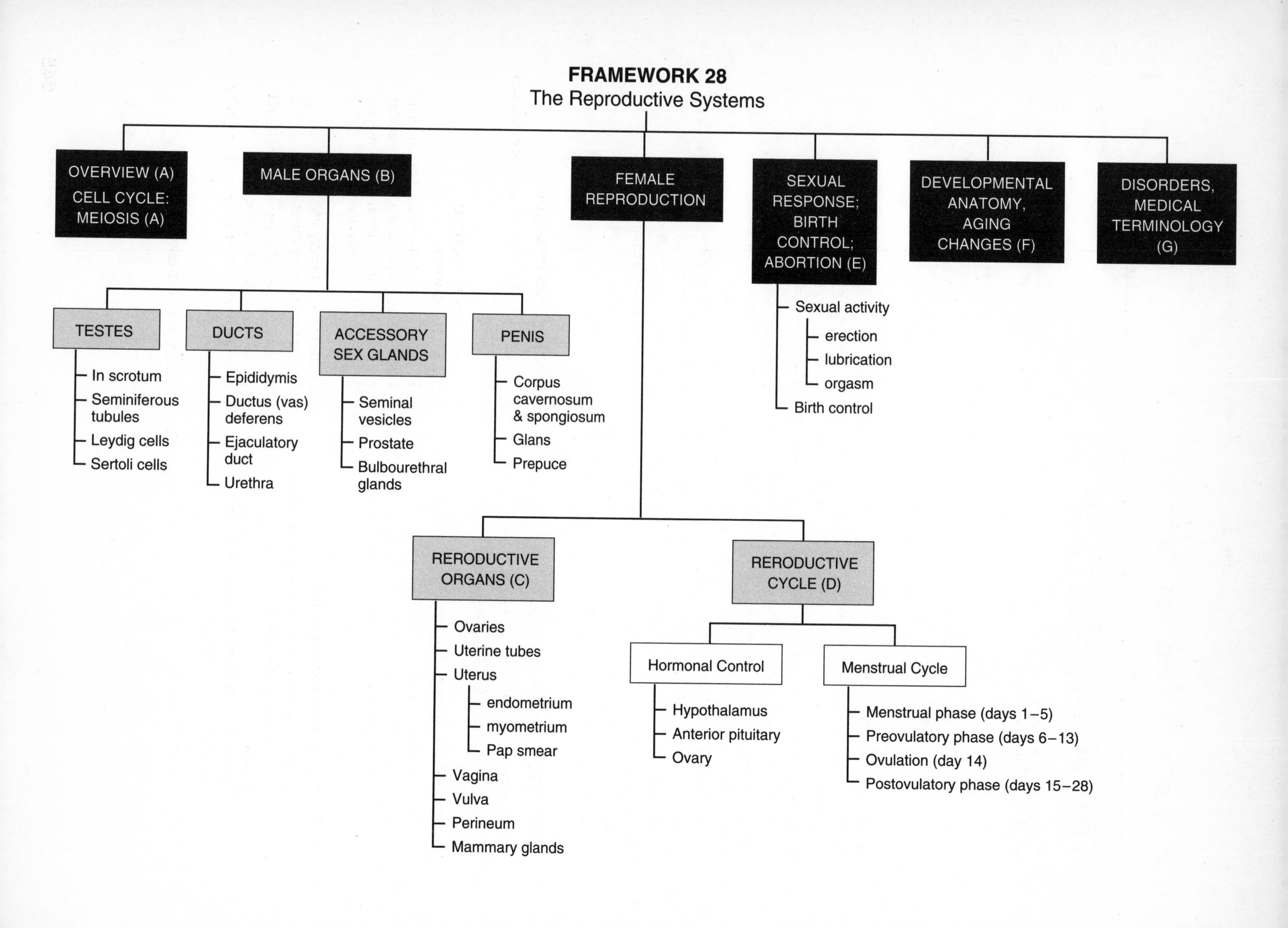
FRAMEWORK 28
The Reproductive Systems
OVERVIEW (A) CELL CYCLE: MEIOSIS (A)
MALE ORGANS (B)
FEMALE REPRODUCTION
SEXUAL RESPONSE; BIRTH CONTROL; ABORTION (E)
DEVELOPMENTAL ANATOMY, AGING CHANGES (F)
DISORDERS, MEDICAL TERMINOLOGY (G)
TESTES
In scrotum
Seminiferous tubules
Leydig cells
Sertoli cells
DUCTS
Epididymis
Ductus (vas) deferens
Ejaculatory duct
Urethra
ACCESSORY SEX GLANDS
Seminal vesicles
Prostate
Bulbourethral glands
PENIS
Corpus cavernosum & spongiosum
Glans
Prepuce
Sexual activity
erection
lubrication
orgasm
Birth control
RERODUCTIVE ORGANS (C)
Ovaries
Uterine tubes
Uterus
endometrium
myometrium
Pap smear
Vagina
Vulva
Perineum
Mammary glands
RERODUCTIVE CYCLE (D)
Hormonal Control
Hypothalamus
Anterior pituitary
Ovary
Menstrual Cycle
Menstrual phase (days 1–5)
Preovulatory phase (days 6–13)
Ovulation (day 14)
Postovulatory phase (days 15–28)

CHAPTER 28

The Reproductive Systems

Unit V introduces the systems in females and males that provide for continuity of the human species, not simply homeostasis of individuals. The reproductive systems produce sperm and ova that may unite and develop to form offspring. In order for sperm and ova to reach a site for potential union, they must each pass through duct systems and be nourished and protected by secretions. External genitalia include organs responsive to stimuli and capable of sexual activity. Hormones play significant roles in the development and maintenance of the reproductive systems. The variations of hormones in ovarian and uterine (menstrual) cycles are considered in detail, along with methods of birth control. As we have done for all other systems, we discuss development of the reproductive system and normal aging changes.

As you begin your study of reproduction, carefully examine the Chapter 28 Topic Outline and Objectives. Check off each objective as you complete it. The Chapter 28 Framework provides an organizational preview of the reproductive systems. Be sure to refer to the Framework frequently and note relationships among key terms in each section.

TOPIC OUTLINE AND OBJECTIVES

A. Overview; the cell cycle in gonads

1. Describe the process of meiosis.

B. The male reproductive system

1. Describe the location, structure, and functions of the organs of the male reproductive system.
2. Discuss the process of spermatogenesis in the testes.

C. The female reproductive system

1. Describe the location, structure, and functions of the organs of the female reproductive system.
2. Discuss the process of the oogenesis in the ovaries.

D. The female reproductive cycle

1. Compare the major events of the ovarian and uterine cycles.

E. The human sexual response; birth control; abortion

1. Describe the similarities and differences in the sexual responses of males and females.
2. Compare the various kinds of birth control methods and their effectiveness.

F. Developmental anatomy and aging of the reproductive systems

1. Describe the development of the male and female reproductive systems.
2. Describe the effects of aging on the reproductive systems.

G. Disorders, medical terminology

WORDBYTES

Now become familiar with the language of this chapter by studying each wordbyte, its meaning, and an example of its use within a term. After you study the entire list, self-check your understanding by writing the meaning of each wordbyte on the line. As you continue through the *Learning Guide,* identify (and fill in) additional terms that contain the same wordbyte.

Wordbyte	Self-check	Meaning	Example(s)
a-	____________	without	*a*menorrhea
acro-	____________	top	*acro*some
andro-	____________	man	*andro*gen
-arche	____________	beginning	men*arche*
cervix-	____________	neck	*cervix* of uterus
corp-	____________	body	*corp*us cavernosum
crypt-	____________	hidden	*crypt*orchidism
dys-	____________	difficult, painful	*dys*menorrhea
ejacul-	____________	to spill	*ejacul*ation
-fer	____________	to carry	semini*fer*ous tubules
-genesis	____________	formation	spermato*genesis*
-gonia	____________	offspring	oo*gonia*
gyneco-	____________	woman	*gyneco*logy
hapl-	____________	single	*hapl*oid
hyster(o)-	____________	uterus	*hyster*ectomy
labia	____________	lips	*labia* majora
mamm-	____________	breast	*mamm*ogram
mast-	____________	breast	*mast*ectomy
meio-	____________	smaller	*meio*sis
men-	____________	month	*men*opause, *men*ses
-metrium	____________	uterus	endo*metrium*
mito	____________	thread	*mito*sis
oo-, ov-	____________	egg	*oo*genesis, *ov*um
orchid-	____________	testis	*orchid*ectomy

rete	______________	network	*rete* testis
-rrhea-	______________	a flow	dysmeno*rrhea*
salping-	______________	tube, trumpet	*salping*itis
troph-	______________	nutrition	*troph*oblast

CHECKPOINT

A. Overview; the cell cycle in gonads (pages 1012–1014)

A1. Define each of these groups of reproductive organs.

a. Gonad

b. Ducts

c. Accessory glands

d. Supporting structures

✓ **A2.** As you go through the chapter, note which structures of the reproductive systems are included in each of the above categories:

a. Female:

b. Male:

✓ **A3.** Do the following exercise about reproductive cell division.

a. Meiosis is a special type of nuclear division that occurs only in ______________ and is one process in the production of sex cells or ______________. Fusion of gametes during the process of ______________ results in formation of a ______________.

b. Cells that begin the process of meiosis are ovarian or testicular cells that have been undergoing mitosis up to this point. Each cell initially contains ________ chromosomes, which is the *(diploid or 2n? haploid or n?)* number for humans. These chromosomes consist of ________ homologous pairs.

c. One pair is known as the *(autosomes? sex chromosomes?)*. In females, this pair consists of *(two X chromosomes? one X and one Y chromosome?)*. The remaining 22 pairs are

known as ____________________.

d. By the time that cells complete meiosis, they will contain *(46? 23?)* chromosomes, the *(2n? n?)* number.

e. Meiosis differs from mitosis in that meiosis involves *(one? two? three?)* divisions with replication of DNA only *(once? twice?)*. The first division is known as the

____________________ division, whereas the second division is the

____________________ division.

f. Replication of DNA occurs *(before? after?)* reduction division, resulting in doubling of

chromosome number in humans from 46 chromosomes to ____________________ "chromatids." (Chromatid is a name given to a chromosome while the chromosome is attached to its replicated counterpart, that is, in the doubled state.) Doubled or paired chromatids are held together by means of a *(centromere? centrosome?)*.

g. A unique event that occurs in Prophase I is the gathering of chromosomes in homo-

logous pairs, a process called ____________________. Note that this process *(does? does not?)* occur during mitosis. The four chromatids of the homologous pair are known as

a ____________________. The proximity of chromatids permits the exchange or

____________________ of DNA between maternal and paternal chromatids. What is the advantage of crossing over to sexual reproduction?

h. Another factor that increases possibilities for variety among sperm or eggs you can produce is the random (or chance) assortment of chromosomes as they line up on either side of the equator in Metaphase *(I? II?)*.

i. Division I reduces the chromosome number to __________ chromatids. Division II

results in the separation of doubled chromatids by splitting of ____________________

so that resulting cells contain only __________ chromosomes.

j. Meiosis is only one part of the process of forming sperm or eggs. Each original diploid cell in a testis will lead to production of *(1? 2? 4?)* mature sperm, while each diploid ovary cell produces *(1? 2? 4?)* mature egg(s) plus three small cells known as

____________________.

B. The male reproductive system (pages 1014–1027)

✓ **B1.** Describe the scrotum and testes in this exercise.

a. Arrange coverings around testes from most superficial to deepest:

______ ______ ______

D.	**Dartos: smooth muscle that wrinkles skin**
TA.	**Tunica albuginea: dense fibrous tissue**
TV.	**Tunica vaginalis: derived from peritoneum**

b. The temperature in the scrotum is about 2–3°C *(higher? lower?)* than the temperature in the abdominal cavity. Of what significance is this?

c. During fetal life, testes develop in the *(scrotum? posterior of the abdomen?)*.

Undescended testes is a condition known as ________________. If not corrected,

this condition places the male at 30 to 50 times greater risk for ________________,

as well as increased risk of ________________.

d. Skeletal muscles that elevate testes closer to the warmth of the pelvic cavity are known as *(cremaster? dartos?)* muscles.

✓ **B2.** Discuss spermatogenesis in this exercise.

a. Spermatozoa are one type of gamete; name the other type: ________________

Gametes are *(haploid? diploid?)*. Human gametes contain ________ chromosomes.

b. Fusion of gametes produces a cell called a ________________. This cell, and all cells of the organism derived from it, contain the *(haploid? diploid?)* number of chromosomes, written as *(n? 2n?)*. These chromosomes, received from both sperm and

ovum, are said to exist in ________________ pairs.

c. Gametogenesis assures production of haploid gametes from otherwise diploid

humans. In males, this process, called ________________-genesis, occurs in the

________________.

✓ **B3.** Refer to Figure LG 28.1 and do the following Checkpoint. Use the following list of cells for answers.

Early spermatid	**Secondary spermatocyte**
Leydig cell	**Spermatogonium**
Late spermatid	**Spermatozoon (sperm)**
Primary spermatocyte	**Sertoli cell**

a. Label cells A and B and 1–6.

b. Mark cells 1–6 according to whether they are diploid (2n) or haploid (n).

c. Meiosis is occurring in cells numbered ________ and ________. Color these meiotic cells according to color ovals on the figure. Which cell undergoes reduction division

(meiosis I)? ________ Which undergoes equatorial division (meiosis II)? ________

d. From this figure, you can conclude that as cells develop and mature, they move *(toward? away from?)* the lumen of the tubule.

e. Which cells are in the process of spermiogenesis? ________ and ________

f. Which cells secrete testosterone? ________

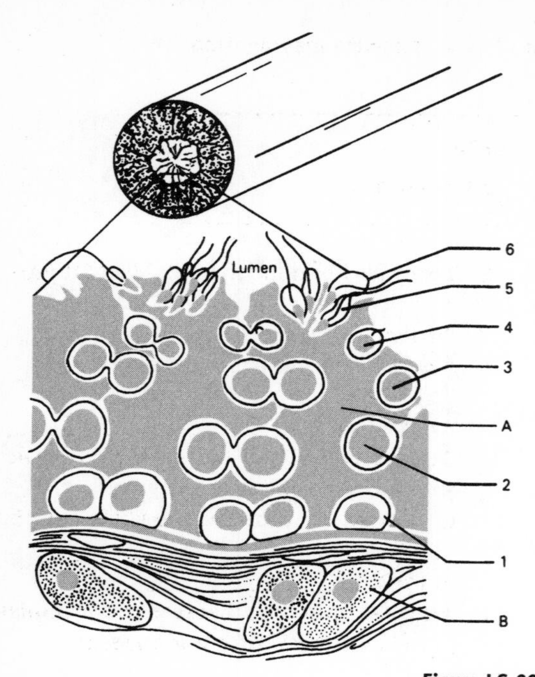

O Primary spermatocyte
O Secondary spermatocyte

Figure LG 28.1 Diagram of a portion of a cross section of a seminiferous tubule showing the stages of spermatogenesis. Color and label as directed in Checkpoint B3.

g. Which cells produce nutrients for developing sperm, phagocytose excess spermatid cytoplasm, form the blood-testis barrier, and produce androgen-binding protein (ABP) and the hormone inhibin? ________

h. Which cells lie dormant near the basement membrane until they divide mitotically beginning at puberty? They also serve as a reservoir of stem cells available for spermatogenesis throughout life. ________

B4. Explain how cytoplasmic bridges between cells formed during spermatogenesis may facilitate survival of Y-bearing sperm and therefore permit generation of male offspring.

✓ **B5.** Describe spermatozoa (sperm) in this Checkpoint.

a. Number produced in a normal male: ____________________/day

b. Life expectancy of sperm once inside female reproductive tract: ________ hours

c. Portion of sperm containing nucleus: ____________________

d. Part of sperm containing mitochondria: ____________________

e. Function of acrosome:

f. Function of tail: ____________________

✓ **B6.** Match the hormones listed in the box with descriptions below. Answers may be used more than once.

DHT	**FSH**	**GnRH**
Inhibin	**LH**	**Testosterone**

a. Released by the hypothalamus: stimulates release of two anterior pituitary hormones:

____

b. Inhibits release of GnRF: ________________

c. Co-stimulate Sertoli cells to secrete ABP, which keeps testosterone near cells undergoing spermatogenesis (two hormones):

____

d. Stimulates release of testosterone from Leydig cells: ________________

e. Principal male hormone and precursor of dihydrotestosterone (DHT):

____

f. Testicular hormone that inhibits FSH production: ________________

g. Before birth, stimulates development of external genitals: ________________

h. At puberty, facilitate development of male sexual organs and secondary sex characteristics; contribute to libido (two hormones): ________________

i. Anabolic hormones, because they stimulate protein synthesis (two hormones):

____

B7. List four functions of the androgens testosterone and DHT.

✓ **B8.** Refer to Figure LG 28.2. Structures in the male reproductive system are numbered in order along the pathway that sperm take from their site of origin to the point where they exit from the body. Match the structures in the figure (using numbers 1 to 9) with descriptions below.

________ a. Ejaculatory duct

________ b. Epididymis

________ c. Prostatic urethra

________ d. Spongy (cavernous) urethra

________ e. Membranous urethra

________ f. Ductus (vas) deferens (portion within the abdomen)

________ g. Ductus (vas) deferens (portion within the scrotum)

________ h. Urethral orifice

________ i. Testis

✓ **B9.** Which structures numbered 1–9 in Figure LG 28.2 are paired? Circle their numbers:

1 2 3 4 5 6 7 8 9

Which structures are single? Write their numbers: ________

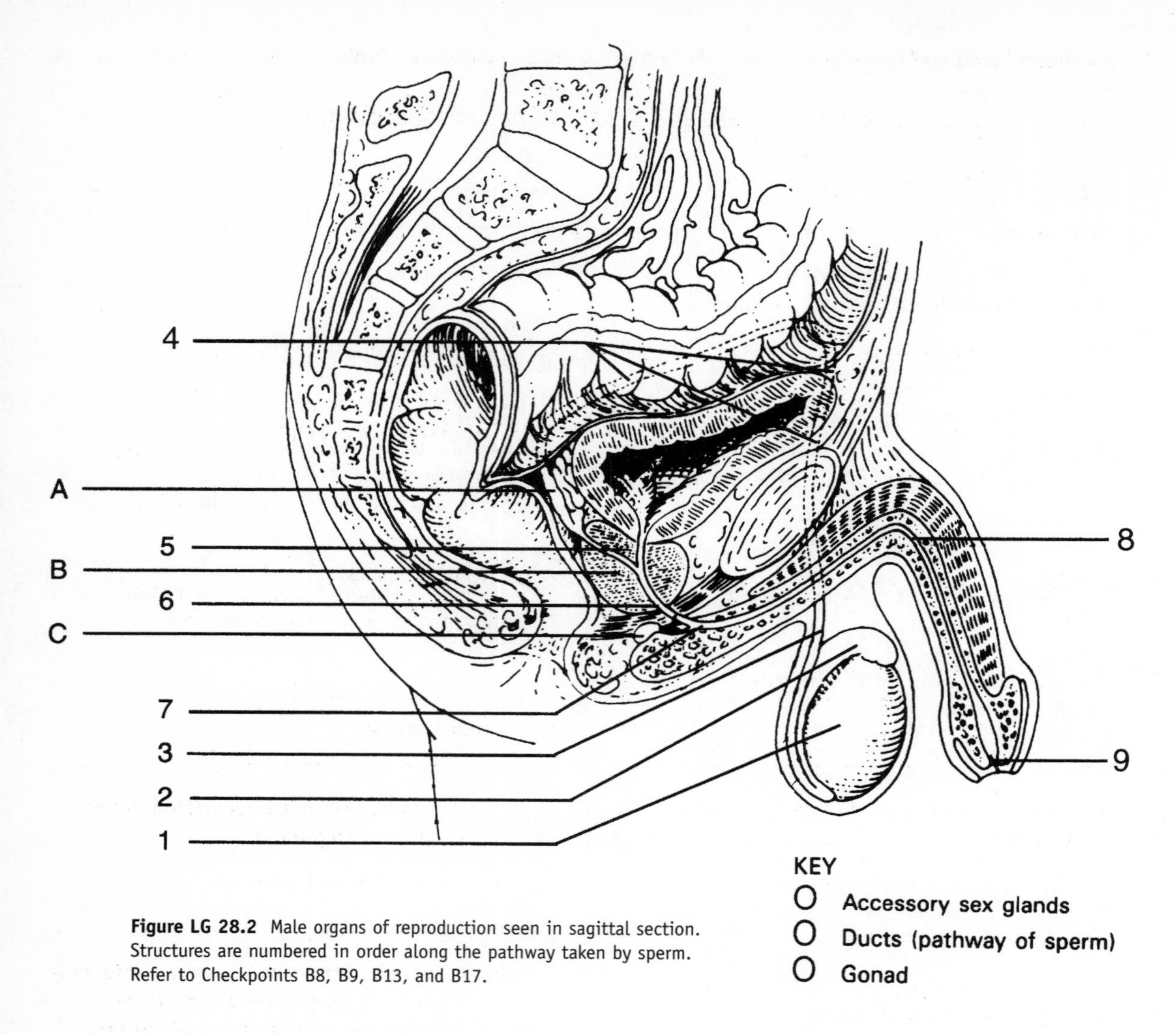

Figure LG 28.2 Male organs of reproduction seen in sagittal section. Structures are numbered in order along the pathway taken by sperm. Refer to Checkpoints B8, B9, B13, and B17.

✓ **B10.** Refer to Figure 28.5 (page 1019) and describe the male reproductive duct system in this Checkpoint.

a. Arrange these reproductive structures in correct order according to the pathway of sperm: ________ ________ ________ ________ ________

BE.	**Body of the epididymis**	**ST.**	**Straight tubules of testes**
HE.	**Head of the epididymis**	**TE.**	**Tail of the epididymmis**
RT.	**Rete testes**		

b. List several functions of the epididymis.

c. The ductus deferens is about ________ cm (________ inches) long.

d. It is located *(entirely? partially?)* in the abdomen. It enters the abdomen via the

____________________. Protrusion of abdominal contents through this weakened

area is known as ____________________.

e. Sperm *(are? are not?)* stored in the ductus (vas) deferens.

f. What structures besides the vas compose the spermatic cord?

g. A vasectomy is performed at a point along the section of the vas that is within the

____________________. Must the peritoneal cavity be entered during this procedure? *(Yes? No?)* Will the procedure affect testosterone level and sexual desire of the male? *(Yes? Not normally?)* Why?

✓ **B11.** Which portion of the male urethra is longest? *(Prostatic? Membranous? Spongy or cavernous?)*

✓ **B12.** Match the accessory sex glands with their descriptions.

B. Bulbourethral glands	**P. Prostate gland**	**S. Seminal vesicles**

________ a. Paired pouches posterior to the urinary bladder

________ b. Structures that empty secretions (along with contents of vas deferens) into ejaculatory duct

________ c. Single doughnut-shaped gland located inferior to the bladder; surrounds and empties secretions into urethra

________ d. Contribute about 60% of seminal fluid

________ e. Pair of pea-sized glands located in the urogenital diaphragm

________ f. Forms 25% of semen

________ g. Produces PSA

________ h. Fluid produced here is alkaline, and contains fructose, prostaglandins (that increase sperm motility), and clotting proteins

✓ **B13.** Color parts of the male reproductive system on Figure LG 28.2 according to color code ovals. Label the *seminal vesicle, prostate gland,* and *bulbourethral gland* on the figure.

✓ **B14.** *A clinical correlation.* Explain the clinical significance of the shape and location of the prostate gland. (*Hint*: See page 1053.)

✓ **B15.** Complete the following exercise about semen.

a. The average amount per ejaculation is ________ mL.

b. The average range of number of sperm is ____________________/mL. When the count falls below ____________________/mL, the male is likely to be sterile. Only one sperm fertilizes the ovum. Explain why such a high sperm count is required for fertility.

c. The pH of semen is slightly *(acid? alkaline?)*. State the advantage of this fact.

d. What is the function of *seminalplasmin*?

✓ **B16.** Refer to Figure 28.12 in the text and do this exercise.

a. The ____________________, or foreskin, is a covering over the ____________________ of the penis. Surgical removal of the foreskin is known as ____________________.

b. The names *corpus* ____________________ and *corpus* ____________________ indicate that these bodies of tissue in the penis contain spaces that distend in the presence of excess ____________________. *(Sympathetic? Parasympathetic?)* nerves stimulate vasodilation of the blood vessels within the penis. As a result, blood is prohibited from leaving the penis because *(arteries? veins?)* are compressed. This temporary state is known as the *(erect? flaccid?)* state.

c. The urethra passes through the corpus ____________________. The urethra functions in the transport of ____________________ and ____________________. During ejaculation, what prevents sperm from entering the urinary bladder and urine from entering the urethra?

B17. Label these parts of the penis on Figure LG 28.2: *prepuce, glans, corpus cavernosum, corpus spongiosum,* and *bulb of the penis.*

C. The female reproductive system (pages 1028–1041)

✓ **C1.** Refer to Figure LG 28.3 and match the structures numbered 1–5 with their descriptions. Note that structures are numbered in order along the pathway taken by ova from the site of formation to the point of exit from the body.

________ a. Uterus (body)

________ b. Uterus (cervix)

________ c. Ovary

________ d. Uterine (Fallopian) tube

________ e. Vagina

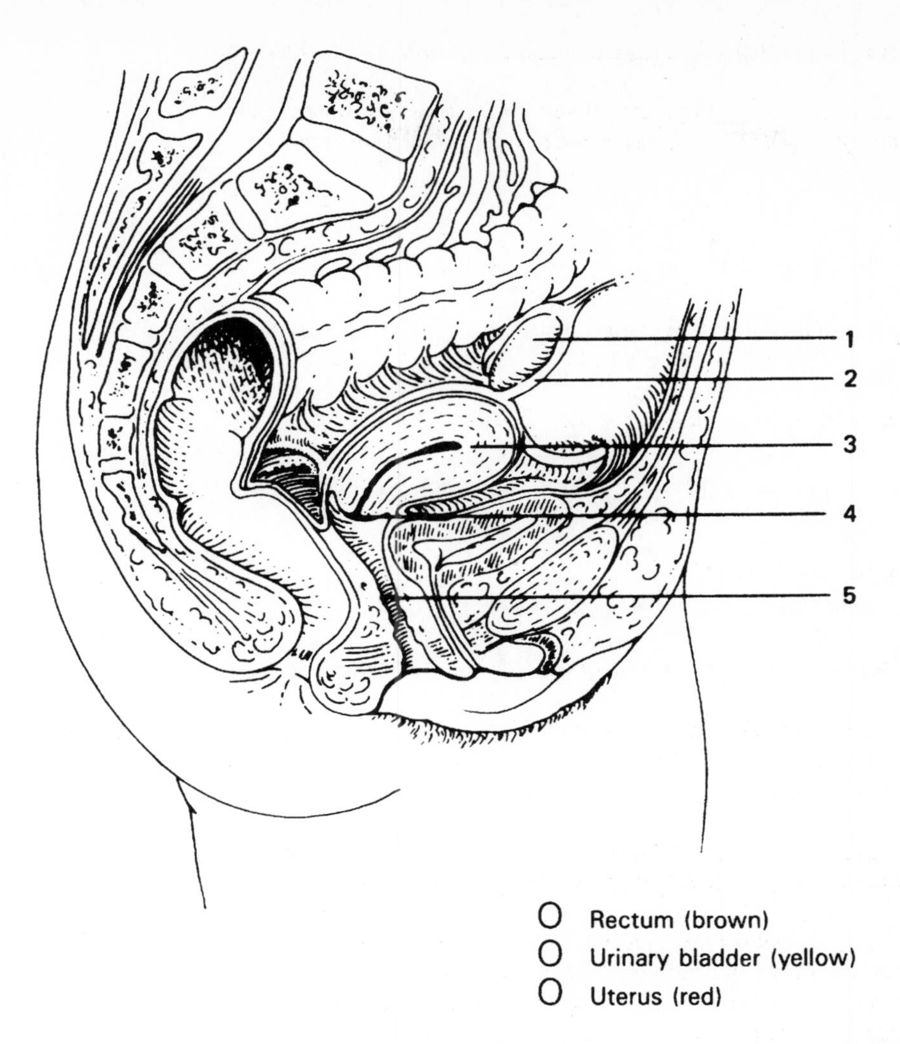

O Rectum (brown)
O Urinary bladder (yellow)
O Uterus (red)

Figure LG 28.3 Female organs of reproduction. Structures are numbered in order in the pathway taken by ova. Refer to Checkpoints C1, C2, C7, C9, and C11.

✓ **C2.** Indicate which of the five structures in Figure LG 28.3 are paired. Circle the numbers.

1 2 3 4 5

Which structures are singular? Write their numbers: ________

✓ **C3.** Describe the ovaries in this Checkpoint.

a. The two ovaries most resemble:
   A. Almonds (unshelled)
   B. Peas
   C. Pears
   D. Doughnuts

b. Position of the ovaries is maintained by *(ligaments? muscles?)*. Name two or more of these structures.

c. Arrange in correct sequence the layers of ovaries from most superficial to deepest:

_______ _______ _______

GE. Germinal epithelium OC. Ovarian cortex OM. Ovarian medulla

d. Arrange in sequence according to chronological appearance, from first to develop to

last. _______ _______ _______ _______ _______

CA. Corpus albicans
CH. Corpus hermorrhagicum
CL. Corpus luteum
MF. Mature (Graafian) follicle
OF. Ovarian follicle

✓ **C4.** Color the four cells on Figure LG 28.4 according to the color code ovals.

✓ **C5.** Describe exactly what is expelled during ovulation. Place an asterisk ([*]) next to that cell on Figure LG 28.4.

What determines whether the second meiotic (equatorial) division will occur?

✓ **C6.** Complete this exercise about the uterine tubes.

a. Uterine tubes are also known as ________________________ tubes. They are about

________ cm (________ inches) long.

b. Arrange these parts of the uterine tubes in order from most medial to most lateral:

______ ______ ______

A. Ampulla B. Infundibulum and fimbriae C. Isthmus

c. What structural features of the uterine tube help to draw the ovum into the tube and propel it toward the uterus?

d. List two functions of uterine tubes.

e. Define *ectopic pregnancy* and list possible causes of this condition. (*Hint*: See page 1067.)

C7. Color the pelvic organs on Figure LG 28.3. Use the color code ovals on the figure.

✓ **C8.** Answer these questions about uterine position.

a. The organ that lies anterior and inferior to the uterus is the ____________________.

The ____________________ lies posterior to it.

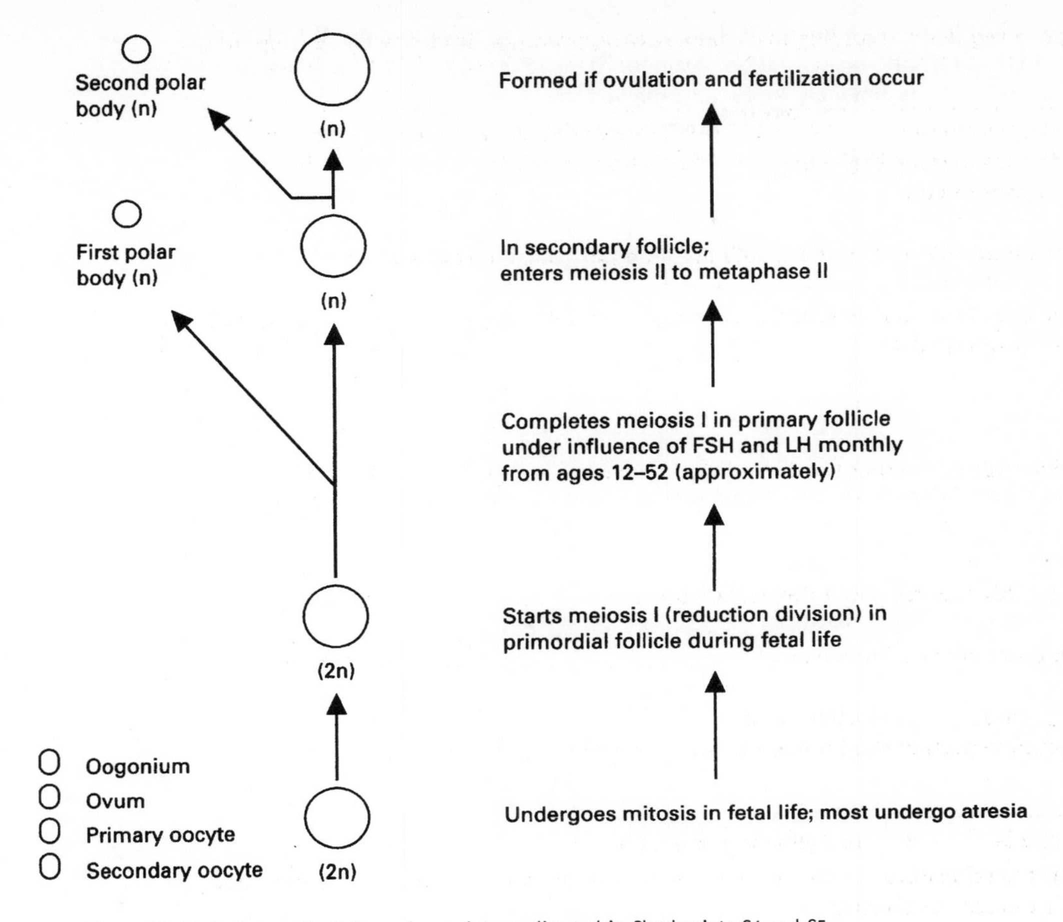

Figure LG 28.4 Oogenesis. Color and complete as directed in Checkpoints C4 and C5.

b. The rectouterine pouch lies *(anterior? posterior?)* to the uterus, whereas the

_________________-uterine pouch lies between the urinary bladder and the uterus.

c. The fundus of the uterus is normally tipped *(anteriorly? posteriorly?)*. If it is malpositioned posteriorly, it would be *(anteflexed? retroflexed?)*.

d. The ___________________ ligaments attach the uterus to the sacrum. The

___________________ ligaments pass through the inguinal canal and anchor into

external genitalia (labia majora). The ___________________ ligaments are broad, thin sheets of peritoneum extending laterally from the uterus.

e. Important ligaments in prevention of drooping (prolapse) of the uterus are the

___________________ ligaments, which attach the base of the uterus laterally to the pelvic wall. In third degree prolapse, the *(cervix of the? entire?)* uterus is ouside the vagina.

✓ **C9.** Refer to Figure LG 28.3 and do this exercise about layers of the wall of the uterus.

a. Most of the uterus consists of ___________________-metrium. This layer is *(smooth muscle? epithelium?)*. It should appear red in the figure (as directed above).

b. Now color the peritoneal covering over the uterus green. This is a *(mucous? serous?)*

membrane known as the ___________________.

c. The innermost layer of the uterus is the ___________________-metrium. Which portion of it is shed during menstruation? Stratum *(basalis? functionalis?)* Which arteries supply the stratum functionalis? *(Spiral? Straight?)*

d. The innermost layer of the cervix produces mucus that is *(more? less?)* viscous around the time of ovulation. This characteristic *(impedes? enhances?)* sperm passage through the cervical canal.

C10. Surgical removal of the uterus is known as a ___________________. Contrast the radical and subtotal forms of this surgery.

✓ **C11.** Refer to Figure LG 28.3. In the normal position, at what angle does the uterus join the vagina? ___________________

C12. Describe these vaginal structures.

a. Fornix

b. Rugae

c. Hymen

✓ **C13.** The pH of vaginal mucosa is *(acid? alkaline?)*. What is the clinical significance of this fact?

✓ **C14.** Refer to Figure LG 28.5 and do the following exercise.

a. Label these perineal structures: *anus, clitoris, labia major, labia minor, mons pubis, urethra,* and *vagina.*

b. Draw the borders of the perineum, using landmarks labeled on the diagram. Draw a line between the right and left ischial tuberosities. Label the triangles that are formed.

In which triangle are all of the female external genitalia located? ___________________

c. Draw a dotted line where an *episiotomy* incision would be performed for childbirth.

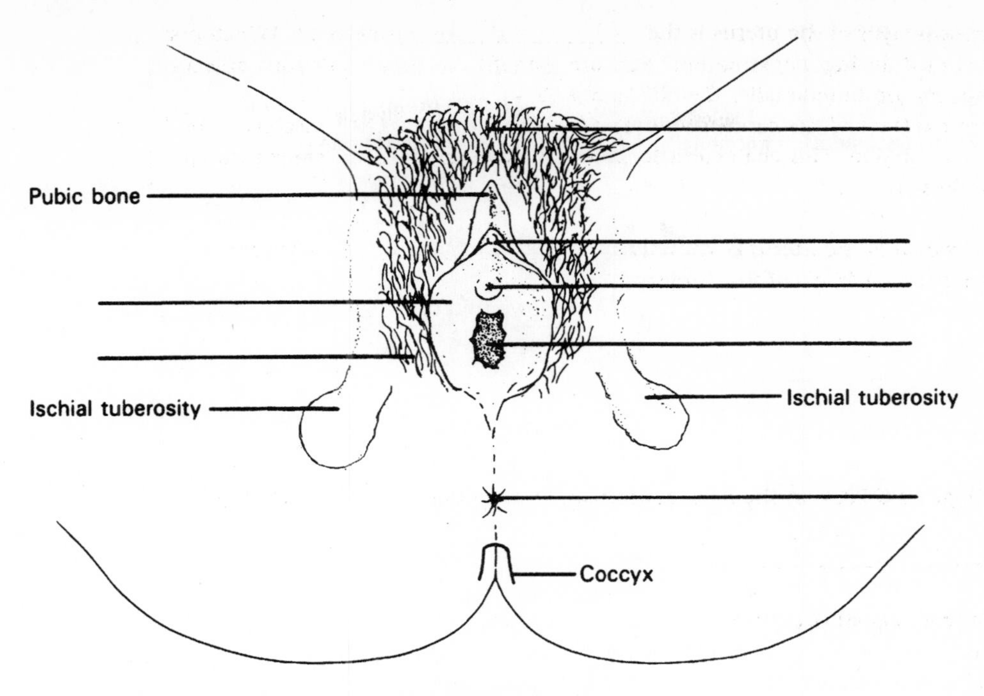

Figure LG 28.5 Perineum. Label as directed in Checkpoint C14.

✓ **C15.** Describe mammary glands in this Checkpoint.

Each breast consists of 15–20 ____________________ composed of lobules. Milk-

secreting glands, known as ____________________, empty into a duct system that terminates at the *(areola? nipple?)*.

✓ **C16.** Identify homologous reproductive structures by completing this table. (*Hint*: refer to text Table 28.2.)

Female structure	**Male structure**
a.	a. Scrotum
b. Paraurethral glands	b.
c.	c. Glans penis
d. Bulb of the vestibule	d.

D. The female reproductive cycle (pages 1041–1046)

D1. Contrast *menstrual cycle* and *ovarian cycle*.

✓ **D2.** Select hormones produced by each of the endocrine glands listed below.

E.	**Estrogens**	**LH.**	**Luteinizing hormone**
FSH.	**Follicle-stimulating hormone**	**P.**	**Progesterone**
GnRH.	**Gonadotropin-releasing hormone**	**R.**	**Relaxin**
I.	**Inhibin**		

Hypothalamus: ____________________ Anterior pituitary: ____________________

Ovary: ____________________

✓ **D3.** Now select the hormones listed in C2 that fit descriptions below.

_______ a. Stimulates release of FSH

_______ b. Stimulates release of LH

_______ c. Inhibits release of FSH

_______ d. Inhibits release of LH

_______ e. Stimulates development and maintenance of uterus, breasts, and secondary sex characteristics; enhances protein anabolism

_______ f. Prepares endometrium for implantation by fertilized ovum

_______ g. Dilates cervix for labor and delivery

_______ h. β-Estradiol, estrone, and estriol are forms of this hormone

_______ i. May lower blood cholesterol level

_______ j. Secreted by follicle cells

_______ k. Secreted by corpus luteum cells

✓ **D4.** Refer to Figure LG 28.6, and check your understanding of events in female cycles in this checkpoint.

a. The average duration of the menstrual cycle is _______ days. The menstrual phase usually lasts about _______ days. During this period about _______ mL of blood, cells, mucus, and other fluid exits from the uterus in menstrual flow. On Figure LG 28.6d, color red the endometrium in the menstrual phase, and note the changes in its thickness. Also write *menses* on the line in that phase on Figure LG 28.6d.

b. Following the menstrual phase is the ____________________ phase. During both of these phases ____________________ develop in the ovary. Usually only _______ matures each month, becoming the dominant follicle, and later the ____________________ (Graafian) follicle. The mature follicle contains a developing ovum that has reached metaphase of meiosis *(I? II?)* of oogenesis.

c. Follicle development occurs under the influence of the hypothalamic hormone _______, which regulates the anterior pituitary hormones _______ and later _______.

d. What are the two main functions of follicles?

e. How do estrogens affect the endometrium?

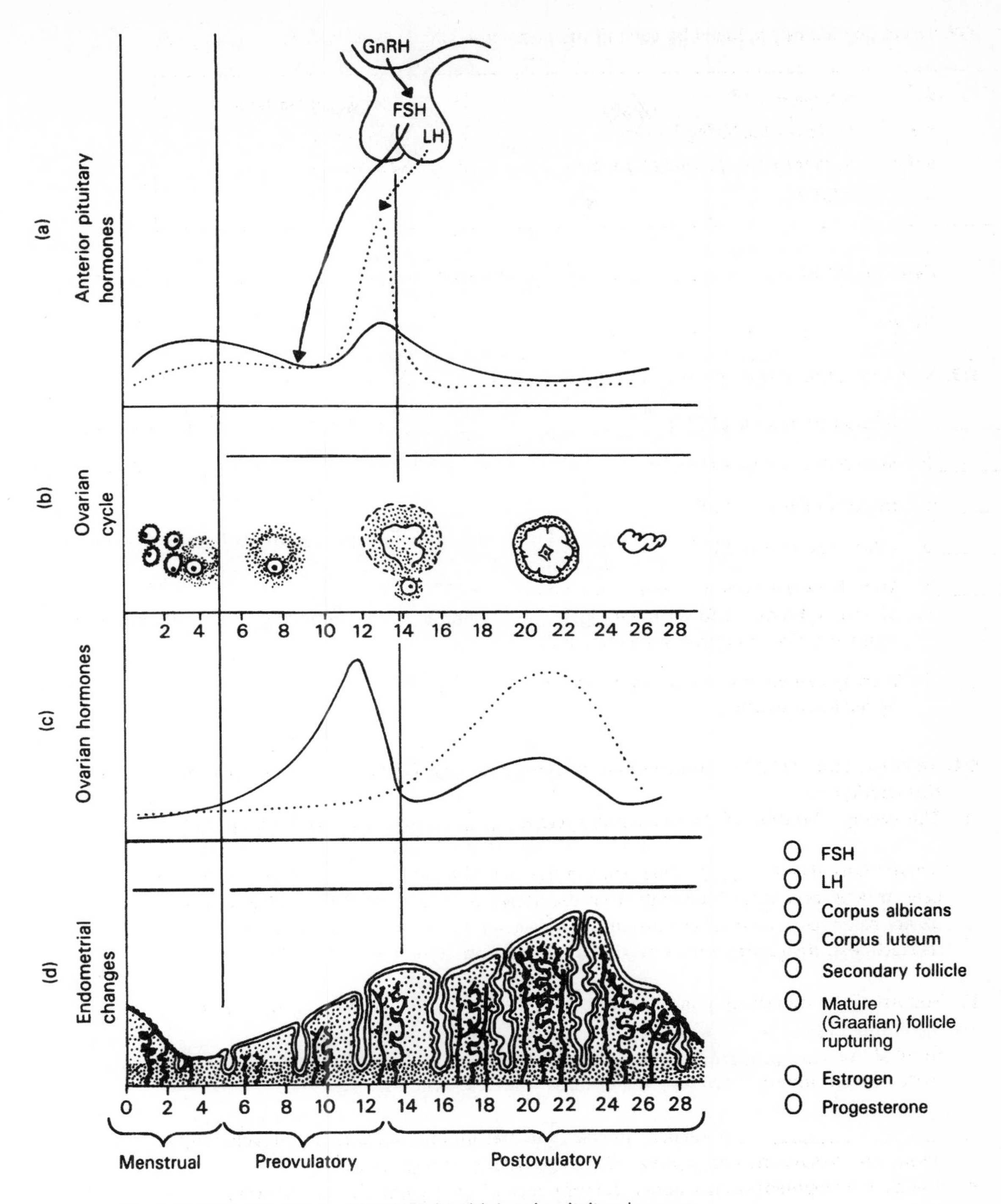

Figure LG 28.6 Correlations of female cycles. (a) Anterior pituitary hormones. (b) Ovarian changes: follicles and corpus luteum. (c) Ovarian hormones. (d) Uterine (endometrial) changes. Color and label as directed in Checkpoint D4.

For this reason, the preovulatory phase is also known as the ______________________ phase. Because follicles reach their peaks during this phase, it is also known as the

______________________ phase. Write those labels on Figures LG 28.6d and 28.6b.

f. A moderate increase in estrogen production by the growing follicles initiates a *(positive? negative?)* feedback mechanism that *(increases? decreases?)* FSH level. It is this change in FSH level plus secretion of the hormone ___________________ by follicles (See Checkpoint D3j) that causes atresia of the remaining partially developed follicles. Atresia means *(degeneration? regeneration?).*

g. The principle of this negative feedback effect is utilized in oral contraceptives (See text page 1048). By starting to take "pills" (consisting of target hormones estrogen plus chemicals similar to progesterone) early in the cycle (day 5), and continuing until day 25, a woman

will maintain a very low level of the tropic hormones ___________________. Without these hormones, follicles and ova will not develop, and so the woman will not

___________________.

h. The second anterior pituitary hormone released during the cycle is ________. The surge of this hormone results from the *(high? low?)* level of estrogens present just at the end of the preovulatory phase. This is an example of one of the very few *(positive? negative?)* feedback mechanisms in the human body.

i. The LH surge causes the release of the ovum (actually secondary oocyte), the event

known as ___________________. The LH surge is the basis for a home test for *(ovulation? pregnancy?)*. Color all stages of follicle development through ovulation on Figure LG 28.6b according to the color code ovals.

j. Following ovulation is the ___________________ phase. It lasts until day ________

of a 28-day cycle. Under the influence of tropic hormone ________, follicle cells are

changed into the corpus ___________________. These cells secrete two hormones:

___________________ and ___________________. Both prepare the endometrium

for ___________________. What preparatory changes occur?

k. What effect do rising levels of target hormones estrogens and progesterone have upon the GnRH and LH?

l. One function of LH is to form and maintain the corpus luteum. So as LH ________-creases, the corpus luteum disintegrates within 2 weeks, forming a white scar on the

ovary known as the corpus ___________________. Color these two final stages of the original follicle on Figure LG 28.6b. Also write alternative names for the postovulatory phase on lines in Figure LG 28.6b and d.

m. The corpus luteum had been secreting _________________ and _________________. With the demise of the corpus luteum, the levels of these hormones rapidly *(increase? decrease?)*. Because these hormones were maintaining the endometrium, endometrial tissue now deteriorates and will be shed during the next

___________________ phase. Declining progesterone especially leads to menses as

this condition stimulates release of ___________________ that *(constrict? dilate?)* endometrial blood vessels.

n. If fertilization should occur, estrogens and progesterone are needed to maintain the endometrial lining. The ____________________ around the developing embryo secretes a hormone named ____. It functions much like LH in that it maintains the corpus ____________________, even though LH has been inhibited (step *k* above). The corpus luteum continues to secrete estrogens and progesterone for several months until the placenta itself can secrete sufficient amounts. Incidentally, hCG is present in the urine of pregnant women only and so is routinely used to detect ____________________.

o. Review hormonal changes by coloring blood levels of hormones in Figure LG 28.6a and c, according to the color code ovals. Then note uterine changes by coloring red the remainder of the endometrium in Figure LG 28.6d.

✓ **D5.** To check your understanding of the events of these cycles, list letters of the major events in chronological order beginning on day 1 of the cycle. To do this exercise it may help to list on a separate piece of paper days 1 to 28 and write each event next to the approximate day on which it occurs.

_____ _____ _____ _____ _____ _____ _____ _____ _____ _____ _____

A. Ovulation occurs and may cause mittelschmerz
B. Estrogen and progesterone levels drop
C. FSH begins to stimulate follicles
D. LH surge is stimulated by high estrogen levels
E. High levels of estrogens and progesterone inhibit LH
F. Estrogens are secreted for the first time during the cycle
G. Corpus luteum degenerates and becomes corpus albicans
H. Endometrium begins to deteriorate, leading to next month's menses
I. Follicle is converted into corpus hemorrhagicum and then into corpus luteum
J. Rising level of estrogen inhibits FSH secretion
K. Corpus luteum secretes estrogens and progesterone

✓ **D6.** Write terms that mean:

a. Painful menstruation: ____________________.

b. Absence of monthly menstrual flow: ____________________.

✓ **D7.** List the three conditions that are part of the *female athletic triad.*

Explain how osteoporosis is associated with amenorrhea.

E. The human sexual response; birth control; abortion (pages 1046–1050)

E1. Define *sexual intercourse* or *coitus.*

✓ **E2.** Answer these questions about the process.

a. *(Sympathetic? Parasympathetic?)* impulses from the *(brainstem? sacral cord?)* cause vasodilation of arteries of the penis or the vagina, leading to the *(flaccid? erect?)* state.

b. *(Sympathetic? Parasympathetic?)* impulses cause sperm to be propelled into the urethra of the male. Expulsion from the urethra to the exterior of the body is called

_______________________.

E3. Compare and contrast the phases of sexual activity in females and males by completing this table.

Phase	Female	Male
a. Erection		
b. Lubrication		
c. Orgasm		

✓ **E4.** Describe the significance of *transudation* in female sexual response.

E5. Discuss *erectile dysfunction* in this Checkpoint.

a. Erectile dysfunction (also known as _______________________). List several causes of this condition.

b. Nitric oxide causes *(contraction? relaxation?)* of smooth muscles in arteries of the penis.

The result of release of nitric oxide is ________-crease of blood flow into the penis, leading to the *(erect? flaccid state?)*.

c. Many cases of erectile dysfunction are associated with nitric oxide insufficiency. Name a medication used to treat this disorder by enhancing effects of nitric oxide.

____

✓ **E6.** Match the methods of birth control listed in the box with descriptions below.

A.	**Abstinence**	**R.**	**Rhythm**
CI.	**Coitus interruptus**	**RU 486.**	**RU 486**
Con.	**Condom**	**S.**	**Spermicides**
D.	**Diaphragm**	**T.**	**Tubal ligation**
IUD.	**Intrauterine device**	**Vag.**	**Vaginal ring**
N.	**Norplant**	**Vas.**	**Vasectomy**
OC.	**Oral contraceptive**		

_______ a. Foams, jellies, and creams that kill sperm

_______ b. "The pill," a combination of progesterone and estrogens, causing decrease in GnRH, FSH, and LH so ovulation does not occur

_______ c. Removal of a portion of the ductus (vas) deferens

_______ d. Tying off of the uterine tubes

_______ e. A natural method of birth control involving abstinence during the period when ovulation is most likely

_______ f. A mechanical method of birth control in which a dome-shaped structure is placed over the cervix; accompanied by use of a jelly or cream

_______ g. Copper T 380 A placed in the uterus by a physician; prevents implantation

_______ h. A rubber sheath over the penis

_______ i. Withdrawal

_______ j. Progestin-containing capsules placed under the skin for up to 5 years

_______ k. Competitive inhibitor of progesterone; a chemical that induces miscarriage

_______ l. The 100% reliable method of avoiding pregnancy by avoidance of sexual intercourse

_______ m. Worn inside the vagina for three weeks each month, where it releases progestin and possibly estrogen

E7. Review the mechanism of action or oral contraceptives (Checkpoint D4g). Then contrast benefits and risks of use of "the pill."

E8. Contrast *spontaneous abortion* with *induced abortion.*

F. Developmental anatomy and aging of the reproductive systems (pages 1050–1053)

✓ **F1.** Name the reproductive structures (if any) that develop from the following embryonic ducts.

a. Mesonephric duct in females: ______________________________

b. Mesonephric duct in males: ______________________________

c. Paramesonephric (Müller's) duct in females: ______________________________

d. Paramesonephric (Müller's) duct in males: ______________________________

F2. Describe the SRY gene

F3. Review homologous structures in Checkpoint C16.

F4. Contrast these terms: *menarche/menopause*

F5. Describe normal aging changes that involve the reproductive system in:

a. Females

b. Males

F6. List five or more common manifestations of benign prostatic hyperplasia (BPH).

G. Disorders, medical terminology (pages 1053–1056)

✓ **G1.** Identify two tests that screen for prostate cancer.

G2. List six symptoms of *premenstrual syndrome (PMS).*

________________________ ________________________

________________________ ________________________

________________________ ________________________

✓ **G3.** *A clinical correlation.* Mrs. Rodriguez, a 55-year-old woman, tells the nurse that she is concerned about having breast cancer. Do this exercise about the nurse's conversation with Mrs. Rodriguez.

a. The nurse asks some questions to evaluate Mrs. Rodriguez's risk of breast cancer. Which of the following responses indicate an increased risk of breast cancer?
   A. "My older sister had cancer when she was 40 and she had the breast removed."
   B. "I've never had any kind of cancer myself."
   C. "I have two daughters; they were born when I was 36 and 38."
   D. "I don't smoke and I never did."
   E. "I do need to cut back on fatty foods; I know I eat too many of them."

b. The nurse tells the client that the American Cancer Society recommends three methods of early detection of breast cancer. These are listed below. On the line next to each of these, write the frequency that the nurse is likely to suggest to Mrs. Rodriguez.
   1. Self breast exam (SBE) ____________________
   2. Mammogram ____________________
   3. Breast exam by physician ____________________

✓ **G4.** Oophorectomy is removal of a(n) ____________________, whereas ____________________ refers to removal of a uterine (Fallopian) tube.

G5. Define *sexually transmitted disease* (STD).

G6. Describe the main characteristics of these sexually transmitted diseases by completing the table.

Disease	Causative Agent	Main Symptoms
a. Gonorrhea		
b.	*Treponema pallidum*	
c.	Type II herpes simplex virus	
d. Chlamydia		

✓ **G7.** Select disorders in the box that fit descriptions below. One answer will be used twice.

A.	**Amenorrhea**	**E.**	**Endometriosis**
Can.	**Candidiasis**	**G.**	**Gonorrhea**
Chl.	**Chlamydia**	**H.**	**Hypospadias**
CC.	**Cervical cancer**	**S.**	**Syphilis**
D.	**Dysmenorrhea**		

_______ a. Painful menstruation, partly due to contractions of uterine muscle

_______ b. Possible cause of infection and possible blindness in newborns if bacteria transmitted to eyes during birth

_______ c. Spreading of uterine lining into abdominopelvic cavity via uterine tubes

_______ d. Caused by bacterium *Treponema pallidum;* may involve many systems in tertiary stage

_______ e. Absence of menstrual periods

_______ f. May be detected by a Pap smear; is linked to genital warts (human papilloma virus = HPV)

_______ g. The most prevalent STD in the U.S., leading to 20,000 cases of sterility each year

_______ h. Leading cause of PID (pelvic inflammatory disease)

_______ i. Displaced urethral opening in males

_______ j. Caused by yeast; most common form of vaginal infection

ANSWERS TO SELECTED CHECKPOINTS: CHAPTER 28

A2. Female: (a) Ovaries. (b) Uterine (Fallopian) tubes, uterus, and vagina. (c) Greater vestibular (Bartholin's) glands. (d) External genitalia, breasts. Male: (a) Testes. (b) Seminiferous tubules, epididymis, vas deferens, ejaculatory duct, urethra. (c) Prostate, seminal vesicles, bulbourethral glands. (d) Scrotum, penis.

A3. (a) Gonads (ovaries or testes), gametes; fertilization, zygote. (b) 46, diploid or 2n; 23. (c) Sex chromosomes; two X chromosomes; autosomes. (d) 23, n. (e) Two, once; reduction, equatorial. (f) Before, "46 × 2" (or 46 doubled); centromere. (g) Synapsis; does not; tetrad; crossing-over; permits exchange of DNA between chromatids and greatly increases possibilities for variety among sperm or eggs and thus offspring. (h) 1. (i) "23 × 2" (or 23 doubled chromatids); centromeres, 23. (j) 4, 1, polar bodies.

B1. (a) D TV TA. (b) Lower; lower temperature is required for normal sperm production. (c) Posterior of the abdomen; cryptorchidism; testicular cancer; sterility. (d) Cremaster.

B2. (a) Ova; haploid; 23. (b) Zygote; diploid, 2n; homologous. (c) Spermato, seminiferous tubules of testes.

B3. (a-b) See Figure LG 28.lA. (c)2, 3; 2; 3. (d) Toward. (e) 4, 5. (f) B. (g) A. (h) 1.

B5. (a) 300 million. (b) 48. (c) Head. (d) Midpiece. (e) A specialized lysosome that releases enzymes that help the sperm to penetrate the secondary oocyte. (f) Propels sperm.

B6. (a) GnRF. (b) Testosterone. (c) FSH and testosterone. (d) LH. (e) Testosterone. (f) Inhibin. (g) DHT. (h-i) Testosterone and DHT.

B8. (a) 5. (b) 2. (c) 6. (d) 8. (e) 7. (f) 4. (g) 3.(h) 9. (i) 1.

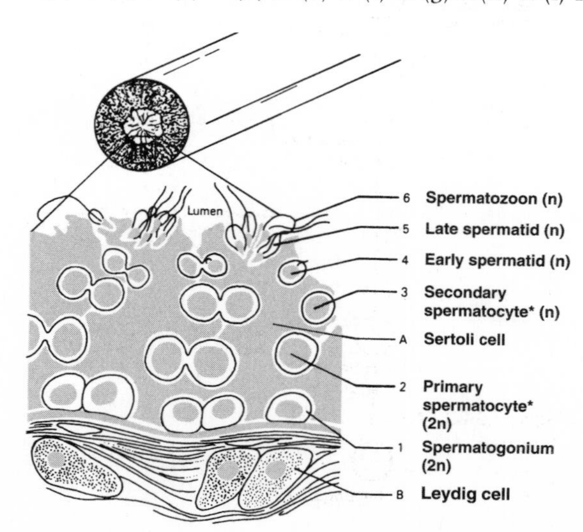

Figure LG 28.1A Diagram of a portion of a cross section of a seminiferous tubule showing the stages of spermatogenesis.

B9. Paired: 1–5; singular: 6–9.

B10. (a) ST RT HE BE TE. (b) Site where sperm mature and acquire motility as well as the ability to fertilize an ovum. (c) 45, (18). (d) Partially; inguinal canal and rings; inguinal hernia. (e) Are. (f) Testicular artery, veins, lymphatics, nerves, and cremaster muscle. (g) Scrotum; no: not normally; hormones exit through testicular veins, which are not cut during vasectomy.

B11. Spongy or cavernous.

B12. (a) S. (b) S. (c) P. (d) S. (e) B. (f) P. (g) P. (h) S.

B13. Accessory sex glands: A (seminal vesicles), B (prostate). C (bulbourethral glands); ducts, 2–9; gonad, 1.

B14. Because it surrounds the urethra (like a donut), an enlarged prostate can cause painful and difficult urination (dysuria) and bladder infection (cystitis) due to incomplete voiding.

B15. 2.5–5 ml. (b) 50–150 million; 20 million; a high sperm count increases the likelihood of fertilization, because only a small percentage of sperm ever reaches the secondary oocyte; in addition, a large number of sperm are required to release sufficient enzyme to digest the barrier around the secondary oocyte. (c) Alkaline; protects sperm against acidity of the male urethra and the female vagina. (d) Serves as an antibiotic that destroys bacteria.

B16. (a) Prepuce, glans; circumcision. (b) Cavernosum, spongiosum, blood; parasympathetic; veins; erect. (c) Spongiosum; urine, sperm; sphincter action at base of bladder.

C1. (a) 3. (b) 4. (c) 1. (d) 2. (e) 5.

C2. Paired: 1–2; Singular: 3–5.

C3. (a) A. (b) Ligaments; mesovarian, ovarian, and suspensory ligament. (c) GE OC OM. (d) OF MF CL CH CA.

C4.

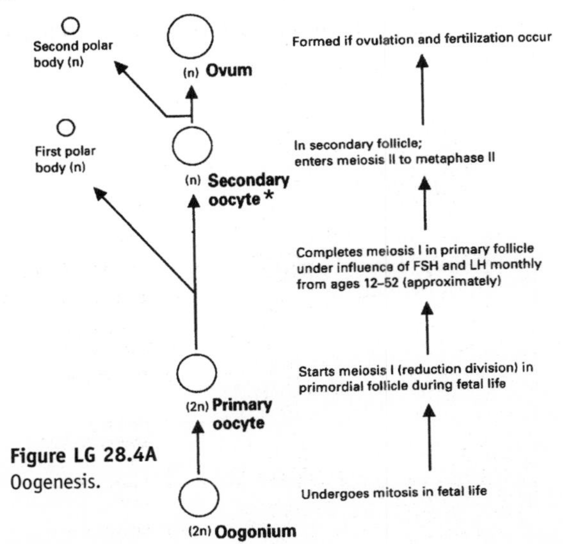

Figure LG 28.4A Oogenesis.

C5. The secondary oocyte (see Figure LG 28.4A) along with the first polar body and some surrounding supporting cells (corona radiata) are expelled. Meiosis II is completed only if the secondary oocyte is fertilized by a spermatazoon.

C6. (a) Fallopian; 10 (4). (b) C A B. (c) Currents produced by movements of fimbriae, ciliary action of columnar epithelial cells lining tubes, and peristalsis of the muscularis layer of the uterine tube wall. (d) Passageway for secondary oocyte and sperm toward each other and site of fertilization. (e) Pregnancy following implantation at any location other than in the body of the uterus, for example, tubal pregnancy, or pregnancy on ovary, intestine, or in cervix of the uterus.

C8. (a) Bladder; rectum. (b) Posterior, vesico. (c) Anteriorly; retroflexed. (d) Uterosacral; round; broad. (e) Cardinal; entire.

C9. (a) Myo; smooth muscle. (b) Serous, perimetrium. (c) Endo; functionalis; spiral. (d) Less; enhances.

C11. Close to 90° angle.

C13. Acid; retards microbial growth but may harm sperm cells.

C14. (a and c) See Figure LG 28.5A. (b) Urogenital triangle.

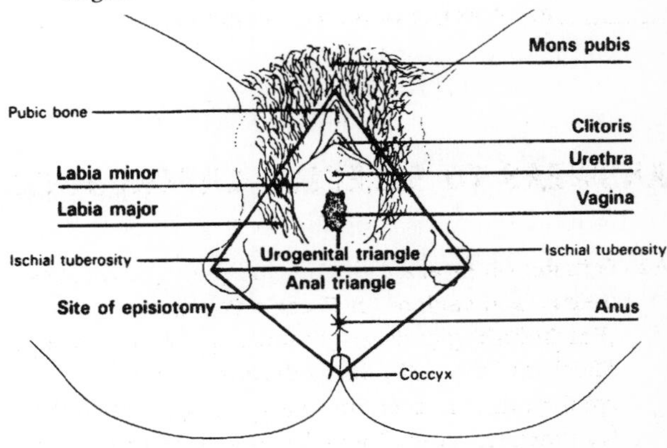

Figure LG 28.5A Perineum.

C15. Lobes; alveoli; nipple.

C16. (a) Labia majora. (b) Prostate. (c) Clitoris. (d) Corpus spongiosum penis and bulb of penis.

D2. Hypothalamus: GnRH; anterior pituitary: FSH, LH; Ovary: E, I, P, R.

D3. (a–b) GnRH. (c) I (E and P indirectly because inhibit GnRH). (d) I (E and P indirectly because they inhibit GnRH). (e) E. (f) P. (g) R. (h) E. (i) E. (j) E, I (and small amount of P). (k) E, P, R, and I.

D4. (a) 28; 5; 50–150; see Figure LG 28.6A. (b) Preovulatory; secondary follicles; 1; mature; II. (c) GnRH, FSH, LH. (d) Secretion of estrogens and development of the ovum. (e) Estrogens stimulate replacement of the sloughed off stratum functionalis by growing (proliferating) new cells; proliferative; follicular. See Figure LG 28.6A. (f) Negative, decreases; inhibin, degeneration. (g) FSH and LH; ovulate. (h) LH; high; positive. (i) Ovulation; ovulation. See Figure LG 28.6A. (j) Postovulatory; 28; LH, luteum; progesterone, estrogens; implantation;

thickening by growth and secretion of glands and increased retention of fluid. (k) Inhibition of these hormones. (1) De; albicans. See Figure LG 28.6A. (m) Progesterone, estrogens; decrease; menstrual; prostaglandins, constrict. (n) Placenta, human chorionic gonadotropin (hCG); luteum; pregnancy. (o) See Figure LG 28.6A.

D5. C F J D A I K E G B H.

D6. (a) Dysmenorrhea. (b) Amenorrhea.

D7. Eating disorder, amenorrhea, and premature osteoporosis; decreased estrogen (that causes amenorrhea) leads to the diminished capacity of bones to retain calcium.

E2. (a) Parasympathetic, sacral cord, erect. (b) Sympathetic; ejaculation.

E4. The vaginal wall lacks glands; most lubrication during sexual response is provided by engorgement of the vaginal wall and its capillaries which then causes liquid to seep out of blood vessels (transudation).

E6. (a) S. (b) OC. (c) V. (d) T. (e) R. (f) D. (g) IUD. (h) Con. (i) CI. (j) N. (k) RU 486. (l) A. (m) Vag.

F1. (a) None. (b) Ducts of testes and epididymides, ejaculatory ducts and seminal vesicles. (c) Uterus and vagina. (d) None.

G1. Digital rectal exam and blood level of PSA (prostate-specific antigen).

G3. (a) A, C, and E. (b) 1: monthly; 2 and 3: annually or more often based on the decision made by physician and client.

G4. Ovary, salpingectomy.

G7. (a) D. (b) G. (c) E. (d) S. (e) A. (f) CC. (g) Chl. (h) Chl. (i) H. (j) Can.

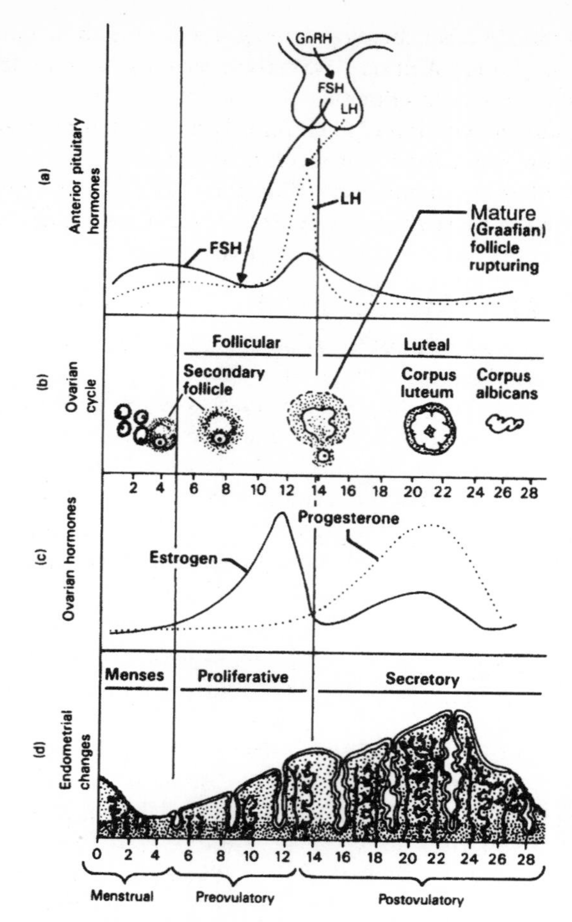

Figure LG 28.6A Correlations of female cycles. (a) Anterior pituitary hormones. (b) Ovarian changes: follicles and corpus luteum. (c) Ovarian hormones. (d) Uterine (endometrial) changes. Refer to Checkpoint D4.

MORE CRITICAL THINKING: CHAPTER 28

1. Explain how gonads function as both endocrine glands and exocrine glands.
2. Contrast the processes of oogenesis and spermatogenesis.
3. Briefly contrast the ovarian cycle with the menstrual cycle.
4. Contrast the roles of each of the following chemicals or structures in the ovarian and menstrual cycles: (a) GnRH and FSH; (b) estrogens and progesterone; (c) follicle and corpus luteum.
5. Tell exactly what specific hormonal change (increase or decrease of which hormone) accounts for each of the following: (a) ovulation; (b) change of the follicle to the corpus luteum; (c) demise of the corpus luteum in a nonpregnant woman; (d) maintenance of the corpus luteum for several months in a pregnant woman; (e) menstruation.
6. Identify the phase that is the "follicular, proliferative, estrogenic phase" of the menstrual cycle.

7. Tell exactly what chemical is tested for in over-the-counter tests for: (a) ovulation; (b) pregnancy. And state which specific hormones are found in oral contraceptives and in Depo-Provera injections.
8. Identify which form of birth control you think is both most effective with the least harmful side effects. State your rationale.
9. Describe the one major physiological change that normally accounts for both erection in the male and engorgement of female genitalia during sexual response.

MASTERY TEST: CHAPTER 28

Questions 1–10: Arrange the answers in correct sequence.

_____ _____ _____ 1. From most external to deepest:
A. Endometrium
B. Serosa
C. Myometrium

_____ _____ _____ 2. Portions of the male urethra, from proximal (closest to bladder) to distal (closest to outside of the body):
A. Prostatic
B. Membranous
C. Spongy

_____ _____ _____ 3. Thickness of the endometrium, from thinnest to thickest:
A. On day 5 of cycle
B. On day 13 of cycle
C. On day 23 of cycle

_____ _____ _____ 4. From anterior to posterior:
A. Uterus and vagina
B. Bladder and urethra
C. Rectum and anus

_____ _____ _____ _____ 5. Order in which hormones begin to increase level, from day 1 of menstrual cycle:
A. FSH
B. LH
C. Progesterone
D. Estrogen

_____ _____ _____ _____ 6. Order of events in the female monthly cycle, beginning with day 1:
A. Ovulation
B. Formation of follicle
C. Menstruation
D. Formation of corpus luteum

_____ _____ _____ _____ 7. From anterior to posterior:
A. Anus
B. Vaginal orifice
C. Clitoris
D. Urethral orifice

_____ _____ _____ _____ 8. Pathway of milk in breasts:
A. Alveoli
B. Secondary tubules
C. Mammary ducts
D. Lactiferous ducts

_____ _____ _____ _____ _____ 9. Pathway of sperm:
A. Ejaculatory duct
B. Testis
C. Urethra
D. Ductus (vas) deferens
E. Epididymis

_____ _____ _____ _____ _____ _____ 10. Pathway of sperm entering female reproductive system:
A. Uterine cavity
B. Cervical canal
C. External os
D. Internal os
E. Uterine tube
F. Vagina

Questions 11–12: Circle the letter preceding the one best answer to each question.

11. Eighty to 90% of seminal fluid (semen) is secreted by the combined secretions of:
A. Prostate and epididymis
B. Seminal vesicles and prostate
C. Seminal vesicles and seminiferous tubules
D. Seminiferous tubules and epididymis
E. Bulbourethral glands and prostate

12. In the normal male, there are two of each of the following structures *except:*
A. Testis
B. Seminal vesicle
C. Prostate
D. Ductus deferens
E. Epididymis
F. Ejaculatory duct

Questions 13–15: Circle letters preceding all correct answers to each question.

13. Which of the following are produced by the testes?
A. Spermatozoa
B. Testosterone
C. Inhibin
D. GnRH
E. FSH
F. LH

14. Which of the following are produced by the ovaries and then *leave* the ovaries?
A. Follicle
B. Secondary oocyte
C. Corpus luteum
D. Corpus albicans
E. Estrogen
F. Progesterone

15. Which of the following are functions of LH?
A. Begin the development of the follicle
B. Stimulate change of follicle cells into corpus luteum cells.
C. Stimulate release of secondary oocyte (ovulation).
D. Stimulate corpus luteum cells to secrete estrogens and progesterone.
E. Stimulate release of GnRH.

Questions 16–20: Circle T (true) or F (false). If the statement is false, change the underlined word or phrase so that the statement is correct.

T F 16. Spermatogonia and oogonia both continue to divide throughout a person's lifetime to produce new primary spermatocytes or primary oocytes.

T F 17. The menstrual cycle refers to a series of changes in the uterus, whereas the ovarian cycle is a series of changes in the ovary.

T F 18. Meiosis is a part of both oogenesis and spermatogenesis.

T F 19. Each oogonium beginning oogenesis produces four mature ova.

T F 20. An increased level of a target hormone (such as estrogen) will inhibit release of its related releasing factor and tropic hormone (GnRH and FSH).

Questions 21–25: Fill-ins. Complete each sentence with the word or phrase that best fits.

_______________ 21. _______ is a term that means undescended testes.

_______________ 22. Oral contraceptives consist of the hormones _______ and _______.

_______________ 23. The most common STD in the United States is currently _______.

_______________ 24. Menstrual flow occurs as result of a sudden _______-crease in the hormones _______ and _______.

_______________ 25. Three major functions of estrogens are development and maintenance of _______, protein _______, and lowering of blood _______ level.

ANSWERS TO MASTERY TEST: CHAPTER 28

Arrange

1. B C A
2. A B C
3. A B C
4. B A C
5. A D B C
6. C B A D
7. C D B A
8. A B C D
9. B E D A C
10. F C B D A E

Multiple Choice

11. B
12. C

Multiple Answers

13. A, B, C
14. B, E, F
15. B, C, D

True–False

16. F. Spermatogonia; primary spermatocytes (Oogonia do not divide in a female after her birth)
17. T
18. T
19. F. One mature ovum and three polar bodies
20. T

Fill-ins

21. Cryptorchidism
22. Progesterone and estrogens
23. Chlamydia
24. De, progesterone and estrogens
25. Female reproductive structures and secondary sex characteristics, anabolism, cholesterol

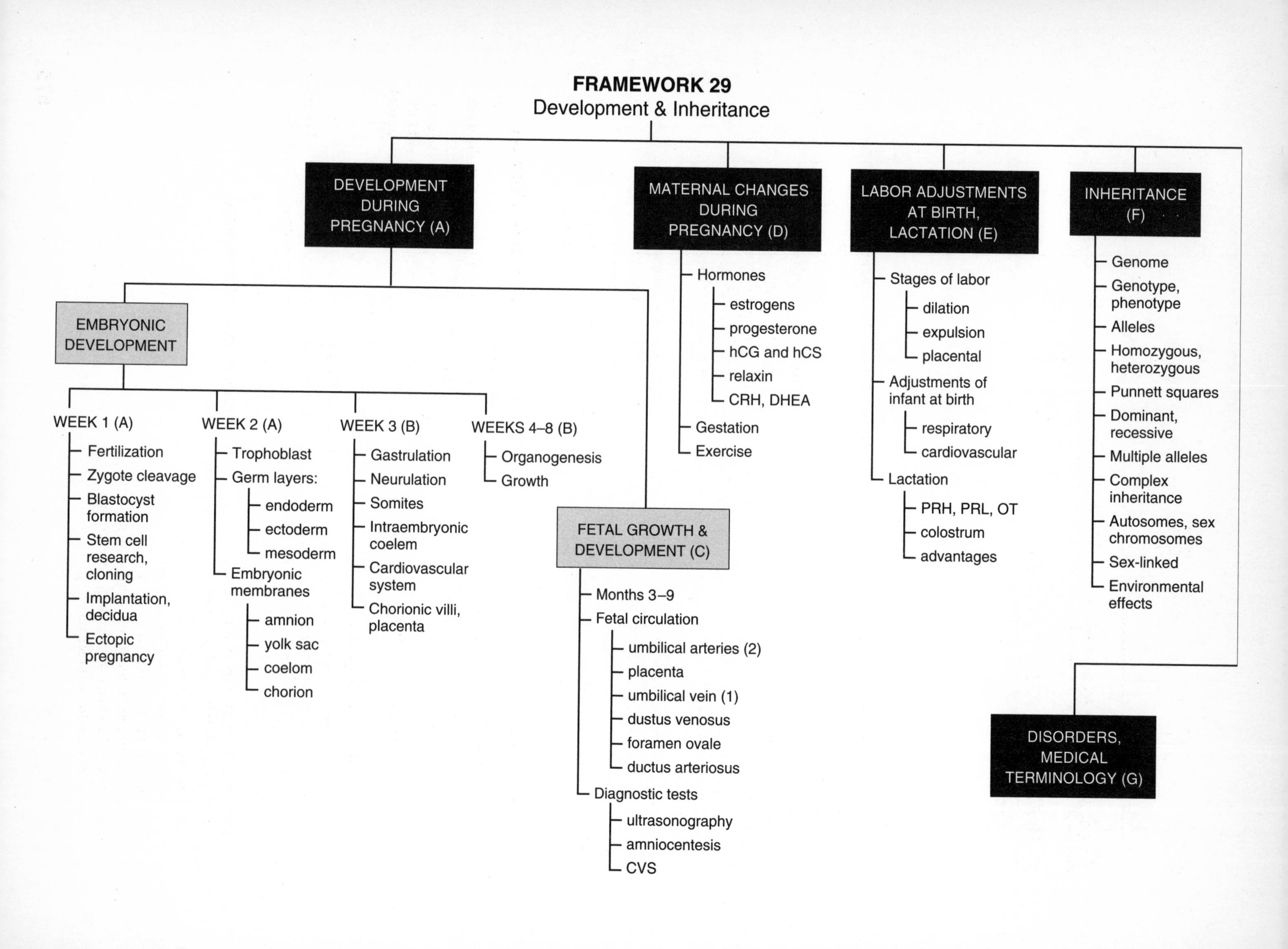

FRAMEWORK 29
Development & Inheritance
DEVELOPMENT DURING PREGNANCY (A)
EMBRYONIC DEVELOPMENT
WEEK 1 (A)
Fertilization
Zygote cleavage
Blastocyst formation
Stem cell research, cloning
Implantation, decidua
Ectopic pregnancy
WEEK 2 (A)
Trophoblast
Germ layers:
endoderm
ectoderm
mesoderm
Embryonic membranes
amnion
yolk sac
coelom
chorion
WEEK 3 (B)
Gastrulation
Neurulation
Somites
Intraembryonic coelem
Cardiovascular system
Chorionic villi, placenta
WEEKS 4–8 (B)
Organogenesis
Growth
FETAL GROWTH & DEVELOPMENT (C)
Months 3–9
Fetal circulation
umbilical arteries (2)
placenta
umbilical vein (1)
dustus venosus
foramen ovale
ductus arteriosus
Diagnostic tests
ultrasonography
amniocentesis
CVS
MATERNAL CHANGES DURING PREGNANCY (D)
Hormones
estrogens
progesterone
hCG and hCS
relaxin
CRH, DHEA
Gestation
Exercise
LABOR ADJUSTMENTS AT BIRTH, LACTATION (E)
Stages of labor
dilation
expulsion
placental
Adjustments of infant at birth
respiratory
cardiovascular
Lactation
PRH, PRL, OT
colostrum
advantages
INHERITANCE (F)
Genome
Genotype, phenotype
Alleles
Homozygous, heterozygous
Punnett squares
Dominant, recessive
Multiple alleles
Complex inheritance
Autosomes, sex chromosomes
Sex-linked
Environmental effects
DISORDERS, MEDICAL TERMINOLOGY (G)

CHAPTER 29

Development and Inheritance

Twenty-eight chapters ago you began a tour of the human body, starting with the structural framework and varieties of buildings (chemicals and cells) and moving on to the global communication systems (nervous and endocrine), transportation routes (cardiovascular), and refuse removal systems (respiratory, digestive, and urinary). This last chapter synthesizes all the parts of the tour: A new individual, with all those body systems, is conceived and develops during a nine-month pregnancy. Adjustments are made at birth for survival in an independent (outside of mother) world. But parental ties are still evident in inherited features, signs of intrauterine care (family's avoidance of alcohol or cigarettes), and provision of nourishment, as in lactation.

As you begin your study of development and inheritance, carefully examine the Chapter 29 Topic Outline and Objectives. Check off each objective as you complete it. The Chapter 29 Framework provides an organizational preview of development and inheritance. Be sure to refer to the Framework frequently and note relationships among key terms in each section. And congratulations on having completed your first tour through the wonders of the human body.

TOPIC OUTLINE AND OBJECTIVES

A. Embryonic period: first and second weeks

1. Explain the major developmental events that occur during the embryonic period.

B. Embryonic period: third through eighth weeks

C. Fetal period; prenatal diagnostic tests

1. Define the fetal period and outline the major events of the fetal period.
2. Describe the procedures for fetal ultrasonography, amniocentesis, and chorionic villi sampling.

D. Maternal changes during pregnancy: hormones; exercise and pregnancy

1. Describe the sources and functions of the hormones secreted during pregnancy.
2. Describe the hormonal, anatomical, and physiological changes in the mother during pregnancy.
3. Explain the effects of pregnancy on exercise, and of exercise on pregnancy.

E. Labor; adjustments at birth; physiology of lactation

1. Explain the events associated with the three stages of labor.
2. Explain the respiratory and cardiovascular adjustments that occur in an infant at birth.
3. Discuss the physiology and hormonal control of lactation.

F. Inheritance

1. Define inheritance, and explain the inheritance of dominant, recessive, complex, and sex-linked traits.

G. Disorders, medical terminology

WORDBYTES

Now become familiar with the language of this chapter by studying each wordbyte, its meaning, and an example of its use within a term. After you study the entire list, self-check your understanding by writing the meaning of each wordbyte on the line. As you continue through the *Learning Guide,* identify (and fill in) additional terms that contain the same wordbyte.

Wordbyte	Self-check	Meaning	Example(s)
angio-	____________	vessel	*angio*genesis
-blasto	____________	germ, to form	*blasto*mere, *blasto*cyst
-cele	____________	hollow	blasto*cele*
-centesis	____________	puncture to remove fluid	amnio*centesis*
corona	____________	crown	*corona* radiata, *corona*ry
dys-	____________	difficult, painful	*dys*tocia
ecto-	____________	outside of	*ecto*derm, *ecto*pic
endo-	____________	inside of	*endo*derm
fertil-	____________	fruitful	*fertil*ization
lact-	____________	milk	*lact*ation
-mere	____________	part	blasto*mere*
meso-	____________	middle	*meso*derm
morula	____________	mulberry	*morula*
pellucida	____________	allowing passage of light	zona *pellucida*
soma	____________	body	*soma*tic mesoderm
syn-	____________	together	*syn*gamy
toc-	____________	birth	dys*toc*ia, oxy*toc*in
-troph	____________	nourish	syncytio*troph*oblast
zygosis-	____________	a joining	*zygo*te; di*zygo*tic

CHECKPOINTS

A. Embryonic period: first and second weeks (pages 1063–1070)

✓ **A1.** Answer these questions about Paula who is seven weeks pregnant.

a. Her developing baby is currently a(n) *(embryo? fetus?)*. When will her baby be classified as a fetus? ____________

b. During the *(first? third?)* trimester, most fetal growth occurs; initial development of all major organs develops during the ____________ trimester.

c. Paula conceived her baby on March first. Her baby is due *(36? 38? 40?)* weeks after March first, or specifically, in late *(Novermber? December?)*. The four weeks following the baby's birth are known as the ________-natal period.

✓ **A2.** Do this exercise about events associated with fertilization.

a. How many sperm are ordinarily introduced into the vagina during sexual intercourse? ____________________ About how many reach the area where the developing ovum is located? ____________________. Sperm are likely to be viable for about *(6? 24? 48?)* hours after they are deposited in the vagina.

b. The name given to the developing ovum at the time of fertilization is the ____________________. This cell is viable for *(6? 24? 48?)* hours after ovulation. List three or more mechanisms in the female and/or male that favor the meeting of sperm with oocyte so that fertilization can occur.

Female	**Male**

c. How long must sperm be in the female reproductive tract before they can fertilize the female gamete? Several *(minutes? hours? days?)* Describe the changes sperm undergo during this time?

These changes in the sperm are known as ____________________.

d. Name the usual site of fertilization. ____.
What do the corona radiata and zona pellucida have to do with fertilization?

e. What triggers the *acrosomal reaction?*

How does this reaction help the sperm to penetrate the oocyte?

f. How many sperm normally penetrate the developing egg? ________ What is the name of this event? ____________________ What prevents further sperm from entering the female gamete (polyspermy)?

✓ **A3.** Describe roles of the following structures in fertilization.

a. Upon fertilization, the secondary oocyte completes meiosis *(I? II?)*, forming the fertilized ovum and releasing a ____________________ body.

b. A male pronucleus present in a *(secondary spermatocyte? sperm?)* combines with a female pronucleus from a *(secondary oocyte? ovum?)*. Each pronucleus contains the *(n? 2n?)* number of chromosomes which is *(23? 46?)*.

c. The product of the fusion of the two haploid pronuclei is known as the ____________________ containing ________ chromosomes.

✓ **A4.** Contrast types of multiple births by choosing the answer that best fits each description.

DTwin. Dizygotic twins	**MTwin. Monozygotic twins**

________ a. Also known as fraternal twins

________ b. Two infants derived from a single fertilized ovum that divides once

________ c. Two infants derived from two secondary oocytes each fertilized by a different sperm

A5. Complete this checkpoint about the next events following formation of the zygote.

a. Rapid mitotic divisions of the zygote are called ____________________. The first of these divisions is completed by about *(6? 30? 48?)* hours after fertilization. By the end of day 3 there are ________ cells, each of which is *(the same size as? progressively smaller than?)* previous cells.

b. A solid sphere is formed called a ____________________ and is composed of cells known as ____________________. This sphere is *(about the same size as? considerably larger than?)* the original zygote. The morula moves through the uterine tube and reaches the uterine cavity by about day ________.

c. The next stage is the ____________________. State two functions of uterine milk derived from endometrial glands.

d. Draw a diagram of a blastocyst. Label: *blastocyst cavity, inner cell mass,* and *trophoblast.* Color red the cells that will become the embryo. Color blue the cells that will become part of the placenta. (Refer to Figure 29.3b in the text, for help.)

✓ **A6.** Check your understanding of stem cells and cloning by completing this Checkpoint.

a. Which type of stem cell has the potential to form any cell in the entire organism, as well as placental cells? *(Multi? Pluri? Toti?)*-potent stem cells. Inner cell mass cells that form virtually all types of cells in the embryo (but not placental membranes) are known as ____________________-potent stem cells. Myeloid cells that can form a number of types of blood cells and keratinocytes that form skin cells are ____________________ -potent stem cells.

b. Name several sources of pluripotent stem cells and describe their theraputic use.

c. Explain how embryonic cloning is designed for therapeutic use.

✓ **A7.** Describe the process of implantation in this Checkpoint.

a. The blastocyst remains free in the uterine cavity for about _______ days.

Implantation is likely to occur about _______ days after ovulation (or _______ days after fertilization). Note that the developing baby is implanted in the uterine wall before the mother "misses" the first day of her menstrual period.

b. Typically, the blastocyst implants in the endometrial lining of the *(body? cervix? fundus?)* of the uterus. List three or more sites where implantation could occur but which would not support a full term pregnancy—a condition known as an ____________________ pregnancy. List several factors that increase risk for ectopic pregnancy.

c. Following implantation, the endometrium is known as the ________________________. Which portion of the decidua seperates the developing embryo from the uterine cavity? Decidua *(basalis? capsularis? parietalis?)*. (*Hint*: refer to text Figure 29.4.)

✓ **A8.** Recall from Checkpoint A5d that the blastocyst contains two portions: the ________________________ and the ________________________. Describe these parts in the exercise below.

a. About the time of implantation, the trophoblast forms two layers that will both become part of the *(amnion? chorion?)* portion of the placenta. Which trophoblast layer secretes enzymes that allow the blastocyst to implant? *(Cyto? Syncytio?)*-trophoblast. Name the hormone secreted by the trophoblast: ________________________ Describe the function of this hormone.

b. Inner cell mass cells also divide into two layers, the ________________________ and the ________________________. Together, these form the ________________________ embryonic disc.

c. The epiblast develops into the *(amnion? chorion?)*, commonly known as the "________________________ of ________________________." The amnion surrounds fluid known as ________________ fluid. What is the source of this fluid? ________________________ List functions of this fluid.

✓ **A9.** Refer to Figure 29.5 and identify developmental structures that fit related descriptions below.

CP.	**Coagulation plug**	**LN.**	**Lacunar networks**
ECM.	**Exocoelemic membrane**	**YS.**	**Yolk sac**
EEM.	**Extraembryonic mesoderm**		

________ a. Cells from the hypoblast that migrate to form a thin inner surface of the blastocyst wall on day 8 after fertilization

________ b. Formed by the hypoblast and ECM on day 9 after fertilization; located on one side of the bilaminar disc with the amniotic cavity on the other

________ c. Although relatively small in humans, serves as a shock absorber, provides nutrients during weeks 2 and 3, and is the source of blood cells from weeks 3–6; also forms part of GI tract and primordial sperm and egg cells

________ d. A temporary seal over the site of implantation

________ e. Small spaces that form within the syncytiotrophoblast and fuse by day 12; serve as the site where maternal blood will surround fetal vessels

________ f. Together with the trophoblast layers, forms the chorion, the thickest placental membrane

B. Embryonic period: third through eighth weeks (pages 1070–1079)

✓ **B1.** Describe developmental steps during week 3 following fertilization.

a. The process of formation of primary germ layers is known as ________________;

it is completed by the end of week _______ following fertilization. During the process the bilaminar disc is converted into a *(3? 4? 6?)*-layer disc consisting of the *(3? 4? 6?)* primary germ layers.

b. The first sign of gastrulation is the __________________ streak from posterior to anterior along the *(dorsal? ventral?)* surface of the embryo. Of what significance is the primitive streak?

c. The process of __________________ involves migration of epiblast cells below the primitive streak. Some of these cells displace the hypoblast and form the *(ecto? endo? meso?)*-derm primary germ layer. These *(loosely? tightly?)* packed cells will form the

linings of __________________ and __________________ tracts.

d. Tightly packed cells remaining in the epiblast form the *(ecto? endo? meso?)*-derm, whereas loosely packed *(epithelial? mesenchyme?)* cells form the mesoderm.

B2. Each of the primary germ layers is involved with formation of both fetal organs and fetal membranes. Name the major structures derived from each primary germ layer: endoderm, mesoderm, and ectoderm. Consult Table 29.1 in your text.

Endoderm	Mesoderm	Ectoderm

✓ **B3.** *For extra review.* Do this matching exercise on primary germ layer derivatives.

Ecto. Ectoderm	**Endo.** Endoderm	**Meso.** Mesoderm

________ a. Epithelial lining of all of digestive, respiratory, and genitourinary tracts except near openings to the exterior of the body

________ b. Epidermis of skin, epithelial lining of entrances to the body (such as mouth, nose, and anus), hair, nails

________ c. All of the skeletal system (bone, cartilage, joint cavities)

________ d. All skeletal and cardiac muscle; most smooth muscle

________ e. Blood and all blood and lymphatic vessels

________ f. Entire nervous system, including posterior pituitary

________ g. Thyroid, parathyroid, thymus, and pancreas

✓ **B4.** Decribe further events during weeks 3 and 4 of gestation in this exercise.

a. What functions are carried out by the notochord? It induces formation of the ____________________ and actually forms the *(annulus fibrosus? nucleus pulposus?)* of intervertebral discs.

b. What structures are formed from the region known as the oropharyngeal membrane?

c. What structures are formed from the region known as the cloacal membrane?

d. Arrange in correct sequence the events in neurulation. Formation of:

________ ________ ________

NF. Neural fold NP. Neural plate NT. Neural tube

e. Name structures formed from neural crest cells.

f. Look back at the table in LG Chapter 14 Checkpoint E2, page 279, and review development of parts of the brain.

g. Define these terms:

1. Neural tube defects (NTF)

2. Anencephaly

✓ **B5.** Describe development of somites in this exercise.

a. Somites form from *(ecto? endo? meso?)*-derm lateral to the notochord and neural tube.

By the end of week 5, a total of ________________ somites are present.

b. Which part of the somite develops into most skeletal muscles? *(Dermatome? Myotome? Sclerotome?)* Which develops into skin? ________________ Which forms bone of vertebrae? ________________

✓ **B6.** Indicate what body parts will be formed by the following structures:

IEC. Intraembryonic coelem	SPM. Splanchnic mesoderm	SOM. Somatic mesoderm

_______ a. Heart and connective tissue of respiratory and digestive organs

_______ b. Visceral layer of serous membranes over organs

_______ c. Parietal layer of serous membranes over organs

_______ d. Pleural, pericardial, and peritoneal regions of the ventral cavity

_______ e. Bones, ligaments, and dermis of skin

✓ **B7.** Discuss formation of the cardiovascular system in this Checkpoint.

a. Explain why development of blood vessels is needed early in embryonic development.

b. Describe roles of angioblasts in this process.

c. Describe roles of endocardial tubes.

✓ **B8.** Further describe the placenta in this exercise.

a. The placenta has the shape of a ________________ embedded in the wall of the uterus. Its fetal portion is the ________________ fetal membrane. Fingerlike projections, known as ________________, grow into the uterine lining. Consequently maternal and fetal blood are *(allowed to mix? brought into close proximity but do not mix?)*. Fetal blood in umbilical vessels is bathed by maternal blood in ________________ spaces.

b. The maternal aspect of the placenta is derived from the *(chorion? decidua basalis?)*.

This is a portion of the ______________________-metrium of the uterus.

c. What is the ultimate fate of the placenta?

✓ **B9.** *A clinical correlation.* Following a birth, vessels of the umbilical cord are carefully examined. A total of ________ vessels should be found, surrounded by a mucous jelly-like connective tissue. The vessels should include *(1? 2?)* umbilical artery(-ies) and *(1? 2?)*-umbilical vein(s). Lack of a vessel may indicate a congenital malformation.

✓ **B10.** Circle T (true) or F (false) next to each statement. Correct the underlined portion of false statements.

T F a. Chrionic villi are formed from cytotrophoblast tissue derived from the zygote rather than from maternal endometrial tissue.

T F b. Umbilical artery blood contains a higher level of oxygen than umbilical vein blood.

T F c. No viruses or antibodies can pass across the placenta.

T F d. In the condition known as placenta previa, the placenta implants into the fundus of the uterus.

B11. List five functions of the placenta.

✓ **B12.** Discuss processes of organogenesis in this Checkpoint.

a. By the end of week *(3? 5? 8?)*, all major organs are present, at least in rudimentary form.

b. Explain two roles played by blood vessels in organogenesis.

c. What does the term *embryonic folding* mean?

As a result, the embryo is *(flat? C-shaped?)* (Figure 29.14c).

d. The stomodeum will become the *(mouth? anus?)*. When the cloacal membrane ruptures during week 7, the ______________________ and ______________________ openings are created.

e. *(Three? Five? Ten?)* pairs of pharyngeal (branchial) arches form (Figure 29.14c). Which primary germ layers form these? ______________________ Between the arches are the pharyngeal ______________________. In addition, *(4? 5? 6?)* pairs of endoderm-lined pharyngeal pouches form.

f. What does the otic placode form (Figure 29.14d)? ________________

What about the lens placode? ________________

g. During which week do limb buds appear? *(4? 6? 8?)* Which limb buds appear first? *(Upper [arm]? Lower [leg]?)* By the end of week 4, a tail *(is? is not?)* found in the human embryo (Figure 29.14c).

C. Fetal period; prenatal diagnostic tests (pages 1080–1085)

✓ **C1.** After you study Table 29.2 in your text, write the number of the nine embryonic and fetal months when each of the following developmental events occurs.

________ a. Heart starts to beat.

________ b. Heart beat can be detected.

________ c. The embryo folds; chorionic villi develop.

________ d. Eyes are fully developed and wideset with eyelids closed.

________ e. External ears develop and are low set (Figure 29.14g).

________ f. Gender is distinguishable from external genitalia.

________ g. Eyes and ears move to their final positions.

________ h. Eyes are open.

________ i. Fetal movements are often felt by the mother (quickening).

________ j. Limb buds develop (Figure 29.14c).

________ k. Limb buds become distinct as arms and legs; digits develop; ossification of limb buds begins (Figure 29.14e).

________ l. Fine hairs (lanugo) and a cheesy covering (vernix caseosa) are found on skin.

________ m. Fetus has become capable of survival this month if given adequate care.

________ n. Testes begin to descend in scrotum of male

________ o. Body fat is about 16% of total body weight (which is about 7 lb).

✓ **C2.** Complete this exercise about prenatal diagnostic tests.

a. Amniocentesis involves withdrawal of ________________ fluid. This fluid is constantly recycled through the fetus (by drinking and urination); it *(does? does not?)* contain fetal cells and products of fetal metabolism. For what purposes is amniotic fluid collected and examined?

Amniocentesis is performed at about ________________ weeks of gestation.

b. *(Amniocentesis? Ultrasound?)* is by far the most common test for determining true fetal age if the date of conception is uncertain. The technique *(is also? is not?)* used for diagnostic purposes. Fetal ultrasonography involves *(use of a transducer on? a puncture into?)* the abdomen. For this procedure the mother's urinary bladder should be *(full? empty?)*.

c. CVS (meaning c ________________ v________________ s________________) is a diagnostic test that can be performed as early as *(4? 8? 16? 22?)* weeks of pregnancy. It *(must? can? does not?)* involve penetration of the uterine cavity. Cells of

the ________________ layer of the placenta embedded in the uterus are studied. These *(are? are not?)* derived from the same sperm and egg that united to form the fetus. Therefore this procedure, like amniocentesis, can be used to diagnose

________________.

d. The three tests just described *(are? are not?)* routinely used to determine gender of the fetus. Explain.

e. Explain how the AFP test is used to detect fetal abnormalities.

D. Maternal changes during pregnancy; exercise and pregnancy (pages 1085–1088)

✓ **D1.** Complete this exercise about the hormones of pregnancy.

a. During pregnancy, the level of progesterone and estrogens must remain *(high? low?)* in order to support the endometrial lining. During the first three or four months or so, these hormones are produced principally by the ________________ located in the ________________ under the influence of the hormone hCG produced by the ________________. (It might be helpful to review Checkpoint D4 in Chapter 28.) High levels of *(estrogens? progesterone?)* are especially critical in preventing premature contraction of the myometrium.

b. Human chorionic gonadotropin (hCG) reaches its peak between the ________ and ________ month of pregnancy. Because it is present in blood, it will be filtered into ________________, where it can readily be detected as an indication of pregnancy as early as ________ days after fertilization. Sometimes the urine of a woman who is not pregnant might erroneously indicate that she is pregnant. List several factors that could account for such a "false pregnancy test."

c. In addition, the chorion produces the hormone hCS; these initials stand for ________________ chorionic ________________. List two functions of this hormone.

Peak levels of hCS occur at about week *(9? 20? 32?)* of gestation.

d. The newly found chorionic hormone CRH, if secreted in high levels early in pregnancy, is associated with *(premature? late?)* delivery. Is this hormone produced in people who are not pregnant?

e. ____________________ is a hormone that relaxes pelvic joints and helps dilate the cervix near the time of birth. This hormone is produced by the ____________________ and ____________________.

✓ **D2.** Complete this exercise about maternal adaptations during pregnancy. Describe normal changes that occur.

a. Pulse, stroke volume, and cardiac output all _______-creases.

b. Blood volume _______-creases by about _______%

c. Tidal volume _______-creases about _______%

d. Gastrointestinal (GI) motility _______-creases, which may cause *(diarrhea? constipation?)*.

e. Pressure upon the inferior vena cava _______ -creases, which may lead to ____________________.

f. Increase in weight of the uterus from _______ grams in nonpregnant state to about _______ grams at term.

✓ **D3.** About 10–15% of all pregnant women in the United States experience PIH, which stands for p________________-i________________ h________________.

This condition apparently results from impaired function of ____________________.
Write three or more signs or symptoms of PIH.

D4. Defend or dispute this statement: "All women who are pregnant or lactating should avoid exercise because of its deleterious effects."

E. Labor; adjustments at birth; physiology of lactation (pages 1088–1093)

E1. Contrast these terms: *labor/parturition*

✓ **E2.** Do the following hormones stimulate (*S*) or inhibit (*I*) uterine contractions during the birth process?

_______ a. Progesterone

_______ b. Estrogens

_______ c. Oxytocin

✓ **E3.** *For extra review,* discuss regulation of hormones associated with labor.

a. During the last six months of gestation, both progesterone and estrogens are produced in large amounts by the ________________. The name progesterone indicates that this hormone is "pro" gestation, for one reason, because it *(stimulates? inhibits?)* uterine contractions, thereby continuing the gestation period.

b. For birth, known as ________________, to occur, levels of estrogen must *(surpass? become less than?)* levels of progesterone. This change in estrogens results from increasing levels of CRH made by the *(fetal hypothalamus? placenta?)*. This hormone causes the fetal anterior pituitary to secrete *(ACTH? DHEA?)* which stimulates the release of cortisol and the androgen ________________ from the fetal adrenal *(cortex? medulla?)*.

c. DHEA is then converted by the placenta into *(estrogen? progesterone?)*. This hormone causes mother's uterine muscle to become more receptive to *(oxytocin? relaxin?)*.

d. Explain how oxytocin promotes labor and delivery.

Oxytocin is involved in a *(negative? positive?)* feedback loop. Explain.

e. The hormone relaxin assists in parturition by *(relaxing? tightening?)* the pubic symphysis. Relaxin and estrogen together lead to *(softening? tightening?)* of the cervix.

✓ **E4.** What is the "show" produced at the time of birth?

Is the "show" associated with true labor or false labor?

✓ **E5.** Identify descriptions of the three phases of labor (*first, second,* or *third*).

_______ a. Stage of expulsion: from complete cervical dilation through delivery of the baby

_______ b. Time after the delivery of the baby until the placenta ("afterbirth") is expelled; the placental stage

_______ c. Time from onset of labor to complete dilation of the cervix; the stage of dilation

E6. *A clinical correlation.* Explain more about labor and delivery in this Checkpoint.

a. The average duration of labor is about _____ hours, whereas the duration for first labor and delivery is about _____ hours.

b. The term ________________ means painful or difficult labor. In a breech presentation, which parts of the body enter the birth canal first? ________________ If necessary, a C-section or ________________ section will be required for such births. Was this procedure named such because Julius Caesar was born by C-section? *(Yes? No?)* Explain.

c. Defend or dispute this statement: "Longer (such as 45 weeks) is better for the fetus when it comes to length of pregnancy."

d. The stress of labor and delivery causes the fetus to initiate "fight-or-flight" responses in which the fetal adrenal medullae release their hormones. Explain how these chemicals help the fetus through the birth process.

E7. Define these terms related to birth and the postnatal period.

a. *Involution*

b. *Lochia*

c. *Puerperium*

E8. What adjustments must the newborn make at birth in attempting to cope with the new environment? Describe adjustments of the following.

a. Respiratory system (What serves as a stimulus for the respiratory center of the medulla?)

The respiratory rate of a newborn is usually about *(one-third? three times?)* that of an adult.

b. Heart and blood vessels (*For extra review* go over changes in fetal circulation, Chapter 21, Table LG 21.1, page LG 468.)

c. Blood cell production

✓ **E9.** A premature infant ("preemie") is considered to be one who weighs less than

________ grams (________ pounds ________ ounces) at birth. These infants are at high

risk for RDS (or ____________________).

✓ **E10.** Complete this exercise about lactation.

a. The major hormone promoting lactation is ____________________, which is secreted by

the ____________________. During pregnancy the level of this hormone increases somewhat, but its effectiveness is limited while estrogens and progesterone are *(high? low?)*.

b. Following birth and the loss of the placenta, a major source of estrogens and progesterone is gone, so effectiveness of prolactin *(increases? decreases?)* dramatically.

c. The sucking action of the newborn facilitates lactation in two ways. What are they?

d. What hormone stimulates milk ejection (let-down)? ____________________

e. The time period from the start of a baby's suckling at the breast to the delivery of milk

to the baby is about ________-seconds. Although afferent impulses (from breast to hypothalamus) require only *(a fraction of a second? 45 seconds?)*, passage of the oxytocin

through the ____________________ to the breasts requires about ________ seconds.

E11. Contrast *colostrum* with *maternal milk* with respect to time of appearance, nutritional content, and presence of maternal antibodies.

E12. Defend or dispute this statement: "Breast-feeding is an effective, reliable form of birth control."

E13. List five or more advantages offered by breast-feeding.

F. Inheritance (pages 1093–1098)

F1. Define these terms.

a. *Inheritance*

b. *Genetic counseling*

c. *Homologues (homologous chromosomes)*

F2. Compare and contrast.

a. *Genotype/phenotype*

b. *Dominant allele/recessive allele*

c. *Homozygous/heterozygous*

✓ **F3.** Answer these questions about genes controlling PKU. (See Figure 29.20 in the text.)

a. The dominant gene for PKU is represented by *(P? p?)*. The dominant gene is for the *(normal? PKU?)* condition.

b. The letters PP are an example of a genetic makeup or *(genotype? phenotype?)*. A person with such a genotype is said to be *(homozygous dominant? homozygous recessive? heterozygous?)*. The phenotype of that individual would *(be normal? be normal, but serve as a "carrier" of the PKU gene? have PKU?)*.

c. The genotype for a heterozygous individual is ________________. The phenotype is

________________.

d. Use a Punnett square to determine possible genotypes of offspring when both parents are Pp.

✓ **F4.** Complete this exercise about inherited disorders.

a. Most genes give rise to *(a different phenotype depending on? the same phenotype regardless of?)* whether mother or father gave the abnormal genes. This phenomenon is known

as genomic ________________. However, exceptions do occur as in Angelman syndrome in which the gene of an abnormal trait is inherited from the *(father? mother?)*, and Prader-Willi syndrome in which the gene for an abnormal trait is inherited from the *(father? mother?)*.

b. What is the meaning of the term *aneuploid*?

One example is *trisomy 21* in which the individual has *(2? 3? 4?)* of chromosome 21 in each cell. Trisomy is caused by nondisjunction in virtually all cases of

________________ syndrome. Define *nondisjunction*.

c. Refer to Table 29.3 and identify whether the genes that cause the following disorders are dominant (D) or recessive (R). Circle the correct answer in each case.

1.	Polydactyly:	D	R
2.	Familial hypercholesterolemia:	D	R
3.	Cystic fibrosis:	D	R
4.	Near-sightedness:	D	R
5.	Albinism:	D	R

d. Timothy who has the genotype for the autosomal dominant disorder, Huntington's disease (HD), must have received the faulty gene from *(one? both?)* parent(s). Stephen who has the autosomal recessive disorder cystic fibrosis (CF) must have received the faulty gene from *(one? both?)* parent(s).

e. Sickle cell anemia (SCA) is a condition that is controlled by *(complete? incomplete?)* dominance. Match the correct genotype below with the related description. Choose from these genotypes:

A. Homozygous recessive	**C. Heterozygous**
B. Homozygous dominant	

1. Genotype present in persons who have SCA *(Hb^SHb^S)*: ________
2. Genotype present in persons who have SCA *(Hb^AHb^S)*: ________

✓ **F5.** Discuss inheritance of blood types in this Checkpoint.

a. List the three possible alleles for inheritance of the ABO group.

__________________ __________________ __________________ Which of these is/are codominant? __________________ List the possible genotypes of persons who produce A antigens on red blood cells. __________________ Describe the phenotype of persons with the 00 (or *ii*) genotype.

b. ABO blood group is an example of *(complex? multiple allele? sex-linked?)* inheritance.

✓ **F6.** Complete this exercise about more complex inheritance.

a. Most inherited traits are controlled by *(a single? many?)* gene(s), a pattern known as __________________ inheritance.

b. Characteristics governed by two or more genes and also environmental factors are said to have __________________ inheritance. List four or more characteristics controlled by complex inheritance.

c. Explain why neural tube defects (NTDs) are classified as being governed by complex inheritance.

d. Jonathan's genotype for skin color is *AABBCC*; Jackie's is *aaBbcc*. Which is likely to have darker skin? *(Jonathan? Jackie?)* Jonathan and Jackie are *(P? F_1? F_2?)* generation, whereas their daughter Keisha is the *(P? F_1? F_2?)* generation. Keisha's genotype for skin color is most likely to be *(AABBCC? AaBBCc? AaBbcc?)*.

F7. Contrast *autosome* with *sex chromosome*.

✓ **F8.** Answer these questions about sex inheritance.

a. All *(sperm? secondary oocytes?)* contain the X chromosome. About half of the sperm produced by a male contain the _______ chromosome and half contain the _______.

b. Every cell (except those forming ova) in a normal female contains *(XX? XY?)* genes on sex chromosomes. All male cells (except those forming sperm) are said to be *(XX? XY?)*.

c. Who determines the sex of the child? *(Mother? Father?)* Show this by using a Punnett square.

d. Name the gene on the Y chromosome that determines that a developing individual will be male. ____________________

✓ **F9.** Complete this exercise about sex-linked inheritance.

a. Sex chromosomes contain other genes besides those determining sex of an individual. Such traits are called ____________________ traits. Y chromosomes are shorter than X chromosomes and lack some genes, In fact, the Y chromosome contains only _______ genes. One of these is the gene controlling ability to ____________________ red from green colors. Thus this ability is controlled entirely by the _______ gene.

b. Write the genotype for females of each type: color-blind, ____________________; carrier, ____________________; normal, ____________________.

c. Now write possible genotypes for males: color-blind, ____________________; normal, ____________________.

d. Determine the results of a mating between a color-blind male and a normal female.

e. Now determine the results of a mating between a normal male and a carrier female.

f. Color blindness and other X-linked, recessive traits are much more common in *(males? females?)*.

F10. Discuss *fragile X syndrome*.

List several examples of other sex-linked traits.

F11. Describe the work of researcher Mary Lyons that elucidated the function of the Barr body and the process of lyonization.

✓ **F12.** Complete this exercise about potential hazards to the embryo and fetus. Match the terms in the box with the related descriptions.

C. Cigarette smoking	FAS. Fetal alcohol syndrome	T. Teratogen

_______ a. An agent or influence that causes defects in the developing embryo

_______ b. May cause defects such as small head, facial irregularities, defective heart, retardation

_______ c. Linked to a variety of abnormalities such as lower infant birth weight, higher mortality rate, GI disturbances, and SIDS

G. Disorders, medical terminology (pages 1099–1100)

G1. Describe the following aspects of in vitro fertilization (IVF).

a. In what year was the first live birth as a result of this procedure? *(1962? 1970? 1978? 1986?)*

b. Briefly describe this procedure.

c. Contrast these two forms of IVF: *embryo transfer* and *GIFT*

G2. Describe how the conditions of being underweight or overweight may affect reproductive hormones and fertility.

✓ **G3.** Fill in the term (*impotence* or *infertility*) that best fits the description. Then list several causes of each condition.

a. Inability to fertilize an ovum: ____________________.

b. Inability to attain and maintain an erection: ____________________.

G4. Each of the following disorders involves a particular chromosome. Identify which chromosome is involved and then write a brief description of characteristics of individuals born with this disorder.

a. Down syndrome involves chromosome ________.

b. Klinefelter's syndrome involves chromosome ________.

c. Turner's syndrome involves chromosome ________.

G5. Write the technical term for morning sickness and describe who is likely to be affected by this condition.

ANSWERS TO SELECTED CHECKPOINTS: CHAPTER 29

A1. (a) Embryo; in one week (at eight weeks of gestation). (b) Third; first. (c) 38; November (22nd); neo.

A2. (a) Approximately 200 million; less than 1%; 48. (b) Secondary oocyte; 24; mechanisms in the female: peristaltic contractions of uterine tubes, uterine contractions promoted by prostaglandins in semen; mechanisms in the male: presence of flagella on sperm. (c) Hours; more vigorous lashing of flagella, and changes that enable plasma membranes of sperm and oocyte to better fuse; capacitation. (d) Uterine (Fallopian) tube; they are the coverings over the oocyte that sperm must penetrate. (e) ZP3, a chemical in the zona pellucida, attracts sperm and triggers this reaction; proteolytic enzymes released from the sperm's acrosome clear a path for the sperm through the zona pellucida. (f) One; syngamy; sperm penetration of the secondary oocyte triggers depolarization of the oocyte (fast block), then release of calcium ions from the oocyte that lead to inactivation of ZP3 and hardening of the zona pellucida so it is impenetrable to sperm (slow block).

A3. (a) II, polar. (b) Sperm, ovum; n, 23. (c) Zygote, 46.

A4. (a) DTwin. (b) MTwin. (c) DTwin.

A5. (a) Cleavage; 30; 16, progressively smaller than. (b) Morula, blastomeres; about the same size as; 4 or 5. (c) Blastocyst; nourishes the developing individual and seeps between cells to form the fluid filling the blastocyst cavity. (d) See Figure 29.3b.

A6. (a) Toti; pluri; multi. (b–c) See text page 1066.

A7. (a) 2; 7. (6). (b) Body or fundus; ectopic; most often the uterine tube, but possibly in the cervix, on an ovary, or on other abdominopelvic structures; tubal (or other) scarring from previous infection or pelvic inflammatory disease, previous tubal pregnancy, abnormal tubal anatomy, and smoking. (c) Decidua; capsularis.

A8. Trophoblast and inner cell mass. (a) Chorion; syncytio; HCG; much like LH, HCG maintains the corpus luteum until placental secretion of estrogens and progesterone eliminates the need for production of these hormones by the corpus luteum; and HCG is the chemical tested for in home pregnancy tests. (b) Hypoblast (primitive endoderm) and epiblast (primitive ectoderm); bilaminar. (c) Amnion, bag (of) waters; amniotic; filtrate of mother's blood (and later, fetal urine); shock absorber for developing baby; see also page 1070.

A9. (a) ECM. (b-c) YS. (d) CP. (e) LN. (f) EEM.

B1. (a) Gastrulation; 3; 3, 3. (b) Primitive, dorsal; it establishes head and tail, right and left. (d) Invagination; endo; tightly, digestive (and) respiratory. (d) Ecto, mesenchyme.

B3. (a) Endo. (b) Ecto. (c) Meso. (d) Meso. (e) Meso. (f) Ecto. (g) Endo.

B4. (a) Vertebral bodies, nucleus pulposus. (b) Upper gastrointestinal tract (mouth to pharynx). (c) Anus and inferior openings of urinary and reproductive tracts. (d) NP NF NT. (e) See page 1073. (f) See LG pages 279 and 282. (g) See text page 1073.

B5. (a) Meso; 42–44. (b) Myotome; dermatome; schlerotome.

B6. (a) SPM. (b) SPM. (c) SOM. (d) IEC. (e) SOM.

B7. (a) To provide nourishment because there is so little yolk in the yolk sac for rapidly dividing embryonic cells. (b) These cells form blood islands that develop into blood vessels. (c) They fuse to form a single primitive heart tube that bends into an S-shape and begins to beat (as the heart) within the third week.

B8. (a) Flattened cake (or thick pancake); chorion; chorionic villi; brought into close proximity but do not mix; intervillous. (b) Decidua basalis; endo. (c) It is discarded from mother at birth (as the "afterbirth"); separation of the decidua from the remainder of endometrium accounts for bleeding during and following birth.

B9. 3; 2; 1.

B10. (a) T. (b) F. Lower. (c) F. Many can. (d) F. Cervix or lower body so the placenta is delivered before the baby, a situation that can lead to serious hypoxia.

B12. (a) 8. (b) Bring oxygen and nutrients and remove wastes; signal cell-to-cell interactions that are required for organogenesis. (c) Conversion from a bilaminar disc to a 3-dimensional structure composed of the three primary germ layers; C-shaped. (d) Mouth; urogenital and anal. (e) Five; all three; clefts; 4. (f) The ear; forms the eye. (g) 4; upper [arm]; is.

C1. (a) 1. (b) 3. (c) 1. (d) 3. (e) 3. (f) 3. (g) 4. (h) 7. (i) 5. (j) 1. (k) 2. (l) 5. (m) 7. (n) 7. (o) 9.

C2. (a) Amniotic; does; detecting suspected genetic abnormalities; 14–18. (b) Ultrasound; is also; use of a transducer on; full. (c) Chorionic villi sampling, 8; can; chorion; are; genetic and chromosomal defects. (d) Are not; they offer some risk of spontaneous abortion. (e) Alpha-fetoprotein is normally produced by the fetus and passes into mother's blood stream up until week 16. But after that time, high level of AFP is predictive of NTD defects such as spina bifida or anencephaly.

D1. (a) High; corpus luteum, ovary, chorion portion of the placenta; progesterone. (b) First, third; urine, 8; excess protein or blood in her urine, ectopic production of hCG by uterine cancer cells, or presence of certain diuretics or hormones in her urine. (c) Human (chorionic) somatomammotropin; stimulates development of breast tissue and alters metabolism during pregnancy (see page 1085 in the text); 32. (d) Premature; yes, made by the hypothalamus, it is the releasing hormone for ACTH. (e) Relaxin; placenta, ovaries.

D2. (a) In. (b) In, 30–50. (c) In, 30–40. (d) De, constipation. (e) In, varicose veins. (f) 60–80, 900–1200.

D3. Pregnancy-induced hypertension; kidneys; hypertension, fluid retention leading to edema with excessive weight gain, headaches and blurred vision, and proteinuria.

E2. (a) I. (b) S. (c) S.

E3. (a) Chorion of the placenta; inhibits. (b) Parturition, surpass; placenta; ACTH, DHEA, cortex. (c) Estrogen; oxytocin. (d) Oxytocin stimulates uterine contractions which force the baby into the cervix. Positive; cervical stretching triggers nerve impulses to the hypothalamus where more oxytocin is released. (e) Relaxing; softening.

E4. A discharge of blood-containing mucus that is present in the cervical canal during labor; true labor.

E5. (a) Second. (b) Third. (c) First.

E9. 2500 grams (5 pounds. 8 ounces); respiratory distress syndrome.

E10. (a) Prolactin (PRL), anterior pituitary; high. (b) Increases. (c) Maintains prolactin level by inhibiting PIH and stimulating release of PRH; stimulates release of oxytocin. (d) Oxytocin. (e) 30–60; a fraction of a second, bloodstream, 30–60.

F3. (a) *P*; normal. (b) Genotype; homozygous dominant; be normal. (c) *Pp*; normal but a carrier of PKU gene. (d) 25% *PP*, 50% *Pp*, and 25% *pp*.

F4. (a) The same phenotype regardless of; imprinting; mother, father. (b) Having chromosome number other than normal (which is 46 for humans); 3; Down; an error in meisosis (or rarely) mitosis in which homologous (or sister) chromatids fail to separate at anaphase. (c1) D. (c2) D. (c3) R. (c4) D. (c5) R. (d) One; both. (e) Incomplete; (e1) A; (e2) C.

F5. (a) I^A, I^B, and *i*; I^A and I^B; I^AI^A or I^Ai; produced neither A nor B antigens on red blood cells. (b) Multiple allele.

F6. Many, polygenic; complex; skin color, eye color, height, or body build. (c) Genes as well as maternal folic acid deficiency contribute to these disorders. (d) Jonathan; P, F_1; *AaBBCc*.

F8. (a) Secondary oocytes; X, Y. (b) XX; XY. (c) Father, as shown in the Punnett square (see Figure 29.25 in the text), because only the male can contribute the Y chromosome. (d) SRY (sex-determining region of the Y chromosome).

F9. (a) Sex-linked; 231; differentiate; X. (b) X^cX^c, X^CX^c, X^CX^C. (c) X^cY, X^CY. (d) All females are carriers and all males are normal. (e) Of females, 50% are normal and 50% are carriers; of males, 50% are normal and 50% are color-blind. (f) Males.

F12. (a) T. (b) FAS. (c) C.

G3. (a) Infertility (or sterility) (b) Impotence (erectile dysfunction).

MORE CRITICAL THINKING: CHAPTER 29

1. Summarize three major events in the embryonic period following implantation.
2. Describe the layers that form the placenta and outline its protective functions.
3. Contrast the three primary germ layers and the types of tissues and organs each will form.
4. Contrast use of amniocentesis and chorionic villi sampling (CVS). Describe when during gestation each procedure is likely to be performed, what is sampled, and how each test gives information about potential genetic anomalies.
5. Describe effects of the following maternal or fetal hormones upon labor and parturition: progesterone, estrogens, oxytocin, and relaxin (maternal); cortisol and epinephrine (fetal).
6. Describe the roles of the *fetal* adrenal gland in assisting the baby through labor and delivery. Be sure to include both cortical and medullary hormones.
7. Discuss effects of embryonic or fetal environment upon complex inheritance traits. Include effects of cigarettes, alcohol and other drugs, and radiation.

MASTERY TEST: CHAPTER 29

Questions 1–2: Arrange the answers in correct sequence.

______ ______ ______ ______ 1. From most superficial to deepest (closest to embryo):
A. Amnion
B. Amniotic cavity
C. Chorion
D. Decidua

______ ______ ______ ______ ______ ______ 2. Stages in development:
A. Morula
B. Blastocyst
C. Zygote
D. Fetus
E. Embryo

Questions 3–11: Circle the letter preceding the one best answer to each question.

3. During early months of pregnancy, hCG is produced by the ______ and it mimics the action of ______.
A. Follicle; progesterone
B. Corpus luteum; FSH
C. Chorion of placenta; LH
D. Amnion of placenta: estrogens

4. These fetal characteristics—weight of 8 ounces to 1 pound, length of 10 to 12 inches, lanugo covers skin, and brown fat is forming—are associated with which month of fetal life?
A. Third month
B. Fourth month
C. Fifth month
D. Sixth month

5. All of the following structures are developed from mesoderm *except:*
A. Aorta
B. Biceps muscle
C. Heart
D. Humerus
E. Sciatic nerve
F. Neutrophil

6. Which part of the structures surrounding the fetus is developed from maternal cells (not from fetal)?
A. Allantois
B. Yolk sac
C. Chorion
D. Decidua
E. Amnion

7. All of the following events occur during the first month of embryonic development *except:*
A. Endoderm, mesoderm, and ectoderm are formed.
B. Amnion and chorion are formed.
C. Limb buds develop.
D. Ossification begins.
E. The heart begins to beat.

8. Implantation of a developing individual (blastocyst stage) usually occurs about ______ after fertilization.
A. Three weeks
B. One week
C. One day
D. Seven hours
E. Seven minutes

9. The high level of estrogens present during pregnancy is responsible for all of the following *except:*
A. Stimulates release of prolactin
B. Prevents ovulation
C. Maintains endometrium
D. Prevents menstruation
E. Stimulates uterine contractions

10. Which of the following is a term that refers to discharge from the birth canal during the days following birth?
A. Autosome
B. Cautery
C. Lochia
D. PKU

11. Formation of the three primary germ layers is known as:
A. Implantation
B. Blastocyst formation
C. Gastrulation
D. Morula formation

Questions 12–20: Circle T (true) or F (false). If the statement is false, change the underlined word or phrase so that the statement is correct.

T F 12. Oxytocin and progesterone both stimulate uterine contractions.

T F 13. Of the primary germ layers, only the ectoderm forms part of a fetal membrane.

T F 14. Maternal blood mixes with fetal blood in the placenta.

T F 15. Amniocentesis is a procedure in which amniotic fluid is withdrawn for examination at about 8–10 weeks during the gestation period.

T F 16. Stage 2 of labor refers to the period during which the "afterbirth" is expelled.

T F 17. Maternal pulse rate, blood volume, and tidal volume all normally increase during pregnancy.

T F 18. The decidua is part of the endoderm of the fetus.

T F 19. Inner cell mass cells of the blastocyst form much of the chorion of the placenta.

T F 20. A phenotype is a chemical or other agent that causes physical defects in a developing embryo.

Questions 21–25: Fill-ins. Complete each sentence with the word or phrase that best fits.

__________________ 21. The newborn infant generally has a respiratory rate of ________, pulse of ________, and a WBC count that may reach ________, all values that are ________ than those of an adult.

__________________ 22. As a prerequisite for fertilization, sperm must remain in the female reproductive tract for at least several hours so that ________________ can occur.

__________________ 23. hCG is a hormone made by the ________; its level peaks at about the end of the ________ month of pregnancy, and up to this time it can be used to detect ________.

__________________ 24. Determine the probable genotypes of children of a couple in which the man has hemophilia and the woman is normal. ________

__________________ 25. This is the ________ mastery test question in this book.

ANSWERS TO MASTERY TEST: CHAPTER 29

Arrange

1. D C A B
2. C A B E D

Multiple Choice

3. C
4. C
5. E
6. D
7. D
8. B
9. A
10. C
11. C

True–False

12. F. Oxytocin but not progesterone stimulates.
13. F. All three layers form parts of fetal membranes: ectoderm (trophoblast) forms part of chorion; mesoderm forms parts of the chorion and the allantois; endoderm lines the yolk sac.
14. F. Does not mix but comes close to fetal blood
15. F. Amniocentesis, 14–18
16. F. Fetus
17. T
18. F. Endometrium of the uterus
19. F. Trophoblast
20. F. Teratogen

Fill-ins

21. 45, 120–160, up to 45,000, higher
22. Capacitation and dissolving of the covering over the ovum by secretion of acrosomal enzymes of sperm
23. Chorion of the placenta, second, pregnancy
24. All male children normal; all female children carriers
25. Last or final. Congratulations! Hope you learned a great deal.